Phytoremediation

Biotechnological Strategies for Promoting Invigorating Environs

Phytoremediation

Biotechnological Strategies for Promoting Invigorating Environs

Edited by

Rouf Ahmad Bhat
Department of Environmental Science, Sri Pratap College, Cluster University Srinagar, Srinagar, India

Fernanda Maria Policarpo Tonelli
Pitágoras College, Campus Divinópolis, Divinópolis, Brazil

Gowhar Hamid Dar
Department of Environmental Science, Sri Pratap College, Cluster University Srinagar, Srinagar, India

Khalid Rehman Hakeem
Department of Biological Sciences, Faculty of Science, King Abdulaziz University, Jeddah, Saudi Arabia

ELSEVIER
ACADEMIC PRESS
An imprint of Elsevier
elsevier.com/books-and-journals

Academic Press is an imprint of Elsevier
125 London Wall, London EC2Y 5AS, United Kingdom
525 B Street, Suite 1650, San Diego, CA 92101, United States
50 Hampshire Street, 5th Floor, Cambridge, MA 02139, United States
The Boulevard, Langford Lane, Kidlington, Oxford OX5 1GB, United Kingdom

British Library Cataloguing-in-Publication Data
A catalogue record for this book is available from the British Library

Library of Congress Cataloging-in-Publication Data
A catalog record for this book is available from the Library of Congress

ISBN: 978-0-323-89874-4

Publisher: Mica Haley
Acquisitions Editor: Emma Hayes
Editorial Project Manager: Lena Sparks
Production Project Manager: Omer Mukthar
Cover Designer: Greg Harris

Typeset by Aptara, New Delhi, India

Contents

Contributors

Ismail Abiola Adebayo Microbiology and Immunology Department, Faculty of Biomedical Sciences, Kampala International University-Western Campus, Ishaka-Busheyi, Uganda

Azeez Omoniyi Adeoye Department of Anatomy, Faculty of Biomedical Sciences, Kampala International University-Western Campus, Ishaka-Busheyi, Uganda

Adam Moyosore Afodun Department of Anatomy, Faculty of Biomedical Sciences, Kampala International University-Western Campus, Ishaka-Busheyi, Uganda,

Moniem B. Ahad Department of Zoology, Central University of Kashmir, Ganderbal, Jammú and Kashmir

Kamoldeen Abiodun Ajijolakewu Microbiology Department, Faculty of Life Sciences, University of Ilorin, Ilorin, Nigeria

Angel Josabad Alonso-Castro Department of Pharmacy, University of Guanajuato Guanajuato, Mexico; Department of Pharmacy, Division of Natural and Exact Sciences, University of Guanajuato, Guanajuato, Mexico

Alice Amoding Department of Agricultural Production, Makerere University, Kampala, Uganda

Sagar Arya School of Life and Environmental Sciences, Deakin University, Victoria, Australia; TERI-Deakin Nanobiotechnology Centre, The Energy and Resources Institute, Gurugram, Haryana, India

Tabinda Athar Institute of Soil and Environmental Sciences, University of Agriculture, Faisalabad, Pakistan

Muhammad Ashar Ayub Institute of Soil and Environmental Sciences, University of Agriculture, Faisalabad, Pakistan

Arpita Banerjee School of Biotechnology, Kalinga Institute of Industrial Technology, Bhubaneswar, Odisha

Monika Bansal School of Agricultural Biotechnology, Punjab Agricultural University, Ludhiana, Punjab, India

Twaha Ateenyi Basamba Department of Agricultural Production, Makerere University, Kampala, Uganda

Ashish Bedi School of Life and Environmental Sciences, Deakin University, Victoria, Australia; TERI-Deakin Nanobiotechnology Centre, The Energy and Resources Institute, Gurugram, Haryana, India

Isha Bedi Saint Ambrose University, Davenport, Iowa, USA

Sashidhar R. Beedu Department of Biochemistry, University College of Science, Osmania University, Hyderabad, Telangana, India

Rouf Ahmad Bhat Department of Environmental Science, Sri Pratap College, Cluster University Srinagar, Srinagar, India

Sasha Cardozo School of Biotechnology, Kalinga Institute of Industrial Technology, Bhubaneswar, Odisha

Candy Carranza-Álvarez Faculty of Professional Studies Zona Huasteca, UASLP, San Luis Potosí., Mexico; Faculty of Professional Studies Zona Huasteca and Center for Research and Postgraduate Studies of the Faculty of Chemical Sciences, Autonomous University of San Luis Potosí, Cd.Valles, S.L.P., Mexico

Ramón Fernando García-De La Cruz Research and Postgraduate Center of the Faculty of Chemical Sciences, Universidad Autónoma de San Luis Potosí, S.L.P., Mexico

Gowhar Hamid Dar Department of Environmental Science, Sri Pratap College, Cluster University Srinagar, Srinagar, India

Xiaorong Fan State Key Laboratory of Crop Genetics and Germplasm Enhancement, Nanjing Agricultural University, Nanjing, China; Key Laboratory of Plant Nutrition and Fertilization in Lower-Middle Reaches of the Yangtze River, Ministry of Agriculture, Nanjing Agricultural University, Nanjing, China

Iqra Farooq Division of Floriculture and Landscape Architecture, Sher-e-Kashmir, University of Agricultural Sciences and Technology of Kashmir, Srinagar, India

Zia Ur Rahman Farooqi Institute of Soil and Environmental Sciences, University of Agriculture, Faisalabad, Pakistan

Nafeesa Farooq Khan Department of Botany, School of Biological Sciences, University of Kashmir, Srinagar, India

Saksham Garg Department of Biotechnology, Delhi Technological University, Delhi, India

Muhammad Imran Ghani College of natural resource and environment Northwest A&F University, Yangling shaanxi, China

Mir Z. Gul Department of Biochemistry, University College of Science, Osmania University, Hyderabad, Telangana, India

Charu Gupta School of Studies in Microbiology, Jiwaji University, Gwalior, Madhya Pradesh

Mahendra K. Gupta Microbiology Research Lab., School of Studies in Botany, Jiwaji University, Gwalior, Madhya Pradesh

Aukib Habib Microbiology Research Lab., School of Studies in Botany, Jiwaji University, Gwalior, Madhya Pradesh

Younis Ahmad Hajam Department of Biosciences, Division Zoology, Career Point University, Hamirpur, Himachal Pradesh, India

Khalid Rehman Hakeem Department of Biological Sciences, Faculty of Science, King Abdulaziz University, Jeddah, Saudi Arabia

Insha Hameed School of Studies in Environmental Science, Jiwaji University, Gwalior, Madhya Pradesh

Humaira Hussain Department of Botany, University of Kashmir, Srinagar, Jammu and Kashmir

Muhammad Mahroz Hussain Institute of Soil and Environmental Sciences, University of Agriculture, Faisalabad, Pakistan

Predrag Ilic PSRI Institute for Protection and Ecology of the Republic of Srpska, Banja Luka, Bosnia and Herzegovina

Rashid Iqbal Department of Agronomy, Faculty of Agriculture and environmental sciences, The Islamia university of Bahawalpur, Pakistan

Akanksha Jaiswar TERI-Deakin Nanobiotechnology Centre, The Energy and Resources Institute, Gurugram, Haryana, India

Muatasim Jan Centre of Research for Ethnobotany, Government Model Science College, Jiwaji University, Gwalior, India

Rahul Singh Jasrotia Department of Biological Science, Florida State University, Tallahassee, Florida, USA

Rajdeep Jaswal National Agri Food Biotechnology Institute (NABI), Mohali, Punjab, India

Manmeet Kaur Department of Microbiology, Punjab Agricultural University, Ludhiana, Punjab, India

Navneet Kaur Division of Biochemistry, FBSc, Sher-e-Kashmir University of Agricultural Sciences and Technology of Jammu, Jammu, India

Iqra F. Khan Division of Fruit Science, Faculty of Horticulture, Sher-e-Kashmir University of Agricultural Sciences and Technology of Kashmir, Srinagar, India

Satya Prakash Khuntia School of Biotechnology, Kalinga Institute of Industrial Technology, Bhubaneswar, Odisha

Md. Khursheed Kazimi International Laboratory, Muscat, Oman

Rajesh Kumar Department of Biosciences, Division Zoology, Career Point University, Hamirpur, Himachal Pradesh, India

Moline Severino Lemos Department of Cell Biology, Institute of Biological Sciences, Federal University of Minas Gerais, Belo Horizonte, Brazil

Mengyun Liu College of natural resource and environment Northwest A&F University, Yangling shaanxi, China

Bisma Malik University Centre for Research and Development (UCRD), Chandigarh University, Mohali, Punjab, India

Sweeta Manhas Division of Biochemistry, FBSc, Sher-e-Kashmir University of Agricultural Sciences and Technology of Jammu, Jammu, India

Javid Manzoor Department of Environmental Science, JJT University, Rajasthan, India

Tawseef A. Mir Centre of Research for Ethnobotany, Government Model Science College, Jiwaji University, Gwalior, India

Ghulam Murtaza Institute of Soil and Environmental Sciences, University of Agriculture, Faisalabad, Pakistan

Umair Mustafa Department of Soil Science, Faculty of Agriculture and environmental sciences, The Islamia university of, Bahawalpur, Pakistan

Misbah Naz State Key Laboratory of Crop Genetics and Germplasm Enhancement, Nanjing Agricultural University, Nanjing, China

Núbia Alexandre de Melo Nunes Department of Biochemistry and Immunology, Institute of Biological Sciences, Federal University of Minas Gerais, Belo Horizonte, Brazil

Anjali Pathak Microbiology Research Lab., School of Studies in Botany, Jiwaji University, Gwalior, Madhya Pradesh

Tanveer Bilal Pirzadah University Centre for Research and Development (UCRD), Chandigarh University, Mohali, Punjab, India

Amauri Ponce-Hernández Faculty of Chemical Sciences, UASLP, San Luis Potosí, Mexico; Research and Postgraduate Center of the Faculty of Chemical Sciences, Universidad Autónoma de San Luis Potosí, S.L.P., Mexico

Paola Lucero Pérez Faculty of Professional Studies Zona Huasteca, UASLP, San Luis Potosí, Mexico

Ayesha Abdul Qadir Institute of Soil and Environmental Sciences, University of Agriculture, Faisalabad, Pakistan

Mir S. Rabani Microbiology Research Lab., School of Studies in Botany, Jiwaji University, Gwalior, Madhya Pradesh

Shilpa Raina Division of Biochemistry, FBSc, Sher-e-Kashmir University of Agricultural Sciences and Technology of Jammu, Jammu, India

Raksha Rani Department of Biosciences, Division Zoology, Career Point University, Hamirpur, Himachal Pradesh, India

Seema Rashid Department of Zoology, Amar Singh College, Cluster University, Srinagar, India

Summia Rehman Department of Botany, University of Kashmir, India

Sobia Riaz Institute of Soil and Environmental Sciences, University of Agriculture, Faisalabad, Pakistan

Umair Riaz Soil and water Testing Laboratory for Research, Agriculture Department, Government of Punjab, Bahawalpur, Pakistan

Arpita Roy Department of Biotechnology, School of Engineering & Technology, Sharda University, Greater Noida, India

Jane Alexander Ruley Department of Agricultural Production, Makerere University, Kampala, Uganda; Department of Agricultural Sciences, University of Juba, Juba, South-Sudan

Karuna Rupula Department of Biochemistry, University College of Science, Osmania University, Hyderabad, Telangana, India

Muhammad Fahad Sardar Institute of Environment and Sustainable Development in Agriculture, Chinese Academy of Agricultural Sciences, Beijing, China

Mohammad Sarraf Department of Horticulture Science, Shiraz Branch, Islamic Azad University, Shiraz, Iran

Riya Sharma Department of Biosciences, Division Zoology, Career Point University, Hamirpur, Himachal Pradesh, India

Shivani Sharma Department of Microbiology, Punjab Agricultural University, Ludhiana, Punjab, India

Sheikh Mansoor Division of Biochemistry, FBSc, Sher-e-Kashmir University of Agricultural Sciences and Technology of Jammu, Jammu, India

Harpreet Singh Sodhi Department of Microbiology, Punjab Agricultural University, Ludhiana, Punjab, India

Hiralal Sonawane Department of Botany, Ramkrishna More College of Arts Commerce and Science, Pune, India

Fernanda Maria Policarpo Tonelli Department of Cell Biology, Institute of Biological Sciences, Federal University of Minas Gerais, Belo Horizonte, Brazil

Flávia Cristina Policarpo Tonelli Department of Biochemistry, Federal University of São João del Rei, Divinópolis, Brazil

Shivani Tripathi Microbiology Research Lab., School of Studies in Botany, Jiwaji University, Gwalior, Madhya Pradesh

John Baptist Tumuhairwe Department of Agricultural Production, Makerere University, Kampala, Uganda

Anjani Kumar Upadhyay Heredity Lifesciences Pvt. Ltd., Bhubaneswar, Odisha

Bilal A. Wani Plant Reproductive Biology, Genetic Diversity and Phytochemistry Researh Lab., Department of Botany, University of Kashmir, Srinagar Jammu and Kashmir

Khursheed Ahmad Wani Department of Environmental Science, Govt Degree College, Bijbehara, India

Shabir H. Wani Mountain Research Center for Field Crops, Sher-e-Kashmir University of Agricultural Sciences and Technology of Kashmir, Khudwani, Jammu and Kashmir, India

About the Editors

Dr. Rouf Ahmad Bhat (PhD) received his doctorate at the Sher-e-Kashmir University of Agricultural Sciences and Technology Kashmir (Division of Environmental Science) and presently working in the Department of School Education, Government of Jammu and Kashmir. Dr. Bhat has been teaching graduate and postgraduate students of environmental sciences for the past three years. He is an author of more than 50 research articles and 40 book chapters, and has published more than 28 books with international publishers (Springer, Elsevier, CRC Press Taylor and Francis, Apple Academic Press, John Wiley, and IGI Global). He specializes in limnology, toxicology, phytochemistry, and phytoremediation. Dr. Bhat has presented and participated in numerous state, national, and international conferences, seminars, workshops, and symposium. Additionally, he has worked as an associate environmental expert at the World Bank funded Flood Recovery Project and on the environmental support staff in the Asian Development Bank (ADB) funded development projects. He has received many awards, appreciations, and recognition for his services to the science of water testing and air and noise analysis. He has served as an editorial board member and reviewer for reputed international journals. Dr. Bhat is still writing and experimenting with diverse capacities of plants for use in aquatic pollution remediation.

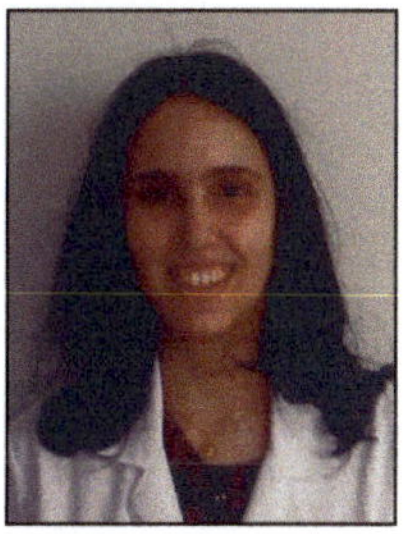

Dr. Fernanda Maria Policarpo Tonelli specializes in molecular biology and has been studying biotechnological topics such as gene delivery approaches (using engineered viral particles and nanomaterials) aiming towards transgenesis. She has taught topics related to biochemistry and molecular biology to graduate students. Dr. Tonelli has authored 12 scientific articles and more than 20 book chapters from international publishers and has reviewed various articles/book proposals. She has presented and participated in numerous national and international conferences and has also had the opportunity to contribute to the organization of various scientific events. Dr Tonelli has also dedicated herself to promote science and technology, cofunding an ONG with this purpose, and also joining a scientific divulgation group composed only of women. She has had her efforts as a researcher recognized by various awards (including *For Women in Science Brazil-L'Oreal/UNESCO/ABC* and *Under30 Brazil – Forbes*) and certificates of merit.

Dr. Gowhar Hamid Dar (PhD) is an assistant professor of environmental science at Sri Pratap College, Cluster University Srinagar, Department of Higher Education (Jammu and Kashmir). He received a PhD in environmental science with specialization in environmental microbiology (fish microbiology, fish pathology, industrial microbiology, taxonomy, and limnology). He has been teaching postgraduate and graduate students for the past many years at Postgraduate Department of Environmental Science, Sri Pratap College, Cluster University Srinagar. He has published more than 45 research articles in international and national journals of repute and several books with international publishers to his credit. Moreover, he supervises several students for the completion of degrees. Previously, he finished his PhD program entitled "Profiling and effect of pathogenic bacteria in fishes of Jammu and Kashmir State and correlation of disease incidence to the trophic status of water bodies" from the Department of the Environmental Science University of Kashmir, India and M. Phil in Environmental Science on "Isolation, Identification and characterization of some *Penicillium* isolates for alpha-amylase activity" from the same department. He has been working on the isolation, identification, and characterization of microbes, their pathogenic behavior, and also the impact of pollution on development of diseases in fish fauna for the past several years. He has received many awards and appreciations for his services towards science and development. Additionally, he also acts as a member of various research and academic committees.

Dr. Khalid Rehman Hakeem (PhD, FRSB) is a professor at King Abdulaziz University, Jeddah, Saudi Arabia. After completed his PhD (botany; specialization in plant ecophysiology and molecular biology) from Jamia Hamdard, New Delhi, India in 2011, he worked as an assistant professor at the University of Kashmir, Srinagar, India for a short period. Later he joined Universiti Putra Malaysia, Selangor, Malaysia and worked there as a postdoctorate fellow in 2012 and fellow researcher (associate Professor) from 2013-2016 respectively. He joined King Abdulaziz University in August 2016 and was recently promoted to professorship.

Dr. Hakeem has more than twelve years of teaching and research experience in plant ecophysiology, biotechnology, and Molecular biology, medicinal plant research, and plant-microbe-soil interactions as well as in environmental studies. He is the recipient of several fellowships at both national and international levels. He has recently been elected as a fellow of the Royal Society of Biology, London. Prof. Hakeem has served as a visiting scientist at Fatih Universiti, Istanbul, Turkey as well as at Jinan University, Guangzhou, China. Currently, he is involved with a number of international research projects with different Government. organizations.

So far, Dr. Hakeem has authored and edited more than 60 books with international publishers, including Springer Nature, Academic Press (Elsevier), CRC Press, etc. He has also to his credit 121 research publications in peer reviewed international journals and 65 book chapters in edited volumes with international publishers.

At present, Dr Hakeem is serving as editorial board member and reviewer of several high impact international scientific journals from publishers Elsevier, Springer Nature, Cambridge, Taylor and Francis, and John Wiley.

Foreword

The world's vicious evils include the three Ps: pollution, poverty, and population. Of these, the first is the most serious because of its causing an imminent threat to the life on Earth. Environmental pollution in the form of water, soil, and air contamination is a consequence of non eco-friendly anthropogenic actions occuring ever since mankind started satisfying its greed, ravaging the environment as a consequence. The damage caused to the global ecosystem has led to the present-day environmental crisis that needs to be immediately redressed and rectified. Various strategies have been devised for remediation of damaged environs. These include biological remediation; among which phytoremediation - the use of plants to promote sustained ecosystem restoration— is an in situ, low-cost, eco-friendly, widely accepted, and easily applicable large-scale remediation approach. A green technology, it involves absorption and uptake of inorganic and/or organic pollutants, including various heavy metals, and the sequestration of accumulated toxicants; thus, helping clean the polluted sites.

Although the term phytoremediation has been coined rather recently (Chaney, 1983), tolerance of plants to heavy metals was already known in the 16th century. Its science has, however, evolved in the recent decades due to advances in biotechnology and it is now an interdisciplinary technology having greatly benefited from microbiology, molecular biology, cell signaling, nanotechnology, and the '-omics' techniques such as genomics, transcriptomics, proteomics, and metabolomics. Nowadays, phytoremediation is even used in *phytomining*—extraction of high-economic-value metals such as Ni, Co, and Au accumulated by appropriate plants at particular polluted sites. As such, compilation of useful scientific information about the current topic is very relevant!

This book, edited by Dr. Rouf Ahmad Bhat, Dr. Fernanda Policarpo Tonelli, Dr. Gowhar Hamid Dar, and Dr. Khalid Rehman Hakeem, provides a broad account of phytoremediation as an economically feasible and ecofriendly strategy to clean up the polluted environment on a sustained basis. It presents an elaborate review of the evolution of phytoremediation vis-à-vis biotechnological development leading to various innovations that have greatly augmented pollutant remediation in different environs. Spread over 23 chapters under 2 parts, it is written by many authors from different countries. Part I is spread over 12 chapters: chapter 1 reviews the historical development of plant-based remediation; chapter 2 deals with potentially toxic elements and phytoremediation, highlighting the latter as a cost-effective and eco-friendly approach to dealing with nonbiodegradable pollutants; mechanisms of phytoremediation are reviewed in chapter 3, with emphasis on phytoaccumulation, phytodegradation, phytohydraulic control, phytostabilization, phytovolatilization, rhizofiltration, and rhizodegradation; chapter 4 is on phytoremediation at molecular level, making out that genetic modification and '-omics' studies have paved the way to enhance the uptake and tolerance capacities in plants, thus boosting phytoremediation efficiency; chapter 5 discusses microbial-assisted phytoremediation, highlighting that bacteria and fungi assist in stress alleviation, pollutant degradation, and in adequate plant growth through a 'mutually beneficial' interaction; nanophytoremediation for soil contamination is the subject matter of chapter 6, elaborating on the merger of nanotechnology and phytotechnology as an emerging approach to improve phytoremediation in tarnished sites; chapters 7 and 8 dwell on phytoremediation vis-à-vis biomass and chemical amendments; chapter 9 deals with the role of '-omics' innovations in the field, highlighting 'omics' data sets as tools for developing genetic methods useful in devising phytoremediation protocols; these methods are elaborated upon in chapter 10, highlighting the use of genetically modified plants (GMPs) for efficient phytoremediation; the insertion of transgene(s) into target-plant DNA (transgenesis) can be site-directed through techniques such as CRISPR-Cas, as revealed in chapter 11; chapter 12 spells out benefits of phytoremediation, as well as its limitations with emphasis on heavy-metal remediation.

Part II covers 11 chapters, devoted to aspects of phytoremediation in specific environs: soil & water (chapter 13); and specific contaminants that are of major concern worldwide— petroleum hydrocarbon soil contaminants (chapter 14), ecotoxicity of Nickel (chapter 15), pesticide contaminants (chapter 16), organic pollutant explosives (chapter 17), dye contaminants (chapter 18), pollutants from pharmaceutical wastes (chapter 19), and persistent organic pollutants (POPs, chapter 20). Chapter 21 reviews the role of biotechnological innovations in optimizing phytoremediation efficiency; heavy-metal pollution, with severe toxicity to humans, is tackled using phytoremediation protocols based on genetically engineered plants as dealt with in chapter 22. This book concludes with a review on the feasibility of phytoremediation from the economic standpoint (chapter 23).

Environmental pollution is a serious crisis mankind is currently faced with. With each passing day, it is increasing the world over, rendering nature increasingly imbalanced and life on Earth unsustainable. Its cumulative upshot has culminated in the form of global warming that, in turn, takes a heavy toll on biodiversity, accelerating its vulnerability to extinction. This book provides an elaborate review on fundamental and applied aspects of phytoremediation from a biotechnological viewpoint, and would be useful to realize the vital role of this promising science in combating the threat posed to humanity by environmental pollution.

Professor Ghulam Hassan Dar
Director and Dean, School of Biosciences and Biotechnology, Baba Ghulam Shah Badshah University, Rajouri, Jammu and Kashmir, India.

Preface

Industrialization and modernization promoted through human actions, without regard to sustainability, has brought serious consequences to ecosystems worldwide. Inorganic and organic environmental pollutants disrupt the balance that previously existed, causing abiotic resources to become toxic to living beings and able to cause their intoxication and even death. As environmental pollution is a major concern, diverse strategies have been developed aiming to mitigate or eliminate the dangers related to environmental pollution. Phytoremediation is an ecofriendly, efficient, low-cost, and well-accepted strategy based on plant species to promote environs restoration.

There are plant species naturally able to deal with environmental contaminants, but due to biotechnological advances species that naturally do not phytoremediate can be turned into organisms able to remove and/or degrade pollutants efficiently.

This book is an attempt to offer readers (from undergraduate students to researchers) extensive and comprehensive knowledge on phytoremediation from a biotechnological point-of-view: from the first strategies proposed to its recent characteristic and status as a very relevant approach to fight pollution harmful consequences. The 23 chapters address themes such as phytoremediation's history, phytoremediation of different types of environments and pollutants from diverse chemical nature, opportunities, challenges, benefits and limitations related to the technique, main mechanisms from a macroscopic level, microscopic level mechanisms that allow remediation to occur, assisted phytoremediation promoted by microbes, nanomaterials and organic and inorganic amendments, omics as a source of targets for phytoremediation's efficiency improvements, the role of omics on innovations on the field, genetically modified plants to optimized remediation by using non-targeted and targeted DNA engineering strategies, and economic feasibility of phytoremediation.

We are extremely grateful to all authors who have contributed chapters in this book and to Elsevier for their generous cooperation and for publishing this book.

Editors
Dr. Rouf Ahmad Bhat
Dr. Fernanda Maria Policarpo Tonelli
Dr. Gowhar Hamid Dar
Dr. Khalid Rehman Hakeem

About the Book

This publication contains 23 book chapters addressing linked themes that explore aspects of phytoremediation and aim to, together; provide readers with a biotechnological point-of-view of the field. From historical aspects to modern hot-topics such as omics technologies, the themes were discussed in a way to allow the understanding of basic knowledge to more complex subjects. Phytoremediation strategies to deal with different types of pollutants (inorganic, such as heavy metals, and organic, such as petroleum hydrocarbon, pesticides, explosives, dyes, and pharmaceutical wastes) in different types of environs (water or soil), as well as the untapped potential of using cutting-edge technology to find targets for genetic engineering through techniques optimized to avoid unwanted effects. This publication aims to offer a comprehensive overview of this environmentally sound and sustainable alternative based on plants to restore polluted environs, which is an important step towards sustainable development.

Chapter 1

The history of phytoremediation

Fernanda Maria Policarpo Tonelli[a], **Rouf Ahmad Bhat**[b], **Gowhar Hamid Dar**[b], **Khalid Rehman Hakeem**[c]

[a]*Department of Cell Biology, Institute of Biological Sciences, Federal University of Minas Gerais, Belo Horizonte, Brazil,* [b]*Department of Environmental Science, Sri Pratap College, Cluster University Srinagar, Srinagar, India,* [c]*Department of Biological Sciences, Faculty of Science, King Abdulaziz University, Jeddah, Saudi Arabia*

1.1 Introduction

Anthropogenic pollution is as old as humanity development of abilities that differ human being from animals. In Bronze Age, for example, archeologists have reported Balkan's existing lead (Pb) pollution (Longman et al., 2018). However, as activities are at an increasing rate performed not considering environmental preservation and restoration, the pollution issue is becoming more dangerous.

Organic and inorganic pollutants in soil, air, and water are concerning problems humanity faces as a consequence of centuries of failure to prioritize sustainability principles when developing activities such as industrial, farming, mining, waste disposal, and military (Peuke and Rennenberg, 2005).

Waste disposal in landfills, for example, is highly threatening once it pollutes soil, water, and air. Direct soil pollution, gas emissions, leachate production (that may pollute surface water and groundwater) are some negative consequences. This waste disposal method appears to be simple and possessing low cost at first sight but the damages provoked are enormous to ecosystems (Kim and Owens, 2011).

Among main environmental inorganic pollutants are toxic metals (which can be divided into heavy metals and radionuclides) (Salt et al., 1995), and among organic ones we can highlight xenobiotics such as aliphatic and polycyclic aromatic hydrocarbons, pharmaceuticals, pesticides, and dyes (Tonelli and Tonelli, 2020).

From organic environmental pollutants, the xenobiotics polychlorinated biphenyls (PCBs) in acute exposure are little toxic to humans; however, this persistent contaminants in chronic exposure are highly dangerous being able to induce cancer development and to disrupt the reproductive, immune, nervous, and endocrine systems (Pieper and Seeger, 2008).

In 21st century, for example, societies are threat by the risk of lack of drinking water due to the low quality of surface water previous to treatment and

Phytoremediation. DOI: https://doi.org/10.1016/B978-0-323-89874-4.00018-2

depletion of groundwater resources (Sharma et al., 2016; Kurwadkar, 2019). Human beings are experiencing diverse adverse health effects due to pollution exposure: from acute effects to DNA damage and life-threatening diseases such as dementia and cancer (Fatima et al, 2018; Singh et al., 2018; Ali et al., 2019).

Therefore, it is necessary that measures aiming remediation of environments and their restoration be taken as soon as possible. This chapter will be dedicated to reviewing the evolutionary history of one of these measures that is based on the use of plants to promote a process known as phytoremediation. This strategy was first developed to remediate heavy metal's pollution and is considered a noninvasive, eco-friend, cost-effective, and well-accepted technology (Peuke and Rennenberg, 2005). It has been evolving as interdisciplinary technology to remediate, in autosustained way, polluted environs (Fig. 1.1) (Padmavathiamma and Li, 2007; Khan, 2020; Palansooriya et al., 2020; Shah and Daverey, 2020).

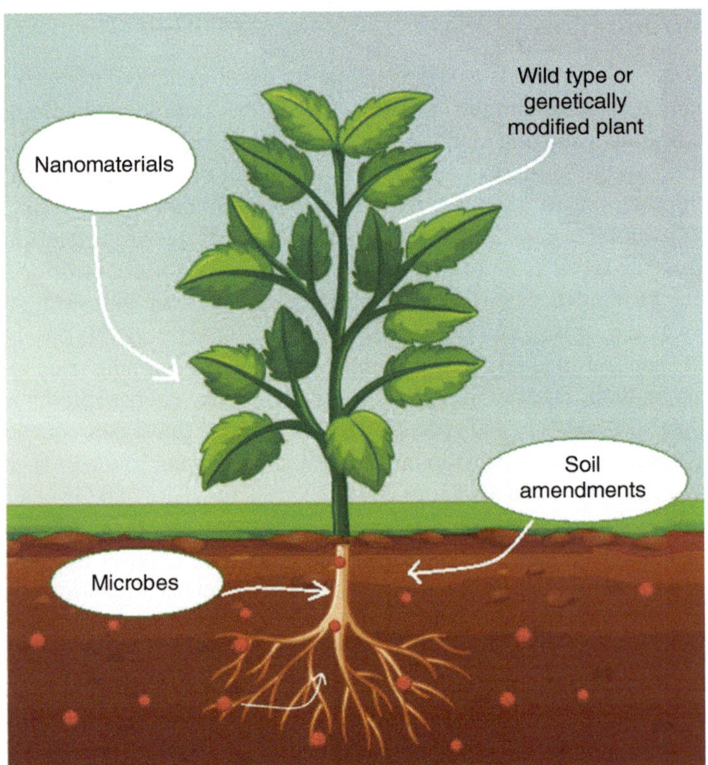

FIG. 1.1 Phytoremediation is an interdisciplinary technology that can benefit, for example, from molecular biology to generate plants with improved capacity to remediate polluted areas; from nanotechnology through nanomaterials capable of enhancing plants' tolerance to pollutants and remediation capacity; from soil amendments to adjust soil characteristics for improved phytoremediation; from microbiology for protocols of optimized microbe-assisted phytoremediation.

1.1.1 Plants and pollutants before the 1980s

Human interest in studying the interaction of plants with pollutants was already present in the 16th century when species tolerant to heavy metals started to be elucidated (Stephenson and Black, 2014). However, environmental pollution was not yet an issue of worldwide concern.

From 1750 to 1840 this scenario started to change for the UK. The First Industrial Revolution, through machinery development, changed not only the economy but also people's lifestyle: consumption habits, means of transportation's use, among other aspects (Mohajan, 2019). Large amounts of coal (mainly) and coke were burned in British industries resulting especially in air pollution, that negatively impacted not only contaminated cities attractiveness to people but also the employees' health impacting productivity: which concerned entrepreneurs (Hanlon, 2016).

From 1860 to 1914 the Second Industrial Revolution took place achieving mainly the United States and despite all the technological benefits it brought (electricity, development of chemical industry, internal combustion engine, etc), it also enhanced the number and types of pollution sources, turning environmental pollution into a more serious threat to humans. Fossil fuels enhanced greenhouse gas emissions, biodiversity reduction started to be observed, pesticides were progressively becoming a more common reality in agriculture (Mohajan, 2013; Mohajan, 2020).

In 1907 the capacity of *Agrobacterium tumefaciens* to induce crown gall tumors in plants (mainly dicotyledonous) was reported (Binns and Campbell, 2001) but its use to plant transgenesis was too far to become true.

In 1914 the World War I started, ending in 1918 severely impacting not only humans but all living forms mainly in France. Forests were converted into non-vegetal-area containing trenches and roads or monocultures, and soils ended up contaminated by heavy metals and explosives (Heiderscheidt, 2018).

In the 1930s, it was already known that some plants grown in the presence of persistent contaminants, such as selenium, could end up uptaking these substances at a high level (Beath et al., 1937). The Second World War was initiated in this decade and was responsible for more deaths, and pollution from nuclear attacks also become a serious problem to be handled by human beings, specially Japanese. The shocking effects of the war on living forms (Lawrence et al., 2015) had an impact to transform the mentality of citizens regarding the environment, and political modifications started to be required (Dukakis, 1996). However, environmental pollution continued to increase worldwide. For example, after 1945, with the Second World War ended, the necessity to increase agricultural production, in order to attend the necessities of the growing population, considerably enhanced pesticide's use: a concerning group of environmental contaminants (Gavrilescu, 2005).

In the 1940s and 1950s, different studies were developed to understand better how pollution may impact plants (Thomas, 1951). Scientists started to observe, in 1944, nocive damages to some plant's leaves induced by polluted air (smog)

(Haagen-Smit et al., 1952). In the 1950s, important progresses were made, such as comprehending that the damage induced by pollution on plants was a reality and it not always was apparent to human eyes: the absence of leaf injury was not a proof of lack of impairments; decrease in plant's growth rate without apparent impairment on plant's appearance may occur as a consequence of exposure to pollutants (Bleasdale, 1952). It also became highlighted by researchers the fact that different plant species react differently to pollution's exposure and that different contaminants are able to cause different symptoms (that sometimes allow their chemical nature's identification) on a plant's organism (Middleton, 1956).

In the 1960s, scientists were able to apply ion exchange and spectrophotometric methods to determine Fe, Cu, Mn, and Zn in vegetal samples (Bagliano et al., 1968) and to study the influence of pollutants on plants' metabolism. The concerning presence of fluoride as an air pollutant, for example, motivated analysis on its effect on vegetal species *Avena sativa L.* The contaminant proved to be able to interfere with cell wall structure (by reducing ^{14}C incorporation in cellulose fibers) and, as consequence, to inhibit growth in a more evident way when present in higher concentrations (Ordin and Skoe, 1963). Systems to facilitate to analyze plants' response to pollutants were being developed; vegetal's response to the contaminants nitrogen oxides, ozone, and hydrogen fluoride, as same as photosynthesis and respiration rates, transpiration index, temperature, and total level of oxidants could be recorded through data recording systems facilitating analysis (Ivie et al., 1963). In an attempt to understand how plants could interact with contaminants, removing them from polluted spots, researchers noticed that some plant species were able to transform the pollutant generating products that differed from one species to another. When it comes to selenium ions, for example, *Astragalus racemosus* could convert it into the volatile substance dimethyl diselenide (Evans et al., 1968) and alfalfa generated dimethyl selenide from Se ions (Lewis et al., 1966) through a process now referred to as phytovolatilization.

In the 1970s, plant species capable to remove pollutants from environs continued to be explored and scientists noticed that some plant species were able to deal well with more than one pollutant. *Aeolanthus biformifolius*, for example, proved to be able to remediate by accumulating high levels of both Cu and Co (Brooks et al., 1978).

Parameters influencing on plant's capacity to uptake pollutants also received attention. Corn, for example, could uptake Cd; however, soils with low pH and high organic matter content impaired the process (Street et al., 1977). In this decade, concerning of regulatory agencies regarding pollutants' threats to human life resulted in the avoidance or banning of some substances' use. In the late 70s PCBs, for example, were banned in a lot of countries due to their persistence in polluted areas and toxicity (Van Aken and Bhalla, 2011): PCBs can affect human's heath causing carcinogenic and or noncarcinogenic effects (Pawełczyk, 2013).

However, stop using or reduce the use of most environmental pollutants does not avoid the damage they may cause once they can be, as PCBs and heavy metals (Kim and Owens, 2011), persistent contaminants: lasting and accumulating for a long time in polluted environs representing threats and doing damage to food webs nearby for a long period of time (Gómez-Gutiérrez et al., 2007).

In the late 1970s, researchers continued to study plant's capacity to deal with polluted areas uptaking pollutants (for example, *Sebertia acuminate*'s ability to accumulate nickel (Jain et al., 1989)) and started to propose protocols to remediate contaminated environs based on plants, in order to restore ecosystems' balance. The strategy was not called phytoremediation yet, but plants were being used, for example, to remediate PCB and landfill leachate (Gómez-Gutiérrez et al., 2007; Jones et al., 2006). Heavy metals such as copper and cobalt also received attention; it was observed that *Haumaniastrum* species could efficiently uptake these inorganic pollutants (Brooks, 1977). Regarding heavy metals' pollution, it was observed that some plants could accumulate very high amounts of them dealing well with the stress induced by the pollutants; these plant species were called, for the first time, as hyperaccumulators (Brooks et al., 1977).

In order to improve protocol's efficiency, factors influencing pollutants' uptake (such as if the remediation needs to be performed in soil or water, local pH, the pollutant's form, the plant's species) were receiving increasing attention. For example, rice's capacity of uptaking the radionuclide ^{241}Am proved to be larger in water than in soil samples; it also proved that the environmental pollution caused by some contaminants, such as this, is a risk to human health once the contaminant can be captured by crops, entering the food chain with little or no difficulties (Adriano, 1979).

In the '70s, researchers also tried to advance the comprehension of important molecules involved in tolerant plant's response to pollutants' stress. *Silene vulgaris,* for example, was able to increase leaves' malic acid content as it may chelate the pollutant Zn in the pollutant's presence (Mathys, 1977).

An important step toward plant's transgenesis promoted by *A. tumafaciens* was also performed in this decade: the Ti plasmid present on virulent bacteria was discovered as being the tumor-inducing plasmid that after infection could persist in host cells as T-DNA – a fragment of it inserted into plants' DNA in a stable way to last for cell generations (Zaenen et al., 1974; Chilton et al., 1977).

1.2 The 1980s and the phytoremediation

In the 1980s, researchers continued to evaluate plant's response to pollutants' presence. *Deschampsia caespitosa's* zinc tolerant ecotypes proved to be able to increase citric acid content after exposure to the pollutant (Thurman and Rankin, 1982) making the vegetal able to accumulate Zn at an increasing rate as days advanced (Godbold et al., 1984).

The term phytoremediation was coined by Chaney (1983) being composed by the prefix "phyto" (Greek term for plant) and the suffix "remediation" that comes from the Latin word "remedium" which means to restore, to clean (Cunningham and Ow, 1996; Vasavi et al., 2010). It refers to plant-based strategies to remove pollutants from the environment or restrict their mobility, avoiding them to do damage, directly or indirectly, to living forms. Plant is an interesting organism to be used in remediation protocols once it consists in an auto sustained and eco-friendly system (that is able to use simple sources as solar energy and CO_2 to survive) capable of dealing well with a large variety of contaminants (Van Aken et al., 2010).

Phytoremediation involves different mechanisms to deal with pollutants; mechanisms that use plants as only living beings involved (for example, phyto-volatilization and phytoextraction), but it can also apply other living forms such as microbes on phytomicroremediation strategies (Farraji et al., 2020).

There are plant species that are naturally more suitable for phytoremediation protocols. In the 80s they received special attention from researchers and some of them even applied to patent certificates; Utsunamyia, for example, developed a method to use plants to extract the metals present in contaminated soil and applied to obtain the patent (Fulekar, 2010; Thakur et al., 2016).

In the evolutionary process these organisms prone to remediate were selected by being exposed to pollutants and allelochemicals, evolving to present detoxification mechanisms and an enhanced capacity to manage the stress caused by these substances (Singer et al., 2003). Metalliferous soils' plants, for example, were selected by being able to survive in places containing toxic metals without presenting symptoms of toxicity, that could have being caused by pollutants' presence. As time advanced species developed the ability to accumulate high amounts of contaminants and to act to inhibit their nocive effects on plant's organism (Reeves and Brooks, 1983; Baker and Brooks, 1989).

Different species, for example, were studied to phytoremediate heavy metals (Table 1.1), such as *Eichhornia crassipes* L. (Kay et al., 1984), *Hydrocotyle*

TABLE 1.1 Important plant species studied on the 80s to access their capacity to phytoremediate.

Plant species	Contaminant	Reference
Thlaspi caerulescens	Zn	Reeves and Brooks (1983)
Alyxia rubricaulis	Mn	Brooks et al. (1981)
Eichhornia crassipes L.	Pb	Kay et al (1984)
Hydrocotyle umbellata L.	Cu, Pb	Dierberg et al. (1987)
Lemna minor L.	Hg	Mo et al. (1989)
Lemna minor L.	Fe, Cu	Mo et al. (1989)
Alyssum euboeum	Ni	Morrison et al. (1980)
Azolla pinnata	Fe, Cu	Jain et al. (1989)
Thlaspi caerulescens	Cd	Baker and Walker (1989)
Ipomea alpine	Cu	Baker and Walker (1989)

umbellata L. (Dierberg et al., 1987), *Azolla pinnata* (Jain et al., 1989), *Lemna minor* L. (Mo et al., 1989), *Ipomea alpine* (Baker and Walker, 1989), *Thlaspi caerulescens* (Reeves and Brooks, 1983) and *Alyxia rubricaulis* (Brooks et al., 1981). Possible factors that could influence remediation capacity were explored: effects of pH, copper, and humic acid in *L. minor*'s Hg remediation and the influence of Fe presence during Cu remediation performed by *H. umbellata* were investigated (Jain et al., 1989).

Even when dealing with different pollutants that present similar chemical nature (for example, different inorganic contaminants), plants can react differently to their presence. *Eichhornia crassipes* L., for example, was studied to deal with Pb, Cd, and Cu pollution on wastewater. A study lasting 6 weeks revealed that Pb was well tolerated by the plant being unable to impair its growth. However, Cd and Cu were toxic inducing chlorosis, impairing growth, and suppressing development of new roots (Kay et al., 1984).

Sometimes, to contribute to reducing pollutant's toxicity to living beings, plants do not need to uptake or transform them; stabilization of these substances may be sufficient to avoid, for example, their uptake by living beings and/or the spread of pollution. Some research groups dedicated to explore influences on stabilization process and notice that humified organic matter and lime could raise soil's pH favoring metals' immobilization and restoration/revegetation of contaminated environs (Williamson and Johnson 1981; Kuo et al., 1985).

In this decade phytomining was also defined by Chaney (Chaney et al., 1981a, b) as an additional phytoremediation's positive possibility. It refers to prospecting among biomass of phytoremediation plants high economic value phytoextracted metals, such as Ni, Co, Au (Chaney et al., 2018). It started to be performed centuries ago and continues until the present day (Meagher, 2000). As harvestable biomass and shoot metals' concentration influence the process's success, species able to hyperaccumulate metals are preferable. Field experiments revealed that *Odontarrhena chalcidica* capacity to extract Ni from the soil is significantly higher than the one from *Noccaea goesingensis*, being the first species more interesting for Ni phytomining (Rosenkranz et al., 2019).

The first successful attempts of plant transgenesis using *A. tumefaciens*, highly common used methodology until today, were performed in the 1980s. Researchers tried to insert foreign genes to be expressed in Ti-plasmid but they failed to be expressed by plants. However, in 1983 it became possible: using the promoter of nopaline synthase gene, the coding sequence of octopine synthase gene and chloramphenicol acetyltransferase gene. The construction could be transferred to plant cells through *A. tumefaciens*, being expressed by target plants (Herrera-Estrella et al., 1983). However, as genetic modification of plants is performed not only through this methodology, other techniques were also being proposed and developed. Graminaceous monocots maize, that was refractory to gene delivery performed by *A. tumefaciens*, revealed to be prone to suffer electroporation and accept foreign DNA to be expressed; however, in a transient way (Fromm et al., 1986).

1.2.1 Phytoremediation from the 1990s to 2020

Developed some years before the beginning of the 1990s and applied in increasing number of plant genetic modification protocols until the beginning of the 21st century (when *A. tumefasciens* strategy became the favorite plant genetic engineering strategy), biolistics consists in high-velocity microprojectiles to deliver material (f.e. transgenes) to intact cells and tissues (Sanford, 1990). The *mt1* gene from mouse (that codes for metallothionein) in a chloroplast transformation vector was delivered through particle bombardment to tobacco plants enhancing their capacity to accumulate mercury, and converting them into suitable organisms for Hg phytoremediation (Ruiz et al., 2011).

In the 1990s, phytoextraction experiments of Zn and Cd were conducted for the first time by Baker et al. (1991). This type of experiment performed in the field is crucial to allow researchers to evaluate other important influences on the success of phytoremediation strategies. When performed in landfill areas field experiments revealed that the restoration of soil quality is necessary to be performed first in order to achieve an enhanced result. Phytoremediation maintains physical structure and biological activity of remediated soils (Baker et al., 1994). Soil amendments are commonly used to improve soil quality enhancing soil's aggregation, porosity, and moisture content ameliorating its physical properties (Kimand and Owens, 2011). Field studies continue to be performed nowadays. A long-term (18 months) phytoextraction's efficiency study using *Celosia argentea* to deal with Cd contamination in the presence and absence of soil amendments is an interesting example. In their absence the phytoremediation could occur; however, its efficiency was increased by ammonium chloride, *Bacillus megaterium* and citric acid, enhancing the quality of farmland soil contaminated by mine tailings (Yu et al., 2020).

Organic pollutants received increasing attention as years passed. Prairie grasses were assayed to access their capacity to degrade polycyclic aromatic hydrocarbons polluting soil samples (Aprill and Sims, 1990). Soil contaminated by pentachlorophenol could be 36% uptaken by wheatgrass (*Agropyron desertorum*) accelerating contaminant's removal (Ferro et al., 1994). Poplar trees proved to be able to uptake the pesticide atrazine and promote its degradation in rhizosphere without presenting visible negative effects due to remediation process (Burken and Schnoor, 1996). Poplars revealed to be able to uptake the pollutant trichloroethylene, transpire part of the pollutant, and degrade part of it in trichloroethanol, trichloroacetic acid, and dichloracetic acid being suitable to remediate this contaminant *in situ* (Newman et al., 1997). *Beta vulgaris* intact cells could rapidly metabolize xenobiotic nitroglycerin into glycerol dinitrate and then glycerol mononitrate showing potential to promote phytoremediation of sites polluted by this nitrate ester (Goel et al., 1997). The aquatic plant *Myriophyllum spicatum* as same as *Catharanthus roseus* could uptake and transform 2,4,6-trinitrotoluene (Hughes et al., 1997). In the 2000s, similar studies continued to be performed and are still being performed by different research groups; trichloroethylene present in groundwater, for example, proved also to be prone to suffer uptake in phytoremediation process promoted by poplars (Orchard et al., 2009) and hydrocarbon-contaminated soils could be phytoremediated by *Oryza longistaminata, Sorghum arundinaceum, Tithonia diversifolia,* and *Hyparrhenia rufa*; enhances in process's success could be achieved by cow dung that allowed plants to develop better (Ruley et al., 2020).

Studies on the joint use of plants and microbes to achieved improved and faster degradation of organic compounds continued to be developed as promising strategy to assisted phytoremediation through plant-microbe systems (Walton and Anderson, 1992).

Different transgenic plants developed to phytoremediate environmental pollutants were being developed since the 1990s. When it comes to metallic pollution, strategies such as expressing metal chelators molecules or metal transporters were being explored. *Pisum sativum*'s *PsMTA* gene (coding metallothionein-like protein) made transgenic *A. thaliana* accumulate more Cu than wild-type plants (Evans et al., 1992). *Arabidopsis thaliana* could express mercuric ion reductase (MerA) from bacteria after genetic modification, becoming able to transform the pollutant Hg^{2+} into a less toxic form (Hg^0) phytoremediating mercury pollution; the genetically modified plants (GMPs) became also resistant to Au^{3+} toxic levels (Rugh et al., 1996). In order to generate *Nicotiana tabacum* and *Carica papaya* able to tolerate being exposed to aluminum soluble toxic form, these organisms were engineered to overproduct citrate enhancing tolerance (De la Fuente et al., 1997). The *CUP1* yeast metallothionein gene, when used to genetically modify *Brassica oleracea*, offered a Cd-tolerant transgenic plant also able to phytoremediate accumulating large amounts of the pollutant (Hasegawa et al., 1997). *Brassica juncea* could be genetically modified to present increased accumulation and tolerance to Cd through overexpression of glutathione synthetase (Zhu et al., 1999a) or overexpression of glutamylcysteine synthetase (Zhu et al., 1999b). The DNA sequence coding calmodulin-binding protein CBP4 from *N. tabacum* in autotransgenic organisms revealed to be responsible for improved tolerance to Ni^{2+} and enhanced Pb^{2+} accumulation (Arazi et al., 1999). Monomethylmercuric chloride and phenylmercuric acetate pollution could be phytoremediated by *Arabidopsis thaliana* engineered to produce bacterial MerB protein (Bizily et al., 1999). A system containing bacterial organomercurial lyase genes *merA* and *merB* when expressed in *A. thaliana* allowed efficient methylmercury phytoremediation (Bizily et al., 2000). Autotransgenic *A. thaliana* overexpressing Nramp1 protein presented increased resistance to Fe at toxic concentration (Curie et al., 2000). *Nicotiana tabacum* genetically modified to express *A. thaliana*'s antiporter CAX2 uptook more Ca^{2+}, Cd^{2+}, and Mn^{2+} and tolerated better high Mn^{2+} levels (Hirschi et al., 2000). The coding sequence for *Triticum aestivum*'s phytochelatin synthase PCS1 was used to genetically modify *Nicotiana glauca* increasing plant's tolerance to Pb and Cd: to be explored in phytoremediation protocols (Gisbert et al., 2003). Transgenic tobacco plants expressing *Pseudomonas fluorescens copC* gene could tolerate Cu-induced stress accumulating high levels of this pollutant

(Pérez-Palacios et al., 2017). *Pseudomonas putida* genes *CzcB* or *CzcA* allowed transgenic tobacco plants to enhance root accumulation of cadmium (Nesler et al., 2017). *Pascopyrum smithii* could be engineered to remediate explosives: 2,4,6-trinitrotoluene (when expressing bacterial gene *nfsI*) and hexahydro-1,3,5-trinitro-1,3,5-triazine (when expressing bacterial genes *xplA* and *xplB*) (Zhang et al., 2018).

At the beginning of the 2000s phytoremediation's approach to urban soils started to be developed (Jusselme et al., 2020) and this kind of protocols continue to be developed nowadays (Capuana, 2020).

Arabdopsis thaliana was the first plant species to have its genome sequenced (Arabidopsis Genome Initiative, 2000; Boivin et al., 2004) increasing researchers' interest in sequencing the genome from other plant species. It also made possible the advances in understanding gene's functions to find suitable candidates for different genetic improvements, such as phytoremediation capacity.

Omics evolution has been facilitating the finding of candidate genes for genome engineering to phytoremediation. Transcriptomics, metabolomics, and proteomics, for example, allow the study of RNA, metabolites, and proteins from plants under stress induced by pollutants and in the absence of them, identifying alterations that could be explored to enhance remediation (Singh et al., 2016). Studying, for example, genes expressed in *Lolium multiflorum* under Cd presence and in the absence of this metallic pollutant led researchers to overexpress *LmAUX1* gene (upregulated in this species under Cd exposure) in *A. thaliana* confirming gene's role as enhancer of Cd accumulation (Hu et al., 2020). The informative potential of each omic platform is enormous and when these platform are combined in multi-omics protocols the potential can be enormously enhanced to provide scientists with a holist view on plants' response to stressful situations (Jia et al., 2016).

Since the late 1950s, nanotechnology field has developed materials in nanoscale, carbon-based, and non-carbon-based, with diverse applications (Drexler, 2004; Weiss et al., 2006; Zhang et al., 2008). The term "nanotechnology" was introduced and defined by Norio Taniguchi in 1974 to refer to these materials (Taniguchi, 1974). Nanotechnology field has continuously advanced to provide low-cost and efficient solutions in different areas including Environmental Science, offering remediation protocols for different pollutants (Lofrano et al., 2017). Among these protocols are the joint use of nanomaterials and plants in a nano-assisted phytoremediation strategy, enhancing the potential that vegetal species can offer. Salicylic acid nanoparticles, for example, can improve the stress tolerance of *Isatis cappadocica* and plant's biomass under As presence enhancing phytoremediation's potential (Souri et al., 2017). Issues related to the safety of nanomaterials are still being discussed such as toxicologic aspects, risk of more environmental pollution generation and food-chain entrance. Green synthesis protocols are already being proposed dedicating more attention to sustainability (Nunes et al., 2019).

Site-directed genome editing technologies, since the late 1990s, provided scientist with strategies to choose the integration site of transgenes avoiding some undesirable effects of random integration (Jaiswal et al., 2019). Zinc finger nucleases (Kim et al., 1996) and then transcription activator-like effector nuclease (Christian et al., 2010) attracted most attention until clustered regularly interspaced short palindromic repeats (CRISPR)/Cas system elucidation and development to transgenesis (Jinek et al., 2012). After its elucidation CRISPR became the preferable technique to targeted genome engineering, possessing an enormous potential for the future of phytoremediation field (Basharat et al., 2018). CRISPR/Cas system possess a guide RNA directing the nuclease to the desired integration site on target DNA; this guide RNA recognizes DNA based on simple Watson–Crick base pairing, which does not require: sophisticated protein engineering necessary for transcription activator-like effector nuclease protein repeats to recognize nucleotide sequences on DNA (Holkers et al., 2013) or Zinc finger nucleases difficult design (Sander et al., 2011). CRISPR/Cas system was already successfully used to study plant genes' function and improve desirable characteristics mainly through mutations (Xing et al., 2014; Ma et al., 2015; Wang et al., 2015; Xie et al., 2015; Didovyk et al., 2016; Khatodia et al., 2016; Richter et al., 2018). However, GMPs to phytoremediation is still a promising perspective that researchers wait to come true in a near future.

1.3 Conclusion

Phytoremediation field has been developing together with increasing human ecological consciousness to offer plant-based protocols (assisted or not by strategies from other fields of expertise) to remediate polluted spots, restoring environs. This technology's evolution suffered influence from different areas such as molecular biology, soil sciences, nanotechnology, microbiology, and bioinformatics through the years aiming to offer even more efficiency, at the lower possible cost, and with the maximum safety possible, solutions to be applied to help to reestablish ecosystems balance worldwide.

1.4 Future perspectives

Phytoremediation, in addition to all the benefits that have already been achieved through plant-based protocols to remediate environmental pollution, is still a field of science with enormous potential to be explored.

It is expected that in the near future the number of studies will increase, generating GMPs optimized for phytoremediation of different pollutants, thanks to the candidate genes that have been identified (and will still continue to be identified) through research involving omic technologies.

Further advances in the development of omic platforms and integrated bioinformatics tools are also expected to be achieved in systems biology studies, involving multiomic approaches.

Site-directed genetic modification tools are also expected to be used in different plant species for more efficient generation of organisms with enhanced phytoremediation's capacity.

Ethical and safety discussions regarding the use of GMPs in field trials and the number of field trials aimed at phytoremediation are also expected to increase in number in the future, as same as efficient and safe multidisciplinary remediation protocols (involving, for example, nanomaterials and microbes).

References

Adriano, D.C., 1979. Factors affecting the availability of americium-241 to the rice plant. J. Agric. Food Chem. 27, 1369–1375.

Ali, H., Khan, E., Ilahi, I., 2019. Environmental chemistry and ecotoxicology of hazardous heavy metals: environmental persistence, toxicity, and bioaccumulation. J.Chem. 2019, 1–14.

Aprill, W., Sims, R.C., 1990. Evaluation of the use of prairie grasses for stimulating polycyclic aromatic hydrocarbon treatment in soil. Chemosphere 20, 253–265.

Arabidopsis Genome Initiative, 2000. Analysis of the genome sequence of the flowering plant Arabidopsis thaliana. Nature 408, 796–815.

Arazi, T., Sunkar, R., Kaplan, B., Fromm, H., 1999. A tobacco plasma membrane calmodulin-binding transporter confers Ni^{2+} tolerance and Pb^{2+} hypersensitivity in transgenic plants. Plant J. 20, 171–182.

Bagliano, G., Rossi, L.C., Ferraris, M.M., 1968. Determination of oligoelements in biologic samples. 1. Application of ion exchange and spectrophotometric methods for the determination of Fe, Cu, Mn and Zn in vegetal samples. Minerva Fisiconucl. 12, 163–167.

Baker, A.J.M., Brooks, R.R, 1989. Terrestrial higher plants which hyper accumulate metallic elements—review of their distribution, ecology, and phytochemistry. Biorecovery 1, 81–126.

Baker, A.J.M., McGrath, S.P., Sidoli, C.M.D., Reeves, R.D, 1994. The possibility of in situ heavy metal decontamination of polluted soils using crops of metal-accumulating plants. Resour., Conserv. Recy. 11, 41–49.

Baker, A.J.M., Reeves, R.D., McGrath, S.P, 1991. In situ decontamination of heavy metal polluted soils using crops of metal-accumulating plants—a feasibility study. In: Hinchee, R.L., Olfenbuttel, R.F. (Eds.), In situ Bioreclamation. Butterworth-Heinemann, Boston, MA, pp. 600–605.

Baker, A.J.M., Walker, P.L, 1989. Ecophysiology of metal uptake by tolerant plants. In: Shaw, A.J. (Ed.), Heavy Metal Tolerance in Plants: Evolutionary Aspects. CRC, Boca Raton, FL, pp. 155–177.

Basharat, Z., Novo, L.A.B., Yasmin, A, 2018. Genome editing weds CRISPR: what is in it for phytoremediation? Plants (Basel) 7, 51.

Beath, O.A., Eppsom, H.F., Gilbert, G.S., 1937. Selenium distribution in and seasonal variation of vegetation type occurring on seleniferous soils. J. Amer. Pharm. Assoc. 26, 394–405.

Binns, A., Campbell, A., 2001. Agrobacterium tumefaciens-mediated transformation of plant cellsEncyclopedia of Life Sciences. Nature Publishing Group, London, pp. 1–6.

Bizily, S.P., Rugh, C.L., Summers, A.O., Meagher, R.B., 1999. Phytoremediation of methylmercury pollution: merB expression in Arabidopsis thaliana confers resistance to organomercurials. Proc. Nat. Acad. Sci. USA 96, 6808–6813.

Bizily, S.P., Rugh, C.L., Meagher, R.B., 2000. Phytodetoxification of hazardous organomercurials by genetically engineered plants. Nat. Biotechnol. 18, 213–217.

Bleasdale, J.K.A, 1952. Atmospheric pollution and plant growth. Nature 169, 376–377.

Boivin, K., Acarkan, A., Mbulu, R.S., Clarenz, O., Schmidt, R., 2004. The Arabidopsis genome sequence as a tool for genome analysis in brassicaceae. A comparison of the arabidopsis and *Capsella rubella* genomes. Plant Physiol. 135, 735–744.

Brooks, R.R., 1977. Copper and cobalt uptake by haumaniastrum species. Plant Soil 48, 541–545.

Brooks, R.R., Lee, J., Reeves, R.D., Jaffrré, T., 1977. Detection of nickeliferous rocks by analysis of herbarium specimens of indicator plants. J. Geochem. Explor. 7, 49–57.

Brooks, R.R., Morrison, R.S., Reeves, R.D., Malaisse, F., 1978. Copper and cobalt in African species of Aeollanthus Mart. (Plectranthinae, Labiatae). Plant Soil 50, 503–507.

Brooks, R.R., Thow, J.M., Veillon, J., Jaffre, T., 1981. Studies on manganese-accumulating Alyxia from New Caledonia. Taxon 30, 420–423.

Burken, J.G., Schnoor, J.L., 1996. Phytoremediation: plant uptake of atrazine and role of root exudates. J. Environ. Eng. 122, 958–963.

Capuana, M., 2020. A review of the performance of woody and herbaceous ornamental plants for phytoremediation in urban areas. iForest —Biogeosci. Forest. 13, 139–151.

Chaney, R.L., Baker, A.J.M., Morel, J.L., 2018. The long road to developing agromining/phytomining. In: Van der Ent, A., Echevarria, G., Baker, A.J.M., Morel, J.L. (Eds.), Agromining: Farming for Metals. Springer, Cham, Baviera, pp. 1–17.

Chaney, R.L., Hornick, S.B., Sikora, L.J., 1981. Review and preliminary studies of industrial land treatment practices, Proceedings of the 7th Annual Research Symposium on Land Disposal of Municipal Solid and Hazardous Waste and Resource Recovery. Philadelphia, PA NTIS Document No. PH81-173882.

Chaney, R. L., Kaufman, D. D., Hornick, S. B., Parr, J. F., Sikora, L. J., Burge, W. D., et al. (1981b). Review of information relevant to land treatment of hazardous wastes. Report to US-EPA Solid and Hazardous Waste Research Division, 476.

Chaney, R.L., 1983. Plant uptake of inorganic waste constituents. In: Parr, J.F.E.A. (Ed.), Land treatment of Hazardous Wastes. Noyes Data Corp, Park Ridge, IL, pp. 50–76.

Chilton, M.D., Drummond, M.H., Merlo, D.J., Sciaky, D., Montoya, A.L., Gordon, M.P., et al., 1977. Stable incorporation of plasmid DNA into higher plant cells: the molecular basis of crown gall tumorigenesis. Cell 11, 263–271.

Christian, M., Cermak, T., Doyle, E.L., Schmidt, C., Zhang, F., Hummel, A., et al., 2010. Targeting DNA double-strand breaks with TAL effector nucleases. Genetics 186, 757–761.

Cunningham, S.D., Ow, D.W., 1996. Promises and pros-pects of phytoremediation. Plant physiology 110, 715–719.

Curie, C., Alonso, J.M., Le Jean, M., Ecker, J.R., Briat, J.F., 2000. Involvement of Nramp1 from Arabidopsis thaliana in iron transport. Biochem. J. 347, 749–755.

De la Fuente, J.M., Ramírez-Rodríguez, B., Cabrera-Ponce, J.L., Herrera-Estrella, L., 1997. Aluminum tolerance in transgenic plants by alteration of citrate synthesis. Science 276, 1566–1568.

Didovyk, A., Borek, B., Tsimring, L., Hasty, J., 2016. Transcriptional regulation with CRISPR-Cas9: principles, advances, and applications. Curr. Opinion Biotechnol. 40, 177–184.

Dierberg, F.E., DeBusk, T.A., Goulet, N.A., 1987. Removal of copper and lead using a thin film technique. In: Reddy, K.B., Smith, W.H. (Eds.), Aquatic Plants for Water Treatment and Resource Recovery. Magnolia Publishing Inc, Miami Beach, FL, pp. 497–504.

Drexler, K.E., 2004. Nanotechnology: from feynman to funding. Belgian Sci., Technol. Soc. 24, 21–27.

Dukakis, M.S., 1996. Environmental politics in post-World War II America. Resour., Conserv. Recycl. 18, 5–9.

Evans, C.S., Asher, C., Johnson, C.M., 1968. Isolation of dimethyl diselenide and other volatile selenium compounds from Astragalus racemosus (Pursh.). Aust. J. Biol. Sci. 21, 13–20.

Evans, K.M., Gatehouse, J.A., Lindsay, W.P., Shi, J., Tommey, A.M., Robinson, N.J., 1992. Expression of the pea metallothionein-like gene PsMTA in *Escherichia coli* and *Arabidopsis thaliana* and analysis of trace metal ion accumulation: implications for gene PsMTA function. Plant Mol. Biol. 20, 1019–1028.

Farraji, H., Robinson, B., Mohajeri, P., Abedi, T., 2020. Phytoremediation: green technology for improving aquatic and terrestrial environments. Nippon J. Environ. Sci. 1, 1002.

Fatima, S.A., Hamid, A., Yaqub, G., Javed, A., Akram, H., 2018. Detection of volatile organic compounds in blood of farmers and their general health and safety profile. Nat. Environ. Pollut. Technol. 17, 657–660.

Ferro, A.M., Sims, R.C., Bugbee, B., 1994. Hycrest crested wheatgrass accelerates the degradation of pentachlorophenol in soil. J. Environ. Qual. 23, 272–279.

Fromm, M.E., Taylor, L.P., Walbot, V., 1986. Stable transformation of maize after gene transfer by electroporation. Nature 319, 791–793.

Fulekar, M.H., 2010. Environmental Biotechnology. Taylor & Francis Group and CRC Press, New York, NY.

Gavrilescu, M., 2005. Fate of pesticides in the environment. Eng. Life Sci. 5, 497–526.

Gisbert, C., Ros, R., De Haro, A., Walker, D.J., Bernal, M.P., Serrano, R., et al., 2003. A plant genetically modified that accumulates Pb is especially promising for phytoremediation. Biochem. Biophys. Res. Commun. 303, 440–445.

Godbold, D.L., Horst, W.J., Collins, J.C., Thurman, D.A., Marschner, H., 1984. Accumulation of zinc and organic acids in roots of zinc tolerant and non-tolerant ecotypes of *Deschampsia caespitosa*. J. Plant Physiol. 116, 59–69.

Goel, A., Kumar, G., Payne, G.F., Dube, S.K., 1997. Plant cell biodegradation of a xenobiotic nitrate ester, nitroglycerin. Nat. Biotechnol. 15, 174–177.

Gómez-Gutiérrez, A., Garnacho, E., Bayona, J.M., Albaigés, J., 2007. Screening ecological risk assessment of persistent organic pollutants in Mediterranean sea sediments. Environ. Int. 33, 867–876.

Haagen-Smit, A.J., Darley, E.F., Zaitlin, M., Hull, H., Noble, W., 1952. Investigation on injury to plants from air pollution in the Los Angeles area. Plant Physiol. 27, 18–34.

Hanlon, W.W., 2016. Coal smoke and the costs of the industrial revolution. Nat. Bur. Econ. Res., 22921.

Hasegawa, I., Terada, E., Sunairi, M., Wakita, H., Shinmachi, F., Noguchi, A., et al., 1997. Genetic improvement of heavy metal tolerance in plants by transfer of the yeast metallothionein gene (CUP1). Plant Soil 196, 277–281.

Heiderscheidt, D., 2018. The Impact of World War one on the Forests and Soils of Europe. Ursidae: Undergrad. Res. J. Univ. Northern Colorado 7, 1–16.

Herrera-Estrella, L., Depicker, A., Van Montagu, M., Schell, J., 1983. Expression of chimaeric genes transferred into plant cells using a Ti- plasmidderived vector. Nature 303, 209–213.

Hirschi, K.D., Korenkov, V.D., Wilganowski, N.L., Wagner, G.J., 2000. Expression of Arabidopsis CAX2 in tobacco. Altered metal accumulation and increased manganese tolerance. Plant Physiol. 124, 125–133.

Holkers, M., Maggio, I., Liu, J., Janssen, J.M., Miselli, F., Mussolino, C., Recchia, A., et al., 2013. Differential integrity of TALE nuclease genes following adenoviral and lentiviral vector gene transfer into human cells. Nuc. Acids Res., 41 (5), e63.

Hu, Z., Zhang, Y., He, Y., Cao, Q., Zhang, T., Lou, L., et al., 2020. Full-length transcriptome assembly of Italian ryegrass root integrated with RNA-Seq to identify genes in response to plant cadmium stress. Int. J. Mol. Sci. 21, 1067.

Hughes, J.B., Shanks, J., Vanderford, M., Lauritzen, J., Bhadra, R., 1997. Transformation of TNT by aquatic plants and plant tissue cultures. Environ. Sci. Technol. 31, 266–271.

Ivie, J.O., Thomas, M.D., Taylor, O.C., Thompson, C.R., Dugger Jr, W.M., Richards, B.L, 1963. Recording the response of plants to various air pollutants. J. Air Pollut. Control Assoc. 13, 355–359.

Jain, S.K., Vasudevan, P., Jha, N.K., 1989. Removal of some heavy metals from polluted water by aquatic plants: Studies on duckweed and water velvet. Biol. Wastes 28 (2), 115–126.

Jaiswal, S., Singh, D.K., Shukla, P., 2019. Gene editing and systems biology tools for pesticide bioremediation: a review. Front. Microbiol. 10, 87.

Jia, X., Sun, C., Zuo, Y., Li, G., Li, G., Ren, L., et al., 2016. Integrating transcriptomics and metabolomics to characterise the response of Astragalus membranaceus Bge. var. mongolicus (Bge.) to progressive drought stress. BMC Genom. 17, 188.

Jinek, M., Chylinski, K., Fonfara, I., Hauer, M., Doudna, J.A., Charpentier, E., 2012. A programmable dual-RNA-guided DNA endonuclease in adaptative bacterial immunity. Science 337, 816–821.

Jones, D.L., Williamson, K.L., Owen, A.G., 2006. Phytoremediation of landfill leachate. Waste Manage. 26, 825–837.

Jusselme, M.D., Bousserrhine, N., Abbad-Andaloussi, S., Brondeau, F., Balland-Bolou-Bi, C., 2020. Phytoremediation: An Ecological Solution for Decontamination of Polluted Urban Soils [Online First]. IntechOpen, London, UK, p. 93621.

Kay, S.H., Haller, W.T., Garrard, L.A., 1984. Effect of heavy metals on water hyacinths [Eichhornia crassipes (Mart.) Solms]. Aquatic Toxicol. 5, 117–128.

Khan, A.G., 2020. Promises and potential of in situ nano-phytoremediation strategy to mycorrhizo-remediate heavy metal contaminated soils using non-food bioenergy crops (Vetiver zizinoides & Cannabis sativa). Int. J. Phytoremed. 9, IPS2019.

Khatodia, S., Bhatotia, K., Passricha, N., Khurana, S.M.P., Tuteja, N, 2016. The CRISPR/Cas genome-editing tool: application in improvement of crops. Front. Plant Sci. 7, 506.

Kim, K.R., Owens, G., 2010. Potential for enhanced phytoremediation of landfills using biosolids—a review. J. Environ. Manage. 91, 791–797.

Kim, Y.G., Cha, J., Chandrasegaran, S., 1996. Hybrid restriction enzymes: zinc finger fusions to Fok I cleavage domain. Proc. Nat. Acad. Sci. USA 93, 1156–1160.

Kuo, S., Jellum, E.J., Baker, A.S., 1985. Effects of soil type, liming, and sludge application on zinc and cadmium availability to Swiss chard. Soil Sci. 139, 122–130.

Kurwadkar, S., 2019. Occurrence and distribution of organic and inorganic pollutants in groundwater. Water Environ. Res. 91, 1001–1008.

Lawrence, M.J., Stemberger, H.L.J., Zolderdo, A.J., Struthers, D.P., Cooke, S.J, 2015. The effects of modern war and military activities on biodiversity and the environment. Environ. Rev. 23, 0039.

Lewis, B.G., Johnson, C.M., Delwiche, C.C., 1966. Release of volatile selenium compounds by plants: collection procedures and preliminary observations. J. Agric. Food Chem. 14, 638–640.

Lofrano, G., Libralato, G., Brown, J., 2017. Nanotechnologies for Environmental Remediation Applications and Implications. Springer, Cham, p. 325.

Longman, J., Veres, D., Finsinger, W., Ersek, V., 2018. Exceptionally high levels of lead pollution in the Balkans from the Early Bronze Age to the industrial revolution. Proc. Nat. Acad. Sci. USA 115, E5661–E5668.

Ma, X., Zhang, Q., Zhu, Q., Liu, W., Chen, Y., Qiu, R., et al., 2015. A robust CRISPR/Cas9 system for convenient, high-efficiency multiplex genome editing in monocot and dicot plants. Mol. Plant 8, 1274e1284.

Mathys, W., 1977. The role of malate, oxalate, and mustard oil glucosides in the evolution of zinc-resistance in herbage plants. Physiol. Plant. 40, 130–136.

Meagher, R.B., 2000. Phytoremediation of toxic elemental and organic pollutants. Curr. Opinion Plant Biol. 3, 153–162.

Middleton, J.T., 1956. Response of plants to air pollution. J. Air Pollut. Control Assoc. 6, 7–50.

Mo, S.C., Choi, D.S., Robinson, J.W., 1989. Uptake of mercury from aqueous solution by duckweed: the effect of pH, copper, and humic acid. J. Environ. Health 24, 135–146.

Mohajan, H.K., 2013. Global Greenhouse Gas Emissions and Climate Change. Lambert Academic Publishing, Berlin.

Mohajan, H.K., 2019. The first industrial revolution: creation of a new global human era. J. Soc. Sci. Human. 5, 377–387.

Mohajan, H.K., 2020. The second industrial revolution has brought modern social and economic developments. J. Soc. Sci. Human. 6, 1–14.

Morrison, R.S., Brooks, R.R., Reeves, R.D., 1980. Nickel uptake by Alyssum sp. Plant Sci. Lett. 17, 451–457.

Nesler, A., DalCorso, G., Fasani, E., Manara, A., Di Sansebastiano, G.P., Argese, E., et al., 2017. Functional components of the bacterial CzcCBA efflux system reduce cadmium uptake and accumulation in transgenic tobacco plants. New Biotechnol. 35, 54–61.

Newman, L.A., Strand, S.E., Choe, N., Duffy, J., Ekuan, G., et al., 1997. Uptake and biotransformation of trichloroethylene by hybrid poplars. Environ. Sci. Technol. 31, 1062–1067.

Nunes, D., Pimentel, A., Santos, L., Barquinha, P., Pereira, L., Fortunato, E., et al., 2019. Introduction. In: Nunes, D., Pimentel, A., Santos, L., Barquinha, P., Pereira, L., Fortunato, E. et al (Eds.), Metal Oxide Nanostructures. Elsevier, New York, NY, p. 328.

Orchard, B.J., Doucette, W.J., Chard, J.K., Bugbee, B., 2009. Uptake of trichloroethylene by hybrid poplar trees grown hydroponically in flow-through plant growth chambers. Environ. Toxicol. Chem. 19, 895–903.

Ordin, L., Skoe, B.P., 1963. Inhibition of metabolism in avena coleoptile tissue by fluoride. Plant Physiol. 38, 416–421.

Padmavathiamma, P.K., Li, L.Y., 2007. Phytoremediation technology: hyper-accumulation metals in plants. Water, Air, Soil Pollut. 184, 105–126.

Palansooriya, K.N., Shaheen, S.M., Chen, S.S., Tsang, D.C.W., Hashimoto, Y., Hou, D., et al., 2020. Environment international review article soil amendments for immobilization of potentially toxic elements in contaminated soils: a critical review. Environ. Int. 134, 105046.

Pawełczyk, A., 2013. Assessment of health risk associated with persistent organic pollutants in water. Environ. Monit. Assess. 185, 497–508.

Pérez-Palacios, P., Agostini, E., Ibáñez, S.G., Talano, M.A., Rodríguez-Llorente, I.D., Caviedes, M.A., et al., 2017. Removal of copper from aqueous solutions by rhizofiltration using genetically modified hairy roots expressing a bacterial Cu-binding protein. Environ. Technol. 28, 2877–2888.

Peuke, A.D., Rennenberg, H., 2005. Phytoremediation. EMBO Rep. 6, 497–501.

Pieper, D.H., Seeger, M., 2008. Bacterial metabolism of polychlorinated biphenyls. J. Mol. Microbiol. Biotechnol. 15, 121–138.

Reeves, R.D., Brooks, R.R., 1983. Hyperaccumulation of lead and zinc by two metallophytes from a mining area of Central Europe. Environ. Pollut. Ser. A 31, 277–287.

Richter, J., Watson, J.M., Stasnik, P., Borowska, M., Neuhold, J., Berger, M., et al., 2018. Multiplex mutagenesis of four clustered CrRLK1L with CRISPR/Cas9 exposes their growth regulatory roles in response to metal ions. Sci. Rep. 8, 12182.

Rosenkranz, T., Hipfinger, C., Ridard, C., Puschenreiter, M., 2019. A nickel phytomining field trial using Odontarrhena chalcidica and Noccaea goesingensis on an Austrian serpentine soil. J. Environ. Manage. 242, 522–528.

Rugh, C.L., Wilde, H.D., Stack, N.M., Thompson, D.M., Summers, A.O., Meagher, R.B., 1996. Mercuric ion reduction and resistance in transgenic Arabidopsis thaliana plants expressing a modified bacterial merA gene. Proc. Nat. Acad. Sci. USA 93, 3182–3187.

Ruiz, O.N., Alvarez, D., Torres, C., Roman, L., Daniell, H., 2011. Metallothionein expression in chloroplasts enhances mercury accumulation and phytoremediation capability. Plant Biotechnol. J. 9, 609–617.

Ruley, J.A., Amoding, A., Tumuhairwe, J.B., Basamba, T.A., Opolot, E., Oryem-Origa, H., 2020. Enhancing the phytoremediation of hydrocarbon-contaminated soils in the Sudd Wetlands, South Sudan, using organic manure. Appl. Environ. Soil Sci., 2020.

Salt, D.E., Blaylock, M., Kumar, N.P.B.A., Dushenkov, V., Ensley, B.D., Chet, I., et al., 1995. Phytoremediation: a novel strategy for the removal of toxic metals from the environment using plants. Nat. Biotechnol. 13, 468–474.

Sander, J.D., Dahlborg, E.J., Goodwin, M.J., Cade, L., Zhang, F., Cifuentes, D., Curtin, S.J., Blackburn, J.S., Thibodeau-Beganny, S., Qi, Y., Pierick, C.J., Hoffman, E., Maeder, M.L., Khayter, C., Reyon, D., Dobbs, D., Langenau, D.M., Stupar, R.M., Giraldez, A.J., Voytas, D.F., Peterson, R.T., Yeh, J.R.J., Joung, J.K, 2011. Selection-free zinc-finger-nuclease engineering by context-dependent assembly (CoDA). Nat. Methods 8, 67–69.

Sanford, J.C., 1990. Biolistic plant transformation. Physiol. Plant. 79, 206–209.

Shah, V., Daverey, A., 2020. Environmental technology & innovation phytoremediation: a multidisciplinary approach to clean up heavy metal contaminated soil. Environ. Technol. Inno. 18, 100774.

Sharma, A.K., Tiwari, R.K., Gaur, M.S., 2016. Nanophotocatalytic UV degradation system for organophosphorus pesticides in water samples and analysis by Kubista model. Arab. J. Chem. 9, 1755–1764.

Singer, A.C., Crowley, D.E., Thompson, I.P., 2003. Secondary plant metabolites in phytoremediation and biotransformation. Trends Biotechnol. 21, 123–130.

Singh, N.S., Sharma, R., Parween, T., Patanjali, P.K., 2018. Pesticide contamination and human health risk factor. In: Mohammad, O., Mohammad, Z.K., Iqbal, M.I.I. (Eds.), Modern Age Environmental Problems and their Remediation. Springer, New York, NY, pp. 49–68.

Singh, S., Parihar, P., Singh, R., Singh, V.P., Prasad, S.M., 2016. Heavy metal tolerance in plants: role of transcriptomics, proteomics, metabolomics and ionomics. Frontiers in Plant Science 6, 1143.

Souri, Z., Karimi, N., Sarmadi, M., Rostami, E., 2017. Salicylic acid nanoparticles (SANPs) improve growth and phytoremediation efficiency of Isatis cappadocica Desv., under As stress. IET Nanobiotechnol. 1, 650–655.

Stephenson, C., Black, C.R., 2014. One step forward, two steps back: the evolution of phytoremediation into commercial technologies. Biosci. Horiz. 7, 1–15.

Street, J.J., Lindsay, W.L., Sabey, B.R., 1977. Solubility and plant uptake by cadmium in soils amended with cadmium and sewage sludge. J. Environ. Qual. 6, 72–77.

Taniguchi, N., 1974. On the basic concept of 'nano-technology', Proceedings of the International Conference on Production Engineering. Tokyo 1974, Part II.

Thakur, S., Singh, L., Wahid, Z.A., Siddiqui, M.F., Atnaw, M.S., Din, M.F.M., 2016. Plant-driven removal of heavy metals from soil: uptake, translocation, tolerance mechanism, challenges, and future perspectives. Environmental Monitoring and Assessment 188 (4), 206.

Thomas, M.D., 1951. Gas damage to plants. Ann. Rev. Plant Physiol. 2, 293–322.

Thurman, D.A., Rankin, J.L., 1982. The role of organic acids in zinc tolerance in Deschampsia caespitosa. New Phytol. 91, 629–635.

Tonelli, F.C.P., Tonelli, F.M.P, 2020. Concerns and threats of xenobiotics on aquatic ecosystems. In: Bhat, R.A., Hakeem, K.R., Al-Saud, N.B.S. (Eds.), Bioremediation and Biotechnology Vol 3: Persistent and Recalcitrant Toxic Substances. Springer, New York, NY, p. 360.

Van Aken, B., Correa, P.A., Schnoor, J.L., 2010. Phytoremediation of polychlorinated biphenyls: new trends and promises. Environ. Sci. Technol. 44, 2767–2776.

Van Aken, B., Bhalla, R., 2011. Microbial degradation of polychlorinated biphenyls. In: Moo-Young, M. (Ed.), Comprehensive BiotechnologySecond Edition. Academic Press, New York, NY: vol. 6, pp. 151–166 .

Vasavi, A., Usha, R., Swamy, P.M., 2010. Phytoremediation - an overview review. J. Ind. Pollut. Control 26, 83–88.

Wang, C., Shen, L., Fu, Y., Yan, C., Wang, K., 2015. A simple CRISPR/Cas9 system for multiplex genome editing in rice. J. Genet. Genomics 42, 703.

Walton, B.T., Anderson, T.A., 1992. Plantmicrobe treatment systems for toxic waste. Curr. Opinion Biotechnol. 3, 267–270.

Weiss, J., Takhistov, P., McClements, D.J., 2006. Functional materials in food nanotechnology. J. Food Sci. 71, R107–R116.

Williamson, A., Johnson, M.S., 1981. Reclamation of metalliferous mine wastes. In: Lepp, N.W. (Ed.), Effect of Heavy Metal Pollution on Plants. Applied Science Publishers, Essex: vol. 2, pp. 185–212 .

Xie, K., Minkenberg, B., Yang, Y., 2015. Boosting CRISPR/Cas9 multiplex editing capability with the endogenous tRNA-processing system. Proc. Nat. Acad. Sci. USA 112, 3570e3575.

Xing, H.L., Dong, L., Wang, Z.P., Zhang, H.Y., Han, C.Y., Liu, B., et al., 2014. A CRISPR/Cas9 toolkit for multiplex genome editing in plants. BMC Plant Biol. 14, 327.

Yu, G., Jiang, P., Fu, X., Liu, J., Sunahara, G.I., Chen, Z., et al., 2020. Environmental pollution phytoextraction of cadmium-contaminated soil by Celosia argentea Linn.: a long-term field study. Environ.Pollut. 266, 115408.

Zaenen, I., Van Larebeke, N., Teuchy, H., Van Montagu, M., Schell, J., 1974. Supercoiled circular DNA in crown-gall inducing Agrobacterium strains. J. Mol. Biol. 86, 109–127.

Zhang, L., Rylott, E.L., Bruce, N.C., Strand, S.E., 2018. Genetic modification of western wheatgrass (*Pascopyrum smithii*) for the phytoremediation of RDX and TNT. Planta 249, 1007–1015.

Zhang, Y., Yang, M., Portney, N.G., Cui, D., Budak, G., Ozbay, E., et al., 2008. Zeta potential: a surface electrical characteristic to probe the interaction of nanoparticles with normal and cancer human breast epithelial cells. Biomed. Microdev. 10, 321–328.

Zhu, Y., Pilon-Smits, E.A.H., Jouanin, L., Terry, N, 1999a. Overexpression of glutathione synthetase in Brassica juncea enhances cadmium tolerance and accumulation. Plant Physiol. 119, 73–79.

Zhu, Y., Pilon-Smits, E.A.H., Tarun, A., Weber, S.U., Jouanin, L., Terry, N, 1999b. Cadmium tolerance and accumulation in Indian mustard is enhanced by overexpressing -glutamylcysteine synthetase. Plant Physiol. 121, 1169–1177.

Chapter 2

Potentially toxic elements and phytoremediation: Opportunities and challenges

Zia Ur Rahman Farooqi[a], Muhammad Mahroz Hussain[a], Muhammad Ashar Ayub[a], Ayesha Abdul Qadir[a], Predrag Ilic[b]

[a]*Institute of Soil and Environmental Sciences, University of Agriculture, Faisalabad, Pakistan,* [b]*PSRI Institute for Protection and Ecology of the Republic of Srpska, Banja Luka, Bosnia and Herzegovina*

2.1 Introduction

Heavy metal or possibly toxic element(s) (PTEs) is the common name assigned to an element group with a density greater than 6 g/cm^3 (Rai et al., 2019). Although these components are ubiquitous on the surface of the Earth, their concentration and distribution range from less than 1000 ppm to a few ppb in soil and water, excluding manganese (Mn), which varies in the range 20–10,000 ppm (Sarwar et al., 2017). About 2–250 mg/kg copper (Cu) is present in the soil but due to anthropogenic activities its range reportedly increases up to 50–900 mg/kg. At extremely high concentrations, PTE-contaminated soils may have certain elements. Likewise, the range of mercury (Hg) level in the soil varies from 0.01 to 0.5 ppm but can be detected at high levels ranging from 100 to 500 ppm, which is 1000 to 10,000 times greater than plant critical level (Ashraf et al., 2019).

The normal soil deposition of PTEs is harmless to living organisms, with the exception of soil contaminated with PTEs. However, big industries like oil mining and processing, along with agricultural operations, have tremendously enhanced the level of certain elements in the soil that exceed the toxicity critical limit. We can assume from the minimum accessible information that such an improvement in the concentration of PTEs can represent an increase in supply that can be transferred to the living cell membrane. The movement of PTEs in the soil is affected by many factors, including soil properties, metal species, water solubility, and soil frequency of humic substances. Therefore the chemicals present in the soil and groundwater will result in water contamination, which in a given ecosystem will eventually impact all living organisms (Cristaldi et al., 2017). Due to the PTEs pollution, different health risks to humans, plants, and soils are evident from the previous literature and summarized in Fig. 2.1.

Phytoremediation. DOI: https://doi.org/10.1016/B978-0-323-89874-4.00020-0

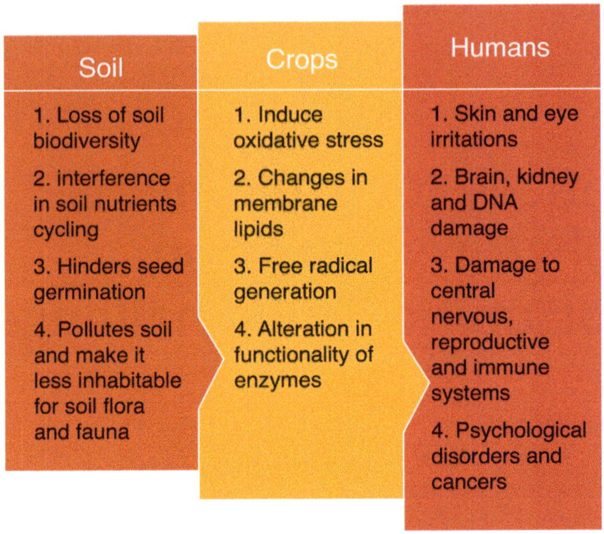

FIG. 2.1 The risks of potentially toxic elements on soil, crops, and human health.

While modern technologies such as chemical treatment or mechanical separation, isolation, and containment, are useful for infected field cleaning, they are generally costly, labor-intensive, and distort the soil's properties (Fig. 2.2). Nowadays, the plants aided metal extraction (phytoremediation) now emerged as a substitute to removing surplus heavy metals from soil and water. On the basis of the contaminant fate or the implicated pathways, phytoremediation can

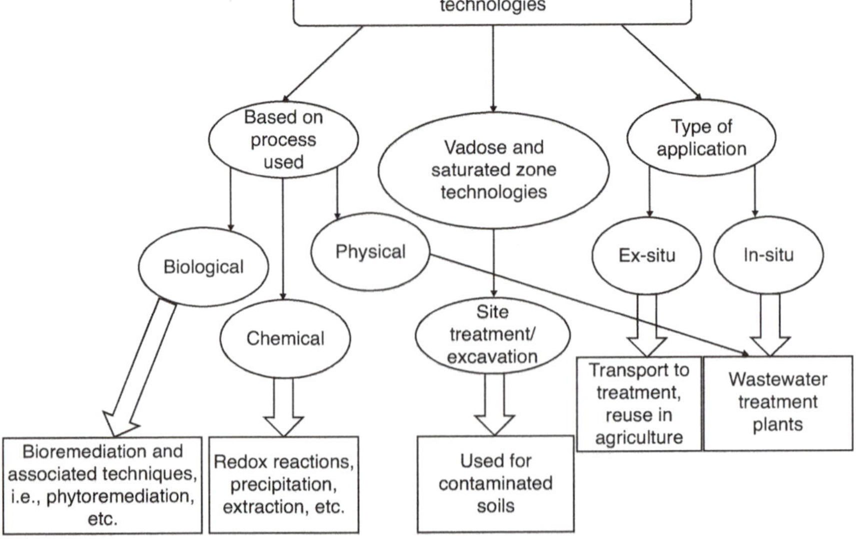

FIG. 2.2 All possible remediation technologies for potentially toxic elements.

be graded. As far as fate is concerned, the definition involves oxidation, extraction, containment, or a combination of these; and as far as the processes are concerned, phytoremediation can be described as plant tissue accumulation, extraction, pollutant degradation, root-level immobilization, volatilization, and, finally, control of erosion and infiltration. While the process is carried out in water-based cultures, extraction is labeled rhizofiltration. Separation, isolation and immobilization, and volatilization are the techniques used in the decontamination of heavy metals and other toxic materials. Hg and selenium are volatilized by certain plant species (Se) (Peng et al., 2018; Masindi and Muedi, 2018).

2.2 Phytoremediation of potentially toxic elements-polluted water

The usage of plants (phytoremediation) in remediation field (decontamination of organic and inorganic) from polluted land and water is an innovative step. This technique focuses on plants grown hydroponically that have been shown to succeed in the extraction of PTEs from solution. Plants like Eichhornia crassipes, and >8000 mg Cu/L when produced with 5 mg/L of these PTEs, have shown the great ability to accumulate >6000 mg/L of Cd and Pb in the entire plant. The ability to consume Cd, Cu, Ni, and Zn, i.e., the water fern (*Azolla filiculoides*) has been shown by another herb. *A. Filiculoids* subjected to PTEs (Cd, Cu, Ni, and Zn) were eliminated by approximately 10,000, 9000, 9000, and 6500 mg/L of Cd, Cu, Ni, and Zn, respectively. This ensures that it is possible to successfully carry out this water cleaning operation. Certain species of aquatic plants, including Nelumbo nucifera can drained Cr quickly from wastewater. Under laboratory conditions, this plant accumulate up to 3000 mg/kg Cr. Contrarily, many plants are sensitive to this metal toxicity. While the heavy metal buildup exhibited by these organisms (mg of metal kg^{-1} of tissue) is strong, productivity is poor in absolute terms, as biomass and slow-growing roots have low growth rates. Scientists have attempted to practice quick growing and higher biomass-producing plant species to cope with this dissatisfaction (Sela et al., 1989). Indian mustard (*Brassica juncea*) can effectively accumulate Zn and Pb under hydroponic conditions. Kentucky bluegrass and colonial bentgrass also accrue >150 mg/g Pb from a solution comprising 300 mg Pb/L (Niu et al., 2007).

To remove contaminants, ponds are also used. Higher plants having aerial canopy more than 50% (submerged or above water) were used to grow in these water bodies for remediation purposes. Plant organisms like tule (Scirpus) and cattail (Typha) have been shown to be able to extract a number of PTEs. Several researchers built/planned a special wetland to measure the removal of contaminants and found that after a duration of 150 days, the water was efficiently decontaminated using *Cyperus alternifolius* (inflow chamber) and *Villarsia exaltata* (outflow chamber). Other tests have shown that wetlands use wetland species to successfully consume PTEs with a bioconcentration factor of 1225 and 2567, respectively. A macro-algae, Microspora, was considered capable of extracting 97% of the aqueous fluid comprising 39.4 mg/L of Pb, while 76% of Pb and 82% of Ni were extracted from a solution containing 10 mg Pb/L and 5 mg Ni/L. Latest experiments have shown that mesquite (Prosopis spp.), a desert hydroponic plant species, has the capacity to extract Cr from the developing medium and convert it into less toxic form (Afzal et al., 2019; Ijaz et al., 2016).

2.3 Toxicity of potentially toxic elements to plants

The contamination of the environment by PTEs has become a major problem on earth. Their improved flexibility has led to the introduction into the world of these components. With the growing industrialization and deterioration of natural biogeochemical systems, the issue of PTE contamination is getting more and more serious. They are essentially nonbiodegradable, unlike biological substances, and thus accrue in the atmosphere. The accretion of heavy metals in soils and waterways sets a danger to biodiversity and to ecosystem security. In the soil, the toxicological impact on soil microbes caused by PTEs contributes to a decline in their numbers and actions. PTEs are categorized as vital and nonessential as far as their function in biological processes is concerned. The living organism also needs some metals to complete their biological functions, these heavy metals are named as critical heavy metals or essential PTEs. These are Fe, Cu, Mn, Ni, and Zn. On the other hand, Cd, Cr, As, Hg, and Pb are toxic for the plant in all cases thus are called nonessential heavy metals. There are detrimental health consequences of heavy metal concentrations over threshold limits since they conflict with the regular functioning of living structures. Via natural and anthropogenic causes, heavy metals collapse into the climate. Mineral weathering, flooding, and earthquake activity are the most significant natural influences, whereas anthropogenic sources involve mining, electroplating, smelting, and chemical fertilizer use. In addition, atmospheric accumulation, agricultural biosolids, industrial discharge, sludge runoff, etc., were also categorized under anthropogenic causes (Shahid et al., 2017; El-Meihy et al., 2019).

By modifying their biochemistry, potentially dangerous components regulate the development and output of plants. These PTEs disturb the metabolism and morphology of plants. For instance, Cr, Ni, and Pb have been observed to affect the development of chlorophyll, whereas Cu, Cd, Cr, Pb, and Ni decrease the growth of plants. Other studies have shown that thylakoid surface region, electron transport, and enzymatic behavior are impaired by Pb, Cr, Cu, Ni, and Hg or decreased chlorophyll production. Plant species, however, have evolved such techniques that enable them to develop in metalliferous soils and flourish. Some herbs classified as metallophytes may be recognized as containing metals. The most recognized metallophytes across the globe are several types of Thlaspi. In some European countries, these organisms are reported to grow in soils infected with Cd, Pb, Zn, and Ni. Another prominent metallophyte plant genus, *Silene vulgaris* (Dubey et al., 2018; Mustafa and Komatsu, 2016).

2.4 Effects of potentially toxic elements on human health

Potentially harmful elements have detrimental impacts on human well-being and thus, specific consideration should be given to the food chain in regard to their exposure. Many of their types are destructive and at very low amounts, can even cause unfavorable consequences and serious complications. Free radicals produced by heavy metals species induce oxidative stress. This stress provides the reason for enhancing reactive oxygen species production in plant which may overcome the intrinsic antioxidant protection of the cell and contribute to cell injury or death (Fryzova et al., 2017). Furthermore, metals in pigments or enzymes that impair their work may be substituted. With respect to their toxicity, Hg, Cd, Pb, As, Cu, Zn, Sn, and Cr are the most problematic heavy metals. Of these, Hg, Cd, Pb, and As are noncritical heavy metals, while Zn and Cu (trace elements) are major heavy metals (Liao et al., 2018). Several health-related issues were also connected with these PTEs. Human health effects of PTEs on human health are explained in Table 2.1.

TABLE 2.1 Harmful effects of PTEs on human health.

PTEs	Harmful effects	References
Al	Alzheimer's disease, kidney diseases	Xu et al. (2018)
As	Interference in oxidative phosphorylation, essential cellular processes, and ATP synthesis	Costa (2019)
Cd	Causes renal failure and anemia, mutagenic and teratogenic, endocrine disruptor and carcinogenic	Fatima et al. (2019)
Cr	Loss of hair, mutagen, hematotoxic, kidney and liver damage, cardiovascular disease, impaired reproduction system, genotoxic, and carcinogenic	Pavesi and Moreira (2020)
Cu	Anemia, brain hemorrhage, kidney, and liver damage	Kalita et al. (2018)
Hg	Fatigue, depression, anxiety, insomnia, hair loss, memory loss, tremors, ulcer, brain damage, vision disturbances, kidney, and lung disorders	Andreoli and Sprovieri (2017)
Ni	Dermatitis, lungs cancer, hair loss. pulmonary toxic, hematotoxic, neurotoxic, reproductive toxic, genotoxic, nephrotoxic, hepatotoxic, and immunotoxic	Das et al. (2018)
Pb	Renal failure, cardiovascular disease. loss of short-term memory, coordination problems, reduced intelligence, learning disabilities, and impaired development	Batool et al. (2017)
Zn	Dizziness and fatigue	Wu et al. (2019)

2.5 Cleanup of potentially toxic elements contaminated soils

The environmental heavy metal is increasing year after year, i.e., the 700 km^2 region of Campine in Belgium and the Netherlands is diffusely saturated with ambient precipitation of Cd, Zn, and Pb. A total area of 2.88×10^6 ha of destroyed land is generated in China alone as a consequence of mining, and an additional estimated area of 46,700 ha of contaminated land is produced annually (Verma and Jaiswal, 2016). Because of extreme contaminants, these devastated fields almost entirely neglect vegetation and inevitably causing significant soil erosion and off-site pollution. In order to mitigate their influence on biodiversity, cleaning up heavy metal-contaminated soils is also necessary. With regard to costs and technological problems, this is a daunting task. For this reason, various human, chemical, and biological methods have been used so far (Fig. 2.2). In situ vitrification, soil incineration, drilling and landfill, soil cleaning, solidification, soil flushing, and electro-kinetic phase recovery are the typical methods of remediation. Physical and chemical approaches usually suffer from restrictions such as high expense, hard work, irreversible adjustments in surface land, and degradation of the existing microflora of the soil. Chemical processes can trigger problems with secondary pollution as well. Study is also required to improve, reliable, and cost and environmentally friendly remediation methods for heavy metal-polluted soil decontamination. Phytoremediation, regarded as a potential green response to the PTEs contamination problem, is one such innovative strategy (Shiomi, 2018).

2.6 Phytoremediation: a green technology

The modern clean-up approaches include either bulk drainage or in situ toxin remediation for depletion, mitigation, or immobilization purposes. Soil extraction and transport involves costly equipment, disturbs the environment, and is not generally accessed by the group. In-situ technology, on the other hand, relies on the cost of washing kg^{-1}, the selectivity of the treatment agent, and the viability of recovering from the surface and subsurface the chemicals used to cure the soil and the loaded material. With the usage of acids or chelating agents, the properties of the soil and its biological components may shift, and the ionic power, flow rate, and form of contaminant might not be stable or influenced. In addition, directly following surgery, soil cannot be included (Cristaldi et al., 2020).

An environment-friendly, cost-effective, and pollution-proof approach of PTEs remediation is phytoremediation. It does not create any mass nor be the cause of metal addition in spite of that metal extraction can be done. Moreover, the remediating medium is in safe and functional condition just after the application of treatment. High metal concentration can limit the growth potential of plant, but this can be overcome by the use of genetic engineering and metal absorption enhancers (Bauddh et al., 2017).

Phytoremediation essentially applies to the usage of plants and associated soil bacteria to decrease the volume of toxins in the atmosphere or their adverse consequences. It may be used both for the decontamination of heavy metals and radionuclides and for the prevention of organic contaminants. In general, plants manage the toxins without impacting topsoil, thus retaining their usefulness and fertility. They will increase soil productivity with enhanced organic matter inputs. Phytoremediation is a fusion of two words: Phyto: Greek word; meaning plant and remedium: Latin word; meaning to fix or eradicate an evil. Green plants have a large potential to capture atmospheric toxins and use multiple techniques to detoxify them (Romika and Kang 2018). This is comparatively recent approach, with scientific trials mainly performed during the last two decades. The concept is aesthetically pleasing, and it is normally embraced by the public. It is ideal for usage in very wide areas where other methods to remediation are not cost-effective or practicable. Compared to other remediation options, phytoremediation can cost as little as 5% of substitutive clean-up technologies in terms of performance. Deforestation and metal leaching are often avoided by the formation of plants on contaminated soils. On economic basis, the purpose of phytoremediation of contaminated land can be threefold: (1) the phytostabilization; (2) the phytoextraction of market-value metals such as Ni, titanium, and silver; (3) the sustainable management of land where phytoextraction ameliorate soil fertility steadily in order to generate higher market-value crops. Furthermore, for both phytoremediation and energy generation, rapidly increasing and high-biomass generating plants like Jatropha, Poplar, and Willow may be used (Singh et al., 2019).

2.7 Strategies in phytoremediation

Phytoremediation strategies include the following (Table 2.2).

TABLE 2.2 Overview of techniques used in phytoremediation.

Technique	Description	References
Phytoextraction	Pollutants accumulation in plant biomass	Pajević et al. (2016)
Phytofiltration	Sequestration in plants through roots	Sandhi et al. (2018)
Phytostabilization	Limiting the contaminants mobility in soil using by plants	Zeng et al. (2018)
Phytovolatilization	Transformation of pollutants to volatile form and consequently release to the atmosphere	Arya et al. (2017)
Phytodegradation	Degradation of contaminants using plants and their enzymes	Chlebek and Hupert-Kocurek (2019)
Rhizodegradation	Rhizospheric degradation of pollutants rhizobacteria	Allamin et al. (2020)

2.7.1 Phytoextraction

Phytoextraction is the ingestion by plant roots of toxins from soil or water and their transition to and deposition, i.e., shoots, in shallow biomass. Effective phytoextraction occurs when most of the plant uptake metal, accumulate in plant shoot through biochemical mechanisms as extraction of metal from the root is usually not feasible.

2.7.2 Phytofiltration

Decontamination of pollutants from polluted seawater or waste from fields, phytofiltration occurs. Rhizofiltration, blastofiltration, and caulofiltration are the use of plant roots, seedlings, and excised plant shoots (caulis latin = shoot) may be an alternative to phytofiltration. The toxins are removed or adsorbed in phytofiltration and hence their flow through underwater waters is reduced.

2.7.3 Phytostabilization

The usage of such plants for stabilizing toxins from contaminated fields is phytostabilization or phytoimmobilization. This approach is used to regulate the volatility and bioavailability of toxins in the environment, thus keeping them from entering the food chain or expanding to groundwater. Via root sorption, complexation, precipitation, or metal valence reduction in the rhizosphere, plants can immobilize PTEs in soils. In toxicity, metals with several valences vary. Plants capably transform hazardous metals to a comparatively less risky condition by eliminating special redox enzymes and decreasing potential metal stress and injury. For starters, it is commonly studied to reduce Cr to Cr, the latter being both less mobile and less poisonous. Phytostabilization limits heavy metal biota deposition and eliminates leaching into underground waters. However, since the PTE metals remain in the soil, phytostabilization is not a permanent solution; it is restricted only to their movement. It is effectively a technique of treatment for the stabilization of extremely harmful pollutants.

2.7.4 Phytovolatilization

Phytovolatilization is the absorption, transformation to a volatile state, and eventual release into the environment of contaminants from the soil by plants. For organic pollutants and certain PTEs, including Hg and selenium (Se), this method may be used. Its implementation though is limited by the reality that the contaminant is not entirely extracted. It is necessary to move only one component (soil) to another (atmosphere) from which it can be redeposited. The most argumentative of the phytoremediation methods is phytovolatilization.

2.7.5 Phytodegradation

phytodegradation is the enzyme-catalyzed oxidation by plants of organic pollutants as it is not dependent on rhizospheric microorganisms. Through their metabolic processes, plants can uptake and detoxify organic xenobiotics from contaminated conditions. Through this standpoint, the biosphere will treat green plants as green liver. As PTEs are nonbiodegradable, phytodegradation is restricted to the elimination of organic toxins. Scientists have recently shown concern in researching the phytodegradation of many organic contaminants, including insecticides and synthetic herbicides. The use of genetically engineered plants (e.g. transgenic poplars) for this reason has been documented in several studies.

2.7.6 Rhizodegradation

Rhizodegradation applies to microorganisms in the rhizosphere breaking down organic materials in the soil. The rhizosphere spreads approximately 1 mm around the root and is under the influence of the vine. The elevated numbers and biochemical pathways of microbes are likely to be the main cause of increased pollutant degradation in the rhizosphere. About 10–100 times higher microbial activity was reported in the rhizosphere due to the plant's secreted carbohydrates, amino acids, and flavonoids. Nutrient-containing exudates that release from plant roots offer sources of carbon and nitrogen to the soil microbes. Plants also contain such enzymes that are capable of removing organic compounds in soils, in addition to the secretion of organic substrates to facilitate the growth and activities of rhizospheric microorganisms (Sarwar et al., 2017; García-Sánchez et al., 2018; Malaviya et al., 2020).

2.8 Phytoextraction of potentially toxic elements: natural vs artificial phytoremediation

The principal and most helpful form of phytoremediation for extracting PTEs and metalloids from contaminated soils, sediments, or water is phytoextraction. For industrial launch, it is the most thrilling. The success of phytoextraction determines by many aspects, such as soil bioavailability of PTEs, soil properties, heavy metal speciation, and the plant species involved (Wang et al., 2017).

Two key variables determined the phytoextraction potential of a plant form, i.e., shoot metal concentration and shoot biomass. Phytoextraction of PTEs was done in two ways: (1) growing plant that are hyperaccumulators, i.e., plant that might produce low above-ground biomass but can have high metals accumulation ability, (2) growing plants may accumulate less metal concentration but grow rapidly and are high biomass producer like Brassica juncea (Indian mustard). The use of nonaccumulators, on the other hand, would produce a low in metal, large-volume biomass that would be inexpensive for the treatment of metal recovery and thus costly for safe disposal (García-Sánchez et al., 2018).

For heavy metals, plants with multiple harvests production in a single growth period (such as Trifolium spp.) may have a high phytoextraction capacity. Due to their fast growth rate, better adaptability to stressful environments and high biomass, grasses are more suitable for phytoextraction than shrubs or trees. Several studies have also explored the use of crops for phytoextraction PTEs, such as maize and barley. In this case, several crops are required in order to reduce the exposure of PTEs to acceptable standards. Phytoextraction of PTEs has a drawback of food chain contamination thus the field crops that in direct animal and human food consumption are not recommended for this purpose. Soil chemical composition and sorption properties are affected by metal mobility and bioavailability. The bioavailability of soil PTEs is a crucial factor influencing phytoextraction efficiency of target PTEs. Bioavailability of Pb is low in soil. Only a fraction of soil metal is generally bioavailable for plant absorption. The extreme binding of PTEs to soil particles or precipitation allows the insolubility of a substantial fraction of soil PTEs and is therefore effectively inaccessible to the uptake of plants. There may be three groups in terms of soil bioavailability of PTEs: readily bioavailable (As, Se, Cd, Ni, Zn, Cu); relatively bioavailable (Co, Mn, Fe); and least bioavailable (Pb, Cr, U). However, certain processes for the solubilization of PTEs in the soil have been developed by plants. In the rhizosphere, plant roots secrete metal-mobilizing compounds termed phytosiderophores. Roots may secrete H^+ ions and improve the dissolution of PTEs to acidify the rhizosphere. Cations of H^+ ions adsorbed to soil particles can be displaced by PTEs. Typically, the pH of the soil in the rhizosphere can be decreased by one to two units above that of the bulk soil by root exudates. By facilitating desorption, decreased soil pH improves the concentration of PTEs in the solution. In addition, the bioavailability of PTEs in soils may be greatly improved by rhizospheric microorganisms (primarily bacteria and mycorrhizal fungi). Labile metal pools and root absorption may be improved through interactions with microbial siderophores (Shmaefsky, 2020a).

Phytoextraction of PTEs in two forms, natural and induced, must be carried out. For the removal of PTEs under natural conditions, plants are used in natural or constant phytoextraction, i.e., no soil modification is made. Various chelating agents such as EDTA, citric acid, elemental sulphur, and ammonium sulphate are added to the soil during induced or chelate-assisted phytoextraction to enhance the bioavailability of PTEs in the soil for plant uptake. With PTEs in the soil, the chelates grow water-soluble complexes and assist in their desorption from soil particles. Decreasing soil pH also enhances the bioavailability of heavy metals. Such chemical treatments might cause problems with secondary pollution.

Synthetic-chelate, EDTA, is nonbiodegradable, can be leached into groundwater sources, somehow toxic to plants, and having an additional impact on the environment thus when performing mediated phytoextraction, proper care should also be taken. However, since it has a natural past and is readily biodegraded in the soil, the use of citric acid as a chelating agent could be appealing. In addition, citric acid is nontoxic to plants, so plant growth does not decrease (Shmaefsky, 2020b).

2.8.1 Metallophytes

In soils rich in PTEs, metallophytes are plants that are uniquely acclimatize and thrive. The primary sites for heavy metal tolerant plants are soils, called metalliferous or orogenic soils, where ores are outcropped. Over thousands of years, metal sensitivity to a surplus of numerous metals has determined the production of metal tolerance in metallophytes under local environmental conditions. Mining activities have ruined metal-enriched habitats and are now reducing them, thereby altering the metallophyte niche. Botanical curiosities encompass metallophytes. Within the plant family Brassicaceae, these species are concentrated. Their use, either alone or in combination with microorganisms, is an enticing phenomenon for phytoremediation of soils contaminated with PTEs. Three classes are known as metallophytes: metal excluders, metal detectors, and metal hyperaccumulators (Midhat et al., 2019).

2.8.2 Excluders

Metal excluders in their roots accumulate PTEs from the substrate but limit their transfer and entry into their aerial sections. These plants have a weak potential for metal absorption but can be useful aimed phytostabilization (Awa and Hadibarata, 2020).

2.8.3 Indicators

In their aerial elements, potentially harmful elements are accumulated by metal markers. These plants typically reflect heavy metal levels in the medium, as the name implies (Ahmad et al., 2019).

2.8.4 Hyperaccumulators

Hyperaccumulators are plants which in quantities well above those contained in the soil or in neighboring nonaccumulating plants, can accumulate PTEs in their aboveground tissues. It is possible to define hyperaccumulators as a special and extreme case of a broader category of accumulators. The metals they accumulate are hyper tolerant in the shoots. The criteria for hyperaccumulators have not been defined in any empirical way. Hyperaccumulators have however been defined by specific authors or research groups. In order to characterize plants for Ni concentrations >1000 mg/kg dry weight, the term "hyperaccumulator" was first coined. Scientists have tried to define Ni hyperaccumulation more precisely because a Ni hyperaccumulator is a plant in which at least 1000 mg/kg of Ni concentration has been reported in at least one specimen of any overground tissue in its natural habitat that develops in dry matter. Overground tissue may be viewed only as plant foliage for the determination of hyperaccumulator status. The phrase "growing in its natural habitat" suggests that metal hyperaccumulation must be carried out by hyperaccumulators while maintaining adequate protection to retain a population that is self-sustaining. Hyperaccumulators accumulate in their shoots >100 mg/kg Cd dry weight or >1000 mg/kg Cu, Pb and Ni dry weight or >10,000 mg/kg Zn and Mn dry weight while growing on soils rich in metal. It is a fact that the hyperaccumulation requirements commonly used for certain metals are overly conservative and suggest reducing the requirement for hyperaccumulation for some metals. For various metals and metalloids in dry vegetation with plants growing in their natural environment, the following concentration criteria are proposed: 100 mg/kg for Cd, Se, and Ti; 300 mg/kg for Cu, Cr, and Co; 1000 mg/kg for As, Ni, and Pb; 10,000 mg/kg for Mn; 3000 mg/kg for Zn (Yan et al., 2019).

The discovery of more effective hyperaccumulators species for PTEs is the initial step toward successful remediation. Hydroponics, metal spiked soils, and chemically acidified soils in lab experiments cannot give comprehensive estimation of hyperaccumulation. These experiments alone may not be considered to be capable of defining a genus as a hyperaccumulator. They say that it is important to monitor natural populations. Literature reveals that metal-hyperaccumulators and more than 300 Ni-hyperaccumulators have been classified as over 400 plant organisms. Several metal-accumulating species exist in the Brassicaceae family. Hyperaccumulator plants also include *Thlaspi caerulescens* and *Alyssum bertolonii*. The best-known metal hyperaccumulator is arguably *T. caerulescens*. For Cd, Zn, and Ni, this fungus is a hyperaccumulator (Narayanan et al., 2020; Prabakaran et al., 2019).

2.9 Phytoremediation of potentially toxic elements using constructed wetlands

Constructed wetlands have been used for effluent and drainage water clean-up. In remediating PTEs and numerous water quality issues, they have a technically viable and cost-effective technology and have proven to be reliable and efficient. For the removal of PTEs from wastewater, the aquatic macrophytes are suitable because of their accelerated growth, high biomass production, and a comparatively higher pollutant absorption capacity. Owing to direct contact with polluted water, they perform enhanced purification. Various floating, developing, and submerged marine organisms are utilized in constructed wetlands. For edge growing in constructed wetlands, poplars (Populus spp.) and willows (Salix spp.) may be included. Aquatic underwater plants accumulate metals from their origins, whereas their entire bodies collect metals from submerged plants. Water hyacinths (Eichhornia crassipes) in constructed wetlands have been used for phytoremediation of PTEs. With a well-developed fibrous root system and substantial biomass, it is a rapidly growing, floating herb. It often easily adapts and plays a major role in the accumulation of metals in water in different marine environments.

Likewise, water lettuce (Pistia stratiotes) was reported best phytoremediator for Mn decontamination in water. Ample wetland growth, large canopy coverage, and easy harvesting are the additional benefits with its growing. Another Aquatic Phytoremediation nominee is Azolla (Hussain et al., 2018; Farzi et al., 2017).

2.10 Mechanism of potentially toxic elements uptake, translocation, and remediation

By means of soil solution, plants draw heavy metals into their roots. From this point of entry these ions either deposit in roots or transported to above-ground parts of the plants through vascular bundles, where they are also accumulated in vacuoles. This cellular organelle has low metabolic activity. One of the procedures to remove surplus metal ions from the cytosol is strong metal sequestration in the vacuole, which can reduce their association with cellular metabolic processes. The resistance process in metal hyperaccumulators is a part of the compartmentalization of complex metals in vacuoles. There are five fundamental aspects of the whole mechanism of heavy metal phytoextraction: heavy metals bioavailability in the soil, roots metal ions absorption, the uptake metal translocation within plant (roots to the aerial tissues), the metal sequestration in plant tissues, and the tolerance of metals. Resistance of metal toxicity is the most important factor in phytoremediation. Cell wall binding, vacuole accretion, and chelating by the incorporation of metal-binding peptides and the creation of metal complexes are processes regulating heavy metal resistance in plant cells (Kumar et al., 2017; Yang and Shen, 2020).

A range of molecules monitor and regulate the ultimate path of heavy metals from soil solution to the vacuoles. Some molecules, in their complexation and eventual sequestration, are active in the cross-membrane transfer of PTEs. For metal ion uptake, plasma membrane of root cell maintained specialized transporters (channel proteins) or H+ coupled carrier. For instance, the zinc–iron permease family of transporters contributes to the absorption of Zn and Fe. Another family of proteins is natural resistance-associated macrophage that assist in transporting divalent metal ions. The metal ions, essential or nonessential, due to comparable oxidation states and ionic radii can enter into the root by the same transmembrane transporters after the competition. The lack of selectivity at transmembrane point of entry and concentration gradient are the possible reasons behind nonessential heavy metals uptake into plant cells. Amino acids and organic acids are worked as ligands in their molecules for the chelation of heavy metal ions due to the presence of donor atoms (N, S, and O) (Raza et al., 2020; Awa and Hadibarata, 2020).

For the phytoremediation of poisonous and risky heavy metals as well as for the phytoremediation of expensive heavy metals (Pd, Pt, and Au), hyperaccumulators can be used. phytoremediation with the use of hyperaccumulators may result in the manufacture of a bio-ore of certain commercial value. The volume of heavy metals which hyperaccumulators remove from the soil is a function of the concentration of tissue metals multiplied by the amount of biomass produced. Several plants have a normal hyperaccumulation capacity for particular heavy metals. These plants are referred to as natural hyperaccumulators of origin. On the other hand, biotechnological approaches can increase the deposition capacity of certain species for certain heavy metals through their genetic modification. These genetically engineered plants have demonstrated positive phytoremediation outcomes for some heavy metals. However, since the bio-safety of genetically modified organisms remains doubtful for some environmental scientists, the marketing of such crops remains a global issue (Ebrahimbabaie et al., 2020; Muthusaravanan et al., 2018).

2.11 Phytoremediation and future prospects

Phytoremediation, as discussed earlier, is a comparatively recent approach in remediation arena. Most of the phytoremediation-based experiments were conducted in laboratory and greenhouse, and only a few studies have been conducted in the real world to assess the effectiveness of phytoremediation. The field's actual effects vary from those in laboratory or greenhouse environments, as the field is a real environment in which multiple variables play a role at the same time. Differences in nutrients, humidity, temperature, precipitation, plant diseases, herbivory, uneven contaminant distribution, soil chemistry, soil pH, and soil structure are the factors that can influence phytoremediation in the field. For true estimation of the efficacy of this technology, field testation, keeping plant species, and target metal in mind is necessary before commercialization.

These characteristics can be selected either by conventional breeding methods or by using modern hybridization methods such as protoplast fusion or by regulating gene expression in transgenic plants since the discovery of desirable traits in natural hyperaccumulators. In order to identify genes coding for the hyperaccumulation of different heavy metals in plants, analysis is underway. The evolution of phytoremediation "superbug" plants allows certain genes to be identified and successfully converted into other viable plants. Transgenic plants secrete metal selective ligands into the rhizosphere that could specifically solubilize phytoremediation target elements. Therefore, several attractive features may be integrated into a single plant genus, which will assist efficiently. However, in order to prevent any accidental damage to the biosphere at present or in the future, gene expression, and interaction must be investigated reliably and carefully. Researchers may be able to precisely tune the phytoremediation mechanism by recognizing the coordination chemistry of metals inside plant tissues. Phytoremediation is seen as a green technique of remediation with an anticipated high potential, considering the many obstacles.

2.12 Conclusions and recommendations

As pollution of soils and waterways with PTEs is a major environmental issue, successful methods of remediation are therefore necessary. There are major disadvantages to physical and chemical methods for the cleaning and amelioration of PTEs polluted soils, such as high prices, irreversible changes in surface resources, deterioration of existing soil microflora, and a reason of adjacent pollution problems. Phytoremediation is a safer, by comparison, approach to the problem. Phytoremediation is a solar-driven process with strong public support that is economically sustainable and ecologically responsible. It is a relatively modern technology that continues to be tested. But with this technique, soil chemistry, microbiology, plant genetics, and ecology, as well as environmental engineering, are strongly intertwined. Research is ongoing to identify native plants for the phytoremediation of selective PTEs and to determine the impact of various parameters on phytoremediation performance, as it takes time for exotic plant species to adapt to the external environment. In addition, studies are being conducted to genetically alter certain suitable plants for enhanced heavy metal phytoremediation and other xenobiotics. The different proteins that are involved in cross-membrane transport and vacuolar PTEs sequestration are also being studied in order to identify and define them. In such molecular science, advances and milestones will help greatly to understand the mechanism and improve the efficacy of phytoremediation. Phytomining, an environmentally friendly plant-based processing of metals that can be used for the production of metals and low-grade minerals, would also be facilitated by improved knowledge of the ingestion of heavy metals by soil plants. In the future, PTEs phytoextraction is anticipated to be an economically viable method of PTEs phytoremediation and phytomining.

List of abbreviations

As Arsenic.
Cr Chromium.
Cd Cadmium.
Cu Copper.
EDTA Ethylenediaminetetraacetic acid.
ha Hectare.
Hg Mercury.
Mn Manganese.
N Nitrogen.
Ni Nickel.
O Oxygen.
Pb Lead.
Ppm Parts per million.
PTEs Potentially toxic elements.
S Sulfur.
Sn Tin.
Zn Zinc.

References

Afzal, M., Arslan, M., Müller, J.A., Shabir, G., Islam, E., Tahseen, R., et al., 2019. Floating treatment wetlands as a suitable option for large-scale wastewater treatment. Nat. Sustain. 2 (9), 863–871.

Ahmad, Z., Khan, S.M., Ali, M.I., Fatima, N., Ali, S., 2019. Pollution indicandum and marble waste polluted ecosystem; role of selected indicator plants in phytoremediation and determination of pollution zones. J. Clean. Prod. 236, 117709.

Allamin, I.A., Halmi, M.I.E., Yasid, N.A., Ahmad, S.A., Abdullah, S.R.S., Shukor, Y, 2020. Rhizodegradation of petroleum oily sludge-contaminated soil using Cajanus cajan increases the diversity of soil microbial community. Sci. Rep. 10 (1), 1–11.

Andreoli, V., Sprovieri, F., 2017. Genetic aspects of susceptibility to mercury toxicity: an overview. Int. J. Environ. Res. Public Health 14 (1), 93.

Arya, S., Devi, S., Angrish, R., Singal, I., Rani, K., 2017. Soil reclamation through phytoextraction and phytovolatilizationVolatiles and Food Security: Springer, Singapore, pp. 25–43.

Ashraf, S., Ali, Q., Zahir, Z.A., Ashraf, S., Asghar, H.N., 2019. Phytoremediation: environmentally sustainable way for reclamation of heavy metal polluted soils. Ecotoxicol. Environ. Safety 174, 714–727.

Awa, S.H., Hadibarata, T., 2020. Removal of heavy metals in contaminated soil by phytoremediation mechanism: a review. Water, Air, Soil Pollut. 231 (2), 47.

Batool, Z., Yousafzai, N.A., Murad, M.S., Shahid, S., Iqbal, A., 2017. Lead toxicity and evaluation of oxidative stress in humans. PSM Biol. Res. 2 (2), 79–82.

Bauddh, K., Singh, B., Korstad, J., 2017. Phytoremediation Potential of Bioenergy Plants. Springer, Singapore.

Chlebek, D., Hupert-Kocurek, K., 2019. Endophytic bacteria in the phytodegradation of persistent organic pollutants. Advanc. Microbiol. 58 (1), 70–79.

Costa, M., 2019. Review of arsenic toxicity, speciation and polyadenylation of canonical histones. Toxicol. Appl. Pharmacol. 375, 1–4.

Cristaldi, A., Conti, G.O., Jho, E.H., Zuccarello, P., Grasso, A., Copat, C., et al., 2017. Phytoremediation of contaminated soils by heavy metals and PAHs. A brief review. Environ. Technol. Innov. 8, 309–326.

Cristaldi, A., Copat, C., Conti, G.O., Zuccarello, P., Grasso, A., Ferrante, M., 2020. PhytoremediationThe Handbook of Environmental Remediation, London, United Kingdom, 268–298.

Das, K.K., Reddy, R.C., Bagoji, I.B., Das, S., Bagali, S., Mullur, L., et al., 2018. Primary concept of nickel toxicity—an overview. J. Basic Clin. Physiol. Pharmacol. 30 (2), 141–152.

Dubey, S., Shri, M., Gupta, A., Rani, V., Chakrabarty, D., 2018. Toxicity and detoxification of heavy metals during plant growth and metabolism. Environ. Chem. Lett. 16 (4), 1169–1192.

Ebrahimbabaie, P., Meeinkuirt, W., Pichtel, J., 2020. Phytoremediation of engineered nanoparticles using aquatic plants: mechanisms and practical feasibility. J. Environ. Sci.

El-Meihy, R.M., Abou-Aly, H.E., Youssef, A.M., Tewfike, T.A., El-Alkshar, E.A., 2019. Efficiency of heavy metals-tolerant plant growth promoting bacteria for alleviating heavy metals toxicity on sorghum. Environ. Exp. Bot. 162, 295–301.

Farzi, A., Borghei, S.M., Vossoughi, M., 2017. The use of halophytic plants for salt phytoremediation in constructed wetlands. Int. J. Phytoremed. 19 (7), 643–650.

Fatima, G., Raza, A.M., Hadi, N., Nigam, N., Mahdi, A.A., 2019. Cadmium in human diseases: it's more than just a mere metal. Ind. J. Clin. Biochem., 1–8.

Fryzova, R., Pohanka, M., Martinkova, P., Cihlarova, H., Brtnicky, M., Hladky, J., Kynicky, J., et al., 2017. Oxidative stress and heavy metals in plants. Rev. Environ. Contam. Toxicol. 245, 129–156.

García-Sánchez, M., Košnář, Z., Mercl, F., Aranda, E., Tlustoš, P., 2018. A comparative study to evaluate natural attenuation, mycoaugmentation, phytoremediation, and microbial-assisted phytoremediation strategies for the bioremediation of an aged PAH-polluted soil. Ecotoxicol. Environ. Safety 147, 165–174.

Hussain, Z., Arslan, M., Malik, M.H., Mohsin, M., Iqbal, S., Afzal, M., 2018. Integrated perspectives on the use of bacterial endophytes in horizontal flow constructed wetlands for the treatment of liquid textile effluent: phytoremediation advances in the field. J. Environ. Manage. 224, 387–395.

Ijaz, A., Iqbal, Z., Afzal, M., 2016. Remediation of sewage and industrial effluent using bacterially assisted floating treatment wetlands vegetated with Typha domingensis. Water Sci. Technol. 74 (9), 2192–2201.

Kalita, J., Kumar, V., Misra, U.K., Bora, H.K., 2018. Memory and learning dysfunction following copper toxicity: biochemical and immunohistochemical basis. Mol. Neurobiol. 55 (5), 3800–3811.

Kumar, S.S., Kadier, A., Malyan, S.K., Ahmad, A., Bishnoi, N.R., 2017. Phytoremediation and rhizoremediation: uptake, mobilization and sequestration of heavy metals by plantsPlant-microbe Interactions in Agro-ecological Perspectives: Springer, Singapore, pp. 367–394.

Liao, F., Wang, G., Shi, Z., Huang, X., Xu, F., Xu, Q., et al., 2018. Distributions, sources, and species of heavy metals/trace elements in shallow groundwater around the Poyang Lake, East China. Exp. Health 10 (4), 211–227.

Malaviya, P., Singh, A., Anderson, T.A., 2020. Aquatic phytoremediation strategies for chromium removal. Rev. Environ. Sci. Bio/Technol., 1–48.

Masindi, V., Muedi, K.L., 2018. Environmental contamination by heavy metals. Heavy Metals 10, 115–132.

Midhat, L., Ouazzani, N., Hejjaj, A., Ouhammou, A., Mandi, L., 2019. Accumulation of heavy metals in metallophytes from three mining sites (Southern Centre Morocco) and evaluation of their phytoremediation potential. Ecotoxicol. Environ. Safety 169, 150–160.

Mustafa, G., Komatsu, S., 2016. Toxicity of heavy metals and metal-containing nanoparticles on plants. Biochimica et Biophysica Acta (BBA)-Proteins Proteomics 1864 (8), 932–944.

Muthusaravanan, S., Sivarajasekar, N., Vivek, J., Paramasivan, T., Naushad, M., Prakashmaran, J., et al., 2018. Phytoremediation of heavy metals: mechanisms, methods and enhancements. Environ. Chem. Lett. 16 (4), 1339–1359.

Narayanan, M., Devarajan, N., Kandasamy, G., Kandasamy, S., Shanmuganathan, R., Pugazhendhi, A., 2020. Phytoremediation competence of short-term crops on magnesite mine tailing. Chemosphere, 270, 128641.

Niu, Z.-X., Sun, L.-n., Sun, T.-h., Li, Y.-s., Hong, W, 2007. Evaluation of phytoextracting cadmium and lead by sunflower, ricinus, alfalfa and mustard in hydroponic culture. J. Environ. Sci. 19 (8), 961–967.

Pajević, S., Borišev, M., Nikolić, N., Arsenov, D.D., Orlović, S., Župunski, M., 2016. Phytoextraction of heavy metals by fast-growing trees: a reviewPhytoremediation. Springer, Switzerland, pp. 29–64.

Pavesi, T., Moreira, J.C., 2020. Mechanisms and individuality in chromium toxicity in humans. J.Appl. Toxicol 40(9), 1183–1197.

Peng, W., Li, X., Xiao, S., Fan, W., 2018. Review of remediation technologies for sediments contaminated by heavy metals. J. Soils Sediments 18 (4), 1701–1719.

Prabakaran, K., Li, J., Anandkumar, A., Leng, Z., Zou, C.B., Du, D., 2019. Managing environmental contamination through phytoremediation by invasive plants: a review. Ecol. Eng. 138, 28–37.

Rai, P.K., Lee, S.S., Zhang, M., Tsang, Y.F., Kim, K.-H., 2019. Heavy metals in food crops: health risks, fate, mechanisms, and management. Environ.Int. 125, 365–385.

Raza, A., Habib, M., Kakavand, S.N., Zahid, Z., Zahra, N., Sharif, R., et al., 2020. Phytoremediation of cadmium: physiological, biochemical, and molecular mechanisms. Biology 9 (7), 177.

Romika, C., Kang, H. (2018). Phytoremediation and physiological effects of mixed heavy metals on poplar hybrids. Heavy metals. IntechOpen, UK. pp. 183–200.

Sandhi, A., Landberg, T., Greger, M., 2018. Phytofiltration of arsenic by aquatic moss (Warnstorfia fluitans). Environ. Pollut. 237, 1098–1105.

Sarwar, N., Imran, M., Shaheen, M.R., Ishaque, W., Kamran, M.A., Matloob, A., et al., 2017. Phytoremediation strategies for soils contaminated with heavy metals: modifications and future perspectives. Chemosphere 171, 710–721.

Sela, M., Garty, J., TEL-OR, E., 1989. The accumulation and the effect of heavy metals on the water fern Azolla filiculoides. New Phytol. 112 (1), 7–12.

Shahid, M., Dumat, C., Khalid, S., Schreck, E., Xiong, T., Niazi, N.K., 2017. Foliar heavy metal uptake, toxicity and detoxification in plants: a comparison of foliar and root metal uptake. J. Haz. Mater. 325, 36–58.

Shiomi, N., 2018. Advances in Bioremediation and Phytoremediation. BoD–Books on Demand, London, UK.

Shmaefsky, B.R., 2020a. Phytoremediation. Springer, Springer, Cham.

Shmaefsky, B.R., 2020b. Principles of Phytoremediation. Phytoremediation. Springer, Springer, Cham., pp. 1–26.

Singh, R., Jha, A.B., Misra, A.N., Sharma, P., 2019. Adaption mechanisms in plants under heavy metal stress conditions during phytoremediation Phytomanagement of Polluted Sites. Elsevier, India, pp. 329–360.

Verma, J.P., Jaiswal, D.K., 2016. Book review: advances in biodegradation and bioremediation of industrial waste. Front. Microbiol. 6, 1555.

Wang, L., Ji, B., Hu, Y., Liu, R., Sun, W., 2017. A review on in situ phytoremediation of mine tailings. Chemosphere 184, 594–600.

Wu, B., Wu, J., Liu, S., Shen, Z., Chen, L., Zhang, X.-X., et al., 2019. Combined effects of graphene oxide and zinc oxide nanoparticle on human A549 cells: bioavailability, toxicity and mechanisms. Environ. Sci.: Nano 6 (2), 635–645.

Xu, L., Zhang, W., Liu, X., Zhang, C., Wang, P., Zhao, X., 2018. Circulatory levels of toxic metals (aluminum, cadmium, mercury, lead) in patients with Alzheimer's disease: a quantitative meta-analysis and systematic review. J. Alzheimer's Dis. 62 (1), 361–372.

Yan, H., Gao, Y., Wu, L., Wang, L., Zhang, T., Dai, C., et al., 2019. Potential use of the Pteris vittata arsenic hyperaccumulation-regulation network for phytoremediation. J. Haz. Mater. 368, 386–396.

Yang, Y., Shen, Q., 2020. Phytoremediation of cadmium-contaminated wetland soil with Typha latifolia L. and the underlying mechanisms involved in the heavy-metal uptake and removal. Environ. Sci. Pollut. Res. 27 (5), 4905–4916.

Zeng, P., Guo, Z., Cao, X., Xiao, X., Liu, Y., Shi, L., 2018. Phytostabilization potential of ornamental plants grown in soil contaminated with cadmium. Int. J. Phytoremed. 20 (4), 311–320.

Chapter 3

Mechanisms of phytoremediation

Flávia Cristina Policarpo Tonelli[a], Fernanda Maria Policarpo Tonelli[b], Moline Severino Lemos[b], Núbia Alexandre de Melo Nunes[c]
[a]*Department of Biochemistry, Federal University of São João del Rei, Divinópolis, Brazil,* [b]*Department of Cell Biology, Institute of Biological Sciences, Federal University of Minas Gerais, Belo Horizonte, Brazil,* [c]*Department of Biochemistry and Immunology, Institute of Biological Sciences, Federal University of Minas Gerais, Belo Horizonte, Brazil*

3.1 Introduction

Air, water, and soil pollution main sources are human activities performed ignoring sustainability principles. Urbanization and industrialization at hectic pace severely impact ecosystems (Chaoua et al., 2019; Zheng et al., 2020). Wastes from industries (containing, for example, pharmaceuticals, heavy metals, and dyes), dumpsites generating leachate, mine exploitation, harbor activities, offshore oil/petroleum extraction, pesticides use are relevant, and dangerous sources of organic and/or inorganic pollutants that may negatively affect living beings in a severe way (Loizidou and Kapetanios, 1993; Jafari et al., 2013; Barrios-Estrada et al., 2018; Carpenter, 2018; Bilal et al., 2019; Mendes et al., 2019; Pesantes et al., 2019; Pu et al., 2019; Rosculete et al., 2019; Vázquez-Luna and Cuevas-Díaz 2019; Zhang et al., 2019; Sorte et al., 2020). Pesticides, for example, can damage organs' function (especially brain, causing neurological diseases) and damage cell's DNA inducing cancer (Fatima et al., 2018; Singh et al., 2018; Vidart d'Egurbide Bagazgoïtia et al., 2018).

Negative effects from environmental pollution can affect living beings directly and indirectly. Air pollution (high level of $PM_{2.5}$ concentration) could, for example, increase the risk of young children (lower than 5 years old) in Vietnam suffer from an acute lower respiratory infection, requiring hospital admission (Luong et al., 2020). However, it can also favor that other harmful events achieve huge proportions. In the pandemic situation, world is facing in 2020, high level of Severe Acute Respiratory Syndrome CoronaVirus 2 lethality, for example, meets in atmospheric pollution in Northern Italy a relevant cofactor (Conticini et al., 2020).

Phytoremediation. DOI: https://doi.org/10.1016/B978-0-323-89874-4.00023-6

Among environs' contaminants the persistent ones are of major concern once they are generally stable substances that may last for a long time on polluted areas or contaminated organisms (once they are not easily degraded), threatening generations (Parrilli et al., 2010). Perfluorooctane sulfonate is a toxic persistent pollutant capable, for example, to alter gut microbiome from mice in contact with it (inducing DNA damage in bacteria), and also to disrupt liver metabolic pathways (Zhang et al., 2020). A large number of organic pollutants can be classified as persistent such as legacy brominated flame retardants, chlorinated paraffins, organochlorine pesticides, polychlorinated biphenyls, and perfluorinated compounds. In China, for example, high levels of short-chain chlorinated paraffins and 1,1-dichloro-2,2-bis (p-chlorophenyl) ethylene (p,p'-DDE) were find in breast milk, posing as severe risk to newborns' health (Hu et al., 2021). So, some organic substances can also bioaccumulate in living beings reaching high levels. Pharmaceuticals, for example, can increase their concentrations in each level of food chain poisoning organisms like zooplankton (Xie et al., 2017), algae (Vannini et al., 2011), mussels (Maruya et al., 2014), and fish (Du et al., 2012) after being incorrectly discharged in wastewaters. Inorganic pollutants such as heavy metals (for example, Hg, Cr, Zn, Cd) can also exhibit bioaccumulation capacity and persist in polluted areas being highly toxic even in small concentrations, affecting not only the health of living beings but also negatively impacting soil quality and agricultural production (Ojuederie and Babalola, 2017; Gupta et al., 2016; Chaturvedi et al., 2015; Ndeddy Aka and Babalola, 2016). They may cause from poisoning to death, being related to serious diseases, such as dementias and cancer (Jaishankar et al., 2014; Jacob et al., 2018; Ali et al., 2019). Common sources of heavy metals are mine exploitation, paints, electronic waste, pesticides, waste oil, some personal care products, pharmaceuticals and photographic chemicals (Tahiri et al., 2017).

In order to deal with environmental pollutants aiming their removal from contaminated areas and/or drastic reduction of their negative effects, there are different remediation strategies such as using nonliving substances such as nanomaterials (Jain and Vaya, 2017) and biochar (Varjani et al., 2019), living beings like microbes (Gupta et al., 2019) and plants (Panesaret al., 2019), or both (Dominguez et al., 2020).

Classic physicochemical technologies used to remediate (such as plasma photocatalysis (Baldissarelli et al., 2019)) may present some disadvantages such as high cost to be performed in a large scale (Ye et al., 2017) and these technologies can also be not eco-friendly generating more waste (Dasgupta et al., 2015; Albering et al., 2016; Bilal et al., 2018).

In order to try to offer a more advantageous protocol to remediate, bioremediation field developed proposing methods that use biological structures to restore contaminated environs (Strong and Burgess, 2008). Among these methods, phytoremediation stands out as a sustainable strategy that uses plants (and in some mechanisms also the microbes associated with plants' roots) to deal with soil, water, and air pollution (Lee et al., 2020; He et al., 2020).

This chapter will explore different mechanisms by which plant species can remediate and restore contaminated environs.

3.2 Phytoremediation: plants promoting bioremediation

By using plants, it is possible to allow *in-situ* and, through some mechanisms, *ex-situ* remediation of environmental contamination. This advantageous strategy of dealing with pollutants attenuating or eliminating their nocive effects generally requires low initial investments, offers an auto-sustained process that is eco-friendly and well accepted by scientific as safe and feasible strategy (Muthusaravanan et al., 2018). Plants considered good candidates to perform phytoremediation generally possess some characteristics such as deep roots, fast growth, and high biomass: being capable of storing a wide range of pollutants such as heavy metals in its aerial organs and offering a root system interacting with living beings to increase rhizosphere's secretions to the decomposition of contaminants (Rostami and Azhdarpoor, 2019). The efficiency of phytoremediation can also be influenced by abiotic factors or biotic aspects.

When it comes to biotic factors, plant species can be used alone, combined with other vegetal species, or together with different living beings, such as microbes in an attempt to improve processes' efficiency. The association between plant species *Eupatorium odoratum* and *Eichhornia crassipes*, for example, has proven to be advantageous to remediate an organic pollutant. In low concentration, the extract from *E. odoratum* helps to improve phytoremediation attenuating formaldehyde toxic action over *E. crassipes* during the process (Gong et al., 2018). Rhizobacteria and plant endophytes that induce plant growth are also interesting tools to improved phytoremediation by microbe-assistance, especially when dealing with environs polluted by different contaminants (mixed pollution) (He et al., 2020).

Among abiotic factors that can influence phytoremediation are pollutants composition (single contaminant or mixed pollution), presence of chelators or other substances such as the ones that can act as soil amendments and environment's pH (Radziemska et al., 2019; Rostami and Azhdarpoor, 2019; Xu et al., 2019, Yu et al., 2019; He et al., 2020). Biochar, for example, is an interesting assistant in some phytoremediation protocols once this product of biomass pyrolysis may sequester carbon (Woolf et al., 2010), regulate soil pH, and reduce the stress contaminants can cause to plants species remediating. It can, for example, help *Ricinus communis* to deal with oxidative stress and other toxic effects of Pb, increasing plant's tolerance to these metals during phytoremediation; by doing that, it can improve the process' efficiency and also improve plant growth by enhancing nutrient intake and diminishing harmful effects that the contaminant may cause (Kiran and Prasad, 2019).

The mechanism by which the environmental restoration will be performed (mechanism of phytoremediation) depends on plant species, contaminant's chemical nature, and environ to be remediated (Panesar et al., 2019). The main mechanisms are phytoaccumulation, phytodegradation, phytohydraulic control, phytoremediation, phytostabilization, phytovolatilization, rhizofiltration, and rhizodegradation.

When it comes to contaminants' chemical natures, as previously discussed, pollutants can be organic (such as fuels, dyes and pesticides) or inorganic (such as nitrates, phosphates, radioactive isotopes, and heavy metals). Mechanisms of phytoremediation that involve sequestration of pollutants or their volatilization are more suitable to remediate inorganic pollutants, and mechanisms involving substances metabolization/degradation are commonly used to organic contaminants (Nwoko, 2010).

Consequently, as different cellular machinery is necessary to perform different remediation mechanisms, there are plant species more suitable to remediate each group of pollutants or both. And depending on the specific characteristics of plant species, they can be able to remediate better water and/or soil. Water hyacinth (*E. crassipes*) can remediate aqueous environs containing organic structures such as dyes, but can also remediate inorganic substances such as phosphate, heavy metals, and radioactive isotopes (Jayaweera and Kasturiarachchi, 2004; Nie et al., 2015; Thapa et al., 2016; Priya and Selvan, 2017).

3.3 Mechanisms of phytoremediation

Remediation of environmental pollution promoted in a natural way and at low costs (Rajakaruna et al., 2006; Mahajan and Kaushal, 2018) by plants can occur, as previously mentioned, through different mechanisms (Figs 3.1–3.7). In this section, phytoaccumulation, phytodegradation, phytohydraulic control, phytostabilization, phytovolatilization, rhizofiltration, and rhizodegradation will be explored.

3.3.1 Phytoaccumulation

Phytoaccumulation is a mechanism of phytoremediation that is also referred to as phytoextraction once it extracts pollutants from contaminated areas. This mechanism is performed to remediate especially inorganic pollutants (for example, heavy metals) and does not involve any kind of contaminants' transformation (Kamal et al., 2004). It removes them, without any kind of reaction, primarily from soil, and concentrates them in plant parts (mainly shoots but it can also be leaves) (Ahalya and Ramachandra, 2006; Rashid et al., 2014; Muthusaravanan et al., 2018).

As represented in Fig. 3.1, the contaminant is captured by the vegetal species among other substances the plant generally extracts, using roots, from soil or water, and transported using plants' vessels to the plant part in which it will be stored. An experiment involving 10 wetland plant species (*Acorus tatarinowii, Alocasia cucullata, Cyperus alternifolius, Echinodorus amazonicus,*

FIG. 3.1 Phytoremediation through phytoaccumulation mechanism.

Echinodorus baothii, Hydrocotyle vulgaris, Scirpus triqueter, Veronica serpyllifolia, Eleocharis geniculate, and *Panicum repens*) concluded that metal phytoaccumulation capacity is directly related to radial oxygen loss (that enables flooded and anoxic areas tolerance and detoxication process) rates (Yang et al., 2017).

The natural capacity of plants to promote this kind of remediation depends upon bioavailability of the pollutant in rhizosphere, how well a plant can deal with the abiotic stress generated by contaminants' increasing concentration, plant capacity to store large amounts of the pollutant, the rate through which roots can uptake the contaminant and the transportation capacity inside plants' organism (Etim, 2012).

A group of plants in which it is possible to find examples of phytoaccumulation is the one from hyperaccumulators, capable of remediating high concentration of pollutants accumulating them inside the plant. There are species that are able to store more than 1000 μg of cooper to 1g of dry plant. *Calandula officinalis,* for example, can deal well with the abiotic stress-induced by Cu ions at 300 mg/kg of soil samples, by increasing the production of antioxidant enzymes like superoxide dismutase and catalase. It can accumulate Cu up to concentrations of 4675 μg/g in leafs and 3995 μg/g in roots (Goswami and Das, 2016). Hyperaccumulator grass species such as *Vetiveria zizanioides* (lately named *Chrysopogon zizanioides*), *Penisetum purpurium* and *Chloris guyana* can efficiently remediate leachate generated by solid waste phytoaccumulating Mn, Cu, and Zn (Hassan et al., 2020). *V. zizanioides* can also phytoaccumulate polycyclic aromatic hydrocarbons remediating diesel contaminated soil (Un Nisa and Rashid, 2015). The tin-hyperaccumulator (Ashraf et al., 2011) *Cyperus Rotundus* is a grass species that can efficiently extract Cd and Cr from polluted soil presenting bioconcentration factors of 44.18 and 4.42, respectively, for these pollutants accumulated in plant body (Subhasini and Swamy, 2014). *Brassica juncea* can enhance its level of the antioxidants to tolerate better the presence of the contaminants Fe, Cr, Mn, and Zn phytoaccumulating them in the presence of amendments of tannery sludge (Singh and Sinha, 2005). The Cd-hyperaccumulator *Centella asiatica* (Liu et al., 2016) can accumulate up to 1353.0 mg of cadmium per kg whereas *E. crassipes* possess the capacity to accumulate 1147.5 mg/kg in roots, both by phytoaccumulation, remediating, for example, Cd-polluted wastewater (Mokhtar et al., 2011). The hyperaccumulator *Pelargonium hortoum* (Orroño and Lavado, 2009) can efficiently phytoaccumulate Pb ions, and this capacity may be enhanced by the presence of some amendments such as citric acid (Gul et al., 2020).

However, not only hyperaccumulators can phytoaccumulate metals; *Arabidopsis thaliana,* for example, by using a metabolite derived from cysteine (methyl cysteinate) can phytoaccumulate cesium in roots. On root's surface or inside the cells methyl cysteinate can bind to Cs enhancing storage (Adams et al., 2017).

3.3.2 Phytohydraulic control

As same as in phytoaccumulation, in this mechanism, the contaminants are not degraded. Phytohydraulic contol or phytohydraulics consists in dealing with pollutants by hydrology, influencing on groundwater's cycle/movement (Awasthi et al., 2020). It is an indirect phytoremediation mechanisms commonly performed by trees and that aims to deal with the pollution present in groundwater (Khalifa and Alkhalf, 2018). The plants work as natural pumps, taking water from groundwater to use or return to the environment through transpiration (Fig. 3.2) followed by condensation and raining. If there is pollution present in this water, roots take up contributing to limit contaminants' migration, also reducing or eliminating its presence (Ahlfeld and Heidari, 1994; Muthusaravanan et al., 2018).

Sometimes plant species can perform both phytohydraulics and phytotransformation, in this situation helping to drastically reduce the threaten posed by the pollutants. *Liquidambar styraciflua, Salix nigra,* and wetland plants *Allenrolfea occidentalis*, *Polygonum punctatum*, *Nymphaea odorata*, *Spirodela polyrhiza* can not only uptake perchlorates, together with water, through the roots, but also phytotransformate them into simpler and smaller metabolites (Susarla et al., 2000).

FIG. 3.2 Phytoremediation through phytohydraulic control.

Some plant's characteristics are considered advantageous to allow them to perform this kind of phytoremediation. Large root mass and high transpiration rate make plants capable of removing pollutants in a large amount and avoid their reinfiltration/leaching back to the water table (Etim, 2012).

Two genres deserve to be highlighted when discussing phytohydraulic control: *Populus* (poplars) and *Salix* (willows): phreatophytic trees that can extract large amounts of shallow groundwater. High-transpiring plants from *Populus* genre, such as *Populus nigra* L. possess genotypes (for example, the Spanish) that present low water use efficiency together with a high transpiration rate and large root system (Hirsh et al., 2003; Bogeat-Triboulot et al., 2019). *Salix nigra* proved to be able to capture shallow contaminated groundwater in a place that once was a landfill in Canada, removing pollutants (Danielescu et al., 2020).

3.3.3 Phytodegradation

Phytodegradation is a phytoremediation mechanism also known as phytotransformation as it involves not only contaminants uptake but also modification. It is commonly performed to remediate organic contaminants (for example, polycyclic aromatic hydrocarbons and polychlorinated biphenyls), transforming them by reactions into less complex metabolites (Newman and Reynolds, 2004)

FIG. 3.3 Phytoremediation through phytodegradation mechanism.

(Fig. 3.3). These metabolites can be released back to the environment in less or non-nocive form, or be used by the plant to grow or perform other metabolic activity (Awasthi t al., 2020); (Trap et al., 2005).

Process' efficiency can suffer the influence from diverse factors such as contaminants' concentration on the polluted area to be remediated (that may influence uptake rate), plant phytochemical's properties, and contaminants hydrophobicity (that directly impacts uptake efficiency) (Etim, 2012).

The explosive trinitrotoluene could not only be uptaken by *Myriophyllum aquaticum* plants but also transformed allowing the recovery of 30% for products in aqueous medium. The remaining products were distributed inside the plant, mainly roots, to be destined (Vanderford et al., 1997).

The phosphorus pesticide ethion could be phytotransformed by water hyacinth (*Eichhornia crassipes*); the pollutant uptake was performed 69% by phytodegradation mechanism decreasing pollutants' presence to 91% in shoots and up to 81% in roots after 1 week in the absence of the contaminant (Xia and Ma, 2006).

Scirpus grossus could phytoremediate diesel's contaminated water degrading petroleum hydrocarbons up to 81.5% to a concentration of 0.1% diesel (v/v) and 66.6% to 0.25% pollutant's concentration after 72 days (Al-Baldawi et al., 2015).

Phragmites australis could phytodegrade the pharmaceutical Ibuprofen (a persistent and biologically active emerging contaminant); the cytochrome P450 monooxygenase enzyme could convert the pollutant into hydroxyl-Ibuprofen, 1,2-dihydroxy-Ibuprofen, carboxy Ibuprofen, and glucopyranosyloxy-hydroxy-Ibuprofen (He et al., 2017). Degradation of organic contaminants is generally performed by a set of enzymes in several steps. Cytochrome P450 monooxygenase enzymes play a vital role on these pollutants' phytoremediation through hydroxylation in a biochemical process that uses NADPH to activate molecular oxygen to degradation occur (Rostami and Azhdarpoor, 2019).

Other emerging and concerning pollutants such as surfactants can also be remediated through phytodegradation. The anionic surfactant sodium dodecyl sulfate could be efficiently removed from water by *E. crassipes* and converted into low-molecular-weight products. The presence of *Chromolaena odorata* L. extract enhanced specially root degradation of the pollutant, involving the enzyme ascorbate peroxidase (Gong et al., 2019).

Some microbes associated with plant roots species can offer considerable enhancement on process' efficiency. There are authors from phytoremediation field that refer to this as phytodegradation (Pawlik et al., 2020), but there are researchers that will use the term rhizodegradation (Maqbool et al., 2012), as in this chapter.

3.3.4 Rhizodegradation

This remediation mechanism is similar to phytodegradation and involves not only plant species but also microbes (Fig. 3.4). Mainly organic contaminants

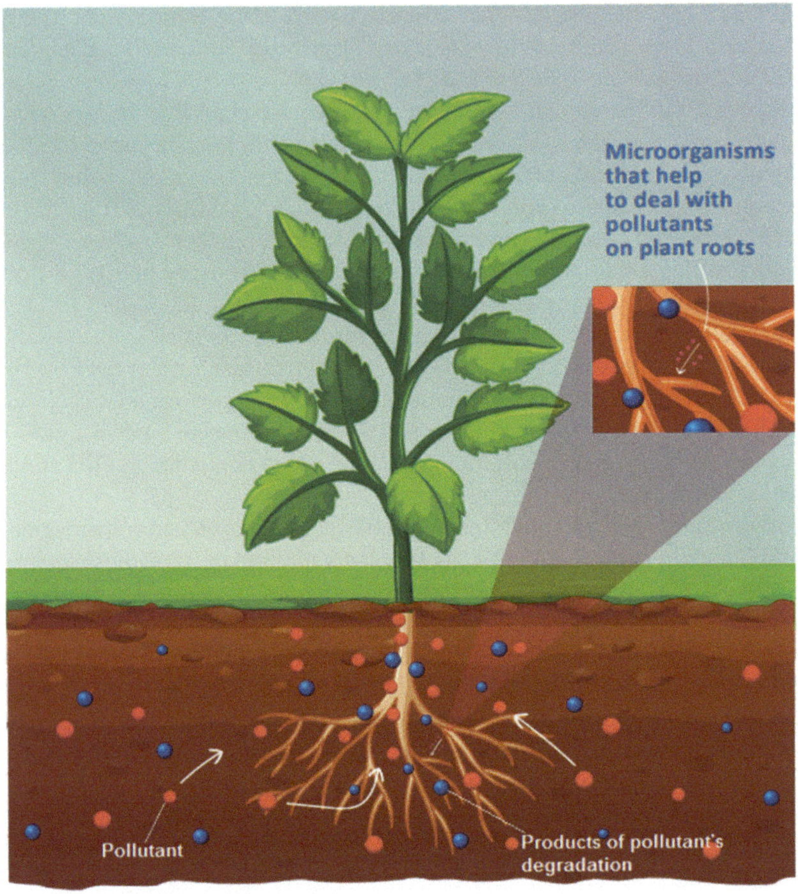

FIG. 3.4 Phytoremediation through rhizodegradation.

can be degraded through this process that involves microorganisms (commonly plant growth-promoting rhizobacteria) associated with the plant's roots. These microorganisms are important to optimize degradation and plants can stimuli their growth by releasing metabolites, such as growth factors, in rhizosphere (Dominguez et al., 2020).

The plant species *Sesbania cannabina* could remediate petroleum-polluted soil through rhizodegradation in the presence of indigenous bacteria in rhizosphere (for example *Rhizobium radiobacter* and *Delta proteobacterium*) reducing total petroleum hydrocarbon concentration from 2541 mg/kg to 679 mg/kg (Maqbool et al., 2012).

Medicago sativa (a N-symbiotic plant) could rhyzodegrade n-alkanes from contaminated soil in a rapid way in the presence of *Bacillus altitudinis*, offering degradation more efficient than the one achieved in the presence of *Bacillus cereus* or *Stenotrophomonas maltophilia* (Bano et al., 2015).

Hylotelephium spectabile could improve its capacity to degrade petroleum hydrocarbons (enhancing its copy number of genes related to petroleum hydrocarbons' degradation up to 14.44 times) in the presence of high salt tolerance microorganisms *Alcanivorax* and *Bacteroidetes* (Cheng et al., 2019).

Sorghum x drummondii in hydroponic culture could promote rhizodegradation of polycyclic aromatic hydrocarbons in an optimized manner in the presence of *Sphingobium sp.* in a way independent of plant's age and culture conditions (Dominguez et al., 2020).

Lolium perenne could remediate soil polluted by petroleum hydrocarbon suffering influence of microorganisms. Preinoculation of the caryopses followed by soil inoculation with *Rhodococcus erythropolis* 5WK and *Rhizobium sp.* 10WK offered a high rate of pollutant's removal, surpassing plant's capacity alone, and the results obtained after direct soil inoculation (Pawlik et al., 2020).

Petroleum oily sludge-spiked soil could be remediated by *Cajanus cajan* in association to a bacterial phylum composed by *Actinobacteria, Proteobacteria, Bacteroidetes, Acidobacteria, Firmicutes, Gemmatimonadetes, Saccharibacteria, Chloroflexi*, and *Verrucomicrobia*. The removal rate reached 92% to 1%(w/w) polluted sample. However, the polluted increasing concentration impaired the process reaching 47.3% removal to a 5%(w/w) concentration (Allamin et al., 2020).

3.3.5 Phytostabilization

Phytostabilization mechanism is very useful to diminish the risk of former mine areas, army bases, and battlefields. It depends on contaminant's uptake to promote pollutant's damaging effects reduction by reducing its bioavailability through immobilization (migration limitation) in rhizosphere (Fig. 3.5) (Berti and Cunningham, 2000; Mendez and Maier, 2008; Muthusaravanan et al., 2018).

Senna multijuga could, in later development, phytostabilize soil polluted by copper. Peat addition did not improve the remediation capacity but could improve morphological parameters of the plant, acting as soil conditioner (De Marco et al., 2017).

Festuca rubra could remediate postmilitary areas phytostabilizing heavy metals as Cu, Ni, and Cd. The addition of the amendment limestone, improved soil pH, optimizing the remediation process (Radziemska et al., 2017).

Cadmium contaminated soil could be remediated through phytostabilization by *Osmanthus fragrans, Ligustrum vicaryi L., Loropetalum chinense var. rubrum*, and *Euonymus japonicas cv. Aureo-mar* dealing with contents of pollutant, respectively, of 15.76, 19.09, 20.59, and 32.91 mg/kg (Zeng et al., 2018).

Contaminants Cd, Cu, Pb, and S could be phytostabilized by *Erica australis* (a plant species that can survive well in extreme acidic condition) through deposition in root cortex also improving vegetation's restoration by less tolerant plant species such as *Nerium oleander L.* (Monaci et al., 2019).

FIG. 3.5 Phytoremediation through phytostabilization.

Miscanthus × *giganteus* proved to be able to phytostabilize Cd and Hg from contaminated soil (Zgorelec et al., 2020).

The phytostabilization's mechanism performed by some plant species may benefit not only from soil inorganic amendments presence as previously reported (Radziemska et al., 2017). The product of biomass' pyrolysis, known as biochar (Saletnik et al., 2019) can also contribute to pollutants' immobilization (for example, capturing metals such as Cd and Zn) facilitating and optimizing remediation promoted by corn plants, as same as manure-based compost (Sigua et al., 2019). *Boehmeria nivea* L. could phytostabilize Cd and Pb from polluted mine soils and process' efficiency could be improved by simultaneous application not only of amendment (Fe-Si-Ca material) but also of organic fertilizer and coconut shell biochar (Lan et al., 2020).

Although phytostabilization does not depend upon microorganisms, it can also contribute to plants' capacity to remediate through this mechanism. *Funneliformis mosseae* (fungi species) could contribute to Pb immobilization and helped to reduce the toxicity it may induce on plant roots enhancing the quality of Pb phytostabilization promoted by *Robinia pseudoacacia* (Huang et al., 2019).

3.3.6 Phytovolatilization

Phytovolatilization is a controversial phytoremediation mechanism to be employed to restore polluted environs as it may cause secondary pollution. It is interesting to evaluate if the pollutant will not be converted into an equally or more toxic substance. In this process plants remove pollutants uptaking them, commonly metabolize them into less toxic form and then release the contaminants as volatile substances to atmosphere (Fig. 3.6). The return of

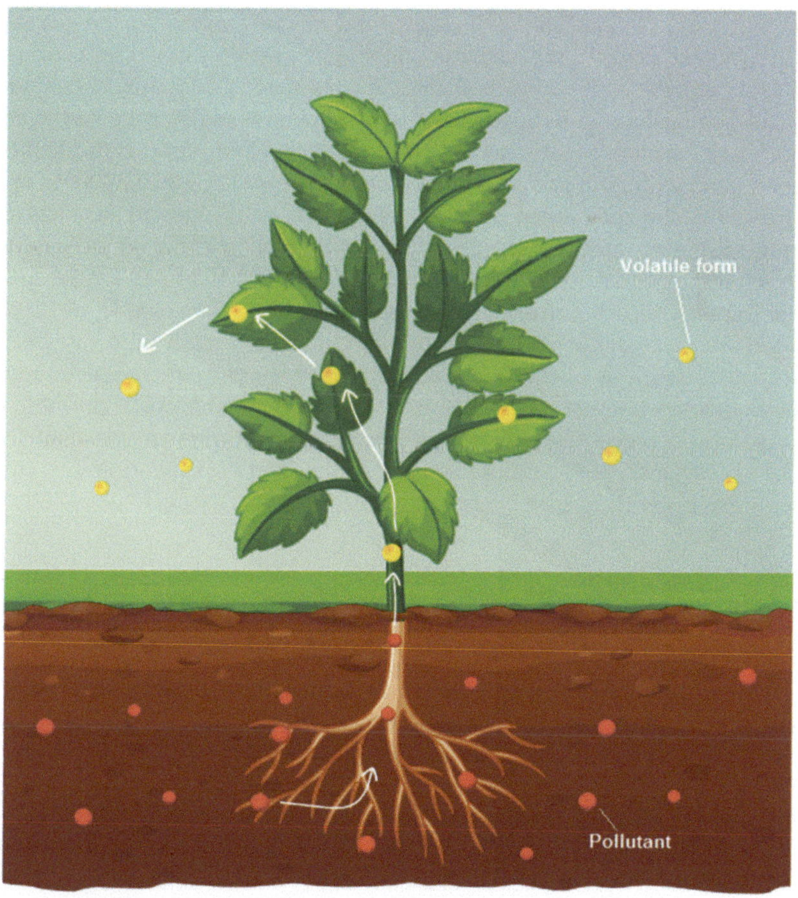

FIG. 3.6 Phytoremediation through phytovolatilization.

remediation's products to atmosphere allows the phytovolatilization classification into: direct—the return occurs from leaves or stems, or indirect—the return occurs from roots (Limmer and Burken, 2016).

Pteris vittata, for example, could remediate mercury and arsenic pollution through this mechanism. However, it is not an eco-friendly procedure once mercuric oxide and dimethyl selenide produced as final products are highly toxic substances (Sakakibara et al., 2010).

Some highly popular species for their enormous capacity to uptake pollutants (for example the ones from *Populus* and *Salix* genre) and species capable to generate volatile products from heavy metals (for example, *Brassica juncea* and *A. thaliana*) are good candidates for this kind of remediation protocol (Pulford and Watson, 2003; Ghosh and Singh, 2005). *Brassica juncea* L. Czern. (the Indian mustard) could phytovolatilize, mainly from roots, mercury: increasing the amount released by plant linearly as a function of pollutant's concentration. However, the vapor of Hg is not a desirable product to be released on atmosphere (Moreno et al., 2008). *Populus deltoides* and *Populus nigra* are suitable plant species (poplar trees) to remediate perchloroethylene; they could transform the pollutant reducing by 99% the total mass of chlorinated ethenes in contaminated water. When dealing with the pollutant volatilization could be triggered in spite of not being the main procedure (that consisted in metabolization of the contaminant into trichloroethylene and cis-dichloroethylene) (James et al., 2009).

However, the mentioned species are not the only ones to perform this phytoremediation mechanism. *Eucalyptus sideroxylon* could phytovolatilize trichloroethylene present on the polluted soil of in Travis and Fairchild Air Force Bases, California, from plant's leaves and roots (Doucette et al., 2013). *Arundo donax* L. could phytovolatilize arsenic pollution at high rates being able to deal with this contaminant eliminating 75% of its total concentration through this mechanism (Guarino et al., 2020). Rice plants could promote the phytovolatilization of 2,4-dibromophenol and 2,4-dibromoanisole contributing to 41.7% and 6.78% of these contaminants' total volatilization into atmosphere (Zhang et al., 2020).

3.3.7 Rhizofiltration

Rhizofiltration is a phytoremediation mechanism similar to phytoaccumulation differentiating mainly on the type of polluted environ preferentially remediated. While phytoaccumulation is a mechanism of remediating mainly soil, rhizofiltration is performed to remediate surface or groundwater, especially contaminated by metals (Ahalya and Ramachandra, 2006). A plant rhizofiltrating uptakes the pollutant through the roots being able to store it on this plant's part (Fig. 3.7) (Panesar et al., 2019). Contaminants do not need to be conducted to shoots and *ex situ* remediation is also possible through this mechanism (Rastogi and Nandal, 2020). Researchers grow vegetal species to apply on this kind of remediation's protocol hydroponically (Mikheev et al., 2017; Tiwari et al., 2019).

FIG. 3.7 Phytoremediation through rhizofiltration.

Helianthus annuus could rhizofiltrate uranium from polluted water samples (groundwater or water contaminated in the laboratory) reducing, in a day time, 80% of the element's concentration in a way that at the end of the process the remaining pollutant level was below drinking water limit. *Phaseolus vulgaris*, as same as *H. annuus*, presented removal capacity of the pollutant using the roots larger than 500 mg/kg and *P. vulgaris* remediated 60% to 80% of the contaminant (Lee and Yang, 2010).

Uranium polluted environs could also be remediated by *Phragmites australis* and Fe plaque assistance could allow achievement of better results in rhizofiltration protocols by enhancing uranium retention on root (Wang and Dudel, 2017). NH_4^+ dependent reductive milieu as same as enhanced plant's transpiration this contaminant's rhizofiltration could be enhanced differently from NO_3^- rich milieu that benefited pollutant's retention on roots (Wang and Dudel, 2018).

Barley, oat, pea, corn, and timothy in bioplateau, using foam plastic as substrate, could remediated radionuclides such as ^{137}Cs from polluted water. The system offered decrease of 80% in radioactivity in 24 hours, 90% in 48 hours, and 97% in 5 days: mainly due to timothy's (*Phleum pretense*) promoted rhizofiltration (Mikheev et al., 2017).

Pistia stratiotes proved to be a phytoremediator to rhizofiltration of polluted Egyptian wetlands. It possesses an interesting potential of rhizofiltration for Cr, Pb, and Cu and 10 times higher for Fe (Galal et al., 2018).

A. donax constructed wetland could deal with Cu pollution rhizofiltrating the pollutant offering in 30 days remaining concentration of 3.5% of the initial Cu in water (Oustriere et al., 2017).

P. australis could be used in rhizofiltration system to deal with pollution related to nitrogen species in warmer months. Nitrates were removed up to 60.4% by plants while in systems without *P. australis* only 45.4% was denitrified (Sikhosana et al., 2020).

3.4 Transgenic plants for improved phytoremediation

As previously discussed in this chapter, phytoremediation can be performed through different mechanisms by different plant species naturally capable of dealing with environmental pollution. However, not only plants that are naturally able to remediate polluted areas can be used in phytoremediation protocols. Plants can be genetically modified to acquire this capacity and plants that are already able to phytoremediate can have this capacity enhanced by genome modifications (Prasad, 2019) (Table 3.1). A common type of DNA alteration is to improve the plant's capacity to deal with the stress generated by the pollutant that can end up limiting the success of remediation (Tiwari and Lata, 2018). Strategies of genetic modification to generate a plant optimized to phytoremediate commonly involves genes related to (1) functionalization/chemical modifications of contaminants through oxidations, reductions, and hydrolysis; (2) conjugation of pollutants with sugars, glutathione and amino acids; (3) pollutants' transport/compartmentalization/sequestration (Rai et al., 2019).

DNA sequence used to generate a transgenic plant can divide the organisms generated into two groups: (1) autotransgenic organisms (the DNA sequence used to generate the transgenic comes from the DNA of an organism from the same species) generally designed to overexpress a previously existent sequence accentuating a previously existing characteristic (Reis et al., 2006). For example, *Solanum tuberosum* engineered to overexpress DREB transcription factors to tolerate and improve performance dealing with Cd pollution (Charfeddine et al., 2017); (2) allotransgenic organism (the DNA sequence inserted on target's genome comes from a different species) generally designed to confer new characteristic to transgenic organism (Ozyigit et al., 2020). For example,

TABLE 3.1 Examples of genetic engineered plants designed to phytoremediation.

DNA sequence source	Transgene	Genetically modified organism	Pollutant to be remediated	References
Escherichia cloacae	*Onr*	*N. tabacum*	Nitroglycerin	French et al. (1999)
Homo sapiens	*CYP2E1*	*Arabidopsis thaliana*	Trichloroethylene and ethylene dibromide	Doty et al. (2000)
Astragalus bisulcatus	Selenocysteine methyltransferase gene	*Brassica juncea*	Se	LeDuc et al. (2004)
Pseudomonas sp.	Atrazine chlorohydrolase gene	*Medicago sativa* and *N. tabacum*	Atrazine	Wang et al. (2005)
Escherichia coli	Nitroreductase gene	*Arabidopsis thaliana*	2,4,6-trinitrotoluene	Kurumata et al., (2005)
Homo sapiens	*CYP1A1, CYP2B6* and *CYP2C19*	*Oryza sativa*	Atrazine, metolachlor, and simazine	Kawahigashi et al., (2006)
Enterobacter cloacae	*NfsI*	*N. tabacum*	2,4,6-trinitrotoluene	Hannink et al. (2007)
Sedum alfredii	MT2 gene	*Nicotiana tabacum*	Cu	Zhang et al. (2014)
Rhodococcus rhodochorus	*XplA, XplB* genes	*Arabidopsis thaliana*	1,3,5-trinitro-1,3,5-triazine	Jackson et al. (2007)
Pseudomonas	Genes from naphthalene dioxygenase system	*Arabidopsis* and rice plants	Naphthalene and phenanthrene	Peng et al. (2014)
Metagenomic library was constructed from polluted soil sample	2,3-dihydroxybiphenyl-1,2-dioxygenase gene	*Medicago sativa*	Polychlorinated biphenyls and 2,4-dichlorophenol	Wang et al., (2015)
Pseudomonas fluorescens	*copC*	*N. tabacum*	Cu	Pérez-Palacios et al. (2017)
Enterobacter cloacae	*nfsI*	*Pascopyrum smithii*	2,4,6-trinitrotoluene	Zhang et al. (2018)
Rhodococcus rhodochrous	*xplA* and *xplB*	*Pascopyrum smithii*	Hexahydro-1,3,5-trinitro-1,3,5-triazine	Zhang et al. (2018)

Nicotiana tabacum engineered to express *CzcB* or *CzcA* genes from bacterium *Pseudomonas putida* increased its capacity to accumulate the inorganic pollutant Cd in the roots (Nesler et al., 2017).

The first transgenic plants were developed for the remediation of heavy metals (Misra and Gedama, 1989; Rugh et al., 1996). Years later, transgenesis in plants also aimed organic compounds' remediation such as explosives and halogenated organic compounds (Mackova et al., 1998; Doty et al., 2000; Eapen et al., 2007; Macek et al., 2008).

Different strategies can be used to deliver transgenes to plants to generate organisms optimized to phytoremediate. Biolistics, for example, allowed the delivery of transgenes through microprojectile bombardment to different plant species (Ismagul et al., 2018). *N. tabacum* through biolistic could be genetically modified to overexpress the coding sequence of glutathione S-transferase I from *Zea mays* becoming an organism capable to phytoremediate chloroacetanilide herbicides (Karavangeli et al., 2005). However, the most popular methodology uses organisms from *Agrobacterium* gender (mainly *Agrobacterium tumefaciens*): plant pathogens commonly found on soil containing the Ti-plasmid on which transgene to be delivered can be inserted in T-DNA region to be integrated on the DNA of target organism (Cunningham et al., 2018; Thompson et al., 2020). There are also protocols that are designed to be performed using both technologies to generate plants with more potential to phytoremediate; for example, the photosynthesizing roots of *Trapa natans* could be generated as hairy roots organism (suitable for phytoremediation) through *Agrobacterium* and biolistic transformation (Mikhaylova et al., 2020).

3.5 Conclusion

Phytoaccumulation, phytodegradation, phytohydraulic control, phytostabilization, phytovolatilization, rhizofiltration, and rhizodegradation are important phytoremediation mechanisms that allow restoration of polluted environs. They present efficiencies to deal with contaminants that vary with pollutant chemical nature, plant species chosen, and type of sample to be remediated, and the results can be optimized by conjugating phytoremediation with other strategies such as biochar use. There are plant species naturally capable to be used on remediation protocols and recombinant DNA technology also allows the generation of genetic modification in plants especially designing them to efficiently perform phytoremediation.

3.6 Futures perspectives

It is expected that innovative phytoremediation protocols continue to be developed especially aiming to improve process' efficiency. For example, it is known that some heavy metals can reduce photosynthesis rates and impair plant nutrition and other essential enzymatic activities. Protocols of phytoremediation that allow plants to hyperaccumulate heavy metals or deal with high amounts of them without suffering from that, are desirable.

There are also welcome new protocols to avoid incomplete remediation and to avoid secondary pollution, such as air pollution as a consequence of phytovolatilization.

Advances in molecular biology field in order to make DNA editing more precise and safe are also expected to enhance the quality of genetically modified plants generated to phytoremediate polluted areas. Safety issues should receive increasing attention especially when it comes to field experiments: that are expected to increase in number and extension of performance.

References

Adams, E., Miyazaki, T., Hayaishi-Satoh, A., Han, M., Kusano, M., Khandelia, H., et al., 2017. A novel role for methyl cysteinate, a cysteine derivative, in cesium accumulation in *Arabidopsis thaliana*. Sci. Rep. 7, 43170.

Ahalya, N., Ramachandra, T.V., 2006. Phytoremediation: processes and Mechanisms. J. Ecobiol. 18, 33–38.

Ahlfeld, D.P., Heidari, M., 1994. Applications of optimal hydraulic control to ground-water systems. J. Water Res. Planning Manage. 120, 350–365.

Al-Baldawi, I.A., Abdullah, S.R.S., Anuar, N., Suja, F., Mushrifah, I, 2015. Phytodegradation of total petroleum hydrocarbon (TPH) in diesel-contaminated water using *Scirpus grossus*. Ecol. Eng. 74, 463–473.

Albering, H.J., Rila, J.P., Moonen, E.J., Hoogewerff, J.A., Kleinjans, J.C., 2016. Human health risk assessment in relation to environmental pollution of two artificial freshwater lakes in the Netherlands. J. Res. Med. Sci. 107, 27–35.

Ali, H., Khan, E., Ilahi, I., 2019. Environmental chemistry and ecotoxicology of hazardous heavy metals: environmental persistence, toxicity, and bioaccumulation. J. Chem. 2019, 1–14.

Allamin, I.A., Halmi, M.I.E., Yasid, N.A., Ahmad, S.A., Abdullah, S.R.S., Shukor, Y, 2020. Rhizodegradation of petroleum oily sludge-contaminated soil using Cajanus cajan increases the diversity of soil microbial community. Sci. Rep. 10, 4094.

Ashraf, M.A., Maah, M.J., Yusoff, I., 2011. Heavy metals accumulation in plants growing in ex tin mining catchment. Int. J. Environ. Sci. Technol. 8, 401–416.

Awasthi, M.K., Guo, D., Awacthi, S.K., Wang, Q., Chen, H., Liu, T., et al., 2020. Recent advances in phytoremediation of toxic metals from contaminated sites: a road map to a safer environment. In: Bharagava, R.N., Saxena, G. (Eds.), Biorremediaiton of Industrial Waste for Environmental Safety. Springer, Singapore, pp. 77–112.

Bagazgoïtia, Vidart d'Egurbide, N., Bailey, D., H., Orsi, L., Lacour, B., Guerrini-Rousseau, L., Bertozzi, A.I., et al., 2018. Maternal residential pesticide useduring pregnancy and risk of malignant childhood brain tumors: a pooledanalysis of the ESCALE and ESTELLE studies (SFCE). Int. J. Cancer 142, 489–497.

Baldissarelli, D.P., Vargas, G.D.L.P., Korf, E.P., Galon, L., Kaufmann, C., Santos, J.B, 2019. Remediation of soils contaminated by pesticides using physicochemical processes: a brief review. Planta Daninha 37, e019184975.

Bano, A., Shahzad, A., Siddiqui, S., 2015. Rhizodegradation of hydrocarbon from oily sludge. J. Bioremed. Biodegr. 6, 289–295.

Barrios-Estrada, C., de Jesús Rostro-Alanis, M., Muñoz-Gutiérrez, B.D., Iqbal, H.M., Kannan, S., Parra-Saldivar, R., 2018. Emergent contaminants: endocrine disruptors and their laccase-assisted degradation–a review. Sci. Total Environ. 612, 1516–1531.

Berti, W.R., Cunningham, S.D., 2000. Phytostabilization of metals. In: Raskin, I., Ensley, B.D. (Eds.), Phytoremediation of Toxic Metals: Using Plants to Clean Up the Environment. John Wiley & Sons, Hoboken, New Jersey, pp. 71–88.

Bilal, M., Adeel, M., Rasheed, T., Zhao, Y., Iqbal, H.M., 2019. Emerging contaminants of high concern and their enzyme-assisted biodegradation–a review. Environ. Int, 124, 336–353.

Bilal, M., Rasheed, T., Sosa-Hernández, J., Raza, A., Nabeel, F., Iqbal, H., 2018. Biosorption: an interplay between marine algae and potentially toxic elements—a review. Marine Drugs 16, 65–70.

Bogeat-Triboulot, M.B., Buré, C., Gerardin, T., Chuste, P.A., Le Thiec, D., Hummel, I., et al., 2019. Additive effects of high growth rate and low transpiration rate drive differences in whole plant transpiration efficiency among black poplar genotypes. Environ. Exper. Botany 166, 1–11.

Carpenter, A., 2018. Oil pollution in the North Sea: the impact of governance measures on oil pollution over several decades. Hydrobiologia 2018, 1–19.

Chaoua, S., Boussaa, S., El Gharmali, A., Boumezzough, A., 2019. Impact of irrigation with wastewater on accumulation of heavy metals in soil and crops in the region of Marrakech in Morocco. J. Saudi Soc. Agri. Sci. 18, 429–436.

Charfeddine, M., Charfeddine, S., Bouaziz, D., Messaoud, R.B., Bouzid, R.G., 2017. The effect of cadmium on transgenic potato (*Solanumtuberosum*) plants overexpressing the StDREB transcription factors. Plant Cell Tissue Organ Culture 128, 521–541.

Chaturvedi, A.D., Pal, D., Penta, S., Kumar, A., 2015. Ecotoxic heavy metals transformation by bacteria and fungi in aquatic ecosystem. World J. Microbiol. Biotechnol. 31, 1595–1603.

Cheng, L., Zhou, Q., Yu, B., 2019. Responses and roles of roots, microbes, and degrading genes in rhizosphere during phytoremediation of petroleum hydrocarbons contaminated soil. Int. J. Phytoremed. 21, 1161–1169.

Conticini, E., Frediani, B., Carob, D., 2020. Can atmospheric pollution be considered a co-factor in extremely high level of SARS-CoV-2 lethality in Northern Italy? Environ. Pollut. 261, 114465.

Cunningham, F.J., Goh, N.S., Demirer, G.S., Matos, J.L., Landry, M.P., 2018. Nanoparticle-mediated delivery towards advancing plant genetic engineering. Trends Biotechnol. 36, 882–897.

Danielescu, S., Stempvoort, D.R.V., Bickerton, G., Roy, J.W, 2020. Use of mature willows (Salix nigra) for hydraulic control of landfill-impacted groundwater in a temperate climate. J. Environ. Manage. 272, 111106.

Dasgupta, J., Sikder, J., Chakraborty, S., Curcio, S., Drioli, E., 2015. Remediation of textile effluents by membrane based treatment techniques: a state of the art review. J. Environ. Manage. 147, 55–72.

De Marco, R., da Silva, R.F., Da Ros, C.O., Vanzam, M., Boeno, D., 2017. Senna multijuga and peat in phytostabilization of copper in contaminated soil. Revista Brasileira de Engenharia Agrícola e Ambiental 21, 421–426.

Dominguez, J.J.A., Inoue, C., Chien, M.F, 2020. Hydroponic approach to assess rhizodegradation by sudangrass (*Sorghum x drummondii*) reveals pH- and plant age-dependent variability in bacterial degradation of polycyclic aromatic hydrocarbons (PAHs). J. Haz. Mater. 387, 121695.

Doty, S.L., Shang, Q.T., Wilson, A.M., Moore, A.L., Newman, L.A., Strand, S.E., et al., 2000. Enhanced metabolism of halogenated hydrocarbons in transgenic plants contain mammalian P450 2E1. Proc. Nat. Acad. Sci. USA 97, 6287–6291.

Doucette, W., Klein, H., Chard, J., Dupont, R., Plaehn, W., Bugbee, B., 2013. Volatilization of trichloroethylene from trees and soil: measurement and scaling approaches. Environ. Sci. Technol. 47, 5813–5820.

Du, B., Perez-Hurtado, P., Brooks, B.W., Chambliss, C.K., 2012. Evaluation of an isotope dilution liquid chromatography tandem mass spectrometry method for pharmaceuticals in fish. J. Chromatogr. A 1253, 177–183.

Eapen, S., Singh, S., D'Souza, S.F, 2007. Advances in development of transgenic plants for remediation of xenobiotic pollutants. Biotechnol. Adv. 25, 442–451.

Etim, E.E., 2012. Phytoremediation and Its Mechanisms: a review. Int. J. Environ. Bioenergy 2, 120–136.

Fatima, S.A., Hamid, A., Yaqub, G., Javed, A., Akram, H., 2018. Detection of volatile organic compounds in blood of farmers and their general health and safety profile. Nat. Environ. Pollut. Technol. 17, 657–660.

French, C.J., Rosser, S.J., Davies, G.J., Nicklin, S., Bruce, N.C., 1999. Biodegradation of explosives by transgenic plants expressing pentaerythritol tetranitrate reductase. Nat. Biotechnol. 17, 491–494.

Galal, T.M., Eid, E.M., Dakhil, M.A., Hassan, L.M., 2018. Bioaccumulation and rhizofiltration potential of *Pistia stratiotes* L. for mitigating water pollution in the Egyptian wetlands. Int. J. Phytoremed. 20, 440–447.

Ghosh, M., Singh, S.P., 2005. A review on phytoremediation of heavy metals and utilization of its byproducts. Appl. Ecol. Environ. Res. 3, 1–18.

Gong, Y., Chen, J., Pu, R., 2019. The enhanced removal and phytodegradation of sodium dodecyl sulfate (SDS) in wastewater using controllable water hyacinth. Int. J. Phytoremed. 2019, 1–10.

Gong, Y., Zhou, X., Ma, X., Chen, J., 2018. Sustainable removal of formaldehyde using controllable water hyacinth. J. Clean. Prod. 181, 1–7.

Goswami, S., Das, S., 2016. Copper phytoremediation potential of *Calandula officinalis* L. and the role of antioxidant enzymes in metal tolerance. Ecotoxicol. Environ. Safety 126, 211–218.

Guarino, F., Miranda, A., Castiglione, S., Cicatelli, A., 2020. Arsenic phytovolatilization and epigenetic modifications in *Arundo donax* L. assisted by a PGPR consortium. Chemosphere 251, 126310.

Gul, I., Manzoor, M., Kallerhoff, J., Arshad, M., 2020. Enhanced phytoremediation of lead by soil applied organic and inorganic amendments: Pb phytoavailability, accumulation and metal recovery. Chemosphere 258, 127405.

Gupta, A., Joia, J., Sood, A., Sood, R., Sidhu, C., Kaur, G., 2016. Microbes as potential tool for remediation of heavy metals: A review. J. Microbial Biochem. Technol. 8, 364–372.

Gupta, J., Rathour, R., Singh, R., Thakur, I.S., 2019. Production and characterization ofextracellular polymeric substance (EPS) generated by carbofuran degrading strain Cupriavidus sp. ISTL7. Biores. Technol. 282, 417–424.

Hannink, N.K., Subramanian, M., Rosser, S.J., Basran, A., Murray, J.A.H., Shanks, J.V., et al., 2007. Enhanced transformation of TNT by tobacco plants expressing a bacterial nitroreductase. Int. J. Phytoremed. 9, 385–401.

Hassan, M.M., Haleem, N., Baig, M.A., Jamal, Y., 2020. Phytoaccumulation of heavy metals from municipal solid waste leachate using different grasses under hydroponic condition. Sci. Rep. 10, 15802.

He, W., Megharaj, M., Wu, C.Y., Subashchandrabose, S.R., Dai, C.C., 2020. Endophyte-assisted phytoremediation: mechanisms and current application strategies for soil mixed pollutants. Crit. Rev. Biotechnol. 40, 31–45.

He, Y., Langenhoff, A.A.M., Sutton, N.B., Rijnaarts, H.H.M., Blokland, M.H., Chen, F., et al., 2017. Metabolism of Ibuprofen by Phragmites australis: uptake and phytodegradation. Env. Sci. Technol. 51, 4576–4584.

Hirsh, S.R., Compton, H.R., Matey, D.H., Wrobel, J.G., Schneider, W.H., 2003. Five-year pilot study: aberdeen proving ground, Maryland. In: McCutcheon, S.C., Schnoor, J.L. (Eds.), Phytoremediation: Transformation and Control of Contaminants. John Wiley & Sons, Hoboken, NJ, pp. 635–659.

Huang, L., Chen, D., Zhang, H., Song, Y., Chen, H., Tang, M., 2019. *Funneliformis mosseae* enhances root development and Pb Phytostabilization in *Robinia* pseudoacacia in Pb-Contaminated Soil. Front. Microbiol. 10, 2591.

Hu, L., Luo, D., Wang, L., Yu, M., Zhao, S., Wang, Y., et al., 2021. Levels and profiles of persistent organic pollutants in breast milk in China and their potential health risks to breastfed infants: a review. Sci. Total Environ. 753, 142028.

Ismagul, A., Yang, N., Maltseva, E., Iskakova, G., Mazonka, I., Skiba, Y., et al., 2018. A biolistic method for high-throughput production of transgenic wheat plants with single gene insertions. BMC Plant Biol. 18, 135.

Jackson, E.G., Rylott, E.L., Fournier, D., Hawari, J., Bruce, N.C., 2007. Exploring the biochemical properties and remediation applications of the unusual explosive-degrading P450 system XplA/B. Proc. Nat. Acad. Sci. USA 104, 16822–16827.

Jacob, J.M., Karthik, C., Saratale, R.G., Kumar, S.S., Prabakar, D., Kadirvelu, K., et al., 2018. Biological approaches to tackle heavy metal pollution: a survey of literature. J. Environ. Manage. 217, 56–70.

Jafari, M., Danesh, Y.R., Goltapeh, E.M., Varma, A., 2013. Bioremediation and genetically modified organisms. In: Goltapeh, E., Danesh, Y., Varma, A.M. (Eds.), Fungi as Bioremediators. Springer, Berlin, pp. 433–450.

Jain, A., Vaya, D., 2017. Photocatalytic activity of TiO_2 nanomaterial. J. Chilean Chem. Soc. 62, 3683–3690.

Jaishankar, M., Tseten, T., Anbalagan, N., Mathew, B.B., Beeregowda, K.N., 2014. Toxicity, mechanism and health effects of some heavy metals. Interdiscip. Toxicol. 7, 60–72.

James, C.A., Xin, G., Doty, S.L., Muiznieks, I., Newman, L., Strand, S.E., 2009. A mass balance study of the phytoremediation of perchloroethylene-contaminated groundwater. Environ. Pollut. 157, 2564–2569.

Jayaweera, M., Kasturiarachchi, J., 2004. Removal of Nitrogen and Phosphorus from industrial wastewaters by phytoremediation using water hyacinth (*Eichhornia crassipes* (Mart.) Solms). Water Sci. Technol. 50, 217–225.

Kamal, M., Ghaly, A.E., Mahmoud, N., CoteCôté, R., 2004. Phytoaccumulation of heavy metals by aquatic plants. Environ. Int. 29, 1029–1039.

Karavangeli, M., Labrou, N.E., Clonis, Y.D., Tsaftaris, A., 2005. Development of transgenic tobacco plants overexpressing maize glutathione S-transferase I for chloroacetanilide herbicides phytoremediation. Biomol. Eng. 22, 121–128.

Kawahigashi, H., Hirose, S., Ohkawa, H., Ohkawa, Y., 2006. Phytoremediation of herbicide atrazine and metolachlor by transgenic rice plants expressing human CYP1A1, CYP2B6 and CYP2C19. J. Agr. Food Chem. 54, 2985–2991.

Khalifa, F.K., Alkhalf, M.I., 2018. Phytoremediation as a cleansing tool for nanoparticles and pharmaceutical wastes toxicity. In: Ansari, A.A., Gill, S.S., Gill, R., Lanza, G.R., Newman, L. (Eds.), Phytoremediation: management of environmental contaminants—volume 6. Springer, New York, NY, pp. 283–294.

Kiran, B.R., Prasad, M.N.V, 2019. Biochar and rice husk ash assisted phytoremediation potentials of *Ricinus communis* L. for lead-spiked soils Author links open overlay panel. Ecotoxicol. Environ. Safety 183, 109574.

Kurumata, M., Takahashi, M., Sakamoto, A., Ramos, J.L., Nepovim, A., Vanek, T., et al., 2005. Tolerance to, and uptake and degradation of 2,4,6-trinitrotoluene (TNT) are enhanced by the expression of a bacterial nitroreductase gene in *Arabidopsis thaliana*. Zeitschrift Naturforschung C 60, 272–278.

Lan, M.M., Liu, C., Liu, S.J., Qiu, R.L., Tang, Y.T., 2020. Phytostabilization of Cd and Pb in highly polluted farmland soils using ramie and amendments. Int. J. Environ. Res. Public Health 17, 1661.

LeDuc, D.L., Tarun, A.S., Montes-Bayon, M., Meija, J., Malit, M.F., Wu, C.P., et al., 2004. Overexpression of selenocysteine methyltransferase in *Arabidopsis* and indian mustard increases selenium tolerance and accumulation. Plant Physiol. 135, 377–383.

Lee, B.X.Y., Hadibarata, T., Yuniarto, A, 2020. Phytoremediation mechanisms in air pollution control: a review. Water, Air, Soil Pollut. 231, 437.

Lee, M., Yang, M., 2010. Rhizofiltration using sunflower (*Helianthus annuus* L.) and bean (*Phaseolus vulgaris* L. var. vulgaris) to remediate uranium contaminated groundwater. J. Haz. Mater. 173, 589–596.

Limmer, M., Burken, J., 2016. Phytovolatilization of Organic Contaminants. Environ. Sci. Technol. 50, 6632–6643.

Liu, K., Zhou, Z., Yu, F., Chen, M., Chen, C., Zhu, J., et al., 2016. A newly found cadmium hyperaccumulator—Centella asiatica linn. Fresenius Environ. Bull. 25, 2668–2675.

Loizidou, M., Kapetanios, E.G., 1993. Effect of leachate from landfills on underground water quality. Sci. Total Environ. 128, 69–81.

Luong, L.T.M., Dang, T.N., Huong, N.T.T., Phung, D., Tran, L.K., Dung, D.V., et al., 2020. Particulate air pollution in Ho Chi Minh city and risk of hospital admission for acute lower respiratory infection (ALRI) among young children. Environ. Pollut. 257, 113424.

Macek, T., Kotrba, P., Svatos, A., Novakova, M., Demnerova, K., Mackova, M., 2008. Novel roles for genetically modified plants in environmental protection. Trends Biotechnol. 26, 146–152.

Mackova, M., Macek, T., Kucerova, P., Burkhard, J., Trisk, J., Demnerova, K., 1998. Plant tissue cultures in model studies of transformation of polychlorinated biphenyls. Chem. Papers 52, 599–600.

Mahajan, P., Kaushal, J., 2018. Role of phytoremediation in reducing cadmium toxicity in soil and water. J. Toxicol. 2018, 1–16.

Maqbool, F., Wang, Z., Xu, Y., Zhao, J., Gao, D., Zhao, Y.G., et al., 2012. Rhizodegradation of petroleum hydrocarbons by Sesbania cannabina in bioaugmented soil with free and immobilized consortium. J. Haz. Mater. 237-238, 262–269.

Maruya, KA, Dodder, NG, Weisberg, SB, Gregorio, D, Bishop, JS, Klosterhaus, S, et al., 2014. The Mussel Watch California pilot study on contaminants of emerging concern (CECs): synthesis and next steps. Mar. Pollut. Bull. 81, 355–363.

Mendes, K.F., Régo, A.P.J., Takeshita, V., Tornisielo, V.L, 2019. Water Resource pollution by herbicide residues. IntechOpen 2019, 85159.

Mendez, M.O., Maier, R.M., 2008. Phytostabilization of mine tailings in arid and semiarid environments—an emerging remediation technology. Environ. Health Perspect. 116, 278–283.

Mikhaylova, E.V., Artyukhin, A., Musin, K., Panfilova, M., Gumerova, G., Kuluev, B., 2020. The first report on the induction of hairy roots in *Trapa natans*, a unique aquatic plant with photosynthesizing roots. Plant Cell, Tissue Organ Culture 2020. doi:10.1007/s11240-020-01963-7.

Mikheev, A.N., Lapan, O.V., Madzhd, S.M., 2017. Experimental foundations of a new method for rhizofiltration treatment of aqueous ecosystems from 137Cs. J. Water Chem. Technol. 39, 245–249.

Misra, S., Gedamu, L., 1989. Heavy metal tolerant transgenic Brassica napus L. and Nicotiana tabacum L. plants. Theor. Appl. Genet. 78, 161–168.

Mokhtar, H., Morad, N., Fizri, F.F.A, 2011. Phytoaccumulation of copper from aqueous solutions using *Eichhornia crassipes* and *Centella asiatica*. Int. J. Environ. Sci. Dev. 2, 205–210.

Monaci, F., Trigueros, D., Mingorance, M.D., Rossini-Oliva, S., 2019. Phytostabilization potential of Erica australis L. and Nerium oleander L.: a comparative study in the Riotinto mining area (SW Spain). Environ. Geochem. Health 2019, 1–16.

Moreno, F.N., Anderson, C.W.N., Stewart, R.B., Robinson, B.H, 2008. Phytofiltration of mercury-contaminated water: volatilisation and plant-accumulation aspects. Environ. Exp. Bot. 62, 78–85.

Muthusaravanan, S., Sivarajasekar, N., Vivek, J.S., Paramasivan, T., Naushad, M., Prakashmaran, J., et al., 2018. Phytoremediation of heavy metals: mechanisms, methods and enhancements. Environ. Chem. Lett. 16, 1339–1359.

Ndeddy Aka, R.J., Babalola, O.O., 2016. Effect of bacterial inoculation of strains of *Pseudomonas aeruginosa, Alcaligenes feacalis* and *Bacillus subtilis* on germination, growth and heavy metal (Cd, Cr, and Ni) uptake of *Brassica juncea*. Int. J. Phytoremed. 18, 200–209.

Nesler, A., DalCorso, G., Fasani, E., Manara, A., Di Sansebastiano, G.P., Argese, E., et al., 2017. Functional components of the bacterial CzcCBA efflux system reduce cadmium uptake and accumulation in transgenic tobacco plants. New Biotechnol. 35, 54–61.

Newman, L.A., Reynolds, C.M., 2004. Phytodegradation of organic compounds. Current Opinion Biotechnol. 15, 225–230.

Nie, X., Dong, F., Ding, D.X., Liu, N., Li, G.Y., Zhang, D., et al., 2015. Ability of Pistia stratiotes L. and Eichhornia crassipes for remediation of uranium-contaminated waste water. Yuanzineng Kexue Jishu/Atomic Energy Sci. Technol. 49, 1946–1953.

Nwoko, C.O., 2010. Trends in phytoremediation of toxic elemental and organic pollutants. Afr. J. Biotechnol. 9, 6010–6016.

Ojuederie, O.B., Babalola, O.O., 2017. Microbial and plant-assisted bioremediation of heavy metal polluted environments: a review. Int. J. Environ. Res. Public Health 14, 1504.

Orroño, D.I., Lavado, R.S., 2009. Heavy metal accumulation in Pelargonium hortorum: effects on growth and development. Phyton 78, 75–82.

Oustriere, N., Marchand, L., Roulet, E., Mench, M., 2017. Rhizofiltration of a Bordeaux mixture effluent in pilot-scale constructed wetland using *Arundo donax* L. coupled with potential Cu-ecocatalyst production. Ecol. Eng. 105, 296–305.

Ozyigit, I.I., Can, H., Dogan, I., 2020. Phytoremediation using genetically engineered plants to remove metals: a review. Environ. Chem. Lett. doi:10.1007/s10311-020-01095-6.

Panesar, A.S., Kumar, A., Kalpana, 2019. Phytoremediation: an ecofriendly tool for in-situ remediation of contaminated soil. J. Pharm. Phytochem. 8 (1), 311–316.

Parrilli, E., Papa, R., Tutino, M.L., Sannia, G., 2010. Engineering of a psychrophilic bacterium for the bioremediation of aromatic compounds. Bioeng. Bugs 1, 213–216.

Pawlik, M., Płociniczak, T., Thijs, S., Pintelon, I., Vangronsveld, J., Zofia Piotrowska-Seget, Z., 2020. Comparison of two inoculation methods of endophytic bacteria to enhance phytodegradation efficacy of an aged petroleum hydrocarbons polluted soil. Agronomy 10, 1196.

Peng, R.H., Fu, X.Y., Zhao, W., Tian, Y.S., Zhu, B., Han, H.J., et al., 2014. Phytoremediation of phenanthrene by transgenic plants transformed with a naphthalene dioxygenase system from Pseudomonas. Environ. Sci. Technol. 48, 12824–12832.

Pesantes, A.A., Carpio, E.P., Vitvar, T., López, M.M.M., Menéndez-Aguado, J.M, 2019. A multi-index analysis approach to heavy metal pollution assessment in river sediments in the Ponce Enríquez Area, Ecuador. Water 11, 590–596.

Prasad, M.N.V, 2019. Transgenic Plant Technology for Remediation of Toxic Metals and Metalloids. Academic Press, New York, NY, p. 564.

Priya, E.S., Selvan, P.S., 2017. Water hyacinth (*Eichhornia crassipes*)—an efficient and economic adsorbent for textile effluent treatment—a review. Arabian J. Chem. 10, S3548–S3558.

Pulford, I.D., Watson, C., 2003. Phytoremediation of heavy metalcontaminated land by trees: A review. Environ. Int. 29, 529–540.

Pu, Q., Sun, J.Q., Zhang, F.F., Wen, X.Y., Liu, W.H., Huang, C.M., 2019. Effects of copper mining on heavy metal contamination in a rice agrosystem. Acta Geochimica 2019, 1–21.

Pérez-Palacios, P., Agostini, E., Ibáñez, S.G., Talano, M.A., Rodríguez-Llorente, I.D., Caviedes, M.A., et al., 2017. Removal of copper from aqueous solutions by rhizofiltration using genetically modified hairy roots expressing a bacterial Cu-binding protein. Environ. Technol. 28, 2877–2888.

Radziemska, M., Bęś, A., Gusiatin, Z.M., Cerdà, A., Mazur, Z., Jeznach, J., et al., 2019. The combined effect of phytostabilization and different amendments on remediation of soils from post-military areas. Sci. Total Environ. 688, 37–45.

Radziemska, M., Vaverková, M.D., Baryła, A., 2017. Phytostabilization management strategy for stabilizing trace elements in contaminated soils. Int. J. Environ. Res. Public Health 14, 958.

Rai, P.K., Kim, K.H., Lee, S.S., Lee, J.H., 2019. Molecular mechanisms in phytoremediation of environmental contaminants and prospects of engineered transgenic plants/microbes. Sc. Total Environ. 705, 135858.

Rajakaruna, N., Tompkins, K.M., Pavicevic, P.G., 2006. Phytoremediation: an affordable green technology for the clean-up of metalcontaminated sites in Sri Lanka. Ceylon J. Sci. (Biol. Sci.) 35, 25–39.

Rashid, A., Mahmood, T., Mehmood, F., Khalid, A., Saba, B., Batool, A., 2014. Phytoaccumulation, competitive adsorption and evaluation of chelators-metal interaction in lettuce plant. Environ. Eng. Manage. J. 13, 2583–2592.

Rastogi, M., Nandal, M., 2020. Toxic metals in industrial wastewaters and phytoremediation using aquatic macrophytes for environmental pollution control: an eco remedial approach. In: Bharagava, R.N., Saxena, G. (Eds.), Biorremediaiton of Industrial Waste for Environmental Safety. Springer, Singapore, pp. 257–282.

Reis, L.F.L., Van Sluys, M.A., Garratt, R.C., Pereira, H.M., Teixeira, M.M, 2006. GMOs: building the future on the basis of past experience. Anais da Academia Brasileira de Ciências 78, 667–686.

Rosculete, C.A., Bonciu, E., Rosculete, E., Olaru, L.A., 2019. Determination of the environmental pollution potential of some herbicides by the assessment of cytotoxic and genotoxic effects on *Allium cepa*. Int. J. Environ. Res. Public Health 16, 75–85.

Rostami, S., Azhdarpoor, A., 2019. The application of plant growth regulators to improve phytoremediation of contaminated soils: a review. Chemosphere 220, 818–827.

Rugh, C.L., Wilde, D., Stack, N.M., Thompson, D.M., Summer, A.O., Meagher, R.B., 1996. Mercuric ion reduction and resistance in transgenic Arabidopsis thaliana plants expressing a modified bacterial merA gene. Proc. Nat. Acad. Sci. USA 93, 3182–3187.

Sakakibara, M., Watanabe, A., Inoue, M., Sano, S., Kaise, T., 2010. Phytoextraction and phytovolatilization of arsenic from As-contaminated soils by *Pteris vittata*. Proc. Ann. Int. Conf. soils, Sed., Water Energy 12, 26.

Saletnik, B., Zaguła, G., Bajcar, M., Tarapatskyy, M., Bobula, G., Puchalski, C., 2019. Biochar as a multifunctional component of the environment—a review. Appl. Sci. 9, 1139.

Sigua, G.C., Novak, J.M., Watts, D.W., Ippolito, J.A., Ducey, T.F., Johnson, M.G., et al., 2019. Phytostabilization of Zn and Cd in mine soil using corn in combination with biochars and manure-based compost. Environments 6, 1–19.

Sikhosana, M.L.M., Botha, A., Mpenyane-Monyatsi, L., Coetzee, M.A.A, 2020. Evaluating the effect of seasonal temperature changes on the efficiency of a rhizofiltration system in nitrogen removal from urban runoff. J. Environ. Manage. 274, 111192.

Singh, N.S., Sharma, R., Parween, T., Patanjali, P.K., 2018. Pesticide contamination and human health risk factor. In: Mohammad, O., Mohammad, Z.K., Iqbal, M.I.I. (Eds.), Modern Age Environmental Problems and their Remediation. Springer, New York, NY, pp. 49–68.

Singh, S., Sinha, S., 2005. Accumulation of metals and its effects in *Brassica juncea* (L.) Czern. (cv. Rohini) grown on various amendments of tannery waste. Ecotoxicol. Environ. Safety 62, 118–127.

Sorte, S., Rodrigues, V., Borrego, C., Monteiro, A., 2020. Impact of harbour activities on local air quality: a review. Environ. Pollut. 257, 113542.

Strong, P.J., Burgess, J.E., 2008. Treatment methods for wine-related ad distillery wastewaters: a review. Bioremed. J. 12, 7087–7095.

Subhasini, V., Swamy, A.V.V.S, 2014. Phytoremediation of cadmium and chromium contaminated soils by *Cyperus Rotundus* L. Am. Int. J. Res. Sci., Technol., Eng. Math. 6, 97–101.

Susarla, S., Bacchus, S.T., Harvey, G., McCutcheon, S.C., 2000. Phytotransformations of perchlorate contaminated waters. Environ. Technol. 21, 1055–1065.

Tahiri, A.A., Laziri, F., Yachaoui, Y., Allaoui, A.E., Tahiri, A.H., 2017. Heavy metals leached from the waste from the landfill in the city of Meknes, and their impact on groundwater. J. Mater. Environ. Sci. 8, 1004–1014.

Thapa, G., Das, D., Gunupuru, L.R., Tang, B., 2016. Endurance assessment of *Eichhornia crassipes* (Mart.) Solms, in heavy metal contaminated site–a case study. Cogent Environ. Sci. 2, 1215280.

Thompson, M.G., Moore, W.M., Hummel, N.F.C., Pearson, A.N., Barnum, C.R., Scheller, H.V., et al., 2020. *Agrobacterium tumefaciens*: a bacterium primed for synthetic biology. BioDes. Res. 2020, 1–16.

Tiwari, J., Ankit, Sweta, Kumar, S., Korstad, J., Bauddh, K, 2019. Ecorestoration of polluted aquatic ecosystems through rhizofiltration. In: Pandey, V.C., Bauddh, K. (Eds.), Phytomanagement of Polluted Sites. Elsevier, New York, NY, pp. 179–201.

Tiwari, S., Lata, C., 2018. Heavy metal stress, signaling, and tolerance due to plant-associated microbes: an overview. Front. Plant Sci. 9, 452.

Trap, S., Kohler, A., Larsen, L.C., Zambrano, K.C., Karlson, U., 2005. Phytotoxicity of fresh and weathered diesel and gasoline to willow and poplar trees. J. Soil Sed. 1, 71–76.

Un Nisa, W., Rashid, A, 2015. Potential of vetiver (*Vetiveria Zizanioides* L.) grass in removing selected pahs from diesel contaminated soil. Pak. J. Bot. 47, 291–296.

Vanderford, M., Shanks, J.V., Hughes, J.B., 1997. Phytotransformation of trinitrotoluene (TNT) and distribution of metabolic products in *Myriophyllum aquaticum*. Biotechnol. Lett. 19, 277–280.

Vannini, C., Domingo, G., Marsoni, M., De Mattia, F., Labra, M., Castiglioni, S., et al., 2011. Effects of a complex mixture of therapeutic drugs on unicellular algae *Pseudokirchneriella subcapitata*. Aquat. Toxicol. 101, 459–465.

Varjani, S., Kumar, G., Rene, E.R., 2019. Developments in biochar application for pesticide remediation: current knowledge and future research directions. J. Environ. Manage. 232, 505–513.

Vázquez-Luna, D., Cuevas-Díaz, M.C., 2019. Soil Contamination and Alternatives for Sustainable Development. IntechOpen, London.

Wang, L., Samac, D.A., Shapir, N., Wackett, L.P., Vance, C.P., Olszewski, N.E., et al., 2005. Biodegradation of atrazine in transgenic plants expressing a modified bacterial atrazine chlorohydrolase (atzA) gene. Plant Biotechnol. J. 3, 475–486.

Wang, W., Dudel, E.G., 2017. Fe plaque-related aquatic uranium retention via rhizofiltration along a redox-state gradient in a natural *Phragmites australis* Trin ex Steud. wetland. Environ. Sci. Pollut. Res. 24, 12185–12194.

Wang, W., Dudel, E.G., 2018. Nitrogen species coupled with transpiration enhance Fe plaque assisted aquatic uranium removal via rhizofiltration of *Phragmites australis* Trin ex Steud. J. Environ. Radioact. 181, 138–146.

Wang, Y., Ren, H., Pan, H., Liu, J., Zhang, L., 2015. Enhanced tolerance and remediation to mixed contaminates of PCBs and 2,4-DCP by transgenic alfalfa plants expressing the 2,3-dihydroxybiphenyl-1,2-dioxygenase. J. Haz. Mater. 286, 269–275.

Woolf, D., Amonette, J.E., Street-Perrott, F.A., Lehmann, J., Joseph, S., 2010. Sustainable biochar to mitigate global climate change. Nat. Commun. 1, 56.

Xia, H., Ma, X., 2006. Phytoremediation of ethion by water hyacinth (*Eichhornia crassipes*) from water. Bioresour. Technol. 97, 1050–1054.

Xie, Z., Lu, G., Yan, Z., Liu, J., Wang, P., Wang, Y., 2017. Bioaccumulation and trophic transfer of pharmaceuticals in food webs from a large freshwater lake. Environ. Pollut. 222, 356–366.

Xu, L., Xing, X., Liang, J., Peng, J., Zhou, J., 2019. In situ phytoremediation of copper and cadmium in a co-contaminated soil and its biological and physical effects. RSC Adv. 9, 993–1003.

Yang, J., Zheng, G., Yang, J., Wan, X., Song, B., Cai, W., et al., 2017. Phytoaccumulation of heavy metals (Pb, Zn, and Cd) by 10 wetland plant species under different hydrological regimes. Ecol. Eng. 107, 56–64.

Ye, S., Zeng, G., Wu, H., Zhang, C., Dai, J., Liang, J., et al., 2017. Biological technologies for the remediation of co-contaminated soil. Crit. Rev. Biotechnol. 37, 1062–1076.

Yu, F., Li, Y., Li, F., Zhou, Z., Chen, C., Liang, X., et al., 2019. Nitrogen fertilizers promote plant growth and assist in manganese (Mn) accumulation by *Polygonum pubescens* Blume cultured in Mn tailings soil. Int. J. Phytoremed. 21, 1225–1233.

Zeng, P., Guo, Z., Cao, X., Xiao, X., Liu, Y., Shi, L., 2018. Phytostabilization potential of ornamental plants grown in soil contaminated with cadmium. Int. J. Phytoremed. 20, 311–320.

Zgorelec, Z., Bilandzija, N., Knez, K., Galic, M., Zuzul, S., 2020. Cadmium and Mercury phytostabilization from soil using Miscanthus × giganteus. Sci. Rep. 10, 6685.

Zhang, J., Zhang, M., Tian, S., Lu, L., Shohag, M.J.I., Yang, X, 2014. Metallothionein 2 (SaMT2) from *Sedum alfredii* Hance confers increased Cd tolerance and accumulation in yeast and tobacco. PLoS One 9, e102750.

Zhang, L., Rimal, B., Nichols, R.G., Tian, Y., Smith, P.B., Hatzakis, E., et al., 2020. Perfluorooctane sulfonate alters gut microbiota-host metabolic homeostasis in mice. Toxicology 431, 152365.

Zhang, L., Rylott, E.L., Bruce, N.C., Strand, S.E., 2018. Genetic modification of western wheatgrass (*Pascopyrum smithii*) for the phytoremediation of RDX and TNT. Planta 249, 1007–1015.

Zhang, Q., Kong, W., Wei, L., Wang, Y., Luo, Y., Wang, P., et al., 2020. Uptake, phytovolatilization, and interconversion of 2,4-dibromophenol and 2,4-dibromoanisole in rice plants. Environ. Int. 142, 105888.

Zhang, Q., Yu, R., Fu, S., Wu, Z., Chen, H.Y.H., Liu, H, 2019. Spatial heterogeneity of heavy metal contamination in soils and plants in Hefei, China. Sci. Rep. 1049, 2045–2051.

Zheng, H., Zhang, C., Liu, B., Liu, G., Zhao, M., Xu, G., et al., 2020. Biochar for water and soil remediation: production, characterization, and application. In: Jiang, G., Li, X. (Eds.), A New Paradigm for Environmental Chemistry and Toxicology. Springer, Singapore, pp. 153–196.

Chapter 4

Phytoremediation at Molecular Level

Sheikh Mansoor[a], Nafeesa Farooq Khan[b], Iqra Farooq[c], Navneet Kaur[a], Sweeta Manhas[a], Shilpa Raina[a], Iqra F. Khan[d]

[a]Division of Biochemistry, FBSc, Sher-e-Kashmir University of Agricultural Sciences and Technology of Jammu, Jammu, India, [b]Department of Botany, School of Biological Sciences, University of Kashmir, Srinagar, India, [c]Division of Floriculture and Landscape Architecture, Sher-e-Kashmir University of Agricultural Sciences and Technology of Kashmir, Srinagar, India, [d]Division of Fruit Science, Faculty of Horticulture, Sher-e-Kashmir University of Agricultural Sciences and Technology of Kashmir, Srinagar, India

All authors have equal contribution.

4.1 Introduction

Owing to numerous anthropogenic causes, such as industrialization, mining and intensive agriculture, environmental contamination has risen dramatically. These factors lead to the accumulation of various inorganic (nitrates, phosphates, heavy metal, and radionuclides) (Arthur et al., 2005) and other compounds such as volatile halogenated hydro-carbons, also pesticides and herbicides like atrazine, hexa-chlorocyclohexane and 2, 4-dichlorophenoxyacetic acid (Mauriz et al., 2006). Heavy metal (H.M) transmission into the food chain can lead to metal toxicity, thereby endangering animal and human health (Mann et al., 2011). Similarly volatile halogenated hydrocarbons, for example, tetrachloroethylene have been classified as suspected human carcinogens (Casas Ferreira et al., 2014). The conventional physicochemical strategies for decontamination of soils like soil excavation, washing, pumping and vapor extraction (Khan et al., 2018) are expensive, labor-intensive, and non-eco-friendly (Pandey and Bajpai, 2019). In addition, such strategies produce adverse changes in the composition of the soil. An eco-friendly, low labor intensive and economical alternative to these methods is phytoremediation. The process is a renewable technique focused on solar energy that uses green plants and their endophytes to eliminate organic and inorganic chemical pollutants (Pilon-Smits,2005; Feng et al., 2017). This technology could be used for soil, water, and air mitigation (Rostami and Azhdarpoor, 2019). Phytoremediation has been extensively used in the elimination of metals, contaminants, petroleum, explosive materials, polycyclic aromatic hydrocarbons (PAH's), and environmental toxins such

Phytoremediation. DOI: https://doi.org/10.1016/B978-0-323-89874-4.00011-X

as CO_2, SO_2, NOx, halogenated volatile hydrocarbons, and ozone (Meagher, 2000; Agarwal et al., 2019). The application of growth promoters such as Gibberellins (GA3), Indole-3-Acetic Acid (IAA), and 6-Benzyl-aminopurine (6-BA) improves phytoremediation quality through increased plant biomass and resistance to toxins (Hadi et al., 2010; Chen et al., 2020).

4.2 Phytoremediation techniques

Plants can use one or more of the following types of phytoremediation techniques (Table 4.1; Fig. 4.1), due to the type of contaminant and the habitat to be remedied.

4.2.1 Phytoextraction

Phytoextraction is the separation of metals from polluted soil and water by the absorption and eventual storage of the toxins into the harvestable areas of the plant part above ground. In general, main groups of higher plants most often used for heavy metal phytoextraction are (1) high-biomass producing and rapid growing plants, (2) genetically modified plants. Around 25% identified hyperaccumulators occur in family *Brassicaceae* (Suman et al., 2018). Efficient phytoextraction requires both metal hyperaccumulation and good biomass yield. Harvested plant produce/biomass is mostly processed for metal retrieval. The practice is called phytomining (Ashraf et al., 2019; Tangahu et al, 2011). Chelators such as citric acid, EDTA, and humic acid improve heavy metals' solubility and bioavailability (Turan and Angin, 2004). Limited bioavailability of organic compounds in the soil is a primary restriction for their effective phytoextraction (Alkorta and Garbisu, 2001).

4.2.2 Phytovolatilization

Phytovolatilization is documented for a lot of inorganic as well as organic pollutants. The instable contaminants are first absorbed by the plants and then removed by releasing into the atmosphere through stem or leaf. It has been confirmed that poplar trees volatilize 90% of the absorbed trichloroethane (McGrath and Zhao, 2003). Several inorganic compounds including Se, As, and Hg can also be volatilized from plants after converting them into less toxic volatile forms (Limmer and Burken, 2016; Arya et al., 2017). There is no need of harvesting the plants or their disposal in phytovolatilization. Volatilization can occur in two different ways:

Direct phytovolatilization: Instable organic compounds via transpiration can be vaporized from stems or foliage. Hydrophobic pollutants may be diffused along all hydrophobic suberizing barrier or cut-in barrier (Limmer and Burken, 2016).

Indirect phytovolatilization: Volatilization of organic impurities from the subsurface soil as a side benefit of plant root activities (Chandra et al., 2017).

TABLE 4.1 List of studies and plants used for phytoremediation techniques.

S.No	Phytoremediation technique	Contaminant	Plant used	References
1.	Phytoextraction	Hg, As, Cd, Cu, Pb, Zn, Pb	*Brassica juncea*	Franchi et al. (2017)
			Thlaspi caerulescens	Schwartz et al. (2003)
			S. acutifolia	Mleczek et al. (2017)
			Glycine max	Aransiola et al. (2013)
2.	Phytovolatilization	2,4-dibromophenol	*Oryza sativa*	Zhang et al. (2020)
		Hg	*Nicotiana tabacum*	Rahman et al. (2008)
		Se	*Brassica napus*	Alder (1996)
3.	Phytostabilization	Cu, Cd, Pb	*Lupinus luteus*	Dary et al. (2010)
4.	Rhizofiltration	Uranium	*Helianthus annuus, Phaseolus vulgaris*	Lee and Yang. (2010)
		Pb	*Carex pendula*	Yadav et al. 2011)
		Cd	*Pistia stratiotes*	Veseley et al. 2011)
5.	Phytodegradation	Ethion	*Eichhornia crassipes*	Xia and Ma 2006)
		Pyrene	*Festuca arundinacea*	Chen et al. 2003)
		Trichloroethylene	*Populusdeltoides, Plantanus* L.	Strycharz and Newman. 2009)

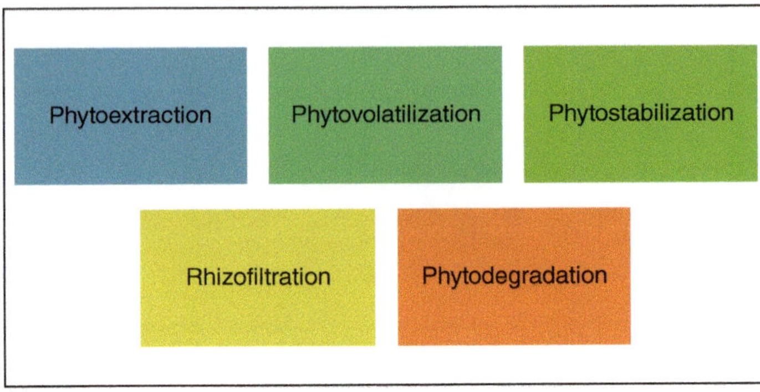

FIG. 4.1 Different types of phytoremediation processes used by plants.

4.2.3 Phytostabilization

The approach is suitable for the elimination of metal toxins, such as Pb, Cd, Cr, Cu, As, and Zn (Jadia and Fulekar 2008). Phytostabilization refers to the contaminant restriction within metal-resistant plants by their rhizosphere. The decreased mobility and bioavailability of pollutants restricts their migration into the ground water and prevents their access to the polluted loam surface, thus reducing direct contact with animals or humans and exposure to wind and water (Bolan et al., 2011). The disposal of contaminated biomass is not required in this method of phytoremediation (Branzini and Zubillaga, 2013). Phytostabilization can occur through sequestration and absorption into the root tissues; deposition metal toxins in rhizosphere; reduction of metal valence, and/or adsorption onto roots (Yan et al., 2020). Certain soil microorganisms are capable to reduce the free movement of metal ions inside soil, for example, *Xanthomonas maltophyla* catalytically reduces Cr^{6+} to Cr^{3+} which remains significantly less toxic with low mobility (Fig. 4.2) (Lasat, 2002).

4.2.4 Rhizofiltration

Rhizofiltration is used mainly for the disposal of polluted surface water, sewage, and waste water containing low amounts of heavy metals or radionuclides. It involves the adsorption and precipitation of the metal pollutants onto the roots or absorption by the roots, therefore plants with large root systems are preferred (Rezania, 2016). The plants used are initially raised hydroponically and later transferred to the contaminated water source after having developed an extensive root area. Such plants are harvested when the roots become saturated with contaminants (Adiloglu, 2018; Etim, 2012). In this method, both terrestrial and aquatic plants are used however terrestrial plants are preferred owing to their fibrous and long roots. Plants other than hyperaccumulators can be used for rhizofiltration since the heavy metals do not translocate into the shoots.

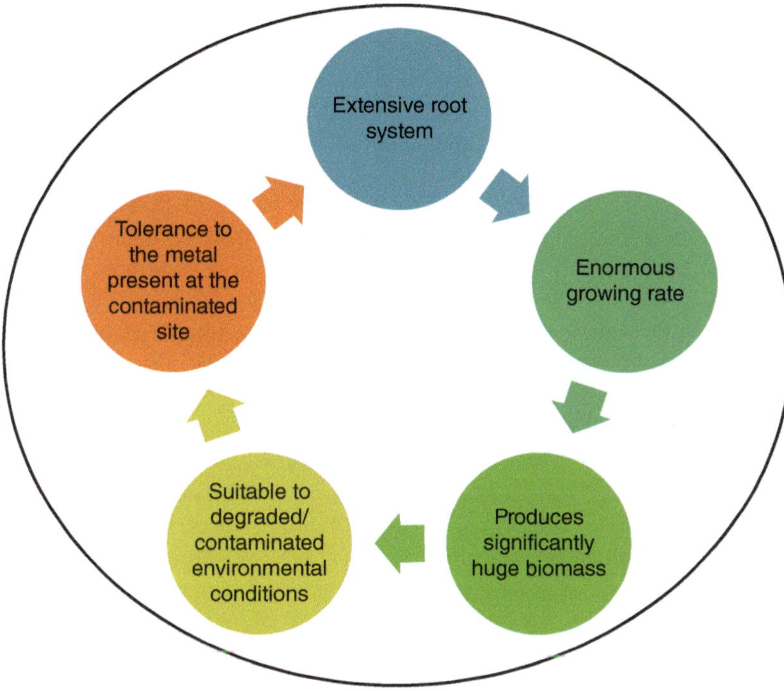

FIG. 4.2 The required characteristics of the plants used for phytostabilization.

4.2.5 Phytodegradation

Phytodegradation is characterized mostly by plant enzyme produced by the root or by metabolic pathways within plant cells, as a part of disintegration or decomposition of organic pollutants. The contaminants are degraded into harmless products (Cluis, 2004). Cytochrome P450s, dehalogenase peroxidases, phosphatases, peroxygenases, and nitroreductases are some of the plant key enzyme taking part in phytodegradation (Qixing et al., 2011). Endophytic bacteria associated with plants promote plant growth and improve phytodegradation performance by promoting the full breakdown of organic compounds into CO_2 and H_2O. For example, Bacillus sp. SBER3 associated with its host plant *Populus deltoides* degrades naphthalene, benzene, toluene, and xylene (Chlebek and Hupert-Kocurek, 2019). Phytodegradation is not applicable to metals since metals are not biodegradable (Prieto et al., 2018).

4.3 Genetic basis and identified target genes in phytoremediation

As it has been acknowledged in recent years that plants act as the potential source of remediation of polluted environment, there has been intensive research on new economical plants and the genetics behind the mechanism of phytoremediation. The responses of the plants to the contaminated environments for instance contamination by heavy metals involve ecological, physiological, molecular, and genetic factors. The tendency of plants to hyperaccumulate toxic heavy metals influences these variables. Many plant species like *Brassica* (Selvam and Wong, 2008), *Arabidopsis* (Saraswat and Rai, 2011), *Helianthus* (Boonyapookana et al., 2005), etc., have been studied for their tendency to accumulate heavy metals. Because of the high biomass level and propensity to absorb contaminants such as radioactive metals, identification of these plant species is of utmost importance (Sharma, 2011). Metal tolerance capacity in many species is due to the absorption through active or passive osmosis (Pilon-Smits et al., 2009). Also, the photosynthetic pigments like chlorophyll and porphyrins pose a direct effect on metal tolerance (Vajpayee et al., 2001; El Sheekh, 1992).

The plants responsible for bioremediation possess the genetically inherited traits of toxin tolerance which include various pathways varying from transcriptome, i.e., variation in the expression of genes and protein product accumulation that are responsible for the degradation of pollutants (Dixit, 2015). siRNA's are considered to be active in controlling gene expression, thereby gaining control over many processes such as disease resistance and heavy metal tolerance in plants. siRNA and miRNA that aid in the RNA silencing and RISC's (RNA induced silencing complex) is formed in reaction to heavy metal stressor. RISCs form the complexes with metals, causing post-transcription processes in target RNA and target DNA methylation. This post-transcriptional gene expression is being used to boost the relationship between plant and plant growth, encouraging microorganisms to promote polluted soil bioremediation (Basu et al., 2018). To comprehend the genetic basis for tolerance and accumulation, certain plant species like *Arabidopsis thaliana* and *Thlaspi caerulescens* have been used as model plants that can be conveniently crossed with other plants aiding in observation of the segregation trend. Genetic studies have confirmed that tolerance to heavy metals is regulated by few genes and tolerance to one metal is independent of its accumulation (Marmiroli et al., 2006). The genes are known for encoding metal transportation commonly belong to large gene families. One of the gene families is the family of ATP-binding cassettes (ABC), a large gene family that codes for ion modulation and detoxification process proteins (Martinoia, 2002). In the transgenic plant tonoplast, these proteins encoded by the ABC family were observed to translocate and also improved metal sequestration in vacuolar lumen (Song et al., 2014). Manipulation in phytoextraction and phytostabilization of metals can be improved this way. Metal tolerance/transport proteins (MTP) encoded by the genes of cation diffusion facilitator family are also reported to remove unnecessary cytoplasm ions. Overexpression of cation diffusion facilitator gene is reported to sequestrate metal in the vacuoles by chelating those metals (Fig. 4.3) (Koźmińska et al., 2018). Phytoremediation of organic contaminants and hazardous herbicides was reported to be conducted by endophytic bacteria like Pseudomonas and genes encoding for these actions

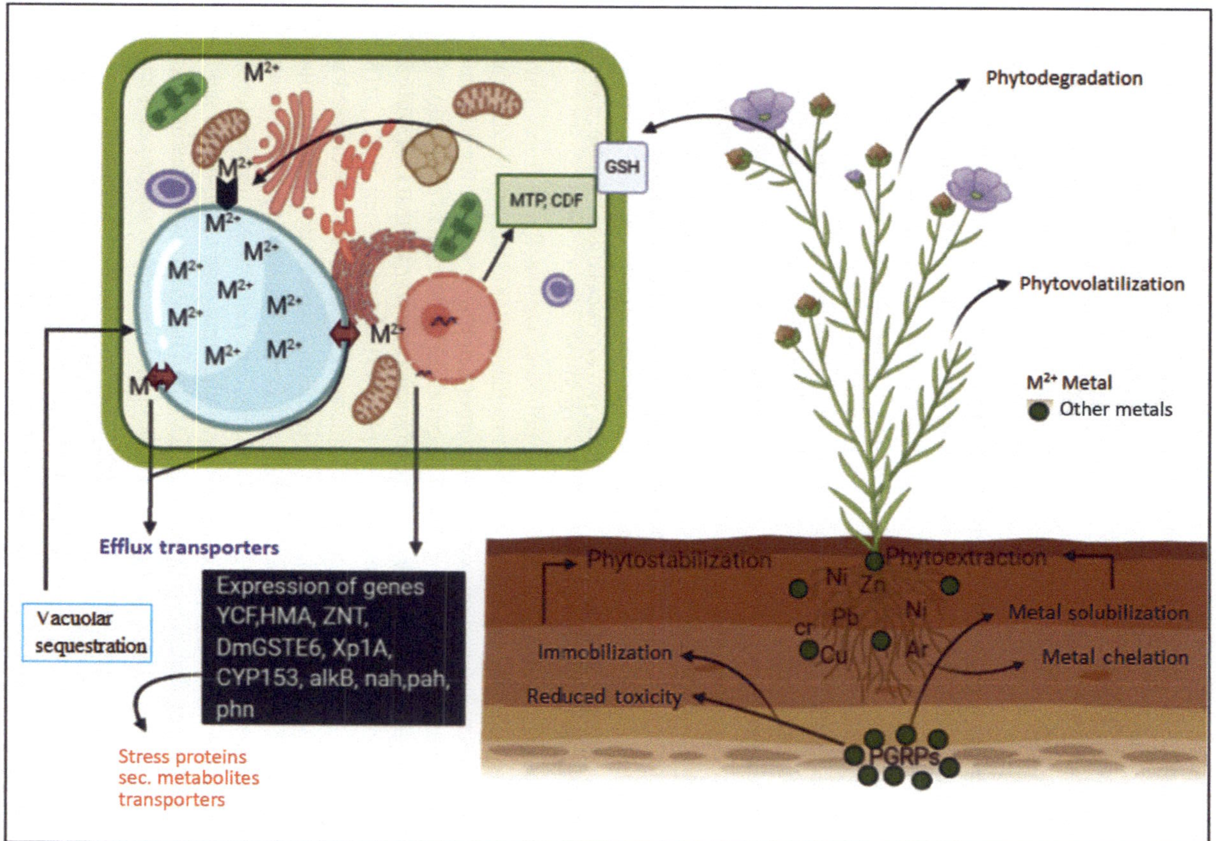

FIG. 4.3 It represents the molecular mechanism of phytoremdiation. Plant growth promoting rhhizobias help in reducing toxicity, solubilisation, but inside cell different genes, i.e., YCF, CYP153, HMA, etc., are involved in production of stress-related proteins, molecules/metabolites (antioxidants). Further, some genes are overexpressed during high metal uptake and involve Gulthione system, vacuolar sequestration, and membrane transporters.

were present in extra chromosomal DNA (Hussain et al., 2018; Chauhan et al., 2008). Bioremediation of phenolics like salicylic acid and catechol was also reported to be facilitated by endophytic bacterial strains like Rhizobium (Teng et al., 2015).

In another research, *Arabidopsis thaliana* plant has reported eight genes that code for the family transporting heavy metal Type 1B. Among these, three of the transporters, HMA 2, 3, and 4 are tightly linked with one another having identical sequence to that of heavy metal transporters in prokaryotes (Hussain et al., 2004). The representative of a group of metal transporter genes, ZIP gene family were found to transport several cations of Cadmium, Iron, Zinc, and Manganese (Kupper and Kochain, 2010). A membrane protein in *Thlaspi* (TgMTP1) is known to influence Zinc hyperaccumulation on a significant level. The high levels of TgMTP1 in vacuolar membranes play a major part in zinc accumulation and uptake (Gustin et al., 2009). Identification of several genes in these gene families has been done through the yeast mutant complementation and sequence homology with well-characterized yeast genes. The mechanisms and processes of some transporters are well characterized using expression systems in oocytes of *E. coli* and yeast. It has been observed that several endophytic bacteria like *Pseudomonas, Stenotrophomonas*, and *Pantoea* have a principal part in phytoremediation of organic contaminants like petroleum hydrocarbons. In this regard, few genes have been identified for the tolerance of this stress in plants, e.g., CYP153, alkB, nah, pah, and phn genes (Mitter et al., 2019). Some of the target genes identified to be coding for phytoremediation have been represented in Table 4.2.

Techniques like tissue culture, callus cultures, suspension cultures have been used to study the phytoremediation mechanisms and also to predict the plant-environment interaction with improvement in the design of these plants by means of transgenics (Doran 2009). The genetic engineering in plants aiming expression of transporters, factors of transcription and enzymes that are involved in metal absorption can result in improved phytoextraction (Fasani et al., 2018; Li et al., 2015). The growth-encouraging microorganisms is now been used to decontaminate soils. Recent study on genes taking part in phytoremediation can be applied for manipulation of associated growth-stimulating microorganism's to enrich bioremediation of polluted environment. mRNA expression is regulated by the RNA elements called riboswitches that bind to the ligand binding to the aptamer domain (Fürtig et al., 2020). Thus, riboswitch technology offers direct administration of ligands without requiring additional proteins and these can be successfully used to control the gene expression as per requirement (Whittmann and Suess, 2012). Ribozymes are even well-known to take to a controlling effect on the expression of genes (Wilkinson and Been, 2005). These techniques can be used to manipulate the plant environment interaction to increase phytoremediation efficacy. Additionally, a popular method to improve phytoremediation is the introduction of those genes whose overexpression will lead to enhanced metal sesquestration (Mani and Kumar, 2014; Das et al., 2016).

TABLE 4.2 Examples of identified genes and their bioremediation functions.

S. No.	Identified gene	*Host*	Function	References
1.	HMA2, HMA3, HMA4	*Arabidopsis thaliana*	Transportation of Cd	Wonget al. (2009)
2.	ZNT_1, ZNT_5	*Thlaspi*	Increased tolerance of Cd, Fe, Mn, Zn	Sarma (2011)
3.	MTP1	*Arabidopsis halleri*	Transportation of Zn^{2+}/H^+	Kawachiet al. (2008)
4.	merA, merB	*Bacterial*	Catalysis removal of Hg (II)	Hong et al. (2010)
5.	ArsC	*E.coli*	Reduces Arsenate to arsenite	DeMel et al. (2004)
6.	γ,ECS	*E. coli*	Reduces Arsenate	Zhu et al. (2009)
7.	SMTA	*A. bisulcatus*	Degrades Se	Krämer et al. (2005)
8.	YCF-1	*S. cerevisiae*	Pb and Cd phytoremediation	Goel et al. (2009)
9.	ohb	*P. aeruginosa*	Encodes enzymes to metabolize chlorobenzoic acid	Romanov et al. (1994)
10.	FcbB	*Arthobacter*	Degrades 2.4-dinitrotoulene	Guo et al. (2019)
11.	MerR	*Shigella flexneri*	Resistance of Hg(II)	Yoon (1994)
12.	Cytochrome $P450_{CAM}$	*P. putida*	Oxidizes Hexane and 3-methylpentane	Bell et al. (2003)
13.	bpH operon	*Burkholderia*	Degrades polychlorinated bipheryls	Agar (2002)
14.	dszA/B/C	*Rhodococcus*	Eliminates Sulphur without hampering fuel content	Basu et al. (2018)
15.	vgb	*Vitreoscilla*	Production of useful products and improved growth	Stark et al. (2011)
16.	pnp operon	*Deltia acidovorans*	Degrades an organophosphorous compound, Paraxon	Kuhad et al. (2004)
17.	copC	*A. thalliana*	Hyperaccumulation of Cu in shoots	Pérez-Palacios et al. (2017)

(*continued*)

TABLE 4.2 (Cont'd)

S. No.	Identified gene	*Host*	Function	References
18.	ABC	*S. cerevisiae, Populus*	Increases accumulation of Pb in roots	Shiet al. (2019).
19.	ThMT3	*Tamarix indica*	Enhances Cd, Zn, and Cu tolerance	Hasan et al. (2017)
20.	OYE3	*Saccharomyces*	TNT phytoremediation	You et al. (2015)
21.	DmGSTE6	*Drosophila, Arabidopsis*	RDX leaching	Tzafestas et al. (2016)
22.	Xp1A	*Rhodococcus*	RDX degradation	Kumar et al. (2012)
23.	alkB	*Pseudomonas*	Phytoremediation of petroleum products	Iqbal et al. (2019)

Gene merA and merB were introduced to *Arabidopsis* and poplar and these species were reported to have noteworthy tolerance toward mercuric compounds (Bizily et al., 2000). Yet another study of transgenic *Arabidopsis* introduced to selenocysteine methyltransferase (SMTA) isolated from Astragalus which is known to be Selenium hyperaccumulator, accumulated methyl-selenocysteine and was observed to contain Se concentrations up to eight folds (Ellis et al., 2004). In a similar fashion, several transgenic grasses like Agrostis and Panicum incorporated with xp1A and xp1B vectors have shown positive eliminated RDX in contaminated groundwater and even in soils (Zhang et al., 2017). Besides, glutathiome transferase (DmGSTE6) expressed in *Arabidopsis* helped in phytoremediation of explosives like TNT (Tzafestas et al., 2017). Clustered Normal Interspaced Palindromic Repeats **(CRISPR)** can be used as a method to inculcate phytoremediation-related characteristics in the genome editing discipline (Basharat et al., 2018). More recently, OsNramp5, carrier gene associated with metals, was observed to be regulated by CRISPR technique in rice which unfastened new platforms with CRISPR in the regulation of metal transport (Peng et al., 2017; Tang et al., 2017).

4.4 Omics and phytoremediation

Several plant species contribute to bioremediation in natural ways, essentially in pollutant accumulation, in the process of degradation via microbial stimulation, volatilization, and stabilization. Before the arrival of omics technologies, plants stood extra discernible to science than microscopic groups. The genetics behind phytoremediation came into light with new omics approaches which revealed molecular variations among apparently similar individuals. Identification of genes involved in phytoremediation has been made simpler by genomic approaches. To understand numerous genetic, metabolic, and hormonal pathways underlying soil-mediated phytoremediation, omics approaches can be employed. For instance, increasing mercury tolerance in plants was laid down when bacterial mercury transformation genes mer A and mer B were encoded and characterized (Muhammad et al., 2016). Similarly, three genomic regions were located with 42% of variance in plant zinc tolerance through QTL analysis of Zinc (Zn)resistant*A. halleri* and Zn-intolerant A. *lyrata* backcross progenies (Willams et al., 2007).

To examine metal contaminated soils for potential phytomediator candidate's approaches like transcriptomics and metabolomics play an important role. The use of transcriptomics to identify genes in *Brassica juncea* that are involved actively in phytoremediation of sulphur and chromium contaminated environments is well studied (Schiavon et al., 2012). Greatest advantages of "omics" in this field would be the availability of ever-growing banks of accessible data. A shift in the microbial activities over a period of time can be predicted by extracting insights from available data (Gonzalez et al., 2012). The study of "omics" in relation to phytoremediation was mainly conducted on plants subjected to metal pollutants. Proving that plants play a part in promoting organic-degrading microorganisms amid their important direct functions in the aggregation, degradation, and volatilization of organic molecules (Brentner et al., 2010).

Omics can be applied in two different ways—descriptive omics: which collects a large database involving the processes of interest and a monitoring of biological communities. Secondly, manipulative omics: using high throughput biological and chemical tools perturbing biological system by adding oxidoreductases, use of fertilizers, soil physical disturbances, etc., to create new ones that are modified allowing beneficial species to flourish, eliminating the undesirable microorganisms (Helbling et al. 2015).

Application of metagenomics, using SIP coupled metagenomics application to contaminated soils to remove contaminants such as hydrocarbons, diesel, mineral oil in a culture-independent way has been used for the determination of hydrocarbon metabolizing strains (Rodgers Vieira et al., 2015). Similarly, the role of root exudates in hydrocarbon degradation has been illustrated by SIP-labeled substrates (Sofie et al., 2017). In phytoremediation study, a significant interest in plasmid DNA was also seen as rhizosphere plants and polluted soils as the key site for plasmid exchange (Chen et al., 2018). In a research, plasmid DNA metagenomic sequencing results from municipal waste water treatments depicted a disproportionate abundance of heavy metal resistance genes when compared to the entire genome. The involvement of genes involved in the degradation and transformation of organic compounds in plasmids was discovered in the same study (Balcom et al., 2018). Proteomics has also been used to identify genes responsible for pollutant-induced stress and reveal plant responses. For example, the use of proteomics to identify changes in Populus caused by cadmium revealed that the presence of cadmium in tissues directly as well as indirectly affected the plant proteome, and the plant response differed between cambium and leaves tissues. The indirect effect can be attributed to reduced metabolic rates due to systemic toxicity (Durand et al., 2010). To find the active genes prevalent in the plume, a Geo Chip functional gene microarray was used to find hydrocarbon degradation genes with a comparatively higher abundance in the plume (Lu et al., 2011).

Integrated omics when applied along with traditional methods may help in determining metaorganism phytoremediation properties that can be targeted such as studying root exudation patterns targeting plantome properties, relationships between expression, activity, translation, and expression properties of a metaorganism can be targeted via microbiome studies. Similarly, interactome studies may reveal undetected information on promotion and suppression of plant growth by microorganisms (Terrence et al., 2014).

Enhancing predictability and community modification and enhancing predictability: one of the obstacles in studying natural communities of microorganisms is yet to be cultured microbial communities that form almost 99% of the microbial taxa. Using cultural independent approaches, these can be described. Analysis of the composition of microcommunities and their behavior analysis can be conducted using Omics technology to retain fingerprints of abiotic and biotic factors. The methodologies to omics are used on different scales specific to phytoremediation (Pham and Kim, 2012).

A combination of multiple omics techniques in the future studies possibly provides additional comprehensive understanding to involved protein, gene, and metabolic networks and their plant-level interface and environmental-based variations. Despite targeting microorganisms or plants, some steps need to be taken at the metaorganism level in order to improve phytoremediation systems. While "omics" studies are highly economical, the data needs finally to be supported by high throughput screening besides isolation of major microbial characteristics like growth rate, which aid the determination of microbial activity that must be targeted in phytoremediation even though it might not be naturally dominant (Blomberg, 2011).

4.5 Physiological and molecular cross-talk involved in phytoremediation

On exposure to an excess of heavy metals, the biochemical and physiomolecular mechanisms in plants get activated to overcome the stressors. The collaborations amongst vegetation and microbes should be explored for better adaptation in addition to survival of plants in metalliferous conditions (Mahmood, 2020). To understand the mechanism of metal tolerance of a plant is a prerequisite for introducing plants suitable for the phytoremediation at metal contaminated sites. Among all the tolerance mechanisms involved, root secretion of low molecular weight organic acids (LMWOAs) is among few strategies in plants that help them overcome the high concentration of metal by excluding metals and metalloids. Chelation of metal ions in the rhizosphere/apoplast region, thus preventing them from entering the symplast region of the cell (Magdziak et al., 2011). Many of the LMWOAs on secretion can lead to microorganism growth and dissolve unsolvable or moderately solvable minerals (P, Fe, Zn) (Choudhary and Varma 2016; Tu et al., 2004; Li et al., 2013). Some of the studies revealed that under metal stress the LMWOAs identical to oxalic acid, malic, citric, and succinic acid secreted by plants, play a noteworthy part in mitigating the metal toxicity to plants (Evangelou et al., 2006; Yuan et al., 2007; Choudhary and Varma 2016).

Various signaling molecules are required for developing communication between plants and microbes, thereby sustaining their growth. The signals from plants remain well-known in microorganisms. These signals and volatiles in turn stimulate the alterations in plant physiology. Flavonoids are the signaling components played vital part in plant-microbe interactions including arbuscular mycorrhizal fungi (AMF) spore germination, hyphal development, differentiation and root colonization (Badri et al., 2009; Mandal et al., 2010). Flavonoids had a clear impact in both the AMF genus and species on early symbiotic hyphae development in the primary phase of AMF and plant interaction (Choudhary and Varma 2016). During the primary establishment of AMF, flavonoids in root were observed in moderate levels and their level increased in later phase of root colonization. There are many different ways like promoting plant biomass, adjusting metal availability in soil by changing pH, or facilitating its transport from soil to root by which microbes can elevate the process of phytoremediation (Choudhary and Varma 2016;Ma et al., 2011, 2013;Rajkumar et al., 2012). In phytoremediation host-plant growth-stimulating microbes provide several benefits above the use of chemical products because the metabolites provided by useful microbes in the rhizosphere have lower toxicity (Rajkumar et al., 2012). Many of the investigations revealed that metal-resistant microbes have the ability to ameliorate plant development and reduce toxic effects of metals, thus aiding phytoremediation. AMF particularly from metalliferous sites have the ability to enhance both growth rate of plants (Choudhary and Varma 2016;Orłowska et al., 2013) and nutrient assimilation (Choudhary and Varma 2016;Guo et al., 2013). A previous study demonstrated that during greenhouse effect, sunflower grown in three concentrations of Cd through successful AMF colonization via *Rhizophagus irregularis* leads to increased rate of phytoextraction of Cd. On the other hand, by treating with *Funneliformis mosseae*, significantly increased phytostabilization was noticed for both Cadmium and Zinc (Hassan et al., 2013). The main component recognized in phytoremediation is the suitable microbial-root exudates interconnection which includes signaling in plant-microbe and heavy metal versus plant-microbe collaborations (Badri et al., 2009; Segura et al., 2009). During metal stress, allelopathy and detoxification are the chief functions performed by root exudates for better adaptation and survival of host plants. In allelopathic condition, both plants and rhizospheric microbial growth get affected whereas the latter process involves chelation, alteration/inactivation, and adsorption of metals. Previous studies reported that two of the LMWOAs, i.e., citric-acid and oxalic-acid from *Echinocholo acrus-galli* increasing the rate of transportation and bioaccumulation of Pb, Cu, and Cd. The overall results therefore suggest the immense role of organic acids as natural chelating agents for enhancing the phytoextraction. On the other hand, few of the root exudate components either do not affect the metal availability or negatively regulate the mobility in metals (Zhao et al., 2001). Additionally, formation of stable metal complexes has been observed in the seedlings (ecto- and non-mycorrhizal) of Scots pine by the action of oxalate secreted via metal immobilization (Johansson et al., 2008). Free living microbes such as PGDB, fungi affect the biochemical composition of root exudates by releasing molecules like volatile organic compounds (VOC), nod factors, Myc factors, microbe associated molecular patterns, exopolysaccharides and thereby affect plant physiology (Choudhary and Varma 2016;Goh et al., 2013). Further, two volatile organic compounds namely acetoin and 2,3-butanediol develop the connection thereby activating the plant resistance and growth mechanism via colonizing the nutrients (S & Fe) of poor soil (Choudhary and Varma 2016;Bailly and Weisskop, 2012), a common practice in phytoremediation. Many previous studies demonstrated that VOCs play a key role on most PGPMs by acting as bioprotectants, biopesticides, and phytostimulators via ISR i.e., induced systemic resistance (Ryu et al., 2004), antibiotic (Trivedi and Pandey, 2008), and activating hormonal network (Zhang et al., 2008) respectively. Consequently, all these factors ameliorate plant growth, a basic prerequisite for successful phytoremediation.

In plants progress has been made in molecular aspect to heavy metals, more specifically in respect to herbaceous plants such as *Arabidopsis thaliana, Arabidopsis halleri,* and *T. caerulescens* (Kaushik, 2015; Thapa et al., 2012; Meyer, 2012). In case of *A. thaliana*, three of the genes namely OPR1 (oxophytodienoate reductase), OPR2, OPR3 were reported to be enhanced on exposure to TNT (trinitrotoluene, an explosive) (Beynon et al., 2009). Further the biochemical characterization demonstrated two OPR1 and all OPR2 lines showed high tolerance to TNT. Later on, a target gene was identified for phytoremediation of some chemical contaminants like RDX and TNT via microarray (Rao et al., 2009). Some of the genes such as oxidoreductases, cytochrome P450s, transferases and other unknown proteins were differentially expressed. One more researcher discussed cross-talk between Fe homeostasis and Zn tolerance at FRD3 MATE transporter locus (Pineau et al., 2013). In two *Populus nigra* L., from northern and southern Italy with 58-861 and Poli genotypes respectively, both physiological and molecular responses were evaluated in response to Cd stress (Qi et al., 2015). Overall studies have displayed that the latter, i.e., Poli genotype was comparatively more tolerant to Cd stress and glutathione pathway was involved in both the genotypes. Moreover, the transcript analysis of *Populus canescens* was also performed in response to Cd which revealed a co-regulation network, formed by 48% of differentially regulated transcripts and thereby 48 hub genes actively participated in crosstalk among different biological processes (Luo et al., 2013). One more promising plant employed for the phytoremediation of heavy metals is *Brassica juncea*. In response to Cd exposure, its root proteome was analyzed and it was found that the different enzymes (methionine sulfoxide reductase, 2-nitropropane dioxygenase, O-acetylserine sulfhydrylase, and glutathione-S-transferase) and glutathione-conjugated transporter were essential components in Cd hyperaccumulation and resistance (Alvarez et al., 2009).

4.6 Future directives and goals of phytoremediation

Phytoremediation is an incredibly simple procedure, a very specific staff and proprietary facility are not required. It refers to the remediation of large-scale environments where most traditional approaches are both highly wasteful and expensive. It is a relatively recent method for plants in the ecosystem or/and human agriculture to extract dangerous contaminants from the environment, which is a part of bioremediation (Leguizamo et al., 2017). Due to the active interdisciplinary collaboration of plant organic chemists, molecular biologists, soil physicists, agronomists, natural designers, and government and state regulations, phytoremediation is becoming feasible. It is an effective and commonly used method of environmental cleaning focused on the volatilization, stabilization, oxidation, or elimination of toxins by plants and their related microorganisms (Zhou et al., 2019). In the course of the most recent couple of years, phytoremediation processes have been carefully researched and deemed to be a sustainable approach for the regeneration of nutrients and the removal of harmful pollutants, such as antibiotics (Cheng et al., 2018), HMs, clothing, dyes, pesticide, hormone, petroleum, landfill leachate, explosives plus even poisonous gases (Hu et al., 2020; Smits, 2005). The system of phytoremediation is powered by solar energy and can create natural habitats for rehabilitated areas, which are also esthetically important in the treatment of contaminants (Markou et al., 2018). Government agencies and companies are becoming increasingly common in terms of cost-effective and restricted support for environmental governance. In-situ remediation of heavy metal-contaminated soils is now appealing owing to low costs and environmental friendliness. Phytoremediation is viewed as an available, financially savvy, eco-accommodating, and maintainable elective answer for the future; however, the presentation of this cycle should be additionally improved (Corami, 2020). Further steps need to be taken to improve the productivity and stability of phytoremediation processes across different approaches to ensure improved performance of environmental protection processes. Furthermore, it should be noted that there is a plant harvesting process that can be handled automatically in the treatment system for harvesting unwanted plants. This mostly leads to a decline in the rate of toxins released by dying plants, it also decreases the cost of labor in plant remediation projects. Countless situations, such as light intensity, stomatological conductivity, temperature, and microscopic organisms, influence purification aspect in plants during the remediation process (Wei et al., 2020). This is important to encourage plants to be used to preserve vegetation and the climate. Generally, plants chosen for phytoremediation was able to focus on either a wide variety of or particular toxins and could have a greater development rate, high biomass, strong fibrous roots, stress-free to control, greater resistance to impurities, quick to develop and yield (Hu et al., 2020; Ekperusi et al., 2019). The most important element of phytoremediation is the discovery of appropriate plants. In remediating heavy metals from the polluted site, aquatic plants play very colorful roles with the same ease as other hyperaccumulator plants. In the eradication of toxic substances, the use of aquatic plants in biomagnification and in biosorption may be successful (Kushwaha et al., 2015). Plant ecology is complex, so careful selection of species is expected to correct and rebuild unique habitats effectively. In particular, it should be noted that the genetic modification of plants for phytoremediation by the use of CRISPR is a fruitful endeavor to increase the potential use of hydrophytes (Basu et al., 2018). Efforts to grow more adaptable and promising plants using gene editing with a surprisingly high outlook are the key artefacts for the future. Moreover, the success of phytoremediation relies on plant-specific metabolism, including air absorption and photosynthesis, diversity of soil microorganisms, and intake of heavy metals. Currently, as nanomaterials are paired with biotechnology and compost is used to extract contaminants, the effect of nanomaterials and compost on plant microorganisms and their own environmental impact should also be taken into account (Wei et al., 2020). Although the use of nanotechnology and green waste promotes the return to the environment of plants, a large dosage can have detrimental effect on plant growth. In order to increase the feasibility of phytoremediation in environmental restoration, further research is required to study the effect of different types of catalysts on the effectiveness of phytoremediation. Phytoremediation has gained significant interest recently and there are a growing number of studies indicating that it should become a technology.

References

Abbasian, F., Lockington, R., Megharaj, M., Naidu, R., 2016. A review on the genetics of aliphatic and aromatic hydrocarbon degradation. Appl. Biochem. Biotechnol. 178 (2), 224–250.

Adiloğlu, S., 2017. Heavy metal removal with phytoremediationAdvances in Bioremediation and Phytoremediation. IntechOpen.

Adiloğlu, S., Eryılmaz Açıkgöz, F., Solmaz, Y., Çaktü, E., Adiloğlu, A., 2018. Effect of vermicompost on the growth and yield of lettuce plant (Lactuca sativa L. var. crispa). International Journal of Plant & Soil Science 21 (1), 1–5.

Adler, T., 1996. Botanical cleanup crews. Sci. News (Washington, DC) 150 (3), 42–43.

Agar, N.Y.R., 2002. Doctoral dissertation. Concordia University.

Agarwal, P., Sarkar, M., Chakraborty, B., Banerjee, T., 2019. Phytoremediation of air pollutants: prospects and challengesPhytomanagement of Polluted Sites. Elsevier, pp. 221–241.

Alkorta, I., Garbisu, C., 2001. Phytoremediation of organic contaminants in soils. Bioresour. Technol. 79 (3), 273–276.

Alvarez, S, Berla, B, Sheffield, J, Cahoon, RE, Jez, JM, et al., 2009. Comprehensive analysis of the Brassica juncea root proteome in response to cadmium exposure by complementary proteomic approaches. Proteomics 9, 2419–2431.

Aransiola, S.A., Ijah, U.J.J., Abioye, O.P., 2013. Phytoremediation of lead polluted soil by Glycine max L. Appl. Environ. Soil Sci. 2013.

Arthur, E.L., Rice, P.J., Rice, P.J., Anderson, T.A., Baladi, S.M., Henderson, K.L., Coats, J.R., 2005. Phytoremediation—an overview. Crit. Rev. Plant Sci. 24 (2), 109–122.

Arya, S.S., Devi, S., Angrish, R., Singal, I., Rani, K., 2017. Soil reclamation through phytoextraction and phytovolatilizationVolatiles and Food Security. Springer, Singapore, pp. 25–43.

Ashraf, S., Ali, Q., Zahir, Z.A., Ashraf, S., Asghar, H.N., 2019. Phytoremediation: environmentally sustainable way for reclamation of heavy metal polluted soils. Ecotoxicol. Environ. Safety 174, 714–727.

Badri, D.V., Weir, T.L., vanderLelie, D., Vivanco, J.M., 2009. Rhizosphere chemical dialogues: plant-microbe interactions. Curr. Opin. Biotechnol. 20, 642–650. doi:10.1016/j.copbio.2009.09.014.

Bailly, A., Weisskopf, L., 2012. The modulating effect of bacterial volatiles on plant growth: current knowledge and future challenges. Plant Signal.Behav 7, 79–85. doi:10.4161/psb.7.1.18418.

Balcom, I.N., Driscoll, H., Vincent, J., Leduc, M., 2018. Metagenomic analysis of an ecological wastewater treatment plant's microbial communities and their potential to metabolize pharmaceuticals. F1000Research 5, 1881.

Basharat, Z., Novo, L.A., Yasmin, A., 2018. Genome editing weds CRISPR: what is in it for phytoremediation? Plants 7 (3), 51.

Basu, S., Rabara, R.C., Negi, S., Shukla, P., 2018. Engineering PGPMOs through gene editing and systems biology: a solution for phytoremediation? Trends Biotechnol. 36 (5), 499–510.

Basu, S., Rabara, R.C., Negi, S., Shukla, P., 2018. Engineering PGPMOs through gene editing and systems biology: a solution for phytoremediation? Trends Biotechnol. 36 (5), 499–510.

Beynon, E.R., Symons, Z.C., Jackson, R.G., Lorenz, A., Rylott, E.L., et al., 2009. The role of oxophytodienoate reductases in the detoxification of the explosive 2, 4, 6-trinitrotoluene by Arabidopsis. Plant Physiol. 151, 253–261.

Bizily, S.P., Rugh, C.L., Meagher, R.B., 2000. Phytodetoxification of hazardous organomercurials by genetically engineered plants. Nat. Biotechnol. 18 (2), 213–217.

Blomberg, A., 2011. Measuring growth rate in high-throughput growth phenotyping. Curr. Opin. Biotechnol. 22, 94–102.

Bolan, N.S., Park, J.H., Robinson, B., Naidu, R., Huh, K.Y., 2011. Phytostabilization. Adv. Agro., 145–204.

Boonyapookana, B., Parkpian, P., Techapinyawat, S., DeLaune, R.D., Jugsujinda, A., 2005. Phytoaccumulation of lead by sunflower (Helianthus annuus), tobacco (Nicotiana tabacum), and vetiver (Vetiveria zizanioides). J. Environ. Sci. Health 40 (1), 117–137.

Branzini, A., Zubillaga, M.S., 2013. Phytostabilization as soil remediation strategyPlant-Based Remediation Processes. Springer, Berlin, pp. 177–198.

Brentner, L.B., et al., 2010. Localization of hexahydro-1,3,5-trinitro- 1,3,5-triazine (RDX) and 2,4,6-trinitrotoluene (TNT) in poplar and switchgrass plants using phosphor imager autoradiography. Environ. Pollut. 158, 470–475.

Buhari Muhammad, L., Sulaiman Babura, R., Vyas, N.L, Badaru, S., Harisu Umar, Y., 2016. Role of biotechnology in phytoremediation. J. Bioremed. Biodeg. 7, 302.

Casas Ferreira, A.M., Fernandez Laespada, M.E., Perez Pavon, J.L., Moreno Cordero, B., 2014. Development of an environmentally friendly methodological approach to determine chlorinated hydrocarbons and chlorobenzenes in soils. Green Chem. Lett. Rev. 7 (1), 50–59.

Chandra, R., Dubey, N.K., Kumar, V., 2017. Phytoremediation of Environmental Pollutants. CRC Press.

Chauhan, A., Oakeshott, J.G., Jain, R.K., 2008. Bacterial metabolism of polycyclic aromatic hydrocarbons: strategies for bioremediation. Ind. J. Microbiol. 48 (1), 95–113.

Chen, Y.C., Banks, M.K., Schwab, A.P., 2003. Pyrene degradation in the rhizosphere of tall fescue (Festuca arundinacea) and switchgrass (Panicum virgatum L. Environ. Sci. Technol. 37, 5778.

Chen, L., Long, C., Wang, D., Yang, J., 2020. Phytoremediation of cadmium (Cd) and uranium (U) contaminated soils by Brassica juncea L. enhanced with exogenous application of plant growth regulators. Chemosphere 242, 125112.

Chen, Y., Zhi, D., Sun, C., Huang, A., Wu, S., Yao, B., Tang, Y., Zhou, Y., 2021. Electrokinetic techniques, their enhancement techniques and composite techniques with other processes for persistent organic pollutants remediation in soil: A review. Journal of Industrial and Engineering Chemistry.

Cheng, D.L., Ngo, H.H., Guo, W.S., Liu, Y.W., Zhou, J.L., Chang, S.W., Nguyen, D.D., Bui, X.T., Zhang, X.B., 2018. Bioprocessing for elimination antibiotics and hormones from swine wastewater. Sci. Total Environ. 621, 1664–1682.

Chlebek, D., Hupert-Kocurek, K., 2019. Endophytic bacteria in the phytodegradation of persistent organic pollutants. Adv. Microbiol. 58 (1), 70–79.

Choudhary, D.K., Varma, A. eds., 2016. Microbial-mediated induced systemic resistance in plants. Springer: New York.

Cluis, C., 2004. Junk-greedy greens: phytoremediation as a new option for soil decontamination. BioTeach. J. 2 (6), 1–67.

Corami, A., 2020. Phytoremediation impacts on water productivity. Water Productivity Journal, 1 (2), 71-81.

Dary, M., Chamber-Pérez, M.A., Palomares, A.J., Pajuelo, E., 2010. "In situ" phytostabilisation of heavy metal polluted soils using Lupinus luteus inoculated with metal resistant plant-growth promoting rhizobacteria. J. Haz. Mater. 177 (1-3), 323–330.

Das, N., Bhattacharya, S., Maiti, M.K., 2016. Enhanced cadmium accumulation and tolerance in transgenic tobacco overexpressing rice metal tolerance protein gene OsMTP1 is promising for phytoremediation. Plant Physiol. Biochem. 105, 297–309.

DeMel, S., Shi, J., Martin, P., Rosen, B.P., Edwards, B.F., 2004. Arginine 60 in the ArsC arsenate reductase of E. coli plasmid R773 determines the chemical nature of the bound As (III) product. Prot. Sci. 13 (9), 2330–2340.

Dixit, R., Malaviya, D., Pandiyan, K., Singh, U.B., Sahu, A., Shukla, R., Singh, B.P., Rai, J.P., Sharma, P.K., Lade, H., Paul, D., 2015. Bioremediation of heavy metals from soil and aquatic environment: an overview of principles and criteria of fundamental processes. Sustainability, 7 (2), 2189-2212.

Doran, P.M., 2009. Application of plant tissue cultures in phytoremediation research: incentives and limitations. Biotechnol. Bioeng. 103 (1), 60–76.

Durand, T.C., et al., 2010. Acute metal stress in Populus tremula _ P. alba (717-1B4 genotype): leaf and cambial proteome changes induced by cadmium2+. Proteomics 10, 349–368.

Ekperusi, A.O., Sikoki, F.D., Nwachukwu, E.O., 2019. Application of common duckweed (Lemna minor) in phytoremediation of chemicals in the environment: state and future perspective. Chemosphere 223, 285–309.

Ellis, D.R., Sors, T.G., Brunk, D.G., Albrecht, C., Orser, C., Lahner, B., Salt, D.E., 2004. Production of Se-methylselenocysteine in transgenic plants expressing selenocysteine methyltransferase. BMC Plant Biol. 4 (1), 1.

Etim, E.E., 2012. Phytoremediation and its mechanisms: a review. Int. J. Environ. Bioenergy 2 (3), 120–136.

Evangelou, M.W., Ebel, M., Schaeffer, A., 2006. Evaluation of the effect of small organic acids on phytoextraction of Cu and Pb from soil with tobacco Nicotiana tabacum. Chemosphere 63, 996–1004. doi:10.1016/j.chemosphere.2005.08.042.

Fasani, E., Manara, A., Martini, F., Furini, A., DalCorso, G., 2018. The potential of genetic engineering of plants for the remediation of soils contaminated with heavy metals. Plant, CellEnviron. 41 (5), 1201–1232.

Feng, N.X., Yu, J., Zhao, H.M., Cheng, Y.T., Mo, C.H., Cai, Q.Y., …, Wong, M.H., 2017. Efficient phytoremediation of organic contaminants in soils using plant–endophyte partnerships. Sci. Total Environ. 583, 352–368.

Franchi, E., Rolli, E., Marasco, R., Agazzi, G., Borin, S., Cosmina, P., …, Petruzzelli, G., 2017. Phytoremediation of a multi contaminated soil: mercury and arsenic phytoextraction assisted by mobilizing agent and plant growth promoting bacteria. J. Soils Sed. 17 (5), 1224–1236.

Fürtig, B., Oberhauser, E.M., Zetzsche, H., Klötzner, D.P., Heckel, A., Schwalbe, H., 2020. Refolding through a linear transition state enables fast temperature adaptation of a translational riboswitch. Biochemistry 59 (10), 1081–1086.

Goel, S., Malik, J.A., Nayyar, H., 2009. Molecular approach for phytoremediation of metal-contaminated sites. Arch. Agro. Soil Sci. 55 (4), 451–475.

Goh, H.H., Sloan, J., Malinowski, R., Fleming, A., 2013. Variable expansion expression in Arabidopsis leads to different growth responses. J. PlantPhysiol. 171, 329–339. doi:10.1016/j.jplph.2013.09.009.

Gonzalez, A., et al., 2012. Characterizing microbial communities through space and time. Curr. Opin. Biotechnol. 23, 431–436.

Guo, X., Xie, C., Wang, L., Li, Q., Wang, Y., 2019. Biodegradation of persistent environmental pollutants by Arthrobacter sp. Environ. Sci. Pollut. Res. 26 (9), 8429–8443.

Guo, W., Zhao, R.X., Zhao, W.J., Fu, R.Y., Guo, J.Y., Bi, N., et al., 2013. Effects of arbuscular mycorrhizal fungi on maize (Zeamays L.) and sorghum (Sorghum bicolor L. Moench) grown in rare earth elements of mine tailings. Appl. Soil Ecol 72, 85–92. doi:10.1016/j.apsoil.2013.06.001.

Gustin, J.L., Loureiro, M.E., Kim, D., Na, G., Tikhonova, M., Salt, D.E., 2009. MTP1-dependent Zn sequestration into shoot vacuoles suggests dual roles in Zn tolerance and accumulation in Zn-hyperaccumulating plants. Plant J. 57 (6), 1116–1127.

Hadi, F., Bano, A., Fuller, M.P., 2010. The improved phytoextraction of lead (Pb) and the growth of maize (Zea mays L.): the role of plant growth regulators (GA3 and IAA) and EDTA alone and in combinations. Chemosphere 80 (4), 457–462.

Hasan, M., Cheng, Y., Kanwar, M.K., Chu, X.Y., Ahammed, G.J., Qi, Z.Y., 2017. Responses of plant proteins to heavy metal stress—a review. Front. Plant Sci. 8, 1492.

Hassan, S.E., Hijri, M., St-Arnaud, M., 2013. Effect of arbuscular mycorrhizal fungi on trace metal uptake by sunflower plants grown on cadmium contaminated soil. New Biotechnol. 30, 780–787. doi:10.1016/j.nbt.2013.07.002.

Helbling, D.E., Johnson, D.R., Lee, T.K., Scheidegger, A., Fenner, K., 2015. A framework for establishing predictive relationships between specific bacterial 16S rRNA sequence abundances and biotransformation rates. Water Res. 70, 471–484.

Hong, B., Nauss, R., Harwood, I.M., Miller, S.M., 2010. Direct measurement of mercury (II) removal from organomercurial lyase (MerB) by tryptophan fluorescence: NmerA domain of coevolved γ-proteobacterial mercuric ion reductase (MerA) is more efficient than MerA catalytic core or glutathione. Biochemistry 49 (37), 8187–8196.

Hu, H., Li, X., Wu, S., Yang, C., 2020. Sustainable livestock wastewater treatment via phytoremediation: Current status and future perspectives. Biores. Technol. 315, 123809.

Hu, H., Li, X., Wu, S., Yang, C., 2020. Sustainable livestock wastewater treatment via phytoremediation: current status and future prespectives. Biores. Technol. 20, 31081.

Hussain, D., Haydon, M.J., Wang, Y., Wong, E., Sherson, S.M., Young, J., …, Cobbett, C.S., 2004. P-type ATPase heavy metal transporters with roles in essential zinc homeostasis in Arabidopsis. Plant cell 16 (5), 1327–1339.

Hussain, I., Aleti, G., Naidu, R., Puschenreiter, M., Mahmood, Q., Rahman, M.M, et al., 2018. Microbe and plant assisted-remediation of organic xenobiotics and its enhancement by genetically modified organisms and recombinant technology: a review. Sci. Total Environ. 628, 1582–1599.

Chen, Y., Zhi, J., Zhang, H., Li, J., Zhao, Q., 2018. Transcriptome analysis of Phytolacca americana L. in response to cadmium stress. PLoS One 13 (6), e0199721.

Iqbal, A., Mukherjee, M., Rashid, J., Khan, S.A., Ali, M.A., Arshad, M., 2019. Development of plant-microbe phytoremediation system for petroleum hydrocarbon degradation: an insight from alkb gene expression and phytotoxicity analysis. Sci. Total Environ. 671, 696–704.

Jadia, C.D., Fulekar, M.H., 2008. Phytotoxicity and remediation of heavy metals by fibrous root grass (sorghum). J. Appl. Biosci. 10 (1), 491–499.

Johansson, E.M., Fransson, P.M.A., Finlay, R.D., vanHees, P.A.W, 2008. Quantitative analysis of exudates froms oil-living basidiomycetes in pure culture as a response to lead, cadmium and arsenic stress. Soil Biol. Biochem. 40, 2225–2236. doi:10.1016/j.soilbio.2008.04.016.

Kaushik, G., 2015. Applied Environmental Biotechnology: Present Scenario and Future Trends. Springer.

Kawachi, M., Kobae, Y., Mimura, T., Maeshima, M., 2008. Deletion of a histidine-rich loop of AtMTP1, a vacuolar Zn_{2+}/H_{+} antiporter of Arabidopsis thaliana, stimulates the transport activity. J. Biol. Chem. 283 (13), 8374–8383.

Khan, N.T., Jameel, N., Khan, M.J., 2018. A brief overview of contaminated soil remediation methods. BioTechnol. Ind. J. Res. 14 (4), 168.

Koźmińska, A., Wiszniewska, A., Hanus-Fajerska, E., Muszyńska, E., 2018. Recent strategies of increasing metal tolerance and phytoremediation potential using genetic transformation of plants. Plant Biotechnol. Rep. 12 (1), 1–14.

Krämer, U., 2005. Phytoremediation: novel approaches to cleaning up polluted soils. Curr. Opinion Biotechnol. 16 (2), 133–141.

Kuhad, R.C., Johri, A.K., Singh, A., Ward, O.P., 2004. Bioremediation of pesticide-contaminated soilsApplied Bioremediation and Phytoremediation. Springer, Berlin, pp. 35–54.

Kumar, S., Jin, M., Weemhoff, J.L., 2012. Cytochrome P450-mediated phytoremediation using transgenic plants: a need for engineered cytochrome P450 enzymes. J. Pet. Environ. Biotechnol. 3 (5).

Küpper, H., Kochian, L.V., 2010. Transcriptional regulation of metal transport genes and mineral nutrition during acclimatization to cadmium and zinc in the Cd/Zn hyperaccumulator, Thlaspi caerulescens (Ganges population). New Phytol. 185 (1), 114–129.

Kushwaha, A., Rani, R., Kumar, S., Gautam, A., 2015. Heavy metal detoxification and tolerance mechanisms in plants: implications for phytoremediation. Environ. Rev. 24 (1), 39–51.

Lasat, M.M., 2002. Phytoextraction of toxic metals: a review of biological mechanisms. J. Environ. Qual. 31 (1), 109–120.

Lee, M., Yang, M., 2010. Rhizofiltration using sunflower (Helianthus annuus L.) and bean (Phaseolus vulgaris L. var. vulgaris) to remediate uranium contaminated groundwater. J. Haz. Mater. 173 (1-3), 589–596.

Leguizamo, M.A.O., Gómez, W.D.F., Sarmiento, M.C.G, 2017. Native herbaceous plant species with potential use in phytoremediation of heavy metals, spotlight on wetlands-A review. Chemosphere 168, 1230–1247.

Li, D., Xu, X., Hu, X., Liu, Q., Wang, Z., Zhang, H., et al., 2015. Genome-wide analysis and heavy metal-induced expression profiling of the HMA gene family in Populus trichocarpa. Front. Plant Sci. 6, 1149.

Li, T., Tao, Q., Liang, C., Shohag, M.J., Yang, X., Sparks, D.L., 2013. Complexation with dissolved organic matter and mobility control of heavy metals in the rhizosphere of hyperaccumulator Sedum alfredii. Environ.Pollut. 182, 248–255. doi:10.1016/j.envpol.2013.07.025.

Limmer, M., Burken, J., 2016. Phytovolatilization of organic contaminants. Environ. Sci. Technol. 50 (13), 6632–6643.

Lu, Z., Deng, Y., Van Nostrand, J.D., He, Z., Voordeckers, J., Zhou, A., et al., 2011. Microbial gene functions enriched in the Deepwater Horizon deep-sea oil plume. ISME J. 6, 451–460.

Luo, Z.B., He, J., Li, H., Luo, J., Ma, C., et al., 2013. A transcriptomic network underlies microstructural and physiological responses to cadmium in Populus x canescens. Plant Physiol. 162, 424–439.

Ma, Y., Prasad, M.N.V., Rajkumar, M., Freitas, H., 2011a. Plant growth promoting rhizobacteria and endophytes accelerate phytoremediation of metalliferous soils. Biotechnol.Adv 29, 248–258. doi:10.1016/j.biotechadv.2010.12.001.

Ma, Y., Rajkumar, M., Luo, Y., Freitas, H., 2013. Phytoextraction of heavy metal polluted soils using Sedum plumbizincicola inoculated with metal mobilizing Phyllobacteriummyrsinacearum RC6b. Chemosphere 93, 1386–1392. doi:10.1016/j.chemosphere.2013.06.077.

Magdziak, Z., Kozlowska, M., Kaczmarek, Z., Mleczek, M., Chadzinikolau, T., Drzewiecka, K., Golinski, P., 2011. Influence of Ca/Mg ratio on phytoextraction properties of Salix viminalis. II. Secretion of low molecular weight organic acids to the rhizosphere. Ecotoxicol. Environ. safety 74 (1), 33–40.

Mahmood, T., Khalid, S., Abdullah, M., Ahmed, Z., Shah, M.K.N., Ghafoor, A., Du, X., 2020. Insights into drought stress signaling in plants and the molecular genetic basis of cotton drought tolerance. Cells 9 (1), 105.

Mandal, S.M., Chakraborty, D., Dey, S., 2010. Phenolic acids act as signaling molecules inplant-microbe symbioses. Plant Signal.Behav. 5, 359–368. doi:10.4161/psb.5.4.10871.

Mani, D., Kumar, C., 2014. Biotechnological advances in bioremediation of heavy metals contaminated ecosystems: an overview with special reference to phytoremediation. Int. J. Environ. Sci. Technol. 11 (3), 843–872.

Mann, R.M., Vijver, M.G., Peijnenburg, W.J.G.M, 2011. Metals and metalloids in terrestrial systems: Bioaccumulation, biomagnification and subsequent adverse effectsEcologica l impacts of toxic chemicals. Bentham Science Publishers, pp. 43–62.

Markou, G., Wang, L., Ye, J.F., Unc, A., 2018. Using agro-industrial wastes for the cultivation of microalgae and duckweeds: Contamination risks and biomass safety concerns. Biotechnol. Adv. 36, 1238–1254.

Marmiroli, N., Marmiroli, M., Maestri, E., 2006. Phytoremediation and phytotechnologies: a review for the present and the futureSoil and water pollution monitoring, protection and remediation. Springer, Dordrecht, pp. 403–416.

Martinoia, E., Klein, M., Geisler, M., Bovet, L., Forestier, C., Kolukisaoglu, U., …, Schulz, B., 2002. Multifunctionality of plant ABC transporters–more than just detoxifiers. Planta 214 (3), 345–355.

Mauriz, E., Calle, A., Montoya, A., Lechuga, L.M., 2006. Determination of environmental organic pollutants with a portable optical immunosensor. Talanta 69 (2), 359–364.

McGrath, S.P., Zhao, F.J., 2003. Phytoextraction of metals and metalloids from contaminated soils. Current opinion in biotechnology 14 (3), 277–282.

Meagher, Richard B., 2000. Phytoremediation of toxic elemental and organic pollutants. Current opinion in plant biology 3 (2), 153–162.

Meyer, C., Verbruggen, N., 2012. The use of the model species Arabidopsis halleri towards phytoextraction of cadmium polluted soils. New biotechnolo 30, 9–14.

Mitter, E.K., Kataoka, R., de Freitas, J.R., Germida, J.J., 2019. Potential use of endophytic root bacteria and host plants to degrade hydrocarbons. Int. J. Phytoremed. 21 (9), 928–938.

Mleczek, M., Rutkowski, P., Goliński, P., Kaczmarek, Z., Szentner, K., Waliszewska, B., Szczukowski, S., 2017. Biological diversity of Salix taxa in Cu, Pb and Zn phytoextraction from soil. Int. J. Phytoremed. 19 (2), 121–132.

Orłowska, E., Przybyłowicz, W., Orlowski, D., Mongwaketsi, N.P., Turnau, K., Mesjasz-Przybyłowicz, J., 2013. Mycorrhizal colonization affects the elemental distribution in roots of Ni-hyperaccumulator Berkheyacoddii Roessler. Environ. Pollut. 175, 100–109. doi:10.1016/j.envpol.2012.12.028.

Pandey, V.C., Bajpai, O., 2019. Phytoremediation: from theory toward practicePhytomanagement of Polluted Sites. Elsevier, pp. 1–49.

Peng, J.S., Wang, Y.J., Ding, G., Ma, H.L., Zhang, Y.J., Gong, J.M., 2017. A pivotal role of cell wall in cadmium accumulation in the Crassulaceae hyperaccumulator Sedum plumbizincicola. Mol. Plant 10 (5), 771–774.

Pérez-Palacios, P., Agostini, E., Ibáñez, S.G., Talano, M.A., Rodríguez-Llorente, I.D., Caviedes, M.A., Pajuelo, E., 2017. Removal of copper from aqueous solutions by rhizofiltration using genetically modified hairy roots expressing a bacterial Cu-binding protein. Environ. Technol. 38 (22), 2877–2888.

Pham, V.H.T., Kim, J, 2012. Cultivation of unculturable soil bacteria. Trends Biotechnol. 30, 475–484.

Pilon-Smits, E., 2005. Phytoremediation. Annu. Rev. Plant Biol. 56, 15–39.

Pilon-Smits, E.A., Quinn, C.F., Tapken, W., Malagoli, M., Schiavon, M., 2009. Physiological functions of beneficial elements. Curr. Opinion Plant Biol. 12 (3), 267–274.

Pineau, C., Loubet, S., Lefoulon, C., Chalies, C., Fizames, C., et al., 2013. Natural variation at the FRD3 MATE transporter locus reveals cross-talk between Fe homeostasis and Zn tolerance in Arabidopsis thaliana. PLoS Gen. 8, 1–8.

Prieto, M.J., Acevedo, S.O.A., Prieto, G.F., Nallely, T.G, 2018. Phytoremediation of soils contaminated with heavy metals. Biodiv. Int. J. 2 (4), 362–376.

Qi, W., Zhang, L., Wang, L., Xu, H., Jin, Q., Jiao, Z., 2015. Pretreatment with low-dose gamma irradiation enhances tolerance to the stress of cadmium and lead in Arabidopsis thaliana seedlings. Ecotoxicol. Environ. Safety 115, 243–249.

Qixing, Z., Zhang, C., Zhineng, Z., Weitao, L., 2011. Ecological remediation of hydrocarbon contaminated soils with weed plant. J. Resour. Ecol. 2 (2), 97–105.

Quan, M., Liu, X., Xiao, L., Chen, P., Song, F., Lu, W., Song, Y., Zhang, D., 2021. Transcriptome analysis and association mapping reveal the genetic regulatory network response to cadmium stress in Populus tomentosa. J. Exp. Bot. 72 (2), 576–591.

Rahman, R.A., Abou-Shanab, R.A., Moawad, H., 2008. Mercury detoxification using genetic engineered Nicotiana tabacum. Glob NEST J 10, 432–438.

Rajkumar, M., Sandhya, S., Prasad, M.N.V., Freitas, H, 2012. Perspectives of plant-associated microbes in heavy metal phytoremediation. Biotechnol.Adv 30, 1562–1574. doi:10.1016/j.biotechadv.2012.04.011.

Rao, M.R., Halfhill, M.D., Abercrombie, L.G., Ranjan, P., Abercrombie, J.M., et al., 2009. Phytoremediation and phytosensing of chemical contaminants, RDX and TNT: identification of the required target genes. Funct. Integr. Gen. 9, 537–547.

Rezania, S., Taib, S.M., Din, M.F.M., Dahalan, F.A., Kamyab, H., 2016. Comprehensive review on phytotechnology: heavy metals removal by diverse aquatic plants species from wastewater. J. Hazard. Mater., 318, 587-599.

Rodgers-Vieira, E.A., Zhang, Z., Adrion, A.C., Gold, A., Aitken, M.D., 2015. Identification of anthraquinone-degrading bacteria in soil contaminated with polycyclic aromatic hydrocarbons. Appl. Environ. Microbiol. 81 (11), 3775–3781.

Romanov, V., Hausinger, R.P., 1994. Pseudomonas aeruginosa 142 uses a three-component ortho-halobenzoate 1, 2-dioxygenase for metabolism of 2, 4-dichloro-and 2-chlorobenzoate. J. Bacteriol. 176 (11), 3368–3374.

Rostami, S., Azhdarpoor, A., 2019. The application of plant growth regulators to improve phytoremediation of contaminated soils: a review. Chemosphere 220, 818–827.

Ryu, C.M., Farag, M.A., Hu, C.H., Reddy, M.S., Kloepper, J.W., Paré, P.W., 2004. Bacterial volatiles induce systemic resistance in Arabidopsis. Plant Physiol 134, 1017–1026. doi:10.1104/pp.103.026583.

Saraswat, S., Rai, J.P.N, 2011. Complexation and detoxification of Zn and Cd in metal accumulating plants. Rev. Environ. Sci. Bio/Technol. 10 (4), 327–339.

Sarma, H., 2011. Metal hyperaccumulation in plants: a review focusing on phytoremediation technology. J. Environ. Sci. Technol. 4 (2), 118–138.

Sarma, H., 2011. Metal hyperaccumulation in plants: a review focusing on phytoremediation technology. J. Environ. Sci. Technol. 4 (2), 118–138.

Schiavon, M., et al., 2012. Transcriptome profiling of genes differentially modulated by sulfur and chromium identifies potential targets for phytoremediation and reveals a complex S–Cr interplay on sulfate transport regulation in B. juncea. J. Hazard. Mater. 239, 192–205 40.

Schwartz, C., Echevarria, G., Morel, J.L., 2003. Phytoextraction of cadmium with Thlaspi caerulescens. Plant Soil 249 (1), 27–35.

Segura, A., Rodríguez-Conde, S., Ramos, C., Ramos, J.L., 2009. Bacterial responses and interactions with plants during rhizo remediation. Microb. Biotechnol. 2, 452–464. doi:10.1111/j.1751-7915.2009.00113.x.

Selvam, A., 2008. Phytochelatin systhesis and cadmium uptake of Brassica napus. Environmental technology 29 (7), 765–773.

Selvam, A., Wong, J.W.C., 2009. Cadmium uptake potential of Brassica napus cocropped with Brassica parachinensis and Zea mays. J. Haz. Mater. 167 (1-3), 170–178.

Shi, W.G., Liu, W., Yu, W., Zhang, Y., Ding, S., Li, H., Mrak, T., Kraigher, H., Luo, Z.B., 2019. Abscisic acid enhances lead translocation from the roots to the leaves and alleviates its toxicity in Populus× canescens. J. Haz. Mater. 362, 275–285.

Smits, P.E., 2005. Phytoremediation. Annu. Rev. Plant Biol. 56, 15–39.

Sofie, T., Wouter, S., Nele, W., Jaco, V., 2017. Phytoremediation: State-of-the-art and a key role for the plant microbiome in future trends and research prospects. Int. J. Phytoremed. 19 (1), 23–38

Song, W.Y., Park, J., Eisenach, C., Maeshima, M., Lee, Y., Martinoia, E., 2014. ABC transporters and heavy metalsPlant ABC Transporters. Springer, Cham, pp. 1–17.

Stark, B.C., Dikshit, K.L., Pagilla, K.R., 2011. Recent advances in understanding the structure, function, and biotechnological usefulness of the hemoglobin from the bacterium Vitreoscilla. Biotechnol. Lett. 33 (9), 1705.

Strycharz, S., Newman, L., 2009. Use of native plants for remediation of trichloroethylene: I. deciduous trees. Int. J. Phytoremed. 11 (2), 150–170.

Suman, J., Uhlik, O., Viktorova, J., Macek, T., 2018. Phytoextraction of heavy metals: a promising tool for clean-up of polluted environment? Front. Plant Sci. 9, 1476.

Tang, L., Mao, B., Li, Y., Lv, Q., Zhang, L., Chen, C., …, Pan, Y., 2017. Knockout of OsNramp5 using the CRISPR/Cas9 system produces low Cd-accumulating indica rice without compromising yield. Sci. Rep. 7 (1), 1–12.

Tangahu, B.V., Sheikh Abdullah, S.R., Basri, H., Idris, M., Anuar, N., Mukhlisin, M., 2011. A review on heavy metals (As, Pb, and Hg) uptake by plants through phytoremediation. Int. J. Chem. Eng. 2011.

Teng, Y., Wang, X., Li, L., Li, Z., Luo, Y., 2015. Rhizobia and their bio-partners as novel drivers for functional remediation in contaminated soils. Frontiers in plant science, 6, p. 32.

Terrence, H.B., Simon, J., Fre´de´ ric, E.P., Etienne, Y., 2014. Increasing phytoremediation efficiency and reliability using novel omics approaches. Trends Biotechnol. 32, 5.

Thapa, G., Sadhukhan, A., Panda, S.K., Sahoo, L., 2012. Molecular mechanistic model of plant heavy metal tolerance. Biometals 25, 489–505.

Trivedi, P., Pandey, A., 2008. Plant growth promotion abilities and formulation of Bacillus megaterium strain B388(MTCC6521) isolated from a temperate Himalayan location. Ind. J.Microbiol 48, 342–347. doi:10.1007/s12088-008-0042-1.

Tu, S.X., Ma, L.Q., Luongo, T., 2004. Root exudates and arsenic accumulation in arsenic hyperaccumulating Pteris vittata and non-hyperaccumulating Nephrolepis exaltata. Plant and Soil 258, 9e19.

Turan, M., Angin, I., 2004. Organic chelate assisted phytoextraction of B, Cd, Mo and Pb from contaminated soils using two agricultural crop species. Acta Agr. Scand., B-Soil Plant Sci. 54 (4), 221–231.

Tzafestas, K., 2016. Doctoral dissertation. University of York.

Tzafestas, K., Razalan, M.M., Gyulev, I., Mazari, A.M., Mannervik, B., Rylott, E.L., Bruce, N.C., 2017. Expression of a Drosophila glutathione transferase in Arabidopsis confers the ability to detoxify the environmental pollutant, and explosive, 2, 4, 6-trinitrotoluene. New Phytol, 214 (1), 294–303.

Vajpayee, P., Rai, U.N., Ali, M.B., Tripathi, R.D., Yadav, V., Sinha, S., Singh, S.N., 2001. Chromium-induced physiologic changes in Vallisneria spiralis L. and its role in phytoremediation of tannery effluent. Bull. Environ. Contam. Toxicol. 67 (2), 246–256.

Veselý, T., Tlustoš, P., Száková, J., 2011. The use of water lettuce (Pistia stratiotes L.) for rhizofiltration of a highly polluted solution by cadmium and lead. Int. J. Phytoremed. 13 (9), 859–872.

Wei, Z., Van Le, Q., Peng, W., Yang, Y., Yang, H., Gu, H., …, Sonne, C., 2020. A review on phytoremediation of contaminants in air, water and soil. J. Haz. Mater. 403, 123658.

Wilkinson, S.R., Been, M.D., 2005. A pseudoknot in the 3′ non-core region of the glmS ribozyme enhances self-cleavage activity. RNA 11 (12), 1788–1794.

Willems, G., et al., 2007. The genetic basis of zinc tolerance in the metallophyte Arabidopsis halleri ssp halleri (Brassicaceae): an analysis of quantitative trait loci. Genetics 176, 659–674.

Wittmann, A., Suess, B., 2012. Engineered riboswitches: expanding researchers' toolbox with synthetic RNA regulators. FEBS letters, 586 (15), 2076–2083.

Wong, C.K.E., Jarvis, R.S., Sherson, S.M., Cobbett, C.S, 2009. Functional analysis of the heavy metal binding domains of the Zn/Cd-transporting ATPase, HMA2, in Arabidopsis thaliana. New Phytol. 181 (1), 79–88.

Xia, H., Ma, X., 2006. Phytoremediation of ethion by water hyacinth (Eichhornia crassipes) from water. Bioresour. Technol. 97 (8), 1050–1054.

Yadav, B.K., Siebel, M.A., van Bruggen, J.J., 2011. Rhizofiltration of a heavy metal (lead) containing wastewater using the wetland plant Carex pendula. Clean–Soil, Air, Water, 39 (5), 467–474.

Yan, A., Wang, Y., Tan, S.N., Yusof, M.L.M., Ghosh, S., Chen, Z, 2020. Phytoremediation: a promising approach for revegetation of heavy metal-polluted land. Front. Plant Sci., 11.

Yoon, K.P., 1994. Overproduction and Operator DNA-Protein Blotting of R100 Mutant MerR from Shigella flexneri. J. Microbiol. Biotechnol. 4 (4), 250–255.

You, S.H., Zhu, B., Han, H.J., Wang, B., Peng, R.H., Yao, Q.H., 2015. Phytoremediation of 2, 4, 6-trinitrotoluene by Arabidopsis plants expressing a NAD (P) H-flavin nitroreductase from Enterobacter cloacae. Plant Biotechnol. Rep. 9 (6), 417–430.

Yuan, S., Xi, Z., Jiang, Y., Wan, J., Wu, C., Zheng, Z., et al., 2007. Desorption of copper and cadmium from soils enhanced by organic acids. Chemosphere 68, 1289–1297. doi:10.1016/j.chemosphere.2007.01.046.

Zhang, L., Routsong, R., Nguyen, Q., Rylott, E.L., Bruce, N.C., Strand, S.E. (2017). Expression in grasses of multiple transgenes for degradation of munitions compounds on live-fire training ranges. Plant biotechnology journal, 15 (5), 624–633.

Zhang, Q., Kong, W., Wei, L., Wang, Y., Luo, Y., Wang, P., …, Jiang, G., 2020. Uptake, phytovolatilization, and interconversion of 2, 4-dibromophenol and 2, 4-dibromoanisole in rice plants. Environ. Int. 142, 105888.

Zhang, H., Xie, X., Kim, M.S., Kornyeyev, D.A., Holaday, S., Paré, P.W., 2008. Soil bacteria augment Arabidopsis photosynthesis by decreasing glucose sensing and abscisic acid levels in planta. Plant J 56, 264–273. doi:10.1111/j.1365-313X.2008.03593.x.

Zhao, F.J., Hamon, R.E., McLaughlin, M.J., 2001. Rootexudatesofthe hyperaccumulator Thlaspicaerulescens do not enhance metal mobilization. New Phytol 151, 613–620. doi:10.1046/j.0028-646x.2001.00213.x.

Zhou, Q., Li, X., Lin, Y., Yang, C.P., Tang, W.C., Wu, S.H., Li, D.H., Lou, W., 2019. Effects of copper ions on removal of nutrients from swine wastewater and on release of dissolved organic matter in duckweed systems. Water Res 158, 171–181.

Zhu, Y.G., Rosen, B.P., 2009. Perspectives for genetic engineering for the phytoremediation of arsenic-contaminated environments: from imagination to reality? Curr. Opinion Biotechnol. 20 (2), 220–224.

Chapter 5

Microbial-assisted phytoremediation

Mir S. Rabani[a], Insha Hameed[b], Tawseef A. Mir[c], Bilal A. wani[d], Mahendra K. Gupta[a], Aukib Habib[a], Muatasim Jan[c], Humaira Hussain[e], Shivani Tripathi[a], Anjali Pathak[a], Moniem B. Ahad[f], Charu Gupta[g]

[a]*Microbiology Research Lab., School of Studies in Botany, Jiwaji University, Gwalior, Madhya Pradesh,* [b]*School of Studies in Environmental Science, Jiwaji University, Gwalior, Madhya Pradesh,* [c]*Centre of Research for Ethnobotany, Government Model Science College, Jiwaji University, Gwalior, India,* [d]*Plant Reproductive Biology, Genetic Diversity and Phytochemistry Researh Lab., Department of Botany, University of Kashmir, Srinagar, Jammu and Kashmir,* [e]*Department of Botany, University of Kashmir, Srinagar, Jammu and Kashmir,* [f]*Department of Zoology, Central University of Kashmir, Ganderbal, Jammu and Kashmir,* [g]*School of Studies in Microbiology, Jiwaji University, Gwalior, Madhya Pradesh*

5.1 Introduction

Rapid industrialization has led to discharge of the high concentrations of toxic substances directly or indirectly into the ecosystem. These toxic substances include organic and inorganic pollutants that have become a major concern across the globe. Soil has become a repository for a number of pollutants released from natural (Supaka et al., 2001) and anthropogenic sources. Majority of these pollutants are released by anthropogenic activities, including industrial, agricultural, and domestic sources, which lead to the deterioration of soil and other ecosystems. As a result of which, there is an upsurge in the polluted areas, which needs to be eliminated or mitigated. A bioremediation technique namely, microbial-assisted phytoremediation is considered as a promising technology to reduce or mitigate the contaminants from the environment. In addition to their ability of increasing plant growth, microbes play a pivotal role in phytoremediation contaminated soil (Doty et al., 2017; Motesharezadeh et al., 2017; Stepniewska and Kuzniar, 2013). Toxic pollutant compounds are lethal and persist in the environment for longer periods and have been found to pose significant threat to various organisms and environment (Nagajyoti et al., 2010; Abdel-Shafy and Mansour, 2016; Ahemad, 2019).

Plants can degrade hydrocarbons by removing or immobilizing them, and by the enhancement of microbe-based degradation (Megharaj et al., 2011). Another group of pollutants, which is hazardous to mankind and animals, are persistant

Phytoremediation. DOI: https://doi.org/10.1016/B978-0-323-89874-4.00006-6

organic compounds (POPs). POPs are carbon containing compounds and are not easily degraded naturally (Weber et al., 2011). The stability of these organic compounds increases with the increase in number of chlorine atoms. Thus, the highly halogenated compounds are not easily degraded (Wikoff et al., 2012). Further, POPs are transported to long distances because of their insoluble and semi-volatile nature, and therefore have worldwide distribution (Wang, 2012). Due to their higher stability and long-distance transportation, POPs are found in air and water, reach into the soils, and then enter the food chain (Hao et al., 2014). They are bioaccumulated and biomagnified after tranfering from different trophic levels. Bioaccumulation of POPs in living organisms was reported for the first in 1970 (Lie et al., 2003). Similarly, plants have the property to eliminate metals from polluted environment by accumulating these pollutants into their body (Baker and Brooks, 1989; Berti and Cunningham, 2000). Studies have indicated a variety of properties of microbes living in rhizospheric zone of the plant body. They release some organic compounds, which can increase the heavy metal bioavailability. Although phytoremediation is considered as a promising and low-cost approach for the degradation of organic and inorganic compounds compared to other technologies, but this process takes more time. However, the phytoremediation may not be effective when there are mixed or combined pollutants, as the contaminants may interact with each other or with plants and the rhizospheric zone (Almeida et al., 2009).

Microbial-assisted remediation is a potential technique for the remediation of mixed or combined pollutants. Both plants and microbes can handle stressed conditions and play a pivotal role in the remediation of mixed pollutants in highly polluted soils. Degradation rate of organic contaminants is dependent on potential microorganisms. Heavy metal contaminants show their impact on microorganisms and their activities negatively, therby affecting the rate of phytoremediation as well (Chigbo, 2013). Different investigations have revealed that gram-negative microbes could induce innate plant growth promoting (PGP) traits like phytohormones, siderophores, and chelating substances. Further, with their evolutionary adaptations these microbes enable and enhance the biodegradation of pollutants. However, the capacity for bacterial-mediated phytoremediation has been undervalued till date.

Plants interact with pollutants such as organic and inorganic pollutants from contaminated environments in various ways, including phytoextraction, phytodegradation, phytostabilization, rhizofiltration, and phytovolatilization (Fig. 5.1).

5.2 Mechanisms of microbe-assisted phytoremediation of environmental pollutants

There are various mechanisms by which microbes in the different environs can improve the pollutant elimination. They are broadly classified into two types, direct and indirect mechanism.

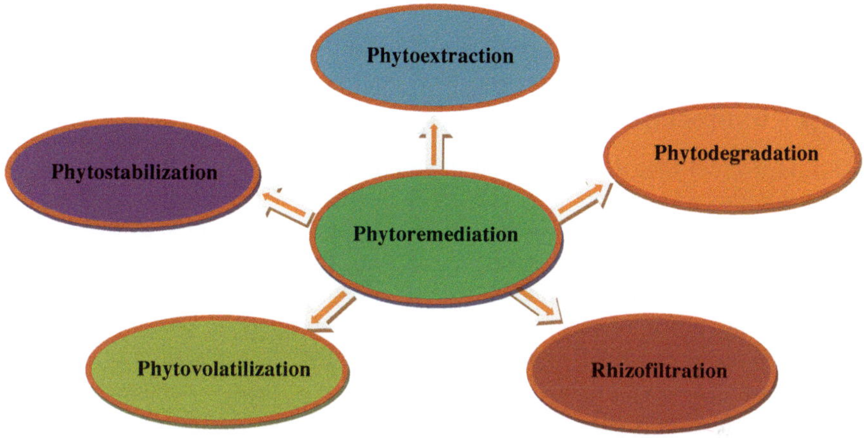

FIG. 5.1 Different phtoremediation techniques used for remediation of contaminated sites.

5.2.1 Direct mechanism

The addition or association of microbes in the process of phytoremediation enhances the rate of phytoremediation. Microorganisms have the capability to release H^+ through transmembrane H^+ ATPase, hence maintains the potential of membrane (Gadd, 2004). Autotropic leaching in many cases is performed by acidophilic bacteria, wherby assimilates CO_2 and energy is produced from various inorganic pollutants (Rawlings, 1997).

Another group of metabolites, which have the capability to enhance pollutant mobility and stimulation of phytoremediation are biosurfactants (Rajkumar et al., 2012). Microbial biosurfactants have the capability to form complex with various contaminants in the soil. As the bond between the contaminant and the biosurfactant is much stronger in comparison to the bond between the contaminant and the soil particles, these contaminants are absdorbed from the soil by decreasing the interfacial tension, due to which the solubility and the bioavailibilty of these pollutants is increased in the soil (Banat, 1995).

Another important mechanism in this regard is the use of siderophores. These are organic molecules having high affinity toward the chelation of inorganic pollutants like Fe^{3+} (Miethke and Marahiel, 2007). By this process, the iron molecule is released from the minerals or organic matter and uptake of this iron is enhanced when its concentration is limited (Li et al., 2012). Generally, the bacterial siderophores are catecholates (i.e. enterobactin), some are carboxylates (i.e. rhizobactin) and hydroxamates (i.e. ferrioxamine B) (Matzanke, 1991) (Schalk et al., 2011). Furthermore, fungi mostly release ferrichrome, caprogen, or triacetylfusarinine commonly called hydroxamates (Haselwandter, 1995). The association of microorganisms with plants help in the immobilization of various types of pollutants in the soil and enhances the rate of phytostabilization of these contaminants (Wenzel, 2008). Most of the attention in the process of bioaccumulation of pollutants is provided to acuumulation in cellular biomass and precipitation (Fein et al., 2001; He et al., 2000).

5.2.2 Indirect mechanism

As it is evident that the successful phytoremdiation is dependant on the pollutant tolerance by plants and capability of plants to produce high biomass. Furthermore, a considerable attention is being given to plant growth promoting fungi (PGPF) and plant growth promoting bacteria (PGPB) (Bent, 2006). Varied number of mechanisms, which are associated are mostly dependent on the taxonomy of the microbes and their metabolic properties. Mehanisms which have been found and confirmed by various scientific communities involve phytohormone and enzyme syenthesis (indole acetic acid [IAA]), enhanced nutrient uptake (such as nitrogen fixation and phosphorus immobilization), biotic and abiotic (such as pathogen attack and salinity) stress tolerance (Ma et al., 2016b). Many of the physiological and biochemical functions of the plants are regulated by IAA, one of the most important phytohormone (Glick, 2012). It has the capability to induce root growth and development, reduces the salt stress, participates in the plant pathogen interactions, and is mainly involved in the stimulation and proliferation of lateral roots in the plant species. It increases the surface area and induces the absorption of contaminants by the plant roots (Lugtenberg and Kamilova, 2009). Most of the phytohormones are produced by bacteria and about 80% of the rhizospheric strains biosynthesize IAA (Spaepen and Vanderleyden, 2011). Similarly, fungi are also believed to release some compounds, which are similar to certain phytohormones like auxins, gibberllins, and ethylene (Chanclud and Morel, 2016). In occurrence of greater concentration of contaminants present in the soils, the plant growth is reduced to higher levels and the absorption of these contaminants is reduced. Microorganisms associated with plants thereby, enhance the contaminant and nutrient uptake by plant roots in such stressful conditions (Rajkumar et al., 2012). Occurrence of siderophores may also stimulate the plant growth in polluted areas by involvement in maintaining levels of IAA and by decreasing generation of reactive oxygen species, by oxidative stress reduction, by the peroxidase activity stimulation, and by the control of phytopathogens through chelation (Rajkumar et al., 2009).

5.3 Phytoremediation of inorganic pollutants

The inorganic pollutants are the pollutants, which are hazardous to all the living organisms or to the environment. The natural sources of the inorganic pollutants are many, including dried water droplets from atmosphere, volcano dust particles, soil erosion, weathering of rocks, and most of the forest fires. The degradation of the animals after death also adds to the inorganic pollutant concentration in the ecosystem. In water, the increased number of inorganic pollutants can be attributed to both the natural and anthropogenic sources (Agarwal et al., 2009).

Besides having some of the beneficial effects, higher conccentrations of these inorganic components (mercury, lead, and cadmium) are hazardous to the living environment. These inorganic components are intentionally added to some industrial products such as cosmetics, battries, circuit boards, switches, and some pigments through which they make way to enter the environment. These pollutants are released in the environment mostly by industries, transportation, and power generation. After their release, the pollutants are taken to other places with the the help of wind, water, and other carriers depending upon the form of the pollutant (Agarwal et al., 2009).

Generally, by name, the inorganic pollutant is the substance that is foreign to the environment. However, if the chemical is native to the environment and when its concentration increases more than the standard value, it is called as the pollutant and causes pollution. When the concentration of these pollutants increases, they cause instability to the environment, and the remediation of such pollutants becomes the priority for different stakeholders (Ali et al., 2013). Phytoremediation has been found as an accepted technique to be followed to achieve the successful remediation of such pollutants. To carry out the phytormediation, several hyperaccumulator plant species have been identified. These hyperaccumulator plant species are limited by slow growth, low biomass, etc. (Maestri, 2011). Besides, several plants that are not hyperaccumulators have been identified for remediating inorganic pollutants (Solhi et al., 2005; Yu et al., 2006). Lasat (2002) tested some 30 plants species for about 4 weeks in hydroponics having low amounts of cadmium (Cd), copper (Cu), and zinc (Zn) to identify the inorganic pollutant accumulating capacity of plants. He observed that certain species of *Brassica,* such as, *Brassica juncea, Brassica napus, Brassica rapa*, enhanced the accumulation of Zn and Cd to some extent. Till date, about 400 plant species have been reported to be used in degradation of inorganic pollutants (Brooks, 1998; Baker et al., 1994).

5.3.1 Bacterial-assisted phytoremediation of inorganic pollutants

From last few decades, the plant-microbe interactions are being studied enormously. These informations on microbial interactions with plants in the rhizospheric zone of plants are proposed as remediation techniques (Kavamura and Esposito, 2010). Different plant species release some secreations (Antoun and Kloepper, 2001) that attract microbes by serving as a source of microbial nutrition. Their presence has some neutral, lethal, or valuable effects on plant growth. Those bacterial strains, which intensively colonize roots are termed as plant growth promoting rhizobacteria (PGPR). These bacterial srains are found to be very useful in terms of reducing the environmental stress such as salinity and floods (Saleem et al., 2007). PGPR promotes the growth of the plant by direct or indirect methods. Heavy metalloids have been identified to produce stress in plants and reduce the growth of plant species. 1-aminocyclopropane-1-carboxylate (ACC) deaminase producing bacteria reduce ethylene stress levels and promote the growth of plant species even in polluted areas (Gerhardt et al., 2006). Alternatively, siderophore producing bacteria could play a pivotal role remediation of contaminated environments. Furthermore, siderophores play a great role in complexing toxic metals and enhancing their mobility in soils (Rajkumar et al., 2010). PGPRs have been used in phytoremediation of arsenic (As) contaminated soils as well. Rhizospheric bacterial strains are worthy appreciable in case of remediation of metals and metalloids because they are capable of changing the metal bioavailability through the adjustment of soil pH, chelator release, induced redox changes, etc. (Kidd et al., 2009; Rajkumar et al., 2012). Some other microorganisms are also able to accumulate etals and reduce them to lower redox states (Wang et al., 2017) Table 5.1.

5.3.2 Fungal-assisted phytoremediation of inorganic pollutants

The properties of plant species such as growth, productivity, and phytoremediation may contribute to microbiota. Fungi are found to be more metal resistant than bacterial species (Deng and Cao, 2017; Kidd et al., 2009). They effectively reveal microsites of the soil due to the limited diameter of the mycelia, are not accessible to plant roots, protect the roots from direct metal contact, and impede the transport of metals via increased soil hydrophobicity (Deng and Cao, 2017; Wenzel, 2009). Fungi are among the energetic components of nature for degradation of pollutant materials and important agent of food web in soil (Rhodes, 2012). In addition, the capacity of forming prolonged networks of mycelia makes them suited for bioremediation. In situations of bacterial failure, fungi may prove helpful as an alternate or a beneficial match (Deng and Cao, 2017; Harms et al., 2013). Extended networks of hyphae can also contribute to soil aggregate stabilization (Leung et al., 2013). *Trichoderma, Aspergillus*, and arbuscular mycorrhizal fungi can improve metal phytoremediation (Schneider et al., 2016) because they are highly capable of immobilizing toxic metals by insoluble oxalate formation, biosorption, or melanin-like polymer chelation. The degradation of endosulfan was evaluated for *Aspergillus niger, Aspergillus terreus, Cladosporium oxysporum, Mucor thermohyalospora, Fusarium ventricosum, Phanerochaetechrysosporium*, and *Trichoderma harzianum* (Bhalerao and Puranik, 2007). Some of the strains of fungi were reported during transformation to conduct esterification, dehydrogenation, hydroxylation, and dioxygenation (Bhalerao and Puranik, 2007). A few zygomycetes and *Aspergillus* sp. fungi demonstrate the participation of nonligninolytic polychlorinated biphenyl (PCB) degradation enzymes (Tigini et al., 2009). Vascular plants, in fact, have with them a high variety of fungi. Some nonpathogenic mutualistic fungi (AM fungi) are also able to colonize plant root system. Most herbs and different tree families are associated with AM fungi, while EM fungi are associated with woody plants only (Deng and Cao, 2017; Mandyam and Jumpponen, 2005). Mycorhizza promote soil nutrient absorption and help translocate nutrients

TABLE 5.1 Role of bacteria in decontamination and plant growth promotion.

Bacterial strains	Host plant	Contaminants	Beneficial effects	References
Kocuria sp. LC2 and LC5, *Enterobacter* sp. LC1, LC4, and LC6, and *Kosakonia* sp. LC7	*Solanum nigrum*	As	IAA, P-solubilization	Mukherjee et al., (2018)
Mesorhizobium Amorphae	*Robinia pseudoacacia*	Cu, Zn, Cr	IAA-induced stress tolerance	Hao et al., (2013)
Burkholderia sp. J62	*Lycopersicon esculentum* and Zea mays	Pb, Cd	P-solubilization, IAA, siderophores, ACC deaminase	Jiang et al., (2008)
Photobacterium sp.	*Phragmites australis*	Hg	IAA, mercury reductase activity	Mathew et al., (2015)
Brevibacillus brevi	*Trifolium repens*	Cd, Pb, Ni	IAA production	Vivas et al., (2006)
Arthrobacter sp. LKS02, *Chryseobacterium* sp. LKS04, *Flavobacterium* sp. LKS03, *Serratia marcescens* LKR01	*Solanum nigrum* L.	Cd, Cu, Pb, Zn	Production of IAA, siderophores, ACC deaminase, P-solubilization	Luo et al., (2011)
Acaulospora, *Rhizophagus*, *Glomus*	*Vetiveria zizanioides* *Ricinus communi*	Pb	Higher tendency for Pb absorbed by the roots to be transported to the shoots	Schneider et al., (2016)

(continued)

TABLE 5.1 (Cont'd)

Bacterial strains	Host plant	Contaminants	Beneficial effects	References
Bacillus sp. MN3-4	*Alnus firma* and *Brassica napus*	Pb, Cd, Zn, Ni, and Cu	Production of IAA and siderophores	Shin et al., (2012)
Rhizophagus intraradices	*Oryza sativa*	Cd	Significant effects on root biomass, straw, and root Cd concentration	Luo et al., (2011)
Rhizophagus intraradices, Funneliformis mosseae,	*Solanum nigrum*	Cd	Significantly enhanced shoot biomass and Cd shoot concentration	Li et al., (2012)
Scleroderma citrinum, Amanita muscaria, Lactarius rufus	*Pinus sylvestris L.*	Zn, Cd, Pb	Reduced translocation of Zn, Cd, or Pb from roots to shoots in pine seedlings	Krupa and Kozdrój, (2007)
Actinobacterium	*Salix caprea*	Cd and Zn	Production of siderophores and ACCD	Kuffner et al., (2010)

ACC, 1-aminocyclopropane-1-carboxylate; ACCD, α-aminocyclopropane-1-carboxylic acid deaminase; As, arsenic, Cd, cadmium; Cu, copper; Hg, mercury; IAA, indole acetic acid; Ni, nickel; Pb, lead; Zn, zinc.

to host plant species, eliminate hazardous metalloids and enhance microbial activity in soils that helps in contaminant degradation (He et al., 2014; Sarwar et al., 2017; Deng and Cao, 2017; Li et al., 2018; Lu and Lu, 2015). Micorbes commonly found in natural and agricultural soils are the AM fungi of Glomeromycota (He et al., 2014; Mohammad and Mittra, 2013). AM fungi are widely spread and commonly found with terrestrial plant species, form symbiotic associations (Leung et al., 2010; Leung et al., 2010). Due to this, they form an essential portion of soil microbiome and offer benefits to their hosts (He et al., 2014), in particular phosphorus; they also boost the acquisition of water and enhance pathogen resistance (Chaturvedi et al., 2018; Qiao et al., 2015). Therefore, mycorrhizal plants spend fewer resources than nonmycorrhizal plants in the root system extension and are tolerant toward drought stress and pathogenic outbreaks (Coninx et al., 2017). Mycorrhizal fungi prevent the increase of pathogenic agents by acidifying rhizospheric zone and by the development of certain antibiotics (Coninx et al., 2017; Fortin et al., 2009). AM fungi can show a higher resistance to herbivore attacks due to jasmonic acid (Coninx et al., 2017; Fortin et al., 2009). It was found that the content of osmolytes in *B. juncea* increased using AM fungi (Sarwat et al., 2016). This association enhances phytoremdiation by increasing the bioavailability of pollutants or by decreasing the toxicity of contaminants (Bhargava et al., 2012; Coninx et al., 2017; Cabral et al., 2015; Colpaert et al., 2011). But it is evident that the efficiency is based on the fungal component (Coninx et al., 2017; Cabral et al., 2015).

5.3.3 Remeidation of inorganic contaminants through root uptake and accumulation

Due to urbanization and industrialization, large amount of chemicals are released in the environment intentionally or unintentionally from the past few decades. Among them some of the chemicals are essential and are needed for the proper functioning of the environment, whereas few are not used up and get accumulated in the environment (Wang et al., 2018). These chemicals, which get accumulated in the environment generally involve metals and metalloids. These contaminants are released by burning of fossils fuels, industrial discharges, mining processes, unregulated dumping of munciple wastes, and the use of fertilizers, which then enter the food chain (Tchounwou et al., 2012). But sometimes these pollutants are not accumulated, and these plants develop resistance against these pollutants. This resistance to the pollutants by some of the plant species is because of PGPB (Beneduzi et al., 2012). PGPB have the capability to increase the biomass production as well as the growth of the plant and help in providing resistance toward harmful inorganic contaminants. This involvement of PGPR with that of the plant species is responsible for degradation of inorganic contaminants (Grobelak et al., 2018). This approach of pollutant remediation has achieved a great success in the world today.

5.4 Phytoremediation of organic pollutants

Many wastes and effleunts are being released in the environment by various agricultural and industrial activities (Safe, 1984). Most of the contaminants (organic contaminants) released by these activities are highly hazardous to living beings. One of the problem associated with these organic contaminants is that they persist in the ecosystem for longer periods and prove hazardous to environment and public health (Neff, 1979; Safe, 1984). Their presence of these contaminants in the environment is not the only problem, but also the concentration and their types are a big concern to hold. One of the main organic pollutants responsible for the degradation of environment is the pesticides (EPA, 1984). These pollutants coming from the use of pesticides are found all over the world, both in rural and urban locations as well as in other ecosystems. Additional group of organics present everywhere are the total petroleum hydrocarbons (TPHs) and polycyclic aromatic hydrocarbons (PAHs). TPHs are maily released by the oil refineries during the process of drilling, transportation, and refining. Table 5.2 shows some of the plant species identified for phytoremediation of organic pollutants.

Due to incomplete combustion of the burning wood or peat at very low temperatures, various organic pollutants are released in the air. It is evident that the process of phytoremediation is quite successful in terms of remediation of inorganic pollutants (metallic pollutants, radionuclides, and other inorganic contaminants), and a lot of research has been done to find out the hyperaccumulator species for the purpose of remediation of inorganic pollutants. In fact, most of the publications are in accordance with the inorganic pollutant remediation. But a limited literature is available for the remdiation of organic pollutants (Cunningham et al., 1996). Some of the researchers have identified a few plants, which are the best hyperaccumulators for the organic pollutants, and this process has received an overwhelming response from the researchers and other environmental bodies. The concept of degradation of organic pollutants is based on the idea that vegetative soils are more readily involved in the disappearance of organic pollutants than the nonvegetated bulk soils (Burken and Schnoor, 1996; Cunningham and Berti, 1993).

5.4.1 Bacterial-assisted phytoremediation of organic pollutants

Organic pllutants are degraded by bacteria called plant rhizobacteria (Afzal et al., 2014a; Glick, 2010). In the process, the potent bacterial populations grow near the roots or in the tissues of the plants (Naveed et al., 2014; Sessitch et al., 2005). Bacterial population helps the plant to proliferate higly and make more biomass to absorb a graeter quantity of pollutants, including POPs from the soil (Weyens et al., 2009a). It is reported that the rate of contaminant degradation in this process is greater than the contribution of individual partner (Khan et al., 2013b; Lunney et al., 2004). By the interaction of plant and rhizobacteria in the soil, the abundance of genes of catabolic property in rhizospheric zone

Table 5.2 Some of the plant species identified for phytoremediation of organic pollutants.

Plant species	Contaminant	Technique for measuring the contaminant degradation	Benefits	References
Alfalfa (*Medicago sativa*)	Petroleum hydrocarbons	Soil acetone-hexane extraction was followed by purification and analysis by GC-FID	Alfalfa-*Pseudomonas aeruginosa* association enhanced the petro remediation	Agnello et al., (2016)
Maize (*Zea mays*)	Atrazine	HPLC analysis	Atrazine elimination was increased by electric current assisted phytoremediation with maize	Sánchez et al., (2018)
Scirpus grossus	Petroleum hydrocarbons	Liquid-liquid extraction with dichloromethane followed by GC	Remediation of polluted sites was enhanced	Al-Baldawi et al., (2013)
Achillea millefolium	Petroleum hydrocarbons	A series of extraction, evaporation, and weight measurement and calculation based on formula	Could be for removal of petro-hydrocarbons	Masu et al., (2014)
Melilotus albus	TPHs	GC-MS	Diesel negatively affected both growth and nutritional status of *Melilotus albus* and AMF combination significantly enhanced the degradation of TPH in the rhizosphere	Hernández- Ortega et al., (2012)

(continued)

TABLE 5.2 (Cont'd)

Plant species	Contaminant	Technique for measuring the contaminant degradation	Benefits	References
Helianthus annuus	Persistent organic pollutants	GC-MS	This type of phytoremediation enhanced the remediation of heptachlor, aldrin, heptachlor epoxide, and trans-chlordane	Almeida et al., (2018)
Acorus calamus	PAHs	HPLC	*Acorus calamus* can be used for the remediation of PAH contaminated soils	Jeelani et al., (2017)
Leucanthemum vulgare	TPHs	Gravimetric analysis	There was a great reduction in the levels of crude oil contaminants	Noori et al., (2018)
Brassica napus, Sorghum bicolor, Drummondii	Petroleum hydrocarbons	GC*GC-TOFMS	*Brassica napus* plants reported high resistance towards PAH accumulation	Petrová et al., (2017)
Populus nigra	n-alkanes, PAHs, and PCBs	GC-MS	Phytoremediation of organic compounds was enhanced	Nissim et al., (2018)

AMF, arbuscular mycorrhizal fungi; GC-FID, gas chromatography-flame ionization detection; GC*GC-TOFMS, gas chromatography time-of-flight mass spectrometry; GC-MS, gas chromatography-mass spectrometry; HPLC, high performance liquid chromatography; PAHs, polycyclic aromatic hydrocarbons; PCBs, polychlorinated biphenyl; TPHs, total petroleum hydrocarbons.

increases, and it leads to the enhanced mineralization, degradation, and stabilization of a number of organic pollutants such as POPs (Jaha and Jaha, 2015; Passatore et al., 2014; Yateem, 2013). Plant-bacterial associations improve the production of biosmass of plants and, in turn, help in better colonization of rhizobacteria, which helps in degradation of organic contaminants (Afzal et al., 2013b; Compant et al., 2010; Khan et al., 2013a; Yousaf et al., 2010). In addition, it has also been found that salicylic acid, terpenes, and flavonoids are capable of enhancing the degradation process of different compounds such as POPs (Gilbert and Crowley, 1997, 1998; Hernandez et al., 1997; Koh et al., 2000). A number of research reports have shown the plant rhizobacterial associations (Abhilash et al., 2013; Gerhardt et al., 2009; Jha and Jaha, 2015; Qin et al., 2014). Some of the POPs and PCBs, which are released in the environment are believed to be destroyed by plant-microbe partnerships (Leigh et al., 2002). The destruction of a number of PCBs is increased by the partnership of alfalfa with that of *Pseudomonas* sp. strain F113 (Villacieros et al., 2005). Some of the exudates of of plant species stimulate the rhizospheric microorganisms to carry out degradation process of contaminants, including organic POPs (Donnnelly et al., 1994; Isidorov and Jdanova, 2002; Rao, 1990; Salt et al., 1998). Most of the plants have phenolic compounds in their exudates by which they are able to enhance the proliferation of rhizobacteria to degrade POPs (Fletcher and Hegde, 1995).

The bioavailability of the contaminant in the environment is another factor, which is responsible for the degradation of POPs (Federici et al., 2012). Rhizobacteria are potential organisms to increase the bioavailability of POPs by producing biosurfactants and help in their remediation (Aslund and Zeeb, 2010). However, surfactant release in exudates of roots is believed to be better (Passatore et al., 2014). It is now evident that the plant species and bacteria, which produces biosurfactants can enhance the availability of organic contaminants (Thijs and Vangronsveld, 2015; Fu et al., 2012). Organic contaminants such as POPs and PCBs can be degraded by adjusting apigenin and flavonoids (Narasimhan et al., 2003; Chávez et al., 2004). Joner and Leyval (2003) showed the degradation of phenanthrene by the utilization of plant microbe partnership in *Trifolium* rhizosphere in contaminated areas. The alfalfa inoculation in *Rhizobium meliloti* decreased the PAH concentrations in polluted areas (Teng et al., 2011). Few microbes are believed to improve the root system to degrade PCBs (Van Aken et al., 2010). *Agrobacterium tumefaciens* is believed to work in this regard (Majumder et al., 2016). In addition, plant-breeding approaches are also believed to improve the root system of plants and, in turn, help in degrading various organic contaminants (Vacheron et al., 2013; Federici et al., 2012).

5.4.2 Fungal-assisted degradation of organic pollutants

A number of prokaryotic organisms are used for the remediation of pollutants on a large scale everywhere. Recently, a few attempts have been made for the use of white rot fungi for this process. These types of filamentous fungi are considerd beneficial in comparison to bacteria as they are able to oxide a number of chemicals (Pointing, 2001). Furthermore, they are more tolerant to hazardous compounds than bacteria (Evans and Hedger, 2001). Therefore, these microorganisms are considered as powerful tools for the remediation of soil organic pollutants (Sasek, 2003). Fungi work in the order to change the structural conformation of organic pollutants and make them nontoxic to the environment. These transformed organic pollutants are released to external environment where they are further degraded by bacterial actions (Diez, 2010). In addition, bacteria can respond to the pollutants to a certain level, below which they are not able to act upon (Adenipekun and Lawal, 2012). White rot fungi have the capability to secrete enzymes, which turns the appearance of wood to bleached color, and brown rot fungi degrade cellulose and liginin and makes brown substance. These organisms can cause some cubical cracks and shrinks in wood, which are clearly visible in conifer trees (Stamets, 2005). Some of the chemical compounds such as polychorinated biphenyls and dioxins, pesticides, paper mill effluents, etc., are degraded by the action of fungal components mainly by white rot fungi (Singh, 2006). These fungi degrade several chemicals such as DDT, lindane, chlordane, etc. (Arisoy, 1998). White rot fungi degrade such a great diversity of pollutants because they have ability to tolerate some extracellular enzymes (Kirk and Farrell, 1987; Adenipekun and Lawal, 2012). These fungi degrade a few lignin contaminants (Barr and Aust, 1994). Grow of mycelia in fungi helps it to colonize rapidly and penetration is enhanced by hyphal extension (Reddy and Mathew, 2001). Furthermore, these fungi use lignin and cellulosic materials for nutrition and are capable of degrading a greater range of contaminants (Maloney, 2001). The capability of white rot fungi, *Lentinus subnudus,* to degrade atrazine, heptachlor, and metachlor was checked by Nwachukwu and Osuji (2007), by observing that about 94% of metachlor and heptachlor were degraded with its action and about 78% of atrazine was degraded. Similarly, Purnomo et al., (2014) studied the degradtion of heptachlor and heptachlor epoxide by the action of *Pleurotus ostreatus,* a white rot fungi, and found that 89% and 32% were degraded, respectively. It was found that white rot fungi such as *Phlebia acanthocystis*, *Phlebia brevispora*, and *Phlebia aurea* are capabale of degrading organic contaminants, including diazinon, methomyl, and endosulfan when used in a proper way (Nyakundi et al., 2011; Xiao et al., 2011; Kamei et al., 2011).

5.5 Rhizoremediation

Remediation of organic pollutants by the action of plants and microbes in the synergistic ways happens to be more successful approach in phytoremediation. Rhizoremediation is the process of utilizing microorganisms present in the rhizospheric zone of plant species in the process of phytoremediation (Mosa et al., 2016). The total of about 40% of phytosynthetic products is stored in the soil in the form of sugar, organic acids, and other forms (Gerhardt et al., 2009). The miroorganisms present in the soil surface utilize such materials as the source of energy (Leigh et al., 2002). Plant species can get a number of benefits from these soil microbes (Hontzeas et al., 2004; Kuiper et al., 2004). The first study on rhizosphere is linked to the degradation of pesticides and herbicides, where it has been suggested that the microorganisms tend to degrade the pollutants (Hoagland et al., 1994; Jacobsen, 1997). There are several reports showing the destruction of organic pollutants including PAHs, trichloroethylene, and PCBs (Walton and Anderson, 1990). Some leguminous plant species are appropriate in the process of rhizoremediation (Kuiper et al., 2004). Efficiency of rhizoremediation is dependent on microorganisms to enhance the root colonization (Capdevila et al., 2004; Sibly and Levy, 2004; de Weert et al., 2002). These aromatic compounds are degraded by microbial consortia. These compounds are fortunately similar to that of organic pollutants such as PCBs and PAHs, therby enhancing the chances of making the use of natural processes for remediation of pollutants (Jacobsen, 1997).

5.6 Enzymes and their role in biodegradation of pollutants

The microrganisms are known to secrete extracellular enzymes. Pesticides target these enzymes to carry out their function (Riya and Jagatpati, 2012). Enzymes are the products that degrade pesticides by the process of detoxification (Scott et al., 2008). It is a novel technique of removing toxic compounds by the help of enzymes from polluted sites. This type of technique can be more beneficial than the other chemical techniques. White rot fungi can degrade chemical substances present in the soil and are not degraded by the help of bacterial processes because they produce some extracellular enzymes for degrading organic contaminants. Some of these promising enzymes, which degrade lignin or lignin-like compounds, manganese peroxidase, laccase, and oxidases. In addition, glutathione S-transferase, esterase, and cytochrome Pb450 are three other enzyme families used for the degradation process of pollutants (Bass and Field, 2011). Use of laccase and peroxidases is reported to show good activity in terms of degradation process (Donoso et al., 2008; Pizzul et al., 2009; Kadimaliev et al., 2011). It was shown by Kadimaliev et al. (2011) and Donoso et al. (2008) that lacases and peroxidases are involved in the degradation of tribromophenol by the help of *Trametes versicolor*. As the biochemistry of pesticides is diverse and complex, there is a need of diverse classes of enzymes for their degradation (Van et al., 2003).

5.7 Conclusion

To control the uprise of toxic environmental pollutants, phytoremediation is considered as an emerging technique to be utilized for the restoration of contaminated sites. This technique is totally ecofriendly and cost-effective technique, hence termed as green technology often. It is a suitable alternative to the other expensive approaches used in the degradation of various contaminants. High biomass producing and contaminant tolerant plant species are selected and used in this process of remediation. It has also been reported that plant species very often have associations with diverse microflora in the soil, which are believed to enhance the potential of plants toward the degradation of contaminants. Phytoremediation assisted by microorganisms could yield better results in terms of degradation of contaminants present in contaminated areas. Presence of microbes and physical and chemical properties of contaminants describe the success of microbial-assisted phytoremediation. Besides, these potential benefits, microbial-assisted phytoremediation technique faces some challenges as well. However, the challenges can be reduced and enhancement in the microbial action can be brought by employing genetic engineering techniques.

References

Abdel-Shafy, H.I., Mansour, M.S., 2016. A review on polycyclic aromatic hydrocarbons: source, environmental impact, effect on human health and remediation. Egypt. J. Pet. 25 (1), 107–123.

Abhilash, P.C., Singh, B., Srivastava, P., Schaeffer, A., Singh, N., 2013. Remediation of lindane by Jatropha curcas L: utilization of multipurpose species for rhizoremediation. Biomass Bioenerg. 51, 189–193. doi:https://doi.org/10.1016/j.biombioe.2013.01.028.

Adenipekun, C.O., Lawal, R., 2012. Uses of mushrooms in bioremediation: a review. Biotechnol. Mol. Biol. Rev. 7 (3), 62–68.

Afzal, M., Khan, Q.M., Sessitsch, A., 2014. Endophytic bacteria: prospects and applications for the phytoremediation of organic pollutants. Chemosphere 117, 232–242. doi:https://doi.org/10.1016/j.chemosphere.2014.06.078.

Afzal, M., Yousaf, S., Reichenauer, T.G., Sessitsch, A., 2013. Ecology of alkane-degrading bacteria and their interaction with the plant. In: Bruijn, FJD (Ed.), Molecular Microbial Ecology of the Rhizosphere. John Wiley & Sons, Hoboken, NJ, USA, pp. 975–989.

Agarwal, S.K., 2009. Heavy Metal Pollution (Vol. 4). APH publishing, 161–165.

Agnello, A.C., Huguenot, D., van Hullebusch, E.D., Esposito, G., 2016. Citric acid-and Tween® 80-assisted phytoremediation of a co-contaminated soil: alfalfa (Medicago sativa L.) performance and remediation potential. Environ. Sci. Pollut. Res. 23 (9), 9215–9226.

Ahemad, M., 2019. Remediation of metalliferous soils through the heavy metal resistant plant growth promoting bacteria: para-digms and prospects. Arab. J. Chem. 12 (7), 2014.

Al-Baldawi, I.A., Abdullah, S.R.S., Suja, F., Anuar, N., Mushrifah, I, 2013. Effect of aeration on hydrocarbon phytoremediation capability in pilot sub-surface flow constructed wetland operation. Ecol. Eng. 61 (Part A), 496–500. doi:https://doi.org/10.1016/j.ecoleng.2013.10.017.

Ali, H., Khan, E., Sajad, M.A., 2013. Phytoremediation of heavy metals—concepts and applications. Chemosphere 91 (7), 869–881.

Almeida, C.M.R., Dias, A.C., Mucha, A.P., Bordalo, A.A., Vasconcelos, M.T.S, 2009. Study of the influence of different organic pollutants on Cu accumulation by Halimione portulacoides. Estuar Coast. Shelf Sci. 85 (4), 627–632. doi:10.1016/j.ecss.2009.10.007.

Almeida, M.V.D., Rissato, S.R., Galhiane, M.S., Fernandes, J.R., Lodi, P.C., Campos, M.C.D, 2018. In vitro phytoremediation of persistent organic pollutants by Helianthus annuus L. plants. Quím. Nova 41 (3), 251–257.

Arısoy, M., 1998. Biodegradation of chlorinated organic compounds by white-rot fungi. Bull. Environ. Contam. Toxicol. 60 (6), 872–876.

Åslund, M.W., Zeeb, B.A., 2010. A review of recent research developments into the potential for phytoextraction of persistent organic pollutants (Pops) from weathered, contaminated soil-Application of Phytotechnologies for Cleanup of Industrial, Agricultural, and Wastewater Contamination. Springer, Dordrecht, pp. 35–59.

Baker, A.J., Brooks, R., 1989. Terrestrial higher plants which hyperaccumulate metallic elements. A review of their distribution, ecology and phytochemistry. Biorecovery 1 (2), 81–126.

Baker, A.J.M., McGrath, S.P., Sidoli, C.M.D., Reeves, R.D, 1994. The possibility of in situ heavy metal decontamination of polluted soils using crops of metal-accumulating plants. Resour. Conserv. Recy 11 (1-4), 41–49.

Banat, I.M., 1995. Characterization of biosurfactants and their use in pollution removal–state of the art. Acta Biotechnol. 15 (3), 251–267. doi:10.1002/abio.370150302.

Barr, D.P., Aust, S.D., 1994. Mechanisms white rot fungi use to degrade pollutants. Environ. Sci. Technol. 28 (2), 78A–87A.

Beneduzi, A., Ambrosini, A., Passaglia, L.M.P, 2012. Plant growth-promoting rhizobacteria (PGPR): their potential as antagonists and biocontrol agents. Genet. Mol. Biol. 35 (4), 1044–1051.

Berti, R., Cunningham, S.D., 2000. Phytostabilization of metals. In: Raskin, I., Ensley, B.D. (Eds.), Phytoremediation of Toxic Metals: Using Plants to Clean up the Environment. Wiley, New York, pp. 71–88.

Bhalerao, T.S., Puranik, P.R., 2007. Biodegradation of organochlorine pesticide, endosulfan, by a fungal soil isolate, Aspergillus niger. Int. Biodeter. Biodegrad. 59 (4), 315–321.

Bhargava, A., Carmona, F.F., Bhargava, M., Srivastava, S., 2012. Approaches for enhanced phytoextraction of heavy metals. J. Environ. Manag. 105, 103–120. doi:https://doi.org/10.1016/j.jenvman.2012.04.002.

Brooks, R.R., Chambers, M.F., Nicks, L.J., Robinson, B.H., 1998. Phytomining. Trends Plant Sci. 3 (9), 359–362.

Burken, J.G., & Schnoor, J.L., (1996). Phytoremediation: plant uptake ofatrazine and role of root exudates. J. Environ. Eng. 1222, 958-963.

Cabral, L., Soares, C.R.F.S., Giachini, A.J., Siqueira, J.O, 2015. Arbuscular mycorrhizal fungi in phytoremediation of contaminated areas by trace elements: mechanisms and major benefits of their applications. World J. Microbiol. Biotechnol. 31 (11), 1655–1664.

Capdevila, S., Martínez-Granero, F.M., Sánchez-Contreras, M., Rivilla, R., Martín, M., 2004. Analysis of Pseudomonas fluorescens F113 genes implicated in flagellar filament synthesis and their role in competitive root colonization. Microbiology 150 (11), 3889–3897.

Chaturvedi, R., Favas, P., Pratas, J., Varun, M., Paul, M.S., 2018. Assessment of edibility and effect of arbuscular mycorrhizal fungi on Solanum melongena L. grown under heavy metal (loid) contaminated soil. Ecotoxicol. Environ. Saf. 148, 318–326. https://doi.org/10.1016/j.ecoenv.2017.10.048.

Colpaert, J.V., Wevers, J.H., Krznaric, E., Adriaensen, K., 2011. How metal-tolerant ecotypes of ectomycorrhizal fungi protect plants from heavy metal pollution. Ann. For. Sci. 68 (1), 17–24.

Compant, S., Clément, C., Sessitsch, A., 2010. Plant growth-promoting bacteria in the rhizo-and endosphere of plants: their role, colonization, mechanisms involved and prospects for utilization. Soil Biol. Biochem. 42 (5), 669–678.

Coninx, L., Martinova, V., Rineau, F., 2017. Mycorrhiza-assisted phytoremediationAdvances in Botanical Research, 83. Academic Press, London, UK, pp. 127–188.

Cunningham, S.D., Anderson, T.A., Schwab, A.P., Hsu, F.C., 1996. Phytoremediation of soils contaminated with organic pollutants. Adv. Agron. 56 (1), 55–114.

Cunningham, S.D., Berti, W.R., 1993. Remediation of contaminated soils with green plants: an overview. In Vitro Cell. Dev. Biol.Plant 29 (4), 207–212.

de Weert, S., Vermeiren, H., Mulders, I.H.M., Kuiper, I., Hendrickx, N., Bloemberg, G.V., Vanderleyden, J., De, Mot, &, R., Lugtenberg, B.J.J, 2002. Flagella-driven chemotaxis towards exudate components is an important trait for tomato root colonization by Pseudomonas fluorescens. Mol. Plant-Microbe Interact. 15 (11), 1173–1180.

Deng, W., Lin, D., Yao, K., Yuan, H., Wang, Z., Li, J.…, Hu, X., 2015. Characterization of a novel β-cypermethrin-degrading Aspergillus niger YAT strain and the biochemical degradation pathway of β-cypermethrin. Appl. Microbiol. Biotechnol. 99 (19), 8187–8198.

Deng, Z., Cao, L., 2017. Fungal endophytes and their interactions with plants in phytoremediation: a review. Chemosphere 168, 1100–1106.

Deng, Z., Cao, L., Huang, H., Jiang, X., Wang, W., Shi, Y., Zhang, R., 2011. Characterization of Cd-and Pb-resistant fungal endophyte Mucor sp. CBRF59 isolated from rapes (Brassica chinensis) in a metal-contaminated soil. J. Hazard. Mater. 185 (2-3), 717–724.

Deng, Z., Zhang, R., Shi, Y., Tan, H., Cao, L., 2014. Characterization of Cd-, Pb-, Zn-resistant endophytic Lasiodiplodia sp. MXSF31 from metal accumulating Portulaca oleracea and its potential in promoting the growth of rape in metal-contaminated soils. Environ. Sci. Pollut. Res. 21 (3), 2346–2357.

Diez, M.C., 2010. Biological aspects involved in the degradation of organic pollutants. J. Soil Sci. Plant Nutr. 10 (3), 244–267.

Donnelly, P.K., Hegde, R.S., Fletcher, J.S., 1994. Growth of PCB-degrading bacteria on compounds from photosynthetic plants. Chemosphere 28 (5), 981–988.

Doty, S.L., Freeman, J.L., Cohu, C.M., Burken, J.G., Firrincieli, A., Simon, A.…, Blaylock, M.J., 2017. Enhanced degradation of TCE on a superfund site using endophyte-assisted poplar tree phytoremediation. Environ. Sci. Technol. 51 (17), 10050–10058.

EPA, 1984. Test Methods for Evaluating Solid Waste: Physical/Chemical Methods, 2nd ed. USEPA, EPA-864. Office of Solid Waste and Emergency Response, Washington, DC.

Evans, C.S., Hedger, J.N., 2001, November. Degradation of plant cell wall polymers. In British Mycological Society Symposium Series (Vol. 23), 1–26.

Federici, E., Giubilei, M.A., Covino, S., Zanaroli, G., Fava, F., D'Annibale, A., Petruccioli, M., 2012. Addition of maize stalks and soybean oil to a historically PCB-contaminated soil: effect on degradation performance and indigenous microbiota. New Biotechnol. 30 (1), 69–79.

Fein, J.B., Martin, A.M., Wightman, P.G., 2001. Metal adsorption onto bacterialsurfaces: development of a predictive approach. Geochim.Cosmochim. Ac. 65 (23), 4267–4273. doi:10.1016/S0016-7037(01)00721-9.

Fletcher, J.S., Hegde, R.S., 1995. Release of phenols by perennial plant roots and their potential importance in bioremediation. Chemosphere 31 (4), 3009–3016.

Fortin, J.A., Plenchette, C., Piché, Y., 2009. Mycorrhizas: The new green revolution. Éditions MultiMondes, Canada.

Fu, D., Teng, Y., Luo, Y., Tu, C., Li, S., Li, Z., Christie, P., 2012. Effects of alfalfa and organic fertilizer on benzo [a] pyrene dissipation in an aged contaminated soil. Environ. Sci. Pollut. Res. 19 (5), 1605–1611.

Gadd, G.M., 2004. Microbial influence on metal mobility and application for bioremediation. Geoderma 122 (2-4), 109–119.

Gerhardt, K.E., Huang, X.D., Glick, B.R., Greenberg, B.M., 2009. Phytoremediation and rhizoremediation of organic soil contaminants: potential and challenges. Plant Sci. 176 (1), 20–30.

Gilbert, E.S., Crowley, D.E., 1997. Plant compounds that induce polychlorinated biphenyl biodegradation by Arthrobacter sp. strain B1B. Appl. Environ. Microbiol. 63 (5), 1933–1938.

Gilbert, E.S., Crowley, D.E., 1998. Repeated application of carvone-induced bacteria to enhance biodegradation of polychlorinated biphenyls in soil. Appl. Microbiol. Biotechnol. 50 (4), 489–494.

Glick, B.R., 2010. Using soil bacteria to facilitate phytoremediation. Biotechnol. Adv. 28 (3), 367–374.

Grobelak, A., Kokot, P., Hutchison, D., Grosser, A., Kacprzak, M., 2018. Plant growth-promoting rhizobacteria as an alternative to mineral fertilizers in assisted bioremediation-sustainable land and waste management. J. Environ. Manag. 227, 1–9. doi:10.1016/j.jenvman.2018.08.075.

Hao, Q., Sun, Y.X., Xu, X.R., Yao, Z.W., Wang, Y.S., Zhang, Z.W., Mai, B.X., 2014. Occurrence of persistent organic pollutants in marine fish from the Natuna Island, South China Sea. Mar. Pollut. Bull. 85 (1), 274–279.

Haselwandter, K., 1995. Mycorrhizal fungi: siderophore production. Crit. Rev. Biotechnol. 15 (3-4), 287–291. doi:10.3109/07388559509147414.

He, L., Yang, H., Yu, Z., Tang, J., Xu, L., Chen, X., 2014. Arbuscular mycorrhizal fungal phylogenetic groups differ in affecting host plants along heavy metal levels. J. Environ. Sci. 26 (10), 2034–2040. https://doi.org/10.1016/j.jes.2014.07.013.

He, L.M., Neu, M.P., Vanderberg, L.A., 2000. Bacillus lichenformis γ-glutamyl exopolymer: physicochemical characterization and U (VI) interaction. Environ. Sci. Technol. 34 (9), 1694–1701. doi:10.1021/es991119e.

Hernandez, B.S., Koh, S.C., Chial, M., Focht, D.D., 1997. Terpene-utilizing isolates and their relevance to enhanced biotransformation of polychlorinated biphenyls in soil. Biodegradation 8 (3), 153–158.

Hernández-Ortega, H.A., Alarcón, A., Ferrera-Cerrato, R., Zavaleta-Mancera, H.A., López-Delgado, H.A., Mendoza- López, M.R., 2012. Arbuscular mycorrhizal fungi on growth, nutrient status, and total antioxidant activity of Melilotus albus during phytoremediation of a diesel-contaminated substrate. J. Environ. Manage. 95 Suppl, S319–S324

Hoagland, R.E., Zablotowicz, R.M., Locke, M.A., 1994. Propanil metabolism by rhizosphere microfloraBioremediation Through Rhizosphere Technology. American Chemical Society, Washington, DC, pp. 160–183.

Hontzeas, N., Zoidakis, J., Glick, B.R., Abu-Omar, M.M., 2004. Expression and characterization of 1-aminocyclopropane-1-carboxylate deaminase from the rhizobacterium Pseudomonas putida UW4: a key enzyme in bacterial plant growth promotion. Biochim. Biophys. Acta ProteinsProteom. 1703 (1), 11–19.

Isidorov, V., Jdanova, M., 2002. Volatile organic compounds from leaves litter. Chemosphere 48 (9), 975–979.

Jacobsen, C.S., 1997. Plant protection and rhizosphere colonization of barley by seed inoculated herbicide degrading Burkholderia (Pseudomonas) cepacia DBO1 (pRO101) in 2, 4-D contaminated soil. Plant Soil 189 (1), 139–144.

Jeelani, N., Yang, W., Xu, L., Qiao, Y., An, S., Leng, X., 2017. Phytoremediation potential of Acorus calamus in soils cocontaminated with cadmium and polycyclic aromatic hydrocarbons. Sci. Rep. 7 (1), 8028.

Jha, P., Jha, P.N., 2015. Plant-microbe partnerships for enhanced biodegradation of polychlorinated biphenyls. Plant Microbes Symbiosis: Applied Facets. Springer, New Delhi, pp. 95–110.

Kamei, I., Takagi, K., Kondo, R., 2011. Degradation of endosulfan and endosulfan sulfate by white-rot fungus Trameteshirsuta. J. Wood Sci. 57 (4), 317–322.

Kavamura, V.N., Esposito, E., 2010. Biotechnological strategies applied to the decontamination of soils polluted with heavy metals. Biotechnol. Adv. 28 (1), 61–69.

Khan, S., Afzal, M., Iqbal, S., Khan, Q.M., 2013. Plant–bacteria partnerships for the remediation of hydrocarbon contaminated soils. Chemosphere 90 (4), 1317–1332.

Khan, S., Afzal, M., Iqbal, S., Mirza, M.S., Khan, Q.M., 2013. Inoculum pretreatment affects bacterial survival, activity and catabolic gene expression during phytoremediation of diesel contaminated soil. Chemosphere 91 (5), 663–668.

Kidd, P., Barceló, J., Bernal, M.P., Navari-Izzo, F., Poschenrieder, C., Shilev, S., Monterroso, C., 2009. Trace element behaviour at the root–soil interface: implications in phytoremediation. Environ. Exp. Bot. 67 (1), 243–259.

Kirk, T.K., Farrell, R.L., 1987. Enzymatic" combustion": the microbial degradation of lignin. Ann. Rev. Microbiol. 41 (1), 465–501.

Koh, S.C., Park, Y.I., Koo, Y.M., So, J.S., 2000. Plant terpenes and lignin as natural cosubstrates in biodegradation of polyclorinated biphenyls (PCBs) and polycyclic aromatic hydrocarbons (PAHs). Biotechnol. Bioprocess. Eng. 5 (3), 164–168.

Krupa, P., Kozdrój, J., 2007. Ectomycorrhizal fungi and associated bacteria provide protection against heavy metals in inoculated pine (Pinus Sylvestris L.) seedlings. Water Air Soil Pollut. 182 (1), 83–90. doi:https://doi.org/10.1007/s11270-006-9323-7.

Kuiper, I., Lagendijk, E.L., Bloemberg, G.V., Lugtenberg, B.J., 2004. Rhizoremediation: a beneficial plant-microbe interaction. Mol. Plant Microbe Interact. 17 (1), 6–15.

Lasat, M.M., 2002. Phytoextraction of toxic metals: a review of biological mechanisms. J. Environ. Qual. 31 (1), 109–120. doi:10.2134/jeq2002.1090.

Leigh, M.B., Fletcher, J.S., Fu, X., Schmitz, F.J., 2002. Root turnover: an important source of microbial substrates in rhizosphere remediation of recalcitrant contaminants. Environ. Sci. Technol. 36 (7), 1579–1583.

Leung, H.M., Wu, F.Y., Cheung, K.C., Ye, Z.H., Wong, M.H., 2010. The effect of arbuscular mycorrhizal fungi and phosphate amendement on arsenic uptake, accumulation and growth of Pteris vittata in As-contaminated soil. Int. J. Phytoremediation 12 (4), 384–403.

Leung, H.M., Wu, F.Y., Cheung, K.C., Ye, Z.H., Wong, M.H., 2010. Synergistic effects of arbuscular mycorrhizal fungi and phosphate rock on heavy metal uptake and accumulation by an arsenic hyperaccumulator. J. Hazard. Mater. 181 (1-3), 497–507.

LEUNG, H.M., Zhen-Wen, W.A.N.G., Zhi-Hong, Y.E., Kin-Lam, Y.U.N.G., Xiao-Ling, P.E.N.G., CHEUNG, K.C, 2013. Interactions between arbuscular mycorrhizae and plants in phytoremediation of metal-contaminated soils: a review. Pedosphere 23 (5), 549–563.

Li, X, Li, D, Yana, Z, Ao, Y, 2018. Biosorption and bioaccumulation characteristics of cadmium by plant growth-promoting rhizobacteria. RSC Adv. 8 (54), 30902–30911.

Li, H., Li, X., Xiang, L., Zhao, H.M., Li, Y.W., Cai, Q.Y., Zhu, L., Mo, C.H., Wong, M.H., 2018. Phytoremediation of soil co-contaminated with Cd and BDE-209 using hyperaccumulator enhanced by AM fungi and surfactant. Sci. Total Environ. 613, 447–455. doi:https://doi.org/10.1016/j.scitotenv.2017.09.066.

Li, H.Y., Wie, D.Q., Shen, M., Zhou, Z.P., 2012. Endophytes and their role in phytoremediation. Fungal Divers. 54 (1), 11–18. doi:10.1007/s13225-012-0165-x.

Lie, E., Bernhoft, A., Riget, F., Belikov, S.E., Boltunov, A.N., Derocher, A.E.…, Skaare, J.U., 2003. Geographical distribution of organochlorine pesticides (OCPs) in polar bears (Ursus maritimus) in the Norwegian and Russian Arctic. Sci. Total Environ. 306 (1-3), 159–170.

Lu, Y.F., Lu, M., 2015. Remediation of PAH-contaminated soil by the combination of tall fescue, arbuscular mycorrhizal fungus and epigeic earthworms. J. Hazard.Mater. 285, 535–541. doi:https://doi.org/10.1016/j.jhazmat.2014.07.021.

Lunney, A.I., Zeeb, B.A., Reimer, K.J., 2004. Uptake of weathered DDT in vascular plants: potential for phytoremediation. Environ. Sci. Technol. 38 (22), 6147–6154.

Luo, S., Chen, L., Chen, J., Xiao, X., Xu, T., Wan, Y., Rao, C., Liu, C., Liu, Y., Lai, C., Zeng, G., 2011. Analysis and characterization of cultivable heavy metal-resistant bacterial endophytes isolated from Cd hyperaccumulator Solanum nigrum L. and their potential use for phytoremediation. Chemosphere 85 (7), 1130–1138.

Ma, Y., Oliveira, R.S., Freitas, H., Zhang, C., 2016. Biochemical and molecular mechanisms of plant-microbe-metal interactions: relevance for phytoremediation. Front. Plant Sci. 7, 918. doi:https://doi.org/10.3389/fpls.2016.00918.

Maestri, E., Marmiroli, N., 2011. Transgenic plants for phytoremediation. Int. J. Phytoremediation 13 (suppl1), 264–279.

Maloney, S.E., 2001. Pesticide degradation. In British Mycological Society Symposium Series (Vol. 23), pp.188–223.

Mandyam, K., Jumpponen, A., 2005. Seeking the elusive function of the root-colonising dark septate endophytic fungi. Stud. Mycol. 53, 173–189.

Masu, S., Albulescu, M., Balasescu, L.C., (2014). Assessment on phytoremediation of crude oil polluted soils with Achillea millefolium and total petroleum hydrocarbons removal efficiency. http://hdl.handle.net/123456789/585.

Megharaj, M., Ramakrishnan, B., Venkateswarlu, K., Sethunathan, N., Naidu, R., 2011. Bioremediation approaches for organic pollutants: a critical perspective. Environ. Int. 37 (8), 1362–1375. doi:10.1016/j.envint.2011.06.003.

Miethke, M., Marahiel, M.A., 2007. Siderophore-based iron acquisition and pathogen control. Microbiol. Mol. Biol. Rev. 71 (3), 413–451. doi:10.1128/MMBR.00012 07.

Mohammad, A., Mittra, B., 2013. Effects of inoculation with stress-adapted arbuscular mycorrhizal fungus Glomus deserticola on growth of Solanum melogena L. and Sorghum sudanese Staph. seedlings under salinity and heavy metal stress conditions. Arch. Agron. Soil Sci. 59 (2), 173–183.

Mosa, K.A., Saadoun, I., Kumar, K., Helmy, M., Dhankher, O.P., 2016. Potential biotechnological strategies for the cleanup of heavy metals and metalloids. Front. Plant Sci. 7, 303. doi:https://doi.org/10.3389/fpls.2016.00303.

Motesharezadeh, B., Kamal-poor, S., Alikhani, H.A., Zariee, M., Azimi, S., 2017. Investigating the effects of plant growth promoting bacteria and Glomus Mosseae on cadmium phytoremediation by Eucalyptus camaldulensis L. Pollution 3 (4), 575–588.

Nagajyoti, P.C., Lee, K.D., Sreekanth, T.V.M, 2010. Heavy metals, occurrence and toxicity for plants: a review. Environ. Chem. Lett. 8 (3), 199–216.

Narasimhan, K., Basheer, C., Bajic, V.B., Swarup, S., 2003. Enhancement of plant-microbe interactions using a rhizosphere metabolomics-driven approach and its application in the removal of polychlorinated biphenyls. Plant Physiol. 132 (1), 146–153.

Naveed, M., Mitter, B., Yousaf, S., Pastar, M., Afzal, M., Sessitsch, A., 2014. The endophyte Enterobacter sp. FD17: a maize growth enhancer selected based on rigorous testing of plant beneficial traits and colonization characteristics. Biol. Fertil. Soils 50 (2), 249–262.

Neff, J.M., 1979. Polycyclic aromatic hydrocarbons in the aquatic environments. Appl. Sci. 18 (1), 262.

Nissim, W.G., Cincinelli, A., Martellini, T., Alvisi, L., Palm, E., Mancuso, S., Azzarello, E., 2018. Phytoremediation of sewage sludge contaminated by trace elements and organic compounds. Environ. Res. 164, 356e366. doi:https://doi.org/10.1016/j.envres.2018.03.009.

Noori, A., Zare Maivan, H., Alaie, E., Newman, L.A., 2018. Leucanthemum vulgare Lam. crude oil phytoremediation. Int. J. Phytoremediation 20 (13), 1292e1299.

Nwachukwu, E.O., Osuji, J.O., 2007. Bioremedial degradation of some herbicides by indigenous white rot fungus, Lentinus subnudus. J. Plant Sci. 2, 619–624.

Nyakundi, W.O., Magoma, G., Ochora, J., Nyende, A.B., 2011. Biodegradation of diazinon and methomyl pesticides by white rot fungi from selected horticultural farms in rift valley and central Kenya. J. Appl. Technol. Environ. Sanit. 1 (2), 107–124.

Passatore, L., Rossetti, S., Juwarkar, A.A., Massacci, A., 2014. Phytoremediation and bioremediation of polychlorinated biphenyls (PCBs): state of knowledge and research perspectives. J. Hazard. Mater. 278, 189–202.

Petrová, _S., Rezek, J., Soudek, P., Van_ek, T., 2017. Preliminary study of phytoremediation of brownfield soil contaminated by PAHs. Sci. Total Environ. 599, 572–580. doi:https://doi.org/10.1016/j.scitotenv.2017.04.163.

Pointing, S., 2001. Feasibility of bioremediation by white-rot fungi. Appl. Microbiol. Biotechnol. 57 (1-2), 20–33.

Purnomo, A.S., Putra, S.R., Shimizu, K., Kondo, R., 2014. Biodegradation of heptachlor and heptachlor epoxide-contaminated soils by white-rot fungal inocula. Environ. Sci. Pollut. Res. 21 (19), 11305–11312.

Qiao, Y., Crowley, D., Wang, K., Zhang, H., Li, H., 2015. Effects of biochar and Arbuscular mycorrhizae on bioavailability of potentially toxic elements in an aged contaminated soil. Environ. Pollut. 206, 636–643.

Qin, H., Brookes, P.C., Xu, J., 2014. Cucurbita spp. and Cucumis sativus enhance the dissipation of polychlorinated biphenyl congeners by stimulating soil microbial community development. Environ. Pollut. 184, 306–312.

Rajkumar, M., Sandhya, S., Prasad, M.N.V., Freitas, H, 2012. Perspectives of plant-associated microbes in heavy metal phytoremediation. Biotechnol. Adv. 30 (6), 1562–1574.

Rajkumar, M., Ae, N., Prasad, M.N.V., Freitas, H, 2010. Potential of siderophore-producing bacteria for improving heavy metal phytoextraction. Trends Biotechnol. 28 (3), 142–149.

Rajkumar, M., Sandhya, S., Prasad, M.N.V., Freitas, H, 2012. Perspectives of plant associated microbes in heavy metal phytoremediation. Biotechnol. Adv. 30 (6), 1562–1574. doi:10.1016/j.biotechadv.2012.04.011.

Rao, A.S., 1990. Root flavonoids. Bot. Rev. 56 (1), 1–84.

Rawlings, D.E., 1997. Mesophilic, autotrophic bioleaching bacteria: description, physiology and role. In: Rawlings, D.E. (Ed.), Biomining: theory, microbes and industrial processes. Springer, Berlin-Heidelberg, Germany, pp. 229–245.

Reddy, C.A., Mathew, Z, 2001, November. Bioremediation potential of white rot fungi. In British Mycological Society Symposium Series (Vol. 23), pp. 52–78.

Rhodes, C.J., 2012. Feeding and healing the world: through regenerative agriculture and permaculture. Sci. Prog. 95 (4), 345–446.

Safe, S, 1984. Polychlorinated biphenyls and polybrominated biphenyls: biochemistry, toxicology and mechanisms of action. Crit. Rev. Toxicol. 13 (4), 319.

Salt, D.E.Smith.RD, Raskin, I, 1998. Phytoremediation. Annu. Rev. Plant Physiol. Plant Mol. Biol 49 (1), 643–668.

Sánchez, V., López-Bellido, F.J., Cañizares, P., Rodríguez, L., 2018. Can electrochemistry enhance the removal of organic pollutants by phytoremediation? J. Environ. Manage. 225, 280e287.

Sarwar, N., Imran, M., Shaheen, M.R., Ishaq, W., Kamran, A., Matloob, A., Rehim, A., Hussain, S., 2017. Phytoremediation strategies for soils contaminated with heavy metals: modifications and future perspectives. Chemosphere 171, 710–721.

Sarwat, M., Hashem, A., Ahanger, M.A., Abd_Allah, E.F., Alqarawi, A.A., Alyemeni, M.N., Ahmad, P., Gucel, S., 2016. Mitigation of NaCl stress by arbuscular mycorrhizal fungi through the modulation of osmolytes, antioxidants and secondary metabolites in mustard (Brassica juncea L.) plants. Front. Plant Sci. 7, 869. doi:https://dx.doi.org/10.3389%2Ffpls.2016.00869.

Sasek, V., 2003. Why mycoremediations have not yet come into practice. In: Sasek, V, Glaser, JA, Baveye, P (Eds.), The utilization of bioremediation to reduce soil contaminants: problems and solutions19th edn. Kluwer Academic, Dordrecht, pp. 247–263.

Schalk, I.J., Hannauer, M., Braud, A., 2011. New roles for bacterial siderophores in metal transport and tolerance. Environ. Microbiol. 13 (11), 2844–2854. doi:10.1111/j.1462-2920.2011.02556.x.

Schneider, J., Bundschuh, J., do Nascimento, C.W.A., 2016. Arbuscular mycorrhizal fungi-assisted phytoremediation of a lead-contaminated site. Sci. Total Environ. 572, 86–97 https://doi.org/10.1016/j.scitotenv.2016.07.185.

Sessitsch, A., Coenye, T., Sturz, A.V., Vandamme, P., Barka, E.A., Salles, J.F...., Wang-Pruski, G., 2005. Burkholderiaphytofirmans sp. nov., a novel plant-associated bacterium with plant-beneficial properties. Int. J. Syst. Evol. Microbiol. 55 (3), 1187–1192.

Silby, M.W., Levy, S.B., 2004. Use of in vivo expression technology to identify genes important in growth and survival of Pseudomonas fluorescens Pf0-1 in soil: discovery of expressed sequences with novel genetic organization. J. Bacteriol. 186 (21), 7411–7419.

Singh, H., 2006. Mycoremediation: Fungal Bioremediation. John Wiley & Sons, Hoboken, New Jersey.

Solhi, M., Shareatmadari, H., Hajabbasi, M.A., 2005. Lead and zinc extraction potential of two common crop plants, Helianthus annuus and Brassica napus. Water Air Soil Pollut. 167 (1-4), 59–71.

Stamets, P., 2005. Mycelium Running: How Mushrooms can Help Save the World. Random House Digital, Inc, USA.

Stepniewska, Z, Kuzniar, A., 2013. Endophytic microorganisms- promising applications in bioremediation of greenhouse gases. Appl. Microbiol. Biotechnol 97 (22), 9589–9596.

Supaka, N., Pinphanichakarn, P., Pattaragulwanit, K., Thaniyavarn, S., Omori, T., Juntongjin, K., 2001. Isolation and characterization of a phenanthrene-degrading Sphingomonas sp. strain P2 and its ability to degrade fluoranthene and pyrene via cometabolism. Sci. Asia 27 (2), 21–28.

Tchounwou, P.B., Yedjou, C.G., Patlolla, A.K., Sutton, D.J., 2012. Heavy metal toxicity and the environment. EXS 101, 133–164. doi:https://doi.org/10.1007/978-3-7643-8340-4_6.

Teng, Y., Shen, Y., Luo, Y., Sun, X., Sun, M., Fu, D...., Christie, P., 2011. Influence of Rhizobium meliloti on phytoremediation of polycyclic aromatic hydrocarbons by alfalfa in an aged contaminated soil. J. Hazard. Mater. 186 (2-3), 1271–1276.

Thijs, S., Vangronsveld, J., 2015. RhizoremediationPrinciples of Plant-Microbe Interactions. Springer, Cham, pp. 277–286.

Tigini, V., Prigione, V., Di Toro, S., Fava, F., Varese, G.C., 2009. Isolation and characterisation of polychlorinated biphenyl (PCB) degrading fungi from a historically contaminated soil. Microb. Cell Fact. 8 (1), 5.

Villacieros, M., Whelan, C., Mackova, M., Molgaard, J., Sánchez-Contreras, M., Lloret, J...., Karlson, U., 2005. Polychlorinated biphenyl rhizoremediation by Pseudomonas fluorescens F113 derivatives, using a Sinorhizobiummeliloti nod system to drive bph gene expression. Appl. Environ. Microbiol. 71 (5), 2687–2694.

Walton, B.T., Anderson, T.A., 1990. Microbial degradation of trichloroethylene in the rhizosphere: potential application to biological remediation of waste sites. Appl. Environ. Microbiol. 56 (4), 1012–1016.

Wang, L., Ji, B., Hu, Y., Liu, R., Sun, W., 2017. A review on in situ phytoremediation of mine tailings. Chemosphere 184, 594–600. doi:https://doi.org/10.1016/j.chemosphere.2017.06.025.

Wang, M., Liu, R., Chen, W., et al., 2018. Effects of urbanization on heavy metal accumulation in surface soils, Beijing. J. Environ. Sci. (China) 64, 328–334. doi:10.1016/j.jes.2016.11.026.

Wang, R.Q., 2012. Degradation of persistent organic pollutants mechanism summaryAdvanced Materials Research. Trans Tech Publications Ltd, Switzerland, pp. 620–623.

Weber, R., Watson, A., Forter, M., Oliaei, F., 2011. Persistent organic pollutants and landfills-a review of past experiences and future challenges. Waste Manage. Res. 29 (1), 107–121.

Wenzel, W.W., 2009. Rhizosphere processes and management in plant-assisted bioremediation (phytoremediation) of soils. Plant Soil 321 (1-2), 385–408. doi:10.1007/s11104-008-9686-1.

Weyens, N., van der Lelie, D., Taghavi, S., Newman, L., Vangronsveld, J., 2009. Exploiting plant–microbe partnerships to improve biomass production and remediation. Trends Biotechnol. 27 (10), 591–598.

Xiao, P., Mori, T., Kamei, I., Kiyota, H., Takagi, K., Kondo, R., 2011. Novel metabolic pathways of organochlorine pesticides dieldrin and aldrin by the white rot fungi of the genus Phlebia. Chemosphere 85 (2), 218–224.

Yateem, A., 2013. Rhizoremediation of oil-contaminated sites: a perspective on the Gulf War environmental catastrophe on the State of Kuwait. Environ. Sci. Pollut. Res. 20 (1), 100–107.

Yousaf, S., Ripka, K., Reichenauer, T.G., Andria, V., Afzal, M., Sessitsch, A., 2010. Hydrocarbon degradation and plant colonization by selected bacterial strains isolated from Italian ryegrass and birdsfoot trefoil. J. Appl. Microbiol. 109 (4), 1389–1401.

Yu, X.Z., Zhou, P.H., Yang, Y.M., 2006. The potential for phytoremediation of iron cyanide complex by willows. Ecotoxicology 15 (5), 461–467.

Chapter 6

Nano-phytoremediation for soil contamination: An emerging approach for revitalizing the tarnished resource

Mir Z. Gul, Karuna Rupula, Sashidhar R. Beedu
Department of Biochemistry, University College of Science, Osmania University, Hyderabad, Telangana, India

6.1 Introduction

In an era of modern industrialization, the developmental innovations characterized by technological advances have raised severe environmental challenges bypassing the procedural safeguards of the natural environment (Bennett et al., 2003). As is apparent from the changes in ecosystem processes, an unprecedented number of pollutants of multiple types, in particular, heavy metals (HMs) that occur as a result of accidental or process spillage, disposal of untreated industrial wastes, and sludge application to cultivated fields, have greatly contributed to the degradation of land and water resources. Such pollutants can contaminate aquatic, terrestrial, or atmospheric habitats and trigger deforestation, loss of biodiversity, and soil erosion. Also, most of these waste products are released into the aquatic sources as well as the soil are recognized or suspected of being carcinogenic that eventually have a deleterious impact on public health and present a black image of a complex high-tech industrialized world (Chaves et al., 2016). Examples of such contaminants are metalloids (namely, HMs - arsenic [As], mercury [Hg], zinc [Zn], chromium [Cr], lead [Pb], cadmium [Cd]), volatile organic compounds [VOCs], dioxins, carbon monoxide, chlorofluorocarbons, hydrocarbons, and other particulate matter. Extensive use of agrochemicals: pesticides, herbicides, and other chemical-based products used for agricultural production also contribute to soil contamination. Hence, the reclamation of polluted environments is a major global concern that needs to be tackled at the earliest. The realization that traditional treatment approaches (namely, isolation, pump, and treat and disposal to landfills, soil washing with solubilizing agents, etc.) are not sustainable, has contributed to significant

Phytoremediation. DOI: https://doi.org/10.1016/B978-0-323-89874-4.00014-5

growth in the development of alternative modern remediation processes (Cundy et al., 2008). The outcome of the field technological innovations over the last two decades highlight the gap between published research and the results of the meaningful remediation (Schwitzguébel, 2017). There is a compelling need for an environmentally sustainable approach to the management of certain environmental concerns. Developing innovative and even more successful methods for the remediation of pollution is of paramount importance and will facilitate to protect and re-establish the integrity of natural ecosystems (Petruzzelli et al., 2016; Schwitzguébel, 2017). There is a growing passion for new technologies to facilitate the clean-up of polluted sites and simultaneously bring the cost of these technologies. As a result, several environmental clean-up strategies have been established using chemical and physical remediation, bioremediation, and phytoremediation. Among the different approaches to environmental remediation, considerable interest is drawn toward phytoremediation technologies (green sustainable technologies) (Mahar et al., 2016). Due to its "green strategy," the exploitation of plants for the remediation of environmentally harmful compounds is expanding its scale by attracting a substantial interest among the researchers. The potential in these phytotechnologies is compelling, given that they are environmentally friendly, easy to use, and economical compared to traditional remediation (Petruzzelli et al., 2016). In this chapter, we attempted to provide insights into nanotechnology-based phytoremediation processes to mitigate environmental contaminants. In addition, the complexities of using nanomaterials (NMs) for phytoremediation to identify future strategies in this area are highlighted.

6.2 Phytoremediation

Plants generally have an immense potential to absorb hazardous chemicals/pollutants from the contaminated surrounding habitats and facilitate their decontamination by a variety of strategies. Phytoremediation is an innovative nonintrusive multidisciplinary approach focused on successful, cost-effective, and eco-friendly treatment strategies (solar driven) for the absorption, accumulation, immobilization, or biodegradation of inorganic and organic contaminants present in various environmental matrices, particularly in water and soil by plants and associated soil microorganisms (Vithanage et al., 2012; Paz-Alberto and Sigua, 2013; Petruzzelli et al., 2016). Compared to other remediations options, phytoremediation has minimal operating and service expenses and can cost around 5% of existing clean-up approaches (Prasad, 2003; Aken et al., 2010). Based on economic aspects, the main objective of phytoremediation could be (**1**) plant-based extraction of metals with financial benefits (phytoextraction), (**2**) risk amelioration (phytostabilization), and (**3**) sustainable soil management in which phytoremediation is gradually increasing soil fertility, allowing crops with economic value to grow (Vangronsveld et al., 2009; Mahar et al., 2016). It could be an efficient tool to stabilize, biodegrade/

eliminate a wide range of pollutants, including trace elements (HMs), and other metalloids, recalcitrant organic pollutants (polycyclic aromatic hydrocarbons (PAHs), petroleum hydrocarbons (PHCs), chlorinated solvents, polychlorinated biphenyls (PCBs), radionuclides, polychlorine, etc., in various environmental matrices in an *in-situ* or *ex-situ* set ups (Gerhardt et al., 2017; Ojuederie and Babalola, 2017). Furthermore, phytoremediation tends to support plant growth by bolstering native microflora and soil health, thereby promoting productivity as well as carbon sequestration, as harvested biomass can be used in the production of biofuels (Gomes, 2012; Ali et al., 2013; Afzal et al., 2014; Ullah et al., 2015; Cristaldi et al., 2017; Gerhardt et al., 2017; Luo et al., 2017; Wang et al., 2017). The risk of pollutant dispersion is greatly reduced by phytoremediation and is also suitable for soils or wastewaters carrying mixed and complex contaminants (Lee, 2013; Cristaldi et al., 2017). The mechanism of phytoremediation and its efficiency are influenced by various aspects factors, including the type of contaminants, their bioavailability in soils, physicochemical characteristics in the matrix, and plant species (Truu et al., 2015; Cristaldi et al., 2017). An ideal plant should moreover have a variety of following characteristics for the elimination of pollutants from the soil environment:

1. Accelerated plant growth and heavy biomass production (growth and productivity)
2. Well-developed root system (widely spread as well as highly branched)
3. Ability to withstand or accumulate pollutants (higher toxicity tolerance threshold)
4. Higher potential for a hyperaccumulation of pollutants ideally in the aboveground parts (sink prospect)
5. Simplistic harvesting (based on sink organ usefulness)
6. Non-consumable by animals and humans

There are several species of plants and ecotypes with potency for phytoremediation (phytostabilization and phytoextraction) and absorption of the pollutants from soil, sediments as well as wastewater. The potential of many plant species namely, alfalfa, chinese cabbage, poplar tree, spinach, sunbeam, sunflower, tomato, willow, etc., has been highlighted and many of these are found to be strong phytoremediators to remediate soil pollutants (Mitton et al., 2014). The metallophytes are recognized to be more effective in phytoremediation as they can survive and propagate on metal-contaminated soils (Wójcik et al., 2017; Awa and Hadibarata, 2020). Among the different phytoremediation strategies, especially HM phytoextraction has faced multiple field-scale hurdles, indicating the need for an understanding of the possible interactions between plants, soils, and pollutants before this approach could become prosperous (Bhargava et al., 2012; Ali et al., 2013; Mahar et al., 2016). However, the phytoremediation strategies typically require a long time, often many years to yield any significant results. Moreover, their implementation is constrained by various environmental factors, soil quality, and phytotoxicity of the pollutants (Song et al., 2019).

6.3 The advent of nanotechnology and nano-phytoremediation

The advent of nanotechnology has lately been the focus of comprehensive research and is recognized as an innovative, rapidly evolving technology that extends its application to all domains (Ingle et al., 2014). It plays an important role in the development of technological innovations for the production of new goods, the replacement of existing production facilities, and the reformation of new materials and chemicals with enhanced quality leading to lower energy and material use and decreases environmental damage and environmental remediation (Ingle et al., 2014). The promise of nanotechnology to mitigate environmental damage is progressing rapidly and could stimulate the most transformative changes in the environmental arena. The application of nanotechnology offers a range of possible environmental advantages such as mitigation and revitalization, sensing and detection as well as strategies for the prevention of pollution (Ibrahim et al., 2016; Tripathi et al., 2018; Qian et al., 2020). The enhanced use of nano-based protocols, tools, or environmental cleanup products may begin to emerge from the immediate need for a technology that is safer, economical, and readily accessible while ensuring fast outcomes with no further pressure on the cleanup process in the form of residues and environmental persistence (Guerra et al., 2018; Qian et al., 2020; Usman et al., 2020). The application of nanoparticles (NPs) or nanomaterials (NMs), or nanostructured materials (NSMs), or nanoclusters (NCs), or nanocomposites (NCMPs) are being widely considered for the remediation of soil environments polluted by various toxic substances such as HMs, pesticides, and persistent organic pollutant (POPs), which are mainly non-biodegradable, is being widely considered. The combination of phytoremediation and nanotechnology in reducing the threats due to environmental pollutants is a promising solution due to high efficiency and cost-effectiveness (Gerhardt et al., 2017).

The elimination or mitigation of environmental pollutants with the help of NPs synthesized by algae, fungi, and bacteria is called nano-bioremediation (Yadav et al., 2017). If these pollutants are reduced /removed utilizing the NPs synthesized by/or with higher plants, it is referred to as nano-phytoremediation (Yadav et al., 2017). Nano-phytoremediation is an integrated approach blending nanotechnology and phytoremediation for the removal or remediation of contaminants (namely HMs, organic, and inorganic pollutants, etc.) from the polluted environs with the help of NPs, NMs, NSMs, or NCMPs (Ingle et al., 2014). The nano-phytoremediation approaches are based on the application of NPs/NMs as well as phytoremediation procedures (e.g., phytostabilization, phytodegradation, and phytoextraction). This innovation has shown enormous promise for the remediation of various contaminants in soils and sediments. There are various potential applications of nano-phytoremediation in alleviating soil contamination and these applications are represented in Fig. 6.1.

Many complex organochlorines (OCs) including organochlorine pesticides are notably impervious to plant and microbial disintegration. This limitation

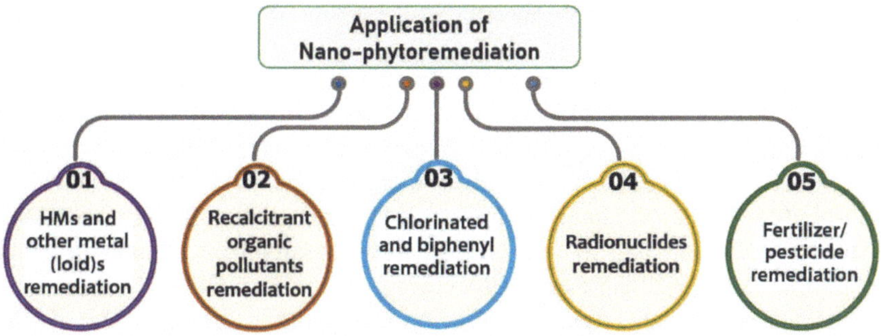

FIG. 6.1 The possible applications of nano-phytoremediation.

could be addressed by a hybrid solution involving nanotechnology and biotechnology; complex OCs would be transformed into simple molecules by nano-encapsulated enzymes, which would, in turn, be swiftly degraded by mutual plant and microbial activities (Singh, 2010). The NMs have been found to potentially affect the nature, behavior, and absorption of contaminants in the phytoremediation processes (Song et al., 2019). Furthermore, the combined use of NMs/NPs together with soil microbiota and plants has been thoroughly explored and envisaged to be superior in the degradation or elimination of HMs, other organic and inorganic substances compared to any other individual technology. The NMs have been recognized as inexpensive, most efficient, and environmentally-friendly substitute materials, both in terms of preservation of resources as well as environmental restoration (Guerra et al., 2018). Certain unique features like nano-size, higher surface area to volume ratio, shape, and reactivity make their chemical behavior entirely different than their bulk counterparts (Yadav et al., 2017). Hence, they gained a great deal of attention for environmental clean-up and as a support medium for various organic and inorganic catalysts (Khan et al., 2019). Advanced NMs, namely, zeolites, carbon nanotubes, magnetic NPs, and biocomposite NMs can catalytically target metal pollutants and other toxins substances to disintegrate them and can eliminate them because of their interaction capabilities. These NMs are capable of absorbing a large flux of pollutants or catalyze reactions at a much rapid pace owing to their higher surface area, thereby minimizing energy consumption during the process of degradation and also help to avoid the release of contaminants (Ingle et al., 2014; Ibrahim et al., 2016). Moreover, NMs bind to roots and are eventually shifted to harvestable portions of the plant that is used for the phytoremediation process. The smaller size of NMs makes it easier for them to penetrate various barriers in the plant systems, therefore, helping to enter specific cellular regions, whilst their larger surface area facilitates their increased absorption as well as target delivery of the desired materials (Ibrahim et al., 2016; Yadav et al., 2017). The potential of NPs to be conjugated with different ligands and controlled surface area to volume ratio by altering their shapes or structures facilitates the development of sensors with greater specificity, sensitivity, and sensitivity.

Nano-phytoremediation generally offers an extensive range of environmental advantages that can be categorized into three areas: (**1**) mitigation and revitalization, (**2**) monitoring and detection, and (**3**) prevention of pollution (Ingle et al., 2014; Guerra et al., 2018). Hence, there may be nano-phytostabilization, nano-phytodegradation, nano-rhizofiltration, nano-phytovolatilization, nano-phytoaccumulation, and so on. The array of possible mechanism of nano-phytoremediation is illustrated in Fig. 6.2.

Recent research has largely centered on increasing the performance of NMs and researchers are focusing on analyzing the mechanisms by which NMs have been acquired and transported to track any toxicological implications on the system. While a series of conventional methods can be used to synthesize NPs, the biological route of synthesizing is desirable due to its expeditious and easy synthesis, regulated toxicity, size regulation, cost-effectiveness, and environmentally friendly approach (Mohanpuria et al., 2008; Ganachari et al., 2019). The green synthesis of NPs by plants is being preferred over the years due to the single-step biosynthesis procedure, the elimination of hazardous material, and the presence of natural capping agents (Ganachari et al., 2019). Biological methodologies of NMs synthesis are rapidly utilized which are relatively cheap, non-toxic, biocompatible, and environmentally benign (Baker et al., 2013;

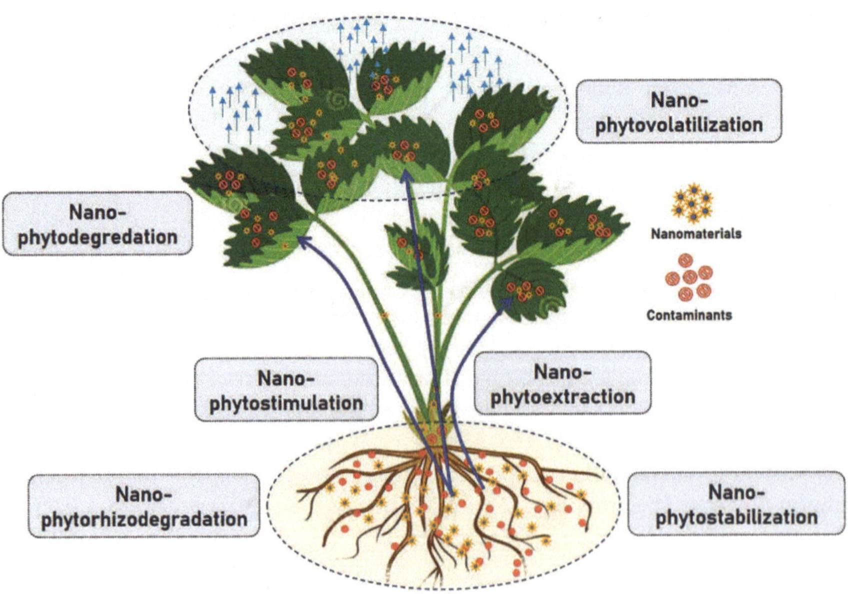

FIG. 6.2 An illustration depicting the array of possible mechanisms of nano-phytoremediation approach.

Das et al., 2017; Khan et al., 2019). The biosynthesis of NMs describes the process that exploits biochemical pathways to produce extracellular or intracellular inorganic metabolites (Mohanpuria et al., 2008; Das et al., 2017). Another approach of synthesizing "smart engineered NMs" to remediate environmental matrices is by using NMs with new coatings or functional groups that could enhance the mobility, stability, and overall performance of NMs/NPs that can become benign once the remediation process is over (von der Kammer et al., 2012; Thatai et al., 2014). The NPs/NMs that are suitable for nano-phytoremediation should possess the following attributes:

1. NPs/NMs should be nontoxic to the plants.
2. They should enhance germination, root-shoot elongation, and biomass of the plants used
3. They should significantly increase the production of phytoenzymes.
4. They should have the potential to stimulate plant growth hormones.
5. They should have ability to bind various contaminants and improved bioavailability for plants.
6. They should also accelerate the process of phytoremediation.

Nano-phytoremediation strategies (NPs and plant-based strategies) utilize either naturally evolving or genetically modified plant species supported by NPs/NMs for the remediation of polluted environments. During the recent past, substantial progress has been made in the development of natural or genetically altered plant species for environmental clean-up processes. The introduction of specified NPs succeeded in substantially improving plant growth and nano-augmentation enhanced phytoremediation potential resulting in considerable decontamination of pollutants from the soil environment. Various NPs have also been identified as "plant growth promoters" because of their potential to improve plant growth hormones and better absorption of contaminants by the plants.

6.4 Soil contamination: a serious concern

Soil is one of the most vital components of the terrestrial ecosystem and a significant life-support system. It modulates essential environmental processes to ensure life on earth, apart from being an extremely valuable resource for plants and food production (Blum, 2005; Amundson et al., 2015). These comprises biogeochemical cycles (carbon cycle, nitrogen fixation cycle, etc.), water cycles, climate change, detoxification of pollutants, biogenic gas regulation, re-establishment of ecosystems, and maintenance and preservation of biodiversity (Smith et al., 2015). Soil health management is important not only for the safe operation of food and agriculture production, but also for maintaining the balance between earth and ecological processes. Furthermore, the sustainable development of modern society needs a healthy and stable soil ecosystem (Lal. 2015). However, in the present times, soil degradation is major environmental issue that poses a significant challenge to agricultural productivity and food security. The chemical contamination by HMs, PAHs, POPs, pesticides have been identified as important factors that contribute heavily to soil degradation. The contaminants from industries particularly HMs deteriorate the soil not only in the vicinity of the source but far off places as well (Petruzzelli et al., 2016; Cristaldi et al., 2017; Jiang et al., 2018; Masindi and Muedi, 2018). Excessive use of pesticides and fertilizers result their accumulation in the soil environs as significant amount of them (~90%) is lost to the atmosphere during their application or as runoff, and subsequently these pollutants find their way into the soil in due course of time. Significant threats to whole ecosystem are known to be environmental hazards with soil pollution. An approximate 33% of global soil resource has adversely affected sequentially because of all the aforementioned factors and further anthropogenic activities (FAO and ITPS, 2015; Tripathi et al., 2017). Over the years, HMs, POPs, and pesticides have polluted the soil to that extent that further agricultural activity on these depleted resources have become impossible (Ali et al., 2013; Luo et al., 2017; Awa and Hadibarata, 2020). Moreover, as these pollutants exhibit bioaccumulation and biomagnification characteristics, contaminated soil is not only inappropriate for food processing, but also presents greater perils to human health via food chain contamination (Ali et al., 2013; Gavrilescu et al., 2015). Most metals/metalloids do not undergo chemical or microbial decomposition or deterioration compared to organic pollutants, which allows the concentrations of these lethal metals to increase incessantly, contributing to long-term soil ecotoxicity (Jadia and Fulekar, 2009; Sarwar et al., 2017). Consequently, most of these polluted and degraded soils are abandoned as they are unfit for crop production, leading to decreased land available for agricultural practices with every passing year. Thus, we face the dual challenge of cultivating more food on an already diminishing land for a large and growing human population, while at the same time, attempting to avoid further harm and preserving the already depleted land. The extensive deterioration to soil and land have not only intensified global concern but has also sparked an immediate need to revive these tarnished resources (Cachada et al., 2012; Li et al., 2018). Therefore, the need of the hour is the prevention together with remediation and regeneration of these polluted soils. These revitalized soils after remediation could successfully then be used to fulfil the food and bioenergy requirements of current population worldwide. There are currently a number of techniques available to confiscate these lethal contaminants from the soil habitats (Chen et al., 2015; Ye et al., 2017). The phytoremediation process has been successfully established and implemented since 1980s in the field studies at the sites polluted with HMs, pesticides, radionuclides, chlorinated solvents, etc. (Paz-Alberto and Sigua, 2013; Petruzzelli et al., 2016; Schwitzguébel, 2017). While phytoremediation has several benefits in the soil remediation, overall remediation efficacy is generally restricted by plant species, contaminant bioavailability, soil properties, and climatic conditions. Factors which influence the chemical absorption and distribution in living plants includes: (**1**) Physical and chemical characteristics of contaminants (e.g., molecular weight, solubility in water, vapor pressure), (**2**) Environmental parameters (e.g., organic matter, soil moisture content, pH, temperature), and (**3**) Plant attributes (e.g., root system and enzymes involved). In view of limiting factors and other constraints, strategies are needed to facilitate the realistic applications of phytoremediation. Several strategies have been employed to enhance phytoremediation processes, which include sustainable agronomic practices, treatment with rhizopheric microorganisms and chemical additives, and also gene manipulation approaches (Gerhardt et al., 2017). The advancement of nanotechnology offers new inspiration and ideas for phytoremediation processes and open new perspectives for the revitalizing contaminated soil habitats (Gong et al., 2018). The application of NMs in plant systems have been documented in several studies in the recent past. Over the last few years, several studies have been performed utilizing NMs to increase the efficacy of phytoremediation. A possible use of combined treatment (phyto-nano-treatment) for contaminated soils could be successfully exploited.

6.5 Nano-phytoremediation of pollutants from contaminated soil environment

Nano-phytoremediation technology is focused on the integration of NMs and phytoremediation methods for the soil decontamination. It is promising approach to integrate NMs into traditional phytoremediation systems. Nano-phytoremediation approach has a wide diverse scope of removal of soil pollutants. This modern remediation approach can easily transform diverse pollutants ranging from HMs to VOCs with high degree of efficiency. There are four main aspects of NM-assisted phytoremediation process: NMs, pollutants, plants, and environmental factors. The schematic representation of interaction of NMs with plants, the influence of environmental factors and type of contaminants in nano-phytoremediation is presented in Fig. 6.3.

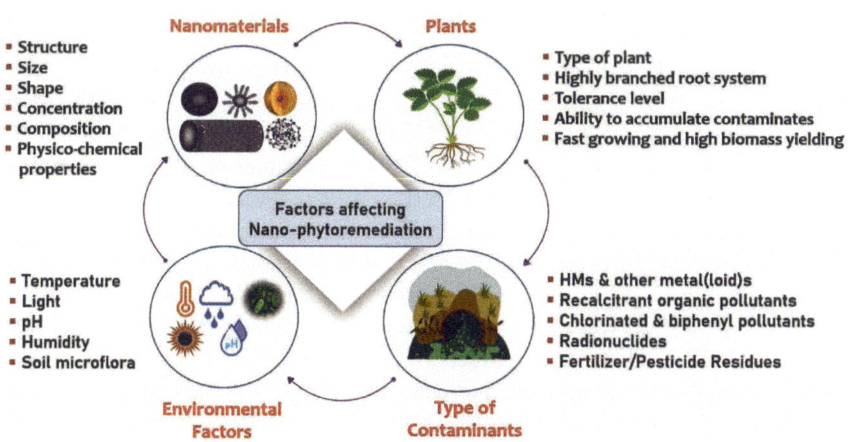

FIG. 6.3 The schematic representation of interaction of nanomaterials with plants, the influence of environmental factors, type of contaminants in nano-phytoremediation.

The application of NMs, mainly because of their outstanding ability to immobilize or absorb metal ions is of great interest in the remediation of soil contaminated with HMs (Ali et al., 2013; Tafazoli et al., 2017; Cai et al., 2019). A significant number of field applications of NMs have been successfully implemented to remediate soil and ground water contaminants (Khin et al., 2012; Cecchin et al., 2017). These NMs may either react directly with the contaminant or facilitate their conversion into less toxic forms in the phytoremediation system, thereby, reducing the burden of pollutant removal (Wang et al., 2016). The NMs can act via absorption or redox reactions for the direct removal of contaminants (Mueller and Nowack, 2010). As in case of phytostabilization, the contaminants, for example, can be immobilized by absorption through carbon nanotubes. Carbon nanotubes have shown outstanding absorption ability toward different organic pollutants (Song et al., 2017; Kang et al., 2018). The nano-zerovalent iron (nZVI), titanium dioxide (TiO_2), multiwalled carbon nanotubes (MWCNTs), fullerenes, zinc oxide (ZnO), bimetallic (Pd/Fe) NPs, and stabilized NPs could rapidly degrade wide variety of organic contaminants such as 2,4-dinitrotoluene, atrazine, chlorpyrifos, lindane, pentachlorophenol, PCBs, pyrene, and trichloroethylene, etc., from polluted soil environments (Das et al., 2015; Guerra et al., 2018; Cai et al., 2019). Due to their remarkable reduction capabilities, extremely small size and their excellent reactivity with wide variety of toxic compounds, the nZVI NPs have become a strategic material of great application potential in environmental remediation, especially in soil clean-up process in the recent past (Xue et al., 2018). nZVI is an effective, low-cost, and environmentally-friendly agent that has been utilized for the mitigation of several organic as well as inorganic contaminants from the soil habitats not only in the laboratory set-up but also in open field conditions. nZVI have been used widely as a reduction agent in recent years (Satapanajaru et al., 2008; Stefaniuk et al., 2016). High reduction capabilities along with higher surface energy of these materials encourage their use in the rapid and efficient use in the elimination of wide variety of contaminants, including HMs. Several studies have been performed for the reduction of organic chlorinated compounds (e.g., PCBs, OC pesticides) as well as for the reduction of high valence toxic metals (e.g., Cr [VI] and U [VI]). A scientific study has shown that nZVI increased the reduction efficiency of Cr (VI) from 14.51% to 86.83% in 2 h. (Singh et al., 2012; Di Palma et al., 2015; El-Temsah et al., 2016; Huang et al., 2016). In another study, authors documented the nZVI efficiently reduced metal availability of As, Cd, Cr, Pb, and Zn in calcareous and multimetal contaminated acidic soils. The availability of HMs like As, Pb, and Cr has plummeted by >82%, whereas the Zn reduction was in the range of 31% to 75%. The minimum reduction (between 13%–42%) was observed for Cd (Gil-Díaz et al., 2017b). It was found that 10% of nZVI substantially reduced the uptake of As by reducing the content of As in available fractions while increasing its quantity in the residual fractions in highly polluted brownfield soils. The plants of Barley (*Hordeum vulgare*) grown on those As-contaminated treated with nZVI demonstrated improved growth, no detrimental effects with regard to nutritive value and also showed least As absorption. In addition, no adverse effects on the physicochemical properties were observed (Gil-Díaz et al., 2017a).

Other NMs extensively explored to remove pollutants from contaminated soil habitats comprise carbon-based NMs, metal-based NMs, phosphate-based NPs, natural mineral NPs, etc. (Liu and Lal, 2012; Cecchin et al., 2017; Baragaño et al., 2020). Owing to their large surface area, carbon nanotubes have shown outstanding absorption ability toward various hydrophobic organic pollutants (Song et al., 2017; Zhang et al., 2019). The OCs loaded with carbon NPs could serve as an excellent carriers of pollutants, particularly organic pollutants, and could drastically alter their fate and transport in the contaminated environments (Hofmann and von der Kammer, 2009). Fullerene NPs are extensively studied for their ability to enhance the phyto-availability of pollutants. C60 fullerene NPs along with *Populus deltoides* (Eastern cottonwood) have been used to facilitate of the removal of trichloroethylene from the polluted soil habitats in the phytoremediation system. Similar findings were reported in related research carried out with redosier dogwood, a brush plant based on similar xylem structures of both plant species (Ma and Wang, 2010). The accumulation and subsequent toxicity of dichlorodiphenyldichloroethylene (p,p'-DDE), a DDT metabolite of the three plants; *Solanum lycopersicum* (tomato), *Glycine max* (soyabean), and *Cucurbita pepo* (zucchini), and was reportedly altered by C60 fullerene NPs. The application of C60 fullerene enhanced the pollutant uptake of plants by 30% to 65% and the effect was mainly noticeable in the roots (De La Torre-Roche et al., 2012). This study further showed that NMs exhibit diverse effects on plant species and their ability for phytoremediation to remediate soil contaminants. Such observations were also confirmed in a separate study in which the application of MWCNTs and C60, four plants: zucchini, soya, tomato, and corn were utilized to degrade chlordane, DDT and its metabolites (DDx). It was shown that based on the type of plant species and concentration of NMs, the pesticide uptake as well as the accumulation ranged between 21% and 80%. Although DDx uptake in tomato and corn was entirely suppressed by C60 application, chlordane accumulation in soyabean and tomato raised to 34.9% (De La Torre-Roche et al., 2013). The effects of amino-functionalized and nonfunctionalized MWCNTs on the coexisting chlordane and DDE accumulation in *Lactuca sativa* (lettuce) were also investigated (Hamdi et al., 2015). Furthermore, the availability of pesticides was greatly affected by the application of various forms of carbon nanotubes (CNTs). Although the nonfunctionalized CNTs reduced the pesticide content by 88% and 70% in root and shoot respectively, the decrease in amino functionalized effect in roots and shoots was only 57% and 23%, respectively. Coexposure of CNTs led to a substantial decrease in the bioavailability of pesticide residues, thus effectively reducing contamination of edible lettuce. Research findings have indicated that iron oxide (IO) NPs function as primary scavenger of a wide variety of hazardous HMs, including As(III), Cd(II), Cr(VI), Cu(II), Pb(II), and Zn(II) (Zhang and Fang, 2010; Roy and Bhattacharya, 2012; Komárek et al., 2015). Thus, environmentalists have used these NPs to remediate these hazardous metals/metalloids from contaminated soils (Gómez-Pastora et al., 2014). Nano TiO_2 degrades organic contaminants like p,p′-DDT, pyrene, phenanthrene, diuron in soil habitats and subsequently brings down the half-lives of these contaminants through photocatalysis, redox reactions, and thermal destruction under irradiation (Makarova et al., 2001; Li et al., 2016). The review of scientific literature has further indicated substantial degradation of pesticide atrazine (2-chloro-4-ethylamino-6-isopropylamino-s-triazine) by means of polyethylene imine copper NPs coated on montmorillonite and sand (Kalidhasan et al., 2017).

Plant biomass and the growing ability of plants are the two fundamental aspects in selecting the right plant species for phytoremediation. Most of the plants used for phytoremediation are also not beneficial because of their minimal biomass and sluggish growth rates resulting from their lower tolerance for contaminants and poor soil conditions. Certain strategies such as inoculation of plant growth promoting rhizobacteria (PGPR), application of growth regulators, and use of genetically modified plants are, therefore, employed in phytoremediation processes to encourage plant growth (Ullah et al., 2015; Wang et al., 2016; Aderholt et al., 2017; Nahar et al., 2017; Yadu et al., 2018).The NMs can accentuate the plant growth in plants by fostering the absorption of nutrients and water, increasing rate of photosynthesis, governing the soil microbial community, and relieving abiotic stress conditions (e.g., drought, salinity, etc.) in *in-situ* and *ex-situ* set-ups, thereby boosting their phytoremediation potential (Singh and Lee, 2016; Souri et al., 2017). In edible plants, alfalfa, soyabean, and corn, following the application of ZnO and cerium dioxide (CeO_2) NPs, an increase in root development system has been reported, envisioning the role of nanotechnology in significantly improving the efficiency of phytoremediation (López-Moreno et al., 2010). An experimental study performed on nano-phytoremediation to assess the influence of nZVI on trinitrotoluene (TNT) absorption and its metabolites by *Panicum maximum* from a TNT-contaminated habitat. The findings imply that the integration of nanotechnology and the process of phytoremediation resulted in significantly greater build-up of TNT in the root system. In presence of nZVI and at an optimal concentration of 500 mg/kg, the absorption of TNT was more significant than without nZVI (Paquin et al., 2004). Bimetallic NPs (Ni/Fe) have also been shown to reduce translocation of polybrominated diphenyl ethers and allied phototoxicity to Chinese cabbage in polluted sites after 3 days of soil culture studies (Wu et al., 2016). The treatment of 20 ppm of Fe-NPs in another study greatly reduced the adverse effects of Cd by enhancing the photosynthetic and antioxidant enzyme activity, Fe uptake, which subsequently increased the plant growth and also minimized the concentration of Cd in plants. Morphological parameters of wheat plants like photosynthetic pigments, biomass of roots/shoots, spike husks, and grains have been improved followed by the exogenous Fe-NPs application (Hussain et al., 2019). Recent scientific studies have reported TiO_2 NPs can boost PGPR efficiency and therefore encourage growth of plants in phytoremediation setups (Timmusk et al., 2018). Experimental studies also revealed that certain NMs like zerovalent Fe NPs (ZVFe-NPs) minimize HM bioavailability in the soils and enhance plant biomass (Tafazoli et al., 2017). However, the mechanisms for posting plant growth of these materials are different. The explicit removal mechanism involved in HMs remediation using ZVFe-NPs are mainly determined by the redox potential of metal pollutants. The metals with similar or even more negative standard redox potential (Zn and Cd) to Fe can be eliminated through adsorption with ZVFe-NPs, whereas the metals with positive standard redox potential (Pb and Ni) both the processes, reduction, and adsorption are needed. Direct elimination of contaminants by NMs decreases phytotoxicity, which is helpful for the plant development. Furthermore, the NMs can interact with plants to improve their pollutant tolerance. As physiological regulators, ZnO NPs were used to reduce the phytotoxicity of Pb and Cd to white popinac during the phytoremediation (Praveen et al., 2018). From the experimental findings, it was implied that these ZnO NPs improved the resistance of plant by regulating genetic expression of enzyme. In other studies on similar lines, Si NPs have been used to alleviate Cr(VI) phytotoxicity in pea plants by limiting the accumulation of Cr and oxidative stress, and also improving antioxidant and photosynthetic activities (Tripathi et al., 2015a).

Furthermore, the pollutant phyto-availability is a critical factor influencing the effectiveness on phytoremediation, particularly for phytoextraction. The plants absorb contaminants in available forms only. The phyto-availability of contaminants relies heavily on the chemical speciation and dissemination of these pollutants in the soil habitats. The maximum phyto-availability of metals contaminants usually exists in exchangeable forms (i.e., dissolved in soil solution), then in mixed forms with oxides, minerals and other organic matter, and the lowest was observed in crystalline state (Sheoran et al., 2016). Additionally, the phyto-availability of contaminants is often influenced by physicochemical parameters and plant physiological properties (Sheoran et al., 2016; Ren et al., 2018). In this direction, several approaches, which include agronomic management (use of fertilizers), inoculation of rhizopheric microorganisms, treatment with chemical additives (chelating agent), and the use of genetic engineering have also been introduced to increase the photo-availability of soil contaminants (Ma et al., 2016; Sheoran et al., 2016; Jacobs et al., 2018). Increasing the phyto-availability of contaminants has been shown to be beneficial in improving the efficacy of phytoremediation process. There are two distinct effects of NMs on the phyto-availability of pollutants in soil. On the one hand, as they penetrate the cells, these NMs may act as a carrier of pollutants, thus, increasing bioavailability (Wild and Jones, 2009; Su et al., 2013). The adsorption of pollutants onto the NMs outside the organism, on the other hand, could decrease the free pollutants, thus, reducing bioavailability (Glomstad et al., 2016). Based on this, the two main criteria for improving the phyto-availability of pollutants with NMs are ought to be fulfilled: (**1**) the NMs can be coupled (primarily by adsorption) with the pollutants and (**2**) the proposed NMs should be phyto-available. Moreover, several other NMs like short CNTs, Fe_3O_4, Si NPs, TiO_2 NPs, and quantum dots (QDs) have been reported to be directly adsorbed by the plants (Wang et al., 2016). This makes it possible for more NMs to be used in future research endeavors to improve the phyto-availability of pollutants in phytoremediation processes.

6.6 Role of nanomaterials on stress tolerance in experimental plants

During the recent past, the various scientific studies have indicated that the use of NMs/NPs can also increase plant stress tolerance *ex-situ* as well as *in-situ* settings, thus, fostering the potential for phytoremediation (Iavicoli et al., 2017; Prasad et al., 2017). In order to overcome the noxious effects of contaminants, particularly HMs, the plant species have evolved multidimensional networks of homeostatic mechanistic strategies that help not only to limit the update and accumulation of these toxic materials but also to control their transportation and inactivation. The NMs play a pivotal role in the regulation of many biochemical and physiological processes in plants under adverse and deleterious stressful conditions (Wu et al., 2019). The interaction of plants with NMs/NPs trigger many morphological and physiological changes and these changes heavily depend on NP type, concentration, speciation, and the plant species involved. The NMs take part in various signal networks regulating main physiological aspects namely, carbon dioxide fixation, nitrogen metabolism, water acquisition, and antioxidant profile, which are an useful strategy for mitigating the adverse consequences of different types of abiotic stress in plants (Tripathi et al., 2015b; Jalil and Ansari, 2019; Wu et al., 2019). The mitigation of HMs/metalloids stress is an imperative form among the different forms of abiotic stress, since this represents a major problem in the present industrial era, and it has been demonstrated that NMs/NPs are extremely efficient in ameliorating the HM-induced plant toxicity in plants (Tripathi et al., 2015b; Du et al., 2017). Several NPs enhance plants' potential to combat stress by enhancing antioxidant potential, whereas the same and/or some new varieties of NMs may reduce the activity of same are different enzymes in other plants introducing oxidative stress. For instance, TiO_2 NPs increased the functions of various antioxidant enzymes like SOD, CAT, APOX, and GPx in spinach (Lei et al., 2008) and SOD, CAT, and GPOX in *Lemna minor* (Song et al., 2012). The scientific literature also describes the mimetic ability of certain NPs in plants with antioxidant enzymes, and this ability has been explored to overcome the oxidative stress in plants. Owing to their minute size and greater surface area, the NMs have great affinity toward HM toxicants and penetrate easily into the contaminated sites. The scientific findings suggested that QDs bind toward metals (Cu and Pb) and reduce their availability and also shield cell walls by functioning as a barrier (Worms et al., 2012). However, if the biophysical barriers are compromised and entry into the cell is enabled, the plants activate a combating phenomenon by accumulating certain metabolites and improve self-protection system of cells to reduce adverse consequences of HM induced stress conditions (Tripathi et al., 2016; Wu et al., 2019). Among the various types of NPs, the role of TiO_2 NPs in plants has been widely investigated (Iavicoli et al., 2017; Faraji and Sepehri, 2018). The better performance of plants regarding physiology, growth, and yield was observed when introducing these NPs under normal as well as stressful conditions. The exogenous treatment of TiO_2 NPs in *Spinacia oleracea* (spinach) curtailed the oxidative stress levels and improved dry mass, biosynthesis of chlorophyll, net photosynthetic rate, and enzymatic activities of Rubisco carboxylase (Zheng et al., 2005; Gao et al., 2006; Lei et al., 2008). Additionally, TiO_2 NPs have also been endorsed to inhibit reactive oxygen free radicals, stimulate antioxidant enzymes, maintain chloroplast structure, and membrane stability of plants (Hong et al., 2005; Lei et al., 2008). The TiO_2 not only mitigate the detrimental effects of pollutants and other plant complications like UV stress, but they also minimize the toxicity of HMs and accelerate the photosynthetic activity as well as growth of the plants (Lei et al., 2008; Singh and Lee, 2016; Faraji and Sepehri, 2018). The treatment of TiO_2 restricts Cd toxicity, accelerates plant growth, relative chlorophyll, moisture content in soyabean exposed to HMs. These NPs greatly minimize the lipid peroxidation and proline content in soyabean plants treated with Cd (Singh and Lee, 2016). Earlier studies on the impact of TiO_2 NPs wheat seedlings treated with Cd have shown that these NPs have great ability to alleviate the adverse effects of Cd toxicity (Faraji and Sepehri, 2018). Additionally, elevated activities of antioxidative enzyme, including ascorbate peroxidases, guaiacol, and catalase have been reported to TiO_2 treatment. The application of other NPs has been found to be effective against HMs toxicity. Nano-hydroxyapatite was found to attenuate the stress levels due to Cd in mustard plants (Li and Huang, 2014). The application of Si NPs for the treatment Cr(VI) stress in plants has been created for better nutrient assimilation, elevated photosynthesis, antioxidant enzymes, and improved growth of the plants. The studies on pea seedlings have demonstrated that Si NPs are efficiently able to ameliorate the phytotoxic effects of Cr(VI) (Tripathi et al., 2015b). This study further provides evidence that the use of Si NPs increases leaf nitrogen content in both normal as well as distressed plants. In cucumber, the exogenous treatment of Si NP maintains K^+/Na^+ ratio under oxidative stress conditions (Alsaeedi et al., 2017). Hence, improving K^+/Na^+ ratio by increasing K^+ and decreasing Na^+ supports a strong cytosolic homeostasis (Parida and Das, 2005) and results in the initiation of many vital enzymes involved in detoxification of reactive oxygen species (Kader and Lindberg, 2010; Siddiqui et al., 2017) in addition to other essential metabolic enzymes.

6.7 Challenges of nano-phytoremediation

The advancement of nanotechnology and bioremediation technologies was followed using NMs to facilitate phytoremediation. However, in realistic implementations, this strategy faces numerous challenges. The most critical challenge is the environmental risk of these NMs in the soil habitats. Most of the NMs are hazardous to plants, animals, and microbial flora of soil ecosystem (Singh, 2016). Moreover, in nano-phytoremediation processes, the phytotoxicity of NMs is of specific concern. Hence, on the one hand, further research to focus on the potential risks of NMs are needed. On the other hand, it is imperative to regulate the use of NMs in phytoremediation processes to make the most of them, but at the same time alleviate their harm. Presently, the application of NMs in phytoremediation processes is in exploration and testing phase, although several promising results have been achieved. Experiences from more applications studies are desired and more research is needed for long-term reliability of the NMs. The principal problems and challenges with respect to nano-phytoremediation are:

1. The phytoremediation studies bolstered by NMs are quite rare.
2. To date, only microcosm experiments have been carried out and more practical studies are needed in future research and better field understanding is essential.
3. Long-term studies are mandated to assess the possible impacts of NMs and changes in the soil profiles during the phytoremediation process.
4. The NMs often aggregate that hampers their mobility. To improve their mobility or bioavalibility, polymer coating or other coatings are critical.
5. Impact and safety assessment of NMs in contaminated soils is also imperative.
6. Sustainable nano-phytoremediation primarily depends on variety of meteorological parameters, so it is important to identify and use environmentally stable NMs.
7. Safe limits of NMs in terms of plant growth and development and the underlying intrinsic factors namely, growth promoters and enzymes need to be understood.

6.8 Conclusion and future perspectives

Phytoremediation is an ecologically sustainable biotechnological approach with many competitive benefits in remediating polluted soils; however, its extensive use is limited by long remediation periods, changing environmental conditions, and phytotoxicity of high concentration pollutants. There have been profound efforts to improve the phytoremediation efficiency. Nanotechnology has revolutionized the available technologies used in various sectors including environmental remediation. In the development of new products to replace current manufacturing processes, nanotechnology has played a significant role, leading to possible environmental and cost benefits. In addition, nanotechnological approaches offers the ability to coordinate and improve manufacturing processes in a more sustainable manner, effectively as close to a zero-emission approach as possible. The review of literature clearly indicated that significant attention is focused on synthesis and potential benefits of NMs in phytoremediation for improving remedial efficiency. The implications of biotechnological tools for environmental cleanup is one of the main characteristics of nano-phytoremediation. It includes three notable features, i.e., (**1**) the usage of green safe and benign NMs, (**2**) provision of removal hazardous compounds from contaminated sites, and (**3**) utilization as ambient variability sensor. The remediation of contaminated sites has been achieved by nano-phytoremediation by its well-known phenomenon, recognized as reactive or adsorptive techniques applied on both *in-situ* and *ex-situ* sites. The former process utilizes HM pollutant removal sequestration, whereas the latter approach involves oxidation of organic pollutants. The implications of these nanobiotechnological strategies are desired for long-term mitigation of environmental pollutants.

NMs can eliminate toxicants, encourage plant growth, and enhance availability of pollutants, therefore, promoting phytoremediation of polluted habitats. Metal oxides and noble metals especially TiO_2, carbon-based nanotubes, and bimetallic forms are diverse biogenic NMs developed and investigated for environmental cleanup. Among these, nZVI and its derivatives have been investigated extensively. The researchers indicated that usefulness of these engineered NPs/NMs in degrading OCs and extracting toxic metals has been promising. The nano-assisted phytoremediation is the most viable and generally favored approach owing to its cost-effectiveness, better performance, and implementation on a wide scale. However, as discussed earlier, nanobiotechnological strategies in phytoremediation process are still in progressive phase, since progressions with respect to the shape, size, and surface chemistry of NMs are yet to optimized. Due to its immense scope nano-phytoremediation is anticipated to rise in near future and play an important role in sustainable development. However, during the process of nano-phytoremediation, the affiliated NMs may find their entry in various strata of environment, and human beings and other animals would certainly be exposed to them through, inhalation, ingestion, injection, or topical contact, etc. Therefore, a comprehensive and in-depth study is recommended to determine the toxicity of these fabricated NMs in living organisms and validate their degree of safety.

Acknowledgments

The author, Mir Zahoor Gul would like to acknowledge the financial support received from the Council of Scientific & Industrial Research (CSIR- HRDG), New Delhi, India under CSIR-Research Associateship Scheme (vide File No. 09/132(0883)/2019-EMR-1).

References

Aderholt, M., Vogelien, D.L., Koether, M., Greipsson, S., 2017. Phytoextraction of contaminated urban soils by Panicum virgatum L. enhanced with application of a plant growth regulator (BAP) and citric acid. Chemosphere 175, 85–96. doi:10.1016/j.chemosphere.2017.02.022.

Afzal, M., Khan, Q.M., Sessitsch, A., 2014. Endophytic bacteria: prospects and applications for the phytoremediation of organic pollutants. Chemosphere 117, 232–242. doi:10.1016/j.chemosphere.2014.06.078.

Aken, B.V., Correa, P.A., Schnoor, J.L., 2010. Phytoremediation of polychlorinated biphenyls: new trends and promises. Environ. Sci. Technol. 44 (8), 2767–2776.

Ali, H., Khan, E., Sajad, M.A., 2013. Phytoremediation of heavy metals-concepts and applications. Chemosphere 91 (7), 869–881.

Alsaeedi, A.H., El-Ramady, H., Alshaal, T., et al., 2017. Engineered silica nanoparticles alleviate the detrimental effects of Na+ stress on germination and growth of common bean (Phaseolus vulgaris). Environ. Sci. Pollut. Res. 24 (27), 21917–21928. doi:10.1007/s11356-017-9847-y.

Amundson, R., Berhe, A.A., Hopmans, J.W., et al., 2015. Soil science. Soil and human security in the 21st century. Science 348 (6235), 1261071. doi:10.1126/science.1261071.

Awa, S.H., Hadibarata, T., 2020. Removal of heavy metals in contaminated soil by phytoremediation mechanism: a review. Water. Air. Soil Pollut. 231, 47. doi:10.1007/s11270-020-4426-0.

Baker, S., Rakshith, D., Kavitha, K.S., et al., 2013. Plants: emerging as nanofactories towards facile route in synthesis of nanoparticles. BioImpacts 3 (3), 111–117. doi:10.5681/bi.2013.012.

Baragaño, D., Forján, R., Welte, L., Gallego, J.L.R., 2020. Nanoremediation of as and metals polluted soils by means of graphene oxide nanoparticles. Sci. Rep. 10 (1). doi:10.1038/s41598-020-58852-4.

Bennett, L.E., Burkhead, J.L., Hale, K.L., et al., 2003. Analysis of transgenic Indian mustard plants for phytoremediation of metal-contaminated mine tailings. J. Environ. Qual. 32 (2), 432–440. doi:10.2134/jeq2003.4320.

Bhargava, A., Carmona, F.F., Bhargava, M., Srivastava, S., 2012. Approaches for enhanced phytoextraction of heavy metals. J. Environ. Manage. 105, 103–120. doi:10.1016/j.jenvman.2012.04.002.

Blum, W.E.H., 2005. Functions of soil for society and the environment. Rev. Environ. Sci. Biotechnol. 4 (3), 75–79. doi:10.1007/s11157-005-2236-x.

Cachada, A., Pato, P., Rocha-Santos, T., et al., 2012. Levels, sources and potential human health risks of organic pollutants in urban soils. Sci. Total Environ. 430, 184–192. doi:10.1016/j.scitotenv.2012.04.075.

Cai, C., Zhao, M., Yu, Z., et al., 2019. Utilization of nanomaterials for in-situ remediation of heavy metal(loid) contaminated sediments: A review. Sci. Total Environ. 662, 205–217. doi:10.1016/j.scitotenv.2019.01.180.

Cecchin, I., Reddy, K.R., Thomé, A., et al., 2017. Nanobioremediation: integration of nanoparticles and bioremediation for sustainable remediation of chlorinated organic contaminants in soils. Int. Biodeterior. Biodegrad. 119, 419–428. doi:10.1016/j.ibiod.2016.09.027.

Chaves, L.C.C., Navoni, J.A., de Morais Ferreira, D., et al., 2016. Water mutagenic potential assessment on a semiarid aquatic ecosystem under influence of heavy metals and natural radioactivity using micronuclei test. Environ. Sci. Pollut. Res. 23 (8), 7572–7581. doi:10.1007/s11356-015-5993-2.

Chen, M., Xu, P., Zeng, G., et al., 2015. Bioremediation of soils contaminated with polycyclic aromatic hydrocarbons, petroleum, pesticides, chlorophenols and heavy metals by composting: applications, microbes and future research needs. Biotechnol. Adv. 33 (6 Pt 1), 745–755. doi:10.1016/j.biotechadv.2015.05.003.

Cristaldi, A., Conti, G.O., Jho, E.H., et al., 2017. Phytoremediation of contaminated soils by heavy metals and PAHs. A brief review. Environ. Technol. Innov. 8, 309–326. doi:10.1016/j.eti.2017.08.002.

Cundy, A.B., Hopkinson, L., Whitby, R.L.D., 2008. Use of iron-based technologies in contaminated land and groundwater remediation: a review. Sci. Total Environ. 400 (1–3), 42–51. doi:10.1016/j.scitotenv.2008.07.002.

Das, R.K., Pachapur, V.L., Lonappan, L., et al., 2017. Biological synthesis of metallic nanoparticles: plants, animals and microbial aspects. Nanotechnol. Environ. Eng. 2, 18. doi:10.1007/s41204-017-0029-4.

Das, S., Sen, B., Debnath, N., 2015. Recent trends in nanomaterials applications in environmental monitoring and remediation. Environ. Sci. Pollut. Res. 22 (23), 18333–18344. doi:10.1007/s11356-015-5491-6.

De La Torre-Roche, R., Hawthorne, J., Deng, Y., et al., 2012. Fullerene-enhanced accumulation of p, p'-DDE in agricultural crop species. Environ. Sci. Technol. 46 (17), 9315–9323. doi:10.1021/es301982w.

De La Torre-Roche, R., Hawthorne, J., Deng, Y., et al., 2013. Multiwalled carbon nanotubes and C60 fullerenes differentially impact the accumulation of weathered pesticides in four agricultural plants. Environ. Sci. Technol. 47 (21), 12539–12547. doi:10.1021/es4034809.

Di Palma, L., Gueye, M.T., Petrucci, E., 2015. Hexavalent chromium reduction in contaminated soil: A comparison between ferrous sulphate and nanoscale zero-valent iron. J. Hazard. Mater. 281, 70–76. doi:10.1016/j.jhazmat.2014.07.058.

Du, W., Tan, W, Peralta-Videa, J.R., et al., 2017. Interaction of metal oxide nanoparticles with higher terrestrial plants: physiological and biochemical aspects. Plant Physiol. Biochem. 110, 210–225. doi:10.1016/j.plaphy.2016.04.024.

El-Temsah, Y.S., Sevcu, A., Bobcikova, K., et al., 2016. DDT degradation efficiency and ecotoxicological effects of two types of nano-sized zero-valent iron (nZVI) in water and soil. Chemosphere 144, 2221–2228. doi:10.1016/j.chemosphere.2015.10.122.

FAO, ITPS, 2015. Status of the World's Soil Resources (SWSR) – Main Report. Food and Agriculture Organization of the United Nations and Intergovernmental Technical Panel on Soils.

Faraji, J., Sepehri, A., 2018. Titanium dioxide nanoparticles and sodium nitroprusside alleviate the adverse effects of cadmium stress on germination and seedling growth of wheat (Triticum aestivum L.). Univ. Sci. 23 (1), 61–87. doi:10.11144/Javeriana.SC23-1.tdna.

Ganachari, S.V., Yaradoddi, J.S., Somappa, S.B., et al., 2019. Green nanotechnology for biomedical, food, and agricultural applicationsHandbook of Ecomaterials. Springer, Cham, Switzerland AG, 2681–2698.

Gao, F., Hong, F., Liu, C., et al., 2006. Mechanism of nano-anatase TiO2 on promoting photosynthetic carbon reaction of spinach: inducing complex of Rubisco-Rubisco activase. Biol. Trace Elem. Res. 111 (1–3), 239–253. doi:10.1385/BTER:111:1:239.

Gavrilescu, M., Demnerová, K., Aamand, J., et al., 2015. Emerging pollutants in the environment: Present and future challenges in biomonitoring, ecological risks and bioremediation. N. Biotechnol. 32 (1), 147–156. doi:10.1016/j.nbt.2014.01.001.

Gerhardt, K.E., Gerwing, P.D., Greenberg, B.M., 2017. Opinion: taking phytoremediation from proven technology to accepted practice. Plant Sci. 256, 170–185. doi:10.1016/j.plantsci.2016.11.016.

Gil-Díaz, M., Alonso, J., Rodríguez-Valdés, E., et al., 2017a. Comparing different commercial zero valent iron nanoparticles to immobilize As and Hg in brownfield soil. Sci. Total Environ. 584–585, 1324–1332. doi:10.1016/j.scitotenv.2017.02.011.

Gil-Díaz, M., Pinilla, P., Alonso, J., Lobo, M.C., 2017b. Viability of a nanoremediation process in single or multi-metal(loid) contaminated soils. J. Hazard. Mater. 321, 812–819. doi:10.1016/j.jhazmat.2016.09.071.

Glomstad, B., Altin, D., Sørensen, L., et al., 2016. Carbon nanotube properties influence adsorption of phenanthrene and subsequent bioavailability and toxicity to Pseudokirchneriella subcapitata. Environ. Sci. Technol. 50 (5), 2660–2668. doi:10.1021/acs.est.5b05177.

Gomes, H.I., 2012. Phytoremediation for bioenergy: challenges and opportunities. Environ. Technol. Rev. 1 (1), 59–66. doi:10.1080/09593330.2012.696715.

Gómez-Pastora, J., Bringas, E., Ortiz, I., 2014. Recent progress and future challenges on the use of high performance magnetic nano-adsorbents in environmental applications. Chem. Eng. J. 256, 187–204. doi:10.1016/j.cej.2014.06.119.

Gong, X., Huang, D., Liu, Y., et al., 2018. Remediation of contaminated soils by biotechnology with nanomaterials: bio-behavior, applications, and perspectives. Crit. Rev. Biotechnol. 38 (3), 455–468.

Guerra, F.D., Attia, M.F., Whitehead, D.C., Alexis, F, 2018. Nanotechnology for environmental remediation: Materials and applications. Molecules 23. doi:10.3390/molecules23071760.

Hamdi, H., De, La., Torre-Roche, R., Hawthorne, J., White, J.C., 2015. Impact of non-functionalized and amino-functionalized multiwall carbon nanotubes on pesticide uptake by lettuce (Lactuca sativa L.). Nanotoxicology 9 (2), 172–180. doi:10.3109/17435390.2014.907456.

Hofmann, T., von der Kammer, F., 2009. Estimating the relevance of engineered carbonaceous nanoparticle facilitated transport of hydrophobic organic contaminants in porous media. Environ. Pollut. 157 (4), 1117–1126. doi:10.1016/j.envpol.2008.10.022.

Hong, F.S., Yang, P., Gao, F.Q., et al., 2005. Effect of nano-anatase TiO_2 on spectral characterization of photosystem II particles from spinach. Chem. Res. Chinese Univ. 21 (2), 196–200.

Huang, D., Xue, W., Zeng, G., et al., 2016. Immobilization of Cd in river sediments by sodium alginate modified nanoscale zero-valent iron: Impact on enzyme activities and microbial community diversity. Water Res. 106, 15–25. doi:10.1016/j.watres.2016.09.050.

Hussain, A., Ali, S., Rizwan, M., et al., 2019. Responses of wheat (Triticum aestivum) plants grown in a Cd contaminated soil to the application of iron oxide nanoparticles. Ecotoxicol. Environ. Saf. 173, 156–164. doi:10.1016/j.ecoenv.2019.01.118.

Iavicoli, I., Leso, V., Beezhold, D.H., Shvedova, A.A., 2017. Nanotechnology in agriculture: opportunities, toxicological implications, and occupational risks. Toxicol. Appl. Pharmacol. 329, 96–111. doi:10.1016/j.taap.2017.05.025.

Ibrahim, R.K., Hayyan, M., AlSaadi, M.A., et al., 2016. Environmental application of nanotechnology: air, soil, and water. Environ. Sci. Pollut. Res. 23 (14), 13754–13788. doi:10.1007/s11356-016-6457-z.

Ingle, A.P., Seabra, A.B., Duran, N., Rai, M., 2014. Nanoremediation: a new and emerging technology for the removal of toxic contaminant from environmentMicrobial Biodegradation and Bioremediation. Elsevier, London, U.K, pp. 233–250.

Jacobs, A., De Brabandere, L., Drouet, T., et al., 2018. Phytoextraction of Cd and Zn with Noccaea caerulescens for urban soil remediation: influence of nitrogen fertilization and planting density. Ecol. Eng. 116, 178–187. doi:10.1016/j.ecoleng.2018.03.007.

Jadia, C.D., Fulekar, M.H., 2009. Phytoremediation of heavy metals: recent techniques. African J. Biotechnol. 8 (6), 921–928.

Jalil, S.U., Ansari, M.I., 2019. Nanoparticles and abiotic stress tolerance in plants: synthesis, action, and signaling mechanismsPlant Signaling Molecules: Role and Regulation under Stressful Environments. Woodhead Publishing, Sawston, U. K., pp. 549–561.

Jiang, D., Zeng, G., Huang, D., et al., 2018. Remediation of contaminated soils by enhanced nanoscale zero valent iron. Environ. Res. 163, 217–227. doi:10.1016/j.envres.2018.01.030.

Kader, M.A., Lindberg, S., 2010. Cytosolic calcium and pH signaling in plants under salinity stress. Plant Signal. Behav. 5 (3), 233–238. doi:10.4161/psb.5.3.10740.

Kalidhasan, S., Dror, I., Berkowitz, B., 2017. Atrazine degradation through PEI-copper nanoparticles deposited onto montmorillonite and sand. Sci. Rep. 7 (1), 1415. doi:10.1038/s41598-017-01429-5.

Kang, J., Duan, X., Wang, C., et al., 2018. Nitrogen-doped bamboo-like carbon nanotubes with Ni encapsulation for persulfate activation to remove emerging contaminants with excellent catalytic stability. Chem. Eng. J. 332, 398–408. doi:10.1016/j.cej.2017.09.102.

Khan, I., Saeed, K., Khan, I., 2019. Nanoparticles: properties, applications and toxicities. Arab. J. Chem. 12 (7), 908–931. doi:10.1016/j.arabjc.2017.05.011.

Khin, M.M., Nair, A.S., Babu, V.J., et al., 2012. A review on nanomaterials for environmental remediation. Energy Environ. Sci. 5 (8), 8075–8109. doi:10.1039/c2ee21818f.

Komárek, M., Koretsky, C.M., Stephen, K.J., et al., 2015. Competitive adsorption of Cd(II), Cr(VI), and Pb(II) onto nanomaghemite: a spectroscopic and modeling approach. Environ. Sci. Technol. 49 (21), 12851–12859. doi:10.1021/acs.est.5b03063.

Lal, R., 2015. Restoring soil quality to mitigate soil degradation. Sustain. 7 (5), 5875–5895. doi:10.3390/su7055875.

Lee, J.H., 2013. An overview of phytoremediation as a potentially promising technology for environmental pollution control. Biotechnol. Bioprocess. Eng. 18, 431–439. doi:10.1007/s12257-013-0193-8.

Lei, Z., Mingyu, S., Xiao, W., et al., 2008. Antioxidant stress is promoted by nano-anatase in spinach chloroplasts under UV-B radiation. Biol. Trace Elem. Res. 121 (1), 69–79. doi:10.1007/s12011-007-8028-0.

Li, G., Sun, G.X., Ren, Y., et al., 2018. Urban soil and human health: a review. Eur. J. Soil Sci. 69 (1), 196–215. doi:10.1111/ejss.12518.

Li, Q., Chen, X., Zhuang, J., Chen, X., 2016. Decontaminating soil organic pollutants with manufactured nanoparticles. Environ. Sci. Pollut. Res. 23 (12), 11533–11548. doi:10.1007/s11356-016-6255-7.

Li, Z., Huang, J., 2014. Effects of nanoparticle hydroxyapatite on growth and antioxidant system in pakchoi (brassica chinensis L.) from cadmium-contaminated soil. J. Nanomater. 2014, 470962. doi:10.1155/2014/470962.

Liu, R., Lal, R., 2012. Nanoenhanced materials for reclamation of mine lands and other degraded soils: a review. J. Nanotechnol. 2012, 461468. doi:10.1155/2012/461468.

López-Moreno, M.L., De, La., Rosa, G., Hernández-Viezcas, J.A., et al., 2010. X-ray absorption spectroscopy (XAS) corroboration of the uptake and storage of ceo2 nanoparticles and assessment of their differential toxicity in four edible plant species. J. Agric. Food Chem. 58 (6), 3689–3693. doi:10.1021/jf904472e.

Luo, J., Cai, L., Qi, S., et al., 2017. A multi-technique phytoremediation approach to purify metals contaminated soil from e-waste recycling site. J. Environ. Manage. 204 (Pt 1), 17–22. doi:10.1016/j.jenvman.2017.08.029.

Ma, X., Wang, C., 2010. Fullerene nanoparticles affect the fate and uptake of trichloroethylene in phytoremediation systems. Environ. Eng. Sci. 27 (11), 989–992. doi:10.1089/ees.2010.0141.

Ma, Y., Rajkumar, M., Zhang, C., Freitas, H., 2016. Beneficial role of bacterial endophytes in heavy metal phytoremediation. J. Environ. Manage. 174, 14–25. doi:10.1016/j.jenvman.2016.02.047.

Mahar, A., Wang, P., Ali, A., et al., 2016. Challenges and opportunities in the phytoremediation of heavy metals contaminated soils: a review. Ecotoxicol. Environ. Saf. 126, 111–121. doi:10.1016/j.ecoenv.2015.12.023.

Makarova, O.V., Rajh, T., Thurnauer, M.C., et al., 2001. Surface modification of TiO_2 nanoparticles for selective photodecomposition of nitrobenzene. ACS Div. Environ. Chem. Prepr. 41 (2), 390–394. doi:10.1021/es001109.

Masindi, V., Muedi, K.L., 2018. Environmental contamination by heavy metalsHeavy Metals. InTechOpen.

Mitton, F.M., Miglioranza, K.S.B., Gonzalez, M., et al., 2014. Assessment of tolerance and efficiency of crop species in the phytoremediation of DDT polluted soils. Ecol. Eng. 71, 501–508. doi:10.1016/j.ecoleng.2014.07.069.

Mohanpuria, P., Rana, N.K., Yadav, S.K., 2008. Biosynthesis of nanoparticles: technological concepts and future applications. J. Nanoparticle Res. 10, 507–517. doi:10.1007/s11051-007-9275-x.

Mueller, N.C., Nowack, B., 2010. Nanoparticles for remediation: solving big problems with little particles. Elements 6 (6), 395–400. doi:10.2113/gselements.6.6.395.

Nahar, N., Rahman, A., Nawani, N.N., et al., 2017. Phytoremediation of arsenic from the contaminated soil using transgenic tobacco plants expressing ACR2 gene of Arabidopsis thaliana. J. Plant Physiol. 218, 121–126. doi:10.1016/j.jplph.2017.08.001.

Ojuederie, O.B., Babalola, O.O., 2017. Microbial and plant-assisted bioremediation of heavy metal polluted environments: a review. Int. J. Environ. Res. Public Health 14 (12), 1504. doi:10.3390/ijerph14121504.

Paquin, D.G., Campbell, S., Li, Q.X., 2004. Phytoremediation in subtropical Hawaii—a review of over 100 plant species. Remediation 14 (2), 127–139. doi:10.1002/rem.20005.

Parida, A.K., Das, A.B., 2005. Salt tolerance and salinity effects on plants: a review. Ecotoxicol. Environ. Saf. 60 (3), 324–349. doi:10.1016/j.ecoenv.2004.06.010.

Paz-Alberto, A.M., Sigua, G.C., 2013. Phytoremediation: a green technology to remove environmental pollutants. Am. J. Clim. Chang. 2, 71–86. doi:10.4236/ajcc.2013.21008.

Petruzzelli, G., Pedron, F., Rosellini, I., Grifoni M., Barbafieri M., 2016. Polycyclic Aromatic Hydrocarbons and Heavy Metal Contaminated Sites: Phytoremediation as a Strategy for Addressing the Complexity of Pollution. In: Ansari A., Gill S., Gill R., Lanza G., Newman L. (eds) Phytoremediation. Springer, Cham. https://doi.org/10.1007/978-3-319-41811-7_5.

Prasad, M.N.V., 2003. Phytoremediation of metal-polluted ecosystems: hype for commercialization. Russ. J. Plant Physiol. 50, 686–701. doi:10.1023/A:1025604627496.

Prasad, R., Bhattacharyya, A., Nguyen, Q.D., 2017. Nanotechnology in sustainable agriculture: recent developments, challenges, and perspectives. Front. Microbiol. 8, 1014. doi:10.3389/fmicb.2017.01014.

Praveen, A., Khan, E., Ngiimei, S., et al., 2018. Iron oxide nanoparticles as nano-adsorbents: a possible way to reduce arsenic phytotoxicity in Indian mustard plant (Brassica juncea L.). J. Plant Growth Regul. 37, 612–624. doi:10.1007/s00344-017-9760-0.

Qian, Y., Qin, C., Chen, M., Lin, S., 2020. Nanotechnology in soil remediation – applications vs. implications. Ecotoxicol. Environ. Saf. 201, 110815. doi:10.1016/j.ecoenv.2020.110815.

Ren, X., Zeng, G., Tang, L., et al., 2018. Sorption, transport and biodegradation – an insight into bioavailability of persistent organic pollutants in soil. Sci. Total Environ. 610–611, 1154–1163. doi:10.1016/j.scitotenv.2017.08.089.

Roy, A., Bhattacharya, J., 2012. Removal of Cu(II), Zn(II) and Pb(II) from water using microwave-assisted synthesized maghemite nanotubes. Chem. Eng. J. 211–212, 493–500. doi:10.1016/j.cej.2012.09.097.

Sarwar, N., Imran, M., Shaheen, M.R., et al., 2017. Phytoremediation strategies for soils contaminated with heavy metals: modifications and future perspectives. Chemosphere 171, 710–721. doi:10.1016/j.chemosphere.2016.12.116.

Satapanajaru, T., Anurakpongsatorn, P., Pengthamkeerati, P., Boparai, H., 2008. Remediation of atrazine-contaminated soil and water by nano zerovalent iron. Water Air. Soil Pollut 192, 349–359. doi:10.1007/s11270-008-9661-8.

Schwitzguébel, J.P., 2017. Phytoremediation of soils contaminated by organic compounds: hype, hope and facts. J. Soils Sediments 17, 1492–1502. doi:10.1007/s11368-015-1253-9.

Sheoran, V., Sheoran, A.S., Poonia, P., 2016. Factors affecting phytoextraction: a review. Pedosphere 26 (2), 148–166. doi:10.1016/S1002-0160(15)60032-7.

Siddiqui, M.N., Mostofa, M.G., Akter, M.M., et al., 2017. Impact of salt-induced toxicity on growth and yield-potential of local wheat cultivars: oxidative stress and ion toxicity are among the major determinants of salt-tolerant capacity. Chemosphere 187, 385–394. doi:10.1016/j.chemosphere.2017.08.078.

Singh, A.K., 2016. Introduction to nanoparticles and nanotoxicology. Engineered Nanoparticles. Academic Press, United States, pp. 1–18.

Singh, B.K., 2010. Exploring microbial diversity for biotechnology: the way forward. Trends Biotechnol 28 (3), 111–116. doi:10.1016/j.tibtech.2009.11.006.

Singh, J., Lee, B.K., 2016. Influence of nano-TiO_2 particles on the bioaccumulation of Cd in soybean plants (Glycine max): a possible mechanism for the removal of Cd from the contaminated soil. J. Environ. Manage. 170, 88–96. doi:10.1016/j.jenvman.2016.01.015.

Singh, R., Misra, V., Singh, R.P., 2012. Removal of Cr(VI) by nanoscale zero-valent iron (nZVI) from soil contaminated with tannery wastes. Bull. Environ. Contam. Toxicol. 88 (2), 210–214. doi:10.1007/s00128-011-0425-6.

Smith P., Cotrufo M.F., Rumpel C., et al (2015) Biogeochemical cycles and biodiversity as key drivers of ecosystem services provided by soils. Soil 1:665–685. doi: 10.5194/soil-1-665-2015.

Song, B., Xu, P., Chen, M., et al., 2019. Using nanomaterials to facilitate the phytoremediation of contaminated soil. Crit. Rev. Environ. Sci. Technol. 49 (9), 791–824. doi:10.1080/10643389.2018.1558891.

Song, B., Zeng, G., Gong, J., et al., 2017. Evaluation methods for assessing effectiveness of in situ remediation of soil and sediment contaminated with organic pollutants and heavy metals. Environ. Int. 105, 43–55. doi:10.1016/j.envint.2017.05.001.

Song, G., Gao, Y., Wu, H., et al., 2012. Physiological effect of anatase TiO_2 nanoparticles on Lemna minor. Environ. Toxicol. Chem. 31 (9), 2147–2152. doi:10.1002/etc.1933.

Souri, Z., Karimi, N., Sarmadi, M., Rostami, E., 2017. Salicylic acid nanoparticles (SANPs) improve growth and phytoremediation efficiency of Isatis cappadocica Desv., under as stress. IET Nanobiotechnol. 11 (6), 650–655. doi:10.1049/iet-nbt.2016.0202.

Stefaniuk, M., Oleszczuk, P., Ok, Y.S., 2016. Review on nano zerovalent iron (nZVI): From synthesis to environmental applications. Chem. Eng. J. 287, 618–632. doi:10.1016/j.cej.2015.11.046.

Su, Y., Yan, X., Pu, Y., et al., 2013. Risks of single-walled carbon nanotubes acting as contaminants-carriers: Potential release of phenanthrene in Japanese medaka (Oryzias latipes). Environ. Sci. Technol. 47 (9), 4704–4710. doi:10.1021/es304479w.

Tafazoli, M., Hojjati, S.M., Biparva, P., et al., 2017. Reduction of soil heavy metal bioavailability by nanoparticles and cellulosic wastes improved the biomass of tree seedlings. J. Plant Nutr. Soil Sci. 180, 683–693. doi:10.1002/jpln.201700204.

Thatai, S., Khurana, P., Boken, J., et al., 2014. Nanoparticles and core-shell nanocomposite based new generation water remediation materials and analytical techniques: a review. Microchem. J. 116, 62–76. doi:10.1016/j.microc.2014.04.001.

Timmusk, S., Seisenbaeva, G., Behers, L., 2018. Titania (TiO_2) nanoparticles enhance the performance of growth-promoting rhizobacteria. Sci. Rep. 8, 617. doi:10.1038/s41598-017-18939-x.

Tripathi, D.K., Singh, S., Singh, V.P., et al., 2016. Silicon nanoparticles more efficiently alleviate arsenate toxicity than silicon in maize cultiver and hybrid differing in arsenate tolerance. Front. Environ. Sci. 4. doi:10.3389/fenvs.2016.00046.

Tripathi, D.K., Singh, V.P., Prasad, S.M., et al., 2015a. Silicon nanoparticles (SiNp) alleviate chromium (VI) phytotoxicity in Pisum sativum (L.) seedlings. Plant Physiol. Biochem. 96, 189–198. doi:10.1016/j.plaphy.2015.07.026.

Tripathi, D.K., Singh, V.P., Prasad, S.M., et al., 2015b. Silicon-mediated alleviation of Cr(VI) toxicity in wheat seedlings as evidenced by chlorophyll florescence, laser induced breakdown spectroscopy and anatomical changes. Ecotoxicol. Environ. Saf. 113, 133–144. doi:10.1016/j.ecoenv.2014.09.029.

Tripathi, S., Sanjeevi, R., Anuradha, J., et al., 2018. Nano-bioremediation: nanotechnology and bioremediationBiostimulation Remediation Technologies for Groundwater Contaminants. IGI Global, United States, pp. 202–219.

Tripathi, V., Edrisi, S.A., Chen, B., et al., 2017. Biotechnological advances for restoring degraded land for sustainable development. Trends Biotechnol 35 (9), 847–859. doi:10.1016/j.tibtech.2017.05.001.

Truu, J., Truu, M., Espenberg, M., et al., 2015. Phytoremediation and plant-assisted bioremediation in soil and treatment wetlands: a review. Open Biotechnol. J. 9, 85–92. doi:10.2174/1874070701509010085.

Ullah, A., Heng, S., Munis, M.F.H., et al., 2015. Phytoremediation of heavy metals assisted by plant growth promoting (PGP) bacteria: a review. Environ. Exp. Bot. 117, 28–40. doi:10.1016/j.envexpbot.2015.05.001.

Usman, M., Farooq, M., Wakeel, A., et al., 2020. Nanotechnology in agriculture: current status, challenges and future opportunities. Sci. Total Environ. 721, 137778. doi:10.1016/j.scitotenv.2020.137778.

Vangronsveld, J., Herzig, R., Weyens, N., et al., 2009. Phytoremediation of contaminated soils and groundwater: Lessons from the field. Environ. Sci. Pollut. Res. 16, 765–794. doi:10.1007/s11356-009-0213-6.

Vithanage, M., Dabrowska, B.B., Mukherjee, A.B., et al., 2012. Arsenic uptake by plants and possible phytoremediation applications: a brief overview. Environ. Chem. Lett. 10, 217–224. doi:10.1007/s10311-011-0349-8.

von der Kammer, F., Ferguson, P.L., Holden, P.A., et al., 2012. Analysis of engineered nanomaterials in complex matrices (environment and biota): General considerations and conceptual case studies. Environ. Toxicol. Chem. 31 (1), 32–49. doi:10.1002/etc.723.

Wang, L., Ji, B., Hu, Y., et al., 2017. A review on in situ phytoremediation of mine tailings. Chemosphere 184, 594–600. doi:10.1016/j.chemosphere.2017.06.025.

Wang, P., Lombi, E., Zhao, F.J., Kopittke, P.M., 2016. Nanotechnology: a new opportunity in plant sciences. Trends Plant Sci. 21 (8), 699–712. doi:10.1016/j.tplants.2016.04.005.

Wild, E., Jones, K.C., 2009. Novel method for the direct visualization of in vivo nanomaterials and chemical interactions in plants. Environ. Sci. Technol. 43 (14), 5290–5294. doi:10.1021/es900065h.

Wójcik, M., Gonnelli, C., Selvi, F., et al., 2017. Metallophytes of serpentine and calamine soils – their unique ecophysiology and potential for phytoremediation. Adv. Bot. Res. 83, 1–42. doi:10.1016/bs.abr.2016.12.002.

Worms, I.A.M., Boltzman, J., Garcia, M., Slaveykova, V.I., 2012. Cell-wall-dependent effect of carboxyl-CdSe/ZnS quantum dots on lead and copper availability to green microalgae. Environ. Pollut. 167, 27–33. doi:10.1016/j.envpol.2012.03.030.

Wu, J., Wang, X., Wang, Q., et al., 2019. Nanomaterials with enzyme-like characteristics (nanozymes): next-generation artificial enzymes (II). Chem. Soc. Rev. 48, 1004–1076. doi:10.1039/C8CS00457A.

Wu, J., Xie, Y., Fang, Z., et al., 2016. Effects of Ni/Fe bimetallic nanoparticles on phytotoxicity and translocation of polybrominated diphenyl ethers in contaminated soil. Chemosphere 162, 235–242. doi:10.1016/j.chemosphere.2016.07.101.

Xue, W., Huang, D., Zeng, G., et al., 2018. Performance and toxicity assessment of nanoscale zero valent iron particles in the remediation of contaminated soil: a review. Chemosphere 210, 1145–1156. doi:10.1016/j.chemosphere.2018.07.118.

Yadav, K.K., Singh, J.K., Gupta, N., Kumar, V., 2017. A review of nanobioremediation technologies for environmental cleanup: a novel biological approach. J. Mater. Environ. Sci. 8 (2), 740–757.

Yadu, B., Chandrakar, V., Korram, J., et al., 2018. Silver nanoparticle modulates gene expressions, glyoxalase system and oxidative stress markers in fluoride stressed Cajanus cajan L. J. Hazard. Mater. 353, 44–52. doi:10.1016/j.jhazmat.2018.03.061.

Ye, S., Zeng, G., Wu, H., et al., 2017. Biological technologies for the remediation of co-contaminated soil. Crit. Rev. Biotechnol. 37 (8), 1062–1076. doi:10.1080/07388551.2017.1304357.

Zhang, L., Fang, M., 2010. Nanomaterials in pollution trace detection and environmental improvement. Nano Today 5 (2), 128–142. doi:10.1016/j.nantod.2010.03.002.

Zhang, Y.N., Niu, Q., Gu, X., et al., 2019. Recent progress on carbon nanomaterials for the electrochemical detection and removal of environmental pollutants. Nanoscale 11, 11992–12014. doi:10.1039/C9NR02935D.

Zheng, L., Hong, F., Lu, S., Liu, C., 2005. Effect of nano-TiO_2 on strength of naturally aged seeds and growth of spinach. Biol. Trace Elem. Res. 104 (1), 83–92. doi:10.1385/BTER:104:1:083.

Chapter 7

Biomass amendments and phytoremediation of environmental pollutants

Ayesha Abdul Qadir[a], Muhammad Fahad Sardar[b], Sobia Riaz[a], Zia Ur Rahman Farooqi[a], Ghulam Murtaza[a]
[a]*Institute of Soil and Environmental Sciences, University of Agriculture, Faisalabad, Pakistan,*
[b]*Institute of Environment and Sustainable Development in Agriculture, Chinese Academy of Agricultural Sciences, Beijing, China*

7.1 Introduction

In the race of development and economic stability, the industry is exponentially growing worldwide, which is the crucial reason for deteriorating pollution indices and soil contamination. To increase per capita income, the farmers' community uses an inappropriate amount of fertilizers and pesticides. This vigorous consumption is eventually polluting nearby soil and water bodies and industrial waste aggravates the problem (Awad et al., 2014; Zeng et al., 2013). World Health Organization reported that contaminated environment such as consumption of polluted water or food had caused several diseases hence mortality of over 12.6 million people globally (W.H.O., 2016). Organic contaminants and heavy metals mainly pollute soil and sediment, and these pollutants are harmful to the environment, human health, and food security (Liu et al., 2017; Sardar et al., 2021; Song et al., 2017; Zhang et al., 2016). Over the last few years, there have been a variety of physical, chemical, biological, and combined technologies to treat the waste and eradicate the problems associated with soil and sediment pollution. Combined technologies include a combination of two or three chemicals, physical or biological treatment approaches. This method of combining different treatment technologies might be the perfect solution to mitigate soil and sediment pollution. Hence, it has been a captivating approach and gained huge attention from scientists globally (Chen et al., 2013; Chen et al., 2016; Chirakkara and Reddy, 2015; Gong et al., 2015).

An increasing demand for knowledge to use or adapt sustainable approaches involving biological entities such as plants. Plants can be the best treatment approach for heavy metals and other organic contaminants because they can

Phytoremediation. DOI: https://doi.org/10.1016/B978-0-323-89874-4.00009-1

decompose or absorb the organic or inorganic pollutants accordingly. This technique's efficiency can be increased by a direct increase with contaminant uptake of plant or by the addition of any organic or inorganic amendment to enhance the bioavailability of pollutants. Pollutant uptake increases by way of the water and nutrient uptake enhance with improving plant growth. Ion competition is another factor that influences the plant uptake of pollutants. Plant biomass can be increased by improving soil properties, and this can be achieved by using some organic amendments or directly adding organic matter (Chirakkara and Reddy, 2015).

This chapter covers the scope, behavior, and mechanisms of biomass amendments towards remediation of organic and inorganic contaminants. In addition, the opportunities for enhanced phytoremediation by applying biomass amendments are also reviewed.

7.2 Scope of biomass amendments

The primary part of phytoremediation is to enhance carbon sequestration (Farooqi et al., 2018; Farooqi et al., 2020), maintain the physical, chemical, and biological characteristics of the polluted soil because any alteration in the soil or rhizosphere chemical properties such as pH, reduction-oxidation potential, and metal speciation or microbiological properties involving the type and activity of mycorrhizae may have a direct impact on its relation with plant or transmission of pollutants from soil to plant. To restore the vegetation cover, it is necessary to enhance the physical and chemical characteristics of soil while remediating the contamination of trace elements. This is done by applying soil organic/biomass amendments such as green manure or other recycled agriculture or industrial leftovers (Fig. 7.1). Meanwhile, all the probable adverse impacts of such

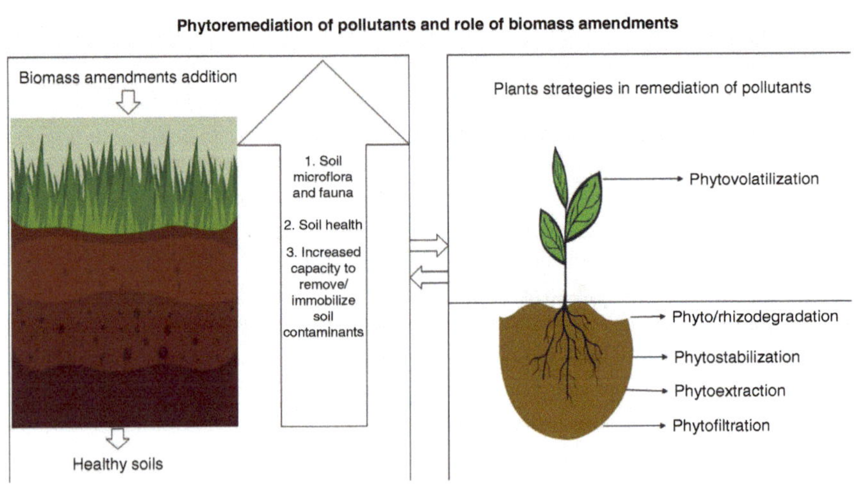

FIG. 7.1 Biomass amendments-assisted phytoremediation of pollutants.

amendments should not be neglected, including the integration of pollutants by other living bodies as well as their mobility.

It was reported that applying any organic or inorganic amendments such as compost, sugarcane bagasse, biosolids, coal fly ash, and lime in the polluted soil enhances the plant efficiency by assisting in plant biomass enhancement, plant subsistence for a longer time, self-proliferation of applied plant raw material as well as increased plant biodiversity (Mench et al., 2010). Application of soil organic amendments augments the soil structure and enhances soil organic matter ratio hence the ability of soil to capture the nutrients and moisture for a long time, which results in plant growth and health improvement. Few soil biomass amendments are known to capture the pollutants by their variety of mechanisms, and for that reason, they are called contaminant immobilizing agents such as agricultural or food byproducts, biochar, animal waste, and compost (Ahmad et al., 2017; Park et al., 2011; Shaheen et al., 2017). Food byproducts from the agriculture sector are being categorized according to the source of food, such as waste from crops and their residues, leftover fruits or vegetables, grains and legumes, distilleries, breweries, waste from food processing units, including the oil industry, sugar, starch as well as sweetmeat industry (Ajila et al., 2012). Crop residues are named as crop wastes as peels, leaves, vines or may contain any other leftover substances after harvesting and processing activities (Ajila et al., 2012). It is reported in various studies that sawdust, rice husk (Singh and Prasad, 2014), sugar beet waste (Madejón et al., 2006), and other agro-industrial residues application reduces the bioavailability of trace elements, hence decreases their toxicity. Whereas waste from vegetal food processing units may produce solid and liquid waste and their wastewater treatment activity generates sludge. These byproducts being used for metal decontamination or remediation were divided into subcategories based on their composition and their efficiency for metal decontamination (Fischer et al., 1998). All the liquid deposits such as effluents from grass and clover silage having aliphatic acids and amino acids were placed in the first group while the solid residues, including blood meal and brewery waste (hot sludge) a having higher amount of proteins or peptides were located in the second group. The third group contains all the solid waste with a bulk quantity of carbohydrates (molasses and potato peel sludge), and the ability to detoxify different metals varies from group to group. Blood meal effectively removes copper (Cu) and Nickel (Ni); Cadmium (Cd) and zinc (Zn) can be removed using grass silage effluent, whereas hydrolyzates having sugar acids are effective to immobilize Cu and lead (Pb).

The use of such organic amendments downgrades the transfer and leaching of pollutants by restoring soil properties. Biomass amendments such as compost and sewage sludge assist in phytoremediation processes by forming stable metal complexes (chemo-phytostabilization), by bounding of metals with the negatively charged functional groups of fulvic and humic acids (immobilization), and by extracting metals through the use of hyperaccumulator plants (phytoextraction). Biomass amendments can be used as an artificial source of biochelators in the soil as these elements can enhance the phytoextraction efficacy by proliferating the mobilization of soil contaminants, mainly heavy metals or metalloids. Agricultural and industrial sectors produce a bulk quantity of waste annually, such as in 2006, 27 countries of the European Union produced 89 Mt of food waste out of which 39% was generated by food manufacturing and processing units (Mirabella et al., 2014). The major categories of biomass amendments are described below.

Processed animal waste: Residues from the livestock industry is full of nutrients and have higher nutrition value hence being utilized by agriculturists, including a variety of animal waste such as swine, beef cattle, dairy as well as farmyard manure, pig/chicken manure, or litters and cow dung. The primary purpose of using animal-derived waste as fertilizer is to enhance the fertility of soil as well as few of such amendments are known to have a reducing effect on mobilization of potentially toxic trace elements within the soil (Bolan et al., 2003).

Biochar: It is a solid resultant of waste from agricultural or forestry operation when burnt at higher temperature by pyrolysis protocol (Zhang et al., 2013). It has significant impacts on soil texture and physicochemical properties as it is a strong surface adsorbent with partially arranged carbon matrix having a greater extent of microporosity, huge surface area, higher pH as well as cation exchange capacity (CEC) (Jiang et al., 2012; Lehmann et al., 2011). Current studies revealed that the biochar application significantly decreases bioavailability and phytotoxicity, resulting in soil pollutants immobilization (Hu et al., 2020; Liu et al., 2020; Wang et al., 2021). By the addition of biochar, metal precipitation occurs along with agronomic effectiveness of soil, including retention and source supplements off nutrients, resulting in enhanced soil fertility (Park et al., 2011), and in the case of sandy or acid soils, plant vegetation plus biomass production remarkably increases (Jeffery et al., 2011; Liu et al., 2013).

7.3 Biosolids/sewage sludge

There are a few differences between biosolids and sewage sludge; hence, both terms could be used identically. Biosolids are the byproducts, having a higher amount of nutrients formed during the wastewater treatment process while treating sewage sludge (Riaz et al., 2020). Appropriately managed and treated biosolids have a very effective role as soil amendments and fertilizers to enhance soil fertility, ultimately improving plant biomass production and effective in decreasing the soil pollutants, that is, potentially toxic trace elements (Riaz et al., 2020). On the other hand, biosolids can be a source of potential contaminants as being a municipal waste. Consequently, biosolids with excellent quality are suggested to use because they may contain comparatively lesser amounts of pathogens and odor, hence fewer heavy metals or metalloid contents (Bolan et al., 2003; Park et al., 2011). Sludge prepared from winery waste is known to have substantial effects when used as an adsorbent for the decontamination of Ni, Cr, and Cu in the aqueous phase (Villaescusa et al., 2004). Winery waste may contain higher concentrations of proteins, phenolic compounds, and carbohydrates with hydroxyl, phosphate, carboxyl, and amino group, which make it more effective for the metal removal (sorption) by binding metal ions (Villaescusa et al., 2004).

7.4 Compost/plant residues

Composting is the saturation and stabilization of any organic residue along with the help of microbial and enzymatic activities that speedily degrades the organic waste by means of mesophilic, thermophilic, and maturation stages, ultimately transforming the waste into mineral ions, CO_2, H_2O, and humic substances (Białobrzewski et al., 2015; Lu et al., 2014; Sardar et al., 2021). The final byproduct of composting comprising of mineral ions, microbes, and humic substances could be a potential source to reduce the mobilization of potentially toxic elements in soil, hence lessens the environmental and other ecological problems linked to these pollutants (Liu et al., 2019; Siedt et al., 2021). It was reported by Huang et al. (2016) that the application of compost material to the soil could significantly decrease the health hazards and environmental issues associated with potentially toxic trace elements. Rice husk is another potential byproduct with efficient soil remediation characteristics. Application of rice husk to the cadmium polluted site resulted in the remarkable reduction of Cd, specially its accumulation in vegetative parts of *Amaranthus caudatus* with a better growth rate and yield (Singh and Prasad, 2014).

7.5 The behavior of biomass amendments in different environments

The application of biomass amendments imparts significant impacts on the soil pH, its composition, and organic acids as well as on its microbial biota. Likewise, scientists are more focused on the rhizosphere because this root zone has intensified microbiological and chemical modification resulting from plant-soil-microbe communications. Hence, any minor or major modifications in soil microbiology or chemistry due to the incorporation of biomass amendments directly impact the mobilization, bioavailability, and transformation of the metals or metalloids (Park et al., 2011).

In a series of research conducted regarding the maintenance of trace compounds within the soil profile, soil organic matter has got much attention because of the transition metals' ability to formulate stable compounds along with ligands, and it is one of the basic aspects in monitoring metal solubility and bioavailability within the plant-soil profile. The qualitative (ligand type) and quantitative characteristics (surface area, CEC) of soil can be controlled by the application of organic matter. This treatment can change metal retention ability as well as modifies the chemistry of soil profile (change in pH and redox potential). Organic matter also controls the chelating and complex agent contents in both soil and aqueous phase. The occurrence of inorganic minerals and other compounds, such as phosphate or aluminum compounds in organic amendments, are held accountable for retaining heavy metals or metalloids (Brown et al., 1998).

Waste from the fresh olive oil industry may have acidic properties and high content of soluble organic complexes, including polyphenols, as stated by Azbar et al. (2004). Thus, it can enhance metal leaching and their bioavailability to both soil and plants by speeding up the metal solubilization in soil (Burgos et al., 2010; Clemente et al., 2007). Chelation of heavy metals by soluble phenols restrict their sorption and form soluble metal complexes, which increase leaching, whereas it has a greater impact on the redox potential of soil (de la Fuente et al., 2008). Metal polluted soil of calcareous nature applied with olive husk responded by enhancing Mn soluble fraction and EDTA-extractable Zn and Pb due to oxidation of water-soluble phenols (de la Fuente et al., 2008).

Contrarily, the complex nature and alkaline properties of waste from the olive oil industry help decrease the solubility and bioavailability of trace elements in polluted areas, ultimately improving the pH of the soil (Moreno-Jiménez et al., 2013; Pardo et al., 2014). Composts from olive oil processing units have a greater content of humic substances, which makes them more viable for interaction with trace elements through different processes such as adsorption, chelation, or retention within the exchange complex.

The metals mobility followed the order Pb > Cu > Cd > Zn in soil that was treated with maize straw, hay, or peat (Hernandez-Soriano and Jimenez-Lopez, 2012). Likewise, used mushroom substrate, silkworm dung, or fly ash could be used alone or with combinations to immobilize the Cd and hence decreasing its availability in the soil for soybean crop (Wang et al., 2014). Additionally, biomass amendments are also helpful in the phytoremediation of explosive materials. The microbial community within the soil profile is held accountable for the degradation of contaminants; hence, some organic materials are applied in order to intensify the process of biological degradation (Feng et al., 2017; Lamichhane et al., 2012).

Molasses, a vital end product of sugar processing units, could be a potential source of minerals, vitamins, and carbon for the soil microbiological community. The royal demolition explosive is usually degraded by the soil bacteria, and their mass volume accelerates by the addition of molasses. Remarkably, without molasses addition, phytoremediation showed no increment in the degradation of royal demolition explosives. Lamichhane et al. (2012) found molasses along with a combination of Guinea grass (a plant used for phytoremediation) showed a higher rate of accumulation of both royal demolition explosive and high-melting explosive. Although, there are no evidence in favor of molasses promoting plant biomass production.

On the other hand, waste residues from both the industrial and agricultural sectors may consist of a notable concentration of trace elements, metal, or toxic compounds that may pose serious environmental or public safety hazards

(McBride, 2003). Consequently, the chemical composition of waste containing such hazardous contaminants should be assessed before their treatment (Sardar et al., 2021b).

7.6 Efficacy of biomass amendments assisting phytoremediation

The resource utilization of municipal and industrial by-products has shown noteworthy results to enhance soil productivity and increase reclamation efficacy (Clemente et al., 2015; Palansooriya et al., 2020). For several hundred years, organic manure has proven its ability to sustain or even enhance agricultural land's physicochemical and biological properties (Mondal et al., 2019). In recent years, it is reported that plants' vegetative and reproductive growth have a positive relation with organic wastes, biosolids, and byproducts of sugar and other industries. Biomass amendments have been used in a consortium to act as phytoremediators, and positive results were found. However, the role of organic matter has a certain level of discrepancy (Chirakkara and Reddy, 2015; de Abreu et al., 2012; Ogundiran et al., 2018).

Palansooriya et al. (2020) reported in a review paper that animal wastes immobilization efficiency for Pb varies from 22% to 99% with an average >60%, whereas biochar immobilized >90% Pb efficiently. Contrastingly, the results obtained for immobilization of Cd were lower than that of Pb and were 60%, 55%, and 15% animal wastes, biochar, and compost, respectively. However, biochar and compost efficiently immobilized Cd and had ≥80 % immobilized rate, while, in the case of Zn, the efficiency of immobilization ranged between 70% and 77%. According to the findings as mentioned earlier, it is reported that most of the soil amendments can be used as immobilizing agents against potentially toxic elements. Along with other amendments, biochar's application enhances the immobilization of Pb, Cd, and Zn contaminated soils. However, special attention must be paid before using certain biomass amendments such as municipal solid waste and biosolids because such amendments could deteriorate the soil quality. Therefore, before their use against toxic elements, a thorough investigation of organic amendments is highly recommended.

Numerous experiments have been conducted to assess the beneficial effects of sugar beet residues against plant-assisted metal extraction. Medina et al. (2006) reported that this lignocellulosic substance, extracted from sugar and fermentation by *Aspergillus niger*, is enriched with polysaccharides and available phosphorus, and its application efficiently increases the phytoextraction of several metals (e.g., Fe, Zn, B, Cd, Cr, and Ni). The phenomenon behind enhanced metal phytoextraction can be partially explained by the elevated microbial community structure in rhizospheres' soil because of microorganisms' ability to counter heavy metals and enhance their availability to crops (Azcón et al., 2009). Moreover, the application of sugar beet residues supports native microbial communities, increases soil aggregates stability, and enhances macronutrients supply (e.g., N, P, and K). The biochemical features of soil are also dependent on foreign additives; thus, the enzymatic activity of soil increases (e.g., hydrolases and oxidoreductases).

Among other biochelators, based on efficiency, composted sewage sludge ranked first. Studies have reported that composted sewage sludge produced noteworthy results during phytoextraction of Cd and Ni. Results showed that sunflower's phytoextraction with composted sludge as an amendment enhances the metal solubilization and ultimately increases its bioavailability (Liphadzi and Kirkham, 2006). In a study, castor oil plant (*Ricinus communis* L.) as phytoextractor with peat and filter cake biomass amendments were investigated. Findings revealed that the addition of peat reduces the required phytoextraction time by 50%. The benefits of using peat might be associated with humic substances and fulvic acid; these substances play a key role in making the complexation of trace metals and significantly affect their bioavailability. Several studies reported that the addition of processed and unprocessed farmyard manure stimulates boron translocation to aerial parts of the plants. Concurrently, phytoextraction of toxic elements is indirectly influenced by pH fluctuation (de Abreu et al., 2012). In another research, *Quercus rubra* L. and *Pinus nigra* Arn. tree species were planted in contaminated soil, and paper industry waste material was also incorporated in it. The enhancement in pH due to paper waste treatment decreased the uptake of Pb and Cd (Doichinova and Velizarova, 2013).

Literature also highlights the drawbacks of few biomass amendments. For example, biochar that is linked with the loss of soil organic matter, accumulation of xenobiotics, and trace metals in the soil (Wardle et al., 2008). The major drawback of biochar is its negative impact on herbicides, which directly reduces crop production (Jones et al., 2011; Lehmann et al., 2011). Moreover, the addition of biochar pollutes the agroecosystem, periurban environment, air quality, and human health (Mara dos Santos Barbosa et al., 2006). Therefore, it is the need of the hour to explore the short-term and long-term behavior of biomass amendments and their environmental complications.

Farmer's community is more concerned about financial contrarians of any amendment; therefore, before selecting and recommending material, policymakers and scientists must focus on its economic outcomes. Previous reports show that natural materials are far better than synthetic agents and are also cost-effective. Meanwhile, due to the lack of environmental regulations in the developing world, environmentally friendly materials must be enforced and encouraged to be used as immobilizing agents.

7.7 Mechanism of biomass amendments assisting phytoremediation

Phytoremediation technique observed various mechanisms that depend on the pollutant type, the bioavailability of the pollutant, and soil properties (Kumar Yadav et al., 2018; Laghlimi et al., 2015). Absorption, transport, and translocation are the main processes involved in metallic pollutant accumulation in plants. The roots of the plants are the key organ for pollutant absorption (Etim, 2012). Mostly the micro and macrobiota of soil associated with the roots play a substantial role in the mobilization of inorganic ions and degradation of organic compounds (Erdei, 2005).

The basic mechanism behind phytoremediation is the tolerance of plants against pollutants toxicity (Ali et al., 2013; Jabeen et al., 2009; Singh et al., 2019). The transportation of metal ions from root to shoot occurs through the xylem transpiration stream (Page et al., 2006). The process involved in heavy metal phytotolerance is the active transport of metal ions into vacuole after binding with cell wall and chelation through the activation of metal-binding peptides and the formation of metallocomplex (Mejáre and Bülow, 2001; Memon and Schröder, 2009). Identified transporters that facilitate metal ions uptake in plants include ZNT1, ZIP1-4, COPT1, IRT1, AtVramp1/3/4, and LCT1 on the plasma membrane-cytosol interface; ZAT, ABC type, AtMRP, CAX2, HMT1, and RAN1 (Shah and Nongkynrih, 2007). The mechanism behind metal detoxification in a plant is the chelation of metallic ions by low molecular weight proteins such as metallothioneins and phytochelatins (Memon and Schröder, 2009). Amino acids, phytochelatins, metallothionein, and glutathione are the metal-binding molecules and plays a significant role in alleviating metal toxicity by making metal chelate complexes (Ali et al., 2013; Singh et al., 2019). Hyperaccumulators have the capability to store a high concentration of heavy metals in the metabolically less active organelle, the vacuole, and thus overcome the toxicity pressure by regulating major metabolic activities (Ali et al., 2013; Leitenmaier and Kupper, 2011). In contrast, the effective metal detoxification occurs by segregating formed metal chelates into subcellular location through transporters (Leitenmaier and Küpper, 2013).

Application of organic amendments like compost or peat reduces the bioavailable fraction of metals in soil by absorption as metal ions form stable complexes with humic substances present in humified organic material (Clemente and Bernal, 2006). Whereas, in the organic pollutant case, organic amendments reduce their toxicity by following two mechanisms: sorption and aiding in biodegradation. Increased soil organic matter due to organic/biomass amendment application provides many different sites for sorption and interaction with other substances. High sorption and less leaching of various organic contaminants was observed in soil with high carbon contents and organically amended soils (Briceño et al., 2007; Cabrera et al., 2014; Wu et al., 2013). However, dissolve organic materials present in applied biomass amendments act as a surfactant, for example, rhamnolipids, which increase the leaching hazard of polar and hydrophobic pesticides by increasing their mobile fraction (Li et al., 2005; Wan et al., 2015). Therefore, the amount of dissolved organic matter present in a biomass amendment is an important factor while considering organically polluted soil remediation (Siedt et al., 2021). Degradation of pesticides can also increase with the application of biomass amendments because of the incorporation of microbes with these amendments' addition and the enhanced activity of indigenous ones due to nutrient availability (Dungan et al., 2003; Getenga, 2003).

Biodegradation and mineralization is the preferable remediation mechanism compared to immobilization by sorption since it finally eliminates the toxic compound. Adsorbed compounds have lesser chances of biodegradation. Whereas it is a reversible process, and any consequent remobilization opens the opportunity for further degradation. Therefore, at first, applied soil organic matter serves as a buffer to absorb pesticides and then gradually releases reversibly bound molecules (Siedt et al., 2021). Petroleum hydrocarbons degradation follows enzyme-specific biodegradation mechanisms. Biosurfactants and emulsifiers produced during aerobic and anaerobic enzymatic activity mediate the degradation process (Ossai et al., 2020; Varjani and Upasani, 2017). Compounds released by several plants, that are, flavonoids and polyphenols, facilitate the enzymatic process in polyaromatic hydrocarbon degrading organisms (Abdullah et al., 2020; Cristaldi et al., 2017).

7.8 Comparison of biomass amendments and other amendments

A comparison of potential soil amendments and efficacy to specific environmental pollutants, that is, toxic metallic, metallic nutrients, and organics, is enlightened in tables 7.1, 7.2, 7.3 for instructive evaluations of soil remediation's practicability.

7.9 Environmental risks and phytoremediated biomass handling

The environmental and commercial constraints of phytoextraction are among major concerns. The disposal of contaminated plant or seedlings requires a well-guided system, which, unfortunately, lacks in developing and underdeveloped countries. Their improper dumping leads to the accumulation of hazardous material in nonpolluted regions. This disposal is the actual cause of the nonpoint source of toxic element pollution in agricultural land. Before discarding, this dangerous biomass must be treated and properly disposed of at an appropriate site to counter such problems to minimize environmental risk. Another alarming issue is the improper biomass handling and low techniques to reduce its volume. For several decades, composting and compaction techniques have been adopted to reduce volume (Sardar et al., 2021). Song and Park (2017) reported that composting could efficiently reduce the biomass of wasted plant material; however, metal-contaminated plant biomass would still require further treatment before clearness. The thermochemical alteration process is another strategy to utilize phytoremediation biomass. Jiang, Lei et al. (2015) gave an economically feasible approach to treat biomass generation and its commercial utilization; the authors reviewed that this technique can be turned into a profit-making industry by commercial production of bio-ore through phyto-mining. This technique has several advantages, and its commercialization can enhance

TABLE 7.1 Amendments and plant species response in the remediation of toxic metallic contaminants.

A type	A dose	Plants	Time	C	Contamination levels	Efficacy (plant accumulation / reduction/immobilization)	References
Groundnut shell biochar	10 g kg^{-1}	*Moringa oleifera*	8 w	Pb	Lead slag contaminated soil	82.6% Pb bioavailable removal Pb uptake: root = 930–1540 mg kg^{-1} Shoot = 420–875 mg kg^{-1}	(Ogundiran et al., 2018)
Wheat straw biochar	1.5%, 3%, 6% (w/w)	*Medicago sativa* L.	90 d	Cd	3 mg kg^{-1}	Cd phytoextraction increased by 207%–64.64%	(Zhang et al., 2019)
Biochar (waste bioenergy industry)	1%, 2.5%, and 5% (w/w)	*Lycopersicon esculentum L.*	9 w	Cr	Tannery waste-polluted soil	87%–97% less Cr accumulation	(Herath et al., 2017)
Sunflower-poultry manure compost	10 g kg^{-1}	*Moringa oleifera*	8 w	Pb	Lead slag contaminated soil	75.5% Pb bioavailable removal in soil Pb uptake: root = 1220–2100 mg kg^{-1} Shoot = 620–1120 mg kg^{-1}	(Ogundiran et al., 2018)
Pig manure compost	5 g kg^{-1}	*Sedum alfredii*	3 m	Cd	31.65 mg kg^{-1}	Accumulation of Cd enhanced by 1.54 to 1.92	(Xiao et al., 2017)
Fermented cow manure	6%, 12%, 25% (w/w)	Sunflower	4 m	Pb	Wastewater contaminated soil	Remediation factor was 0.5–0.9	(Yazdanbakhsh et al., 2020)

(continued)

TABLE 7.1 (Cont'd)

A type	A dose	Plants	Time	C	Contamination levels	Efficacy (plant accumulation / reduction/immobilization)	References
Spent mushroom substrates	75 g kg^{-1}	*Sedum alfredii*	180 d	Cd	25.7 mg kg^{-1}	Cd extraction increased by 92% and 73%	(Yang et al., 2019)
Egg shell	1% (w/w)	*Sorghum bicolor*	15 w	Cd	Sewage effluent irrigated soil	245.1% increase in tissue conc.	(Shaheen et al., 2017)
Bone meal	25 g kg^{-1}	*Lupinus albus L.*	14 w	Pb	702 mg kg^{-1}	74.2% less conc. in shoot	(Houben et al., 2012)
Mussel shell	5% (w/w)	*Zea mays L.*	3 w	Pb	Army firing range soil	71.22% reduce Phytoavailability	(Ahmad et al., 2014)
Bone mill	1% (w/w)	*Sorghum bicolor*	15 w	Cd	Sewage effluent irrigated soil	111.5% increase in tissue conc.	(Shaheen, Shams, Khalifa, El-Dali, & Rinklebe, 2017)
Drinking water treatment residual	1% (w/w)	*Sorghum bicolor*	15 w	Cd	Sewage effluent irrigated soil	129.2% increase in tissue conc.	(Shaheen, Shams, Khalifa, El-Dali, & Rinklebe, 2017)
Urban wastewater treatment plant biosolids	6, 12, 25% (w/w)	Sunflower	4 m	Pb	wastewater contaminated soil	Remediation factor was 0.6-0.9	(Yazdanbakhsh, Alavi, Valadabadi, Karimi, & Karimi, 2020)
Phosphogypsum	1% (w/w)	*Sorghum bicolor*	15 w	Cd	Sewage effluent irrigated soil	121.7% increase in tissue conc.	(Shaheen, Shams, Khalifa, El-Dali, & Rinklebe, 2017)
$CaCO_3$	25 g kg^{-1}	*Lupinus albus L.*	14 w	Cd	33 mg kg^{-1}	82.2% less conc. in shoot	(Houben, Pircar, & Sonnet, 2012)

A, amendments; C, contaminants; $CaCO_3$, calcium carbonate; Cd, cadmium; Conc., concentration; Cr, chromium; Pb, lead.

TABLE 7.2 Amendments and plant species response in the remediation of organic contaminants.

A type	A dose	Plants	Time	C	Contamination levels	Efficacy (plant accumulation / reduction/immobilization)	References
Cow dung	0, 5 and 10 $g\ kg^{-1}$	*Tithonia diversifolia*	120 d	Light crude oil	75 $g\ kg^{-1}$	~85% reduction of total hydrocarbon	(Ruley et al., 2020)
Cow dung	0, 5 and 10 $g\ kg^{-1}$	*Sorghum arundinaceum*	120 d	Light crude oil	75 $g\ kg^{-1}$	~78% reduction of total hydrocarbon	(Ruley et al., 2020)
Cow dung	0, 5 and 10 $g\ kg^{-1}$	*Hyparrhenia rufa*	120 d	Light crude oil	75 $g\ kg^{-1}$	~90% reduction of total hydrocarbon	(Ruley et al., 2020)
Compost	5% w/w	*Lolium perenne*	90 d	Pyrene	50, 100, and 150 $mg\ kg^{-1}$	98%–99% degraded	(Wang et al., 2012)
Compost	5% w/w	*Medicago sativa*	90 d	Pyrene	50, 100, and 150 $mg\ kg^{-1}$	97%–99% degraded	(Wang et al., 2012)
Green garden waste material compost	5% v/v	*Lolium multiflorum*	75 d	Crude oil	3.4% (w/w)	70% degraded	(Hussain et al., 2018)
Green garden waste biochar	5% v/v	*Lolium multiflorum*	75 d	Crude oil	3.4% (w/w)	65% degraded	(Hussain et al., 2018)

A, amendments; C, contaminants.

TABLE 7.3 Amendments and plant species response in the remediation of metallic nutrients contaminants.

A type	A dose	Plants	Time	C	Contamination levels	Efficacy (plant accumulation /reduction/ immobilization)	References
Urban wastewater treatment plant biosolids	6,12, 25% (w/w)	Sunflower	4 m	Zn	wastewater contaminated soil	Remediation factor was 0.9–1.4	(Yazdanbakhsh et al., 2020)
Egg shell	1% (w/w)	*Sorghum bicolor*	15 w	Cu	Sewage effluent irrigated soil	110.6% increase in tissue conc.	(Shaheen et al., 2017)
Egg shell	1% (w/w)	*Sorghum bicolor*	15 w	Mn	Sewage effluent irrigated soil	38.5% decrease in tissue conc.	(Shaheen et al., 2017)
Egg shell	1% (w/w)	*Sorghum bicolor*	15 w	Ni	Sewage effluent irrigated soil	145.5% increase in tissue conc.	(Shaheen et al., 2017)
Fermented cow manure	6, 12, 25% (w/w)	Sunflower	4 m	Zn	wastewater contaminated soil	Remediation factor was 0.6–0.7	(Yazdanbakhsh et al., 2020)
Spent mushroom substrates biochar	30 kg ha^{-1}	*Sedum alfredii*	180 d	Zn	5108 mg kg^{-1}	Zn extraction increased by 12% and 15%	(Yang et al., 2019)
Bone mill	1% (w/w)	*Sorghum bicolor*	15 w	Cu	Sewage effluent applied soil	38.5% increase in tissue conc.	(Shaheen et al., 2017)
Bone mill	1% (w/w)	*Sorghum bicolor*	15 w	Fe	Sewage effluent applied soil	18.5% increase in tissue conc.	(Shaheen et al., 2017)

Bone mill	1% (w/w)	*Sorghum bicolor*	15 w	Mn	Sewage effluent applied soil	30.7% decrease in tissue conc.	(Shaheen et al., 2017)
Sugar beet factory lime	1% (w/w)	*Sorghum bicolor*	15 w	Ni	Sewage effluent applied soil	78.5% increase in tissue conc.	(Shaheen et al., 2017)
Activated charcoal	1% (w/w)	*Sorghum bicolor*	15 w	Mn	Sewage effluent applied soil	40.0% decrease in tissue conc.	(Shaheen et al., 2017)
Activated charcoal	1% (w/w)	*Sorghum bicolor*	15 w	Ni	Sewage effluent applied soil	126.9% increase in tissue conc.	(Shaheen et al., 2017)
Brick factory residual	1% (w/w)	*Sorghum bicolor*	15 w	Fe	Sewage effluent applied soil	91.9% increase in tissue conc.	(Shaheen et al., 2017)
Sulfur	1% (w/w)	*Sorghum bicolor*	15 w	Mn	Sewage effluent applied soil	1941.7% increase in tissue conc.	(Shaheen et al., 2017)
Fermented cow manure	6, 12, 25% (w/w)	Sunflower	4 m	Ni	wastewater contaminated soil	Remediation factor was 0.1–0.2	(Yazdanbakhsh et al., 2020)
Cement bypass kiln dust	1% (w/w)	*Sorghum bicolor*	15 w	Ni	Sewage effluent applied soil	99.2% increase in tissue conc.	(Shaheen et al., 2017)

A, amendments; C, contaminants; Conc., concentration; Cu, copper; Fe, iron; Mn, manganese; Ni, nickel; Zn, zinc.

its adaptability. In recent decades, combustion and gasification methods are the most adopted techniques for the commercial production of electric and thermal energy. Combustion is a simple and easy method, which includes the burning of plant biomass under controlled conditions. Several authors have reported that this method can reduce plant biomass to less than 5%, and final products can be appropriately dumped (Delplanque et al., 2013; Khan et al., 2009). In comparison, thermochemical energy conversion is a more advanced technique and has several economic advantages, for example, fodder and fertilizer production. In the gasification process, a series of steps are included, producing clean combustive gas at high thermal efficiencies (Saxena and Misra, 2010). A mixture of gases, called producer gas and pyro gas, may be combusted for energy production (thermal and electrical). The contaminated biomass can be effectively used in the brick kilns industry, and the production of non-leachable bricks will be a long-term solution for the storage of toxic elements. However, prior to adopting this method, the feasibility must be investigated in detail. Metals and compounds with high economic value and industrial consumption can be extracted from contaminated biomass for processing and producing other useful products. This will be a sustainable approach, and the metallurgic industry will exponentially grow. Although cost and profit ratio may be a major challenge, it can be resolved by policymakers and concerned authorities.

7.10 Conclusion and future strategies

The extensive use of soil amendments in phytoremediation is primarily designed to achieve the objective of pollutant-free soil. Based on the requirements, both inorganic and organic/biomass amendments may be used to increase the bioaccumulation of pollutants. The addition of biomass amendments provides two types of benefits in the phytoremediation of soil contaminants, either organic or inorganic. One is the enhancement of plant biomass by improving soil properties and enriching the soil with nutrients and microbial activities. The increase in plant biomass, especially hyperaccumulators, expresses the potential to accumulate toxic metals from contaminated soil. Secondly, it immobilized most of the bioavailable forms of metallic contaminants due to their high sorption capacity. Moreover, the compound released in root exudates and the humic acid component of the biomass amendment can stabilize toxic pollutants for a longer period. Considering soil organic pollutants, this type of amendments preferentially enhances the microbial activity that plays a significant role in the degradation of organic pollutants such as organochlorine pesticides, poly-aromatic hydrocarbons, phthalate esters pesticides, and hydrocarbons, etc. On the other hand, a contradiction in amendment response forces the scientific community to explore further the short-term and long-term behavior of amendments and their environmental complications. Furthermore, its quantity and application time must be carefully monitored and recorded. Future research should also focus on making better use of microbial diversity for plants and a better understanding of the complex interactions between pollutants in the rhizosphere, plant roots, soil, and microorganisms (bacteria and mycorrhizae). Special attention should also be paid to evaluate the remediation potential of an advanced approach, the use of extract and exudate from various biological material for pollutants decontamination in soil. This technique may be applied as plant-assisted phytoremediation or natural materials in the absence of remediating plant. Both options are believed to be cost-effective and environmentally friendly. Besides this, the development of novel technologies for the disposal of phytoremediator biomass should be the coming hot topic for the scientific community.

References

Abdullah, S.R.S., Al-Baldawi, I.A., Almansoory, A.F., Purwanti, I.F., Al-Sbani, N.H., Sharuddin, S.S.N, 2020. Plant-assisted remediation of hydrocarbons in water and soil: Application, mechanisms, challenges and opportunities. Chemosphere 247 (1), 125932.

Ahmad, M., Lee, S.S., Lee, S.E., Al-Wabel, M.I., Tsang, D.C.W., Ok, Y.S, 2017. Biochar-induced changes in soil properties affected immobilization/mobilization of metals/metalloids in contaminated soils. J. Soils. Sediments 17 (3), 717–730. https://doi.org/10.1007/s11368-015-1339-4.

Ahmad, M., Lee, S.S., Lim, J.E., Lee, S.-E., Cho, J.S., Moon, D.H., et al., 2014. Speciation and phytoavailability of lead and antimony in a small arms range soil amended with mussel shell, cow bone and biochar: EXAFS spectroscopy and chemical extractions. Chemosphere 95, 433–441.

Ajila, C.M., Brar, S.K., Verma, M., Rao, U.J.S.P, 2012. Sustainable solutions for agro processing waste management: an overviewEnvironmental Protection Strategies for Sustainable Development, 1st. ed. Springer, Netherlands, pp. 65–109, 10.1007/978-94-007-1591-2.

Ali, H., Khan, E., Sajad, M.A., 2013. Phytoremediation of heavy metals-concepts and applications. Chemosphere 91 (7), 869–881.

Awad, Y.M., Kim, S.C., Abd El-Azeem, S.A.M., Kim, K.H., Kim, K.R., Kim, K., et al., 2014. Veterinary antibiotics contamination in water, sediment, and soil near a swine manure composting facility. Environ. Earth Sci. 71, 1433–1440. https://doi.org/10.1007/s12665-013-2548-z.

Azbar, N., Bayram, A., Filibeli, A., Muezzinoglu, A., Sengul, F., Ozer, A., 2004. A review of waste management options in olive oil production. Crit. Rev. Environ. Sci. Technol. 34, 209–247.

Azcón, R., Medina, A., Roldán, A., Biró, B., Vivas, A., 2009. Significance of treated agrowaste residue and autochthonous inoculates (Arbuscular mycorrhizal fungi and Bacillus cereus) on bacterial community structure and phytoextraction to remediate soils contaminated with heavy metals. Chemosphere 75, 327–334. https://doi.org/10.1016/j.chemosphere.2008.12.029.

Białobrzewski, I., Mikš-Krajnik, M., Dach, J., Markowski, M., Czekała, W., Głuchowska, K., 2015. Model of the sewage sludge-straw composting process integrating different heat generation capacities of mesophilic and thermophilic microorganisms. Waste Manage. 43, 72–83. https://doi.org/10.1016/j.wasman.2015.05.036.

Bolan, N.S., Adriano, D.C., Duraisamy, P., Mani, A., 2003. Immobilization and phytoavailability of cadmium in variable charge soils. III. Effect of biosolid compost addition. Plant Soil 256, 231–241. https://doi.org/10.1023/A:1026288021059.

Bolan, N.S., Adriano, D.C., Natesan, R., Koo, B.-J., 2003. Effects of Organic Amendments on the Reduction and Phytoavailability of Chromate in Mineral Soil. J. Environ. Qual. 32, 120–128. https://doi.org/10.2134/jeq2003.1200.

Briceño, G., Palma, G., Durán, N., 2007. Influence of organic amendment on the biodegradation and movement of pesticides. Crit. Rev. Environ. Sci. Technol. 37 (3), 233–271.

Brown, S.L., Chaney, R.L., Angle, J.S., Ryan, J.A., 1998. The phytoavailability of cadmium to lettuce in long-term biosolids-amended soils. J. Environ. Qual. 27 (5), 1071–1078.

Burgos, P., Madejón, P., Cabrera, F., Madejón, E., 2010. By-products as amendment to improve biochemical properties of trace element contaminated soils: effects in time. Int. Biodeter. Biodegr. 64 (6), 481–488.

Cabrera, A., Cox, L., Spokas, K., Hermosín, M.C., Cornejo, J., Koskinen, W.C., 2014. Influence of biochar amendments on the sorption–desorption of aminocyclopyrachlor, bentazone and pyraclostrobin pesticides to an agricultural soil. Sci.Total Environ. 470 (1), 438–443.

Chen, S., Chao, L., Sun, L.N., Sun, T.H., 2013. Plant-microorganism combined remediation for sediments contaminated with heavy metals. Adv. Mater. Res. 610 (1), 1223–1228.

Chen, X., Li, H., Liu, X., Zhang, X., Liang, X., He, C., et al., 2016. Combined remediation of pyrene-contaminated soil with a coupled system of persulfate oxidation and phytoremediation with ryegrass. Environ. Sci. Pollut. Res. 23 (20), 20672–20679.

Chirakkara, R.A., Reddy, K.R., 2015. Biomass and chemical amendments for enhanced phytoremediation of mixed contaminated soils. Ecol. Eng. 85, 265–274. https://doi.org/10.1016/j.ecoleng.2015.09.029.

Clemente, R, Paredes, C., Bernal, M.P., 2007. A field experiment investigating the effects of olive husk and cow manure on heavy metal availability in a contaminated calcareous soil from Murcia (Spain). Agriculture. Ecosyst. Environ. 118 (1–4), 319–326.

Clemente, R., Bernal, M.P., 2006. Fractionation of heavy metals and distribution of organic carbon in two contaminated soils amended with humic acids. Chemosphere 64 (8), 1264–1273.

Clemente, R., Pardo, T., Madejón, P., Madejón, E., Bernal, M.P., 2015. Food byproducts as amendments in trace elements contaminated soils. Food Res. Int. 73, 176–189. https://doi.org/10.1016/j.foodres.2015.03.040.

Cristaldi, A., Conti, G.O., Jho, E.H., Zuccarello, P., Grasso, A., Copat, C., et al., 2017. Phytoremediation of contaminated soils by heavy metals and PAHs. A brief review. Environ. Technol. Innov. 8, 309–326. https://doi.org/10.1016/j.eti.2017.08.002.

de Abreu, C.A., Coscione, A.R., Pires, A.M., Paz-Ferreiro, J., 2012. Phytoremediation of a soil contaminated by heavy metals and boron using castor oil plants and organic matter amendments. Journal of Geochemical Exploration 123, 3–7. https://doi.org/10.1016/j.gexplo.2012.04.013.

de la Fuente, C., Clemente, R., Bernal, M.P., 2008. Changes in metal speciation and pH in olive processing waste and sulphur-treated contaminated soil. Ecotoxicol. Environ. Saf. 70 (2), 207–215.

Delplanque, M., Collet, S., Del Gratta, F., Schnuriger, B., Gaucher, R., Robinson, B., et al., 2013. Combustion of Salix used for phytoextraction: The fate of metals and viability of the processes. Biomass Bioenerg 49, 160–170. https://doi.org/10.1016/j.biombioe.2012.12.026.

Doichinova, V., Velizarova, E., 2013. Reuse of paper industry wastes as additives in phytoremediation of heavy metals polluted substrates from the spoil banks of the Kremikovtsi region, Bulgaria. Procedia Environ. Sci. 18, 731–736. https://doi.org/10.1016/j.proenv.2013.04.099.

Dungan, R.S., Gan, J., Yates, S.R., 2003. Accelerated degradation of methyl isothiocyanate in soil. Water Air Soil Pollut. 142 (1), 299–310.

Erdei, L., 2005. Phytoremediation as a program for decontamination of heavy-metal polluted environment. Acta Biol. (Szeged) 49 (1–2), 75–76.

Farooqi, Z.U.R., Sabir, M., Zeeshan, N., Naveed, K., Hussain, M.M, 2018. Enhancing carbon sequestration using organic amendments and agricultural practices. Carbon Capture, Utilization and Sequestration, 1st ed. IntechOpen, London, UK, pp. 17–34.

Farooqi, Z.U.R., Sabir, M., Zia-Ur-Rehman, M., Hussain, M.M, 2020. Mitigation of climate change through carbon sequestration in agricultural soilsClimate Change and Agroforestry Systems: Adaptation and Mitigation Strategies, 1st ed. CRC-Apple academic press, Florida, USA. pp. 87–118.

Feng, N.-X., Yu, J., Zhao, H.-M., Cheng, Y.-T., Mo, C.-H., Cai, Q.-Y., et al., 2017. Efficient phytoremediation of organic contaminants in soils using plant–endophyte partnerships. Sci. Total Environ. 583 (1), 352–368.

Fischer, K., Bipp, H.-P., Riemschneider, P., Leidmann, P., Bieniek, D., Kettrup, A., 1998. Utilization of biomass residues for the remediation of metal-polluted soils. Environ. Sci. Technol. 32 (14), 2154–2161.

Getenga, Z.M., 2003. Enhanced mineralization of atrazine in compost-amended soil in laboratory studies. Bull. Environ. Contam. Toxicol. 71 (5), 933–941.

Gong, X., Xu, X., Gong, Z., Li, X., Jia, C., Guo, M., et al., 2015. Remediation of PAH-contaminated soil at a gas manufacturing plant by a combined two-phase partition system washing and microbial degradation process. Environ. Sci. Pollut. Res. 22 (16), 12001–12010.

Herath, I., Iqbal, M.C.M., Al-Wabel, M.I., Abduljabbar, A., Ahmad, M., Usman, A.R.A., et al., 2017. Bioenergy-derived waste biochar for reducing mobility, bioavailability, and phytotoxicity of chromium in anthropized tannery soil. J. Soils Sediments 17 (3), 731–740.

Hernandez-Soriano, M.C., Jimenez-Lopez, J.C., 2012. Effects of soil water content and organic matter addition on the speciation and bioavailability of heavy metals. Sci. Total Environ. 423, 55–61. https://doi.org/10.1016/j.scitotenv.2012.02.033.

Houben, D., Pircar, J., Sonnet, P., 2012. Heavy metal immobilization by cost-effective amendments in a contaminated soil: effects on metal leaching and phytoavailability. J. Geochem. Explor. 123 (1), 87–94.

Hu, B., Ai, Y., Jin, J., Hayat, T., Alsaedi, A., Zhuang, L., et al., 2020. Efficient elimination of organic and inorganic pollutants by biochar and biochar-based materials. Biochar 2 (1), 47–18 doi:10.1007/s42773-020-00044-4.

Huang, M., Zhu, Y., Li, Z., Huang, B., Luo, N., Liu, C., et al., 2016. Compost as a soil amendment to remediate heavy metal-contaminated agricultural soil: mechanisms, efficacy, problems, and strategies. Water Air Soil Pollut 227, 359. https://doi.org/10.1007/s11270-016-3068-8.

Hussain, F., Hussain, I., Khan, A.H.A., Muhammad, Y.S., Iqbal, M., Soja, G., et al., 2018. Combined application of biochar, compost, and bacterial consortia with Italian ryegrass enhanced phytoremediation of petroleum hydrocarbon contaminated soil. Environ Exp. Bot. 153 (1), 80–88.

Jabeen, R., Ahmad, A., Iqbal, M., 2009. Phytoremediation of heavy metals: physiological and molecular mechanisms. Bot. Rev. 75, 339–364. https://doi.org/10.1007/s12229-009-9036-x.

Jeffery, S., Verheijen, F.G.A., van der Velde, M., Bastos, A.C, 2011. A quantitative review of the effects of biochar application to soils on crop productivity using meta-analysis. Agric. Ecosyst. Environ. 144 (1), 175–187.

Jiang, J., Xu, R., Jiang, T., Li, Z., 2012. Immobilization of Cu (II), Pb (II) and Cd (II) by the addition of rice straw derived biochar to a simulated polluted Ultisol. J. Hazard. Mater. 229, 145–150.

Jiang, Y., Lei, M., Duan, L., Longhurst, P., 2015. Integrating phytoremediation with biomass valorisation and critical element recovery: a UK contaminated land perspective. Biomass Bioenerg 83, 328–339. https://doi.org/10.1016/j.biombioe.2015.10.013.

Jones, D.L., Edwards-Jones, G., Murphy, D.V., 2011. Biochar mediated alterations in herbicide breakdown and leaching in soil. Soil Biol Biochem. 43, 804–813. https://doi.org/10.1016/j.soilbio.2010.12.015.

Khan, A.A., de Jong, W., Jansens, P.J., Spliethoff, H., 2009. Biomass combustion in fluidized bed boilers: Potential problems and remedies. Fuel Process. Technol. 90, 21–50. https://doi.org/10.1016/j.fuproc.2008.07.012.

Kumar Yadav, K., Gupta, N., Kumar, A., Reece, L.M., Singh, N., Rezania, S., et al., 2018. Mechanistic understanding and holistic approach of phytoremediation: A review on application and future prospects. Ecol. Eng. 120, 274–298. https://doi.org/10.1016/j.ecoleng.2018.05.039.

Laghlimi, M., Baghdad, B., El Hadi, H., Bouabdli, A., 2015. Phytoremediation mechanisms of heavy metal contaminated soils: a review. Open J. Ecol. 5 (8), 375.

Lamichhane, K.M., Babcock, R.W., Turnbull, S.J., Schenck, S., 2012. Molasses enhanced phyto and bioremediation treatability study of explosives contaminated Hawaiian soils. J. Hazard. Mater. 243, 334–339. https://doi.org/10.1016/j.jhazmat.2012.10.043.

Lehmann, J., Rillig, M.C., Thies, J., Masiello, C.A., Hockaday, W.C., Crowley, D., 2011. Biochar effects on soil biota - a review. Soil Biol. Biochem. 43, 1812–1836. https://doi.org/10.1016/j.soilbio.2011.04.022.

Leitenmaier, B., Kupper, H., 2011. Cadmium uptake and sequestration kinetics in individual leaf cell protoplasts of the Cd/Zn hyperaccumulator Thlaspi caerulescens. Plant Cell Environ. 34, 208–219. https://doi.org/10.1111/j.1365-3040.2010.02236.x.

Leitenmaier, B., Küpper, H., 2013. Compartmentation and complexation of metals in hyperaccumulator plants. Front. Plant Sci. 4, 374. https://doi.org/10.3389/fpls.2013.00374.

Li, K., Xing, B., Torello, W.A., 2005. Effect of organic fertilizers derived dissolved organic matter on pesticide sorption and leaching. Environ. Pollut. 134, 187–194.

Liphadzi, M.S., Kirkham, M.B., 2006. Chelate-assisted heavy metal removal by sunflower to improve soil with sludge. J. Crop Improv. 16, 153–172. https://doi.org/10.1300/J411v16n01_11.

Liu, G., Meng, J., Huang, Y., Dai, Z., Tang, C., Xu, J., 2020. Effects of carbide slag, lodestone and biochar on the immobilization, plant uptake and translocation of As and Cd in a contaminated paddy soil. Environ. Pollut. 266 (1), 115194.

Liu, L., Wang, S., Guo, X., Wang, H., 2019. Comparison of the effects of different maturity composts on soil nutrient, plant growth and heavy metal mobility in the contaminated soil. J. Environ. Manage. 250 (1), 109525.

Liu, X., Zhang, A., Ji, C., Joseph, S., Bian, R., Li, L., et al., 2013. Biochar's effect on crop productivity and the dependence on experimental conditions—a meta-analysis of literature data. Plant Soil 373 (1), 583–594.

Liu, Y., Zeng, G., Zhong, H., Wang, Z., Liu, Z., Cheng, M., et al., 2017. Effect of rhamnolipid solubilization on hexadecane bioavailability: enhancement or reduction? J. Hazard. Mater. 322, 394–401. https://doi.org/10.1016/j.jhazmat.2016.10.025.

Lu, D., Wang, L., Yan, B., Ou, Y., Guan, J., Bian, Y., et al., 2014. Speciation of Cu and Zn during composting of pig manure amended with rock phosphate. Waste Manage. 34, 1529–1536. https://doi.org/10.1016/j.wasman.2014.04.008.

Madejón, E., De Mora, A.P., Felipe, E., Burgos, P., Cabrera, F., 2006. Soil amendments reduce trace element solubility in a contaminated soil and allow regrowth of natural vegetation. Environ. Pollut. 139 (1), 40–52.

Mara dos Santos Barbosa, J., Ré-Poppi, N., Santiago-Silva, M, 2006. Polycyclic aromatic hydrocarbons from wood pyrolyis in charcoal production furnaces. Environ. Res. 101, 304–311. https://doi.org/10.1016/j.envres.2006.01.005.

McBride, M.B., 2003. Toxic metals in sewage sludge-amended soils: Has promotion of beneficial use discounted the risks? Adv. Environ. Res. 8, 5–19. https://doi.org/10.1016/S1093-0191(02)00141-7.

Medina, A., Vassileva, M., Barea, J.M., Azcón, R., 2006. The growth-enhancement of clover by Aspergillus-treated sugar beet waste and Glomus mosseae inoculation in Zn contaminated soil. Appl. Soil Ecol. 33, 87–98. https://doi.org/10.1016/j.apsoil.2005.08.003.

Mejáre, M., Bülow, L., 2001. Metal-binding proteins and peptides in bioremediation and phytoremediation of heavy metals. Trends Biotechnol. 19, 67–73. https://doi.org/10.1016/S0167-7799(00)01534-1.

Memon, A.R., Schröder, P., 2009. Implications of metal accumulation mechanisms to phytoremediation. Environ. Sci. Pollut. Res. 16, 162–175. https://doi.org/10.1007/s11356-008-0079-z.

Mench, M., Lepp, N., Bert, V., Schwitzguébel, J.-P., Gawronski, S.W., Schröder, P., et al., 2010. Successes and limitations of phytotechnologies at field scale: outcomes, assessment and outlook from COST Action 859. J. Soils Sediments 10 (6), 1039–1070.

Mirabella, N., Castellani, V., Sala, S., 2014. Current options for the valorization of food manufacturing waste: a review. J. Clean. Prod. 65 (1), 28–41.

Mondal, M.M.A., Ahmed, F., Nabi, K.M.E., Al Noor, M.M., Mondal, M.T.R, 2019. Performance of organic manures on the growth and yield of red amaranth (Amaranthus tricolor) and soil properties. Res. Agric. Livest. Fish. 6 (2), 263–269.

Moreno-Jiménez, E., Clemente, R., Mestrot, A., Meharg, A.A., 2013. Arsenic and selenium mobilisation from organic matter treated mine spoil with and without inorganic fertilisation. Environ. Pollut. 173 (1), 238–244.

Ogundiran, M.B., Mekwunyei, N.S., Adejumo, S.A., 2018. Compost and biochar assisted phytoremediation potentials of Moringa oleifera for remediation of lead contaminated soil. J. Environ. Chem. Eng. 6, 2206–2213. https://doi.org/10.1016/j.jece.2018.03.025.

Ossai, I.C., Ahmed, A., Hassan, A., Hamid, F.S., 2020. Remediation of soil and water contaminated with petroleum hydrocarbon: A review. Environ. Technol. Innov. 17, 100526. https://doi.org/10.1016/j.eti.2019.100526.

Page, V., Weisskopf, L., Feller, U., 2006. Heavy metals in white lupin: uptake, root-to-shoot transfer and redistribution within the plant. New Phytol. 171 (2), 329–341.

Palansooriya, K.N., Shaheen, S.M., Chen, S.S., Tsang, D.C.W., Hashimoto, Y., Hou, D., et al., 2020. Soil amendments for immobilization of potentially toxic elements in contaminated soils: A critical review. Environ. Int. 134, 105046. https://doi.org/10.1016/j.envint.2019.105046.

Pardo, T., Bernal, M.P., Clemente, R., 2014. Efficiency of soil organic and inorganic amendments on the remediation of a contaminated mine soil: I. Effects on trace elements and nutrients solubility and leaching risk. Chemosphere 107 (1), 121–128.

Park, J.H., Lamb, D., Paneerselvam, P., Choppala, G., Bolan, N., Chung, J.W., 2011. Role of organic amendments on enhanced bioremediation of heavy metal(loid) contaminated soils. J. Hazard. Mater. 185 (2–3), 549–574. https://doi.org/10.1016/j.jhazmat.2010.09.082.

Riaz, U., Murtaza, G., Saifullah, Farooq, M., Aziz, H., Qadir, A.A., Mehdi, S.M., Qazi, M.A., 2020. Chemical fractionation and risk assessment of trace elements in sewage sludge generated from various states of Pakistan. Environ Sci Pollut Res Int 27 (32), 39742–39752. doi:10.1007/s11356-020-07795-4. 31993896.

Ruley, J.A., Amoding, A., Tumuhairwe, J.B., Basamba, T.A., Opolot, E., Oryem-Origa, H., 2020. Enhancing the phytoremediation of hydrocarbon-contaminated soils in the Sudd Wetlands, South Sudan, using organic manure. Appl. Environ. Soil Sci., 2020. doi:https://doi.org/10.1155/2020/4614286, 4614286.

Sardar, M.F., Zhu, C., Geng, B., Ahmad, H.R., Song, T., Li, H., 2021. The fate of antibiotic resistance genes in cow manure composting: shaped by temperature-controlled composting stages. Bioresour. Technol. 320 (Part B), 124403. doi:https://doi.org/10.1016/j.biortech.2020.124403.

Sardar, M.F., Zhu, C., Geng, B., Huang, Y., Abbasi, B., Zhang, Z., Song, T., Li, H., 2021b. Enhanced control of sulfonamide resistance genes and host bacteria during thermophilic aerobic composting of cow manure. Environ Pollut 275, 116587. doi:10.1016/j.envpol.2021.116587. 33582626.

Saxena, P., Misra, N., 2010. Remediation of heavy metal contaminated tropical landSoil Heavy Metals. Springer, Berlin, Heidelberg, 19, pp. 431–477, https://doi.org/10.1007/978-3-642-02436-8_19.

Shah, K., Nongkynrih, J.M., 2007. Metal hyperaccumulation and bioremediation. Biol. Plant. 51, 618–634. https://doi.org/10.1007/s10535-007-0134-5.

Shaheen, S.M., Shams, M.S., Khalifa, M.R., El-Dali, M.A., Rinklebe, J., 2017. Various soil amendments and environmental wastes affect the (im)mobilization and phytoavailability of potentially toxic elements in a sewage effluent irrigated sandy soil. Ecotoxicol. Environ. Saf. 142, 375–387. https://doi.org/10.1016/j.ecoenv.2017.04.026.

Siedt, M., Schäffer, A., Smith, K.E.C., Nabel, M., Roß-Nickoll, M., van Dongen, J.T, 2021. Comparing straw, compost, and biochar regarding their suitability as agricultural soil amendments to affect soil structure, nutrient leaching, microbial communities, and the fate of pesticides. Sci. Total Environ. 751, 141607. https://doi.org/10.1016/j.scitotenv.2020.141607.

Singh, A., Prasad, S.M., 2014. Effect of agro-industrial waste amendment on Cd uptake in Amaranthus caudatus grown under contaminated soil: an oxidative biomarker response. Ecotoxicol. Environ. Saf. 100 (1), 105–113.

Singh, R., Jha, A.B., Misra, A.N., Sharma, P., 2019. Adaption mechanisms in plants under heavy metal stress conditions during phytoremediationPhytomanagement of Polluted Sites: Market Opportunities in Sustainable Phytoremediation, 1st ed. Elsevier, India, pp. 329–360. https://doi.org/10.1016/B978-0-12-813912-7.00013-2.

Song, B., Zeng, G., Gong, J., Zhang, P., Deng, J., Deng, C., et al., 2017. Effect of multi-walled carbon nanotubes on phytotoxicity of sediments contaminated by phenanthrene and cadmium. Chemosphere 172, 449–458. https://doi.org/10.1016/j.chemosphere.2017.01.032.

Song, U., Park, H., 2017. Importance of biomass management acts and policies after phytoremediation. J. Ecol. Environ. 41 (13). https://doi.org/10.1186/s41610-017-0033-4.

Varjani, S.J., Upasani, V.N., 2017. A new look on factors affecting microbial degradation of petroleum hydrocarbon pollutants. Int. Biodeter. Biodegr. 120 (1), 71–83.

Villaescusa, I., Fiol, N., Martínez, M., Miralles, N., Poch, J., Serarols, J., 2004. Removal of copper and nickel ions from aqueous solutions by grape stalks wastes. Water Res. 38 (4), 992–1002.

W.H.O., 2016. News Release. Geneva. Retrieved from. http://www.who.int/news-room/detail/15-03-2016-an-estimated-12-6-million-deaths-each-year-are-attributable-to-unhealthyenvironments.

Wan, J., Meng, D., Long, T., Ying, R., Ye, M., Zhang, S., et al., 2015. Simultaneous removal of lindane, lead and cadmium from soils by rhamnolipids combined with citric acid. PLoS One 10, e0129978.

Wang, F., Ouyang, W., Hao, F., Lin, C., Song, N., 2014. In situ remediation of cadmium-polluted soil reusing four by-products individually and in combination. J. Soils Sediments 14, 451–461. https://doi.org/10.1007/s11368-013-0827-7.

Wang, J., Shi, L., Zhai, L., Zhang, H., Wang, S., Zou, J., et al., 2021. Analysis of the long-term effectiveness of biochar immobilization remediation on heavy metal contaminated soil and the potential environmental factors weakening the remediation effect: a review. Ecotoxicol. Environ. Saf. 207, 111261. https://doi.org/10.1016/j.ecoenv.2020.111261.

Wang, M.C., Chen, Y.T., Chen, S.H., Chien, S.W.C., Sunkara, S.V, 2012. Phytoremediation of pyrene contaminated soils amended with compost and planted with ryegrass and alfalfa. Chemosphere 87, 217–225.

Wardle, D.A., Nilsson, M.C., Zackrisson, O., 2008. Fire-derived charcoal causes loss of forest humus. Science 320, 629. https://doi.org/10.1126/science.1154960.

Wu, M., Pan, B., Zhang, D., Xiao, D., Li, H., Wang, C., et al., 2013. The sorption of organic contaminants on biochars derived from sediments with high organic carbon content. Chemosphere 90, 782–788.

Xiao, W., Li, D., Ye, X., Xu, H., Yao, G., Wang, J., et al., 2017. Enhancement of Cd phytoextraction by hyperaccumulator Sedum alfredii using electrical field and organic amendments. Environ. Sci. Pollut. Res. 24, 5060–5067.

Yang, W., Wang, S., Ni, W., Rensing, C., Xing, S., 2019. Enhanced Cd-Zn-Pb-contaminated soil phytoextraction by Sedum alfredii and the rhizosphere bacterial community structure and function by applying organic amendments. Plant Soil 444, 101–118. https://doi.org/10.1007/s11104-019-04256-x.

Yazdanbakhsh, A., Alavi, S.N., Valadabadi, S.A., Karimi, F., Karimi, Z., 2020. Heavy metals uptake of salty soils by ornamental sunflower, using cow manure and biosolids: a case study in Alborz city, Iran. Air Soil Water Res. 13. https://doi.org/10.1177/1178622119898460.

Zeng, G., Chen, M., Zeng, Z., 2013. Risks of neonicotinoid pesticides. Science 340 (6139), 1403. https://doi.org/10.1126/science.340.6139.1403-a.

Zhang, C., Lai, C., Zeng, G., Huang, D., Yang, C., Wang, Y., et al., 2016. Efficacy of carbonaceous nanocomposites for sorbing ionizable antibiotic sulfamethazine from aqueous solution. Water Res. 95, 103–112. https://doi.org/10.1016/j.watres.2016.03.014.

Zhang, M., Wang, J., Bai, S.H., Zhang, Y., Teng, Y., Xu, Z., 2019. Assisted phytoremediation of a co-contaminated soil with biochar amendment: Contaminant removals and bacterial community properties. Geoderma 348, 115–123. https://doi.org/10.1016/j.geoderma.2019.04.031.

Zhang, X., Wang, H., He, L., Lu, K., Sarmah, A., Li, J., et al., 2013. Using biochar for remediation of soils contaminated with heavy metals and organic pollutants. Environ. Sci. Pollut. Res. 20, 8472–8483.

Chapter 8

Chemical amendments and phytoremediation

Amauri Ponce-Hernández[a], Paola Lucero Pérez[c], Angel Josabad Alonso-Castro[b], Candy Carranza-Álvarez[c]
[a]*Faculty of Chemical Sciences, UASLP, San Luis Potosí, Mexico,* [b]*Department of Pharmacy, University of Guanajuato, Guanajuato, Mexico,* [c]*Faculty of Professional Studies Zona Huasteca, UASLP, San Luis Potosí, Mexico*

8.1 Introduction

Due to population growth in the world, anthropogenic activities such as industrialization, urbanization, transportation, agriculture, and inappropriate waste disposal methods have increased (Leštan et al. 2008). This has caused an increase in the contamination of environmental resources by heavy metals (HM), becoming a major problem in various parts of the world (Suman et al., 2018). It is well known that HM are not degradable by biological or chemical processes, so they persist in soil or water for a long period (Yan et al., 2020). Due to their persistence, HM are introduced into the food chain by plants and ultimately accumulate in the human body through biomagnification, causing cell malfunction and toxicity (Rehman et *al.*, 2017). For this reason, it is extremely important to apply remediation measures to prevent HM from remaining in the soil, atmosphere, and water. Although there is a wide variety of techniques to clean soil and water contaminated with HM. These physicochemical, mechanical, and biological techniques often have limitations, such as high cost, damage to the soil ecosystem, and the secondary pollutants introduction (DalCorso et al., 2019).

The phytoremediation is a strategy that uses plants to capture, remove, and degrade pollutants *in situ* in soils, surface, and groundwater (McCutcheon and Jørgensen, 2008). It is considered to be an environmentally friendly, safe, and potentially inexpensive technique (Tangahu et al., 2011). In the phytoremediation, there are absorption mechanisms for organic and inorganic compounds. For organic compounds, plants use mechanisms such as rhizofiltration, rhizodegradation, phytostabilization, phytodegradation, and phytovolatilization. For inorganic compounds, they use phytostabilization, rhizofiltration, phytoextraction, phytoaccumulation, and phytovolatilization (Fig. 8.1) (Prasad and De Oliveira Freitas, 2003).

Phytoremediation. DOI: https://doi.org/10.1016/B978-0-323-89874-4.00013-3

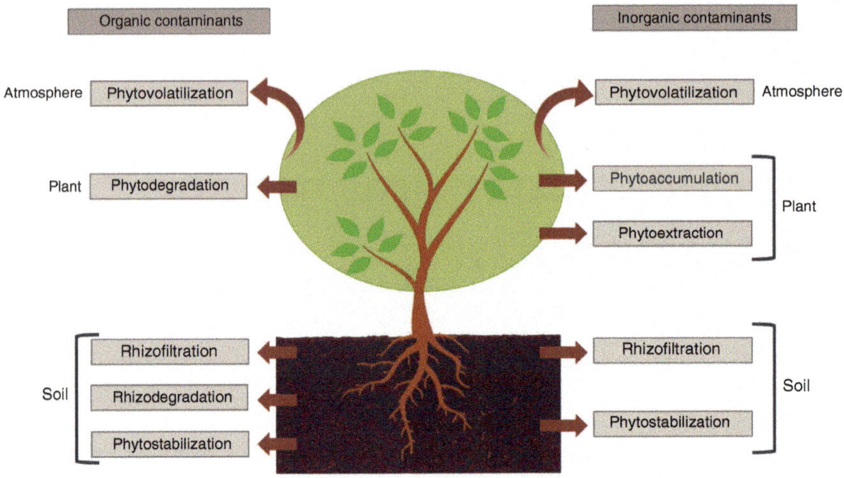

FIG. 8.1 Absorption mechanisms for organic and inorganic pollutants (Tangahu et al., 2011).

However, the absorption of toxic elements is limited by their availability. For this, various chemical chelating compounds have been used that improve the availability of HM, thus facilitating the uptake of metals (Hasan et al., 2019). Some compounds that have been reported to improve HM absorption include citric acid (CA), ethylene diamine tetra acetic acid (EDTA), exogenous glutathione, sulfur (S) compounds, nitric oxide (NO), among others. It is necessary to employ strategies that increase the efficiency of heavy metal phytoremediation. These strategies can be through the use of chelating chemicals. This review describes and discusses the benefits of some chelating chemicals that enhance the absorption of toxic elements.

8.2 Chemical amendments

Some chemical amendments are used to improve phytoremediation processes because they mainly modulate the pH, redox conditions, and the chemical species present in the rhizosphere. Some of the compounds used as chemical amendments are described in Fig. 8.2.

8.2.1 Citric acid

CA is a weak tricarboxylic organic acid, has a low molecular weight, and is excreted by plants through the root in stressful situations caused by HM. It is known that CA is capable of forming complexes with metal ions, thus it can be useful for the fixation, mobilization, and HM phytoextraction (Chen et al., 2003). Some studies have shown that CA has been used as an environmental strategy to improve the HM phytoextraction (Blaylock et al., 1997; Turgut et al., 2003). In accordance with Amir et al. (2020), the addition of CA improves

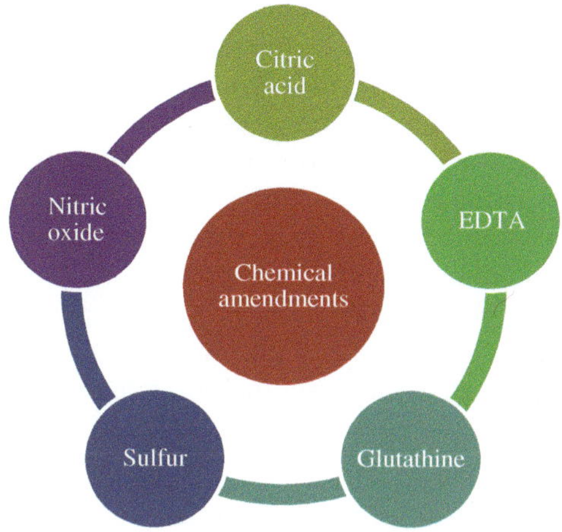

FIG. 8.2 Compounds used as chemical amendments in the phytoremediation.

plant physiology, antioxidant enzymes activity, and electrolyte leakage in *Typha latifolia* plants exposed to stressful conditions caused by metals, such as Pb and Hg. In addition, in this same research, it is mentioned that the accumulation and concentration of these metals increases in different plant tissues (roots, stems, and leaves) with the addition of CA.

On the other hand, Ehsan et al. (2014) indicate in his work that the concentration and accumulation of Cd in roots, stem, and leaves in *Brassica napus* plants are increased with the application of CA, in addition, CA reduced the toxicity of Cd, because the plants presented a reduction in stress oxidative and an increase in its biomass, photosynthetic pigments, and growth (Fig. 8.3). In

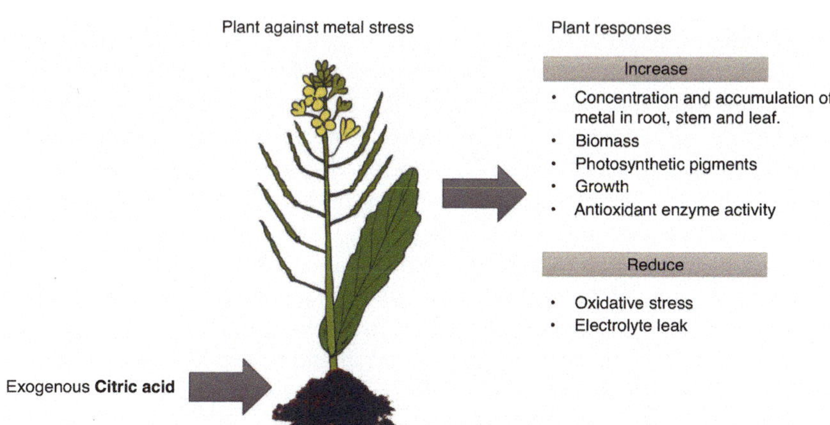

FIG. 8.3 Citric acid effects in mitigating the heavy metal toxicity in plants.

the case of Copper (Cu), studies have shown an increase in the accumulation and absorption in tissues of *B. napus* and *Lemna minor* L. when CA is added to the medium (Zaheer et al., 2015; Sallah-Ud-Din et al., 2017). Furthermore, the toxicity effects caused by Cu in CA-assisted *B. napus* and *L. minor* plants were significantly reduced when compared to control plants. In the case of Nickel (Ni), it has been reported that the addition of CA improves phytoextraction, increasing the concentrations of this metal in leaves, stems, and roots in *Mentha piperita* plants. In addition, a significant improvement was observed in the biochemical and physiological responses of *M. piperita* plants exposed to nickel when AC was applied (Khair et al., 2020). According to previously discussed studies, CA can be considered a useful tool to increase the phytoextraction of metals in soils and water.

8.2.2 Ethylene diamine tetra acetic acid

EDTA is a synthetic agent with a strong chelating capacity for various metal ions (Salt et al., 1998). EDTA is a chelating agent used for phytoremediation. Chelates increase the toxic elements bioavailability, induce greater absorption, and translocation from the roots to the stem and leaves of the plants (Liphadzi et al., 2003). It is well known that EDTA is an effective agent that improves the HM accumulation in plants (Iqbal et al., 2012). The work of Meighan et al. (2011) indicated that in *Helianthus annuus* plants, EDTA promoted the transfer of Cd from the roots to the shoots. In addition, the translocation factor increased up to 100 times by applying a threefold excess of EDTA, which resulted in high concentrations of Cd in plant leaves. Furthermore, the efficiency of the translocation depends mainly on two factors, which are the type of metal and the plant (Evangelou et al., 2007). In some studies, it has been proposed that in hydroponic media, EDTA can chelate Pb and later the complex (EDTA-Pb) is captured, transported, and accumulated in the leaves (Vassil et al., 1998). On the other hand, it has been suggested that free EDTA is absorbed by the roots and immediately forms complexes with metals, which are subsequently transported to the shoots (Evangelou et al., 2007). In the study by Muhammad et al. (2009), it is indicated that the concentrations of HM, such as Cd, Pb, and Cr increase in shoots of *T. latifolia* when EDTA is added. However, the application of EDTA caused adverse effects for plants due to a reduction in height, biomass, and delayed growth of the plants. Some of the negative effects of EDTA that have been reported are that EDTA hardly degrades, which can contribute to environmental damage caused by leaching of metals into groundwater, and also microorganisms on-site can be affected (Araujo do Nacimento and Xing, 2006). In Fig. 8.4, an EDTA-assisted metal absorption mechanism is shown. Finally, Salt et al. (1998) indicate that the chelating agent should be applied at the moment of maximum biomass of the crop, to minimize that the synthetic chelating exceed the metal absorption capacity in the plant.

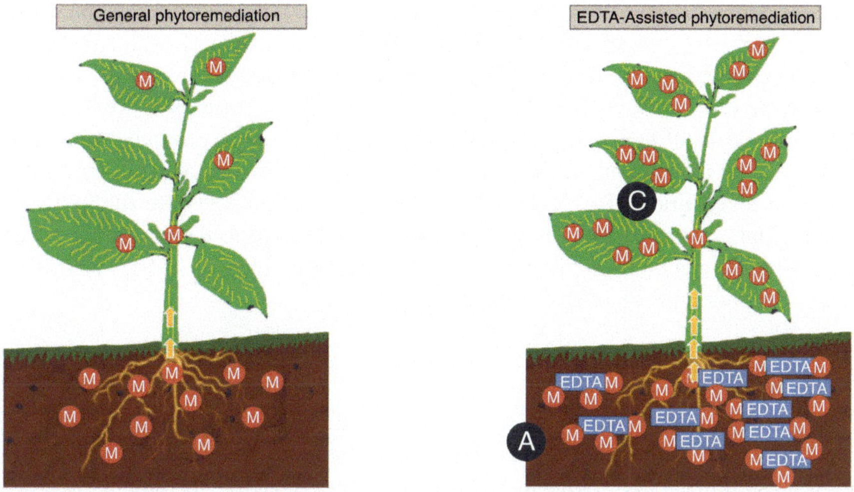

FIG. 8.4 The EDTA Assisted Metal Absorption Mechanism. (A) EDTA increases HM availability in the soil, (B) EDTA improves absorption of metals, and (C) EDTA improves translocation of HM. *EDTA*, Ethylene diamine tetra acetic acid; *HM*, heavy metals (Hasan et al., 2019; Jiang et al., 2019).

8.2.3 Glutathione

Glutathione (GSH) is a compound formed by low molecular weight thiol groups. Glu, Cys, Gly amino acids make up the chemical structure of GHS. It is a precursor compound of phytochelatins and is chelating that transports metals to vacuoles (Cobbett and Goldsbrough, 2002). GHS's role in heavy metal detoxification has had a major impact because the thiol group has a high affinity for metals, which leads to complex formation GSH-metal, reducing the metal toxicity in cells (Seth et al., 2012). Due to the importance of GSH on metal detoxification, various studies have chosen to apply exogenous GSH in plants to improve the relief of metal toxicity and increase the absorption capacity. Qiu et al.'s (2013) study showed that the addition of exogenous GSH significantly alleviated the effect caused by Cr^{6+} on the chloroplasts ultrastructure, on growth and on the nutrients absorption in rice plants. Additionally, exogenous GSH reduced plant stress by increasing the content of secreted organic acids, which are generally released by plants during stressful conditions. GSH has been shown to affect the ions Cr forms at the root and their distribution at the subcellular level in aerial tissue. Ding et al. (2017) indicated that the application of exogenous GSH in *Populus x canescens* plants exposed to Cd^{2+} showed in a better absorption of Cd in the roots, a greater amount of Cd in plants, and a lower amount of H_2O_2 induced by Cd. Furthermore, GSH induced the transcription of positively regulated genes that probably have a role in the capture, bioaccumulation, and detoxification of Cd in the root. Fang et al. (2020) indicate that the addition of exogenous GSH improved the Cd concentration in roots. However, the Cd concentration in the aerial parts and the translocation from roots to shoots was reduced. Furthermore, when exogenous GSH was applied, the concentration of Cd in the root cell wall showed an increase; however, in the soluble fraction the concentration decreased. In the case of Arsenic (As), Jung et al. (2019) reported that the use of exogenous GSH on *Oryza sativa* seedlings exposed to As resulted in a decrease in oxidative stress and increased the antioxidant defense system, the content of antioxidant ascorbic acid, and the translocation of As from the roots to the aerial parts. These studies show that the application of exogenous GSH in plants exposed to metals can alleviate toxicity and even increase the absorption and accumulation of metals. Resulting in a good option for an effective phytoremediation.

8.2.4 Sulfur

S is an essential macronutrient of plant metabolism. Normally, plants capture S in the form of sulfate (SO_4^{-2}). Furthermore, S is part of the chemical structure of some amino acids, antioxidants, vitamins, coenzymes, secondary metabolites, prosthetic groups, and phytochelatins (Zhou et al., 2018). It is known that in plants, S is related to biotic and abiotic stresses and has an important action in the metals and metalloids detoxification. Ding et al. (2019), showed that, by adding S in *Arabidopsis thaliana* previously treated with Cr, the content of metal and malonyldialdehyde was reduced. On the contrary, the chlorophyll and the antioxidant enzymes activity increased when S was added. In the work of Zhong et al. (2011), it is indicated that the addition of S produced a decrease in the As concentration in the root and grain of *B. napus* L.

On the other hand, Cui et al. (2004) evaluated the S activity on the solubility of Pb, Cd, and Zn in soil and its absorption in *Zea mays* plants. The results show that the S application decreases the soil pH and the Zn and Cd solubility. In addition, the concentration of Zn and Cd increased in shoots and roots of *Z. mays* with the application of S. However, the addition of S did not alleviate the negative effects caused by toxic elements on the plant shoot and root biomass. Likewise, Sun et al. (2020) mention that S supplementation decreased soil pH and enzyme activity. For the metals Cd, Cr, Zn, and Cu, it was observed that their availability in soil presented an increase. In this same study, it is indicated that the S application at 10 g/Kg showed an increase in the biomass of the corn plants and an increase in the concentration of Cd in the root and stem. Therefore, it is recommended to use adequate concentrations of S to guarantee positive biomass results and an increase in Cd uptake in plants. According to previously analyzed works, the S application can be used as an economical and suitable tool for metals and metalloids phytoremediation (Sun et al., 2020).

8.2.5 Nitric oxide

NO is a compound that acts in various physiological and biochemical processes in plants. These processes include germination, growth of roots and pollen tubes,

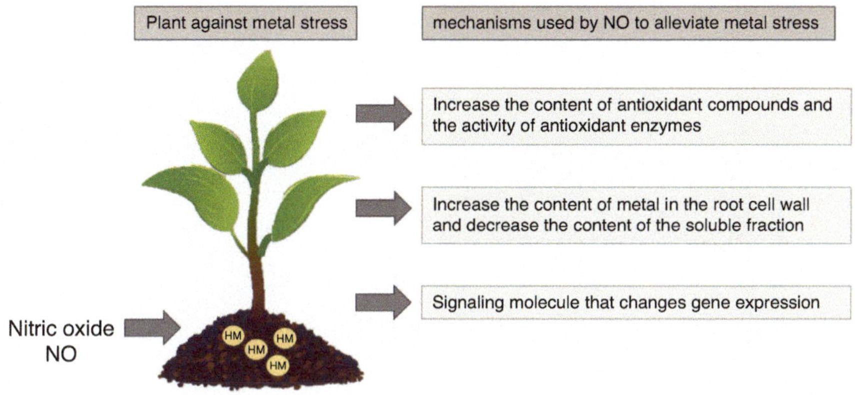

FIG. 8.5 Nitric oxide mechanism to alleviate stress in plants exposed to HM (heavy metals).

flowering, stomatal closure, among others (Wilson et al., 2007). Furthermore, NO is known to be a signaling molecule in plants in response to biotic and abiotic stress (Saxena and Shekhawat, 2013). Currently, the effects of exogenous NO on the toxic elements detoxification in plants have been widely informed. Research indicates that exogenous NO application generally prevents oxidative stress and decreases the metals accumulation in various plant species under metallic stress. Some reports have proposed mechanisms by which NO alleviate the plants stress induced by HM. Some mechanisms of NO reported are (1) NO can increase the content of antioxidant molecules and the action of antioxidant enzymes, (2) NO can increase the content of HM in the root and decrease the content in the soluble fraction, and (3) the NO can have activity as a signaling molecule that produces alterations in gene expression in case of metallic stress (Fig. 8.5) (Xiong et al., 2010). On the other hand, it should be considered that the effect of NO on the responses of plants depends on the concentration, since studies have reported that a high concentration of NO could produce harmful effects and a low concentration produces beneficial effects (Wink and Mitchell, 1998).

8.3 Chemical compounds used to improve the heavy metals phytoremediation

Although there are many studies on the use of plants to remove HM, few studies have been reported on the use of chemical amendments to enhance phytoremediation processes. Table 8.1 presents a review of different chemical compounds used to promote the HM uptake under *in situ*, *in vitro*, and artificial conditions. The information was obtained from electronic sources such as Google Scholar, PubMed, SciELO, Science Direct, Scopus, Springer Link, and Web of Science. Keywords such as phytoremediation, CA, chemical amendments were used to search for scientific works in the previously mentioned databases. Publications within the dates from 2017 to 2020 were considered in this review.

TABLE 8.1 Chemical amendments are used to improve the elements phytoextraction by some plant species.

Plant species used for phytoremediation	*In situ/ In vitro*	Chemical condition	Removed metals	References
Sunflower genotype (Faisalabad Hybrid FH-614)	*In vitro*	Citric acid (0, 2.5, and 5.0 mM)	Cr (0, 5, 10 and 20 mg kg)	Farid et al. (2017)
Brassica napus L. genotype Faisal Canola (RBN-03060)	Artificial solution	Citric acid (0, 5.0 mM)	Ni (0, 100, and 500 μM)	Farid et al. (2017)
Odontarrhena muralis	*In vitro*	Citric acid (0, 5, 10, 20, and 40 mM)	Ni, Co, Cr, Fe, Mn, and Zn	do Nascimento et al. (2020)
Typha latifolia L.	Artificial solution	Citric acid (5 mM)	Ni (0, 100, and 500 μM)	Amir et al. (2020)
Helianthus annuus L.	Artificial solution	Glutamic acid (0 and 2.5 mM)	Ag (0, 100, 250, and 500 μM)	Farid et al. (2018)
Lepidium sativum L.	*In vitro*	Nitrilotriacetic acid (NTA) 5 mg/kg	Hg	Smolińska (2020)
Alternanthera bettzickiana	*In vitro*	Acetic acid (2.5 mM)	Pb (0, 2.5, 5, 7.5 and 10 mM)	Latif et al. (2020)
Solanum nigrum L.	*In vitro*	Liquid amino acid fertilizer (LAAF, waste proteins from hydrolysates of animal carcasses) and polyaspartate (PASP)(5.0 mg/kg) and EDTA (20mM)	Cd and Pb	He et al. (2019)
Sedum alfredii	*In vitro*	S,S-ethylenediamine disuccinic acid (EDDS), citric acid, and oxalic acid (1.0 mM/kg and 2.5 mM/kg)	Cd, Pb and Zn.	Liang et al. (2019)
Helianthus annuus L.	Artificial Solution	EDTA and NTA (0.25, 0.5, 1.0, and 2.0 g kg)	Cu (100, 250, 500 and 1000 mg kg)	Mishra et al. 2020)

Vigna radiata (L.) R. Wilczek	Artificial Solution	EDTA and NTA (0.25, 0.5, 1.0, and 2.0 g kg)	Cu (100, 250, 500 and 1000 mg kg)	Mishra et al. (2020)
Pennisetum glaucum (L.) R. Br.	Artificial Solution	EDTA and NTA (0.25, 0.5, 1.0, and 2.0 g kg)	Cu (100, 250, 500 and 1000 mg kg)	Mishra et al. (2020)
Fagopyrum esculentum	*In situ*	Citric acid (5 and 50 mM)	Pb	Braud et al. (2019)
Fagopyrum esculentum	*In vitro*	Citric acid (5 and 50 mM)	Pb	Braud et al. (2019)
Cichorium intybus L.	*In vitro*	EDTA (5 mM), NTA (5 mM), and citric acid (60 mM)	Cr, Cu, Pb, and Zn	Bursztyn Fuentes et al. (2018)
Ricinus communis L.	*In vitro*	EDTA (5 mM), NTA (5 mM), and citric acid (60 mM)	Cr, Cu, Pb, and Zn	Bursztyn Fuentes et al. (2018)
Brassica juncea, Coss	*In vitro*	EDTA, citric acid and oxalic acid (5 mM and 10 mM)	Cd and Zn	Guo et al. (2019)
Pelargonium hortorum	Artificial solution	EDTA (0–5mM)	Pb (0–1500 mg kg) and Cd (0–150 mg kg)	Gul et al. (2019)
Pelargonium zonale	Artificial solution	EDTA (0–5mM)	Pb (0–1500 mg kg) and Cd (0–150 mg kg)	Gul et al. (2019)
Raphanus sativus L.	*In vitro*	EDTA (5 mM)	Zn and Pb	Chaturvedi et al. 2019)
Brassica oleracea L.	*In vitro*	EDTA (5 mM)	Zn and Pb	Chaturvedi et al. (2019)
Dahlia variabilis	Artificial solution	Sodium dodecyl Sulphate (SDS) (2.0 mM) and EDTA (2.5 mM)	Cd (0, 10, 25, and 100 mg kg) and Pb	Alzahrani et al, (2020)
Lantana Camara	*In vitro*	S, S]-ethylenediaminedisuccinic acid (EDDS) and methy glycinedi-acetic acid (MGDA) (10 mM)	Cu, Zn, Cr and Mn	Pandey and Bhattacharya (2018)
Zea mays L.	*In vitro*	EDTA (0 and 0.5 mM)	Ni (0 and 40 μM)	Tipu et al. (2020)

(continued)

TABLE 8.1 (Cont'd)

Plant species used for phytoremediation	*In situ/In vitro*	Chemical condition	Removed metals	References
Corchorus capsularis L.	Artificial solution	Citric acid and EDTA (3mM)	Cu (80 μM)	Saleem et al. 2020)
Coronopus didymus L.	Artificial solution	EDDS (1,2,3,4 mM)	Ni (30, 50, and 70 mg kg)	Sidhu et al. (2018)
Trigonella foenum-graceum L.	Artificial solution	Ascorbic acid (5mM)	Cu (100, 250 and 500 mM)	Fatima et al. (2020)
Helianthus annuus L.	Artificial solution	Glutamic acid (5mM)	Cr (1, 2, and 5 mg kg)	Farid et al. (2020)
Salix viminalis L.	Artificial solution	Citric acid (20mM/kg)	Cd (3 and 6ppm)	Arsenov et al. (2019)
Salix matsudana Koidz	Artificial solution	Citric acid (20mM/kg)	Cd (3 and 6ppm)	Arsenov et al. (2019)
Salix alba L.	Artificial solution	Citric acid (20mM/kg)	Cd (3 and 6ppm)	Arsenov et al. (2019)
Solanum nigrum	*In vitro*	Indole-3-acetic acid (IAA) (0, 10, 20, or 40 mgL)	Cd	Ran et al. (2020)
Solanum melongena	*In vitro*	Indole-3-acetic acid (IAA) (0, 10, 20, or 40 mgL)	Cd	Ran et al. (2020)
Pelargonium hortorum	*Artificial solution*	EDTA (0–5 mM/kg), citric acid (0–10 mM/kg), ammonium nitrate (0–10 mM/kg), and TiO_2 NPs (0–100 mg kg)	Cd (25–150 mg kg)	Gul et al. (2019)

Credit: Own elaboration.

8.4 Conclusions

The use of some compounds as chemical amendments notably favors phytoremediation processes, by increasing the HM availability in the soil and the potential of plants to remove them. Among the main advantages offered by the use of chemical amendments in plant remediation is: (1) the detoxification of metals in plants, (2) prevention of oxidative stress by toxic elements, (3) modification of soil pH, increased enzyme activity, biomass, photosynthetic pigments, and growth, (4) increased the translocation, (5) increased the availability of heavy metal chelating. The use of chemical amendments can increase the environmental and economic potential of phytoremediation, favoring a greater accumulation of HM, and reducing the toxic effects on plant species.

References

Alzahrani, Y., Alharby, H.F., Hakeem, K.R., Alsamadany, H., 2020. Modulating effect of EDTA and SDS on growth, biochemical parameters and antioxidant defense system of Dahlia variabilis grown under cadmium and lead-induced stress. Notulae Botanicae Horti Agrobotanici Cluj-Napoca 48 (2), 906–923. https://doi.org/10.15835/nbha48211909.

Amir, W., Farid, M., Ishaq, H.K., Farid, S., Zubair, M., Alharby, H.F., Bamagoos, A.A., Rizwan, M., Raza, N., Hakeem, K.R., Ali, S., 2020. Accumulation potential and tolerance response of Typha latifolia L. under citric acid assisted phytoextraction of lead and mercury. Chemosphere 257. https://doi.org/10.1016/j.chemosphere.2020.127247.

Araujo do Nacimento, C.W., Xing, B., 2006. Phytoextraction: a review on enhanced metal availability and plant accumulation. Sci. agric. (Piracicaba, Braz.) (3), 63. http://dx.doi.org/10.1590/S0103-90162006000300014.

Arsenov, D., Nikolić, N., Borišev, M., Župunski, M., Orlović, S., Pilipović, A., Pajević, S., 2019. Greenhouse assessment of citric acid-assisted phytoremediation of cadmium by Willows (Salix spp.)—effect on photosynthetic performances and metal tolerance. Baltic Forest. 25 (2), 203–212. https://doi.org/10.46490/vol25iss2pp203.

Blaylock, M.J., Salt, D.E., Dushekov, O.Z., Gussman, C., Kapulnik, Y., Enley, B.D., Raskin, I., 1997. Enhanced accumulation of Pb in Indian mustard by soil-applied chelating agents. Environ. Sci. Technol. 31, 860–865.

Braud, A.M., Gaudin, P., Hazotte, A., Le Guern, C., Lebeau, T., 2019. Chelate-assisted phytoextraction of lead using Fagopyrum esculentum: laboratory vs. field experiments. Int. J. Phytoremed. 21 (11), 1072–1079. https://doi.org/10.1080/15226514.2019.1606778.

Bursztyn Fuentes, A.L., José, C., de los Ríos, A., do Carmo, L.I., de Iorio, A.F., Rendina, A.E., 2018. Phytoextraction of heavy metals from a multiply contaminated dredged sediment by chicory (Cichorium intybus L.) and castor bean (Ricinus communis L.) enhanced with EDTA, NTA, and citric acid application. Int. J. Phytoremed. 20 (13), 1354–1361. https://doi.org/10.1080/15226514.2018.1524826.

Chaturvedi, R., Favas, P., Pratas, J., Varun, M., Paul, M.S., 2019. EDTA-assisted metal uptake in Raphanus sativus L. and Brassica oleracea L.: assessment of toxicity and food safety. Bull. Environ. Contamination Toxicol. 103 (3), 490–495. https://doi.org/10.1007/s00128-019-02651-9.

Chen, Y.X., Lin, Q., Luo, Y.M., He, Y.F., Zhen, S.J., Yu, Y.L., Tian, G.M., Wong, M.H., 2003. The role of citric acid on the phytoremediation of heavy metal contaminated soil. Chemosphere 50, 807–811.

Cobbett, C., Goldsbrough, P., 2002. Phytochelatins and metallothioneins: roles in heavy metal detoxification and homeostasis. Annu. Rev. Plant Biol. 53, 159–182. doi:10.1146/annurev.arplant.53.100301.135154.

Cui, Y., Dong, Y., Li, H., Wang, Q., 2004. Effect of elemental sulphur on solubility of soil heavy metals and their uptake by maize. Environ. Int. 30 (3), 323–328. https://doi.org/10.1016/S0160-4120(03)00182-X.

DalCorso, G., Fasani, E., Manara, A., Visioli, G., Furini, A., 2019. Heavy metal pollutions: state of the art and innovation in phytoremediation. Int. J. Mol. Sci. 20, 3412. doi:10.3390/ijms20143412.

Ding, G., Jin, Z., Han, Y., Sun, P., Li, G., Li, W., 2019. Mitigation of chromium toxicity in Arabidopsis thaliana by sulfur supplementation. Ecotoxicol. Environ. Safety 182, 109379. https://doi.org/10.1016/j.ecoenv.2019.109379.

Ding, S., Ma, C., Shi, W., Liu, W., Lu, Y., Liu, Q., Luo, Zhi-Bin, 2017. Exogenous glutathione enhances cadmium accumulation and alleviates its toxicity in Populus × canescens. Tree Physiol. 37 (12), 1697–1712. https://doi.org/10.1093/treephys/tpx132.

do Nascimento, C.W.A., Hesterberg, D., Tappero, R., Nicholas, S., da Silva, F.B.V., 2020. Citric acid-assisted accumulation of Ni and other metals by Odontarrhena muralis: implications for phytoextraction and metal foliar distribution assessed by μ-SXRF. Environ. Pollut. 260, 114025. https://doi.org/10.1016/j.envpol.2020.114025.

Ehsan, S., Ali, S., Noureen, S., Mahmood, K., Farid, M., Ishaque, W., Shakoor, M.B., Rizwan, M., 2014. Citric acid assisted phytoremediation of cadmium by Brassica napus L. Ecotoxicol. Environ. Safety 106, 164–172. https://doi.org/10.1016/j.ecoenv.2014.03.007.

Evangelou, M.W.H., Ebel, M., Schaeffer, A, 2007. Chelate assisted phytoextraction of heavy metals from soil. Effect, mechanism, toxicity, and fate of chelating agents. Chemosphere 68, 989–1003. doi:10.1016/j.chemosphere.2007.01.062.

Fang, Z., Hu, Z., Yin, X., Song, G., Cai, Q., 2020. Exogenous glutathione alleviation of Cd toxicity in Italian Ryegrass (Lolium multiflorum) by modulation of the Cd absorption, subcellular distribution, and chemical form. Int. J. Environ. Res. Public Health 17 (21), 8143. https://doi.org/10.3390/ijerph17218143.

Farid, M., Ali, S., Rizwan, M., Ali, Q., Abbas, F., Bukhari, S.A.H., Saeed, R., Wu, L, 2017. Citric acid assisted phytoextraction of chromium by sunflower; morpho-physiological and biochemical alterations in plants. Ecotoxicol. Environ. Safety 145, 90–102. https://doi.org/10.1016/j.ecoenv.2017.07.016.

Farid, M., Ali, S., Rizwan, M., Saeed, R., Tauqeer, H.M., Sallah-Ud-Din, R., Azam, A., Raza, N., 2017. Microwave irradiation and citric acid assisted seed germination and phytoextraction of nickel (Ni) by Brassica napus L.: morpho-physiological and biochemical alterations under Ni stress. Environ. Sci. Pollut. Res. 24 (26), 21050–21064. https://doi.org/10.1007/s11356-017-9751-5.

Farid, M., Ali, S., Zubair, M., Saeed, R., Rizwan, M., Sallah-Ud-Din, R., Azam, A., Ashraf, R., Ashraf, W., 2018. Glutamic acid assisted phyto-management of silver-contaminated soils through sunflower; physiological and biochemical response. Environ. Sci. Pollut. Res. 25 (25), 25390–25400. https://doi.org/10.1007/s11356-018-2508-y.

Farid, M., Farid, S., Zubair, M., Ghani, M.A., Rizwan, M., Ishaq, H.K., Alkahtani, S., Abdel-Daim, M.M., Ali, S., 2020. Glutamic acid-assisted phytomanagement of chromium contaminated soil by sunflower (Helianthus annuus L.): morphophysiological and biochemical alterations. Front. Plant Sci. 11, 1–14. https://doi.org/10.3389/fpls.2020.01297.

Fatima, A., Farid, M., Alharby, H.F., Bamagoos, A.A., Rizwan, M., Ali, S., 2020. Efficacy of fenugreek plant for ascorbic acid assisted phytoextraction of copper (Cu); a detailed study of

Cu induced morpho-physiological and biochemical alterations. Chemosphere 251, 126424. https://doi.org/10.1016/j.chemosphere.2020.126424.

Gul, I., Manzoor, M., Hashim, N., Yaqoob, K., Kallerhoff, J., Arshad, M., 2019. Comparative effectiveness of organic and inorganic amendments on cadmium bioavailability and uptake by Pelargonium hortorum. J. Soils Sed. 19 (5), 2346–2356 https://doi.org/10.1007/s11368-018-2202-1.

Gul, I., Manzoor, M., Silvestre, J., Rizwan, M., Hina, K., Kallerhoff, J., Arshad, M., 2019. EDTA-assisted phytoextraction of lead and cadmium by Pelargonium cultivars grown on spiked soil. Int. J. Phytoremed. 21 (2), 101–110. https://doi.org/10.1080/15226514.2018.1474441.

Guo, D., Ali, A., Ren, C., Du, J., Li, R., Lahori, A.H., Xiao, R., Zhang, Z., Zhang, Z., 2019. EDTA and organic acids assisted phytoextraction of Cd and Zn from a smelter contaminated soil by potherb mustard (Brassica juncea, Coss) and evaluation of its bioindicators. Ecotoxicol. Environ. Safety 167, 396–403. https://doi.org/10.1016/j.ecoenv.2018.10.038.

Hasan, M.M., Uddin, M.M., Ara-Sharmeen, I., Alharby, H.F., Alzahrazi, Y., Hakeem, K.R., Zhang, L., 2019. Assisting phytoremediation of heavy metals using chemical amendments. Plants (Basel) 8 (9), 295. doi:10.3390/plants8090295.

He, X., Zhang, J., Ren, Y., Sun, C., Deng, X., Qian, M., Hu, Z., Li, R., Chen, Y., Shen, Z., Xia, Y., 2019. Polyaspartate and liquid amino acid fertilizer are appropriate alternatives for promoting the phytoextraction of cadmium and lead in Solanum nigrum L. Chemosphere 237, 124483. https://doi.org/10.1016/j.chemosphere.2019.124483.

Iqbal, M., Bakht, J., Shafi, M., Ullah, R., 2012. Effect of heavy metal and EDTA application on heavy metal uptake and gene expression in different Brassica species. Afr. J. Biotechnol. 11 (30), 7649–7658. doi:10.5897/AJB11.3223.

Jiang, M., Liu, S., Li, Y., Li, X., Luo, Z., Song, H., Chen, Q., 2019. EDTA-facilitated toxic tolerance, absorption and translocation and phytoremediation of lead by dwarf bamboos. Ecotoxicol. Environ. Safety 170, 502–512. https://doi.org/10.1016/j.ecoenv.2018.12.020.

Jung, H.I., Kong, M.S., Lee, B.R., Kim, T.H., Chae, M.J, Lee, E.J., Jung, G.B., Lee, C.H., Sung, J.k., Kim, Y.H., 2019. Exogenous glutathione increases arsenic translocation into shoots and alleviates arsenic-induced oxidative stress by sustaining ascorbate–glutathione homeostasis in rice seedlings. Front. Plant Sci. 10, 1089. https://doi.org/10.3389/fpls.2019.01089.

Khair, K.U., Farid, M., Ashraf, U., Zubair, M., Rizwan, M., Farid, S., Ishaq, H.K., Iftikhar, U., Ali, S., 2020. Citric acid enhanced phytoextraction of nickel (Ni) and alleviate Mentha piperita (L.) from Ni-induced physiological and biochemical damages. Environ. Sci. Pollut. Res. 27, 27010–27022. https://doi.org/10.1007/s11356-020-08978-9.

Latif, U., Farid, M., Rizwan, M., Ishaq, H.K., Farid, S., Ali, S., El-Sheikh, M.A., Alyemeni, M.N., Wijaya, L., 2020. Physiological and biochemical response of alternanthera bettzickiana (Regel) G.Nicholson under Acetic Acid Assisted Phytoextraction of Lead. Plants, 9, 1084. https://doi.org/10.3390/plants9091084.

Leštan, D., Luo, C., Li, X., 2008. The use of chelating agents in the remediation of metal-contaminated soils: a review. Environ. Pollut. 153, 3–13. doi:10.1016/j.envpol.2007.11.015.

Liang, Y., Zhou, C., Guo, Z., Huang, Z., Peng, C., Zeng, P., Xiao, X., Xian, Z., 2019. Removal of cadmium, lead, and zinc from multi-metal–contaminated soil using chelate-assisted Sedum alfredii Hance. Environ. Sci. Pollut. Res. 26 (27), 28319–28327. https://doi.org/10.1007/s11356-019-06041-w.

Liphadzi, M.S., Kirkham, M.B., Mankin, K.R., Paulsen, G.M., 2003. EDTA assisted heavy metals uptake by poplar and sunflower grown at a long-term sewage-sludge farm. Plant Soil 257, 171–182.

McCutcheon, S.C., Jørgensen, S.E., 2008. Phytoremediation. Encyclopedia Ecol., 2751–2766. https://doi.org/10.1016/B978-008045405-4.00069-0.

Meighan, M.M., Fenus, T., Karey, E., MacNeil, J., 2011. The impact of EDTA on the rate of accumulation and root/shoot partitioning of cadmium in mature dwarf sunflowers. Chemosphere 83 (11), 1539–1545. https://doi.org/10.1016/j.chemosphere.2011.01.035.

Mishra, S.R., Chandra, R., Prusty, B.A.K, 2020. Chelate-assisted phytoaccumulation: growth of Helianthus annuus L., Vigna radiata (L.) R. Wilczek and Pennisetum glaucum (L.) R. Br. in soil spiked with varied concentrations of copper. Environ. Sci. Pollut. Res. 27 (5), 5074–5084. https://doi.org/10.1007/s11356-019-07257-6.

Muhammad, D., Chen, F., Zhao, J., Zhang, G., Wu, F., 2009. Comparison of EDTA- and citric acid-enhanced phytoextraction of heavy metals in artificially metal contaminated soil by Typha angustifolia. Int. J. Phytoremed. 11 (6), 558–574. doi:10.1080/15226510902717580.

Pandey, S.K., Bhattacharya, T., 2018. Effect of two biodegradable chelates on metals uptake, translocation and biochemical changes of Lantana Camara growing in fly ash amended soil. Int. J. Phytoremed. 20 (3), 214–224. https://doi.org/10.1080/15226514.2017.1365350.

Prasad, M.N.V., De Oliveira Freitas, H.M, 2003. Metal hyperaccumulation in plants—biodiversity prospecting forphytoremediation technology. Electron. J. Biotechnol. 6 (3), 110–146.

Qiu, B., Zeng, F., Cai, S., Wu, X., Haider, S.I., Wu, F., Zhang, G. (2013). Alleviation of chromium toxicity in rice seedlings by applying exogenous glutathione. J Plant Physiol. 170 (8), 772–779. Doi: 10.1016/j.jplph.2013.01.016.

Ran, J., Zheng, W., Wang, H., Wang, H., Li, Q., 2020. Indole-3-acetic acid promotes cadmium (Cd) accumulation in a Cd hyperaccumulator and a non-hyperaccumulator by different physiological responses. Ecotoxicol. Environ. Safety 191, 110213. https://doi.org/10.1016/j.ecoenv.2020.110213.

Rehman, M.Z.U., Rizwan, M., Ali, S., Ok, Y.S., Ishaque, W., Saifullah, et al., 2017. Remediation of heavy metal contaminated soils by using Solanum nigrum: a review. Ecotox. Environ. Safe. 143, 236–248. doi:10.1016/j.ecoenv.2017.05.038.

Saleem, M.H., Ali, S., Rehman, M., Rizwan, M., Kamran, M., Mohamed, I.A.A., khan, Z., Bamagoos, A.A., Alharby, H.F., Hakeem, K.R., Liu, L, 2020. Individual and combined application of EDTA and citric acid assisted phytoextraction of copper using jute (Corchorus capsularis L.) seedlings. Environ. Technol. Innov. 19, 100895. https://doi.org/10.1016/j.eti.2020.100895.

Sallah-Ud-Din, R., Farid, M., Saeed, R., Ali, S., Rizwan, M., Tauqeer, H.M., Bukhari, S.A.H, 2017. Citric acid enhanced the antioxidant defense system and chromium uptake by Lemna minor L. grown in hydroponics under Cr stress. Environ. Sci. Pollut. Res. 24, 17669–17678. https://doi.org/10.1007/s11356-017-9290-0.

Salt, D.E., Smith, R.D., Raskin, I., 1998. Phytoremediation. Annu. Rev. Plant Physiol. Plant Mol. Biol. 49, 643–668.

Salt, D.E., Smith, R.D., Raskin, L., 1998. Phytoremediation. Ann. Rev. Plant Phys. Plant Mol. Biol. 49 (1), 643–668. doi:10.1146/annurev.arplant.49.1.643.

Saxena, I., Shekhawat, G.S., 2013. Nitric oxide (NO) in alleviation of heavy metal induced phytotoxicity and its role in protein nitration. Nitric Oxide 32, 13–20. https://doi.org/10.1016/j.niox.2013.03.004.

Seth, C.S., Remans, T., Keunen, E., Jozefczak, M., Gielen, H., Opdenakker, K., Weyens, N., Vangronsveld, J., Cuypers, A., 2012. Phytoextraction of toxic metals: a central role for glutathione. Plant Cell Environ. 35, 334–346.

Sidhu, G.P.S., Bali, A.S., Singh, H.P., Batish, D.R., Kohli, R.K, 2018. Ethylenediamine disuccinic acid enhanced phytoextraction of nickel from contaminated soils using Coronopus didymus (L.) Sm. Chemosphere 205, 234–243. https://doi.org/10.1016/j.chemosphere.2018.04.106.

Smolińska, B., 2020. The influence of compost and nitrilotriacetic acid on mercury phytoextraction by Lepidium sativum L. J. Chem. Technol. Biotechnol. 95 (4), 950–958. https://doi.org/10.1002/jctb.5970.

Suman, J., Uhlik, O., Viktorova, J., Macek, T., 2018. Phytoextraction of heavy metals: a promising tool for clean-up of polluted environment? Front. Plant Sci. 9, 1476. doi:10.3389/fpls.2018.01476.

Sun, K., Yue, Y., Wen, D., Li, X., Yang, Y., Yang, N., Zhang, H., Chen, N., Wang, K., 2020. Effects of exogenous sulfur on maize (Zea mays L.) growth and Cd accumulation in Cd-contaminated plastic shed soil. Environ. Monit. Assess. 192, 651. https://doi.org/10.1007/s10661-020-08616-4.

Tangahu, B.V., Sheikh Abdullah, S.R., Basri, H., Idris, M., Anuar, N., Mukhlisin, M., 2011. A review on heavy metals (As, Pb, and Hg) uptake by plants through phytoremediation. Int. J. Chem. Eng. 2011, 31. https://doi.org/10.1155/2011/939161.

Tipu, M.I., Ashraf, M.Y., Sarwar, N., Akhtar, M., Shaheen, M.R., Ali, S., Damalas, C.A., 2021. Growth and Physiology of Maize (Zea mays L.) in a Nickel-Contaminated Soil and Phytoremediation Efficiency Using EDTA. J. Plant Growth Reg. 40, 774–786. Doi: 10.1007/s00344-020-10132-1. (Accessed 18 May 2020).

Turgut, C., Pepe, M.K., Cutrightb, T.J. (2003). The effect of EDTA and citric acid on phytoremediation of Cd, Cr, and Ni from soil using Helianthus annuus. Environmental Pollution, 131, 147–154. Doi: 10.1016/j.envpol.2004.01.017.

Vassil, A.D., Kapulnik, Y., Raskin, I., Salt, D.E., 1998. The role of EDTA in lead transport and accumulation in Indian mustard. Plant Physiol. 117, 447–453.

Wilson, I.D., Neill, S.J., Hancock, J.T., 2007. Nitric oxide synthesis and signalling in plants. Plant Cell Environ. 31, 622–631.

Wink, D.A., Mitchell, J.B., 1998. Chemical biology of nitric oxide: insights into regulatory, cytotoxic, and cytoprotective mechanisms of nitric oxide. Free Rad. Biol. Med. 25, 434–456.

Xiong, J., Fu, G., Tao, L., Zhu, C., 2010. Roles of nitric oxide in alleviating heavy metal toxicity in plants. Arch. Biochem. Biophys. 497, 13–20. doi:10.1016/j.abb.2010.02.014.

Yan, A., Wang, Y., Tan, S.N., Mohd, Y.M.L., Ghosh, S., Chen, Z, 2020. Phytoremediation: a promising approach for revegetation of heavy metal-polluted land. Front. Plant Sci. 11, 359. doi:10.3389/fpls.2020.00359.

Zaheer, I.E., Shafaqat Ali, S., Rizwan, M., Farid, M., Shakoor, M.B., Gill, R.A., Najeeb, U., Iqbal, N., Ahmad, R., 2015. Citric acid assisted phytoremediation of copper by Brassica napus L. Ecotoxicol. Environ. Safety 120, 310–317. https://doi.org/10.1016/j.ecoenv.2015.06.020. (Accessed April 29, 2021).

Zhong, L., Hu, C., Tan, Q., Liu, J., Sun, X., 2011. Effects of sulfur application on sulfur and arsenic absorption by rapeseed in arsenic-contaminated soil. Plant Soil Environ. 57 (9), 429–434.

Zhou, J., Hao, M., Liu, Y., Huang, G., Fu, Q., Zhu, J., Hu, H., 2018. Effects of exogenous sulfur on growth and Cd uptake in Chinese cabbage (Brassica campestris spp. pekinensis) in Cd-contaminated soil. Environ. Sci. Pollut. Res. https://doi.org/10.1007/s11356-018-1712-0. (Accessed April 29, 2021).

Chapter 9

Omics and phytoremediation

Md. Khursheed[a], Khursheed Ahmad Wani[b], Summia Rehman[c], Javid Manzoor[d]

[a]*Kazimi International Laboratory, Muscat, Oman,* [b]*Department of Environmental Science, Govt Degree College, Bijbehara, India,* [c]*Department of Botany, University of Kashmir, India,* [d]*Department of Environmental Science, JJT University, Rajasthan, India*

9.1 Introduction

Genomics, proteomics, metabolomics, and glycomics, etc., are branch of science known as omics. The objective of this branch is the classification and quantification of reservoirs of biological molecules to convert them in structure, function, and dynamics of an individual or organism. The research community has well acknowledged the wordy "omes" outside the initial "genome" in scientific literature. "Proteomics" is a well-developed concept for large-scale protein analysis. "Omes" may offer easy shorthand for summing up a field, for example, an interactomics research is easily defined as requiring large-scale examination of gene-gene, protein-protein, or protein-ligand interactions. Rapid industrialization in the past 40 years has caused dramatic increase in soil and water pollution, mainly with heavy metals and chemical contaminants (Chaâbene et al., 2018). The removal methods include not only consistent resources, but the accurate techniques and methods that are cost-effective and possibly can be utilized in mega remediation projects subject to ever-changing circumstances (Grobelak et al., 2019). Phytoremediation methods, including the plant-based approach for the management of contaminated environments have been thoroughly studied in the past.

Phytoremediation utilizes the inherent capacity of plants to remove different pollutants, including the soil, water toxins (Anjum et al., 2014). The primary advantage of this procedure is cost-effectiveness as it has been found to be at least 10 times cheaper relative to conventional methods of environmental cleanup. It is often known as a sustainable activity, because plants will often maintain the polluted region, reducing deforestation and the volume of contaminated soil that might possibly migrate to another distant area (Cevher et al., 2018). In contrast to fungal and bacterial bioremediation, phytoremediation is noticeable and can thus be tracked and constantly monitored. In addition, plants can provide nutrients and activate bacteria in the rhizosphere, which can both

Phytoremediation. DOI: https://doi.org/10.1016/B978-0-323-89874-4.00017-0

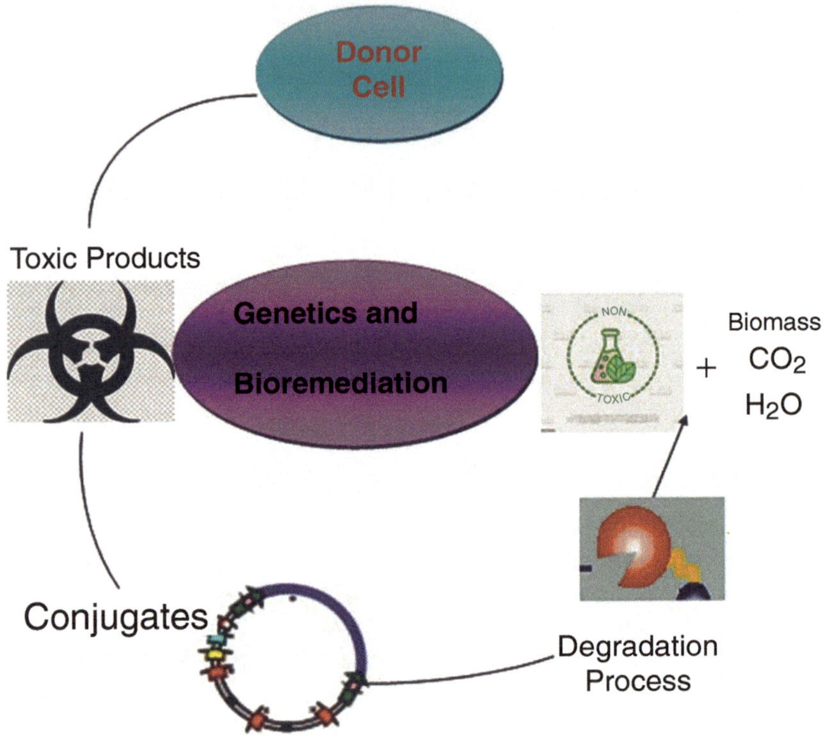

FIG. 9.1 An approach in the involvement of conjugates for degradation process.

increase soil quality and contribute to the entire remediation phase (Song et al., 2016). Phytoremediation in the polluted areas, particularly after mega industrial operations, can provide a new environment for wildlife, insects, Avian fauna, and small mammals (Chen et al., 2018). The diverse dynamics of flora and their related species have a critical role in the breakdown of complex molecular weight polyaromatic hydrocarbons and related molecules (Fig. 9.1).

It is also important to consider the dynamic relationship between the biological, physiological, and molecular processes of microbes, fungi, and plants involved in the natural breakdown of pollutants (Fester et al., 2014). Ecological engineering methods incorporating "myco, phyto, and rhizoremediation" are required to build best approaches for successful cleaning and regeneration of degraded soil habitats. Such type of procedures must consider the chemical biodegradation as an ecosystem property and as a result, relate information of the chemical molecular determinants to govern the transition of environmental drivers for functioning of the ecosystem (Fester et al., 2014; Mitsch, 2012). The microbial biodegradation linked with plants called rhizodegradation also contributes to the effective sustainable remediation (Muratova et al., 2003;

Rentz et al., 2005; Cunningham et al., 1995). Bioremediation systems are also unpredictable and need to be refined since they have been discussed independently through "phytoremediation," "mycoremediation," and "bioremediation" to remediate the degraded areas.

Phytoremediation is a best renewable solution to conventional soil remediation methods such as drilling accompanied by chemical refining. This strategy has supplementary advantages like the biomass collected from polluted areas may be used as biofuel until plants at the onsite will reach their mature stage. This may serve as model organisms for the growth of vegetation, growing local habitats, and leading to the fixation of CO_2 and the reformation of disrupted soils. The performance of phytoremediation is due to synergistic relationships of plants and the nearby environment, in particular microorganisms. Different plants growing in various mediums have the ability to translocate and segregate pollutants as is done by microorganisms in case of organic pollutants.

Next-generation sequencing technology arrived in 2005 that expanded the knowledge of scientific community toward fungus, microbe, and plant microbe interfaces. Broad omics "data sets" are now converted into functional machinery in field of medical science with redaction in sequencing costs that made it easier to apply omics to ecological problems. It is well argued that omics testing of plant-microbe assemblies is likely to lead the next wave of strategies for controlling alien organisms, remediating disturbed sites, and maximizing crop output. In this chapter, different methods at genetic level that are quite useful for removal of pollutants based on plant and microbial interactions to toxins and in general the role of phytoremediation strategies based on omics data sets.

9.2 Genomics

The amalgamation of genomic methods in ecological sciences has contributed to the discovery of several genes implicated in phytoremediation and abiotic stress resistance of plants, including some pollutants (Das et al., 2016). Genomic methods have established genes involved in phytoremediation and plant resistance to different soil toxins. The elevated-performance molecular techniques are particularly relevant for the detection of plant characteristics that rely on the mutual action of several genes. 42% of the difference in plant zinc tolerance has been described by the backcross of F1 hybrids (zinc-resistant) *Arabidopsis halleri* and the zinc-resistant *Arabidopsis lyrata* with *A. Lyrata* (Willems et al., 2007). Similarly, quantitative trait loci (QTLs) for arsenic aggregation have been found in various tissues of recombinant inbreed maize row (Ding et al., 2011). Interestingly, this analysis showed that only single QTL was allocated to tissues, indicating that separate chromosomes are accountable for the aggregation of arsenic in dissimilar tissues of plants. The discovery of possible genes entailed in phytoremediation usually doesn't go through regulated crosses. The interaction mapping analysis of 349 *Arabidopsis thaliana* (wild) accessions aided to classify the heavy metal gene ATPase 3 as the only locus accountable for cadmium (Cd) buildup in foliage (Chao et al., 2012). In view of the heterogeneity of several plant genomes, often involving intense polyploidy, even more comprehensive genomic characterization through plant remediation would be required to define the processes involved in rapid remediation across a number of ecosystems. The computational approach quantitative trait locus analysis correlates two forms of knowledge to phenotypic data and genotypic data based on molecular markers to understand the genetic origin of variance in difficult characteristics (Cox et al., 2017). It should also be remembered that the difficulty of plant chromosomes, which also comprise intense polyploidy, needed thorough genomic categorization through species appropriate for phytoremediation for determining the precise pathways concerned with suitable kind of remediation of the particular pollutant. Activity-focused methods like transcriptomics, proteomics can therefore enable for a detailed analysis of the functionality of particular plant species when exposed to different environmental conditions, along with form and amount of pollutant (Bayçu et al., 2016). Different plants have been identified that are used for removal of pollutants have been documented below (Table 9.1).

There are several reports of the usage of transcriptomics to classify genes involved in plant stress resistance. *Brassica chinensis* while exposed to chromium (Cr) depicted high stress was identified by using these technologies (Zhou et al., 2016). The transcriptome profiling in *Populus* sp. developed by boron exposure has identified multiple genes accountable for boron absorption, transferb and decontamination (Yıldırım and Uylaş, 2016). The maximum inductions were reported for genes encoding, tyrosine aminotransferase, cassette transporters binding ATP, glutathione S transferases and metallochaperones. Most of the chromosomes are active in cell decontamination, for example, transport of contaminants to vacuoles. In addition, several other genes have been found to be strongly upregulated after exposure to boron, including antioxidant and signaling genes (Yıldırım and Uylaş, 2016). As many as 72 metabolic pathways like metabolism of phenylalanine, biosynthesis of phenylpropanoid, fixation of carbon in *Phytolacca americana* has been revealed when exposed to cadmium (Cd). Further, numerous genes linked to Cd resistance, absorption, transport, and accumulation were also established, including a total of 11 areas, eight for nicotianamine synthases, six for aquaporins, four for ZRT/IRT-like proteins, three for ABC transporters, and three for metallothioneins (Chen et al., 2018). Transcriptome review contributed to the determination of usable genes that were upregulated by tension induced by herbicide-atrazine. Similarly, cellular enzymes, finger proteins, antistress, or antidisease proteins that encode for Zn, structural proteins were also found (Song et al., 2016).

The transcriptomic method is also used to explore the impact of thorium (Th) on metabolism of plant. Th is a dangerous actinide metal in nuclear technology and its exposure is primarily triggered by mines or leaks, which presents a major hazard due to its radioactivity. The development of jasmonic and salicylic acid signaling pathways and diverse abiotic and biotic stress responsive

TABLE 9.1 Plants used to remove heavy metals around the world.

Heavy metal	Plant used	Country	Author
Cd	*Brassica napus, Pelargonium*	USA	(Zaier et al., 2010; Arshad et al., 2008)
PAEs	*Medicago osativa*	China	(Ma et al., 2012)
PAEs and PAHs	*Medicago sativa and Tagetes patula*	China	(Fu et al., 2012)
Pb	*Ambrosia artemisiifilia, Viciafaba, Pisum sativum, Phaseolus vulgaris, Vigna unguiculata, Nicotiana tabacum, Lathyrus sativus, Zea mays, Avicennia marina, Sedum alfredii, Allium sativum*	USA	(Huang and Cunningham, 1996; Piechalak et al., 2002; Małecka et al., 2008; Kopittke et al., 2007; Jiang and Liu, 2010.)
Cd	*Pelargonium, Sedum alfredii*	USA, China	(Arshad et al., 2008, Xiong et al., 2004)
Zn, Cd, and Ni	*Thlaspi caerulescens, Sedum alfredii*	China	(Lasat, 1996; Assuncao et al., 2001 Xiong et al., 2004)
Pb, Zn, Cu, Ni, and Cd	*Euphorbia cheiradenia, Scariola orientals, Centaurea virgata, Gundelia tournefortii, and Elaeagnus angustifolia*	Iran	(Chehregani and Malayeri, 2007)
Ni	*Aquilegia bertolonii, Alyssum lesbiacum, Thlaspi goesin-gense*	China	(Kupper et al., 2001)
Cd, Pb, and Zn	*Brassica juncea, Eichhornia crassipes*	India	(Schneider et al., 1999)
Pb, Cd, Ni, and Zn	*Thlaspi caerulescens*	Belgium, India, Italy	(Bert et al., 2002; Prasad, 2005; Maestri et al., 2010)
Ni and Co	*Alyssum bertoloni*	Belgium, India, Italy	(Bert et al., 2002; Prasad, 2005; Maestri et al., 2010)
Zn and Cd	*Arabidopsis halleri*	Belgium, India, Italy	(Bert et al., 2002; Prasad, 2005; Maestri et al., 2010)
Cd and Ni	*T. caerulescens*	Netherland, Italy	(Assuncaoet al., 2001, Maestri et al., 2010)

Cd, cadmium; Co, cobalt; Cu, copper; Ni, nickel; PAEs, phthalic esters; PAHs, polycyclic aromatic hydrocarbons; Pb, lead; Zn, zinc.

genes are due to induced genes. In addition, upregulation of phosphate starvation genes and downregulation of the phytic acid synthesis genes showed that thorium disrupted the absorption or signaling of phosphate. In addition, iron-responsive genes were also greatly affected by the contact. The downregulation of certain aquaporins has demonstrated a serious disruption of water homeostasis (Mazari et al., 2017).

Activity-focused methods like metabolomics, transcriptomics, proteomics make it possible to analyze the involvement of candidate phytoremediators in various environmental and toxin circumstances and have become especially functional in identifying plant activity in metal-polluted locations. Transcriptomics is helpful to locate genes that are dynamically involved in remediation of *Brassica juncea* in sulfur and chromium rich ecosystems (Schiavon et al., 2012; Gao et al., 2013).The transcriptomic analyses between organisms with opposing remediation capabilities can often expose genes accountable for these variations. For instance, an analysis comparing the transcriptomes of *Solanum nigrum* that is considered hyperaccumulator of Cd and *Solanum torvum* a low accumulator of Cd showed higher expression of few chromosomes in *S. Nigrum* like metal carriers, in reaction to the inclusion of Cd (Xu et al., 2012). While most phytoremediation analyses conducted on different plants that were exposed to metal pollutants, plants have essential direct roles in deteriorating, collecting, or evaporating organic molecules along with invigorating natural breakdown. Plants absorb PCBs (Ficko et al., 2011) of 1,3,5-trinitro-1, 3,5-triazine (RDX) (Brentner et al., 2010) and 246-trinitrotoluene (Brentner et al., 2010). Perhaps more surprising is that most of the plants may specifically hydroxylate PCBs (Zhai et al., 2010). While some omics have targeted plant reactions to natural toxins, the gene expression analysis in *Poplar* sp. revealed that five main genes are involved in the decontamination of unknown contaminants (Tanaka et al., 2007). The pollutant-induced stress was studied by using proteomics and metabolomics to classify candidate genes and to track plant reactions (Sergeant et al. 2012; Ariza et al. 2013). The proteomics is helpful to categorize changes stimulated by Cd in *Populus* sp. and has shown that plant responses differ in various tissues like cambium and leaves. The plant proteome is predisposed by Cd in tissues and indirectly by reduced metabolic rates due to general toxicity (Durand et al., 2010). As with microorganisms, potential research incorporating several omics approaches can have a more international observation of genetic material, protein, and metabolite networks, their association at plant level and ecological variance (Villiers et al., 2012).

9.3 Metagenome

The plant-associated microbial interaction like endosphere, rhizosphere, or phyllosphere may be tested either by shotgun sequencing or by amplicon targeting specific genes. The microorganisms in soil, rhizosphere, and endosphere are dissimilar has been illustrated by high-throughput bacterial sequencing 16S rDNA (Bulgarelli et al., 2012; Lundberg et al., 2012). The bacteria inhabiting roots are partially based on plant, soil form and partially on generic plant structural features (Bulgarelli et al., 2012). The population of bacteria and not the shape and type of plant influences the environmental factors has been revealed by sequencing of 16S rDNA amplicons (Rastogi et al., 2012). The extent by which plants are used in the process of phytoremediation is often reliant on the amount of soil contaminant. The fungal community are established by plant phylogeny and the amount of contaminant in the soil (Bell et al., 2014). Similarly, molecular research has shown that arbuscular mycorrhizal composition in rhizosphere relies on the amount of metals (Hassan et al., 2011). This indicates that plant tension may potentially assess the activation of microbial remediation. The wild rye exposed to hydrocarbons were less restrictive as compared to control rye that indicates root exudates can repress natural biodegradation (Phillips et al., 2012; Rentz et al., 2004) and showing that plants may change the degree of biodegradation of the rhizosphere. Mobile DNA, primarily plasmid DNA, is also of importance in phytoremediation, as plant rhizospheres and polluted soils are hot spots for plasmid trade (Van and Bailey, 2002; Top and Springael, 2003). Plasmids are source of genomic creativity in toxic environments and enable microorganisms to respond to novel stress (Sentchilo et al., 2013). The heavy metal tolerant genes are predominant in plasmids has been revealed by metagenomic sequencing of "mobilome" from waste water treatment plants (Sentchilo et al., 2013) and that multiple genes implicated in organic degradation are plasmid-borne (Top and Springael, 2003). Plasmids have a vital role in the modulation of microbial communities. For example, toluene degradation gene plasmids have been successfully added to aerial plant endophytes that seem to lack certain genes and allowed degradation of contaminants that would have been volatilized (Barac et al., 2004; Taghavi et al., 2005). Hence, metagenomics can be employed to search for novel plasmids that will help in delivery of genes to different microorganisms so that transformation efficiency can be measured within the microbial community. The metagenomics can be coupled with (meta) transcriptomics to understand the influence of plants and native microbial communities for plasmid introduction. Study of plant-associated viruses has demonstrated that certain non-pathogenic viruses of uncertain roles inhabit plants (Roossinck, 2012). As the segregation of exceptional metagenome constituents like virome becomes normal, less examined modes of genetic transmission, including viruses and phage-like gene transfer agents (Lang et al., 2012; Williams, 2013) should be explored for possible improvements in microbial community structure. Metagenomic examinations may allocate functional properties to microorganisms and microbial assemblies. Metagenomic observations undertaken through environmental gradients have demonstrated that genetic makeup and taxonomic framework of the microbial population may be linked to its carbon/hydrocarbon-metabolizing behavior (Bell et al., 2013; Fierer et al., 2012). PhyloChip study of metagenomic DNA established unique taxa correlated with soil suppression of plant disease (Mendes et al., 2011).

This illustrates the potential of metagenomics for environmental applications, since microbial genes and taxonomic abundance may be specifically linked to functions like degradation of hydrocarbons and the promotion of plant growth in the determination of specific environmental parameters that benefit target microorganisms. New single-cell segregation and sequencing techniques can establish the genome content of uncultivated strains and can theoretically recognize targets for insulation or field modification in order to better characterize the ability of microorganisms are not yet grown in terms of remediation of toxins or plant growth promotion (Lasken, 2012). Diffusion compartments that enable signals to move from the original ecosystem can aid in microbial cultivation (Pham and Kim, 2012) and the incorporation of plant-derived composites or soil pollutants can help in the insulation of target microbes. Further, clone libraries have bigdata sets of metagenomic inserts may be screened to allocate functional properties to genes (Uchiyama and Miyazaki, 2009) and to test functions for species that remain immune to isolation (Fig. 9.2).

Lignin can be degraded quite easily by ligninolytic fungi can be tested to degrade similar compounds like polycyclic aromatic hydrocarbon (PAHs). The generation of extracellular enzymes like lignin peroxidase manganese (Mn) peroxidase, or flexible peroxidase and laccases may be quite useful in degradation of aromatic compounds. These enzymes have the ability to mineralize several phenolic compounds fully or partially or by taking further help from bacterial population that may later contribute to the creation of enough water-soluble molecules. (Pozdnyakova, 2012; Harms et al., 2011). Mn peroxidase or lignin peroxidase and enzymes, for example, catalyze electron oxidation of PAHs with strong potential ionization accompanied by aromatic cleavage. In-vivo and in-vitro studies revealed that six ringed PAHs can be metabolizable through Mn peroxidase-dependent lipid peroxidation (Bezalel et al., 1996). The fungal cytochrome P450 (CYP) monooxygenases are active in the breakdown of PAHs in addition to the extracellular enzymatic PAH degradation mechanism. The initial step in intracellular phenanthrenic mineralization includes the amalgamation of O_2 by CYP monooxygenase with good capacity to catabolize PAHs with a large number of fused rings (Syed et al., 2010). The entire genome of the model fungi *Phanerochaete chrysosporium* has been identified and is now available to everyone; (Martinez et al., 2004) the existence of approximately 150 putative CYP-encoding genes has been established, several of which are called orphan genes of unknown purpose (Syed et al., 2010). These fungi affect the bacterial soil population and the possible stimulus of PAH depletion is still an unsolved topic. It is very important to remember that mycosphere recognized as the microenvironments having thick fungal hyphae, hosts preferred bacteria and is obviously an significant driver of the large diversity of bacteria inhabited in soil (Zhang et al., 2014). While the exact process involved in the selection of bacteria present in mycosphere is still overlooked, when it is proposed that many bacteria are suited to the mycosphere environment owing to the release of nutrients and signaling of fungi molecules. In addition, fungi mycelia form a thick

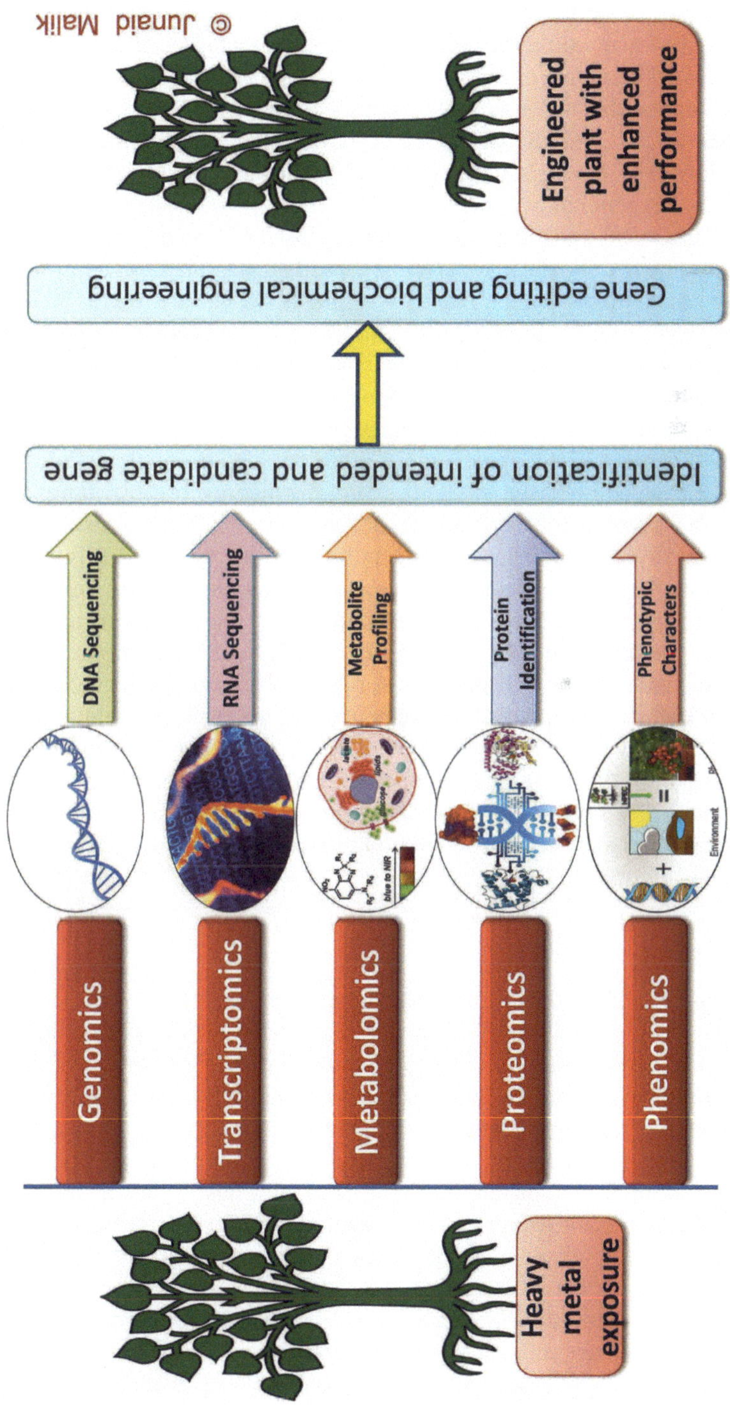

FIG. 9.2 Heavy metal exposure and role of omics, fungi, and bacteria in recycling of polycyclic aromatic hydrocarbons.

and prolonged spatial network of thread-like hyphae in the soil. These mycelia have been shown to communicate with bacteria, (Zhang et al., 2014a), function as efficient dispersal networks for both undirected and chemotactic mobilization of contaminant-degrading bacteria (referred to as "fungal highways") (Kohlmeier et al., 2005; Schamfuß et al., 2013), improve the movement of a broad variety of PAHs as they move during cytoplasmic streaming. Due to their ubiquity and mycelial networks duration fungi has a vital role and show unexploited biochemical and biophysical capacity for PAH biodegradation in polluted environments (Pozdnyakova, 2012). Further, mycorrhizal fungi develop symbiotic association with the majority of plants (Smith and Read, 2008) that are determinants of plant interactions in these complicated environments as they mitigate stress induced by contaminants (Aranda et al., 2013). Unfortunately, mycorrhizal fungi display a dynamic genetic organization since they occur in heterokaryotic forms where the nucleus is genetically distinct or in homokaryotic forms having the same nucleus. In addition, also in single isolates, very high variability can be observed, and several distinct mycorrhizal fungi can be colonized by a single root system. Thus, the usage of "omics" techniques to explain plant mycorrhiza interactions is still in its infancy, but ideally can offer fresh perspectives to decode these dynamic interactions in PAH-polluted settings. Until quite recently, it was challenging to research taxonomic classes and fungi types in a diverse environment comprising DNA from several organisms. However the assessment of certain DNA regions as fungi-specific barcodes by the Fungal Barcoding Consortium (Schoch et al., 2012) should help to increase knowledge of how the interaction of plant fungi in soil populations is involved in the dissipation of dangerous molecules such as PAHs. In addition, a study of the "omics" methods and systems biology used to illustrate the function of mycorrhizal fungi and their association with plants has recently been published (Salvioli and Bonfante, 2013).

9.4 Future prospects

This chapter addressed, in particular, metagenomics and "omics" linked approaches, for degradation of pollutants that enable the detection of the species that are collaborating in field of phytoremediation. Very few confirmations indicate that involvement of plant-fungal-bacterial consortia is helpful in the degradation of complex molecules (Boonchan et al., 2000). It is therefore essential to revise phytoremediation, mycoremediation, bioremediation, and rhizoremediation, taking into consideration either one or a few organisms along with the genetic dataset. Consequently, the remediation of organic xenobiotics at the level of the metaorganism must be justified at the level of the infected field sites (Bell et al., 2014). It is argued that not only genomic and biochemical factors must be taken into account, but also ecological factors must be considered. Applying metagenomics, metatranscriptomics, metabolomics, and other complementary omics innovations to forecast metaorganism/holobiont-scale catabolic capacity in contaminated soils opens up new possibilities to improve the productivity of bio-based remediation. Omics mechanisms have the potential to categorize targeted active native hydrocarbon-degrading microorganisms in the rhizosphere that have been selected by plants (Siciliano et al., 2001). The isolation and usage of these microorganisms as inoculums would be of great benefit for bioaugmentation and stimulation of biotechnology. "Omics"-based awareness of the catabolic ability and metabolite movement may be used in ecological modelling as feedback for estimation of the fate of PAHs in the plant-microbe complex and can also be used for predictive risk assessment. In addition, better information on spatial temporal distribution of PAH biotransformation potential should be accessible in order to better understand and forecast the ecosystem role of PAH biodegradation.

Phytoremediation may be a safe, cost-effective, and socially appropriate technology for working with polluted property. In order to boost its performance, modern omics methods have gained a lot of attention in recent years (Zhang et al., 2018). Our analysis reveals that certain broad data sets of omics have already established particular environmental factors that form the quantity and quality of soil microbial communities. The risk of modifying the microbial makeup by adding plants or altering the development of plants by controlling microbial populations relates not only specifically to phytoremediation purposes, but also to the control of alien organisms by growing crops and biofuels. One of the most important benefits of utilizing omics for such research is its continuously-growing data sets and continuously-decreasing prices. Therefore, translocation of the gathered data and overall information into usable technology relies on the study that utilizes omics analyses for the particular objectives to be reached.

Acknowledgement

The authors are thankful to Dr. Jundaid Malik for his help and advice.

References

Anjum, N.A., Umar, S., Iqbal, M., 2014. Assessment of cadmium accumulation, toxicity, and tolerance in Brassicaceae and Fabaceae plants implications for phytoremediation. Environ. Sci. Pollut. Res. 21 (17), 10286–10293.

Aranda, E., Scervino, J.M., Godoy, P., Reina, R., Ocampo, J.A., Wittich, R.M., García-Romera, I., 2013. Role of arbuscular mycorrhizal fungus Rhizophagus custos in the dissipation of PAHs under root-organ culture conditions. Environ. Pollut. 181, 182–189.

Ariza, J.G., García-Barrera, T., García-Sevillano, M.A., González-Fernández, M., Gómez-Jacinto, V., 2013. Metallomics and metabolomics of plants under environmental stress caused by metalsHeavy Metal Stress in Plants. Springer, Berlin, Heidelberg, pp. 173–201.

Arshad M., Silvestre J., Pinelli E., Kallerhoff J., Kaemmerer M., Tarigo A., Shahid M., Guiresse M., Pradere P., Dumat C. (2008) A field study of lead phytoextraction by various scented Pelargonium cultivars. Chemosphere 71 (11):2187–2192.

Assunção, A., Da Costa Martins, P., De Folter, S., Vooijs, R., Schat, H., Aarts, M., 2001. Elevated expression of metaltransporter genes in three accessions of the metal hyperaccumulator. Thlaspicaerulescens. Plant Cell. Environ. 24 (2), 217–226.

Banitz, T., Johst, K., Wick, L.Y., Schamfuß, S., Harms, H., Frank, K., 2013. Highways versus pipelines: contributions of two fungal transport mechanisms to efficient bioremediation. Environ. Microbiol. Rep. 5 (2), 211–218.

Barac, T., Taghavi, S., Borremans, B., Provoost, A., Oeyen, L., Colpaert, J.V., van der Lelie, D., 2004. Engineered endophytic bacteria improve phytoremediation of water-soluble, volatile, organic pollutants. Nat. Biotechnol. 22 (5), 583–588.

Bayçu, G., Gevrek-Kürüm, N., Moustaka, J., Csatári, I., Rognes, S.E., Moustakas, M., 2016. Cadmium-zinc accumulation and photosystem II responses of Noccaea caerulescens to Cd and Zn exposure. Environ. Sci. Pollut. Res. 24 (3), 2840–2850.

Bell, T.H., Hassan, S.E.D., Lauron-Moreau, A., Al-Otaibi, F., Hijri, M., Yergeau, E., St-Arnaud, M, 2014. Linkage between bacterial and fungal rhizosphere communities in hydrocarbon-contaminated soils is related to plant phylogeny. ISME J. 8 (2), 331–343.

Bell, T.H., Joly, S., Pitre, F.E., Yergeau, E., 2014. Increasing phytoremediation efficiency and reliability using novel omics approaches. Trends Biotechnol. 32 (5), 271–280.

Bell, T.H., Yergeau, E., Maynard, C., Juck, D., Whyte, L.G., Greer, C.W., 2013. Predictable bacterial composition and hydrocarbon degradation in Arctic soils following diesel and nutrient disturbance. ISME J. 7 (6), 1200–1210.

Bert, V., Bonnin, I., Saumitou-Laprade, P., De Lague´rie, P., Petit, D., 2002. Do Arabidopsis halleri from nonmetallicolous populations accumulate zinc and cadmium more effectively than those from metallicolouspopulations? New Phytol. 155 (1), 47–57.

Bezalel, L.E.A., Hadar, Y., Fu, P.P., Freeman, J.P., Cerniglia, C.E, 1996. Metabolism of phenanthrene by the white rot fungus Pleurotus ostreatus. Appl. Environ. Microbiol. 62 (7), 2547–2553.

Boonchan, S., Britz, M.L., Stanley, G.A., 2000. Degradation and mineralization of high-molecular-weight polycyclic aromatic hydrocarbons by defined fungal-bacterial cocultures. Appl. Environ. Microbiol. 66 (3), 1007–1019.

Brentner, L.B., Mukherji, S.T., Walsh, S.A., Schnoor, J.L., 2010. Localization of hexahydro-1, 3, 5-trinitro-1, 3, 5-triazine (RDX) and 2, 4, 6-trinitrotoluene (TNT) in poplar and switchgrass plants using phosphor imager autoradiography. Environ. Pollut. 158 (2), 470–475.

Bulgarelli, D., Rott, M., Schlaeppi, K., van Themaat, E.V.L., Ahmadinejad, N., Assenza, F., Peplies, J., 2012. Revealing structure and assembly cues for Arabidopsis root-inhabiting bacterial microbiota. Nature 488 (7409), 91–95.

Cevher-Keskin, B., Selçukcan-Erol, Ç., Yüksel, B., Ertekin, Ö., Yıldızhan, Y., Onarıcı, S., Memon, A.R., 2018. Comparative transcriptome analysis of Zea mays in response to petroleum hydrocarbon stress. Environ. Sci. Pollut. Res. 25 (32), 32660–32674.

Chaâbene, Z., Rorat, A., Hakim, I.R., Bernard, F., Douglas, G.C., Elleuch, A., Mejdoub, H., 2018. Insight into the expression variation of metal-responsive genes in the seedling of date palm (Phoenix dactylifera). Chemosphere 197, 123–134.

Chao, D.Y., Silva, A., Baxter, I., Huang, Y.S., Nordborg, M., Danku, J...., Salt, D.E., 2012. Genome-wide association studies identify heavy metal ATPase3 as the primary determinant of natural variation in leaf cadmium in Arabidopsis thaliana. PLoS Genet, 8 (9), e1002923.

Chehregani, A., Malayeri B.E., 2007. Removal of heavy metals by native accumulator plants. Int. J. Agric. Biol., 9: 462-465.

Chen, Y., Zhi, J., Zhang, H., Li, J., Zhao, Q., Xu, J., 2018. Correction: Transcriptome analysis of Phytolacca americana L. in response to cadmium stress. PloS one, 13 (6), e0199721.

Cox, A., Venkatachalam, P., Sahi, S., Sharma, N., 2016. Silver and titanium dioxide nanoparticle toxicity in plants: a review of current research. Plant Physiol. Biochem. 107, 147–163.

Cox A., Venkatachalam P., Sahi S., Sharma N., 2017, Silver and titanium dioxide nanoparticle toxicity in plants: A review of current research, Plant Physiology Biochemistry, 107, 147–163.

Cunningham, S.D., Berti, W.R., Huang, J.W., 1995. Phytoremediation of contaminated soils. Trends Biotechnol. 13 (9), 393–397.

Das, N., Bhattacharya, S., Maiti, M.K., 2016. Enhanced cadmium accumulation and tolerance in transgenic tobacco overexpressing rice metal tolerance protein gene OsMTP1 is promising for phytoremediation. Plant Physiol. Biochem. 105, 297–309.

Ding, D., Li, W., Song, G., Qi, H., Liu, J., Tang, J., 2011. Identification of QTLs for arsenic accumulation in maize (Zea mays L.) using a RIL population. PLoS One, 6 (10), e25646.

Durand, T.C., Sergeant, K., Planchon, S., Carpin, S., Label, P., Morabito, D...., Renaut, J., 2010. Acute metal stress in Populus tremula× P. alba (717-1B4 genotype): Leaf and cambial proteome changes induced by cadmium2+. Proteomics, 10 (3), 349–368.

Fester, T., Giebler, J., Wick, L.Y., Schlosser, D., Kästner, M., 2014. Plant–microbe interactions as drivers of ecosystem functions relevant for the biodegradation of organic contaminants. Curr. Opin. Biotechnol. 27, 168–175.

Ficko, S.A., Rutter, A., Zeeb, B.A., 2011. Phytoextraction and uptake patterns of weathered polychlorinated biphenyl–contaminated soils using three perennial weed species. J. Environ. Qual. 40 (6), 1870–1877.

Fierer, N., Lauber, C.L., Ramirez, K.S., Zaneveld, J., Bradford, M.A., Knight, R., 2012. Comparative metagenomic, phylogenetic and physiological analyses of soil microbial communities across nitrogen gradients. ISME J. 6 (5), 1007–1017.

Fu D., Teng Y., Luo Y., Tu C., Li S., Li Z., Christie P. (2012) Effects of alfalfa and organic fertilizer on benzo[a]pyrene dissipation in an aged contaminated soil. Environ. Sci. Pollut. Res. 19 (5):1605–1611.

Furuno, S., Foss, S., Wild, E., Jones, K.C., Semple, K.T., Harms, H., Wick, L.Y., 2012. Mycelia promote active transport and spatial dispersion of polycyclic aromatic hydrocarbons. Environ. Sci. Technol. 46 (10), 5463–5470.

Gao, J., Sun, L., Yang, X., Liu, J.X., 2013. Transcriptomic analysis of cadmium stress response in the heavy metal hyperaccumulator Sedum alfredii Hance. PloS one, 8 (6), e64643.

Grobelak, A., Świątek, J., MurtaŚ, A., Jaskulak, M., 2019. Cadmium-induced oxidative stress in plants, cadmium toxicity, and tolerance in plants: from physiology to remediationCadmium Toxicity and Tolerance in Plants. Academic Press, Cambridge, Massachusetts, pp. 213–231.

Harms, H., Schlosser, D., Wick, L.Y., 2011. Untapped potential: exploiting fungi in bioremediation of hazardous chemicals. Nat. Rev. Microbiol. 9 (3), 177–192.

Hassan, S.E.D., Boon, E.V.A., ST-ARNAUD, M.A.R.C., Hijri, M, 2011. Molecular biodiversity of arbuscular mycorrhizal fungi in trace metal-polluted soils. Mol. Ecol. 20 (16), 3469–3483.

Huang, J.W., Cunningham, S.D., 1996. Lead phytoextraction: species variation in leaduptake and translocation. New Phytol. 134 (1), 75–84.

Jiang, W., Liu, D., 2010. Pb-induced cellular defense system in the root meristematiccells of Allium sativum L. BMC Plant Biol. 10 (1), 40.

Kohlmeier, S., Smits, T.H., Ford, R.M., Keel, C., Harms, H., Wick, L.Y., 2005. Taking the fungal highway: mobilization of pollutant-degrading bacteria by fungi. Environ. Sci. Technol. 39 (12), 4640–4646.

Kopittke, P.M., Asher, C.J., Kopittke, R.A., Menzies, N.W., 2007. Toxic effects of Pb2+on growth of cowpea (Vignaunguiculata). Environ. Pollut. 150 (2), 280–287.

Kupper, H, Lombi, E, Zhao, FJ, Wieshammer, G, McGrath, SP., 2001. Cellular compartmentation of nickel in the hyperaccumulators Alyssum lesbiacum, Alyssum bertolonii and Tulips goesingense. J. Exp. Bot. 52 (365), 2291–2300. doi:10.1093/jexbot/52.365.2291.

Lang, A.S., Zhaxybayeva, O., Beatty, J.T., 2012. Gene transfer agents: phage-like elements of genetic exchange. Nat. Rev. Microbiol. 10 (7), 472–482.

Lasken, R.S., 2012. Genomic sequencing of uncultured microorganisms from single cells. Nat. Rev. Microbiol. 10, 631–640.

Lasat M.M., Baker A.J.M., Kochian L.V. 1996. Physiological characterization of root Zn^{2+} absorption and translocation to shoots in Zn hyperaccumulator and nonaccumulator species of Thlaspi. Plant Physiology 112: 1715–1722.

Liao, H.L., Chen, Y., Bruns, T.D., Peay, K.G., Taylor, J.W., Branco, S…., Vilgalys, R., 2014. Metatranscriptomic analysis of ectomycorrhizal roots reveals genes associated with P iloderma–P inus symbiosis: improved methodologies for assessing gene expression in situ. Environ. Microbiol. 16 (12), 3730–3742.

Lundberg, D.S., Lebeis, S.L., Paredes, S.H., Yourstone, S., Gehring, J., Malfatti, S., Edgar, R.C., 2012. Defining the core Arabidopsis thaliana root microbiome. Nature 488 (7409), 86–90.

Ma, X.F., Tudor, S., Butler, T. et al. (2012). Transgenic expression of phytase and acid phosphatase genes in alfalfa (Medicago sativa) leads to improved phosphate uptake in natural soils. Mol Breeding 30, 377–391. https://doi.org/10.1007/s11032-011-9628-0.

Maestri, E., Marmiroli, M., Visioli, G., Marmiroli, N., 2010. Metal tolerance and hyperaccumulation: costs andtrade-offs between traits and environment. Environ. Exp. Bot. 68 (1), 1–13.

Małecka, A., Piechalak, A., Morkunas, I., Tomaszewska, B., 2008. Accumulation oflead in root cells of Pisumsativum. Acta Physiol. Plant 30 (5), 629–637.

Martinez, D., Larrondo, L.F., Putnam, N., Gelpke, M.D.S., Huang, K., Chapman, J…., Coutinho, P.M, 2004. Genome sequence of the lignocellulose degrading fungus Phanerochaete chrysosporium strain RP78. Nat. Biotechnol. 22 (6), 695–700.

Mazari, K., Landa, P., Přerostová, S., Müller, K., Vaňková, R., Soudek, P., Vaněk, T., 2017. Thorium impact on tobacco root transcriptome. J. Hazard. Mater. 325, 163–169.

Mendes, R., Kruijt, M., De Bruijn, I., Dekkers, E., van der Voort, M., Schneider, J.H…., Raaijmakers, J.M., 2011. Deciphering the rhizosphere microbiome for disease-suppressive bacteria. Science, 332 (6033), 1097–1100.

Mitsch, W.J., 2012. What is ecological engineering? Ecol. Eng. 45, 5–12.

Muratova, A., Hübner, T., Tischer, S., Turkovskaya, O., Möder, M., Kuschk, P., 2003. Plant–rhizosphere-microflora association during phytoremediation of PAH-contaminated soil. Int. J. Phytoremediation, 5 (2), 137–151.

Nahar, N., Rahman, A., Nawani, N.N., Ghosh, S., Mandal, A., 2017. Phytoremediation of arsenic from the contaminated soil using transgenic tobacco plants expressing ACR2 gene of Arabidopsis thaliana. J. Plant Physiol. 218, 121–126.

Nazir, R., Warmink, J.A., Boersma, H., Van Elsas, J.D., 2009. Mechanisms that promote bacterial fitness in fungal-affected soil microhabitats. FEMS Microbiol. Ecol. 71 (2), 169–185.

Pham, V.H., Kim, J., 2012. Cultivation of unculturable soil bacteria. Trends Biotechnol. 30 (9), 475–484.

Phillips, L.A., Greer, C.W., Farrell, R.E., Germida, J.J., 2012. Plant root exudates impact the hydrocarbon degradation potential of a weathered-hydrocarbon contaminated soil. Appl. Soil Ecol. 52, 56–64.

Piechalaka, A., Tomaszewskaa, B., Baralkiewiczb, D., Maleckaa, A., 2002. Accumulation and detoxification of lead ions in legumes. Phytochemistry 60, 153–162.

Pozdnyakova, N.N., 2012. Involvement of the ligninolytic system of white-rot and litter-decomposing fungi in the degradation of polycyclic aromatic hydrocarbons. Biotechnol. Res. Int. 2012, 1–20.

Prasad, M.N.V., 2005. Nickelophilous plants and their significance in phytotechnologies. Braz. J. Plant Physiol. 17 (1), 113–128.

Rastogi, G., Sbodio, A., Tech, J.J., Suslow, T.V., Coaker, G.L., Leveau, J.H., 2012. Leaf microbiota in an agroecosystem: spatiotemporal variation in bacterial community composition on field-grown lettuce. ISME J. 6 (10), 1812–1822.

Rentz, J.A., Alvarez, P.J., Schnoor, J.L., 2004. Repression of Pseudomonas putida phenanthrene-degrading activity by plant root extracts and exudates. Environ. Microbiol. 6 (6), 574–583.

Rentz, J.A., Alvarez, P.J., Schnoor, J.L., 2005. Benzo [a] pyrene co-metabolism in the presence of plant root extracts and exudates: implications for phytoremediation. Environ. Pollut. 136 (3), 477–484.

Roossinck, M.J., 2012. Plant virus metagenomics: biodiversity and ecology. Ann. Rev. Genet. 46, 359–369.

Salvioli, A., Bonfante, P., 2013. Systems biology and "omics" tools: a cooperation for next-generation mycorrhizal studies. Plant Sci. 203, 107–114.

Schamfuß, S., Neu, T.R., van der Meer, J.R., Tecon, R., Harms, H., Wick, L.Y., 2013. Impact of mycelia on the accessibility of fluorene to PAH-degrading bacteria. Environ. Sci. Technol. 47 (13), 6908–6915.

Schiavon, M., Galla, G., Wirtz, M., Pilon-Smits, E.A., Telatin, V., Quaggiotti, S., Malagoli, M., 2012. Transcriptome profiling of genes differentially modulated by sulfur and chromium identifies potential targets for phytoremediation and reveals a complex S–Cr interplay on sulfate transport regulation in B. juncea. J. Hazard. Mater. 239, 192–205.

Schneider, T., Haag-Kerwer, A., Maetz, M., Niecke, M., Povh, B.Rausch, T., and Schüßler, 1999. A. Micro-pixe studies of elemental distribution in Cd-accumulating Brassica Juncea L. Nucl. Instrum. Methods Phys. Res. B 158 (1), 329–334.

Schoch, C.L., Seifert, K.A., Huhndorf, S., Robert, V., Spouge, J.L., Levesque, C.A.…Fungal Barcoding Consortium, 2012. Nuclear ribosomal internal transcribed spacer (ITS) region as a universal DNA barcode marker for Fungi. Proc. Nat.l Acad. Sci. U.S.A, 109 (16), 6241–6246.

Sentchilo, V., Mayer, A.P., Guy, L., Miyazaki, R., Tringe, S.G., Barry, K., Van Der Meer, J.R., 2013. Community-wide plasmid gene mobilization and selection. ISME J. 7 (6), 1173–1186.

Sergeant, K., Renaut, J., Hausman, J.F., 2012. Proteomics as a toolbox to study the metabolic adjustment of trees during exposure to metal trace elementsMetal Toxicity in Plants: Perception, Signaling and Remediation. Springer, Berlin, Heidelberg, pp. 143–164.

Siciliano, S.D., Fortin, N., Mihoc, A., Wisse, G., Labelle, S., Beaumier, D.…, Schwab, P., 2001. Selection of specific endophytic bacterial genotypes by plants in response to soil contamination. Appl. Environ. Microbiol. 67 (6), 2469–2475.

Smith, S., Read, D., 2008. Mycorrhizal Symbiosis, 3rd ed. Academic press, Elsevier, London, UK.

Song, F., Li, J., Fan, X., Zhang, Q., Chang, W., Yang, F., Geng, G., 2016. Transcriptome analysis of Glomus mosseae/Medicago sativa mycorrhiza on atrazine stress. Sci. Rep. 6, 20245.

Syed, K., Doddapaneni, H., Subramanian, V., Lam, Y.W., Yadav, J.S., 2010. Genome-to-function characterization of novel fungal P450 monooxygenases oxidizing polycyclic aromatic hydrocarbons (PAHs). Biochem. Biophys. Res. Comm. 399 (4), 492–497.

Taghavi, S., Barac, T., Greenberg, B., Borremans, B., Vangronsveld, J., van der Lelie, D., 2005. Horizontal gene transfer to endogenous endophytic bacteria from poplar improves phytoremediation of toluene. Appl. Environ. Microbiol. 71 (12), 8500–8505.

Tanaka, S., Brentner, L.B., Merchie, K.M., Schnoor, J.L., Yoon, J.M., Aken, B.V., 2007. Analysis of gene expression in poplar trees (Populus deltoides× nigra, DN34) exposed to the toxic explosive hexahydro-1, 3, 5-trinitro-1, 3, 5-triazine (RDX). Int. J. Phytoremediation, 9 (1), 15–30.

Top, E.M., Springael, D., 2003. The role of mobile genetic elements in bacterial adaptation to xenobiotic organic compounds. Curr. Opin. Biotechnol. 14 (3), 262–269.

Uchiyama, T., Miyazaki, K., 2009. Functional metagenomics for enzyme discovery: challenges to efficient screening. Curr. Opin. Biotechnol. 20 (6), 616–622.

Van Elsas, J.D., Bailey, M.J., 2002. The ecology of transfer of mobile genetic elements. FEMS Microbiol. Ecol. 42 (2), 187–197.

Villiers, F., Hugouvieux, V., Leonhardt, N., Vavasseur, A., Junot, C., Vandenbrouck, Y., Bourguignon, J., 2012. Exploring the plant response to cadmium exposure by transcriptomic, proteomic and metabolomic approaches: potentiality of high-throughput methods, promises of integrative biologyMetal Toxicity in Plants: Perception, Signaling and Remediation. Springer, Berlin, Heidelberg, pp. 119–142.

Willems, G., Dräger, D.B., Courbot, M., Godé, C., Verbruggen, N., Saumitou-Laprade, P., 2007. The genetic basis of zinc tolerance in the metallophyte Arabidopsis halleri ssp. halleri (Brassicaceae): an analysis of quantitative trait loci. Genetics, 176 (1), 659–674.

Williams, S.C.P, 2013. The other microbiome. Proc. Natl. Acad. Sci. U.S.A. 110, 2682–2684.

Xiong, YH, Yang, XE, Ye, ZQ, He, ZL., 2004. Characteristics of cadmium uptake and accumulation by two contrasting ecotypes of sedum alfrediihance. J. Environ. Sci. Heal. A 39 (11–12), 2925–2940.

Xu, J., Sun, J., Du, L., Liu, X., 2012. Comparative transcriptome analysis of cadmium responses in Solanum nigrum and Solanum torvum. New Phytol. 196 (1), 110–124.

Yıldırım, K., Uylaş, S., 2016. Genome-wide transcriptome profiling of black poplar (Populus nigra L.) under boron toxicity revealed candidate genes responsible in boron uptake, transport and detoxification. Plant Physiol. Biochem. 109, 146–155.

Zaier H., Ghnaya T., Ben Rejeb K., Lakhdar A., Rejeb S., Jemal F. (2010) Effects of EDTA on phytoextraction of heavy metals (Zn, Mn and Pb) from sludge-amended soil with Brassica napus. Bioresour Technol 101 (11):3978–3983.

Zhai, G., Lehmler, H.J., Schnoor, J.L., 2010. Identification of hydroxylated metabolites of 3, 3′, 4, 4′-tetrachlorobiphenyl and metabolic pathway in whole poplar plants. Chemosphere, 81 (4), 523–528.

Zhang, L., Rylott, E.L., Bruce, N.C., Strand, S.E., 2018. Genetic modification of western wheatgrass (Pascopyrum smithii) for the phytoremediation of RDX and TNT. Planta, 249 (4), 1007–1015.

Zhang, M., Pereira e Silva, M.D.C., De Mares Maryam, C., van Elsas, J.D, 2014. The mycosphere constitutes an arena for horizontal gene transfer with strong evolutionary implications for bacterial-fungal interactions. FEMS Microbiolo.gy Ecol. 89 (3), 516–526.

Zhang, J.Z., Zhao, K., Ai, X.Y., Hu, C.G. (2014a) Involvements of PCD and changes in gene expression profile during self-pruning of spring shoots in sweet orange (Citrus sinensis). BMC Genomics 15: 892.

Zhou, Q., Guo, J.J., He, C.T., Shen, C., Huang, Y.Y., Chen, J.X., Yang, Z.Y., 2016. Comparative transcriptome analysis between low-and high-cadmium-accumulating genotypes of pakchoi (Brassica chinensis L.) in response to cadmium stress. Environ. Sci, Technol. 50 (12), 6485–6494.

Chapter 10

Recent advancement in plant genetic engineering for efficient phytoremediation

Monika Bansal[a], Shabir H. Wani[b]
[a]*School of Agricultural Biotechnology, Punjab Agricultural University, Ludhiana, Punjab, India,* [b]*Mountain Research Center for Field Crops, Sher-e-Kashmir University of Agricultural Sciences and Technology of Kashmir, Khudwani, Jammu and Kashmir, India*

10.1 Introduction

Metals pollution present in the environment comes from weathering of earth crust, urban runoff, erosion of soil, mining, industrial discharge, effluents, air pollutants, pest and disease control agents that has a detrimental effect on the biosphere and disrupts the ecological equilibrium. The concentrations of contaminant can range from extremely toxic amount in an accidental spill to barely visible concentrations that can be dangerous to human health following long-term exposure. Latest estimates have revealed that more than five million hectares of land is already contaminated due to heavy metal worldwide. Heavy metals if enters in the food chain will be a matter of concern due to the risk to health of human beings (Ghori et al., 2019). Heavy metals are nondegradable and can remain in soil for hundreds of years. Contamination of heavy metals like cobalt (Co), iron (Fe), cadmium (Cd), mercury (Hg), arsenic (As), manganese (Mn), nickel (Ni), and copper (Cu) in agriculture is a matter of concern for global safety. Excessive deposition of metals has adverse effect on life by modifying cellular machinery like proteins, lipids, DNA, and enzymes. The various forms of remediation of polluted areas are chemical, physical, electrical, and biological methods are offered for cleaning polluted areas. The cleanup techniques for these contaminated soils includes physical methods like surface capping and encapsulation, whereas vitrification and electrokinetics come in the sense of electrical approach, further chemical method includes flushing and immobilization of the soil and last but most important is by phytoremediation in the sense of the biological approach (Liu et al., 2018). Another common cleaning approach is by improved bioremediation by inclusion of particular microbial strains with known properties

Phytoremediation. DOI: https://doi.org/10.1016/B978-0-323-89874-4.00001-7

for degrading the pollutant. Collectively, bacteria and fungi may use wide variety of organic molecules for their growth and development. However, in order to be effective in bioremediation using microbes at a given location, certain criteria must be met. This involve the capability of that microorganism to have preferred metabolic functions to thrive in that particular environment for which that requires availability of certain substrates and inducers for activating expression of the necessary enzymes. Mostly pollutants are metabolized by enzymes along with natural substrates, so the substrates must be present in that environment for expression of associated gene. This criterion is sometimes questionable if inducer chemical like phenol behaves as a harmful pollutant. Bioremediation is dependent on the availability of adequate supplies of carbon and energy. Although plants have natural and solar power dependent systems for cleaning of pollutes areas, which leads to their use for phytoremediation purpose (Pilon-Smits et al., 2005; Van Aken et al., 2008). Plants depend upon efficient root system for extracting nutrients and water from soil and groundwater. In plants, enzymes responsible for degradation of pollutants originated from natural defensive systems in response to a diversity of chemicals released by challenging organisms like microbes, plants, and pathogens (Singer et al., 2006). Phtoremediation technology focused on the use of plants for extraction, conversion, and stabilization of the organic and inorganic pollutants found in water, soils, or sediments (Cherian and Oliveira, 2005; Mello-Farias and Chaves, 2008). Phytoremediation is based on benefit of natural and selective uptake capacity of plants root systems for bioaccumulation, translocation, and degradation of pollutants.

10.2 Biochemical and physiological effect of contaminants on plants

The method of phytoremediation can be improved by study and modulation of the biochemical and physiological processes and most importantly genetic control of the target plant. Due to excess quantities of metal in the contaminated soil, plant faces growing number of biochemical and physiological changes like malfunction of photosynthetic machinery (Ozyigit et al., 2016; Sorrentino et al., 2018), poor germination, reduced seeds growth, impaired balance of water-electrolyte, poor rate of transpiration, misbalanced plant nutritional status, and activities of key enzymes associated events are also increasingly disturbed (Latef, 2013; Sarwar et al., 2017; Yadav et al., 2019). When exposed to any stress conditions in plants productions of reactive oxygen species (Dal-Corso, 2012; Shahid et al., 2014) are generated due to oxidative stress arising from impaired cellular reactions affects stress associated genes expression and negatively affects biomolecules, including proteins, lipids, and nucleic acids. There exists chemical similarity between heavy metals and some of plant's essential macronutrients, so binding of these metals impairs nutrition and related metabolic activities in plants. In most of the enzymes, cysteine-rich sites are essential

(Hossain et al., 2012) for their secondary, tertiary, and quaternary folding and stabilizing structure of reaction centers, which play a crucial roles for catalyzing the reactions. When heavy metals (Cd, Hg, As, and Pb) binds to such stabilization centers and restricts activities of enzymes, which further interrupts regular plant development and growth. In plants, metal transporters are needed for the absorption and translocation of metals to specific tissues. The genes coding for metal transporter in plants are responsible for tolerance and accumulation of toxic metal; hence, a great interest had been shown to understand their function and mechanism involved. Due to recent advancement in transcriptomic (Singh et al., 2016) and metabolomics studies, genes coding for regulatory factors and signaling pathways have emerged as modern tools for conferring metal tolerance and metabolism regulation.

10.3 Molecular mechanisms involved during phytoremediation

Huge numbers of plant genes are occupied with function for metal uptake, translocation and sequestration. Transfer of those genes to target plants is a potential technique for plant genetic modification for improving traits for phytoremediation. Transgenic plants are engineered for accumulation of high metal concentrations in harvestable parts. Overexpressing genes leads to improved metal uptake and intracellular targeting. Different genes from different sources are required to be moved to plants in order to improve performance of phytoremediation. These are genes for biodegradable enzymes, genes for improved production of biomass, genes for protein transfer of metals, genes to strengthen chelators and transporters of metal, genes for protein sequestration of metal, genes for increased root depth, and genes to modify oxidation (Fig. 10.1).

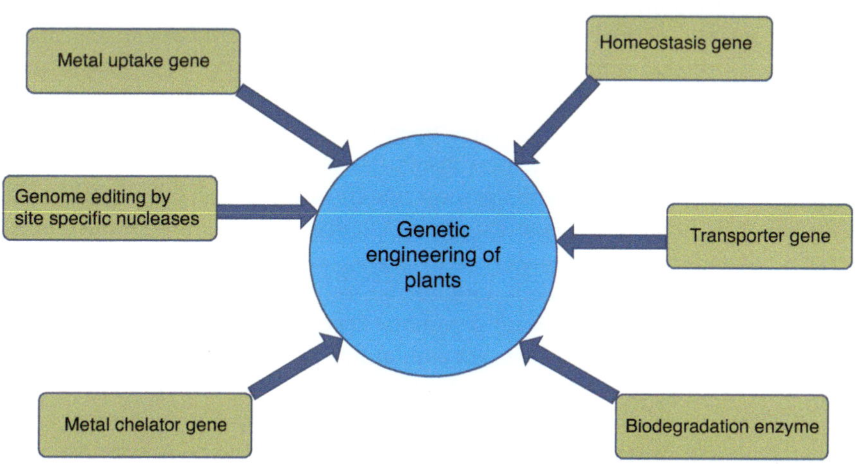

FIG. 10.1 An overview of genetic engineering approaches for phytoremediation.

The experiments in *Arabidopsis* and tobacco are mentioned here because they give fundamental evidence to promote the production of plants suitable for functional use. In this chapter, we are going to discuss few case studies where these genes were used successfully. Transgenic canola varieties have been developed from using the rice gene, OsMyb4, along with the COR15a stress-inducible promoter (Raldugina et al., 2018). The transgenic spring canola plant grown at a high Cu level (150 μM CuSO4) and Zn's (5000 μM ZnSO4) concentrations has been shown to have an extended survival potential (about 15 days) in comparison with wild form. This indicates that the OsMyb4 has a crucial role to play in survival, when plants are exposed to excessive concentration of Cu and Zn. Rui et al. (2016) proposed that the Cd stress in *Vicia sativa* induced the lignin development. These findings were used to generate transgenic *Arabidopsis thaliana* by cloning VsCCoAOMT gene from *V. sativa* (Xia et al., 2018). VsCCoAOMT genes have been found to play main functions that result in increased resistance to heavy metals. A new glutathione synthase enzyme γ-glutamylcysteine synthetase glutathione synthetase gene (StGCS-GS), was revealed in *Streptococcus thermophiles* (Liedschulte et al., 2010). As a result, the coding sequence of the StGCS-GS gene was cloned by Liu et al. (2015) and transferred into sugar beet to assess the possible increase in accumulation of heavy metals. Transgenic sugar beets lines showed increased tendency for retaining Cu, Cu, and Zn ions and exhibits improved glutathione and phytochelatin activities in heavy metal stress conditions (Table 10.1). Sun et al. (2018), utilized *Poplar* for phytoremediation of Hg from soils by using a class of ATP-binding cassette transporter gene. The ATP-binding cassette transporter gene (PtABCC1) was transferred to the *A. thaliana* plants. Transgenic lines showed higher rates of Hg accumulation in transgenic *Arabidopsis* and *Poplar* in comparison with wild plants. Nahar et al. (2017) cloned As reductase 2 gene of the *A. thaliana* for the purpose of phytoremediation. Transformation of 1356 bp AtACR2 open reading frame was performed on leaf desk using *Nicotiana tabacum* leaf disks by using *Agrobacterium* mediated genetic transformation. Transgenics lines were grown in medium-containing As for a duration of 45 days. As levels in the roots of transgenic tobacco have been shown to be considerably higher. Although As levels in transgenic tobacco shoots was found to be very low compared when compared with the roots, the transformation of the ACR2 gene may be a beneficial alternative for phytoremedial purposes. There have been reports where plants like *Populus angustifolia, Nicotiana tabacum,* and *Silene cucubalis* have been genetically engineered to overexpress glutamylcysteine synthetase, resulting in increased accumulation of heavy metal (Fulekar et al., 2008). Transgenic tobacco plants accumulating Cd, Ca, and Mn were reported after transfer of gene CAX-2 (vacuolar transporter) from *A. thaliana* (Hirschi et al., 2000). *A. thaliana* plants, which were resistant to Cu, aluminum (Al), and sodium (Na) were generated with the gene glutathione-S-transferase from tobacco (Ezaki et al., 2000).

TABLE 10.1 Few selected case studies of phytoremediation by transgenic approaches during last 5 years.

Plant species	Gene name	Response/tolerance	References
Arabidopsis thaliana	*AdPCs1-2-3*	Cadmium tolerance	Li et al. (2019)
	PCs1	Cadmium tolerance	Bai et al. (2019)
	PCs1	Cadmium tolerance	Zhang et al. (2018)
	YSL	Cadmium resistance	Chen et al. (2018)
	SaCAD	Cadmium accumulation	Qiu et al. (2018)
	SaNramp6	Cadmium accumulation	Chen et al. (2017)
	IlMT2b	Enhances copper tolerance	Gu et al. (2015)
	PtoHMA5	Improved cadmium transport	Wang et al. (2018)
	AtACR2	Arsenic tolerance	Nahar et al. (2017)
	OsMTP1	Enhanced cadmium accumulation/tolerance	Das et al. (2016)
Oryza sativa	*ricMT*	Copper and cadmium tolerance	Zhang et al. (2017)
Brassica napus	*CKX2 OsPCs5/ OsPCs15*	Cadmium and zinc tolerance	Nehnevajova et al. (2019)
Brassica napus	*OsMyb4*	Copper/zinc tolerance	Raldugina et al. (2018)
Brassica juncea	*AtACBP1 AtACBP4*	Accumulation of lead	Du et al. (2015)

10.4 Genome editing for phytoremediation

Targeted mutagenesis and Site-specific gene insertion/deletion can be achieved by genome editing methods by site specific nucleases like zinc-finger nucleases, transcription activator like effector nucleases, and clustered regularly interspersed, short palindromic repeats (CRISPR/Cas9) (Basu et al., 2018). After introduction of double-strand DNA breaks at target sites by these machineries, DNA repair mechanism is utilized to repair these breaks. Involving either the homologous recombination or nonhomologous end joining, the broken DNAs ends are joined again after insertion or deletions. CRISPR is a revolutionary tool for genome-engineering that allows the improvement of the basic traits of the plant genome. It had proved to be one of the promising approaches in the field of functional genomics (Perez-Pinera et al., 2013). The strength of CRISPR/

Cas9 system is because of its simplicity, efficiency, and versatility. Any of the hardest plant modifications, which is reported to have been made quite hard to achieve earlier can be done easily by using CRISPR/Cas9 technology. The use of genetic engineering techniques for bioremediation shall lead to the production of increased biomass content, strong root system, and hyperaccumulation plants that may be able to grow under minimum specific environmental conditions (Mosa et al., 2016). In general, plant's phytoremediation properties can be improved significantly by the manipulation of gene expression by CRISPR/Cas9 approach to increase overexpression of metal ligands like metallothioneins and phytochelatins along with increased in synthesis of metal transport proteins.

10.5 Future prospects

For the last two decades, genetic engineering techniques have become a field of interest for generating advances phenotypes to fulfill global demands. This approach involves changes in few nucleotides to further deletion of several megabases. The phytoremediation experiments using diverse transgenes have resulted in positive findings to date and more success is expected in future due to the advances in combinational omics technologies. There is an increased amount of sequence and expression data available for the metallophyte community, and further increased number of species capable of genetic transformation. The concern related to use of genetically modified species as the hypothetical danger for human health and the probable dispersion of the transgene in the ecosystem. Phytoremediation by green plants makes environmental harmless and free of pollutants and provides an efficient, non-destructive and cheap method for remediation of contaminated soils. To receive the huge breakthrough in the method of bioremediation to establish a viable and sustainable solution for a clean climate, these approaches can be mapped and incorporated from theoretical to experimental techniques.

References

Basu, S., Rabara, R.C., Negi, S., Shukla, P., 2018. Engineering PGPMOs through gene editing and systems biology: a solution for phytoremediation? Trends Biotechnol. 36 (5), 499–510.

Bai, J., Wang, X., Wang, R., Wang, J., Le, S., Zhao, Y., 2019. Overexpression of three duplicated BnPCS genes enhanced Cd accumulation and translocation in *Arabidopsis thaliana* mutant cad1–3. Bull. Environ. Contam. Toxicol. 102 (1), 146–152.

Chen, H., Zhang, C., Guo, H., Hu, Y., He, Y., Jiang, D., 2018. Overexpression of a *Miscanthus sacchariflorus* yellow stripe-like transporter MsYSL1 enhances resistance of Arabidopsis to cadmium by mediating metal ion reallocation. Plant Growth Regul. 85, 101–111.

Chen, S.S., Han, X.J., Fang, J., Lu, Z.C., Qiu, W.M., Liu, M.Y., Sang, J., Jiang, J., Zhuo, R.Y., 2017. Sedum alfredii sanramp6 metal transporter contributes to cadmium accumulation in transgenic arabidopsis thaliana. Sci. Rep. 7, 13318.

Cherian, S., Oliveira, M.M., 2005. Transgenic plants in phytoremediation: recent advances and new possibilities. Environ. Sci. Technol. 39 (24), 9377–9390.

DalCorso, G., 2012. Heavy metal toxicity in plants. In: Furini, A. (Ed.), Plants and Heavy Metals. Springer, Heidelberg, pp. 1–26.

Das, N., Bhattacharya, S., Maiti, M.K., 2016. Enhanced cadmium accumulation and tolerance in transgenic tobacco overexpressing rice metal tolerance protein gene osmtp1 is promising for phytoremediation. Plant Physiol. Biochem. 105, 297–309.

Du, Z.Y., Chen, M.X., Chen, Q.F., Gu, J.D., Chye, M.L., 2015. Expression of arabidopsis acyl-coa-binding proteins atacbp1 and atacbp4 confers pb(ii) accumulation in brassica juncea roots. Plant Cell Environ. 38 (1), 101–117.

Ezaki, B., Gardner, R.C., Ezaki, Y., Matsumoto, H., 2000. Expression of aluminium induced genes in transgenic Arabidopsis plants can ameliorate aluminium stress and/or oxidative stress. Plant Physiol. 122 (3), 657–665.

Fulekar, M.H., Singh, A., Bhaduri, A.M., 2008. Genetic engineering strategies for enhancing phytoremediation of heavy metals. Afr. J. Biotechnol. 8 (4), 529–535.

Ghori, T., Hayat, M.Q., Imadi, S.R., et al., 2019. Heavy metal stress and responses in plants. Int. J. Environ. Sci. Technol. 16, 1807–1828.

Gu, C.-S., Liu, L.-Q., Deng, Y.-M., Zhu, X.-D., Huang, S.-Z., Lu, X.-Q., 2015. The heterologous expression of the Iris lactea var chinensis type 2 metallothionein IlMT2b gene enhances copper tolerance in Arabidopsis thaliana. Bull. Environ. Contam. Toxicol. 94 (2), 247–253.

Hirschi, K.D., Korenkov, V.D., Wilganowski, N.L., Wagner, G.J., 2000. Expression of Arabidopsis CAX2 in tobacco altered metal accumulation and increased manganese tolerance. Plant Physiol. 124 (1), 125–133.

Hossain, M.A., Piyatida, P., da Silva, J.A.T., Fujita, M., 2012. Molecular mechanism of heavy metal toxicity and tolerance in plants:central role of glutathione in detoxification of reactive oxygen species and methylglyoxal and in heavy metal chelation. J. Bot. Article ID 872875.

Latef, A.A.A., 2013. Growth and some physiological activities of pepper (*Capsicum annuum* L.) in response to cadmium stress and mycorrhizal symbiosis. J. Agric. Sci. Technol. Iran. 15, 1437–1448.

Li, M., Stragliati, L., Bellini, E., Ricci, A., Saba, A., Sanità di Toppi, L., Varotto, C., 2019. Evolution and functional differentiation of recently diverged phytochelatin synthase genes from *Arundo donax* L. J. Exp. Bot. 70 (19), 5391–5405.

Liu, D., An, Z., Mao, Z., Ma, L., Lu, Z., 2015a. Enhanced heavy metal tolerance and accumulation by transgenic sugar beets expressing Streptococcus thermophilus StGCS-GS in the presence of Cd, Zn and Cu alone or in combination. Plos One 10, e0128824.

Liu, L.W., Li, W., Song, W.P., Guo, M.X., 2018. Remediation techniques for heavy metal-contaminated soils: principles and applicability. Sci. Total Environ. 633, 206–219.

Liedschulte, V., Wachter, A., An, Z.G., Rausch, T., 2010. Exploiting plants for glutathione (GSH) production: uncoupling GSH synthesis from cellular controls results in unprecedented GSH accumulation. Plant Biotechnol. J. 8, 807–820.

Mosa, K.A., Saadoun, I., Kumar, K., Helmy, M., Dhankher, O.P., 2016. Potential biotechnological strategies for the cleanup of heavy metals and metalloids. Front. Plant Sci. 7, 303.

Mello-Farias, P.C., Chaves, A.L.S, 2008. Biochemical and molecular aspects of toxic metals phytoremediation using transgenic plants. In: Tiznado Hernandez, M.E., Troncoso Rojas, R., Rivera Domínguez, M.A. (Eds.), Transgenic Approach in Plant Biochemistry and Physiology. Research Signpost, Kerala, India, pp. 253–266.

Nahar, N., Rahman, A., Nawani, N.N., Ghosh, S., Mandal, A., 2017. Phytoremediation of arsenic from the contaminated soil using transgenic tobacco plants expressing ACR2 gene of *Arabidopsis thaliana*. J. Plant Physiol. 218, 121–126.

Nehnevajova, E., Ramireddy, E., Stolz, A., Gerdemann-Knörck, M., Novák, O., Strnad, M., Schmülling, T., 2019. Root enhancement in cytokinin- deficient oilseed rape causes leaf mineral enrichment, increases the chlorophyll concentration under nutrient limitation and enhances the phytoremediation capacity. BMC Plant Biol. 19, 83.

Ozyigit, I.I., Dogan, I., Igdelioglu, S., Filiz, E., Karadeniz, S., Uzunova, Z., 2016. Screening of damage induced by lead (Pb) in rye (*Secale cereale* L.)—a genetic and physiological approach. Biotechnol. Biotechnol. Equip. 30 (3), 489–496.

Perez-Pinera, P., Kocak, D.D., Vockley, C.M., Adler, A.F., Kabadi, A.M., Polstein, L.R., et al., 2013. RNA-guided gene activation by CRISPR-Cas9_based transcription factors. Nat. Methods 10 (10), 973.

Pilon-Smits, E., 2005. Phytoremediation. Ann. Rev. Plant Biol. 56 (15), 39. https://doi.org/10.1146/annurev.arplant.56.032604.144214.

Qiu, W., Song, X., Han, X., Liu, M., Qiao, G., Zhuo, R., 2018. Overexpression of sedum alfredii cinnamyl alcohol dehydrogenase increases the tolerance and accumulation of cadmium in arabidopsis. Environ. Exp. Bot. 155, 566–577.

Raldugina, G.N., et al., 2018. Expression of rice OsMyb4 transcription factor improves tolerance to copper or zinc in canola plants. Biol. Plant 62, 511–520.

Rui, H.Y., Chen, C., Zhang, X.X., Shen, Z.G., Zhang, F.Q., 2016. Cd induced oxidative stress and lignification in the roots of two *Vicia sativa* L. varieties with different Cd tolerances. J. Hazard. Mater. 301, 304–313.

Shahid, M., Pourrut, B., Dumat, C., Nadeem, M., Aslam, M., Pinelli, E., 2014. Heavy-metal-induced reactive oxygen species: phytotoxicity and physicochemical changes in plants. Rev. Environ. Contam. Toxicol. 232, 1–44.

Sarwar, N., et al., 2017. Phytoremediation strategies for soils contaminated with heavy metals: modifications and future perspectives. Chemosphere 171, 710–721.

Singh, S., Parihar, P., Singh, R., Singh, V.P., Prasad, S.M., 2016. Heavy metal tolerance in plants: role of transcriptomics, proteomics, metabolomics, and ionomics. Front. Plant Sci. https://doi.org/10.3389/fpls.2015.01143.

Singer, A., 2006. The chemical ecology of pollutants biodegradation. In: Mackova, M., Dowling, D., Macek, K. (Eds.), Phytoremediation and Rhizoremediation: Theoretical Background. Springer.

Sorrentino, M.C., Capozzi, F., Amitrano, C., Giordano, S., Arena, C., Spagnuolo, V., 2018. Performance of three cardoon cultivars in an industrial heavy metal-contaminated soil: effects on morphology, cytology and photosynthesis. J. Hazard. Mater. 351, 131–137.

Sun, L, et al., 2018. Overexpression of PtABCC1 contributes to mercury tolerance and accumulation in Arabidopsis and poplar. Biochem. Biophys. Res. Commun. 497 (4), 997–1002.

Van Aken, B., 2008. Transgenic plants for phytoremediation: helping nature to clean up environmental pollution. Trends Biotechnol. 26 (5), 225–227.

Wang, X.T., Zhi, J.K., Liu, X.R., Zhang, H., Liu, H.B., Xu, J.C., 2018. Transgenic tobacco plants expressing a P1B-ATPase gene from Populus tomentosa Carr. (PtoHMA5) demonstrate improved cadmium transport. Int. J. Biol. Macromol. 113, 655–661.

Xia, Y., Liu, J., Wang, Y., Zhang, X.X., Shen, Z.G., Hu, Z.B., 2018. Ectopic expression of *Vicia sativa* Caffeoyl-CoA O-methyltransferase (VsCCoAOMT) increases the uptake and tolerance of cadmium in Arabidopsis. Environ. Exp. Bot. 145, 47–53.

Yadav, A., Raj, A., Purchase, D., Ferreira, L.F.R., Saratale, G.D., Bharagava, R.N., 2019. Phytotoxicity, Cytotoxicity and Genotoxicity Evaluation of Organic and Inorganic Pollutants Rich Tannery wastewater from a Common Effluent Treatment Plant (CETP) in Unnao district, India using Vigna radiata and Allium cepa. Chemosphere. 24:324–332.

Zhang, X., Rui, H., Zhang, F., Hu, Z., Xia, Y., Shen, Z., 2018. Overexpression of a functional *Vicia sativa* PCS1 homolog increases cadmium tolerance and phytochelatins synthesis in Arabidopsis. Front. Plant Sci 9, 107.

Chapter 11

Targeted genetic modification technologies: Potential benefits of their future use in Phytoremediation

Hiralal Sonawane[a], Sagar Arya[b,c], Ashish Bedi[b,c], Akanksha Jaiswar[c]
[a]*Department of Botany, Ramkrishna More College of Arts Commerce and Science, Pune, India,* [b]*School of Life and Environmental Sciences, Deakin University, Victoria, Australia,* [c]*TERI-Deakin Nanobiotechnology Centre, The Energy and Resources Institute, Gurugram, Haryana, India*

11.1 Introduction

Phytoremediation is an ecological, environment –friendly, and affordable strategy for exploring the prospects of plants to remediate and restore contaminants and harmful pollutants (Pilon-Smits, 2005). Technically, plants that can absorb/adsorb hazardous compounds and biotransform, bioaccumulate, or biosequester them are used in the process of phytoremediation. There are some additional benefits to this approach, such as mitigation of soil erosion and control of solubilized contaminants by hydraulic activity, altogether representing a sustainable alternative to the pollution generated by anthropogenic activity (Mendez and Maier, 2008; Srivastava, 2016). The rapidly increasing population in the world has significantly impacted agricultural benefits, productivity, and arable land (Abhilash, 2012). To meet the ever-increasing global food requirements, excessive use of pesticides, insecticides, and other synthetic chemical products (fertilizers) in agriculture has resulted in a substantial enhancement in the overall pollution levels in the lithosphere and atmosphere. The heavy use of such chemicals results in the release of toxic organic/inorganic compounds and also heavy metals, which negatively affects the environment (Cheng, 2016; Larsson, 2014). Phytoremediation encompasses different processes employed by plants such as phytoextraction, phytostabilization, rhizofiltration, biotransformation, biosequestration, bioaccumulation, phytodegradation, and phytovolatilization to remove contaminants from soil. The pollutants released from

Phytoremediation. DOI: https://doi.org/10.1016/B978-0-323-89874-4.00007-8

toxic chemicals mainly belong to the polychlorinated biphenyls, the hydrocarbons, petroleum-based hydrocarbons, phthalate esters, polycyclic aromatic nitroaromatic compounds, industrial solvents, trichloroethylene, and the heavy metals. (Deblonde et al., 2011; Perelo, 2010; Dangi et al., 2018). Several plant species have the ability to uptake and sequester carbon products and use them to produce biomass and valuable primary and specialized plant metabolites (Basu et al., 2018., Srivastava, 2016). These processes basically depend on the efficiency of the plant root and shoot system for the uptake and assimilation of metals and other xenobiotics (Callahan et al., 2006; Karenlampi et al., 2000). Apart from this, the selective uptake capabilities of plants, along with the translocating, the bioaccumulating, and the metal biodegrading ability make them a better alternative to human or machine intensive remediation processes (Mello-Farias et al., 2011). Hence, exploiting plants' ability to sequester pollutants can certainly ensure ecological balance, cost reduction, and human/environmental safety.

Some plants involved or studied for the phytoremediation with high contamination neutralizing ability are reviewed and discussed elsewhere (Sudhir et al., 2014; Eevers et al., 2017, Singh et al., 2017). Very few among them can accumulate potentially toxic compounds to the concentration of 50- to 500-fold higher than the normal average plants. These plants are categorized as hyperaccumulators as they can grow and sustain in highly contaminated soils. However, despite such benefits, the plant-based remediation faces certain limitations. Moreover, it is observed that most of the plants are selective toward the spatiotemporal requirements and their ability to work as a phytoremediator is limited to a certain extent. Also, their growth rate and biomass accumulation are inversely affected, that is, they decrease with the enhancement in the concentration of pollutants in their habitat. Therefore, it is essential to resort to the transgenic approaches to develop and enhance the remediation potential and at the same time reduce the specificity of spatiotemporal needs.

The understanding and manipulation of the genetic makeup of the plants has proved to be an effective strategy to increase their ability to grow under various abiotic stress factors, provide them climate-resilient, increase their nutritive properties, and allow them to retain their vigor in different spatiotemporal conditions. Owing to this advantage, the transgenic approaches are heavily relied and therefore used to enhance the remedial capacity of plants. The plant genetic engineering should become an important and mainstream approach to improve the hyperaccumulation of contaminants and their biosequestration in the plants. The information of the processes such as uptake, translocation, and distribution in the metal hyperaccumulators offers opportunities to use molecular biology tools to develop sustainable and economic phytoremediation process in heterologous plants or nonhyperaccumulators (Chaney et al., 1997; Fulekar et al., 2008). However, the plants considered for genetic manipulation to improve the remediating potential should fulfill certain criteria such as high biomass with internal phytoremediation capability and amenable to genetic transformation (Eapen and D'Souza, 2005; Mello-Farias and Chaves, 2008).

Targeted gene editing technique has enormous applicability in the field of plant sciences, as it can produce insertions and deletions at desired genetic locations of the plant. However, gene editing differs from the traditional gene manipulation techniques that do not necessarily incorporate in the foreign DNA (Kumar et al., 2016; Gupta and Shukla, 2017). Multigene engineering approach are involved in the transfer of several genes into suitable plants to remove contaminants and it may resolve the problem of mixed contamination of the polluted sites (Arya et al., 2020a; Lasat et al., 2002). The fundamental knowledge of the plant genes encoding for metal ion transporters and hyperaccumulating inclusion bodies may be useful to expand the capacity of plants to remediate contaminants and toxic pollutants from soil (Yang et al., 2005; Mello-Farias and Chaves, 2008). This knowledge can be harnessed and employed by altering the genes or by genome molecular editing toolboxes such as mega nucleases, zinc finger nucleases, (ZFNs), transcription activator-like effector nucleases (TALENs), and clustered regularly interspaced palindromic repeats-Cas associated protein (CRISPR-Cas) which can be used to modify the plants genome or regulate the gene expression (Foht, 2016; Voytas and Gao, 2014). These techniques can improve the bioremediation processes such as removing and eliminating pollutants, toxic chemicals from soil and degradation of chemical fertilizers and pesticides (Basu et al., 2018; Hussain et al., 2018). The genetic modification in the model plant such as *Arabidopsis thaliana* can serve as a proof-of-concept study to assess the function of genes and sharpen the effectivity of gene editing tools (Arya et al., 2020b), followed by implementing the same in plants involved in the process of phytoremediation.

11.2 Phytoremediators

More than 400 species of the natural hyperaccumulators were identified as phytoremediators and *Thlaspi caerulescens* is the most explored among them (Brown et al., 2015). The plants such as *Festuca rubra*, *Noccaea caerulescens*, *Jasione montana*, *Rumex acetosa*, and *Plantago lanceolata* accumulated metals from the contaminated soil and acted as potential hyperaccumulators (Wiszniewska et al., 2015). Some of the plants studied as genomes of model phytoremediation, such as *N. caerulescens*, *Arabidopsis halleri*, *Pteris vittata*, *Hirschfeldia incana*, *Brassica juncea*, have been fully or partially sequenced (Mandáková et al., 2015, Briskine et al., 2017, Xie et al., 2009). Tolerance to various metals such as Cd, Cu, Fe, Mn, Ni, and Zn, showed by Arabidopsis and tobacco plants improved with the introduction of NAS1 gene (Kim et al., 2005, Pianelli et al., 2005).

In Arabidopsis *AtNRAMP1*, *AtNRAMP3*, and *AtNRAMP4* genes transported in to heavy metal Cd, Mn, and Fe through heterologous expression and play important role in tolerance of metals such as Zn and Cd (Ehrnstorfer et al., 2014; Oomen et al., 2009). In fiber crops, genetic engineering of flax for heavy metals tolerance and accumulation has potential to the improve phytoremediation (Vrbova et al., 2013). In *Linum usitatissimum*, overproduction of Cd-binding peptide by genetic fusion of α-domain of mammalian MT 1 (alpha MT1a) and uidA gene was engineered with cysteine-rich peptides having the ability of high-affinity coordination to heavy metal ions (Coyle et al., 2002; Freisinger, 2008). The genetic engineering of some species such as tobacco, Arabidopsis, and Brassica has improved production of heterologous metal-binding proteins/peptides, such as MTs, thereby increasing the capability of the engineered plants to tolerate and accumulate the heavy metal ions (Bhargava et al., 2012; Kotrba et al., 2009). Transgenic *B. juncea* expressing yeast cadmium factor1 (YCF1), which involved in the metal transport, showed tolerance to Cd and Pb (Bhuiyan et al. 2011). Metallothionein gene MT2b exhibits an increased tolerance and accumulation to Pb even at high concentration. Similarly, the *B. juncea* transformed with the APS and SMT genes that showed the tolerance for Se (LeDuc et al., 2006; Lv et al., 2013). The APS and SMT genes are involved in the synthesis of selenocysteine methyltransferase and ATP sulfurylase. Likewise, the gene from the rice plant encoding for a metal transporter OsNRAMP5 mutated using the CRISPR-Cas system resulted in reduced Cd accumulation in hydroponic condition as well as in Cd-contaminated paddy field. Similarly, it was found that rice OsNRAMP1 expressed in yeast system restored the growth of Fe-sensitive yeast mutant, suggesting its potential role in hyperaccumulation of Cd in rice. The OsNRAMP5 has absorbed and transported in to the metals Cd, Mn, and Fe (Ishimaru et al., 2012; Tang et al., 2017). Therefore, more information on the transporters such as OsNRAMP5 and OsNRAMP1 is vital to use them as a target for increasing phytoremediation efficiency of native as well as heterologous plants. Table 11.1 provides a summary of some notable transgenic plants with increased remediation potential.

In phytoremediators, the biomass production and bioconcentration efficiency are the major parameters. Introduction of new characters to remediate contaminants to either valuable plant metabolites or high biomass is one of the important attributes of phytoremediators. Furthermore, overexpression of the genes encoding for enzymes, proteins, and transporters that are essential in the metal adsorption/absorption, uptake, transport, transformation, accumulation, and sequestration is also an option (Macek et al., 2012). In a quest to enhance phytoremediation, using genome editing tools is essential for targeted genome modification and introduction of novel genes in an accelerated manner. Hence, next we will shed light on such prominent genome editing toolboxes that is used so far in the field of plant sciences to enhance the potential of available phytoremediators.

TABLE 11.1 **Details of some notable examples of transgenic plants involved in development to facilitate and enhance phytoremediation efficiency.**

Source	Gene isolated	Target plant	Phytoremediation	References
Pea	PsMTA	*Arabidopsis thaliana*	Cu accumulation	Evans et al. (1992)
Saccharomyces cerevisiae	*CUP1*	Cauliflower	$CdCl_2$ tolerance	(Hasegawa et al., 1997)
Arabidopsis thaliana	*ZAT1*	*Arabidopsis thaliana*	Zn tolerance	(Van der Zaal et al.,1999)
Arabidopsis thaliana	*AtMHX1, AtCAX2*	Tobacco	Mg and Zn tolerance	(Hirschi et al., 2000; Shaul et al., 1999)
Tobacco	*NtCBP4*	Tobacco	Pb tolerance	(Sunkar et al. 2000
Arabidopsis thaliana	*AtPCS1*	*Arabidopsis thaliana*	Hypersensitivity to $CdCl_2$ and $ZnCl_2$	(Lee et al., 2003)
Wheat	*TaPCS1*	Tobacco	Pb and Cd tolerance	(Gisbert et al., 2003)
Saccharomyces cerevisiae	*CUP1*	Tobacco	Cu tolerance	(Thomas et al., 2003)
Arabidopsis thaliana	*APS* and *CGS*	Indian mustard	Tolerance to Se	(Van Huysen et al., 2003)
Arabidopsis bisulcatus	*SMT*	*Arabidopsis thaliana*	MeSeCys accumulation	(Ellis et al. 2004)
Arabidopsis thaliana	*OASTL*	*Arabidopsis thaliana*	Cd accumulation	(Dominguez-Solis et al., 2004)
Arabidopsis thaliana	*AtPCS1*	*Arabidopsis thaliana*	As and Cd hypersensitivity	(Li et al., 2004)
Arabidopsis thaliana	*PCS1*	*Brassica juncea*	As tolerance	Gasic and Korban, (2007)

(continued)

TABLE 11.1 (Cont'd)

Source	Gene isolated	Target plant	Phytoremediation	References
Pteris vittate	*GRX5*	*Arabidopsis thaliana*	As tolerance	(Sundaram et al., 2009)
Thlaspi goesingense	*TgMTP1*	Yeast	Zn accumulation	(Gustin et al., 2009)
Arabidopsis thaliana	*AtHMA4*	Tobacco	Improve root growth in the presence of Zn, Cd, and Co	(Siemianowski et al., 2011)
Medicago sativa	*HSP23*	*Festuca arundinacea*	As tolerance	(Lee et al., 2012)
Oidiodendron maius	*OmFET*	Yeast	Zn tolerance	(Khouja et al., 2013)
Pteris vittate	*ACR3*	*Arabidopsis thaliana*	As tolerance	(Chen et al., 2013)
Noccaea Caerulescens	*NcNramp1*	*Arabidopsis thaliana*	Cd tolerance	(Milner et al., 2014)
Oryza sativa	*OsHMA3*	*O. sativa*	Cd accumulation and Zn transporter overexpression	(Sasaki et al., 2014)
Cucumis sativus	*CsMTP1, CsMTP4*	Yeast	Enhance Zn and Cd tolerance	(Migocka et al., 2014)
Arabidopsis thaliana	*AtABCC3*	*Saccharomyces cerevisiae*	Cd tolerance	(Brunetti et al., 2015)
Cucumis sativus	*CsHMA5,1* *CsHMA5,2*	Yeast	Cu sequestration	(Migocka et al., 2015)
E. coli	*ECS*	*Populus tremula* and *Populus alba*	Cd remediation	(He et al., 2015)

O. sativa	*GRX_C7, GRX_C2.1*	*Arabidopsis thaliana*	As tolerance	(Verma et al., 2016)
O. sativa	*OsMTP1*	Transgenic tobacco	Cd hyperaccumulation	(Das et al., 2016)
Pseudomonas	*mtL*	*Arabidopsis thaliana* and *O. sativa*	Polyaromatic hydrocarbons Degradation	(Peng et al., 2017)
Saccharomyces cerevisiae	*Pdr5p*	*Arabidopsis thaliana*	Xenobiotics tolerance	(Remy et al., 2017)
B. napus	*BnNRAMP1b*	*B. juncea*	Enhance metal uptake	(Meng et al., 2017)
Sedum plumbizincicola	*SpHMA3*	*Sedum alfredii*	Cd accumulation and tolerance	(Liu et al., 2017)
Pseudomonas putida	*CzcB, CzcA*	Tobacco	Cd accumulation	(Nesler et al., 2017)
Arabidopsis thaliana	*ACR2*	Tobacco	As remediation	(Nahar et al., 2017)
Human	*CYP1A2*	*Arabidopsis thaliana*	Phenylurea herbicide isoproturon degradation	(Azab et al., 2020)
Sphingomonas wittichii DC-6	*cndA*	*Arabidopsis thaliana*	Acetochlor degradation	(Chu et al., 2020)

11.3 Molecular toolboxes for gene/genome editing

Gene/genome editing in plants altogether is considered as a permanent modification in the nuclear or chloroplast DNA at either a specific site or multiple locations. As mentioned earlier, ZFNs, TALENs, and CRISPR-Cas are among in the most important and widely used nucleases for genome editing. Table 11.2 provides some primary characteristics of these endonucleases. However, owing

TABLE 11.2 Salient feature of genome modification toolboxes: ZFNs, TALENs, and CRISPR-Cas.

Characteristics	ZFNs	TALENs	CRISPR-Cas
DNA sequence recognition	9–18 bp	30–40 bp	22 bp + PAM (protospacer adjacent motif) sequence
Mechanism of interaction with target DNA	DNA–protein	DNA–protein	DNA–RNA
DNA cleavage	Coupling with *FokI* nuclease	Coupling with *FokI* nuclease	Innate to Cas
Designing	Challenging, libraries are available for zinc finger motifs with predefined target specificity	Easy, TALE motifs with their specificities are well studied	Easy, sgRNA based on complementarity with the target DNA sequence
Cloning	Need to engineer linkages among juxtaposed zinc finger motifs	Linkages not required and the TALE motif cloning can be performed with Golden gate assembly	Cas vectors are well studied and are commercially available. sgRNA is also delivered into the plant cells as a DNA expression vector or directly as an RNA molecule or preloaded Cas-RNA complex
Mechanism of cleavage	Double stranded breaks by *FokI*	Double stranded breaks by *FokI*	Single-/double-stranded breaks by Cas
Usage and throughput	Low	Low to moderate	High

to the complexities involved in the synthesis and efficient functioning, ZNFs and TALENs have now superseded CRISPR-Cas. Below we have briefly highlighted the functioning of these nucleases and have also provided a detailed use of CRISPR-Cas as a next-generation genome editing toolbox in the field of phytoremediation.

11.3.1 Zinc finger nucleases (ZFNs)

ZFNs were the first to be used for genome editing and are synthesized by using a naturally occurring transcription factor named Zinc finger proteins fused with FokI nucleases (Kim et al., 1996). FokI endonuclease functions as a dimer, and hence, the DNA is cleaved when two ZFNs are juxtaposed on the either side of complementary DNA strands. ZFNs specifically recognize a trinucleotide DNA sequence and if arranged in series, ZFNs can recognize a longer sequence thereby increasing the specificity (Fig. 11.1A). However, it is observed that in a longer repeat of ZFNs, a ZFN can limit the specificity of the neighboring ZFN, which ultimately compromises the overall functioning of the nuclease. Furthermore, synthesizing an array of ZFNs is technically more challenging, time-consuming, and laborious.

11.3.2 Transcription activator-like effector nucleases (TALENs)

TALENs are like ZFNs and comprise fusion proteins belonging to bacterial TALE proteins and FokI endonuclease (Sanjana et al., 2012). Their target specificity is a result of DNA-protein association, where one TALE recognizes a single nucleotide, and therefore, an array of TALEs is needed to target a longer sequence to achieve high specificity (Fig. 11.1B). However, unlike ZFNs the specificity of a single TALE is not affected by its neighboring ones, making the engineering of TALENs much easier.

11.3.3 CRISPR-Cas

CRISPR-Cas is a precision molecular toolbox that can be used for functional elucidation of genetic elements, modify DNA at specific sites, and is labeled as a most effective gene editing technique discovered so far. CRISPR collectively with Cas proteins are known as the CRISPR-Cas system. Cas is a DNA endonuclease guided by RNA to the target location for gene/DNA manipulation (Fig. 11.1C). They are mainly categorized in three types, that is, type I, II, and III. Each system has specific Cas with distinct function based on the organism they originate from (Bortesi and Fischer, 2015; Mahas and Mahfouz, 2018). Among these types, the type II is highly explored, whereas types I and III involve Cas endonucleases that make the pre-crRNAs, this crRNA forms complex with Cas protein that recognizes and cleaves nucleic acid bases complementary to the crRNA (Bortesi and Fischer, 2015; Jinek et al., 2012). The

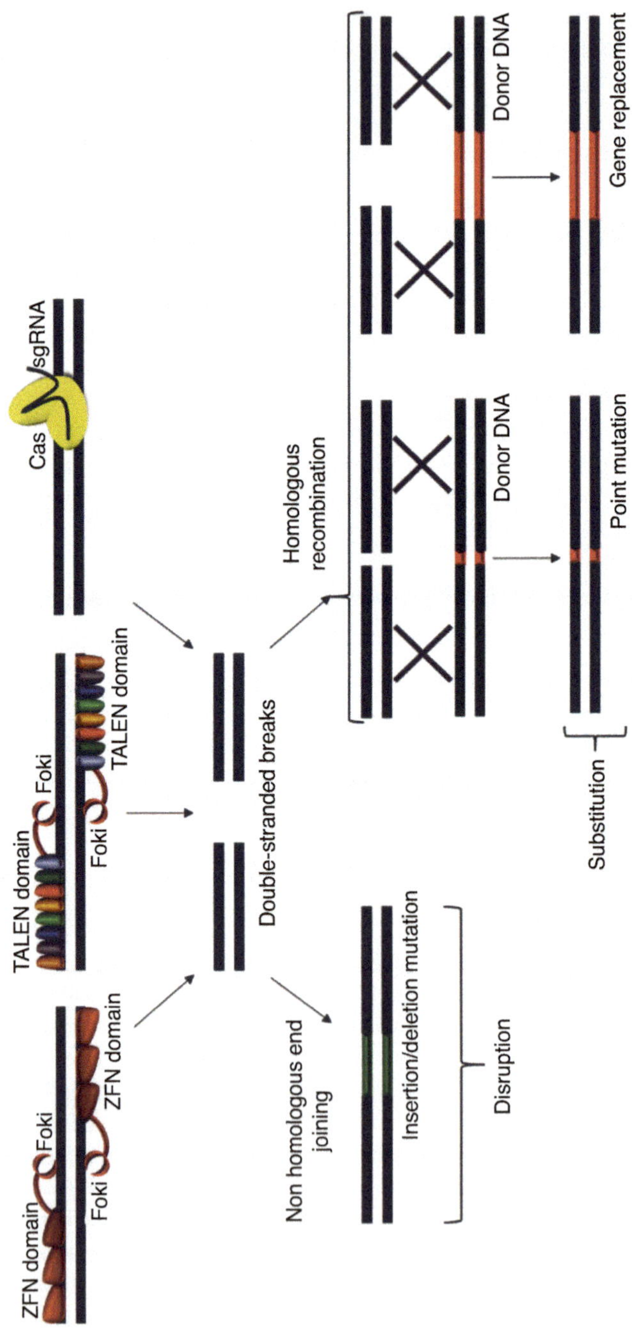

FIG. 11.1 (A) Zinc finger nuclease, (B) Transcription activator-like effector nuclease, (C) CRISPR-Cas.

processing and transcription crRNA is featured by relative construction simplicity along with high functional efficiency (Zhang et al., 2018a, Nemudryi et al., 2014). CRISPR is basically 30–40 bp direct repeat sequence that is separated by a spacer region complementary to the foreign DNA sequence. The gRNA (guide RNA) originated from CRISPR binds specifically to the target DNA site, which constitutes the particularity and specificity of the Cas cleavage (Listgarten et al., 2018; Shah et al., 2018). Thus, the gene or DNA can be altered using CRISPR-Cas by the introduction of double-strand breaks at the target site (Shapiro et al., 2018). For this, a suitable expression system is essential for efficient expression of the CRISPR-Cas sgRNA sequence and ideal promoters for transcription of sgRNA and Cas protein, and effective functionality of the codon-optimized variant of Cas is a prerequisite (Rico et al., 2018).

The changes and improvements that can be made in a plant genome using CRISPR include gene replacements, insertion/deletion mutations, inversions, knockouts, and translocations. These modifications are widely applied and studied in plant stress physiology, plant and microbial biotechnology, plant and alga-based biofuel production, and plant pathology. The modifications have also shown to strengthen the nutrient uptake and improve the bioavailability of metals and xenobiotics for plants through plant-microbe interactions (Abhilash et al., 2012; Saxena et al., 2020). Alteration of plant traits may also modulate their symbiotic relationship or resistance with/to bacteria/fungi, thereby exhibiting beneficial effects on the overall process of phytoremediation (Pandey and Singh, 2019). For instance, CRISPR-Cas mediated introduction of mutation in OsERF922 (rice ERF transcription factor) resulted in enhanced resistance to *Magnaporthe oryzae* and mutations in the ALS2 (acetolactate synthase) gene of maize resulted in tolerance to chlorsulfuron (Svitashev et al., 2015, Liu et al., 2012). These effects are indirectly related to phytoremediation and are essentially not studied to increase the remediation efficiency of these plants. So far it is highly unlikely to find CRISPR-Cas-based editing in the literature that is directed to increase the potential of phytoremediation. However, in Table 11.3 we have listed work that suggests CRISPR-edited plants can have a positive effect on its phytoremediation potential as well as overall plant health. These studies can be used as a proof-of-concept to develop CRISPR-edited plants with enhanced phytoremediation capacity.

11.3.3.1 CRISPR-Cas for genome editing

As stated earlier, genome editing of the plants for phytoremediation using CRISPR systems has the potential to become an important aspect for increasing their remedial capacity (Fig. 11.3). Genomics and transcriptomics data of plants can be utilized to establish CRISPR-Cas for phytoremediation by targeted engineering methods. Several processes such as accumulation, complexation, volatilization, and degradation of pollutants can be significantly enhanced using CRISPR-Cas with various variants of CRISPR-Cas is a certainly a next-generation method that can be used for both discovering genes

TABLE 11.3 Summary of reports that applied CRISPR-Cas to improve the plant health and reduce environmental toxicity.

Plant	Gene	Observation	Remark	Reference
Oryza sativa	*OsNramp5*	*OsNramp5* knockout resulted in low Cd accumulation	Overexpression of *OsNramp5* would increase the bioaccumulation of Cd	(Tang et al., 2017)
Oryza sativa	*OsATX1*	*OsATX1* knockout resulted in increased uptake and accumulation of Cu in shoots	Knockout of OsATX1 homologs in other plants would increase Cu tolerance and remediation	(Zhang et al., 2018b)
Oryza sativa	*OsLCT1*	*OsLCT1* knockout resulted in low Cd accumulation	*OsLCT1 overexpression* would increase the bioaccumulation of Cd	(Songmei et al., 2019)
Tomato	*Slhak5*	*Slhak5* knockout resulted in low Cs (Cesium) accumulation	Overexpression of *Slhak5* would increase the bioaccumulation of Cs	Nieves-Cordones et al., (2020)

involved in phytoremediation and their manipulation as well as regulating the transcriptome and epigenome of the plant (Arya et al., 2020a; Jinek et al., 2012; Basharat et al., 2018). Therefore, genome-edited phytoremediators would have the ability to intensify their uptake and transport of heavy metals (Mosa et al., 2016; Kotrba et al., 2009). Furthermore, we can increase the specificity of the ligands for the metals and the metalloids uptake and the subsequent process of their bioconversion to no or less toxic forms. Also, we can increase the selective permeability of the cell membrane to xenobiotics and toxic volatiles and their subsequent transformation into biocompatible, nonvolatile, and other useful forms by using CRISPR-Cas.

Some of the important genes involved in the uptake and transport mechanism belong to families such as NRAMP, ZIP, HMAs, CPx-type ATPases, Copper transporter, oligopeptide transporters, cation diffusion facilitators, ABC transporters, MATE protein, etc. CPx-ATPases or P-type ATPases are intramembranous cysteine-proline-cysteine/histidine/serine sequences that are highly conserved. They facilitate the influx/efflux of cations across the cell membrane. The P1B-type HMAs (ATPases for heavy metal) functions in metal distribution, translocation, and localization by consuming ATP to pump a spectrum of xenobiotics across the cellular. They are also active in the transport of essential and nonessential metals such as Zn, Cu, Cd, and Pb (Guo et al., 2015). Such broad-spectrum metal-specific transporters can further be improved to accommodate other metals and metal-bound compounds for effective phytoremediation using CRISPR-Cas. The genes coding for transporters with high influx/efflux can be introduced in a target-specific manner using CRISPR-Cas or can be engineered using CRISPR variants.

11.3.3.2 CRISPR-Cas for multiplexing

Boosting the remediation capacity sometimes may be a function of multiple genes rather than one gene. Therefore, a spectrum of genes could be potential targets to simultaneous improve metal/xenobiotic remediation as well as plant growth. Multiplexing with CRISPR-Cas was first reported in the year 2013 (Cong et al., 2013), which has now become a routine in the field of plant genetic engineering. Multiplex genome editing is significant for the efficient expression of multiple gRNAs. Numerous approaches are available to optimize gRNA expressions such as CRISPR-Cas expression cassette configuration, delivery methods, and transcript processing procedures. To edit multiple target sites, multiple gRNA complementary to the target DNA are designed and synthesized, here each gRNA is driven by an individual promoter and/or multiple gRNAs expression are regulated by a single promoter into a single transcript (Xie et al., 2015). As of now, the multiplex gene assemblies are constructed using type II restriction endonuclease (golden gate cloning) or a ligation-independent method (Gibson) (Gibson, 2011; Weber et al., 2011). Although it is observed that either knock in or knocking out of a particular gene enhances the phytoremediation capacity (Table 11.1 and Table 11.3); however, some remediation traits can be attributed to plant gene clusters that are coregulated upon environmental stimuli. For instance, the ALMT1 gene cluster in *Secale cereale* L. controls aluminum tolerance (Collins et al., 2008). Therefore, in some cases, it will be essential to knockout entire or group of genes functioning as a plant gene cluster, and this task will be efficiently carried out using CRISPR-multiplexing technique. Furthermore, like multiplexing, CRISPR-Cas can be modified to achieve modularity and enhanced functionality by attaching it with DNA and histone-altering proteins (Fig. 11.2).

11.3.3.3 CRISPR-Cas for base editing

A very recent update in the CRISPR-Cas-based system includes base editing (BE). The BE system consists of a mutated Cas protein that works as either

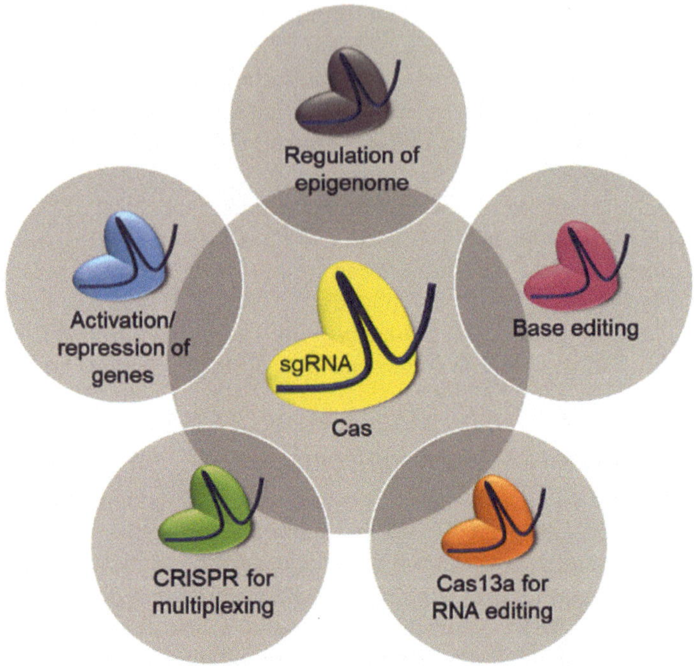

FIG. 11.2 Potential advantages of multiplexing with CRISPR-Cas variants.

dead (dCas; D10A and H840A mutations; inactivated nuclease) or as nickase (nCas; D10A mutation; nuclease with single-strand cleavage activity) conjugated to appropriate nucleoside deaminase, which allows targeted nucleotide conversion. Based on the biochemical characteristics of conjugated nucleoside deaminase, base editors are classified as adenine base editors, cytidine base editors, and RNA base editors. Arya et al. (2020b) have reviewed the properties and functionality base editors. These base editors can be used to engineer proteins and enzymes to increase the uptake capacity of transport proteins as well as substrate specificity. BE combined with molecular and protein docking can be a game changer if used in the field of phytoremediation. Protein and enzyme engineering approaches are widely studied to alter the uptake efficiency of transmembrane proteins and transporters (Bu et al., 2020), as well as the catalytic potential of enzymes involved in the synthesis of high-value plant specialized metabolites (Werner et al., 2019).

11.3.3.4 CRISPR-Cas for transcriptional and epigenome regulation

Recently, CRISPR-Cas is also adapted to regulate the transcription as well as the epigenome in plants. The activation and repression of genes are achieved by

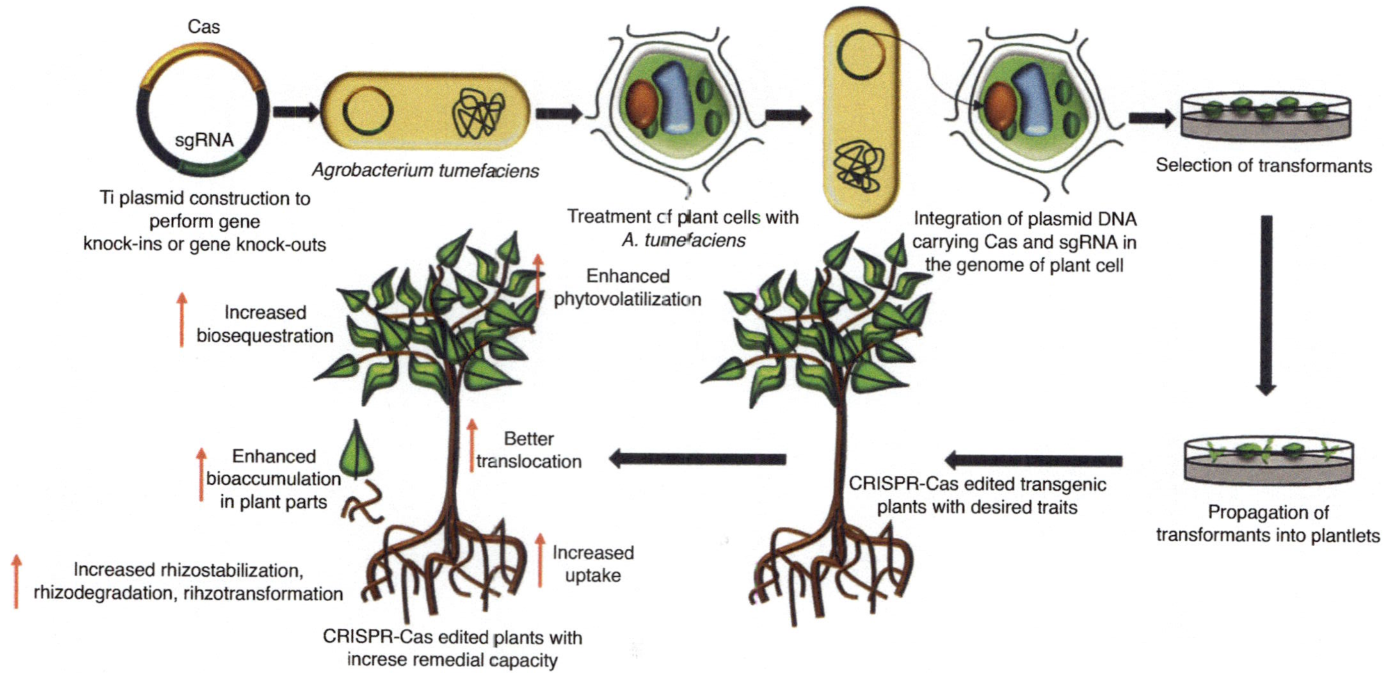

FIG. 11.3 Schematics of gene editing protocol using CRISPR-Cas to generate plants with high remedial capacity.

fusing the dCas with either a gene activator (CRISPRa) or repressor/interference (CRISPRi) (Seth 2016). The activators and repressors comprise transcription factors and epigenetic regulators such as histone acetylases and methylases. These CRISPR variants can be used to study the underlying transcriptional and epigenomic regulation behind the genes involved in phytoremediation. This will help identifying the function of genes as well as epigenome in remediation activity. This system can also be used to activate cryptic genes, which are usually not expressed upon encountering pollutant(s) or contaminant(s)'. The transcriptional activator (named as "TV") has shown >55-fold stronger transcriptional activation potential than the routinely used dCas-VP64 (Li et al., 2017). Although CRISPRi and CRISPRa remain unexplored; however, we assume that their use to understand and increase remediation would revolutionize the field of phytoremediation.

11.4 Conclusion and future perspectives

Phytoremediation process in plants can be markedly enhanced using genetic engineering tools. Over the past few years, CRISPR-Cas-mediated genome editing has emerged as a preferred choice over other molecular tools. Owing to the multiplexing ability and variants available, targeted genome and epigenome editing using CRISPR-Cas will certainly become a regular practice soon. Availability of multiomics data such as genomics, epigenomics, transcriptomics, metabolomics, and proteomics would significantly enhance the identification of new genes involved in metal uptake, transport, and sequestration, as well as in the degradation of hazardous xenobiotics. This will also increase our understanding of functional genomics and accelerate the use of synthetic biology for trait improvement. Moreover, bioinformatics and molecular docking may help to analyze the types of genes, transporters, proteins, and enzymes are vital to the process of phytoremediation. If implemented, CRISPR-edited transgenic plants with improved capacity can become pivotal players for effective detoxification of contaminated sites and will also have substantial cost-benefit compared to their wild counterparts.

References

Abhilash, P.C., et al., 2012. Plant–microbe interactions: novel applications for exploitation in multipurpose remediation technologies. Trends Biotechnol 30, 416–420.

Arya, S.S., Mahto, B.K., Ramkumar, T.R., Lenka, S.K., 2020b. Sharpening gene editing toolbox in Arabidopsis for plants. J. Plant Biochem. Biotechnol. 29, 1–6.

Arya, S.S., Rookes, J.E., Cahill, D.M., Lenka, S.K., 2020a. Next-generation metabolic engineering approaches towards development of plant cell suspension cultures as specialized metabolite producing biofactories. Biotechnol. Adv. 23, 107635.

Azab, E., Hegazy, A.K., Gobouri, A.A., Elkelish, A., 2020. Impact of transgenic *Arabidopsis thaliana* plants on herbicide isoproturon phytoremediation through expressing human cytochrome P450-1A2. Biology 9 (11), 362.

Basharat, Z., Novo, L.A., Yasmin, A., 2018. Genome editing weds CRISPR: what is in it for phytoremediation? Plants 7 (3), 51.

Basu, S., Rabara, R.C., Negi, S., Pratyoosh, S., 2018. Engineering PGPMOs through gene editing and systems biology: a solution for phytoremediation? Trends Biotechnol 36 (5), 499–510.

Bhargava, A., Carmona, F.F., Bhargava, M., Srivastava, S., 2012. Approaches for enhanced phytoextraction of heavy metals. J. Environ. Manage. 105, 103–120.

Bhuiyan, M.S.U., Min, S.R., Jeong, W.J., Sultana, S., Choi, K.S., Lee, Y., et al., 2011. Overexpression of atatm3 in *Brassica juncea* confers enhanced heavy metal tolerance and accumulation. Plant Cell Tiss. Org. Cult. 107, 69–77.

Bortesi, L., Fischer, R., 2015. The CRISPR system for plant genome editing and beyond. Biotechnol. Adv. 33, 41–52.

Briskine, R.V., Paape, T., Shimizu-Inatsugi, R., Nishiyama, T., Akama, S., Sese, J., Shimizu, K.K., 2017. Genome assembly and annotation of *Arabidopsis halleri*, a model for heavy metal hyperaccumulation and evolutionary ecology. Mol. Ecol. Resour. 17, 1025–1036.

Brown, G.R., Hem, V., Katz, K.S., et al., 2015. Gene: a gene-centered information resource at NCBI. Nucleic Acids Res 43 (D1), D36–D42.

Brunetti, P., Zanella, L., Paolis, A.D., Litta, D.D., Cecchetti, V., Falasca, G., et al., 2015. Cadmium-inducible expression of the ABC-type transporter AtABCC3 increases phytochelatin-mediated cadmium tolerance in *Arabidopsis*. J. Exp. Bot. 66 (13), 3815–3829.

Bu, X., Lin, J.Y., Cheng, J., Yang, D., Duan, C.Q., Koffas, M., Yan, G.L., 2020. Engineering endogenous ABC transporter with improving ATP supply and membrane flexibility enhances the secretion of β-carotene in *Saccharomyces cerevisiae*. Biotechnol. Biofuels 13 (1), 1–14.

Callahan, D.L., Baker, A.J., Kolev, S.D., Wedd, A.G., 2006. Metal ion ligands in hyperaccumulating plants. J. Biol. Inorg. Chem. 11 (1), 2–12.

Chaney, R.L., Malik, M., Li, Y.M., Brown, S.L., Angle, J.S., Baker, A.J.M., 1997. Phytoremediation of soil metals. Curr. Opin. Biotechnol. 8, 279–284.

Chen, Y., et al., 2013. Engineering arsenic tolerance and hyper-accumulation in plants for phytoremediation by a PvACR3 trans-genic approach. Environ. Sci. Technol. 47, 9355–9362.

Cheng, Z., 2016. The spatial correlation and interaction between manufacturing agglomeration and environmental pollution. Ecol. Indic. 61, 1024–1032.

Chu, C., Liu, B., Liu, J., He, J., Lv, L., Wang, H., Xie, X., Tao, Q., Chen, Q., 2020. Phytoremediation of acetochlor residue by transgenic Arabidopsis expressing the acetochlor N-dealkylase from *Sphingomonas wittichii* DC-6. Sci. Total Environ. 728, 138687.

Collins, N.C., Shirley, N.J., Saeed, M., Pallotta, M., Gustafson, J.P., 2008. An ALMT1 gene cluster controlling aluminum tolerance at the Alt4 locus of rye (*Secale cereale* L.). Genetics 179 (1), 669–682.

Cong, L., Ran, F.A., Cox, D., Lin, S., Barretto, R., Habib, N., Hs u, P.D., Wu, X., Jiang, W., Marraffini, L.A., Zhang, F., 2013. Multiplex genome engineering using CRISPR/Cas systems. Science 339 (6121), 819–823.

Coyle, P., Philcox, J.C., Carey, L.C., Rofe, A.M., 2002. Metallothionein: the multipurpose protein. Cell. Mol. Life Sci. 59, 627–647.

Dangi, A.K., Babita, S., Russell, T.H., Pratyoosh, S., 2018. Bioremediation through microbes: systems biology and metabolic engineering approach. Crit. Rev. Biotechnol. 39 (1), 79–98.

Das, N., Bhattacharya, S., Maiti, M.K., 2016. Enhanced cadmium accumulation and tolerance in transgenic tobacco overexpressing rice metal tolerance protein gene OsMTP1 is promising for phytoremediation. Plant Physiol. Biochem. 105, 297–309.

Deblonde, T., Cossu-Leguille, C., Hartemann, P., 2011. Emerging pollutants in wastewater: a review of the literature. Int. J. Hyg. Environ. Health 214, 442–448.

Dominguez-Solis, J.R., Lopez-Martin, M.C., Ager, F.J., Ynsa, M.D., Romero, L.C., Gotor, C., 2004. Increased cysteine availability is essential for cadmium tolerance and accumulation in *Arabidopsis thaliana*. Plant Biotechnol. J. 2, 469–476.

Eapen, S., D'Souza, S.F, 2005. Prospects of genetic engineering of plants for phytoremediation of toxic metals. Biotechnol. Adv. 23, 97–114.

Eevers, N., White, J.C., Vangronsveld, J., et al., 2017. Bio- and phytoremediation of pesticide-contaminated environments: a review. Adv. Bot. Res. 83, 277–318.

Ehrnstorfer, I.A., Geertsma, E.R., Pardon, E., Steyaert, J., Dutzler, R., 2014. Crystal structure of a SLC11 (NRAMP) transporter reveals the basis for transition metal ion transport. Nat. Struct. Mol. Biol. 21 (11), 990–996.

Ellis, D.R., Sors, T.G., Brunk, D.G., Albrecht, C., Orser, C., Lahner, B., Wood, K.V., Harris, H.H., Pickering, I.J., Salt, D.E., 2004. Production of Se-methyl seleno cysteine in transgenic plants expressing selenocysteine methyltransferase. BMC Plant Biol 4, 1.

Evans, K.M., Gatehouse, J.A., Lindsay, W.P., Shi, J., Tommey, A.M., Robinson, N.J., 1992. Expression of pea metallothionein-like gene PsMTA function. Plant Mol. Biol. 20, 1019–1028.

Foht, B.P., 2016. Gene editing: new technology, old moral questions. New Atlantis, 3–15.

Freisinger, E., 2008. Plant MTs-long neglected members of the metallothionein superfamily. Dalton. Trans. 47, 6663–6675.

Fulekar, M.H., Singh, A., Bhaduri, A.M., 2008. Genetic engineering strategies for enhancing phytoremediation of heavy metals. Afr. J. Biotechnol. 8 (4), 529–535.

Gasic, K., Korban, S.S., 2007. Transgenic Indian mustard (*Brassica juncea*) plants expressing an *Arabidopsis* phytochelatin synthase (AtPCS1) exhibit enhanced As and Cd tolerance. Plant Mol. Biol. 64, 361–369.

Gibson, D.G., 2011. Enzymatic assembly of overlapping DNA fragmentsMethods in Enzymology498. Academic Press, MA, pp. 349–361.

Gisbert, C., Ros, R., Haro, A.D., Walker, D.J., Bernal, M.P., Serrano, R., Avino, J.N.A., 2003. plant genetically modified that accumulates Pb is especially promising for phytoremediation. Biochem. Biophys. Res. Commun. 303, 440–445.

Guo, J., Green, B.R., Maldonado, M.T., 2015. Sequence analysis and gene expression of potential components of copper transport and homeostasis in *Thalassiosira pseudonana*. Protist 166 (1), 58–77.

Gupta, S.K., Shukla, P., 2017. Gene editing for cell engineering: trends and applications. Crit. Rev. Biotechnol. 37, 672–684.

Gustin, J.L., Loureiro, M.E., Kim, D., Na, G., Tikhonova, M., Salt, D.E., 2009. MTP1-dependent Zn sequestration into shoot vacuoles suggests dual roles in Zn tolerance and accumulation in Zn-hyperaccumulating plants. Plant J 57, 1116–1127.

Hasegalwa I., Terada E., Sunairi M., Wakita H., Shinmachi F., Noguchi A., Nakajima M., Yazaki J., 1997, Genetic improvement of heavy metal tolerance in plants by transfer of the yeast metallothionein gene (CUP1), Plant Soil 196, 277–281.

He, J., Li, H., Ma, C., Zhang, Y., Polle, A., Rennenberg, H., Cheng, X., Luo, Z.B., 2015. Overexpression of bacterial γ-glutamylcysteine synthetase mediates changes in cadmium influx, allocation and detoxification in poplar. New Phytol 205 (1), 240–254.

Hirschi, K.D., Korenkov, V.D., Wilganowski, N.L., Wagner, G.J., 2000. Expression of *Arabidopsis* CAX2 in tobacco: altered metal accumulation and increased manganese tolerance. Plant Physiol 124, 125–133.

Hussain, I., Aleti, G., Naidu, R., Puschenreiter, M., Mahmood, Q., Rahman, M.M., et al., 2018. Microbe and plant assisted-remediation of organic xenobiotics and its enhancement by genetically modified organisms and recombinant technology: a review. Sci. Total Environ. 628, 1582–1599.

Ishimaru, Y., Takahashi, R., Bashir, K., Shimo, H., Senoura, T., Sugimoto, K., et al., 2012. Characterizing the role of rice NRAMP5 in manganese, iron and cadmium transport. Sci. Rep. 2, 286.

Jinek, M., Chylinski, K., Fonfara, I., Hauer, M., Doudna, J.A., Charpentier, E., 2012. A programmable dual-RNA-guided DNA endonuclease in adaptive bacterial immunity. Science 337, 816–821.

Karenlampi, S., Schat, H., Vangronsveld, J., Verkleij, J.A., van der Lelie, D., Mergeay, M., et al., 2000. Genetic engineering in the improvement of plants for phytoremediation of metal polluted soils. Environ. Pollut. 107 (2), 225–231.

Khouja, H.R., Abba, S., Lacercat-Didier, L., Daghino, S., Doillon, D., Richaud, P., et al., 2013. OmZnT1 and OmFET, two metal transporters from the metal-tolerant strain Zn of the ericoid mycorrhizal fungus *Oidiodendron maius*, confer zinc tolerance in yeast. Fungal Genet. Biol. 52, 53–64.

Kim, S., Takahashi, M., Higuchi, K., Tsunoda, K., Nakanishi, H., Yoshimura, E., Mori, S., Nishizawa, N.K., 2005. Increased nicotianamine biosynthesis confers enhanced tolerance of high levels of metals, in particular nickel, to plants. Plant Cell Physiol 46, 1809–1818.

Kim, Y.G., Cha, J., Chandrasegaran, S., 1996. Hybrid restriction enzymes: zinc finger fusions to Fok I cleavage domain. Proc. Natl. Acad. Sci. USA 93 (3), 1156–1160.

Kotrba, P., Najmanova, J., Macek, T., Ruml, T., Mackova, M., 2009. Genetically modified plants in phytoremediation of heavy metal and metalloid soil and sediment pollution. Biotechnol. Adv. 27 (6), 799–810.

Kumar, V., et al., 2016. Recent developments in systems biology and metabolic engineering of plant microbe interactions. Front. Plant Sci. 7, 1421.

Larsson, D.G.J., 2014. Pollution from drug manufacturing: review and perspectives. Philos. Trans. R. Soc. Lond. B. 369, 20130571.

Lasat, M.M., 2002. Phytoextraction of toxic metals: a review of biological mechanisms. J. Environ. Qual. 31, 109–120.

LeDuc, D.L., AbdelSamie, M., Montes-Bayon, M., Wu, C.P., Reisinger, S.J., Terry, N., 2006. Overexpressing both ATP sulfurylase and selenocysteine methyltransferase enhances selenium phytoremediation traits in Indian mustard. Environ. Pollut. 144 (1), 70–76.

Lee, K.W., Choi, G.J., Kim, K.Y., et al., 2012. Transgenic expression of MsHs p23 confers enhanced tolerance to abiotic stresses in tall fescue. Asian-Aust. J. Anim. Sci. 25, 818–823.

Lee, S., Moon, J.S., Ko, T.S., Petros, D., Goldsbrough, P.B., Korban, S.S., 2003. Overexpression of *Arabidopsis* phytochelatin synthase paradoxically leads to hypersensitivity to cadmium stress. Plant Physiol 131 (2), 656–663.

Li Z., Zhang D., Xiong W., et al., 2017, A potent Cas9-derived gene activator for plant and mammalian cells, Nat. Plants 3 (12), 930–936.

Li, Y., Dhankher, O.P., Carreira, L., Lee, D., Chen, A., Schroeder, J.I., Balish, R.S., Meagher, R.B., 2004. Overexpression of phytochelatin synthase in *Arabidopsis* leads to enhanced arsenic tolerance and cadmium hypersensitivity. Plant Cell Physiol 45 (12), 1787–1797.

Listgarten, J., Weinstein, M., Kleinstiver, B.P., Sousa, A.A., Joung, J.K., Crawford, J., et al., 2018. Prediction of off-target activities for the end-to-end design of CRISPR guide RNAs. Nat. Biomed. Eng. 2, 38.

Liu, D., et al., 2012. The rice ERF transcription factor OsERF922 negatively regulates resistance to *Magnaporthe oryzae* and salt tolerance. J. Exp. Bot. 63, 3899–3911.

Liu, H., Zhao, H., Wu, L., Liu, A., Zhao, F.J., Xu, W., 2017. Heavy metal ATPase 3 (HMA3) confers cadmium hypertolerance on the cadmium/zinc hyperaccumulator *Sedum plumbizincicola.* New Phytol 215, 687–698.

Lv, Y., Deng, X., Quan, L., Xia, Y., Shen, Z., 2013. Metallothioneins BcMT1 and BcMT2 from *Brassica campestris* enhance tolerance to cadmium and copper and decrease production of reactive oxygen species in *Arabidopsis thaliana*. Plant Soil 367, 507–519.

Macek, T., Sura-de Jong, M., Kotrba, M., Viktorova, J., Lovecka, P., Fiser, J., et al., 2012. Genetically modified plants designed for phytoremediation of toxic organic and inorganic contaminants. In: Anjum, N.A., Pereira, M.E., Ahmad, I., Duarte, A.C., Umar, S., Khan, N.A. (Eds.), Phytotechnologies: Remediation of Environmental Contaminantsfirst Ed. CRC Press, Raton, FL, pp. 415–426.

Mahas, A., Mahfouz, M., 2018. Engineering virus resistance via CRISPR–Cas systems. Curr. Opin. Virol. 32, 1–8.

Mandáková, T., Singh, V., Krämer, U., Lysak, M.A., 2015. Genome structure of the heavy metal hyperaccumulator *Noccaea caerulescens* and its stability on metalliferous and nonmetalliferous soils. Plant Physiol. 169, 674–689.

Mello-Farias, P.C., Chaves, A.L.S., 2008. Biochemical and molecular aspects of toxic metals phytoremediation using transgenic plants. In: Tiznado-Hernandez, M.E., Troncoso-Rojas, R., Rivera- Domínguez, M.A. (Eds.), Transgenic Approach in Plant Biochemistry and Physiology. Research Signpost, Kerala, India, pp. 253–266.

Mello-Farias, P.C., Chaves, A.L.S., Lencina, C.L., 2011. Transgenic plants for enhanced phytoremediation physiological studies. In: Alvarez, M. (Ed.), Genetic Transformation. In Tech, Rijeka, Croatia.

Mendez, M.O., Maier, R.M., 2008. Phytostabilization of mine tailings in arid and semiarid environments—an emerging remediation technology. Environ. Health Perspect. 116, 278–283.

Meng, J.G., Zhang, X.D., Tan, S.K., Yang, Z.M., 2017. Genome-wide identification of Cd-responsive NRAMP transporter genes and analyzing expression of NRAMP 1 mediated by miR167 in *Brassica napus*. Biometals 30 (6), 917–931.

Migocka, M., Kosieradzka, A., Papierniak, A., Maciaszczyk-Dziubinska, E., Posyniak, E., Garbiec, A., et al., 2014. Two metal-tolerance proteins, MTP1 and MTP4, are involved in Zn homeostasis and Cd sequestration in cucumber cells. J. Exp. Bot. 66 (3), 1001–1015.

Migocka, M., Posyniak, E., Maciaszczyk-Dziubinska, E., Papierniak, A., Kosieradzka, A., 2015. Functional and biochemical characterization of cucumber genes encoding two Copper ATPases CsHMA5.1 and CsHMA5.2. J. Biol. Chem. 290, 15717–15729.

Milner, M.J., Mitani-Ueno, N., Yamaji, N., Yokosho, K., Craft, E., Fei, Z., et al., 2014. Root and shoot transcriptome analysis of two ecotypes of *Noccaea caerulescens*. uncovers the role of NcNramp1 in Cd hyperaccumulation. Plant J 78 (3), 398–410.

Mosa, K.A., Saadoun, I., Kumar, K., Helmy, M., Dhankher, O.P., 2016. Potential biotechnological strategies for the cleanup of heavy metals and metalloids. Front. Plant Sci. 7, 303.

Nahar, N., Rahman, A., Nawani, N.N., Ghosh, S., Mandal, A., 2017. Phytoremediation of arsenic from the contaminated soil using transgenic tobacco plants expressing ACR2 gene of *Arabidopsis thaliana*. J. Plant Physiol. 218, 121–126.

Nemudryi, A.A., Valetdinova, K.R., Medvedev, S.P., Zakian, S.M., 2014. TALEN and CRISPR/Cas genome editing systems: tools of discovery. Acta Nat 6, 19–40.

Nesler, A., dalCorso, G., Fasani, E., Manana, A., Sansbastiano, G.P.D., Argese, E., et al., 2017. Functional components of the bacterial CzcCBA efflux system reduce cadmium uptake and accumulation in transgenic tobacco plants. New Biotech 35, 54–61.

Nieves-Cordones, M., Lara, A., Silva, M., Amo, J., Rodriguez-Sepulveda, P., Rivero, R.M., Martínez, V., Botella, M.A., Rubio, F., 2020. Root high-affinity K+ and Cs+ uptake and plant fertility in tomato plants are dependent on the activity of the high-affinity K+ transporter SlHAK5. Plant Cell Environ 43 (7), 1707–1721.

Oomen, R.J., Wu, J., Lelievre, F., Blanchet, S., Richaud, P., Barbier-Brygoo, H., et al., 2009. Functional characterization of NRAMP3 and NRAMP4 from the metal hyperaccumulator *Thlaspi caerulescens*. New Phytol 181, 637–650.

Pandey, V.C., Singh, V., 2019. Exploring the potential and opportunities of current tools for removal of hazardous materials from environments. In: Pandey, V.C., Bauddh, K (Eds.), Phytomanagement of Polluted Sites. Elsevier, Amsterdam, pp. 501–516.

Peng, J.S., Wang, Y.J., Ding, G., Ma, H.L., Zhang, Y.J., Gong, J.M., 2017. A pivotal role of cell wall in cadmium accumulation in the Crassulaceae hyperaccumulator *Sedum plumbizincicola*. Mol. Plant 10 (5), 771–774.

Perelo, L.W., 2010. Review: in situ and bioremediation of organic pollutants in aquatic sediments. J. Hazard. Mater. 177, 81–91.

Pianelli, K., Mari, S., Marquès, L., Lebrun, M., Czernic, P., 2005. Nicotianamine over-accumulation confers resistance to nickel in *Arabidopsis thaliana*. Transgenic Res 14, 739–748.

Pilon-Smits, E., 2005. Phytoremediation. Annu. Rev. Plant Biol. 56, 15–39.

Remy, E., Nino-Gonzalez, M., Godinho, C.P., Cabrito, T.R., Teixeira, M.C., Sa-Correia, I., et al., 2017. Heterologous expression of the yeast Tpo1p or Pdr5p membrane transporters in *Arabidopsis* confers plant xenobiotic tolerance. Sci. Rep. 7, 4529.

Rico, E., Jeacock, L., Kováˇrová, J., Horn, D., 2018. Inducible high-efficiency CRISPR-Cas9-targeted gene editing and precision base editing in African trypanosomes. Sci. Rep. 8 (1), 1–10.

Sanjana N.E., et al., 2012. A transcription activator-like effector toolbox for genome engineering. Nat. Protoc. (7), 171 192.

Sasaki, A., Yamaji, N., Ma, J.F., 2014. Overexpression of OsHMA3 enhances Cd tolerance and expression of Zn transporter genes in rice. J. Exp. Bot. 65 (20), 6013–6021.

Saxena, P., Singh, N.K., Singh, A.K., Pandey, S., Thanki, A., Yadav, T.C., 2020. Recent advances in phytoremediation using genome engineering CRISPR–Cas9 technology. In: Pandey, V., Singh, V. (Eds.), Recent advances in phytoremediation using genome engineering CRISPR–Cas9 technology. Bioremediation of Pollutants, 125–141.

Seth, K., 2016. Current status of potential applications of repurposed Cas9 for structural and functional genomics of plants. Biochem. Biophys. Res. Commun. 480 (4), 499–507.

Shah, T., Andleeb, T., Lateef, S., Noor, M.A., 2018. Genome editing in plants: advancing crop transformation and overview of tools. Plant Physiol. Biochem. 131, 12–21.

Shapiro, R.S., Chavez, A., Collins, J.J., 2018. CRISPR-based genomic tools for the manipulation of genetically intractable microorganisms. Nat. Rev. Microbiol. 16 (6), 333–339.

Shaul, O., Hilgemann, D.W., de Almeida-Engler, J., van Montagu, M., Inze, D., Galili, G., 1999. Cloning and characterization of a novel Mg^{2+}/H^{+} exchanger. EMBO J. 18, 3973–3980.

Siemianowski, O., Mills, R.F., Williams, L.E., Antosiewicz, D.M., 2011. Expression of the P1B-type ATPase AtHMA4 in tobacco modifies Zn and Cd root to shoot partitioning and metal tolerance. Plant Biotechnol. J. 9, 64–74.

Singh, P., Jain, R., Srivastava, N., et al., 2017. Current and emerging trends in bioremediation of petrochemical waste: a review. Crit. Rev. Environ. Sci. Technol. 14, 203–232.

Songmei, L.I.U., Jie, J.I.A.N.G., Yang, L.I.U., Jun, M.E.N.G., Shouling, X.U., Yuanyuan, T.A.N., Youfa, L.I., Qingyao, S.H.U., Jianzhong, H.U.A.N.G., 2019. Characterization and evaluation of OsLCT1 and OsNramp5 mutants generated through CRISPR/Cas9-mediated mutagenesis for breeding low Cd rice. Rice Sci 26 (2), 88–97.

Srivastava, N., et al., 2016. Phytoremediation of heavy metals contaminated soils through transgenic plants. In: Ansari, A.A. et al (Ed.), Phytoremediation. Springer International Publishing, Switzerland, pp. 345–391.

Sudhir, K.S., Neelam, M., Subba, T.R., et al., 2014. Biofilm-mediated bioremediation of polycyclic aromatic hydrocarbons. Microb. Biodegrad. Bioremed., 203–232.

Sundaram, S., Wu, S., Ma, L.Q., et al., 2009. Expression of a *Pteris vittata* glutaredoxin PvGRX5 in transgenic *Arabidopsis thaliana* increases plant arsenic tolerance and decreases arsenic accumulation in the leaves. Plant Cell Environ 32, 851–858.

Sunkar, R., et al., 2000. Expression of a truncated tobacco NtCBP4 channel in transgenic plants and disruption of the homologous *Arabidopsis* CNGC1 gene confer Pb^{2+} tolerance. Plant J 24, 533–542.

Svitashev, S., et al., 2015. Targeted mutagenesis, precise gene editing, and site-specific gene insertion in maize using Cas9 and guide RNA. Plant Physiol 169, 931–945.

Tang, L., Mao, B., Li, Y., Lv, Q., Zhang, L., Chen, C., He, H., Wang, W., Zeng, X., Shao, Y., Pan, Y., 2017. Knockout of OsNramp5 using the CRISPR/Cas9 system produces low Cd-accumulating indica rice without compromising yield. Sci. Rep. 7 (1), 1–12.

Thomas, J.C., Davies, E.C., Malick, F.K., Endreszl, C., Williams, C.R., Abbas, M., Petrella, S., Swisher, K., Perron, M., Edwards, R., Ostenkowski, P., Urbanczyk, N., Wiesend, W.N., Murray, K.S., 2003. Yeast metallothionein in transgenic tobacco promotes copper uptake from contaminated soils. Biotechnol. Prog. 19 (2), 273–280.

van der Zaal, B.J., Neuteboom, L.W., Pinas, J.E., Chardonnens, A.N., Schat, H., Verkleij, J.A., et al., 1999. Overexpression of a novel *Arabidopsis* gene related to putative zinc-transporter genes from animals can lead to enhanced zinc resistance and accumulation. Plant Physiol 119 (3), 1047–1056.

Van Huysen, T., Abdel-Ghany, S., Hale, K.L., LeDuc, D., Terry, N., Pilon-Smits, E.A., 2003. Overexpression of cystathionine-γ-synthase enhances selenium volatilization in *Brassica juncea*. Planta 218 (1), 71–78.

Verma, P.K., Verma, S., Pande, V., et al., 2016. Overexpression of rice glutaredoxin OsGrx-C7 and OsGrx-C2.1 reduces intracellular arsenic accumulation and increases tolerance in *Arabidopsis thaliana*. Front. Plant Sci. 7, 740.

Voytas, D.F., Gao, C., 2014. Precision genome engineering and agriculture: opportunities and regulatory challenges. PLoS Biol 12 (6), e1001877.

Vrbova, M., Kotrba, P., Horacek, J., Smykal, P., Svabova, L., Vetrovcova, M., et al., 2013. Enhanced accumulation of cadmium in *Linum usitatissimum* L. plants due to overproduction of metallothionein α-domain as a fusion to β- glucuronidase protein. Plant Cell Tiss. Org. Cult. 112, 321–330.

Weber, E., Engler, C., Gruetzner, R., Werner, S., Marillonnet, S., 2011. A modular cloning system for standardized assembly of multigene constructs. PloS one 6 (2), e16765.

Werner, N., Ramirez-Sarmiento, C.A., Agosin, E., 2019. Protein engineering of carotenoid cleavage dioxygenases to optimize β-ionone biosynthesis in yeast cell factories. Food Chem 299, 125089.

Wiszniewska, A., Hanus-Fajerska, E., Smoleń, S., Muszyńska, E., 2015. In vitro selection for lead tolerance in shoot culture of *Daphne* species. Acta Sci. Pol. Hortorum Cultus 14 (1), 129–142.

Xie, K., Minkenberg, B., Yang, Y., 2015. Boosting CRISPR/Cas9 multiplex editing capability with the endogenous tRNA-processing system. Proc. Natl. Acad. Sci. USA 112 (11), 3570–3575.

Xie, Q.-E., Yan, X.-L., Liao, X.-Y., Li, X., 2009. The arsenic hyperaccumulator fern *Pteris vittata* L. Environ. Sci. Technol. 43, 8488–8495.

Yang, X., Jin, X.F., Feng, Y., Islam, E., 2005. Molecular mechanisms and genetic bases of heavy metal tolerance/hyperaccumulation in plants. J. Integr. Plant Biol 47 (9), 1025–1035.

Zhang, C., Konermann, S., Brideau, N.J., Lotfy, P., Novick, S.J., Strutzenberg, T., et al., 2018a. Structural basis for the RNA-guided ribonuclease activity of CRISPR-Cas13d. Cell 175, 212–223.e17.

Zhang, Y., Chen, K., Zhao, F.J., Sun, C., Jin, C., Shi, Y., Sun, Y., Li, Y., Yang, M., Jing, X., Luo, J., 2018. OsATX1 interacts with heavy metal P1B-type ATPases and affects copper transport and distribution. Plant Physiol 178 (1), 329–344.

Chapter 12

Benefits and limitations of phytoremediation: Heavy metal remediation review

Azeez Omoniyi Adeoye[a], Ismail Abiola Adebayo[b], Adam Moyosore Afodun[a], Kamoldeen Abiodun Ajijolakewu[c]

[a]*Department of Anatomy, Faculty of Biomedical Sciences, Kampala International University-Western Campus, Ishaka-Busheyi, Uganda,* [b]*Microbiology and Immunology Department, Faculty of Biomedical Sciences, Kampala International University-Western Campus, Ishaka-Busheyi, Uganda,* [c]*Microbiology Department, Faculty of Life Sciences, University of Ilorin, Ilorin, Nigeria*

12.1 Introduction

Soil and water contamination by hazardous substances remained a global concern fronting researchers. Activities such as waste disposal practices, farming, and human industry have resulted in extensive water and soil pollution. In addition, animate compounds and dense metals beside their deleterious effects have also contributed to the pollution of the ecosystem, which poses serious health risks to both wildlife and human life. However, standard systems of soil remedy are highly exorbitant, together with involvement of soil depot in nominated field without immediate offering of solution but rather postponing the problem. As a result, there is need to search for and provide alternative methods to deal with the problem associated with hazardous contaminants.

During past decades, dire requirement to ascertain another strategy expressed the technical and monetary gain of using both vegetation and the allied microbes for recycling irrigate and ground (Meagher, 2000). The aforementioned alternative means as an ecological technological approach comprises plant utilization to sanitize soils polluted with heavy as well as toxic metals. Recently, a reliable method for the remediation of various soil environments has been developed. Phytoremediation as a modern process involves the convention of plants that have been naturally mutated in addition to microorganisms from the soil to decrease their hazardous or clustering effects within the environment (Ali et al., 2013).

Some groups of greenery known as hyper accumulators' have resistant ability to heavy metals. As the name implies, they have potential of accumulating and transporting soil pollutants to high concentrations (Gratão et al., 2005). The

Phytoremediation. DOI: https://doi.org/10.1016/B978-0-323-89874-4.00002-9

usages of greenery that have been naturally mutated are very feasible method because they are highly modifiable, cost effective, and have the ability to preserve the environment successfully. Besides keeping the environment protected, they also proved beneficial (Glass, 1999). Among their beneficial effects are developed biological diversification, soil protection, sequestration of carbon, proper management of watershed, and varied sources of power and decorative (Dickinson et al., 2009). Apart from benefits previously mentioned, it is highly stable and sustainable as well (Adams et al., 2013).

Phytoremediation as a current and emerging treatment approach for environmental pollution is believed to provide lasting solution to the problems associated with other treatment methods because it is eco-friendly, more feasible, and affordable. However, the efficiency of this green technology has been limited by factors such as anthropogenic pollution. Fair assessment is therefore necessary for limitation associated with this method. This will unveil more knowledge about phytoremediation and provide views for future recommendations.

12.2 Contamination and pollution of soil by heavy metals

In many literature evaluations, the words "contamination" and "pollution" are habitually wrongly interchangeably accustomed. While contamination simply means existence of substances that are virtually destructive over particular concentration or volume, pollution can be defined as contamination that results into detrimental biological effects to inhabitant societies (Dada et al., 2015). Decontamination of soil with heavy metals is an operation that is tough, monetarily challenging, and time consuming (Schaefer and Juliane, 2007). Lead (Pb), thallium (Ti), chromium (Cr), arsenic (As), cadmium (Cd), and mercury (Hg) are referred to as essential heavy metals that constitute the lithosphere. These metals can neither be degraded nor destroyed, but at higher concentrations lead to poisoning in plants. According to Chapman (2007), all pollutants can cause contamination, but not all contaminants can cause pollution. The usage of the word "contamination" is suitable to topics related to bioremediation; this is because the term bioremediation is often used when the degree of contaminants is midway between low and moderate (Schaefer and Juliane, 2007).

Heavy metals can either be grouped as essential or nonessential metals. Metals important for chemical functions in living things are known as essential, basic, or microelement and examples of such include manganese (Mg), copper (Cu), cobalt (Co) and zinc (Zn). For metals that are nonessential, they have functions that are not known in animal physiology; examples are lead (Pb), mercury (Hg), cadmium (Cd), and arsenic (As) (Naidu et al., 2001). Nevertheless, remarkable numbers of nonessential elements possess exhilarating or motivating result, especially when exerted at a concentration that is lower; a good example is Cd that has been reported to produce effects of stimulation in plants such as Barley seedlings (Appenroth, 2010)

12.3 Biological remediation

Bioremediation as a procedure involves hazardous ravage decay or removal from the environment through utilization of living organisms such as plants, microbes, microbial, or plant enzymes. As a rising technology, it may be used concurrently along with other approach be it chemical or physical; this is essential to conclude diverse group treatment of environmental waste products (Singh et al., 2020). It is based on a consolidated strategy that employs microbial communities such as actinomycetes, bacteria, fungi, and earthworms. A promising subset of bioremediation is phytoremediation that uses vegetation or herbs to sequestrate pollutants from soil and water (Marques et al., 2009; Singh and Prasad, 2015). Plant-mediated bioremediation is the most precise technique that utilizes higher living organisms in the ecosystem to eliminate harmful compounds from groundwater, surface water, sediments, contaminated soils, and external atmosphere (Macek et al., 2000).

12.4 Techniques of phytoremediation

Due to advancement in urban and industrial development, there has been substantial rise of heavy metal deposition over decades around the Universe (Suman et al., 2018; Ashraf et al., 2019).Phytoremediation biotechnology as a plant-based method entails the usage of plants to select and eliminate impurities or diminish their soil bioaccessibility (Berti and Cunningham, 2000). There are certain approaches of phytoremediation that have higher capabilities of remediating soils polluted with heavy metals. The mostly widely used strategies for the remediation of heavy metal–polluted soils are phytostabilization, phytoextraction, phytovolatilization, and phytofiltration (Yan et al., 2020).These strategies work in diverse ways for the purpose of rectifying problems caused by polluted soils. According to Halder and Ghosh, the techniques of phytoremediation can be classified mainly into five; phytostabilization, phytoextraction, phytovolatilization, phytofiltration, and phytodegradation (Fig. 12.1). This chapter will highlight the benefits and limitations associated with common remediation techniques mentioned above.

12.4.1 Phytostabilization

The technique involves the application of some sort of plants that have high metallic-tolerant ability for the aim of decreasing the bioavailability of heavy metal in soil, and as a result immobilize them below ground level and thus decline the possibility of these minerals from getting into the food chain (Wong,2003; Marques et al., 2009). *Miscanthus* × *giganteus* (M × G) have been reported to be important energy supplying crops that have the capacity of balancing Cd and Hg effect in soil, and as a result decrease metal accumulation in aboveground biomass (Zgorelec et al., 2020). This method has the advantage of standard soil preservation in a zone where soil has been contaminated with heavy metals through plant growth. Establishment of green vegetation covering

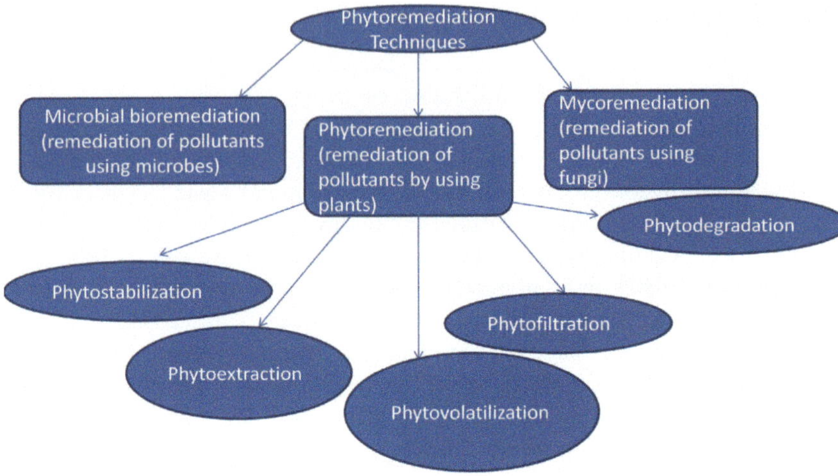

FIG. 12.1 Bioremediation techniques.

the soil also supports hard metals, reduces the groundwater percolation, and hinders dissemination of heavy metals by wind (Vangronsveld et al., 2009). Another advantage is that unsafe biomass distribution is not mandatory when compared to other remediation strategies (Wuana and Okieimen, 2011).

The disadvantage of adopting these strategies for remediating contaminated soil is that there are limited species of plants that can be used; this is because soils with high concentrations of traced elements can result into low crop yield and reduce water-retention capacities of plants. Regarding the vegetative impact associated with this strategy, it is advisable to use native plant for the same purpose. This is because plants inbred possess the capability to adjust to changes caused by contamination and the local climatic conditions (Christou et al., 2017; Radziemska et al., 2018)

12.4.2 Phytoextraction

Another phytoremediation technique that involves the utilization of plants to extract pollutants away from water or soil is phytoextraction. These extracted toxic substances will be translocated and then accumulated on the top of the soil (Salt et al., 1995; Jacob et al., 2018). Contrary to phytostabilization whereby temporary absorption of heavy metals that still remains in the ground-level occurs, phytoextraction is a permanent procedure that removes heavy metals from polluted soil, which are then transported above-ground biomass. The best vital method of phytoremediation that can be used for toxic metals reclamation in recent times is phytoextraction; it is also the most suitable approach for commercial purposes (Ali et al., 2013; Sarwar et al., 2017; Yan et al., 2020). For phytoextraction to be effective, certain factors need to be considered. Factors such as selection of appropriate plants, heavy metal bioavailability, plant performance, and soil nature as well as rhizosphere properties determine the efficiency of phytoextraction as a process (Yan et al., 2020).

12.4.3 Phytovolatilization

Phytovolatilization is a procedure of phytoremediation that uses greenery for absorbing heavy metals from soil, which is then converted from toxic substances to less toxic ones followed by their subsequent release into the air via transportation processes that take place in leaves. This approach of phytoremediation is applicable for detoxifying organic pollutants as well as certain hard metals such as Hg, Se, and As (Mahar et al., 2016). When compared with other strategies of phytoremediation, phytostabilization has the ability of removing and dispersing heavy metal (metalloid) contaminants in form of vaporous combination with the exception of plant gathering and distribution. There is a limitation to this procedure; pollutants are not completely removed from the environment as they are conveyed away from the ground into the air. In the atmosphere, harmful vaporizable contaminants present within are reposited to the soil through water droplets in form of downfall or drizzle (Vangronsveld et al., 2009).

12.4.4 Phytofiltration

Phytofiltration as a strategy of phytoremediation brings about plant usage for eliminating toxic wastes from polluted surface and wastewaters through different parts of plants. The varied forms of phytofiltration are rhizofiltration (when the roots are used), caulofiltration (when the shoots are used), and blastofiltration (when the seeds are used) for the removal of toxic substances (Mesjasz-przybylowicz et al., 2004). It also means the utilization of plant roots to support absorption or adsorption of toxic substances from surface wastewater thereby helps in the prevention of their leaching to underground water (Sarma, 2011). In phytofiltration, unsafe suspended particles in air are diminished via external parts of the plants (Singh and Tripathi, 2007). Direct growth of plants within contaminated aquatic environment revealed this technique in situ but can generally be appropriate by adopting species of plants from marine environment. (Olguin et al., 2012; da Conceicão Gomes,2016).Terrestrial plants can also be used to remediate metals through the process of precipitation, which is achieved through microorganism root biopurifier (Salt et al., 1995). Basically, ideal plants required during rhizofiltration must possess certain characteristics such as elevated biomass formation, dense root system, and must have high tolerant ability to hard metals (Yan et al., 2020). When compared with phytoextraction, phytofiltration has the advantage of showing restricted translocation capability for abstracted metals between roots and shoot (da Conceicão Gomes, 2016).

This is because low translocation of contaminants also reduces their spread to other portion of the vegetation (Kamal et al., 2018).

12.4.5 Phytodegradation

Phytodegradation (or phytotransformation) is another phytoremediation strategy whereby contaminants are chemically broken down through an action of metabolites that occur in plants. Enzymes action produced by vegetation renders related pollutants inoperative in some portion of plants such as shoots and roots (Tangahu et al., 2011).The organic contaminants are changed into less toxic substances that are then incorporated with fibers of plants; this later on promotes development of plant (EPA, 1998). Within the plant rhizosphere or inside the greenery itself, phytotransformation of plant organic components can occur. This approach of remediation is suitable for removing various compounds contaminants from the environment inclusive of groundwater diluents, crude oil, soil odoriferous compounds as well as vaporizable group of chemical molecules in the atmosphere (Newman and Reynolds, 2004). However, this approach usually takes a longer period of time and also requires specific soil characteristics such as deepness, subterranean water accessibility as well as modification of topsoil (Kamal et al., 2018). (Fig. 12.1)

12.5 Advantages of phytoremediation

12.5.1 Eco-friendly

Phytoremediation is thought to be environmentally friendly; it is safer and harmless when compared with some mechanical means of remediation. In 2009, the Environmental Protection Agency (EPA) discovered the growing problems integrated with modern remediation methods whereby more hazards are introduced to the environment than they are cleaning up. As a result, EPA developed Green remediation strategies for the purpose of decreasing energy usage and conservatory gas exudations amid of remediating dump ground with dangerous dissipates (Adams et al., 2013). There should be a limitation to the use of mechanical means of remediation simply because more problems are created than they are solved. The physical remediation procedures greatly alter the soil structure because innumerable layers of residues are isolated and also increase the chances of their exposures to environmental hazards. Moreover, mechanized machines also burn fossil fuel and thus contribute to pollution of the surrounding. Excavation is among the majority of the machine-driven procedure of remediation. Although excavation is straightforward, it can lead to environmental related issues such as washing away of soil, costly depletion of fuel likewise soil disintegration through offsite disposal of dug-up substances, thus rendering the environment unfavorable to both plants and animals, which can adversely affect the human health (EPA, Green Remediation, 2008).

12.5.2 Hyperaccumulators

There are different ways by which phytoremediation can eliminate toxins out of the earth via roots and modified them into neutral or safety compounds. Such plants have the potential of reclamations throughout their whole lifetime and have no requirement to be harvested (Kumar et al., 1995). The most widely used method of eliminating contaminants from soil and eliminate them without causing damage to disposition and fertility of soil is phytoextraction (or phytoaccumulation) (USPA, 2000). Certain groups of plants applicable to phytoextraction are called hyper accumulators. This group of plants has the potential to extract metals and translocate them out of the ground into biomass of above ground level (Kumar et al., 1995). Varieties of plants terrestrial in origin were also reported to suit the function of hyper accumulators because of their capacity to hyper accumulate certain metal from soil. Examples of such plants are *paspalum*, *Apocynum*, *Brassica*, *Thlaspi*, and *Acollanthus* among others (Table 12.1).

12.5.3 Cost effectiveness

Generally, phytoremediation is less expensive when compared to other remediation techniques because it uses plants and natural resources to remove pollutants. The roots of plants fix the soil and help to inhibit movements with regard to pollutants through run off and wind blow dust. Phytoremediation may be executed in situ or ex situ. Standard ex-situ process of phytoremediation includes digging of the ground, off-spot depot, waterlogged of soil

TABLE 12.1 List of plants with hyper accumulating potential.

Species of plants	Parts of plant tissue involved	Resistant metals involved
Solanum nigrum	Roots, stems, and leaves	Cr, Cd, Cu
Thlaspi goesingense	Stem	Ni
Thlaspi caerulescens	Root and stem	Cd, Zn
Elsholtzia splendens	Leaves, stems, and roots	Cu
Commelina communis	Leaves, stems, and roots	Pb
Brassica chinensis	Roots	Pb, Cd
Brassica napus	Roots	Pb
Alnus firma	Roots	Cu, Pb
Alyssum bertolonii	Roots, leaves, and stem	Ni
Arabis hirsuta	Leaves and stems	Zn, Pb

TABLE 12.2 Cost of phytoremediation when compared to others.

Approach	Expenses ($/ton)	Supplementary Effort/Expenditure
Land filling	100–500	Transport/excavation/monitoring
Chemical treatment	100–500	Recycling of contaminants
Vitrification	75–425	Long-term monitoring
Electrokinetic	20–200	Monitoring
Phytoextraction	5–40	Monitoring

as well in situ coating for neutralization; these procedures are additionally costly than those of in situ phytoremediation approach (Pilon-Smits and Freeman, 2006). Techniques of in situ require conventions that are simple to execute in addition to an affordable cost of sustenance (Table 12.2). Due to the fact that in situ procedures are used for the treatment of fairly contaminated soil beyond extensive surfaces, it was regarded and recommended to have considerable ability as a consistent and solar power–driven techniques, provided the plants are carefully selected and there is sufficient agronomic methods to regulate the ability of the traced elements to be transferred correctly to a plant from the soil (Schwitzguébel, 2017).

12.5.4 Limitations

There are problems associated with the use of phytoremediation as a technique of removing contaminants. However, most of the problems depend immensely upon spot of contaminated place, foreign matter nature, and suitability of approach of phytoremediation hired (Rungwa et al., 2013). Although phytoremediation is far less expensive when compared to other remediation technologies; it requires more time to be executed as a strategy unlike chemical, physical, and microbial means. Plants that have reduced biomass turnout with underdeveloped roots lack the probability of facilitating phytoremediation effectively and as a result cannot hinder the percolation of toxins into an aquatic environment (Hemen, 2011).

Reports indicated that appropriate plants must be selected for phytoremediation to be effective. Some of the factors to be considered for selection are immunity ability of plants against toxins, ability to adapt to ecological situation, sufficient plant fertility, reduced rate of substance accumulation by plants, suitability of different soil types as well as resistance against the invasion of foreign species (Kaneez et al., 2017). Factors such as a harsh state of the atmosphere, virulence, as well as total soil status of the polluted land can also affect the performance of phytoremediation technique.

The aforementioned factors can determine the fate of plant survival and their development as they have direct influence of affecting the plants adversely (Dahn et al., 2009).

Another limitation is that there should be specific metal hyper accumulator for multiple metals contaminated soil and water; this necessitates a comprehensive research view before embarking on its utilization to achieve success (Hemen, 2011). Metals must also be in bioavailability state to plants; this is because when metals are strongly attached to organic portions of soil, it will affect its availability to plants for extraction or absorption. Moreover, the metals must not be water soluble in nature; this is because water-soluble metals will not undergo accumulation at the root of plants. Some hard metals previously recognized at the contaminated sites are Pb, Zn, Hg, Cu, As, Cr, Cd, and Ni. These metals availability differ in several locations and directly rely on the derivation of particular contaminants (Lone et al., 2008)

Excessive plant metal absorption can bring about poison in man nourishment through food chain. The toxicity can lead to different types of ailments in humans both persistent and severe; a good example is Zn and Cd that can induce impairment to respiratory and gastrointestinal system; it can also result into severe diseases of the brain, heart, and kidney (Lone et al., 2008). Elevated level of absorption of hard metals by soil will adversely alter the development of crops. Heavy metal toxicity interference, alteration of plant metabolic, physiological as well as biochemical functions have detrimental effects on cellular respiration and chemosynthesis. These can cause deterioration of primary organelles of the cell and thus induce plant death (Garbisu and Alkorta, 2001; Schmidt, 2003; Schwartz et al., 2003).

12.6 Conclusion

Phytoremediation as a technology is established as a result of plant usage that is profitable, feasible, and economically favorable. It works in diverse ways and its main principles include phytoextraction, phytostabilization, phytovolatilization, phytofiltration, and phytodegradation. Heavy metals are important for biochemical functions of both plants and animals, but their higher concentrations can lead to poisons in plants. The toxic substances can reach humans when they enter the food chain and expose humans to serious health problems such as gastrointestinal and respiratory diseases among others. Though the plant-based technology of remediating heavy metals is sustainable, it has some limitations as a technique of remediation when compared to other traditional techniques; it takes longer period and depends on location of site pollution, nature of toxic substances, and suitability of the phytoremediation procedure to be employed. More studies on applicability of phytoremediation need to be considered to achieve maximum result as a means of remediation.

Conflict of Interest

The author declares no interest of conflicts

References

Adams, A., Raman, A., Hodgkins, D., 2013. How do the plants used in phytoremediation in constructed wetlands, a sustainable remediation strategy, perform in heavy-metal-contaminated mine sites? Water Environ. J. 27, 373–386.

Appenroth, K.J., 2010. Definition of heavy metals and their role in biological systems. Soil Heavy Metals. Soil Biology, vol 19.

Ashraf, S., Ali, Q., Zahir, Z.A., Ashraf, S., Asghar, H.N., 2019. Phytoremediation: environmentally sustainable way for reclamation of heavy metal polluted soils. Ecotoxicol, Environ. Safe. *174*, 714–727.

Berti, W.R., Cunningham, S.D., 2000. Phytostabilization of metals. In: Raskin, I., Ensley, B.D. (Eds.), Phytoremediation of Toxic metals; Using Plants to Clean-up the Environment. *John Wiley & Sons, Inc, New York, NY*, pp. 71–88.

Chapman, P.M., 2007. Determining when contamination is pollution-weight of evidence determinations for sediments and effluents. Environ. Int. 33, 492–501.

Christou, A., Theologides, C.P., Costa, C., Kalavrouziotis, I.K., Varnavas, S.P., 2017. Assessment of toxic heavy metals concentrations in soil and wild and cultivated plant species in *Limni* abandoned copper mining site, Cyprus. J. Geochem. Explor 178, 16–22.

da Conceição Gomes, M.A., Hauser-Davis, R.A., de Souza, A.N., Vitória, A.P., 2016. Metal phytoremediation: general strategies, genetically modified plants and applications in metal nanoparticle contamination. Ecotoxicol. Environ. Saf. 134, 133–147.

Dada, E.O., Njoku, K.L., Osuntoki, A.A., Akinola, M.O., 2015. A review of current techniques of in situ psycho-chemical and biological remediation of heavy metals polluted soil. Ethiop. J. Environ. Studies Manag. 8, 606–615.

Dahn, L.T., Truong, R.Mammucari, Tran, T., Foster, N, 2009. *Vetiveria zizanioides*: a choice plant for phytoremediation of heavy metals and organic wastes. Int. J. Phytoremediation 11, 664–691.

Dickinson, N.M., Baker, A.J.M., Doronila, A., Laidlaw, S., Reeves, R.D, 2009. Phytoremediation of inorganics : realism and synergies. Int. J. Phytoremediation 11, 97–114.

Environmental Protection Agency (EPA), 2008. Green Remediation: Incorporating sustainable Environmental practices into Remediation of contaminated sites, EPA 542-R-08-002.

EPA, 1998. A citizen's guide to phytoremediation. U.S. Environmental protection Agency, Office of solid waste and Emergency Response EPA 542-F-98-011.

Garbisu, C, Alkorta, I., 2001. Phytoextraction: a cost effective plant-based technology for the removal of metals from the environment. Bioresour. Technol 77, 229–236.

Glass, D.J., 1999. Economic potential of phytoremediation. In: Raskin, I., Ensley, B.D. (Eds.), Phytoremediation of Toxic Metals: using plants to clean up the Environment. John Wiley & Sons Inc, New York, 15–31.

Gratão, P.L., Prasad, M.N.V., Cardoso, P.F., Lea, P.J, Azevedo, R.A., 2005. Phytoremediation: green technology for the cleanup of toxic metals in the environment. Braz. J. Plant Physiol. 17, 1.

Hemen, S., 2011. Metal hyperaccumulation in plants: a review focusing on phytoremediation Technology. J. Environ. Sci. 4, 118–138.

Jacob, J.M., Karthik, C., Saratale, R.G., Kumar, S.S., Prabakar, D., Kadirvelu, K., 2018. Biological approaches to tackle heavy metal pollution: a survey of literature. J. Environ. Manage. 217, 56–70.

Kamal, U., Mohammed, A.A., Mohammed, H.A., 2018. Phytoremediation: halophytes as promising heavy metal hyper accumulators. In: Saleh, H. (Ed.). Heavy Metals, 12. Intechopen, London, UK, pp. 201–217.

Kaneez, F, Asma, I, Mohammad, N., Muhammad, A., 2017. Plant-bacteria synergism: an innovative approach for the remediation of crude oil-contaminated soils. Soil Environ 36, 93–113.

Kumar, N., Dushenkov, V., Motto, H., Raskin, I., 1995. Phytoextraction: the use of plants to remove heavy metals from soils. Environ. Sci. Technol 29, 1232–1238.

Lone, M.I., He, Z.I., Stoffella, P.J., Yang, X., 2008. Phytoremediation of heavy metal polluted soils and water: progresses and perspectives. J. Zhej. Univ. Sci. B, 210–220.

Macek, T., Mackova, M., Kas, J., 2000. Exploitation of plants for the removal of organics in environmental remediation. Biotechnol. Adv. 18, 23–34.

Mahar, A., Wang, P., Ali, A., Awasthi, M.K., Lahori, A.H., Wang, Q., 2016. Challenges and opportunities in the phytoremediation of heavy metals contaminated soils: a review. Ecotoxicol. Environ. Saf. 126, 111–121.

Meagher, R.B., 2000. Phytoremediation of toxic elemental and organic pollutants. Curr. Opin. Plant Biol. 3, 153–162.

Mesjasz-Przybylowicz, J., Nakonieczny, M., Migula, P., Augustyniak, M., Tarnawska, M., Reimold, W.U., Koeberl, C., Przybylowicz, W., Glowacka, E. (2004). Uptake of cadmium, lead, nickel and zinc from soil and water solutions by the nickel hyperaccumulator Berkheya coddii. Acta Biol. Cracov. Bot. 46, 75–85.

Naidu, R., Krishnamurti, G.S.R., Bolan, N.S., Wenzel. W., Megharaj, M. (2001). Heavy metal interactions in soils and implications for soil microbial biodiversity. In: M.N.V. Prasad (ed.), Metals in the Environment. Analysis by Biodiversity. MarcelDekker Inc., New-York, 401–431.

Newman, L.A., Reynolds, CM., 2004. Phytodegradation of organic compounds. Curr. Opin. Biotechnol. 15, 225–230.

olguín, E.J., Sanchez-Galvan, G, 2012. Heavy metal removal in phytofiltration and phytoremediation: the need to differentiate bioadsorption and bioaccumulation. New Biotechnol. 30, 3–8.

Pilon-Smits, E.A.H., Freeman, J.L, 2006. Environmental clean up using plants: biotechnological advances and ecological considerations. Front. Ecol. Environ. 4, 203–210.

Radziemska, M., Koda, E., Bilgin, A., Vaverkova, M.D., 2018. Concept of aided phytostabilization of contaminated soils in post-industrial areas. Int. J. Environ. Res. Public Health 15 (24).

Rungwa, S., Arpa, G, Sakulas, H, Harakuwe, A, Timi, D, 2013. Phytoremediation-an eco-friendly and sustainable method of heavy metal removal from closed mine Environments in Papua New Guinea. Procedia Earth Planet. Sci. 6, 269–277.

Salt, D.E., Blaylock, M., Kumar, N.P.B.A., Dushenkov, V., Ensley, B.D., Chet, I., 1995. Phytoremediation: a novel strategy for the removal of toxic metals from the environment using plants. Nat. Biotechnol. 13, 468–474.

Sarma, H., 2011. Metal hyperaccumulation in plants: a review focusing on phytoremediation technology. J. Environ. Sci. Technol. 4, 118–138.

Sarwar, N., Imran, M., Shaheen, M.R., Ishaque, W., Kamran, M.A., Matloob, A., 2017. Phytoremediation strategies for soils contaminated with heavy metals: modifications and future perspectives. Chemosphere 171, 710–721.

Schaefer, M., Juliane, F., 2007. The influence of earthworms and organic additives on the biodegradation of oil contaminated soil. Appl. Soil Ecol. 36, 53–62.

Schmidt, U., 2003. Enhancing phytoremediation: the effect of chemical soil manipulation on mobility, plant accumulation, and leaching of heavy metals. J. Environ. Qual 32, 1939–1954.

Schwartz, C, Echevarria, G, Morel, J.L., 2003. Phytoextraction of cadmium with *Thlaspi caerulescens*. Plant Soil 249, 27–35.

Schwitzguébel, J., 2017. Phytoremediation of soils contaminated by organic compounds: hype, hope and facts. J. Soil Sediment. 17, 1492–1502.

Singh, S.N., Tripathi, R.D., 2007. Environmental Bioremediation Technologies46. *Springer Science & Business Media*, Berlin, Germany, pp. 299–300.

Singh, A., Prasad, S.M., 2015. Remediation of heavy metal contaminated ecosystem: an overview on technology advancement. Int. J. Environ. Sci. Technol. 12, 353–366.

Singh, P., Singh, V.K., Singh, R., Borthakur, A., Madhav, S., Ahamad, A., Kumar, A, Pal, D.B., Tiwary, D, Mishra, P.K, 2020. Bioremediation: a sustainable approach for management of environmental contaminants. In: Singh, P., Kumar, A., Borthakur, A. (Eds.), Bioremediation: a sustainable approach for management of environmental contaminants. Abatement of Environmental Pollutants: Trends and Strategies, 1–23.

Suman, J., Uhlik, O., Viktorova, J., Macek, T., 2018. Phytoextraction of heavy metals: a promising tool for clean-up of polluted environment? Front. Plant Sci. 9, 1476.

Tangahu, B.V., Abdullah, S.R.S., Basri, H., Idris, M., Anvar, N., Mukhlisin, M., 2011. A review on heavy metals (As, Pb, and Hg) uptake by plants through phytoremediation. Int. J. Chem. Eng. 31 Article ID 939161.

United States Protection Agency (USPA) 2000. Water Air Soil Pollution 47: 189–215.

Vangronsveld, J., Herzig, R., Weyens, N., Boulet, J., Adriaensen, K., Ruttens, A., 2009. Phytoremediation of contaminated soils and ground water: lessons from the field. Environ. Sci. Pollut. Res. 16, 765–794.

Wong, M.H., 2003. Ecological restoration of mine degraded soils, with emphasis on metal contaminated soils. Chemosphere 50, 775–780.

Wuana, R.A., Okieimen, F.E., 2011. Heavy metals in contaminated soils: a review of sources, chemistry, risks and best available strategies for remediation. ISRN Ecol. 2011, 402647.

Yan, A., Wang, Y., Tan, S.N., Yusof, MLM, Ghosh, S, Chen, Z, 2020. Phytoremediation: a promising approach for revegetation of heavy metal-polluted land. Front. Plant Sci. 11, 359.

Zgorelec, Z., Bilandzija, N., Knez, K., Galic, M., Zuzul, S., 2020. Cadmium and mercury phytostabilization from soil using *Miscanthus* X *giganteus*. Sci. Rep. 10, 6685.

Chapter 13

Phytoremediation of soil and water

Rajdeep Jaswal[a], Ashish Bedi[b,c], Isha Bedi[d], Akanksha Jaiswar[c], Rahul Singh Jasrotia[e]

[a]*National Agri Food Biotechnology Institute (NABI), Mohali, Punjab, India,* [b]*School of Life and Environmental Sciences, Deakin University, Victoria, Australia,* [c]*TERI-Deakin Nanobiotechnology Centre, The Energy and Resources Institute, Gurugram, Haryana, India,* [d]*Saint Ambrose University, Davenport, Iowa, USA,* [e]*Department of Biological Science, Florida State University, Tallahassee, Florida, USA*

13.1 Introduction

All living beings require certain essential entities such as oxygen, food, and liquids to carry out their basic biological process that further sustain their lives. Among these, water and soil are essential components required by all living organisms, including human beings, to metabolize food and provide energy (Fig. 13.1). Water is the first medium where life has been predicted to be originated. It is the primary constituent of every cell of the human body; moreover, it constitutes 60%–70% of the total human body weight (Riveros-Perez and Riveros, 2018). The metabolism of food also produces water molecules; however, the amount produced is insufficient to support various biological processes. Therefore, an adult human being requires 3–5 L of water externally. Water serves several functions in the body, such as building material, water and thermoregulation, lubricant and shock absorber, blood pressure regulation, toxin removal, and several other functions vital for life (Jéquier and Constant, 2010).

The presence of water and oxygen on the planet earth makes it one of the prime locations for life existence (McKay, 2014). Seventy percent of the earth is covered with the water, 96% of the water is present in the sea (Visbeck, 2018). In comparison, the remaining 4% of water is contributed by rivers, lakes, and other freshwater suitable for human consumption (Pimentel et al., 2004).

Similarly, soil is an important resource in an ecological community. It plays an important part in the growth of the plants as helps in the biogeochemical recycling of the nutrients.

Soil is a very diverse heterogeneous assembly of various sized/shaped organic matter in multiple stages of degradation. Several processes comprising

Phytoremediation. DOI: https://doi.org/10.1016/B978-0-323-89874-4.00027-3

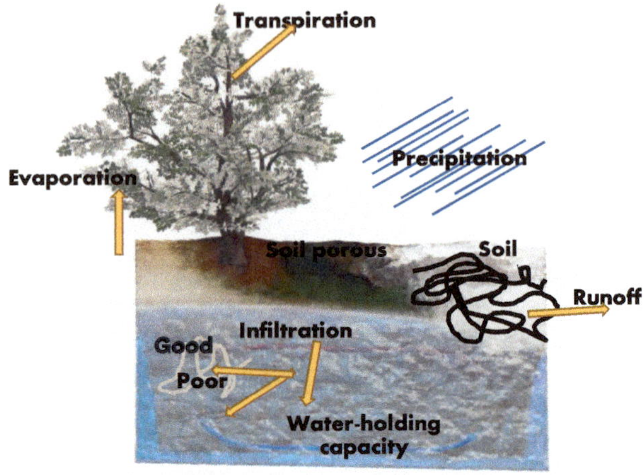

FIG. 13.1 Plant-Soil-Water relationship. *Source: Fig. from Soil, Water & Plant Relationships Lecture note for Soil and Water Management Course Prepared by Dr. ND Nang.*

weathering, biological, topographical and climate changes have given soil its diverse characteristics. Soil constantly exchanges its boundaries between solid, liquid, and gaseous phases imparting it transient properties that serve as a reservoir and a filtration medium breaking down wastes' products. In this bioecological community, there is constant recycling of carbon, nitrogen, and other necessary elements (Yaron et al., 1996). It provides habitat for a plethora of microorganisms, which helps in maintaining biodiversity in an ecosystem. Soil also contributes in equipping us with all types of antibiotics used to fight various infections. It is the basis of the earth's agroecosystems, which replenishes living being with necessities such as food, fiber, and fuel (Yaron et al., 1996).

Degradation of the soil is caused by soil erosion due to continuous spread of urbanization that poses a major threat of our soil resource sustainability. Potentially toxic substances can seep in the ground water causing major health problems for the human population.

The availability of fresh, pure water and good soil quality for agriculture is a major challenge that developing countries face in this era. Moreover, with the increase in the world population, the food and water requirement have increased drastically, putting tremendous pressure on agriculture, promoting more urbanization, more industries establishments, deforestation causing soil erosion, and ultimately leading to soil and water pollution (Calder et al., 2008). The global industrialization process demands increased agro-economic output, which leads to extensive usage of fertilizers, pesticides, crude petroleum and its derivatives, polymers, leading to the addition of assorted range of chemical pollutants and other wastes in the environment.

A study showed that water consumption has increased by sixfold within a century (1900–1995) (Sivakumar, 2011). With this pace, the freshwater resources will be exhausted soon. The conservation of freshwater resources is, therefore, essential to sustain the available water resource.

In addition to freshwater sources such as lakes and ponds, groundwater also constitutes significant human and animal consumption sources. Unfortunately, the level and quality of underground water is lowered due to overuse, and is deteriorating due to industry effluents, anthropogenic activities, overutilization due to increased population (Pimentel et al., 2004, Boretti and Rosa, 2019). The industry effluent makes the remaining underground water contaminated with heavy metals such as arsenic, mercury, lead, cadmium (Masindi and Muedi, 2018, Wuana and Okieimen, 2011). Additionally, environmental changes such as global warming and other deforestation are significantly affecting rainwater accumulation (Ellison et al., 2017).

In the current scenario, due to the poor quality of water and soil, serious health problems are being reported in several studies in developing countries.

It has been estimated that nearly one-third of the world population is consuming groundwater for drinking and food preparation (WHO, 2002) (Boretti and Rosa, 2019). Approximately two million deaths occur due to the consumption of focally contaminated water, causing diarrhea. Several international health care agencies such as WHO, CPCB, ICMR claim that the direct pollutant discharge contaminates approximately 60%–70% of river water quality in India (Giri et al., 2020). In modern world, soil in industrially advanced territories of the world is already polluted to a slight extent. It is estimated by the global Assessment of Soil Degradation (GLASOD) that almost 21.8×10^6 ha of the land in Europe, Asia, Africa, South America are already affected by soil pollution.

The conservation of natural resources must be carried out to avoid natural calamities such as famine, as most agriculture and water consumption are dependent on good soil quality and freshwater resources. Several biological approaches, such as forestation, can be applied to protect the soil and water that will consequently restore groundwater. An effective and cost-effective biological strategy such as phytoremediation, a plant-based conservation and improvement approach, must be introduced to remove the metals from the soil and the water bodies. Several plant species have been reported to act as a hyperaccumulator of various heavy metals such as arsenic, cadmium. These plant species can be grown in contaminated areas to minimize or manage soil and water pollution. In brief, we can say that, phytoremediation is an emanating green technology with higher efficiency and minor environmental impact. A process that uses various types of plants to eliminate, relocate, maintain, and/or eradicate contaminants in the soil and groundwater.

13.2 Water and soil pollution

Water pollution can be defined as any undesired material, mostly toxic chemicals, heavy metals pollutant, organic/inorganic matter, and harmful microorganisms mixed into natural water bodies such as lake river, pond, and sea (Schwarzenbach et al., 2010, Raghav et al., 2019).

Soil pollution is defined as existence of toxic contaminants including organic compounds (hydrocarbons), inorganic ions and salts (phosphates, carbonates, sulfates, nitrates), radioactive and industrial waste, in high concentrations, which pose a risk to the ecosystem (Mishra et al., 2015).

This mixing of the toxic substances causes deterioration of the soil and water quality and renders them unfit for humans and other living organisms. Most of the chemicals are poorly soluble and as a result form a layer on the surface leading to deterioration of the ecosystem of these natural resources or, if consumed, may cause toxicity or even death (Zaghloul et al., 2019).

13.3 Sources of water pollution

There are major sources classified that contribute to water pollution.

13.3.1 Groundwater pollution

Groundwater is one of the most important sources of freshwater for consumption. Agricultural products such as fertilizers, pesticides, waste leaching from landfills or industry pipes, and sewage treatment plants with a leakage are the major groundwater contamination sources (Ashraf et al., 2014, Raman and Narayanan, 2008). Removing contaminants from groundwater could be a very difficult or very costly process (Fig. 13.2).

13.3.2 Surface water

Surface water constitutes >70% of the earth formed by oceans, lakes and rivers. Rivers and other water streams carry different contaminants such as heavy metals, chemicals, organic and inorganic generated by urban/rural areas and industries into the bays and estuaries, ultimately reaching the ocean (Kennish, 1994, Simcock et al., 2016). Synthetic material, predominantly plastic bags, nets, and bottles carried away by the storm, pollute the ocean. Oil spills and leaks are also significant contributors to ocean pollution (Fig. 13.2). (Thompson et al., 2009, Barnes et al., 2009, Gallo et al., 2018).

FIG. 13.2 Major sources of water pollution.

13.4 Various causes of water pollution

13.4.1 Industrial waste

Industries are one of the prime water-polluting factors as they produce a large amount of waste in the form of toxic chemicals and pollutants, which gets absorbed in the oceans (Helmer and Hespanhol, 1997, Tchounwou et al., 2012, Kjellstrom et al., 2006). In addition to several xenobiotics and harmful chemicals, the industry effluents contain several cancer-causing metals such as lead, mercury, sulfur, nitrates, asbestos, and several others (Singh et al., 2017, Akpor et al., 2014). The effluent's introduction changes may change the water's color, the temperature, making it toxic that poses severe hazards to life present in the water (Fig. 13.3, Table 13.1) (Akpor et al., 2014, Edokpayi et al., 2018).

13.4.2 Sewage and wastewater

Daily household activities', sewage, and wastewater also cause significant water pollution if introduced untreated into the clean water streams. The sewage water carries several pathogens and organic matter as typical water pollutants (Pandey et al., 2014, Cabral, 2010, Zheng et al., 2013). The household water is mostly contaminated with fecal matter that predominately contains *Escherichia coli* and other bacterial pathogens and, if consumed, can cause serious health problems (Cabral, 2010, Figueras and Borrego, 2010, Pandey et al., 2014). Additionally, the contaminated water can also be a breeding ground for pathogenic microbes such as malaria. The excess organic matter in water promotes the growth of microorganisms that facilitate decomposition (Joutey et al., 2013). The decomposition of organic matter utilizes oxygen, depleting the dissolved oxygen in the water, thereby generating an anoxygenic environment. Moreover, organic effluents also contain many suspended solids particles, reducing sunlight's penetration for photosynthetic organisms (Fig. 13.3) (Zheng et al., 2013).

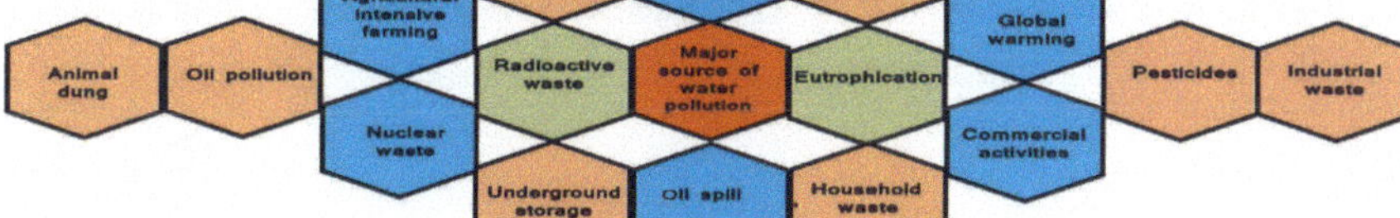

FIG. 13.3 Various causes of water pollution.

TABLE 13.1 List of various contaminants responsible for soil and water pollution, their sources, and health hazards.

Major Contaminant	Sources	Potential health hazards	References
Arsenic[a]	Generated due to natural processes in environment, waste produced by industries, copper, lead, and zinc ore smelting process	May cause toxicity (acute and chronic), in some cases, affects liver and kidney, decreased hemoglobin level	(Ramanjaneyulu and Giri, 2006)
Barium[c]	Occurs naturally in some limestones, sandstones	Heart problems, gastrointestinal, and neuromuscular problems, hypertension, and cardiotoxicity	Kravchenko et al. (2014)
Cadmium[a]	Found in rocks and petroleum. Major source industrial discharge, mining residues	Rise in blood pressure, damage to liver, kidney and may cause anemia. Damages testicular tissue and red blood cells. Harmful to aquatic biota	(Ramanjaneyulu and Giri, 2006)
Chromium[a]	Leaches into groundwater from the mining sites, produced by fossil-fuel use and cement-plant industries	May cause liver and kidney damage, internal hemorrhaging, respiratory problems, dermatitis, and ulcers on the skin when in high concentrations	(Ramanjaneyulu and Giri, 2006)
Copper[a]	Major sources are waste generated due to industrial, domestic activities as well as mining and mineral leaching activities	Contaminated water may cause stomach and intestinal problems, liver, and kidney damage, in some cases, anemia when accumulated in high doses	(Ramanjaneyulu and Giri, 2006)
Cyanide[c]	Produced by steel processing, plastics, and fertilizer producing industries	Consumption of polluted water may cause poisoning or may result in spleen, brain, and liver damage	(Mudder and Botz, 2004)
Lead[b]	Present in batteries, ammunitions, pigment and paints, ceramic coating, hair colors, fishing equipment, burning coal and lead containing gasoline (vehicular exhaust) due to mining and plumbing	Can lead to neurological damage, lower IQ level, impairment of coordination between hand and eye, encephalopathy, weakening of bones, hypertension and kidney damage	(Ramanjaneyulu and Giri, 2006)

Mercury[b]	Present in electrical switches, fluorescent bulbs, and lamps. Some other items include batteries, mercury thermometers, tooth fillings, mining (mainly small scale gold mining), some pesticides, medical waste, and produced during combustion of coal and fuel oil	Can cause damage to central nervous and gastric system, negatively affects development of brain subsequently resulting in reduced IQ, vision-related problem and loss of sense of touch, lead to liver disease as well heart and kidney damage	(Park and Zheng, 2012)
Organic pollutants[a]	Herbicides, house hold sources, byproducts generated from industry and during combustion	Cancer, some of them might cause chronic diseases (hypertension, cardiovascular disease and diabetes), problem associated with reproductive system and developmental, suppression of immune system, interfere with hormones (thyroid and sex steroids)	(Carpenter, 2011), (Burgess, 2013)
Silver[c]	Produced by ore mining and processing, photography	Skin diseases such as argyria	(Drake and Hazelwood, 2005)
Volatile organic compounds[c]	Produced in the manufacturing processing of plastics, pharmaceuticals products, dyes, rubbers, solvents, crude oil, insecticides, inks, varnishes, paints, disinfectants	May cause liver damage, skin problems, anemia, gastrointestinal problems, and nervous system damage or cancer in severe cases	(He et al., 2015)
Microbial contamination[c]	*Escherichia coli* found naturally in the environment, human gut, and soil. However, the presence of *E. coli* is treated as fecal contaminate water and other human pathogenic microbes such as bacteria, viruses, and parasites.	The microbes can cause varied diseases such as dysentery, typhoid, cholera, polio, hepatitis and several other life-threatening diseases	(Ashbolt, 2004)

[a]*Major pollutants in soil and water,*
[b]*Major pollutants in soil,*
[c]*Major pollutants in water.*

13.4.3 Mining activities

The metal extraction also causes water pollution as it produces several harmful chemicals along with raw metals (Carvalho, 2017). These chemicals, when get mixed with water, produce toxic elements rendering water unsuitable for humans' consumption. The mining process also produces metal waste and other elements such as sulfide from the rocks (Fig. 13.3) (Mohapatra and Kirpalani, 2017).

13.4.4 Marine dumping and oil spillage

In several coastal areas, household garbage such as paper, food, plastic, biomass is dumped into the seashores (van Emmerik and Schwarz, 2020). These dumped products are carried along with the tides and enter the sea and adversely affect sea animals and their ecosystem. Oil spills due to petroleum mining and transport also cause huge damage to marine life, killing fishes, birds and other sea organisms (Fig. 13.3) (van Emmerik and Schwarz, 2020, Zabbey and Olsson, 2017).

13.4.5 Fossil fuels

The burning of fossil fuels, such as coal and petroleum oil produces a lot of ash containing suspended particles in the environment. When mixed with water vapors, these particles produce acid rain that degrades the environment (Fig. 13.3) (Casiday and Frey, 1998, Grennfelt et al., 2020).

13.4.6 Use of chemical fertilizers and pesticides in agriculture

To enhance the yields and protection against pathogens, use of synthetic chemicals such as fertilizers and pesticides has increased substantially. The excess chemicals present on the crop or soil get mixed with rain and reach groundwater or water bodies, posing serious damage to aquatic life (Pérez-Lucas et al., 2018).

13.5 Sources of soil pollution

Soil pollution may arise from deliberate sources such as the addition of fertilizers, pesticides and manure which add heavy metals (As, Cd, Mg, Zn, U, Cu) to the soil and are absorbed by the clay mineral lattice work of the soil water depending on the volatility of the active ingredients.

13.5.1 Heavy metals

For the last few decades, the sales of fertilizer and pesticide have skyrocketed. Due to the continuous addition of fertilizers and pesticides, there is severe contamination of urban and agricultural soil, which results in the reduction of growth and yield of the crops and severe heavy metal contamination. Also, fossil fuel combustion has added trace elements (Cd, Zn, As, Sb, Se, Ba, Cu, Mn and V) to the soil as well as to the atmosphere (Table 13.1) (NRCS, 2000).

13.5.2 Hydrocarbons

Hydrocarbon pollutants come from petroleum and mainly comprise straight and branched chain alkanes and aromatic hydrocarbons. The most common one is BTEX compounds consisting of benzene, toluene, ethyl benzyl, and xylene, occurring in plumes in groundwater. Various industrial sites consist of iron pyrites (FeS_2) which leads to high risk of fire and explosion.

Depending on the behavior and density of chemical pollutant in soil/ground water they are grouped as nonaqueous phase liquid (NAPLs), light nonaqueous phase liquids (LNAPLs), and dense nonaqueous phase liquids (DNALPs). Many pesticides derivatives such as Polycyclic aromatic hydrocarbons (PAHs), PCBs, dionines all (organo- micro pollutants) lead to the formation of free Al^{3+} and $Al\ (OH)_2^+$ ions that are toxic to most plants (Alloway and Control, 1996).

13.5.3 Radioactive waste spill and disposal

Decades earlier nuclear accidents at Windscale, United Kingdom (1957) and Chernobyl, Ukraine (1986) resulted in the dispersion of radioactive substance in the environment. The problem associated with radioactive waste is that they remain for a long period in the environment. For example, K present in soil gets replaced by Cs^{137} (half-life: 30 years) and Ca present in biological system is replaced by Sr^{90} (half-life: 29 years). Moreover, accidental leakage of the harmful chemicals during the storage, transportation, or disposal significantly deteriorates the quality of soil and making them unsuitable for cultivation. The pollutants entering into the groundwater make the water undrinkable. The radioactive chemical pollutants also act differently when dumped in the soil making them last for long (Alloway and Control, 1996).

To protect the ecosystem, we should develop imperative tools to estimate the potential risks of human exposure to these chemical pollutants and to establish their threshold concentrations in soil.

13.6 The consequence of water pollution

Water pollution may cause serious threats to humans, including the destruction of the environment. These effects may include the origin of disease outbreak due to contaminated water, destruction of the environment's natural ecosystem, disruption of food chains, degradation of the agricultural system, and many others (Bashir et al., 2020, Tuomisto et al., 2017). A few of them are discussed here. The effect of pollutants also varies depending on their types and source. For example, several heavy metals, dyes, and organic pollutants have been reported to cause cancer (Tuomisto et al., 2017, Masindi and Muedi, 2018).

13.6.1 Destruction of the ecosystem

Every species of plants, animals, birds, and other life forms are dependent on the water to complete their lifecycle and further their survival. The water dependence of all these species directly affects their habitat. For example, the oil spill caused by various ships and oil mining companies is killing countless sea species, including fishes (Saadoun, 2015). The Deep Horizon oil spill in the year 2010 affected more than 82,000 birds, 5000–6000 turtles, approximately 25,900 mammals, and a countless number of fish and invertebrates species (Ainsworth et al., 2018). Similar to oceans, the effluent of several industries discharged in rivers is also destroying their ecosystem.

13.6.2 Diseases

The water contaminated with waste material of industries, human and animal excreta, and untreated sewage is of poor quality and not suitable for drinking due to presence of various pathogenic microbes. Drinking or cooking with such untreated water is the major source of waterborne diseases and infections. In developing countries, the water bodies' fecal contamination and its further consumption without improper treatment can lead to the outbreak of waterborne diseases such as typhoid, cholera, hepatitis A, dysentery, amoebiasis, giardiasis, toxoplasmosis, and several others (Gerba, 2009, Fazal-ur-Rehman, 2019). The World Health Organization claims that approximately 80% of diseases worldwide originate from contaminated water. Moreover, more than 3% of deaths worldwide are due to poor water quality consumption (Prüss-Ustün et al., 2014).

13.6.3 Agriculture

The irrigation of crops directly correlates with the yields. The population increase has also put much pressure on the agriculture field for increased production. The irrigation by poor quality water or water contaminated pathogens will decrease the yield; furthermore, it may deteriorate the nutritional quality and lower the number of grains or promote infection by other plant pathogens (Café-Filho et al., 2019, Islam and Karim, 2019). In addition to this, untreated water containing heavy metals may also accumulate in the water bodies, further creating health hazards if consumed by humans or animals (Jan et al., 2015). The excessive pesticides used in crop protections may also get accumulated in the groundwater or other water bodies. These compounds also accumulate in the animals at each trophic level due to biological magnification (Lushchak et al., 2018).

13.7 Consequences of soil pollution

The consequence of soil pollution is catastrophic for humans, crops, ecosystems as these directly contaminate the ground water that we consume and pollute the air that we breathe. The crops yields would be reduced as they assimilate contaminants in their system, while growing in polluted land and consuming them expose us to concentrated pollutants leading to heavy metal poisoning, allergies, weaken immune system contributing to degrading health.

The world would see increased outbreaks of infectious diseases resulting in higher mortality rate and medical costs. These pollutants would cause reduced comprehensive health of large parts of the human population as various allergies become universal affecting the quality of life of the human race.

13.8 Need for soil and water purification

Water and soil remediation are required so that high quality standards of soil, water, and air can be maintained, which will subsequently benefit agriculture and will be safe for human use as well as for the wild flora and fauna. Some chemicals originating due to human activity do not bound or remain immobilized due to changes in redox potential or acidic conditions of soil and become far more lethal (Gavrilescu and Journal, 2006), (Pisani, 2004). These noxious substances remain in the environment for a long period and overtime get accumulated in high quantities, which is lethal to the organisms present at upper trophic levels of the food chain. Most of these chemicals are toxic and cause mutation, cancer, or interfere with the development of embryo or fetus.

13.8.1 Types of pollutants (organic and inorganic)

The main source of organic pollutants is the drainage water formed by animal waste, household activities, industrial water, and agricultural chemicals such as pesticides, fertilizers, herbicides, etc. (Ashraf et al., 2014). The suspended organic matter contributes to a major share of the organic waste. The main components of organic waste typically include cellulose, lipids, lignin, protein, amino acids (Cutrim et al., 2019). The decomposition of organic matter also produces macro- and micronutrients, such as nitrates and phosphates that also affect aquatic life. The excess of these nutrients promotes excessive plant and algae growth causing detrimental effects on other aquatic life (Weigelhofer et al., 2018).

13.8.2 Inorganic pollutant

The inorganic substances, such as heavy metals, contribute to soil and water pollution as much as organic pollutants do (Häni, 1990). Whether it is high or low, the accumulation of metals has a crucial impact on the environment. Major toxic metals found in contaminated areas are mercury (Hg), lead (Pb), cadmium (Cd), arsenic (As), zinc (Zn), chromium (Cr), copper (Cu), and nickel (Ni). Few of the elements, such as Zn, Cr, Cu, and Ni, are required for the growth and development; however, when in high concentration, these elements may cause the water, soil, and air pollution (Tchounwou et al., 2012). The primary toxic metals, such as As, Cd, Cr, Hg, and Pb, can cause serious health issues, even cancer when accumulated in higher concentrations (Jaishankar et al., 2014). The industrial effluent, agriculture, medical waste, oil refineries, pharmaceutical industries, and household activities are primarily responsible for water and soil pollution (Ferronato and Torretta, 2019).

13.9 Phytoremediation of various types of water pollutants and its mechanism

The plants are widely reported to act as the accumulator of different metals, including various heavy metals responsible for causing water and soil pollution. Studies have shown that the forest can significantly benefit the ecosystem and significantly reduce pollution (Baró et al., 2014, Yang et al., 2005). The water flowing through the plant roots usually slows down; moreover, due to reduced flow rate, various elements such as heavy metals reach the water bodies in a limited amount, as these elements adhere to the soil or taken up by the roots to accumulate in the plants leading to reduced heavy metal pollution (Lone et al., 2008a, Peuke and Rennenberg, 2005). Due to their extensive root system, plants can efficiently accumulate heavy metals and manage water and soil pollution by heavy metals (Chibuike and Obiora, 2014). Moreover, the process is easy to implement and cost-effective also. The utilization of plants to manage water or soil pollution caused by heavy metals and other pollutants is known as phytoremediation. The phytoremediation involves two major pollutant removal strategies; the first involves where the pollutant accumulate in the plant bodies such as root and other harvestable regions, and the second where the pollutants are either degraded/metabolized or immobilized; as a result, unable to reach water bodies or soil to cause pollution (Lone et al., 2008a, Gupta et al., 2020).

Aquatic plant species are known to extract and accumulate several classes of heavy metals and other pollutants without causing excessive growth of their own and deteriorating other aquatic organisms, unlike algae. Utilizing these plants would be a feasible and economic approach for efficiently removing water- and soil-associated pollutants. The aquatic plant species *Phragmites australis* naturally present throughout the world in high water areas. *Phragmites australis* is also reported to tolerate different environmental stresses, including high heavy metal concentrations (Saleem et al., 2019). The *P. australis* is also used to treat industrial wastewater rich in high heavy metals content. Similarly, several other plant species accumulate several heavy metals and pollutants that can act as hyperaccumulators of heavy metals and other pollutants (Saleem et al., 2019). The plant species identified for soil remediation include either plants with high biomass such as willow or those that have low biomass but high metal accumulation ability, such as herbs, *Thlaspi* and *Arabidopsis* (Landberg and Greger, 1996).

13.10 Mechanism of phytoremediation

Phytoremediation involves the use of plants for the restoration of the polluted environment. Phytoremediation can be utilized based on the type of pollutants or strategies for minimizing heavy metal toxicity. For the site where the pollution has to be stabilized the techniques such as phytostabilized is preferred. In contrast, the sites that require a cleaning process, phytoextraction and phytovolatilization are used. In general, three major plant classes can be utilized for phytoremediation of heavy metals and other pollutants present in water: (1) use of natural hyperaccumulators plants, (2) plant species, which are fast-growing and produce high-biomass, (3) genetically modified plants with superior features.

13.10.1 Natural hyperaccumulator

Two major strategies have been proposed by Baker (1981) for defining the natural metal accumulation process in the plants when growing in metal-rich sites. The excluders include most of the plants that accumulate the metals in the roots and further restricting the transfer in other plant parts. In contrast to this, the second class, known as hyperaccumulators, accumulate a very high amount of metals in all plant parts without showing any metal toxicity symptoms (Baker, 1981; Rascio and Navari-Izzo, 2011; Van der Ent et al., 2013). Extensive research is currently being carried out to identify hyperaccumulator plants for different metals (Table 13.2). With the increase in plants' exploration, more species with hyperaccumulation capacity are coming to light. Currently, more than 450 metal hyperaccumulators are being discovered where approximately 25% of species belong to Brassicaceae family (Rascio and Navari-Izzo, 2011, Krzciuk and Gałuszka, 2015; Bhargava et al., 2012). In addition to this, several other families such as *Fabaceae*, *Asteraceae*, *Euphorbiaceae* are known to act as hyperaccumulators (Reeves, 2006, Rascio and Navari-Izzo, 2011). Genus *Alyssum* is reported to contain 48 members of 170 species analyzed. The member of Brassicaceae family *Brassica juncea* is reported to accumulate different heavy metals such as Pb Ni, Cd, Cu, Cr, and Zn when analyzed in a hydroponic medium (Dushenkov et al., 1995). Similarly, the *Eichhornia crassipes,* also known as water hyacinth, accumulate Se, Cr, Cd, Cu in roots exclusively. The *Sesbania drummondii*, a perennial shrub, also accumulate Pb efficiently while maintaining producing sufficient biomass (Sahi et al., 2002).

13.10.2 Fast-growing high-biomass producing nonaccumulators

Several fast-growing deep root plant species such as *Salix* and *Populus* spp. can be used for the phytoremediation/phytoextraction process (Unterbrunner et al., 2007). These species can produce large amounts of biomass in a shorter time. They also

TABLE 13.2 List of various hyperaccumulator plants used for phytoremediation of heavy metals.

S. no.	Metal	Plant species	Family	References
1	Zn, Cd	*Thlaspi caerulescens*	Brassicaceae	(Reeves and Brooks, 1983, Baker and Walker, 1990)
2	Cu	*Ipomea alpine*	Convolvulaceae	(Baker and Walker, 1990)
3	Ni	*Sebertia acuminata*	Sapotaceae	(Jaffre et al., 1976)
4	Co	*Haumaniastrum robertii*	Lamiaceae	(Brooks, 1977)
5	Se	*Astragalus racemosus*	Fabaceae	(Beath et al., 1937)
6	Zn, Cu, Pb, Mn, P	*Arabidopsis thaliana*	Brassicaceae	(Lasat, 2002)
7	Ni	*Thlaspi goesingens*	Brassicaceae	(Krämer et al., 2000)
8	Cd	*Brassica oleracea*	Brassicaceae	(Salt et al., 1995)
9	Zn, Cd	*Arabidopsis halleri*	Brassicaceae	(Reeves, 2000, Cosio et al., 2004)
10	Pb, Zn	*Sonchus asper*	Asteraceae	(Yanqun et al., 2005)
11	Zn, Cd	*Corydails pterygopetala*	Papaveraceae	(Yanqun et al., 2005)
12	Ni	*Alyssum bertolonii*	Brassicaceae	(Li et al., 2003, Chaney, 1999)

13	Se		*Astragalus bisulcatus*	Fabaceae	(Vallini et al., 2005)
14	Ni		*Stackhousia tryonii*	Celastraceae	(Bhatia et al., 2005)
15	Pb		*Hemidesmus indicus*	Apocynaceae	(Sekhar et al., 2005)
16	Cd		*Salsola Kali*	Chenopodiaceae	(de la Rosa et al., 2004)
17	Pb, Zn		*Sedum alfredii*	Crassulaceae	(Li et al., 2005)
18	As		*Pteris Vittata*	Pteridaceae	(Francesconi et al., 2002, Zhang et al., 2004)
19	Cd, Cr, Ni		*Helianthus anus*	Asteraceae	(Turgut et al., 2004)
20	Zinc		*Euphorbia macroclada*	Euphorbiaceae	(Nematian and Kazemeini, 2013)
21	Selenium		*Cardamine hupingshanesis*	Brassicaceae	(El Mehdawi et al., 2011)
22	Copper		*Eichhornia crassipes*	Pontederiaceae	(Mokhtar et al., 2011)
23	Manganese		*Phytolacca americana*	Phytolaccaceae	(Pollard et al., 2009)
24	Chromium		*Phragmites australis,*	Poaceae	(Calheiros et al., 2008)

have a high water uptake capacity, possess a deep root system; moreover, the aerial part is easier to harvest. Studies have discovered that these species preferentially accumulate Zn and Cd in the leaves rather in the other parts of the plants. The accumulated heavy metals can be extracted by harvesting total leaves. In addition to these, plant species with agricultural values and medicinal values such as cotton, peppermint, and lavender are also reported to accumulate significant levels of heavy metals in the harvestable region. This method has been used to produce aromatic cotton oil that is free from heavy metals contamination (Yankov et al., 2000, Vangronsveld et al., 2009).

13.10.3 Genetically modified plant with enhanced accumulation of pollutants/heavy metals

Both classes of plants, the natural hyperaccumulators and high biomass producing nonaccumulators, have several advantages but at the same time some disadvantages as well. The hyperaccumulators produce low biomass, while high biomass producing species have low phytoremediation efficiency. The genetic manipulation of these species for the limiting factors can produce hyperaccumulation of heavy metals and increased biomass production. Several biotechnological approaches are being utilized that can manipulate plants for efficient phytoremediation of heavy metals. The recent biotechnological/molecular biology, including omics studies such as genomics/transcriptomics/metabolomics and proteomics, has been extensively used for elucidating candidates in heavy metals transports tolerance. The strategies implemented involved manipulating metal transporter candidate genes and its uptake process or converting toxic metals into less toxic forms (Bhargava et al., 2012, Kotrba et al., 2009).

The accumulation of heavy metals generally leads to reactive oxygen species (ROS) production, which subsequently causes oxidative stress. Therefore, one of the crucial processes to enhance heavy metal accumulation in plants is to minimize ROS production. The most common strategy to increase heavy metal tolerance is to enhance antioxidant activity (Koźmińska et al., 2018). The oxidative stress can be managed by overexpressing genes involved in antioxidant systems such as superoxide dismutase, peroxidase, and catalase (Yan et al., 2020).

The next step for heavy metal accumulation enhancement is to engineer the plant's transport/uptake or sequestration mechanism. Several gene family members such as metal tolerance proteins, multidrug and toxic compound extrusion (MATE), and the heavy metal ATPase (HMA) family are well characterized to transport heavy metals in plants and thus can be overexpressed in the plant for enhanced metal accumulation (Das et al., 2016, Mani and Kumar, 2014).

It has been reported that genetic engineering leading to the synthesis of metal chelators in plants enhanced the metal assimilation ability of plants (Table 13.3) (Karenlampi et al., 2000; Pilon-Smits and Pilon, 2002; Clemens et al., 2002; Eapen and D'souza, 2005).

13.11 Types of phytoremediation

Various studies have divided the plant-based water and soil pollution management process into the following five groups.

1. *Phytoextraction*: In this process, the plants extract various pollutants, including heavy metals and other pollutants, from the contaminated soil and water bodies and accumulate them into their harvestable bodies (Suman et al., 2018). In this process, plants with very high metal accumulation capacity known as hyperaccumulators are preferably used as these plants naturally accumulate 100–1000 times metals/pollutants compared to nonaccumulators.
2. *Phytodegradation*: In this process, the plants and its associated microbiota degrade the organic pollutants (Newman and Reynolds, 2004).
3. *Rhizofiltration*: In this process plant root system is utilized to absorb, concentrate, and further precipitate the toxic metals from contaminated groundwater or other water bodies. The pollutants from the contaminated waters get accumulated into the plant biomass (Lone et al., 2008b).
4. *Phytostabilization*: in the process, the plants alternately reduce the pollution of water, either by reducing the movements of contaminated or immobilizing it (Bolan et al., 2011).
5. *Phytovolatilization*: The phytovolatilization involves converting the pollutants into a volatile form and releasing them into the environment (Limmer and Burken, 2016).

13.12 Conclusion

Water and soil pollution is a serious environmental issue in all developing countries. Currently, limited attention is being given to solve this problem. Therefore, a sustainable and cost-effective system is required to improve soil and freshwater resources, including groundwater. Plant-based purification processes such as phytoremediation could be a suitable alternative approach for soil and water remediation. This emerging technology can be applied to various organic and inorganic pollutants deposited in the soil, water, and air. The ability of the plant to assimilate pollutants varies among plant species, as plants utilize several mechanisms for ion uptake. In addition to improving the soil and water quality, the process also improves the ecosystem thereby protecting the associated microorganisms. Several plant species such as *P australis,* willow *and Arabidopsis spp.* have been proved to be promising in this field. Implantation of such plant species near contaminated areas and industrial areas can help remove heavy metals and other water pollutants and improve quality of soil and water resources. Problems related to central nervous system, skin, kidney, and liver poisoning are gradually on the rise, which are associated with long-term exposure to these toxins. Thus, phytoremediation is a cost-effective and an efficient method of treating heavy metal polluted soils.

TABLE 13.3 Enhanced uptake of heavy metals in transgenic plants.

Transgenic plant	Gene inserted and source	Heavy metal tolerance/accumulation	References
Brassica juncea	*gshl* (gamma–glutamylcysteine synthetase) from *Escherichia coli*	Cadmium	(Ow, 1996); (Fulekar et al., 2008)
Nicotiana tabacum	*CAX-2* from *Arabidopsis thaliana*	Cadmium, calcium, manganese	(Hirschi et al., 2000)
Arabidopsis thaliana	Glutathione-S-transferase gene (*gst)* from *N. tabacum*	Aluminum, copper, sodium	(Ezaki et al., 2000)
Nicotiana tabacum	*NtCBP4* from *N. Tabacum*	Nickel and lead	(Arazi et al., 1999)
Nicotiana tabacum and *Oryza sativa*	*Ferritin gene* from *Glycine max*	Iron	(Goto et al., 1998); (Goto et al.,1999)
A. thaliana and *N. tabacum*	*merA* from bacteria	Mercury	(Rugh et al., 2000); (Bizily et al., 2000); (Eapwn and D'souza, 2005)
Arabidopsis thaliana	ArsC (arsenate reductase) and € >flflgamma-ECS (glutamylcysteine synthetase)	Arsenic	(Dhankher et al., 2002); (Mello -Farias and Chaves, 2008)

References

Ainsworth, C.H., Paris, C.B., Perlin, N., Dornberger, L.N., Patterson III, W.F., Chancellor, E., Murawski, S., Hollander, D., Daly, K., Romero, I.C., 2018. Impacts of the deepwater horizon oil spill evaluated using an end-to-end ecosystem model. PloS One 13, e0190840.

Akpor, O.B., Ohiobor, G.O., Olaolu, D., 2014. Heavy metal pollutants in wastewater effluents: sources, effects and remediation. Adv. Biosci. Bioeng. 2, 37–43.

Alloway, B.J.P., 1996. Causes, effects & control. Soil Pollution and Land Contamination, 318–339.

Ashbolt, N.J., 2004. Microbial contamination of drinking water and disease outcomes in developing regions. Toxicology 198, 229–238.

Ashraf, M.A., Maah, M.J., Yusoff, I., 2014. Soil contamination, risk assessment and remediation. Environ. Risk Assess. Soil Contamin. 3–56.

Baker, A.J., Walker, PL., 1990. Ecophysiology of metal uptake by tolerant plants. Heavy metal tolerance in plants: evolutionary. Aspects 2, 155–165.

Baker, A.J., 1981. Accumulators and excluders-strategies in the response of plants to heavy metals. J. Plant Nutr. 3 (1-4), 643–654.

Barnes, D.K., Galgani, F., Thompson, R.C., Barlaz, M., 2009. Accumulation and fragmentation of plastic debris in global environments. Philos. Trans. R. Soc. B 364, 1985–1998.

Baró, F., Chaparro, L., Gómez-Baggethun, E., Langemeyer, J., Nowak, D.J., Terradas, J., 2014. Contribution of ecosystem services to air quality and climate change mitigation policies: the case of urban forests in Barcelona, Spain. Ambio. 43, 466–479.

Bashir, I., Lone, F., Bhat, R.A., Mir, S.A., Dar, Z.A., Dar, S.A., 2020. Concerns and threats of contamination on aquatic ecosystems. In: Hakeem1, K.R., Bhat, R.A., Qadri, H. (Eds.), Concerns and threats of contamination on aquatic ecosystems. Bioremediation and Biotechnology: Sustainable Approaches to Pollution Degradation, 1–26.

Beath, O, Eppson, H, Gilbert, C., 1937. Selenium distribution in and seasonal variation of type vegetation occurring on seleniferous soils. J. Am. Pharma. Assoc. 26 (5), 394–405.

Bhargava, A., Carmona, F.F., Bhargava, M., Srivastava, S., 2012. Approaches for enhanced phytoextraction of heavy metals. J. Environ. Manage. 105, 103–120.

Bhatia, NP, Walsh, KB, Baker, AJ., 2005. Detection and quantification of ligands involved in nickel detoxification in a herbaceous Ni hyperaccumulator *Stackhousia tryonii Bailey*. J. Exp. Bot. 56 (415), 1343–1349.

Bizily, S.P., Rugh, C.L., Meagher, R.B., 2000. Phytodetoxification of hazardous organomercurials by genetically engineered plants. Nat. Biotechnol. 18, 213–217.

Bolan, N.S., Park, J.H., Robinson, B., Naidu, R., Huh, K.Y., 2011. Phytostabilization: a green approach to contaminant containment. Advances in Agronomy. Elsevier, Amsterdam.

Boretti, A., Rosa, L., 2019. Reassessing the projections of the world water development report. NPJ Clean Water 2, 1–6.

Brooks, R., 1977. Copper and cobalt uptake by Haumaniastrum species. Plant Soil 48 (2), 541–544.

Burgess L.J.E., 2013. The Effects of Organic Pollutants in Soil on Human Health. EGU2013–979.

Cabral, J.P., 2010. Water microbiology. Bacterial pathogens and water. Int. J. Environ. Res. Public Health 7 (10), 3657–3703.

Café-Filho, A.C., Lopes, C.A., Rossato, M, 2019. Management of plant disease epidemics with irrigation practices. Irrigation in Agroecosystems, 123–142.

Calder, I., Hofer, T., Vermont, S., Warren, P., 2008. Towards a new understanding of forests and water. Unasylva, 229, 3–10.

Calheiros, C.S., Rangel, A.O., Castro, P.M., 2008. The effects of tannery wastewater on the development of different plant species and chromium accumulation in *Phragmites australis*. Arch. Environ. Contamin. Toxicol. 55 (3), 404–414.

Carpenter, D.O.J.R.O.E.H., 2011. Health effects of persistent organic pollutants: the challenge for the Pacific Basin and for the world. Rev. Environ. Health 26, 61–69.

Carvalho, F.P., 2017. Mining industry and sustainable development: time for change. Food Energy Secur. 6, 61–77.

Casiday, R., Frey, R., 1998. Acid rain. Inorganic Reactions Experiment. Washington University http://www. chemistry. wustl. edu/~ edudev/LabTutorials/Water/FreshWater/acidrain.html.

Chaney, R., 1999. Improving metal hyperaccumulator wild plants to develop commercial phytoextraction systems. Approaches and progress. In: Terry, N., Bañuelos, G. (Eds.), Improving metal hyperaccumulator wild plants to develop commercial phytoextraction systems. Approaches and progress. Phytoremediation of Contaminated Soil and Water.

Chibuike, G.U., Obiora, S.C., 2014. Heavy metal polluted soils: effect on plants and bioremediation methods. Appl. Environ. Soil Sci. 2014.

Clemens, S., Palmgren, M.G., Kramer, U., 2002. A long way ahead: understanding and engineering plant metal accumulation. Trends Plant Sci. 7, 309–315.

Cosio, C, Martinoia, E, Keller, C., 2004. Hyperaccumulation of cadmium and zinc in *Thlaspi caerulescens* and *Arabidopsis halleri* at the leaf cellular level. Plant Physiol. 134 (2), 716–725.

Cutrim, F.M., Ramos, E.C., Abreu, M.C., Godinho, A.S., Maciel, A.P., Mendonça, C.J., Cavalcante, K.S., 2019. A study of chemical composition and enzymatic hydrolysis of solid organic waste from agrosilvopastoral systems. J. Braz. Chem. Soc. 30, 1955–1963.

Das, N., Bhattacharya, S., Maiti, M.K., 2016. Enhanced cadmium accumulation and tolerance in transgenic tobacco overexpressing rice metal tolerance protein gene OsMTP1 is promising for phytoremediation. Plant Physiol. Biochem. 105, 297–309.

de la Rosa, G, Peralta-Videa, JR, Montes, M, Parsons, JG, Cano-Aguilera, I, Gardea-Torresdey, JL., 2004. Cadmium uptake and translocation in tumbleweed (Salsola kali), a potential Cd-hyperaccumulator desert plant species: ICP/OES and XAS studies. Chemosphere 55 (9), 1159–1168.

Drake, P.L., Hazelwood, K.J., 2005. Exposure-related health effects of silver and silver compounds: a review. Ann. Occup. Hyg. 49, 575–585.

Dushenkov, V., Kumar, P.N., Motto, H., Raskin, I., 1995. Rhizofiltration: the use of plants to remove heavy metals from aqueous streams. Environ. Sci. Technol. 29, 1239–1245.

Eapen, S., D'souza, S.F, 2005. Prospects of genetic engineering of plants for phytoremediation of toxic metals. Biotechnol. Adv. 23 (2), 97–114.

Edokpayi, J.N., Rogawski, E.T., Kahler, D.M., Hill, C.L., Reynolds, C., Nyathi, E., Smith, J.A., Odiyo, J.O., Samie, A., Bessong, P., 2018. Challenges to sustainable safe drinking water: a case study of water quality and use across seasons in rural communities in Limpopo province. S. Afr. Water 10, 159.

El Mehdawi, AF, Quinn, CF, Pilon-Smits, EA., 2011. Selenium hyperaccumulators facilitate selenium-tolerant neighbors via phytoenrichment and reduced herbivory. Curr. Biol. 21 (17), 1440–1449.

Ellison, D., Morris, C.E., Locatelli, B., Sheil, D., Cohen, J., Murdiyarso, D., Gutierrez, V., Van Noordwijk, M., Creed, I.F., Pokorny, J., 2017. Trees, forests and water: cool insights for a hot world. Glob. Environ. Change 43, 51–61.

Fazal-Ur-Rehman, M., 2019. Polluted water borne diseases: symptoms, causes, treatment and prevention. J. Med. Chem. Sci. 2, 85–91.

Ferronato, N., Torretta, V., 2019. Waste mismanagement in developing countries: a review of global issues. Int. J. Environ. Res. Public Health 16, 1060.

Figueras, M., Borrego, J.J., 2010. New perspectives in monitoring drinking water microbial quality. Int. J. Environ. Res. Public Health 7, 4179–4202.

Francesconi, K, Visoottiviseth, P, Sridokchan, W, Goessler, W., 2002. Arsenic species in an arsenic hyperaccumulating fern, *Pityrogramma calomelanos*: a potential phytoremediator of arsenic-contaminated soils. Sci. Total Environ. 284 (1–3), 27–35.

Gallo, F., Fossi, C., Weber, R., Santillo, D., Sousa, J., Ingram, I., Nadal, A., Romano, D., 2018. Marine litter plastics and microplastics and their toxic chemicals components: the need for urgent preventive measures. Environ. Sci. Eur. 30, 13.

Gavrilescu, M.J.E.E., Journal, M., 2006. Overview of In Situ Remediation Technologies for Sites and Groundwater. Environ. Eng. Manag. J. 5 (1).

Gerba, C.P., 2009. Environmentally transmitted pathogens. In: Maier, R.M., Pepper, I.L., Gerba, C.P. (Eds.), Environmentally transmitted pathogens. Environmental Microbiology.

Giri, A., Bharti, V.K., Kalia, S., Arora, A., Balaje, S., CHAURASIA, O., 2020. A review on water quality and dairy cattle health: a special emphasis on high-altitude region. Appl. Water Sci. 10, 1–16.

Goto, F., Yoshihara, T., Shigemoto, N., Toki, S., Takaiwa, F., 1999. Iron fortification of rice seed by the soybean ferritin gene. Nat. Biotechnol. 17, 282–286.

Grennfelt, P., Engleryd, A., Forsius, M., Hov, Ø., Rodhe, H., Cowling, E., 2020. Acid rain and air pollution: 50 years of progress in environmental science and policy. Ambio 49, 849–864.

Gupta, G., Khan, J., Singh, N.K., 2020. Phytoremediation of Metal-Contaminated Sites. Plant Ecophysiology and Adaptation under Climate Change: Mechanisms and Perspectives II. Springer, Cham.

Häni, H., 1990. The analysis of inorganic and organic pollutants in soil with special regard to their bioavailability. Int. J. Environ. Anal. Chem. 39, 197–208.

He, Z., Li, G., Chen, J., Huang, Y., An, T., Zhang, C., 2015. Pollution characteristics and health risk assessment of volatile organic compounds emitted from different plastic solid waste recycling workshops. Environ. Int. 77, 85–94.

Helmer, R., Hespanhol, I., 1997. Water Pollution Control: A Guide to the Use of Water Quality Management Principles. CRC Press, Boca Raton, FL.

Islam, S.M.F., Karim, Z, 2019. World's Demand for Food and Water: The Consequences of Climate ChangeDesalination-Challenges and Opportunities. IntechOpen, London, UK.

Jaffre, T, Brooks, R, Lee, J, Reeves, R., 1976. Sebertia acuminata: a nickel accumulating plant from New Caledonia. Science 193 (4253), 579–580.

Jaishankar, M., Tseten, T., Anbalagan, N., Mathew, B.B., Beeregowda, K.N., 2014. Toxicity, mechanism and health effects of some heavy metals. Interdiscip. Toxicol. 7, 60–72.

Jan, A.T., Azam, M., Siddiqui, K., Ali, A., Choi, I., Haq, Q.M., 2015. Heavy metals and human health: mechanistic insight into toxicity and counter defense system of antioxidants. Int. J. Mol. Sci. 16, 29592–29630.

Jéquier, E., Constant, F., 2010. Water as an essential nutrient: the physiological basis of hydration. Eur. J. Clin. Nutr. 64, 115–123.

Joutey, N.T., Bahafid, W., Sayel, H., El Ghachtouli, N., 2013. Biodegradation: involved microorganisms and genetically engineered microorganisms. In: Chamy, R., Rosenkranz, F. (Eds.), Biodegradation: involved microorganisms and genetically engineered microorganisms. Agricultural and Biological Sciences: Biodegradation-life of science, 289–320.

Kärenlampi, S., Schat, H., Vangronsveld, J., Verkleij, J.A.C., van der Lelie, D., Mergeay, M., Tervahauta, A.I., 2000. Genetic engineering in the improvement of plants for phytoremediation of metal polluted soils. Environ. Pollut. 107 (2), 225–231.

Kennish, M.J., 1994. Pollution in estuaries and coastal marine waters. J. Coast. Res., 27–49.

Kjellstrom, T., Lodh, M., Mcmichael, T., Ranmuthugala, G., Shrestha, R., Kingsland, S., 2006. Air and Water Pollution: Burden and Strategies for Control. Disease Control Priorities in Developing Countries, Second ed. The International Bank for Reconstruction and Development/ The World Bank.

Kotrba, P., Njamanova, J., Macek, T., RumiL, T., Mackova, M., 2009. Genetically modified plants in phytoremediation of heavy metal and metalloid soil and sediment pollution. Biotechnol. Adv. 27, 799–810.

Koźmińska, A., Wiszniewska, A., Hanus-Fajerska, E., Muszynska, E., 2018. Recent strategies of increasing metal tolerance and phytoremediation potential using genetic transformation of plants. Plant Biotechnol. Rep. 12, 1–14.

Krämer, U., Pickering, I.J., Prince, R.C., Raskin, I., Salt, D.E., 2000. Subcellular localization and speciation of nickel in hyperaccumulator and non-accumulator Thlaspispecies. Plant Physiol. 122 (4), 1343–1354.

Kravchenko, J., Darrah, T.H., Miller, R.K., Lyerly, H.K., Vengosh, A., 2014. A review of the health impacts of barium from natural and anthropogenic exposure. Environ. Geochem. Health 36, 797–814.

Krzciuk, K., Gałuszka, A., 2015. Prospecting for hyperaccumulators of trace elements: a review. Crit. Rev. Biotechnol. 35 (4), 522–532.

Landberg, T., Greger, M.J.A.G., 1996. Differences in uptake and tolerance to heavy metals in Salix from unpolluted and polluted areas. Appl. Geochem. 11, 175–180.

Lasat, M.M., 2002. Phytoextraction of toxic metals: a review of biological mechanisms. J. Environ. Qual. 31 (1), 109–120.

Li, T.-Q., Yang, X.-E., Jin, X.-F., He, Z.-L., Stoffella, P.-J., Hu, Q.-H., 2005. Root responses and metal accumulation in two contrasting ecotypes of *Sedum alfredii* Hance under lead and zinc toxic stress. J. Environ. Sci. Health 40 (5), 1081–1096.

Li, Y.-M., Chaney, R., Brewer, E., Roseberg, R., Angle, J.S., Baker, A., Reeves, R., Nelkin, J., 2003. Development of a technology for commercial phytoextraction of nickel: economic and technical considerations. Plant Soil 249 (1), 107–115.

Limmer, M., Burken, J., 2016. Phytovolatilization of organic contaminants. Environ. Sci. Technol. 50, 6632–6643.

Lone, M.I., He, Z.-L., Stoffella, P.J., Yang, X.-E., 2008. Phytoremediation of heavy metal polluted soils and water: progresses and perspectives. J. Zhej. Univ. Sci. B 9, 210–220.

Lushchak, V.I., Matviishyn, T.M., Husak, V.V., Storey, J.M., Storey, K.B., 2018. Pesticide toxicity: a mechanistic approach. EXCLI J. 17, 1101.

Mani, D., Kumar, C., 2014. Biotechnological advances in bioremediation of heavy metals contaminated ecosystems: an overview with special reference to phytoremediation. Int. J. Environ. Sci. Technol. 11 (3), 843–872.

Masindi, V., Muedi, K.L., 2018. Environmental contamination by heavy metals. Heavy metals 10, 115–132.

Mckay, C.P., 2014. Requirements and limits for life in the context of exoplanets. Proc. Natl. Acad. Sci. USA 111, 12628–12633.

Mello-Farias, P.C., Chaves, A.L.S., 2008. Biochemical and molecular aspects of toxic metals phytoremediation using transgenic plants. In: Tiznado-Hernandez, M.E., Troncoso-Rojas, R., Rivera-Dominguez, M.A. (Eds.), Transgenic approach in plant biochemistry and physiology. Research Signpost, Kerala, India, pp. 253–266.

Mishra, R.K., Mohammad, N., Roychoudhury, N., 2015. Soil pollution: Causes, Effects and Control 3, 20–30.

Mohapatra, D.P., Kirpalani, D.M., 2017. Process effluents and mine tailings: sources, effects and management and role of nanotechnology. Nanotechnol. Environ. Eng. 2, 1.

Mokhtar, H, Morad, N, Fizri, FFA., 2011. Phytoaccumulation of copper from aqueous solutions using *Eichhornia Crassipes* and *Centella Asiatica*. Int. J. Environ. Sci. Dev. 2 (3), 205.

Mudder, T.I., Botz, M.M., 2004. Cyanide and society: a critical review. Eur. J. Miner. Process. Environ. Protect. 4, 62–74.

Nematian, M.A., Kazemeini, F., 2013. Accumulation of Pb, Zn, Cu and Fe in plants and hyperaccumulator choice in Galali iron mine area. Iran. Int. J. Agric. crop Sci. 5 (4), 426.

Newman, L.A., Reynolds, C.M., 2004. Phytodegradation of organic compounds. Curr. Opin. Biotechnol. 15, 225–230.

NRCS, U., 2000. Heavy metal soil contamination. Soil Quality-Urban Technical Note (3).

Pandey, P.K., Kass, P.H., Soupir, M.L., Biswas, S., Singh, V.P., 2014. Contamination of water resources by pathogenic bacteria. Amb. Express 4, 51.

Park, J.-D., Zheng, W., 2012. Human exposure and health effects of inorganic and elemental mercury. J. Preven. Med. Public Health 45, 344.

Pérez-Lucas, G., Vela, N., El Aatik, A., Navarro, S, 2018. Environmental Risk of Groundwater Pollution by Pesticide Leaching Through the Soil Profile. Pesticides-Use and Misuse and their Impact in the Environment. IntechOpen, London, UK.

Peuke, A.D., Rennenberg, H., 2005. Phytoremediation: molecular biology, requirements for application, environmental protection, public attention and feasibility. EMBO Rep. 6, 497–501.

Pilon-Smiths, E., Pilon, M., 2002. Phytoremediation of metals using transgenic plants. Crit. Rev. Plant Sci. 21, 439–456.

Pimentel, D., Berger, B., Filiberto, D., Newton, M., Wolfe, B., Karabinakis, E., Clark, S., Poon, E., Abbett, E., Nandagopal, S., 2004. Water resources: agricultural and environmental issues. Bioscience, 54, 909–918.

Pisani, P., 2004. Soil pollution. Origin, monitoring & remediation. JSTOR.

Pollard, AJ, Stewart, HL, Roberson, CB., 2009. Manganese hyperaccumulation in *Phytolacca americana* L. from the Southeastern United States. Northeastern Nat. 16 (sp5), 155–162.

Prüss-Ustün, A., Bartram, J., Clasen, T., Colford, JR, J., M., Cumming, O., Curtis, V., Bonjour, S., Dangour, A.D., De France, J., Fewtrell, L, 2014. Burden of disease from inadequate water, sanitation and hygiene in low-and middle-income settings: a retrospective analysis of data from 145 countries. Trop. Med. Int. Health 19, 894–905.

Raghav, S., Painuli, R., Kumar, D., 2019. Threats to water: issues and challenges related to ground water and drinking water. In: Naushad, M. (Ed.). A New Generation Material Graphene: Applications in Water Technology, 1–19.

Raman, N., Narayanan, D.S., 2008. Impact of solid waste effect on ground water and soil quality nearer to Pallavaram solid waste landfill site in Chennai. Rasayan J. Chem. 1, 828–836.

Ramanjaneyulu, A.V., Giri, G., 2006. Phytoremediation-A review. Agric. Rev. 27, 216–222.

Rascio, N., Navari-Izzo, F., 2011. Heavy metal hyperaccumulating plants: how and why do they do it? And what makes them so interesting? Plant Sci. 180 (2), 169–181.

Reeves RBrooks, R., 1983. European species of Thlaspi L.(Cruciferae) as indicators of nickel and zinc. J. Geochem. Explor. 18 (3), 275–283.

Reeves, R.D., 2006. Hyperaccumulation of trace elements by plants. In Phytoremediation of metal-contaminated soils. Springer, Dordrecht, pp. 25–52.

Reeves, R.D., 2000. Metal-accumulating plants. In: Raskin, I., Ensley, B.D. (Eds.), Metal-accumulating plants. Phytoremediation of Toxic Metals: Using Plants to Clean up the Environment.

Riveros-Perez, E., Riveros, R., 2018. Water in the human body: an anesthesiologist's perspective on the connection between physicochemical properties of water and physiologic relevance. Ann. Med. Surg. 26, 1–8.

Saadoun, I.M., 2015. Impact of oil spills on marine life. Emerging pollutants in the environment-current and further implications, 75–104.

Sahi, S.V., Bryant, N.L., Sharma, N.C., Singh, S.R., 2002. Characterization of a lead hyperaccumulator shrub, Sesbania drummondii. Environ. Sci. Technol. 36, 4676–4680.

Saleem, H., Arslan, M., Rehman, K., Tahseen, R., Afzal, M., 2019. Phragmites australis—a helophytic grass—can establish successful partnership with phenol-degrading bacteria in a floating treatment wetland. Saudi J. Biol. Sci. 26, 1179–1186.

Salt, DE, Prince, RC, Pickering, IJ, Raskin, I., 1995. Mechanisms of cadmium mobility and accumulation in Indian mustard. Plant Physiol. 109 (4), 1427–1433.

Schwarzenbach, R.P., Egli, T., Hofstetter, T.B., Von Gunten, U., Wehrli, B., 2010. Global water pollution and human health. Annu. Rev. Environ. Resour. 35, 109–136.

Sekhar, KC, Kamala, C, Chary, N, Balaram, V, Garcia, G., 2005. Potential of *Hemidesmus indicus* for phytoextraction of lead from industrially contaminated soils. Chemosphere 58 (4), 507–514.

Simcock, A., Halpern, B., Kirubagaran, R., Hossein, M., Polette, M., Smith, E., Inniss, L., Simcock, A., 2016. Coastal, riverine and atmospheric inputs from land. Inniss L, Simcock A, coordinators. First Global Integrated Marine Assessment. United Nations, pp. 1–93.

Singh, N., Gupta, V.K., Kumar, A., Sharma, B., 2017. Synergistic effects of heavy metals and pesticides in living systems. Front. Chem. 5, 70.

Sivakumar, B., 2011. Water crisis: from conflict to cooperation—an overview. Hydrol. Sci. J. 56, 531–552.
Suman, J., Uhlik, O., Viktorova, J., Macek, T., 2018. Phytoextraction of heavy metals: a promising tool for clean-up of polluted environment? Front. Plant Sci. 9, 1476.
Tchounwou, P.B., Yedjou, C.G., Patlolla, A.K., Sutton, D.J., 2012. Heavy metal toxicity and the environment. Molecular, clinical and environmental toxicology, 133–164.
Thompson, R.C., Moore, C.J., Vom Saal, F.S., SWAN, S.H., 2009. Plastics, the environment and human health: current consensus and future trends. Philos. Trans. R. Soc. B 364, 2153–2166.
Tuomisto, H.L., Scheelbeek, P.F., Chalabi, Z., Green, R., Smith, R.D., Haines, A., Dangour, A.D., 2017. Effects of environmental change on agriculture, nutrition and health: a framework with a focus on fruits and vegetables. Wellcome Open Res. 2, 21.
Turgut, C., Pepe, M.K., Cutright, T.J., 2004. The effect of EDTA and citric acid on phytoremediation of Cd, Cr, and Ni from soil using *Helianthus annuus*. Environ. Pollut. 131 (1), 147–154.
Vallini, G, Di Gregorio, S, Lampis, S., 2005. Rhizosphere-induced selenium precipitation for possible applications in phytoremediation of Se polluted effluents. Z. Naturforsch. C 60 (3-4), 349–356.
Van der Ent, A., Baker, A.J., Reeves, R.D., Pollard, A.J., Schat, H., 2013. Hyperaccumulators of metal and metalloid trace elements: facts and fiction. Plant Soil 362 (1), 319–334.
Van Emmerik, T., Schwarz, A., 2020. Plastic debris in rivers. Wiley Interdiscip. Rev. Water 7, e1398.
Vangronsveld, J., Herzig, R., Weyens, N., Boulet, J., Adriaensen, K., Ruttens, A., Thewys, T., Vassilev, A., Meers, E., Nehnevajova, E., 2009. Phytoremediation of contaminated soils and groundwater: lessons from the field. Environ. Sci. Pollut. Res. 16, 765–794.
Visbeck, M., 2018. Ocean science research is key for a sustainable future. Nat. Commun. 9, 1–4.
Weigelhofer, G., Hein, T., Bondar-Kunze, E., 2018. Phosphorus and nitrogen dynamics in riverine systems: human impacts and management options. In: Schmutz, S., Sendzimir, J. (Eds.). Springer, Cham, pp. 187–202.
Wuana, R.A., Okieimen, F.E., 2011. Heavy metals in contaminated soils: a review of sources, chemistry, risks and best available strategies for remediation. ISRN Ecol. 2011 Article ID 402647.
Yan, A., Wang, Y., Tan, S.N., Yusof, M.L.M., Ghosh, S., Chen, Z., 2020. Phytoremediation: a promising approach for revegetation of heavy metal-polluted land. Front. Plant Sci. 11.
Yang, J., Mcbride, J., Zhou, J., Sun, Z., 2005. The urban forest in Beijing and its role in air pollution reduction. Urban Forest. Urban Green. 3, 65–78.
Yankov, B., Dellibaltova, V., Bojinov, M., 2000. Contents of Cu, Zn, Cd and Pb in the vegetative organs of cotton cultivars from industrially polluted region. Rasteniev dni Nauki 37, 525–531.
Yanqun, Z, Yuan, L, Jianjun, C, Haiyan, C, Li, Q, Schvartz, C., 2005. Hyperaccumulation of Pb, Zn and Cd in herbaceous grown on lead–zinc mining area in Yunnan, China. Environ. Int. 31 (5), 755–762.
Yaron, B., Calvet, R., Prost, R., Prost, R., 1996. Soil Pollution: Processes and Dynamics. Springer Science & Business Media, Berlin, Germany.
Zabbey, N., Olsson, G., 2017. Conflicts–oil exploration and water. Glob. Chall. 1, 1600015.
Zaghloul, A., Saber, M., El-Dewany, C., 2019. Chemical indicators for pollution detection in terrestrial and aquatic ecosystems. Bull. Natl. Res. Cent. 43, 156.
Zhang, W., Cai, Y., Downum, K.R., Ma, L.Q., 2004. Arsenic complexes in the arsenic hyperaccumulator *Pteris vittata* (Chinese brake fern). J. Chromatogr. A 1043 (2), 249–254.
Zheng, C., Zhao, L., Zhou, X., Fu, Z., Li, A., 2013. Treatment technologies for organic wastewater. Water Treat. 11, 250–286.

Chapter 14

Rhizoremediation of petroleum hydrocarbon–contaminated soils: A systematic review of mutualism between phytoremediation species and soil living microorganisms

Jane Alexander Ruley[a,b], Alice Amoding[a], John Baptist Tumuhairwe[a], Twaha Ateenyi Basamba[a]

[a]*Department of Agricultural Production, Makerere University, Kampala, Uganda*, [b]*Department of Agricultural Sciences, University of Juba, Juba, South-Sudan*

14.1 Introduction

In economies of both developed and developing countries, it is orthodox truth that extraction of crude oil is a potent factor for achievement of economic growth and development. Globally, oil is one of the most strategic resources that has enabled oil-extracting countries to realize substantial economic development (Maletic et al., 2013). That is why petroleum oil is has always attracted fierce competition between countries (Xu et al., 2018). The derived energy from crude oil is vital for both domestic and industrial use (Rossiana et al., 2020; Viesser et al., 2020; Yuniati, 2018). The energy is used for heating, transportation, and manufacturing (Adipah, 2019). Indirectly, the major contributions of petroleum oil are reflected in growth of gross domestic product (GDP) from export revenue (of crude oil, refined petroleum products and bi-products) and stimulation of industrial development (specifically, petrochemical industries such as paint and cosmetics, synthetic rubber among others) (Adipah, 2019). In crude oil–rich and crude oil–extracting countries, the direct and indirect contributions mentioned above have led to unprecedented economic growth and development (Koshlaf and Ball, 2017). Collectively, it is therefore eminent to infer that exploitation and utilization of petroleum oil has improved living standards (Ahmad et al., 2020). Cognizant of this, petroleum oil exploitation is a major and sought after planned

Phytoremediation. DOI: https://doi.org/10.1016/B978-0-323-89874-4.00008-X

activity in oil-rich countries leading to increased dependence (Koshlaf and Ball, 2017). In pursuit of the said benefits, a series of activities are carried out. These are concerned with exploration, extraction, processing, and transportation (Rossiana et al., 2020). One of the most undesirable outcomes of this chain of activities is environmental pollution (Koshlaf and Ball, 2017; Rossiana et al., 2020) and this has become a matter of great concern worldwide (Hussain et al., 2018).

Crude oil denotes a complex mixture of both hydrocarbon and nonhydrocarbon constituents (Brown et al., 2017). The hydrocarbon molecules exist as linear, branched, cyclic, and aromatic components; while nonhydrocarbon fractions, alternatively termed resins (carbazoles, thiophenes, and oxygenated hydrocarbons) and asphaltenes, contain nitrogen, sulfur, and oxygen (Brown et al., 2017). In this review, hydrocarbon contamination is used to represent pollution by total petroleum hydrocarbons (TPHs). According to Adipah (2019), TPH is composed of crude oil, gasoline, creosote, and diesel and other petroleum distillates. TPH enters soil through one or a combination of the following channels; leakage of underground storage tanks, oil spills caused by blow out accidents, installation of petrochemical refining equipment, transportation, illicit storage, and as dumped and disposed waste from various industrial processes (Aisien et al., 2015; Resosudarmo et al., 2012; Hussain et al., 2018; Rossiana et al., 2020). Studies (such as Correa-Garcia et al., 2018; Hussain et al., 2018; Viesser et al., 2020; Yu et al., 2020) indicate that TPH exists in different forms. It may, for example, be saturated, aromatic, asphaltene, and colloid. The four forms manifest in two broad families, gasoline range organics (GRO) and diesel range organics (DRO). GRO is composed of small chain alkanes (C_6–C_{10}) such as BTEX, while DRO has long-chain compounds (C_{10}–C_{40}) as well as highly hydrophobic chemicals including PAH (Hussain et al., 2018). At processing sites, distillates take different forms. These include methane, ethane, diesel fuel (petro-diesel), fuel oils, gasoline (petrol), jet fuel, kerosene, and liquefied petroleum gas) (Hussain et al., 2018).

14.2 Impact of soil hydrocarbon contamination on ecosystem and human health

Hydrocarbon contamination greatly affects the quality of ecosystems and human health. Once they mix with soil, they are: (1) adsorbed on the surface of organic soil constituents, (2) get fused within the soil pores, (3) circulate and become mobile, or (4) form a continuous surface layer on top of soil (Ghodeja et al., 2017). Petroleum hydrocarbons pollute soil and aquatic ecosystem (Ghodeja et al., 2017). Altogether, the different forms of petrol hydrocarbons (PHCs) become toxic, mutagenic, or carcinogenic (Viesser et al., 2020) and have affected the quality of ecosystems and human life (Hussain et al., 2018; Malla et al., 2018; Viesser et al., 2020; Zamani et al., 2014; Zangh et al., 2019).

With regard to ecosystems, TPH accumulation affects the health of the inhabiting organisms negatively (Tang et al., 2011). Thus, their toxicity causes

TABLE 14.1 Selected PHC contamination associated heavy metals and their effects.

Heavy metal	Associated human health hazard
Antimony	Ingestion causes a rise in the blood cholesterol levels. Ingestion leads to drop in blood sugar leading type 1 diabetes
Arsenic	Affects the circulatory system. Causes severe skin damage and could induce skin cancer
Barium	Causes a sudden rise in the body blood pressure
Cadmium	Known for causing damage to the kidneys
Chromium	Causes severe skin allergies
Copper	Short-term exposure leads to gastrointestinal problems. Long-term exposure causes severe damage to liver and kidney
Lead	Causes mental development problems in infant and children. Leads to severe kidney problems and high blood pressure in adults.
Mercury (inorganic)	Causes damages to the kidney system
Selenium	Multiple health problems including loss of hair and finger nails, circulatory system failures, and numbness in fingers and toes.

a deleterious impact on the plants as well as affecting the natural state of the oil field and its environs (Yu et al., 2020). The toxicological effect arises from the carcinogenic and mutagenic compounds (Yu et al., 2020). TPH destroy soil structure, downgrade water quality, and scotch growing plants; while they also inhibit possibilities of seed germination (Hassan et al., 2019; Jain et al., 2011). Also, the components pollute drinking water (Yu et al., 2020) and equally cause a decline in the water, air, and soil quality and wastage of nonrenewable resources (Yu et al., 2020).

To humans, in areas of high concentration, the quality of life is threatened (Adipah, 2019). For example, when hydrocarbon contaminants accumulate in food such as vegetables, muscles, and fish, they carry deleterious effects to whoever consumes these foods (Adipah, 2019; Vieser et al., 2020). The toll of their toxicity is also manifested through skin diseases, impaired red blood cells, and damages to the nephron, which have become salient ill health characteristics of the population in many oil-producing and oil-exploiting countries (Adipah, 2019; Viesser et al., 2020). Further, it contains harmful substances such as PAHs, which can cause mutation and cancer (Ahmad et al., 2020). Equally, human life is endangered especially by the increased risks of man's exposure to hydrocarbon compounds through inhalation (Adipah, 2019). The effect is felt when the compounds enter soil directly or indirectly through surface runoff. For example, when toluene is breathed at higher concentrations such as beyond 100 parts per million (100 ppm) for days, the affected individuals or communities report frequent episodes of fatigues, pounding headache, recurrent nausea, eye irritation, and feeling sleepy all the time. For benzene, long periods of exposure, the net effect of inhalation causes grievous damage to the central nervous system and leukemia (Adipah, 2019; Koshlaf and Ball, 2017) (Table 14.1). In addition, in the environments where there are compounds rich in chlorinated solvents, liver diseases become common, kidney diseases multiply, and the entire central nervous system is depressed (Adipah, 2019; Koshlaf and Ball, 2017). If exposures are in extremes, death can occur (Adipah, 2019).

PHC soil contamination has become one of the overarching ecological problems in oil producing and oil-extracting countries worldwide (Hewelke et al., 2018; Zhang et al., 2019; Yu et al., 2020). Once the soil is polluted by hydrocarbon compounds, recovery may take several years (Adipah, 2019). Against this backdrop, this chapter presents an extensive review of rhizoremediation as antidote for TPH contamination but specifically dwells on the mutual contribution of phytoremediation plant species and soil microorganisms in the biodegradation of recalcitrant contaminants. Soil pollution is considered in this review as a prime issue, given the fact that soil serves as the major medium through which hydrocarbon contamination manifests in the ecosystem and humans. Petroleum contaminants in the soil impede the soil microbiological processes, alter the soil chemical composition, and degrade the soil physical properties (Hewelke et al., 2018). The peculiar characteristics of hydrocarbons leading to alteration of the physicochemical properties of soil for example include poor solubility of hydrocarbons in water but high solubility in soil, which causes the organic compounds to sorb to soil particles in the same way as soil organic matter (SOM) (Ghodeja et al., 2017). Consequently, soil aeration and water permeability are impeded especially when the compounds fill soil pores and cause accumulation of acidic metabolites (Yu et al., 2020). Given the high contents of heavy metals in petroleum hydrocarbon contaminants such as produced water (CORDAID, 2014), the soil becomes polluted with the same metals that affect plant growth negatively by making the soil toxic (Ghodeja et al., 2017). On another tone, PHC contaminants lead to a drop in quantity of organic matter content in the soil through increasing organic carbon content and consequently accumulation of humic acid in the soil. The acid alters physicochemical properties of soil as well as scotching biotic components of the entire ecosystem (Ghodeja et al., 2017). As seen earlier, soil contamination is a reason for concern nowadays (Alvim and Pontes, 2018) because it affects human health, vegetation growth, and biological environment (Ahmad et al., 2020). To blur out its ugly side and therefore its landslide challenges, worldwide, there is increasing attention on finding plausible solutions for enhancing soil protection, sustainable use, conservation and remediation of polluted and wasted soil (Rada et al., 2019; Xu et al., 2018). This growing concern for massive remediation of these soils has been catapulted by the sharp increase in human population with more additional mouths to feed, which presents a begging question of how to increase the available land yet it fixed in supply. Also, increased modernization of society is another critical factor calling for restoration of PHC-contaminated soils as agro-based industries with their composite advantages are cost effective mostly when homegrown agricultural products are available.

14.3 Phytoremediation of hydrocarbon-contaminated soils

Remediation refers to the efforts made to remedy soils contaminated with PHC contaminants. It is binary in nature and may be in situ or ex situ. In situ approaches are implemented in the natural environment, while ex situ approaches involve application of artificial treatments in another environment (Ahmad et al., 2020). Remediation thus entails application of treatments or methods that may either be in form of containment, separation, or destruction. These are adopted and implemented to restore contaminated or polluted soil (Chukwunonso et al., 2020). According to Vidonish et al. (2017), a cornucopia of restoration measures and approaches have been selected in different oil-extracting environments to reclaim the contaminated soils. These are either physical, chemical, thermal, electrical, or biological. However, as noted by Ruley et al. (2020), the physical and chemical remediation approaches have proved quite untenable in terms of cost and have secondary contamination effects on the soil once used such as the chemical methods that induce greenhouse effects deleterious to the environment (See Table 14.2). Therefore, quick, nature friendly and cost-effective methods are required to remove and minimize the dangerous effects of crude oil (Ahmad et al., 2020) making the application of the biological approaches as the most cost-effective and sustainable strategies. Most importantly, the biological approaches are safe, desirable, and tenable in financing (Hauptvogl et al., 2020; Kumar et al., 2020; Ruley et al., 2020).

Phytoremediation plant species biodegrade PHC contaminants (Correa-Garcia et al., 2018; Ruley et al., 2020). With this approach, the organic pollutants are metabolized leading to their complete removal, immobilization, or conversion into innocuous products (Viesser et al., 2020). However, many

TABLE 14.2 Cost comparison among the remediation methods for contaminated soil.

Method	Cost (USD/m^3)*	Cost (USD/ton)**
Traditional	100–1200	200–1500
Microremediation	60–200	50–150
Phytoremediation	17–50	10–30

*Schnoor, J. L. (2002). Phytoremediation of soil and groundwater. FGWARTAC Technology Report TE-02-01 (March 2002)

**Frick, C. M., Germida, J. J., & Farrell, R. E. (1999, December). Assessment of phytoremediation as an in-situ technique for cleaning oil-contaminated sites. In *technical seminar on chemical spills* (pp. 105a-124a). Environment Canada; 1998.

studies (Divya and Kumar, 2011; Ite and Ibok, 2019; Ruley et al., 2019) have shown that although phytoremediation induces the said metabolization, large-scale implementation is at times minimized by recalcitrant compounds especially in highly contaminated soil. Chiefly, high concentration causes delayed or failure of seed germination, which inhibits the biological processes of photosynthesis, respiration, and transpiration and consequently stressing conditions for plants leading to low shoot and root biomass (Divya and Kumar, 2011). The loopholes are circumvented through adoption of rhizoremediation that involves the combined action of phytoremediation species and plant growth–promoting rhizobacteria (PGPR) (Divya and Kumar, 2011) as well as a suite of fungi. The mutual action posts effective results compared to use of plants alone (Ite and Ibok, 2019).

14.4 Microorganisms and rhizoremediation of PHC-contaminated soils

In various environments, rhizoremediation has become the most preferred approach for reclamation of PHC-contaminated soils (Hussain et al., 2018). The approach has the potential to restore such soils (Guarino et al., 2019; Liu et al., 2019). The preference of the approach builds on the historical role played as a clean-up technology dating to twentieth century when Stormer (1908) and Sohngen (1913) observed and confirmed that bacterial isolates from PHC-contaminated soil possessed traits that empowered them to degrade PAHs. This provided a formidable ground for many studies to venture into degradation of PAHs (Kuiper et al., 2004) to establish the most amenable and effective ways of restoring PHC-contaminated soil (Malla et al., 2018).

Rhizoremediation involves concurrency of phytoremediation and biostimulation (Ite and Ibok, 2019). In this process, plant roots nurture inhabiting microbes and increase their capacity to scavenge on the PHC organic contaminants (Correa-Garcia et al., 2018; Viesser et al., 2020). The process of rhizoremediation, as the process suggests, occurs in the plant root zones, scientifically known as the rhizosphere (Nadeem et al., 2013). The rhizosphere is defined by Haldar and Sengupta (2015) as a chemically complex zone between the soil and roots of plants. On this reasoning, Bisht et al. (2010) define rhizoremediation as the degradation of PHC contaminants by the bacteria residing in the rhizosphere of plants. Compared to bulk soil, the rhizosphere is regarded as region of ideal survival and thriving of microorganisms. Their concentrations in this region make microbial activity more pronounced in the rhizosphere as compared to bulk soil. It is a unique niche for microbial colonization (Haldar and Sengupta, 2015) which have engineer the clean-up of the PHC-contaminated soil environments (Veerapagu et al., 2019). The rhizosphere supports various strains of microorganisms (Haldar and Sengupta, 2015), a reason it is regarded the most complex ecosystem on the earth: a hotspot fostering millions of microbial cells (Compant et al., 2019; Korenblum et al., 2020). In rhizosphere, there is horizontal gene transfer and abundance of plasmids that biodegrade the PHC contaminants (Correa-Garcia et al., 2018).

Rhizoremediation, a natural and sustainable method of reclaiming PHC-contaminated soil (Yu et al., 2020), involves the mutual action of both the growing plants and microorganisms to breakdown PHC organic compounds in the soil (Ite and Ibok, 2019). According to Correa-Garcia et al. (2018), Ite and Ibok (2019), and Viesser et al. (2020), under this approach, the root zones of the plants harbor, nourish, and catalyze the biodegradability of the resident microorganisms. As a result of reliance on the native biodiversity, studies (such as Guarino et al., 2019; Liu et al., 2019; Obi et al., 2016) have branded rhizoremediation, a nondestructive and environmentally or eco-friendly approach making it a desirable and sustainable remediation technique. For example, it does not generate toxic metabolites (Liu et al., 2019) but rather enhances the churning out of root exudates, siderophores, phytohormones, and phosphatases as products of the symbiosis between plant species and microorganisms (Shukla et al., 2013). This makes rhizoremediation an effective natural remedy for eco-restoration of polluted sites (Shukla et al., 2013) accounting for its popularity as a green technology (Brooijmans et al., 2009; Oyetibo et al., 2016; Prince, 2005). This also proves why it is widely accepted as a remediation strategy in long-term hydrocarbon-contaminated soils (Liu et al., 2019) and a suitable, efficient, and cost-effective method for bioremediation of PHC-contaminated soils.

14.5 Factors affecting the efficiency of soil rhizoremediation

Given the astronomical costs associated, the use of conventional remediation strategies such as chemical engineering (Ruley et al., 2019, 2020; Viesser et al., 2020), and the negative impact on the ecosystems (Ahmad et al., 2020; Ruley et al., 2020), rhizoremediation is a feasible approach. The practice does not harm the environment save for its pocket friendliness when compared to other remediation approaches in both absolute and relative terms. Deserving to mention however is the fact that like other clean-up technologies, rhizoremediation is not a cost-free technology. However, the cost incurred is less in both absolute and relative terms, in comparison with other approaches. For example, it requires site-based installation of vegetation. Moreover, designing the site and an assortment of secondary activities such as reporting, monitoring of the experiments, and expenditure on other operation and maintenance activities lend the approach to be a money-consuming venture though as earlier said, the amount is not as colossal as it is for other remediation approaches. According to Ghodeja et al. (2017), at an advanced level, during the initial phase, the expenditure involves installations for screenhouse experiments. Besides, there is a mandatory requirement of testing the physicochemical properties of the soil as a litmus to guarantee vigorous plant growth at the selected field site (Ghodeja et al., 2017).

The success is pegged to various factors enumerated and exposited as follows:

1. Phytoremediation plant species must be present as they have a key role to play with regard to attracting and supporting microbial abundance and diversity as well as their biodegradation potentials boosted by root exudates (Baetz and Martinoia, 2014). Thus, in any setting where rhizoremediation is carried out, the plants act as promoters for microbial degraders (Correa-Garcia et al., 2018).
2. Rhizoremediation is most active in the surface layers of the soil. Still, the soil layers should not be very compact to allow infiltration and should not be clayey (Correa-Garcia et al., 2018; Ghodeja et al., 2017).
3. Different plant species have different success rates in rhizoremediation. To achieve the desired results, efforts ought to be taken to ensure that plant species with high potential for biodegradation of the PHC contaminants are selected over others (Correa-Garcia et al., 2018).
4. Abundance of organic acids determines the success rate of rhizoremediation. Organic acids exuded from roots can alter the chemistry of the rhizosphere either indirectly through promotion of the growth of indigenous petroleum hydrocarbon–degrading microbes or directly by changing conditions of the soil (e.g., pH) (Ite and Ibok, 2019).
5. The ability of microbes to colonize plants roots is another important factor (Hussain et al., 2019) that can affect microbial drift in the rhizosphere (Baetz and Martinoia, 2014; Lefèvre and Bazylinski, 2013; Martin et al. 2014). The

BOX 14.1 Ideal characteristics required of a plant selected for phytoremediation

Several studies (e.g., Anderson et al.,1993; Salt et al.,1998; Pilon-Smits 2005; Memon and Schröder 2009; Ali et al., 2013; Cook and Hesterberg, 2013) have identified the following 11 characteristics for an effective phytoremediation plant species:

1. The species should have traits for fast growth (Rapid growth).
2. The species should have the ability to withstand toxicity of the hydrocarbons (Tolerance to toxic contaminants).
3. The phytoremediation species should exhibit potential of contaminant accumulation in their roots.
4. Great efficiency in translocating contaminants from the roots to the shoots.
5. Capability to produce large quantities of biomass.
6. Possession of a dense network of root system.
7. Ability to produce high amounts of different degrading enzymes.
8. Ability to withstand the prevailing natural environmental conditions.
9. Capability to resist both pathogens and pests.
10. Ability to repel both grazers and browsers (herbivores).
11. Low cost and low technology are required to cultivate and harvest them.

Numbers 1 to 7 are for handling the contaminant itself, whereas numbers 8 to 11 allure practical implementation of phytoremediation at a contaminated site.

TABLE 14.3 Plant species with proven potential to phytoremediate PHC-contaminated soil.

Common name	Latin name	References
Western wheat grass	*Agropyron smithii*	Greenberg et al.,2010; Zhang et al., 2019; Wang et al.,2019; Mishra and Pandey, 2019.
Side oats grama	*Bouteloua curtipendula*	Herlambang and Suryati, 2018; Nedunuri et al.,2009.
Buffalo grass	*Buchloe dactyloides*	Herlambang and Suryati, 2018; Hutchinson et al., 2003; Zhu et al., 2018.
Bell Rhodes grass	*Chloris gayana*	Keeling and Warren, 2005; Shahsavari et al., 2015; Frick et al.,1999; Farrell and Germida, 2016; Danh et al., 2009; Chibuike and Obiora, 2014.
Tall fescue	*Festuca arundinacea Schreb*	Kim and Cho, 2006; Chekol and Vough, 2001; Huang et al., 2019; Sun et al.,2011; Li et al., 2013; Liu et al.,2013; Edwin-Wosu and Albert, 2010.
Annual ryegrass	*Lolium multiflorum*	Kim and Cho, 2006; Muratova et al., 2008
Alfalfa	*Medicago sativa L.*	Kim and Cho, 2006; Mitton et al., 2016;Mutaova et al., 2008; Ruley et al.,2019; Sun et al., 2011; Li et al., 2013; Liu et al., 2013; Pokethitiyook, 2017; Chekol and Vough, 2001; Robson et al., 2001.
Switchgrass	*Panicum virgatum*	Murphy and Coats, 2011; Shrestha et al., 2019; Guo et al., 2019; Li et al., 2011; Phouthavong-Murphy et al., 2020; Albright III and Coats, 2014; Edwin-Wosu and Albert, 2010.
Poplar trees	*Populus deltoides x nigra*	Kim and Cho, 2006; Doty et al., 2017; Lopez-Echartea et al., 2020; Ancona et al., 2020.
Big bluestem	*Andropogon gerardi*	Thomas et al., 2017; Zhu et al., 2018; Mercier, 2016.
Sorghum	*Sorghum bicolor*	Revathi et al., 2011; Angelova et al., 2011; Al Chami, et al.,2015; Wang et al., 2017; Nawab et al., 2016; Chen et al., 2019; Omara et al., 2019.
Sudan grass	*Sorghum vulgare L.*	Babu et al.,2014; Liu et al., 2020; Angellova et al.,2011; Jia et al., 2014.
Blue grama	*Bouteloua gracilis*	Frick et al.,1999; Wang et al., 2019.

(continued)

TABLE 14.3 (Cont'd)

Common name	Latin name	References
Bermuda grass	*Cynodon dactylon L.*	Basumatary and Bordoloi, 2016; Razmjoo and Adavi, 2012; Huang et al., 2019; Saraeian et al., 2018.
Canada wild-rye	*Elymus canadensis*	Nedunuri et al., 2009; Germida et al., 2002; Sung et al., 2003; Shahsavari et al., 2015.
Arctared red fescue	*Festuca rubra var. Arctared*	Greenberg et al., 2010; Shahsavari et al., 2015; Edwin-Wosu and Albert, 2010; Schenker, 2014.
Duckweed	*Lemna gibba*	Landesman et al., 2010; Patel and Kanungo, 2010; Khataee et al., 2012; Appenroth et al., 2010; Ekperusi et al., 2019; Parra et al., 2012; Saha et al., 2015; Chaudhary and Sharma, 2014.
Perennial ryegrass	*Lolium perenne L.*	Schenker, 2014; Muratova et al., 2008; Shores et al., 2018.
Verde kleingrass	*Panicum coloratum var. Verde*	Dominguez-Rosado and Pichtel, 2004; Wang and Oyaizu, 2009; Ghany et al., 2015.
Winter rye	*Secale cereale L.*	Pokethitiyook, 2017.
Meyer zoysiagrass	*Zoysia japonica var. Meyer*	Gianfreda and Nannipieri, 2001; Hutchinson et al., 2003; Huang et al., 2019; Shahsavari et al., 2015.
Thatching grass	*Hyparrhenia rufa*	Ruley et al., 2019
False sunflower	*Tithonia Diversifolia*	Ruley et al., 2019
Wild rice	*Oryza Longistaminata*	Ruley et al., 2019
Kidney cotton	*Gossypium barbadense*	Ruley et al., 2019

colonization and transmission of endophytes within plants depends on many factors such as the allocation of plant resources and the endophyte ability to colonize plants (Compant et al., 2019).

6. The concentration levels of crude oil play a big role in the success of rhizoremediation in terms of influencing the strains of microorganisms capable of living in stressing conditions. Higher concentrations are recalcitrant especially in areas affected by high molecular weight (HMW) contaminants. HMW are recalcitrant, cause low bioavailability of PHC (Gamerdinger et al., 1997), and are found to cause increased adsorption to SOM (Reddy and Sethunathan, 1983). These defects put microbes at bay by inhibiting microbial access. Besides, in instances where there are high concentrations of hydrocarbon contaminants, root growth is inhibited (Peña-Castro et al., 2006; Ma et al., 2010) and yet root zones or rhizosphere is an enclave for microorganisms (Yu et al., 2020).
7. Where there is excess hydrocarbon contamination of the soil, microbial remediation is largely possible through biostimulation. Biostimulation with

TABLE 14.4 Proven PHC degrading Bacteria genera isolated from soil.

Genera	Type of hydrocarbon	References
Acidovorax	Phenanthrene, anthracene	Benedek et al., 2018; Singleton et al., 2018; Cui et al., 2017.
Alcaligenes	Phenanthrene, fluorine, fluoranthene	Kim and Cho, 2006; Singha et al., 2017; Mahjoubi et al., 2019.
Arthrobacter	(n-alkanes (C10 to C40), benzene, naphthalene, phenanthrene	Abdulrasheed et al., 2020a, Jiang et al., 2017; Abdulrasheed et al., 2020b.
Mycobacterium	2-methylnaphthalene, phenanthrene, pyrene, benzo[a]pyrene, carbazole	Ivanova and Borzenkov, 2018; Wu et al., 2006; Marinescu et al., 2017; Eskandari et al., 2017; Ite, and Ibok, 2019.
Pseudomonas	Phenanthrene, fluoranthene, fluorine, benzo[a]pyrene	Kim and Cho, 2006; Benedek et al., 2018; Marinescu et al., 2017; Eskandari et al., 2017; Ite, and Ibok, 2019.
Rhodococcus	Pyrene and benzo[a] pyrene	Pi et al., 2017; Pi et al., 2019; Cappelletti et al., 2019; Eskandari et al., 2017.
Sphingomonas	Phenanthrene, fluoranthene, anthracene	Zhou et al., 2016; Chaudhary and Kim, 2017; Liu et al., 2016.
Xanthomonas	Carbazole	Gutierrez, 2019; Abdel-Shafy and Mansour, 2018; Ite and Ibok, 2019;
Bacillus	Phenanthrene, xylene, naphthalene, styrene	Kim and Cho 2006; Ite and Ibok, 2019; Wang et al., 2019.

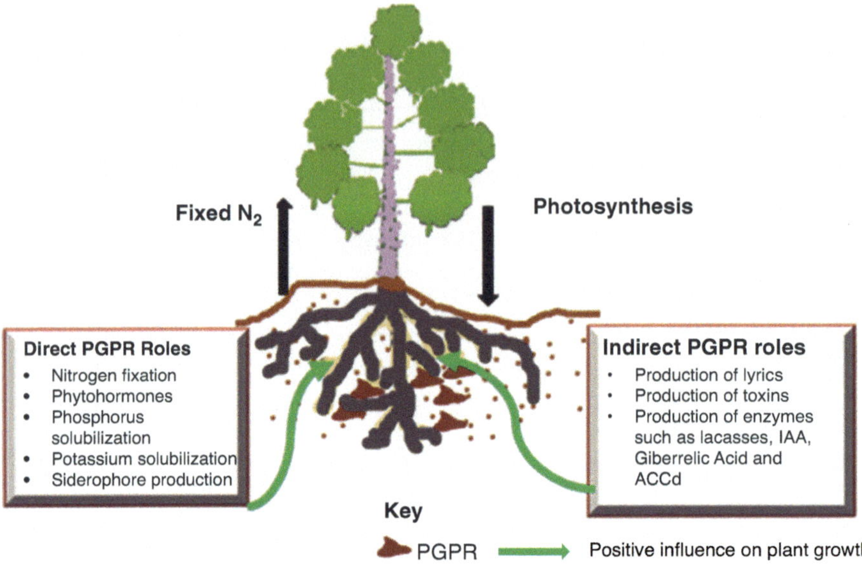

FIG. 14.1 Schematic overview of the roles played by PGPR in remediation of PHC-contaminated soil environments.

cow dung for example restores the lost soil nutrients such as nitrogen, phosphorus, and potassium that are key plant growth–enabling mineral nutrients (Ruley et al., 2020). On the same side of the coin, the amendments are capable of adding high-energy electron acceptors such as oxygen that are highly required by growing plants (Gkorezis et al., 2016). Therefore, it is eminent to infer that biostimulation trebles the rates of bioremediation notably by adding limiting mineral nutrients, improving habitat conditions for microorganisms, and improving soil porosity (Ghodeja et al., 2017; Gkorezis et al., 2016). For example, cow dung improves soil physiochemical characteristics, which enables speedy adaptation of microbes to contaminated soil (Agamuthu et al., 2013). Thus, the now nutrient-rich environment created by the fully established plants acts as a stimulant for microbial activities (Kirk et al., 2005; Lu et al., 2014).

8. The effectiveness of rhizoremediation is dependent on region-specific characteristics. According to Ghodeja et al. (2017), countries in tropics characterized by year-round growth of vegetation are most suited for rhizoremediation projects compared to the temperate lands where seasons such as winter cause excessive deleafing of plants as well as death of some plants.

In sum, with regard to plants, the successful rhizoremediation largely depends on: (1) nature of plant root systems, plants supported by fibrous root systems are known for harboring large numbers of microbial communities; (2) potentialities of the growing plant species to catabolize the PHC contaminants;

TABLE 14.5 Proven PHC-degrading fungi genera isolated from soil.

Genera	Type of hydrocarbon	References
Cunninghamella	Benzo[a]pyrene	Aydin et al., 2017
Penicillium	n-alkanes (C10–C40), benzene, naphthalene, phenanthrene	Aydin et al., 2017; Al-Hawash et al., 2018; Pointing, 2001
Fusarium	n-alkanes (C10–C40), benzene, naphthalene, phenanthrene	Ou-ya et al., 2018; Hesham et al., 2017; Romauld et al., 2019; Azin et al., 2018
Phanerochaete	Benzo[a]pyrene	Taha et al., 2018; Agrawal and Shahi, 2017; Torres-Farradá et al., 2019

and (3) creation of an ideal environment for enhanced mutualism between plants and microorganisms. The plant characteristics or factors for effective phytoremediation are summarized in Box 14.1 and plants proven as best phytoremediation species in Table 14.3. On the tone of microorganisms, rhizoremediation becomes effective when (1) the resident microbes are resilient to unconducive conditions and can ably survive in the rhizosphere, (2) the microorganisms can easily compete for the available nutrients in root exudates, (3) there is abundance of the microorganisms in the rhizosphere, and (4) the ease with which the resident microbes can inhabit and colonize the rhizosphere (Ite and Ibok, 2019). The above determinants work hand in hand to enhance the effective rhizoremediation of hydrocarbon-contaminated soils.

14.6 Characterization of rhizosphere microorganisms

The rhizosphere is a habitat for different categories of microorganisms. Correa-Garcia et al. (2018) found bacteria, fungi, and other minor organisms mainly earthworms and nematodes in rhizosphere. This chapter zeroes on two most commonly researched microorganisms: bacteria and fungi. Bacteria are commonly involved in rhizoremediation projects (Bell et al., 2013; El Amrani et al., 2015). The major reason for their dominance in rhizoremediation is connected to the linkage between bacterial strains and the prevalence of PHC-degradation genes mostly found in the rhizosphere (Gonzalez et al., 2018; Yergeau et al., 2018). The most common bacteria involved in rhizoremediation are called PGPR (Nadeem et al., 2013) (Table 14.4). The PGPR are free-living strains belonging to many bacteria genera such as *Klebsiella, Burkholderia, Gluconacetobacter Bacillus, Erwinia, Pseudomonas, Beijerinckia, Serratia, Enterobacter* and *Flavobacterium* (Nadeem et al., 2013) and are known for varied or diverse metabolic and functional characteristics (Nadeem et al., 2013).

PGPR act directly and indirectly. Directly, they enable solubilization of phosphates, and produce phytohormones as well as siderophores that are key plant nutrients (Divya and Kumar, 2011). Indirectly, PGPR inhibit growth of plant pathogens as well as enable plant to outgrow stressors (Rajkumar et al., 2012; de Zelicourt et al., 2013) (See Fig. 14.1). For example, PGPR produce exopolysaccharides that play a key role in the lowering ethylene production; ethylene triggers stress to growing plants. Further, the PGPR help to regulate the uptake of nutrients by plants and activation of antioxidant enzymes (Nadeem et al., 2013). High concentrations of ethylene cause adverse stress conditions in plants (Correa-Garcia et al., 2018). Deserving mention, siderophore-producing bacteria sequestrate iron in the rhizosphere, which creates unfavorable conditions for the survival of phytopathogens (Divya and Kumar, 2011).

Besides bacteria, mycorrhizal fungi are capable of degrading PHC contaminants in the soil, while they too possess the capability of creating an alluring condition that promotes growth of phytoremediation species or any other plants (Table 14.5). This makes this type of fungi candidates for selection and use in projects mooted to restore contaminated soil (Jambon et al., 2018). Like bacteria, the fungi reside in the rhizosphere (Correa-Garcia et al., 2018). Although they have not been given adequate scholarship, in this chapter, efforts are made to shed some light on arbuscular mycorrhizal fungi (AMF) given their multipronged roles that enhance rhizoremediation (Ahanger et al., 2014; Begum et al., 2019; Correa-Garcia et al., 2018; Zhu et al., 2010a). For example, they secrete enzymes such as

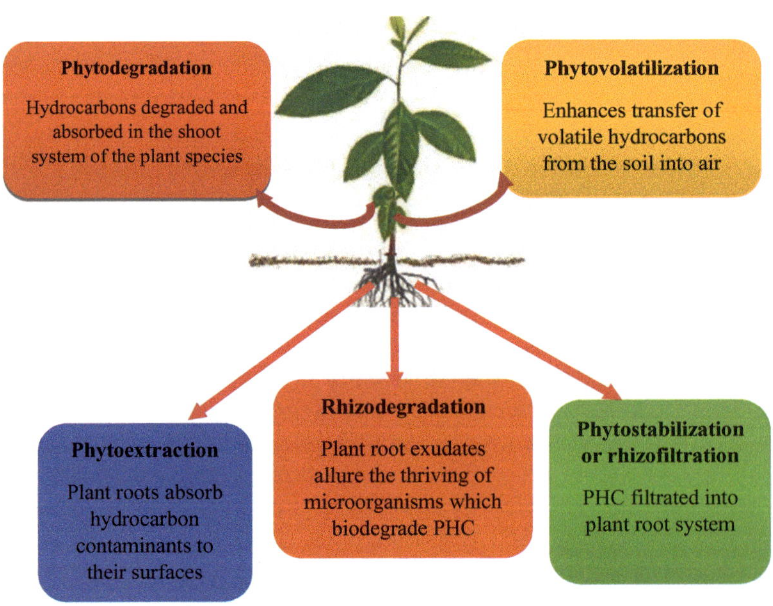

FIG. 14.2 Variants of phytoremediation of PHC-contaminated soil.

laccases (Harms et al., 2011) known for their potential to biodegrade PHC compounds such as lignin (Karigar and Rao, 2011).

14.7 The symbiosis between phytoremediation species and microbes in rhizosphere

The mutualism between plants and microbes breeds synergy where 1 + 1 = 3 and not 2! Suffice to say, the squared or combinatorial effect of both plants and colonizing microbes by far offers greater potential for ecorestoration of the PHC-contaminated soil compared to mere use of phytoremediation plant species alone (Truu et al., 2015). This implies that when phytoremediation species interact with either bacteria or fungi or both in a PHC-polluted soil environment, the potential of successful and efficient rhizoremediation is guaranteed (Jambon et al., 2018). The rhizosphere is the most plausible environment where the said interactions take place (Begum et al., 2019; Nadeem et al., 2013). Our point of argument in this review is rooted in the findings of Jambon et al. (2018), that on the planet earth, there exists about 20,000 plant species that can hardly grow and thrive in the absence of microorganisms, which puts the writing on the wall that the survival of most plant species is significantly dependent on plant–microbe symbiosis. To note also, plant species have forward linkages with microbial abundance given their indispensable role in serving as habitat for microbial communities (Schirawski and Perlin, 2018). For example, the growing plants serve as a primary source of carbon for soil microbes (Ishaq, 2017). In these multifarious interactions, sustainable ties increase on mutualism (Singh et al., 2017) and consequently boost degradation power.

Plants are holobiomes (Harman and Uphoff, 2019) and exist as a visible part of plant–microbe complexes (Harman and Uphoff, 2019). They interact with microbes majorly in three ways (Schirawski and Perlin, 2018). First, plants benefit from microbes especially when the microorganisms catalyze the cycling of major soil nutrients such as carbon, nitrogen, calcium, magnesium, and phosphorus. Second, the microorganisms encourage the fixation of environmental elements such as atmospheric nitrogen and beyond and lead to the decomposition of biological detritus needed for luxuriant plant growth (Ishaq, 2017). Finally, root exudates serve as a medium through which plant–microbial communication takes places and further enables the ecosystem to become responsive to environmental change (Williams and de Vries, 2020). Besides nutrition, Liu et al. (2019) established that by evolving intimate interactions with soil microbes enables plants to become resilient and resistant to many pathogens while also exhibiting tolerance to stress. Second, plants interaction with microbes sometimes results into suffering of the plants. For example, Ke and Wan (2020) acknowledge the harmfulness of the microbial communities with pathogens. In cases when plants become susceptible to pathogens and accumulate the pathogens in their growth organs from the surrounding soil, their growth performance is impeded and some catabolic and metabolic plant processes denatured. Berendsen et al. (2018) call this scenario, negative soil-feedback. These pathogens can be disastrous especially where the plants are growing under monoculture (Berendsen et al., 2018).

Ite and Ibok (2019) contend that phytoremediation plant species have the potential to biodegrade PHC organic compounds either by extracting them from the soil directly, biodegrading the compounds into other forms, volatilizing them, or even vaporization of those that are soluble. The plant species do so through water and mineral uptake, transportation partitioning, transpiration, assimilation, or translocation (Fig. 14.2). This benefits microbes especially those that form intimate associations with the plant host, collectively forming a metaorganism (Berg et al., 2014; Compant et al., 2019; Hassani et al., 2019; Sadhana, 2014) Plant roots are known for secreting chemicals that favor the formation of biofilms that among others increase on the resistance of plants to stresses and infections by pathogens (Compant et al., 2019). Where purposeful capabilities are established in this kind of symbiosis, enhanced plant holobiomes are formed (Harman and Uphoff, 2019).

The composite roles played by phytoremediation plant species in a PHC-contaminated soil environment include: (1) provision of enabling conditions for survival of PHC-degrading microbes leading to increased abundance, (2) squares the plant catabolic processes, (3) enhancement of the bioavailability of recalcitrant and hydrophobic hydrocarbons (Ite and Ibok, 2019). Besides, plants directly interact with the soil microbes as hosts (Korenblum et al., 2020). This facilitative role enhances the survival and multiplication of the soil microbes and in backward linkage, the interaction enables the plants to outgrow all the biotic and abiotic stresses (Korenblum et al., 2020). Deserving mention also, is the fact that roots exude a diverse repertoire of metabolites, including up to 10–50% of the plants' fixed carbon (Korenblum et al., 2020). Thus, the root growth dynamics and exudation processes change the chemical composition of the rhizosphere in space and time.

Growing root systems act as an injection system to spread the bacteria through soil (Harms and Wick, 2006). The roots act as a substitute for tilling the soil to incorporate additives (nutrients) and to improve aeration (Bisht et al., 2010; Shukla et al., 2013). Moreover, the mucigel secreted by root cells, lost root cap cells, the starvation of root cells, or the decay of complete roots provide favorable conditions that enhance the thriving of microbes in the rhizosphere (Bisht et al., 2010; Ite and Ibok, 2019). Some of the microbes stimulated by the root exudates are capable of degrading petroleum hydrocarbons (Correa-García et al., 2018).

The rhizosphere is an ideal environment for hydrocarbon degradation in the following ways; (1) The zone is laden with exudates that stimulate microbial growth and increase their prowess to biodegrade PHC contaminants (Yergeau et al., 2014). (2) Exudates provide a favorable environment for the multiplication and consequently abundance of microbial communities compared to those in bulk soil. (3) The rhizosphere has favorable conditions that enhance horizontal gene transfer among the microbial strains in addition to richness in plasmids that enable the microorganisms to absorb contaminant stresses, become resilient, and gain more potency for degradation of the organic compounds exerted in the soil environment by the PHC (Sentchilo et al., 2013). (4) Some root exudates steer the conversion of organic content in the soil into food for microbes (Gao et al., 2010). Thus, the combined use of native microbes microorganisms and plants enhance the biodegradation and removal of hydrocarbons contaminants from soil-polluted environments cost effectively with reportedly high success rates (Garrido-Sanz et al., 2019; Gkorezis et al., 2016), which makes plant species prominent promoters for microbial degraders of hydrocarbon-contaminated soils (Correa-García et al., 2018).

Phytoremediation species are enabled to grow in PHC-contaminated soils by the action of the resident microorganisms through lessening the toxicity of the soil, enrichment of the contaminated soil with metabolites that induce growth of the plants, fixation of nitrogen, and generally forming the basis for soil nutrient cycling (Gkorezis et al., 2016). Some PGPR strains are also known for secreting phytohormones such as auxins, IAA (indole-3-acaetic acid), cytokines (Patten and Glick, 1996; Spaepen and Vanderleyden, 2011; Raut et al., 2017) as well as gibberellins or gibberellic acid (GA) (Gutiérrez-Mañero et al., 2001) and these are known for promoting plant growth. Confirmation of this relationship is relayed in the work of Napitupulu et al. (2019) and Shahab et al. (2009) who observed that PGPR stimulate root development in plants. Other studies (e.g., Gupta and Chakrabarty, 2013; Napitupulu et al., 2019) have established that availability of gibberellins leads to improved germination of seeds, induced shoot growth, and other plant growth processes such as leaf growth, flowering or blossoming, grain development, and sexual expression or development. Also, according to Tanimoto (2005), compared to auxins, gibberellins promote the normal development of roots. As earlier noted, the roots constitute the rhizosphere, a hot spot for microbes including those that enzymatically attack PHC contaminants consequently converting them into innocuous compounds (Ruley et al., 2020).

Rodriguez and Fraga (1999) observed that certain bacterial strains such as those with capabilities of phosphate solubilization help in the mobilization and availing of nutrients to the growing plants. Also, according to Correa-Garcia et al. (2018), while some bacteria help in the modulation of plant stress genes, others protect plants from environmental stresses by secreting enzymes such as ACC deaminase (ACCd) which helps in biodegrading 1-aminocyclopropane-1-carboxylic acid (ACC). ACC is known for secreting stress ethylene hormone that exposes plants to high levels of stress (Hubbard et al., 2014; Ryu et al., 2004). Consequently, this improves plant growth, further suggesting an increase in capacity to biodegrade PHC contaminants from soil.

Another important aspect of bacteria-plant interaction is observed in root exudation. Root exudates are mineralized by bacteria as growth substrates. According to Ite and Ibok (2019), root exudates are highly rich in metabolites that play a fundamental role in the biodegradation of organic pollutants. Soil indigenous bacteria recycle plant nutrients, maintain soil structure, and play a great role in the conversion of xenobiotics. Generally, the bacteria form symbiotic associations with plants (Ite and Ibok, 2019) whereby the plants provide an alternative carbon source that serves to boost the survival of bacteria under toxic conditions induced by PHC contaminants (Lu et al., 2014). In relation to this, the organic compounds secreted by plant roots increase the solubility and availability of metal nutrient ions (Ite and Ibok, 2019), facilitate plant growth, and eventually enhance biological remediation of organic contaminants (Ite and Ibok, 2019).

AMF-plant interactions, like is the case for bacteria, greatly affect the rate of rhizoremediation. In the root systems of plants, the AMF enhance the formation of vesicles, arbuscules, and hyphae in roots consequently leading to the creation of a hyphal network (Bowles et al., 2016; Correa-Garcia et al., 2018). The hyphal network guarantees plants access to large soil surface areas consequently enhancing plant growth. In a related development, Begum et al. (2019) and Rouphael et al. (2015) consider AMF as biofertilizers given their contribution of inorganic nutrients to the soil, which consequently improves plant growth. Relatedly, the rhizoremediation effect of the AMF presence in the PHC-contaminated soil Besides, it improves the soil quality and mostly its structure and texture (Zou et al., 2016; Thirkell et al., 2017) and also catalyzes the decomposition of SOM (Paterson et al. 2016) leading to improvement of soil fertility and consequent plant growth. AMF also enable plants to fix atmospheric CO_2 (Begum et al., 2019) further enhancing long-term soil fertility and health (Giovanni et al., 2020). In relative and absolute terms therefore, AMF-plant interactions hasten the process of rhizoremediation.

AMF also safeguards plants from fungal pathogens (Jung et al., 2012;Smith and Read,2008 2010). According to Dehgahi et al. (2015), plants affected by fungal pathogens end up wilting. Those that do not succumb to wilting are characteristically liable to leaf chlorosis, vein clearing, and, in some instances, many of such plants are dotted with water-soaked leaf lesions caused by mycelium and toxins secreted by the pathogens. In arable lands famous for crop production, the presence of AMF provides an assurance of sustaining productivity (Gianinazzi et al., 2010). Though we agree with Chen et al. (2018) argument plant-AMF interaction and mostly cereals, vegetables, and fruit trees translate to sustaining of agriculture productivity, this review paper sets the ball rolling by merely providing insights to scholars about carrying out studies to determine the best crop phytoremediation species that are sustained by AMF in oil-contaminated sites. In reviewing the mycorrhizal benefits and their role in sustainable development, Agarwal et al. (2011) established that the association of plants with AMF enhances acquisition of key mineral nutrients such as phosphorus as well as other trace mineral nutrients from the soil. These are limiting nutrients in PHC-contaminated soils and hence provide antecedents for effective remediation of the soil by plant species as plant growth is assured.

Fungi remediate heavy metal–contaminated soils. Evidence is provided by Yang et al. (2015) from their study of two AMF strains; *Funneliformis mosseae* and *Rhizophagus intraradices*. Yang and colleagues established that interaction of the two strains and *Robinia pseudoacacia* L., a leguminous tree translated into a significant degradation of lead from the plant leaves, accumulation of chlorophyll, and increase in plant biomass. Also, in the same plant species, three enzymes were activated by the symbiosis; namely superoxide dismutase (SOD), ascorbate peroxidases (APX), and glutathione peroxidase (GPX). The availability of SOD in phytoremediation species shields them from oxidative stress; APX is involved in the plant physiological processes including protection from stressors and biosynthesis of lignin, a polymer that makes plants stronger with studded cell walls; and GPX is vital for redox balance in plants. Besides the three enzymes, hydrogen peroxide (H_2O_2) and malondialdehyde contents were reduced by the interaction of AMF and the leguminous tree that by implication infers that AMF rid PHC-contaminated soil of lead and henceforth, protecting the growing plants from cellular oxidative damage. This guarantees survival of phytoremediation species in PHC-contaminated soil, given the direct relationship between PHC infiltration into the soil and heavy metal concentrations. Although certain fungi potentially degrade PHC (Kadri et al., 2017), the most effective fungi for rhizoremediation continue to be unknown. In addition, the prominent environmental factors influencing their biodegradation efficiency are not known. Though studies (e.g., Harms et al., 2011) mention factors such as root exudate patterns, the degree of influence has not been reported in any studies. This gap provides a starting point for future microbiological and rhizoremediation studies to assess the potential of various strains of fungi in biodegradation of hydrocarbon contaminants in soil.

14.8 Microbial inoculation for improved rhizoremediation

Although rhizoremediation can succeed in crude oil–contaminated soils riding on the mutual interaction of naturally growing plants and resident microorganisms, the best results are realized through inoculation of the soil with microorganisms that have a greater capacity to biodegrade the PHC contaminants (Divya and Kumar, 2011; Ghodeja et al., 2017). Specifically, microorganism-inoculated plant species are shielded from harmful toxins exerted by PHC contaminants leading to improved growth and consequently rhizoremediation efficacy (Yousaf et al., 2010). The protected plants become resilient to heat, salinity, drought, heavy metals, and extremely high temperatures (Begum et al., 2019) leading to improved plant metabolism. In such physicochemical interactions, the propensity of degrading hydrocarbon contaminants is high. In this review, both the inoculation of bacteria and fungi are explored.

The introduction of bacteria inoculants in PHC-contaminated soil boosts degradation of PHC-contaminated soil (Divya and Kumar, 2011). This can for example, involve genetic engineering of endophytic and rhizospheric bacteria as reported by Liu et al. (2019), Xun et al. (2015), Reed and Glick (2005), Asghar et al. (2017), Gurska et al. (2009), and Khan et al. (2014). Unanimously, the listed studies recommended inoculation of PGPR and mostly the strains with high potential of biodegrading PHC so as to make rhizoremediation successful. Endeavors by Liu et al. (2019) provide confirmatory evidence. Liu and colleagues used an inoculum of *Klebsiella* sp. on Fescue, a flowering plant of the grass family growing in soil artificially contaminated with PHC and observed a high PHC removal rate. Using *Avena sativa* (common oat) and inoculating it with an Acinetobacter PGPR strain in oil-contaminated soil, Xun et al. (2015) established that shoot biomass and plant height significantly improved in comparison with the noninoculated treatments. A similar development was established with the potential of *A. sativa* decontaminating hydrocarbon-polluted soil at a higher rate. Reed and Glick (2005) inoculated *Brassica napus* (rapeseed or canola) with plasmid-transformed bacteria rich in ACC deaminase (*ACCd*) gene and also came up with empirical evidence that inoculation of PHC-degrading bacteria increases the affinity for biodegradation of the contaminants. As an amplification of the studies reported above, the review also appeals to latest studies such as Asghar et al. (2017) who used treatments of *Zea mays*, artificially PHC-contaminated soil with a concentration, 10 g/kg soil. Their study used light crude oil inoculated with a *Bacillus subtilis* strain that produces ACCd. Sixty days after planting, the study reported a proportionately greater increase in dry weight biomass, shoot, and root growths. Collectively, the said plant characteristics scavenged 43% of PHC compared to uninoculated treatments.

In Canada, Gurska et al. (2009) observed significant growth of shoots, roots, and increased plant biomass among native species in a field test in Ontario during a three-year study. Plant growth promoting endophytes were introduced in the screenhouse experiment that caused a significant change in the plant growth characteristics. The survival of plants and consequently their abundance caused a significant reduction in HMW and low molecular weight (LMW) PHC fractions. Besides the recalcitrant HMW, existing LMW fractions. The high concentration of 130 g/kg soil was decimated to 50 g/kg over the study period. This provides a further manifestation that introduction of strains of bacteria inoculants plays a significant role in the biodegradation of soils with high concentrations of PHC and further provides a basis for the improved growth of naturally living phytoremediation species hence leading to successful rhizoremediation.

Khan et al. (2014) introduced inoculated endophyte *Pseudomonas putida* PD1 in soil artificially contaminated with PHC concentration of 100 mg/kg. The introduction of two clones of willow, *Salix purpurea* 94006 and *Salix discolour* S-365 caused biodegradation of the contaminants by a margin of 25%–40% compared to the control. From this result, we therefore recommend inoculation of strains of bacteria and fungi for achievement of better rhizoremediation. This is a great need especially in oil-producing areas where the toll of hydrocarbon contaminants has rendered large chunks of hitherto arable and fertile land as wastelands. The need for introduction of microbial inoculants is further highlighted by Hassan et al. (2014) who observed that in soils with high concentration of PHC, strains of fungi such as AMF can hardly survive. Given this debilitating effect, we concur with Hassan and colleagues that introduction of AMF inoculants resilient to the scotching effects of PHC is critically needed in crude oil–extracting environments. In a related development, Asmelash et al. (2016) recommend use of native inoculum for better rhizoremediation. This is vital because the native inoculum easily adapts to the environment compared to introduction of foreign inoculum. The viability of introducing microbial inoculants in the soil is also mirrored in the works of Joner et al. (2001), who established that in soils contaminated with chrysene and dibenz(a,h)anthracene, treatments with inoculated AMF posted significantly high biodegradation of PHC by AMF and consequently greater reductions in their levels of concentration (chrysene and dibenz(a,h)anthracene) compared to the controls.

Relatedly, in a coal-mining zone, Bi et al. (2018) noted that AMF inoculation improved shoot and root growth of some phytoremediation species on the one hand and soil properties on the other. This provides a ray of hope that in most coal mining centers such as China, India, United States, Germany, Indonesia, and Poland (Dillinger, 2020), cloning of AMF potentially enhances the ecorestoration of the polluted soils. Further circumstantial evidence is provided by Ingraffa et al. (2019) whose study in a faba bean/wheat intercropping system using the ^{15}N stem labeling method established that a high abundance of AMF catalyzes plant uptake of the requisite macro- and micronutrient transfer. Ingraffa and colleagues further used the natural isotopic abundance method to gauge the potential of AMF in brokering biological nitrogen fixation and uptake of other highly needed minerals such as phosphorus, iron, and zinc. The study established a mutual interaction between AMF and the plants (faba bean and wheat) leading to proportionately high growth in root length and density of both faba beans and wheat although the uptake of minerals (phosphorus, iron and zinc) occurred in wheat and not faba beans. This provided confirmatory evidence that AMF in PHC-contaminated soil boosts growth of phytoremediation plant species and consequently, rhizoremediation.

14.9 Conclusion

Rhizoremediation approach cleans up hydrocarbon contaminants in soil and so holds great potential for the restoration of lost land now and in future. It is economical and environmentally sustainable. In this review, we discussed it as an in situ approach greatly relying on the synergy between naturally occurring plants and a host of microbes. In this approach, proper selection of a suitable plant-microbe-soil combination provides a sure way and guarantees that rhizoremediation can become an excellent clean up technology. In addition, biostimulation of the hydrocarbon-contaminated soil with suitable soil amendments is important. Case studies reviewed in this paper have highlighted the viability of the addition of soil amendments such as organic manure in squaring the vitality of both plants and microbes, which, consequently, increases the biodegradation potential manifold. Aside from biostimulation, it has also been established that inoculation guarantees desirable remediation results. Majorly, this review recommends the need for carrying out various bioengineering activities as a means of establishing the best modalities for effective rhizoremediation of PHC-contaminated soils. Besides, extensive research ought to be conducted to establish the conditions under which the most effective phytoremediation plant species can attract the most microorganisms with high PHC degradation potential conditions. Furthermore, given the fact that the symbiotic relationships between phytoremediation species occurs in the soil environment, studies aimed at enriching us with the knowledge about plant-microorganism-soil interaction mechanisms in crude oil–extracting zones is worth undertaking and last but not the least the need for a comparative study about the PHC degradation potential between indigenous microorganisms inhabiting PHC-contaminated soil and recombinant microorganisms. Efforts should further be made to establish the potentials of microbial inoculants in the degradation of PHC-contaminated soil and more so in situations of high toxicity.

Acknowledgments

The authors acknowledge support toward compilation of this Chapter by NORAD through the Sudd project (NORHED Project No. SSD-13/0021) implemented by University of Juba, Makerere University and the Norwegian University of Life Sciences, NMBU. Special gratitude to Dr. Bojoi Moses Tomor and Dr. Henry Busulwa, University of Juba and Makerere University Project Administrators respectively.

References

Abdel-Shafy, H.I., Mansour, M.S., 2018. Microbial degradation of hydrocarbons in the environment: an overview. In: Kumar, V., Kumar, M., Prasad, R. (Eds.), Microbial Action on Hydrocarbons. Springer, Singapore, pp. 353–386.

Abdulrasheed, M., Zakaria, N.N., Roslee, A.F.A., Shukor, M.Y., Zulkharnain, A., Napis, S., et al., 2020b. Biodegradation of diesel oil by cold-adapted bacterial strains of *Arthrobacter* spp. from Antarctica. Antarct. Sci., 1–13.

Abdulrasheed, M., Zulkharnain, A., Zakaria, N.N., Roslee, A.F.A., Khalil, Abdul, K., Napis, S., et al., 2020a. Response surface methodology optimization and kinetics of diesel degradation by a cold-adapted Antarctic bacterium, *Arthrobacter* sp. strain AQ5-05. Sustainability 12 (17), 6966.

Adipah, S., 2019. Introduction of petroleum hydrocarbons contaminants and its human effects. J. Environ. Sci. Public Health 3 (1), 1–9.

Agamuthu, P., Tan, Y.S., Fauziah, S.H., 2013. Bioremediation of hydrocarbon contaminated soil using selected organic wastes. Procedia Environ. Sci. 18, 694–702.

Agrawal, N., Shahi, S.K., 2017. Degradation of polycyclic aromatic hydrocarbon (pyrene) using novel fungal strain *Coriolopsis byrsina* strain APC5. Int. Biodeter. Biodegr. 122, 69–81.

Ahanger, M.A., Hashem, A., Abd-Allah, E.F., Ahmad, P., 2014. Arbuscular mycorrhiza in crop improvement under environmental stress. In: Ahmad, P, Rasool, S. (Eds.), Emerging Technologies and Management of Crop Stress Tolerance. Academic Press, pp. 69–95.

Ahmad, A.A., Muhammad, I., Shah, T., Kalwar, Q., Jianbo, Z., Zeyi, L., et al., 2020. Remediation methods of crude oil contaminated soil. World J. Agric. Soil Sci. 4 (3), 1–8.

Aisien, F.A., Aisien, E.T., Oboh, I.O., 2015. Phytoremediation of petroleum-polluted soils. In: Ansari, A.A., Gill, S.S., Gill, R., Lanza, G.R., Newman, L. (Eds.), Phytoremediation. Springer, Cham, pp. 243–252.

Al Chami, Z., Amer, N., Al Bitar, L., Cavoski, I., 2015. Potential use of *Sorghum bicolor* and *Carthamus tinctorius* in phytoremediation of nickel, lead and zinc. Int. J. Environ. Sci. Technol. 12 (12), 3957–3970.

Albright III, V.C., Coats, J.R, 2014. Disposition of atrazine metabolites following uptake and degradation of atrazine in switchgrass. Int. J. Phytoremediation 16 (1), 62–72.

Al-Hawash, A.B., Dragh, M.A., Li, S., Alhujaily, A., Abbood, H.A., Zhang, X., Ma, F., 2018. Principles of microbial degradation of petroleum hydrocarbons in the environment. Egypt. J. Aquat. Res. 44 (2), 71–76.

Ali, H., Khan, E., Sajad, M.A., 2013. Phytoremediation of heavy metals—concepts and applications. Chemosphere 91 (7), 869–881.

Alvim, G.M., Pontes, P.P., 2018. Aeration and sawdust application effects as structural material in the bioremediation of clayey acid soils contaminated with diesel oil. Int. Soil Water Conserv. Res. 6 (3), 253–260.

Ancona, V., Caracciolo, A.B., Campanale, C., Rascio, I., Grenni, P., Di Lenola, M., et al., 2020. Heavy metal phytoremediation of a poplar clone in a contaminated soil in southern Italy. J. Chem. Technol. Biotechnol. 95 (4), 940–949.

Anderson, T.A., Guthrie, E.A., Walton, B.T., 1993. Bioremediation in the rhizosphere. Environ. Sci. Technol. 27 (13), 2630–2636.

Angelova, V.R., Ivanova, R.V., Delibaltova, V.A., Ivanov, K.I., 2011. Use of sorghum crops for in situ phytoremediation of polluted soils. J. Agric. Sci. Technol. A 1 (5), 693–702.

Appenroth, K.J., Krech, K., Keresztes, A., Fischer, W., Koloczek, H., 2010. Effects of nickel on the chloroplasts of the duckweeds *Spirodela polyrhiza* and *Lemna minor* and their possible use in biomonitoring and phytoremediation. Chemosphere 78 (3), 216–223.

Asghar, H.N., Rafique, H.M., Khan, M.Y., Zahir, Z.A., 2017. Phytoremediation of light crude oil by maize (*Zea mays* L.) bio-augmented with plant growth promoting bacteria. Soil Sediment Contam. 26 (7-8), 749–763.

Asmelash, F., Bekele, T., Birhane, E., 2016. The potential role of arbuscular mycorrhizal fungi in the restoration of degraded lands. Front. Microbiol. 7, 1095. doi:10.3389/fmicb.2016.01095.

Azin, E., Moghimi, H., Heidarytabar, R., 2018. Petroleum degradation, biosurfactant and laccase production by *Fusarium neocosmosporiellum* RH-10: a microcosm study. Soil Sediment Contamin. 27 (4), 329–342.

Baetz, U., Martinoia, E., 2014. Root exudates: the hidden part of plant defense. Trends Plant Sci. 19 (2), 90–98.

Basumatary, B., Bordoloi, S., 2016. Phytoremediation of crude oil-contaminated soil using cynodon dactylon (L.) Pers. In: Ansari, A.A., Gill, S.S., Gill, R., Lanza, G.R., Newman, L. (Eds.), Phytoremediation. Springer, Cham, pp. 41–51.

Begum, N., Qin, C., Ahanger, M.A., Raza, S., Khan, M.I., Ahmed, N., et al., 2019. Role of arbuscular mycorrhizal fungi in plant growth regulation: implications in abiotic stress tolerance. Front. Plant Sci. 10, 1068.

Bell, T., Hassan, S., Lauron-Moreau, A., Al-Otaibi, F., Hijri, M., Yergeau, E., St-Arnaud, Marc., 2013. Linkage between bacterial and fungal rhizosphere communities in hydrocarbon-contaminated soils is related to plant phylogeny. ISME J. 8 (10), 149.

Benedek, T., Szentgyörgyi, F., Szabó, I., Kriszt, B., Révész, F., Radó, J., et al., 2018. Aerobic and oxygen-limited enrichment of BTEX-degrading biofilm bacteria: dominance of Malikia versus Acidovorax species. Environ. Sci. Pollut. Res. 25 (32), 32178–32195.

Berendsen, R.L., Vismans, G., Yu, K., Song, Y., de Jonge, R., Burgman, W.P., et al., 2018. Disease-induced assemblage of a plant-beneficial bacterial consortium. ISME J. 12 (6), 1496–1507.

Berg, G., Mahnert, A., Moissl-Eichinger, C, 2014. Beneficial effects of plant-associated microbes on indoor microbiomes and human health? Front. Microbiol. 5, 15.

Bi, Y., Zhang, Y., Zou, H., 2018. Plant growth and their root development after inoculation of arbuscular mycorrhizal fungi in coalmine subsided areas. Int. J. Coal Sci. Technol. 5 (1), 47–53.

Bisht, S., Pandey, P., Sood, A., Sharma, S., Bisht, N.S., 2010. Biodegradation of naphthalene and anthracene by chemo-tactically active rhizobacteria of Populus deltoides. Braz. J. Microbiol. 41 (4), 922–930.

Bowles, T.M., Barrios-Masias, F.H., Carlisle, E.A., Cavagnaro, T.R., Jackson, L.E., 2016. Effects of arbuscular mycorrhizae on tomato yield, nutrient uptake, water relations, and soil carbon dynamics under deficit irrigation in field conditions. Sci. Total Environ. 566, 1223–1234.

Brooijmans, R.J., Pastink, M.I., Siezen, R.J., 2009. Hydrocarbon-degrading bacteria: the oil-spill clean-up crew. Microb. Biotechnol. 2 (6), 587–594.

Brown, D.M., Bonte, M., Gill, R., Dawick, J., Boogaard, P.J., 2017. Heavy hydrocarbon fate and transport in the environment. Q. J. Eng. Geol. Hydrogeol. 50 (3), 333–346.

Cappelletti, M., Fedi, S., Zannoni, D., 2019. Degradation of alkanes in rhodococcus. In: Alvarez, H.M (Ed.), Biology of Rhodococcus. Springer, Cham, pp. 137–171.

Chaudhary, D.K., Kim, J., 2017. *Sphingomonas olei* sp. nov., with the ability to degrade aliphatic hydrocarbons, isolated from oil-contaminated soil. Int. J. Syst. Evol. Microbiol. 67 (8), 2731–2738.

Chaudhary, E., Sharma, P., 2014. Duckweed plant: a better future option for phytoremediation. Int. J. Emerg. Sci. Eng. 2 (7), 39–41.

Chekol, T., Vough, L.R., 2001. A study of the use of alfalfa (*Medicago sativa* L.) for the phytoremediation of organic contaminants in soil. Remediation 11 (4), 89–101.

Chen, C., Wang, X., Wang, J., 2019. Phytoremediation of cadmium-contaminated soil by *Sorghum bicolor* and the variation of microbial community. Chemosphere 235, 985–994.

Chen, M., Arato, M., Borghi, L., Nouri, E., Reinhardt, D., 2018. Beneficial services of arbuscular mycorrhizal fungi–from ecology to application. Front. Plant Sci. 9, 1270.

Chibuike, G.U., Obiora, S.C., 2014. Heavy metal polluted soils: effect on plants and bioremediation methods. Appl. Environ. Soil Sci. 2014 (1), 1–12.

Chukwunonso, O.I., Ahmed, A., Hassan, A., Hamid, F.S., 2020. Remediation of soil and water contaminated with petroleum hydrocarbon: a review. Environ. Technol. Innov. 17, 100526.

Compant, S., Samad, A., Faist, H., Sessitsch, A., 2019. A review on the plant microbiome: ecology, functions and emerging trends in microbial application. J. Adv. Res. 19, 29–37.

Cook, R.L., Hesterberg, D., 2013. Comparison of trees and grasses for rhizoremediation of petroleum hydrocarbons. Int. J. Phytoremediation 15 (9), 844–860.

Correa-García, S., Pande, P., Séguin, A., St-Arnaud, M., Yergeau, E., 2018. Rhizoremediation of petroleum hydrocarbons: a model system for plant microbiome manipulation. Microb. Biotechnol. 11 (5), 819–832.

Cui, G., Chien, M.F., Suto, K., Inoue, C., 2017. Analysis of stable 1, 2-dichlorobenzene-degrading enrichments and two newly isolated degrading strains, Acidovorax sp. sk40 and Ralstonia sp. sk41. Appl. Microbiol. Biotechnol. 101 (17), 6821–6828.

Danh, L.T., Truong, P., Mammucari, R., Tran, T., Foster, N., 2009. Vetiver grass, *Vetiveria zizanioides*: a choice plant for phytoremediation of heavy metals and organic wastes. Int. J. Phytoremediation 11 (8), 664–691.

de Zelicourt, A., Al-Yousif, M., Hirt, H., 2013. Rhizosphere microbes as essential partners for plant stress tolerance. Mol. Plant 6 (2), 242–245.

Dehgahi, R., Subramaniam, S., Zakaria, L., Joniyas, A., Firouzjahi, F.B., Haghnama, K., et al., 2015. Review of research on fungal pathogen attack and plant defense mechanism against pathogen. Int. J. Sci. Res. Agric. Sci. 2 (8), 197–208.

Dillinger, J., 2020. The Top 10 Coal Producers Worldwide in World Facts. https://www.worldatlas.com/articles/the-top-10-coal-producers-worldwide.html#:~:text=The%20Top%2020%20Coal%20Producers%20In%20The%20World,%20%20481.3%20%2016%20more%20rows%20.

Divya, B., Kumar, M.D., 2011. Plant-microbe interaction with enhanced bioremediation. Res. J. Biotechnol. 6 (1), 72–79.

Dominguez-Rosado, E., Pichtel, J., 2004. Phytoremediation of soil contaminated with used motor oil: II. Greenhouse studies. Environ. Eng. Sci. 21 (2), 169–180.

Doty, S.L., Freeman, J.L., Cohu, C.M., Burken, J.G., Firrincieli, A., Simon, A., 2017. Enhanced degradation of TCE on a superfund site using endophyte-assisted poplar tree phytoremediation. Environ. Sci. Technol. 51 (17), 10050–10058.

Edwin-Wosu, N., Albert, E., 2010. Total petroleum hydrocarbon content (TPH) as an index assessment of macrophytic remediation process of a crude oil contaminated soil environment. J. Appl. Sci. Environ. Manag. 14 (1), 39–42.

Ekperusi, A.O., Sikoki, F.D., Nwachukwu, E.O., 2019. Application of common duckweed (*Lemna minor*) in phytoremediation of chemicals in the environment: state and future perspective. Chemosphere 223, 285–309.

El Amrani, A., Dumas, A.S., Wick, L.Y., Yergeau, E., Berthomé, R., 2015. Omics" insights into PAH degradation toward improved green remediation biotechnologies. Environ. Sci. Technol. 49 (19), 11281–11291.

Farrell, R.E., Germida, J.J., 2016. Phytotechnologies: Plant-based Systems for Remediation of Oil Impacted Soils.

Frick, C.M., Germida, J.J., Farrell, R.E., 1999. Assessment of phytoremediation as an in-situ technique for cleaning oil-contaminated sitesIn Proc. Technical seminar on chemical spills1998. Environment Canada, pp. 105a–124a.

Gamerdinger, A.P., Achin, R.S., Traxler, R.W., 1997. Approximating the impact of sorption on biodegradation kinetics in soil-water systems. Soil Sci. Soc. Am. J. 61 (6), 1618–1626.

Gao, Y., Ren, L., Ling, W., Gong, S., Sun, B., Zhang, Y., 2010. Desorption of phenanthrene and pyrene in soils by root exudates. Bioresour. Technol. 101 (4), 1159–1165.

Garrido-Sanz, D., Redondo-Nieto, M., Guirado, M., Pindado Jiménez, O., Millán, R., Martin, M., et al., 2019. Metagenomic Insights Into The Bacterial Functions Of A Diesel-Degrading Consortium For The Rhizoremediation Of Diesel-Polluted Soil. Genes 10 (6), 456.

Germida, J.J., Frick, C.M., Farrell, R.E., 2002. Phytoremediation of oil-contaminated soilsDevelopments in soil science28. Elsevier, pp. 169–186.

Gianfreda, L., Nannipieri, P., 2001. Basic principles, agents and feasibility of bioremediation of soil polluted by organic compounds. Minerva Biotecnol. 13 (1), 5.

Gianinazzi, S., Gollotte, A., Binet, M.N., van Tuinen, D., Redecker, D., Wipf, D., 2010. Agroecology: the key role of arbuscular mycorrhizas in ecosystem services. Mycorrhiza 20 (8), 519–530.

Giovannini, L., Palla, M., Agnolucci, M., Avio, L., Sbrana, C., Turrini, A., et al., 2020. Arbuscular mycorrhizal fungi and associated microbiota as plant biostimulants: research strategies for the selection of the best performing inocula. Agronomy 10 (1), 106.

Gkorezis, P., Daghio, M., Franzetti, A., Van Hamme, J.D., Sillen, W., Vangronsveld, J., 2016. The interaction between plants and bacteria in the remediation of petroleum hydrocarbons: an environmental perspective. Front. Microbiol. 7, 1836.

Godheja, J., Shekhar, S.K., Modi, D.R., 2017. Bacterial rhizoremediation of petroleum hydrocarbons (PHC). In: Singh, D.P, Singh, H.B., Prabha, R. (Eds.), Plant-Microbe Interactions in Agro-Ecological Perspectives. Springer, Singapore, pp. 495–519.

Gonzalez, E., Pitre, F.E., Pagé, A.P., Marleau, J., Nissim, W.G., St-Arnaud, M., et al., 2018. Trees, fungi and bacteria: tripartite metatranscriptomics of a root microbiome responding to soil contamination. Microbiome 6 (1), 53.

Greenberg, B., Huang, X.D., Gerhardt, K., Yu, X.M., Liddycoat, S., Lu, X., et al., 2010. Successful full-scale deployments of advanced PGPR enhanced phytoremediation systems (PEPS) for decontamination of petroleum and salt impacted soils. https://tools.niehs.nih.gov/srp/1/2010_an_meet/pgpr_enhanced_phytoremediation_systems.pdf.

Guarino, C., Zuzolo, D., Marziano, M., Conte, B., Baiamonte, G., Morra, L., et al., 2019. Investigation and Assessment for an effective approach to the reclamation of polycyclic aromatic hydrocarbon (PAHs) contaminated site: SIN Bagnoli, Italy. Sci. Rep. 9 (1), 1–12.

Guo, Z., Gao, Y., Cao, X., Jiang, W., Liu, X., Liu, Q., et al., 2019. Phytoremediation of Cd and Pb interactive polluted soils by switchgrass (*Panicum virgatum* L.). Int. J. Phytoremediation 21 (14), 1486–1496.

Gupta, R., Chakrabarty, S.K., 2013. Gibberellic acid in plant: still a mystery unresolved. Plant Signal. Behav. 8 (9), e25504.

Gurska, J., Wang, W., Gerhardt, K.E., Khalid, A.M., Isherwood, D.M., Huang, X.D., et al., 2009. Three-year field test of a plant growth promoting rhizobacteria enhanced phytoremediation system at a land farm for treatment of hydrocarbon waste. Environ. Sci. Technol. 43 (12), 4472–4479.

Gutierrez, T., 2019. Aerobic hydrocarbon-degrading *Gammaproteobacteria*: Xanthomonadales. In: McGenity, T.J. (Ed.), Aerobic hydrocarbon-degrading *Gammaproteobacteria*: Xanthomonadales. Taxonomy, Genomics and Ecophysiology of Hydrocarbon-Degrading Microbes, 191–205.

Gutiérrez-Mañero, F.J., Ramos-Solano, B., Probanza, A.N., Mehouachi, J., R. Tadeo, F., Talon, M, 2001. The plant-growth-promoting rhizobacteria *Bacillus pumilus* and *Bacillus licheniformis* produce high amounts of physiologically active gibberellins. Physiol. Plant. 111 (2), 206–211.

Haldar, S., Sengupta, S., 2015. Plant-microbe cross-talk in the rhizosphere: insight and biotechnological potential. Open Microbiol. J. 9, 1.

Harman, G.E., Uphoff, N., 2019. Symbiotic root-endophytic soil microbes improve crop productivity and provide environmental benefits. Scientifica 2019.

Harms, H., Wick, L.Y., 2006. Dispersing pollutant-degrading bacteria in contaminated soil without touching it. Eng. Life Sci. 6 (3), 252–260.

Harms, H., Schlosser, D., Wick, L.Y., 2011. Untapped potential: exploiting fungi in bioremediation of hazardous chemicals. Nat. Rev. Microbiol. 9 (3), 177–192.

Hassan, S.E.D., Bell, T.H., Stefani, F.O., Denis, D., Hijri, M., St-Arnaud, M, 2014. Contrasting the community structure of arbuscular mycorrhizal fungi from hydrocarbon-contaminated and uncontaminated soils following willow (*Salix* spp. L.) planting. PLoS One 9 (7), e102838.

Hassani, M.A., Özkurt, E., Seybold, H., Dagan, T., Stukenbrock, E.H., 2019. Interactions and coadaptation in plant metaorganisms. Annu. Rev. Phytopathol. 57, 483–503.

Hauptvogl, M., Kotrla, M., Prčík, M., Pauková, Ž., Kováčik, M., Lošák, T., 2020. Phytoremediation potential of fast-growing energy plants: challenges and perspectives – a review. Pol. J. Environ. Stud. 29 (1), 505–516. doi:10.15244/pjoes/101621.

Herlambang, A., Suryati, T., 2018. Teknologi Fitoremediasi Untuk Pemulihan Lahan Tercemar Minyak. Prosiding Seminar Nasional dan Konsultasi Teknologi Lingkungan Jakarta.

Hesham, A.E.L., Mohamed, E.A., Mawad, A.M., Elfarash, A., El-Fattah, A., Bahaa, S., et al., 2017. Molecular characterization of *Fusarium solani* degrades a mixture of low and high molecular weight polycyclic aromatic hydrocarbons. Open Biotechnol. J. 11 (1), 27–35.

Hewelke, E., Szatyłowicz, J., Hewelke, P., Gnatowski, T., Aghalarov, R., 2018. The impact of diesel oil pollution on the hydrophobicity and CO_2 efflux of forest soils. Water Air Soil Pollut. 229 (2), 51.

Huang, S., Jiang, S., Liang, J., Chen, M., Shi, Y., 2019. Current knowledge of bermudagrass responses to abiotic stresses. Breeding Sci. 69 (2), 215–226.

Hubbard, M., Germida, J.J., Vujanovic, V., 2014. Fungal endophyte colonization coincides with altered DNA methylation in drought-stressed wheat seedlings. Can. J. Plant Sci. 94 (2), 223–234.

Hussain, I., Puschenreiter, M., Gerhard, S., Sani, S., Khan, W.U., Reichenauer, T.G., 2019. Differentiation between physical and chemical effects of oil presence in freshly spiked soil during rhizoremediation trial. Environ. Sci. Pollut. Res. Int. 26 (18), 18451–18464. doi:10.1007/s11356-019-04819-6.

Hussain, I., Puschenreiter, M., Gerhard, S., Schöftner, P., Yousaf, S., Wang, A., et al., 2018. Rhizoremediation of petroleum hydrocarbon-contaminated soils: improvement opportunities and field applications. Environ. Exp. Bot. 147, 202–219.

Hutchinson, S.L., Schwab, A.P., Banks, M.K., 2003. Biodegradation of petroleum hydrocarbons in the rhizosphere. In: McCutcheon, S.C., Schnoor, J.L (Eds.), Biodegradation of petroleum hydrocarbons in the rhizosphere. Phytoremediation: Transformation and Control of Contaminants, 355-386.

Ingraffia, R., Amato, G., Frenda, A.S., Giambalvo, D., 2019. Impacts of arbuscular mycorrhizal fungi on nutrient uptake, N_2 fixation, N transfer, and growth in a wheat/faba bean intercropping system. PloS One 14 (3), e0213672.

Ishaq, S.L., 2017. Plant-microbial interactions in agriculture and the use of farming systems to improve diversity and productivity. AIMS Microbiol. 3 (2), 335.

Ite, A.E., Ibok, U.J., 2019. Role of plants and microbes in bioremediation of petroleum hydrocarbons contaminated soils. Int J 7 (1), 1–19.

Ivanova, A.E., Borzenkov, I.A., 2018. Aerobic biodegradation of crude oil components by acidophilic mycobacteria. Microbiology 87 (4), 453–462.

Jain, P.K., Gupta, V.K., Gaur, R.K., Lowry, M., Jaroli, D.P., Chauhan, U.K., 2011. Bioremediation of petroleum oil contaminated soil and water. Res. J. Environ. Toxicol. 5 (1), 1–26.

Jambon, I., Thijs, S., Weyens, N., Vangronsveld, J., 2018. Harnessing plant-bacteria-fungi interactions to improve plant growth and degradation of organic pollutants. J. Plant Interact. 13 (1), 119–130.

Jia, C., Hu, S.B., Zhang, C.H., 2014. Phytoremediation of cadmium and pyrene co-polluted soil by Sudan grass (*Sorghum vulgare* L.). J. Agro-Environ. Sci. 33 (6), 1139–1145.

Jiang, Y., Wang, W., Xie, Q., Liu, N., Liu, L., Wang, D., et al., 2017. Plants transfer lipids to sustain colonization by mutualistic mycorrhizal and parasitic fungi. Science 356 (6343), 1172–1175.

Joner, E., Johansen, A., Loibner, A., Cruz, M., Szolar, O., Portal, J., Leyval, C., 2001. Rhizosphere effects on microbial community structure and dissipation and toxicity of polycyclic aromatic hydrocarbons (PAHs) in spiked soil. Environ. Sci. Technol. 35, 2773–2777. doi:10.1021/es000288s.f.

Jung, S.C., Martinez-Medina, A., Lopez-Raez, J.A., Pozo, M.J., 2012. Mycorrhiza-induced resistance and priming of plant defenses. J. Chem. Ecol. 38 (6), 651–664.

Karigar, C.S., Rao, S.S., 2011. Role of microbial enzymes in the bioremediation of pollutants: a review. Enzyme Res. 2011 (7), 805187.

Ke, P.J., Wan, J., 2020. Effects of soil microbes on plant competition: a perspective from modern coexistence theory. Ecol. Monogr. 90 (1), e01391.

Khan, Z., Roman, D., Kintz, T., delas Alas, M., Yap, R., Doty, S., 2014. Degradation, phytoprotection and phytoremediation of phenanthrene by endophyte *Pseudomonas putida*, PD1. Environ. Sci. Technol. 48 (20), 12221–12228.

Khataee, A.R., Movafeghi, A., Torbati, S., Lisar, S.S., Zarei, M., 2012. Phytoremediation potential of duckweed (*Lemna minor* L.) in degradation of CI Acid Blue 92: artificial neural network modeling. Ecotoxicol. Environ. Saf. 80, 291–298.

Kim, J.Y., Cho, K.S., 2006. Bioremediation of oil-contaminated soil using rhizobacteria and plants. Microbiol. Biotechnol. Lett. 34 (3), 185–195.

Kirk, J.L., Klironomos, J.N., Lee, H., Trevors, J.T., 2005. The effects of perennial ryegrass and alfalfa on microbial abundance and diversity in petroleum contaminated soil. Environ. Pollut. 133, 455–465.

Korenblum, E., Dong, Y., Szymanski, J., Panda, S., Jozwiak, A., Massalha, H., et al., 2020. Rhizosphere microbiome mediates systemic root metabolite exudation by root-to-root signalling. Proc. Natl Acad. Sci. USA 117 (7), 3874–3883.

Koshlaf, E., Ball, A.S., 2017. Soil bioremediation approaches for petroleum hydrocarbon polluted environments. AIMS Microbiol. 3 (1), 25.

Kuiper, I., Lagendijk, E.L., Bloemberg, G.V., Lugtenberg, B.J., 2004. Rhizoremediation: a beneficial plant-microbe interaction. Mol. Plant-Microbe Interac. 17 (1), 6–15.

Kumar, S., Kumari, N., Karmakar, S., Singh, R., Behera, M., Rani, A., et al., 2020. Advances in plant–microbe-based remediation approaches for environmental clean-up. In: Bharagava, R.N. (Ed.), Emerging Eco-friendly Green Technologies for Wastewater Treatment. Springer, Singapore, pp. 103–128.

Landesman, L., Fedler, C., Duan, R., 2010. Plant nutrient phytoremediation using duckweed. In: Ansari, A.A, Gill, S (Eds.), Eutrophication: Causes, consequences and Control,. Springer, Dordrecht, pp. 341–354.

Lefèvre, C.T., Bazylinski, D.A., 2013. Ecology, diversity, and evolution of magnetotactic bacteria. Microbiol. Mol. Biol. Rev. 77 (3), 497–526.

Li, C., Wang, Q.H., Xiao, B., Li, Y.F., 2011. Phytoremediation potential of switchgrass (*Panicum virgatum* L.) for Cr-polluted soil, In Proc. 2011 International Symposium on Water Resource and Environmental Protection, 3. pp. 1731–1734, IEEE.

Li, Y., Liang, F., Zhu, Y., Wang, F., 2013. Phytoremediation of a PCB-contaminated soil by alfalfa and tall fescue single and mixed plants cultivation. J. Soils Sediment. 13 (5), 925–931.

Liu, F., Hewezi, T., Lebeis, S.L., Pantalone, V., Grewal, P.S., Staton, M.E., 2019. Soil indigenous microbiome and plant genotypes cooperatively modify soybean rhizosphere microbiome assembly. BMC Microbiol. 19 (1), 1–19.

Liu, S., Guo, C., Liang, X., Wu, F., Dang, Z., 2016. Nonionic surfactants induced changes in cell characteristics and phenanthrene degradation ability of *Sphingomonas* sp. GY2B. Ecotoxicol. Environ. Saf. 129, 210–218.

Liu, W., Sun, J., Ding, L., Luo, Y., Chen, M., Tang, C., 2013. Rhizobacteria (*Pseudomonas* sp. SB) assist phytoremediation of oily-sludge-contaminated soil by tall fescue (*Testuca arundinacea* L.). Plant Soil 371 (1-2), 533–542.

Liu, X., Shen, S., Zhang, X., Chen, X., Jin, R., Li, X., 2020. Effect of enhancers on the phytoremediation of soils polluted by pyrene and Ni using Sudan grass (*Sorghum sudanense* (Piper) Stapf.). Environ. Sci. Pollut. Res., 1–8.

Lopez-Echartea, E., Strejcek, M., Mukherjee, S., Uhlik, O., Yrjälä, K., 2020. Bacterial succession in oil-contaminated soil under phytoremediation with poplars. Chemosphere 243, 125242.

Lu, L., Huggins, T., Jin, S., Zuo, Y., Ren, Z.J., 2014. Microbial metabolism and community structure in response to bio-electrochemically enhanced remediation of petroleum hydrocarbon-contaminated soil. Environ. Sci. Technol. 48 (7), 4021–4029.

Ma, J., Xiao, R., Li, J., Yu, J., Zhang, Y., Chen, L., 2010. Determination of 16 polycyclic aromatic hydrocarbons in environmental water samples by solid-phase extraction using multi-walled carbon nanotubes as adsorbent coupled with gas chromatography–mass spectrometry. J. Chromatogr. A 1217 (34), 5462–5469.

Mahjoubi, M., Aliyu, H., Cappello, S., Naifer, M., Souissi, Y., Cowan, D.A., et al., 2019. The genome of Alcaligenes aquatilis strain BU33N: insights into hydrocarbon degradation capacity. PloS One 14 (9), e0221574.

Maletić, S., Dalmacija, B., & Rončević, S. (2013). Petroleum hydrocarbon biodegradability in soil–Implications for bioremediation. Ed Vladimir Kutcherov, 43.

Malla, M.A., Dubey, A., Yadav, S., Kumar, A., Hashem, A., Abd_Allah, E.F., 2018. Understanding and designing the strategies for the microbe-mediated remediation of environmental contaminants using omics approaches. Front. Microbiol. 9, 1132.

Martin, B.C., George, S.J., Price, C.A., Ryan, M.H., Tibbett, M., 2014. The role of root exuded low molecular weight organic anions in facilitating petroleum hydrocarbon degradation: current knowledge and future directions. Sci. Total Environ. 472, 642–653.

Memon, A.R., Schröder, P., 2009. Implications of metal accumulation mechanisms to phytoremediation. Environ. Sci. Pollut. Res. 16 (2), 162–175.

Mercier, G., 2016. Using Phytotechnology to Remove Total Suspended Sediments and Polycyclic Aromatic Hydrocarbons in the Edwards Aquifer.

Mishra, T., Pandey, V.C., 2019. Phytoremediation of red mud deposits through natural succession. In: Pandey, V.C., Bauddh, K. (Eds.), Phytomanagement of Polluted Sites. Elsevier, pp. 409–424.

Mitton, F.M., Gonzalez, M., Monserrat, J.M., Miglioranza, K.S., 2016. Potential use of edible crops in the phytoremediation of endosulfan residues in soil. Chemosphere 148, 300–306.

Muratova, A.Y., Dmitrieva, T.V., Panchenko, L.V., Turkovskaya, O.V., 2008. Phytoremediation of oil-sludge–contaminated soil. Int. J. Phytoremediation 10 (6), 486–502.

Murphy, I.J., Coats, J.R., 2011. The capacity of switchgrass (*Panicum virgatum*) to degrade atrazine in a phytoremediation setting. Environ. Toxicol. Chem. 30 (3), 715–722.

Nadeem, S.M., Naveed, M., Zahir, Z.A., Asghar, H.N., 2013. Plant–microbe interactions for sustainable agriculture: fundamentals and recent advances. In: Arora, N.K. (Ed.), Plant microbe symbiosis: fundamentals and advances. Springer, New Delhi, pp. 51–103.

Napitupulu, T.P., Kanti, A., Sudiana, I.M., 2019. Evaluation of the environmental factors modulating Indole-3-acetic Acid (IAA) production by *Trichoderma harzianum* InaCC F88, In Proc. IOP Conference Series: Earth and Environmental Science, 308. IOP Publishing, 012060.

Nawab, J., Khan, S., Aamir, M., Shamshad, I., Qamar, Z., Din, I., et al., 2016. Organic amendments impact the availability of heavy metal (loid) s in mine-impacted soil and their phytoremediation by *Penisitum americanum* and *Sorghum bicolor*. Environ. Sci. Pollut. Res. 23 (3), 2381–2390.

Obi, L.U., Atagana, H.I., Adeleke, R.A., 2016. Isolation and characterisation of crude oil sludge degrading bacteria. SpringerPlus 5 (1), 1946.

Omara, T., Adupa, E., Laker, F., Kalukusu, R., Owori, T., 2019. Potential of *Sorghum bicolor* L. (Moench) and the effectiveness of some organic amendments in remediation of petroleum oil-vitiated soils of an automobile repair workshop in Urbanite Kampala. Asian J. Appl. Chem. Res 3 (1), 1–10.

Ou-ya, Z., Sheng-dong, F., Hai-bin, J., Xue-na, Z., Shi, W., Wang, W., et al., 2018. Biodegradation of high molecular weight polycyclic aromatic hydrocarbons mixture by a newly isolated

Fusarium sp. and co-metabolic degradation with starch. Polycycl. Aromat. Compd. 38 (1), 32–41.

Oyetibo, G.O., Miyauchi, K., Huang, Y., Chien, M.F., Ilori, M.O., Amund, O.O., et al., 2016. Biotechnological remedies for the estuarine environment polluted with heavy metals and persistent organic pollutants. Int. Biodeter. Biodegr. 199, 614–625.

Parra, L.M.M., Torres, G., Arenas, A.D., Sánchez, E., Rodríguez, K, 2012. Phytoremediation of low levels of heavy metals using duckweed (*Lemna minor*). In: Ahmad, P., Majeti, P. (Eds.), Abiotic Stress Responses in Plants: Metabolism, Productivity and Sustainability. Springer, New York, NY, pp. 451–463.

Patel, D.K., Kanungo, V.K., 2010. Phytoremediation potential of duckweed (*Lemna minor* L: a tiny aquatic plant) in the removal of pollutants from domestic wastewater with special reference to nutrients. Bioscan 5 (3), 355–358.

Paterson, E., Sim, A., Davidson, J., Daniell, T.J., 2016. Arbuscular mycorrhizal hyphae promote priming of native soil organic matter mineralisation. Plant Soil 408 (1-2), 243–254.

Patten, C.L., Glick, B.R., 1996. Bacterial biosynthesis of indole-3-acetic acid. Can. J. Microbiol. 42 (3), 207–220.

Peña-Castro, J.M., Barrera-Figueroa, B.E., Fernández-Linares, L., Ruiz-Medrano, R., Xoconostle-Cázares, B., 2006. Isolation and identification of up-regulated genes in Bermuda grass roots (*Cynodon dactylon* L.) grown under petroleum hydrocarbon stress. Plant Sci. 170 (4), 724–731.

Phouthavong-Murphy, J.C., Merrill, A.K., Zamule, S., Giacherio, D., Brown, B., Roote, C., et al., 2020. Phytoremediation potential of switchgrass (*Panicum virgatum*), two United States native varieties, to remove bisphenol-A (BPA) from aqueous media. Sci. Rep. 10 (1), 1–10.

Pi, Y., Chen, B., Bao, M., Zhang, B., 2019. Microbial degradation of four dispersed crude oils by *Rhodococcus* sp. evaluated using carbon stable isotope analysis. Journal of Chem. Technol. Biotechnol. 94 (6), 1800–1807.

Pi, Y., Chen, B., Bao, M., Fan, F., Cai, Q., Ze, L., Zhang, B., 2017. Microbial degradation of four crude oil by biosurfactant producing strain *Rhodococcus* sp. Bioresour. Technol. 232, 263–269.

Pilon-Smits, E., 2005. Phytoremediation. Annu. Rev. Plant Biol. 56, 15–39.

Pointing, S., 2001. Feasibility of bioremediation by white-rot fungi. Appl. Microbiol. Biotechnol. 57 (1-2), 20–33.

Pokethitiyook, P., 2017. Phytoremediation of petroleum-contaminated soil in association with soil bacteria. In: Ansari, A.A., Gill, S.S., Gill, R., Lanza, G.R., Newman, L. (Eds.), Phytoremediation. Springer, Cham, pp. 77–99.

Prince, R.C., 2005. The microbiology of marine oil spill bioremediation. In: Ollivier, B., Magot, M. (Eds.), Petroleum microbiology. American Society of Microbiology, pp. 317–335.

Rada, E.C., Andreottola, G., Istrate, I.A., Viotti, P., Conti, F., Magaril, E.R., 2019. Remediation of soil polluted by organic compounds through chemical oxidation and phytoremediation combined with DCT. Int. J. Environ. Res. Public Health 16 (17), 3179.

Rajkumar, M., Sandhya, S., Prasad, M.N.V., Freitas, H, 2012. Perspectives of plant-associated microbes in heavy metal phytoremediation. Biotechnol. Adv. 30 (6), 1562–1574.

Raut, V., Shaikh, I., Naphade, B., Prashar, K., Adhapure, N., 2017. Plant growth promotion using microbial IAA producers in conjunction with azolla: a novel approach. Chem. Biol. Technol. Agric. 4 (1), 1.

Razmjoo, K., Adavi, Z., 2012. Assessment of bermudagrass cultivars for phytoremediation of petroleum contaminated soils. Int. J. Phytoremediation 14 (1), 14–23.

Reddy, B.R., Sethunathan, N., 1983. Mineralization of parathion in the rice rhizosphere. Appl. Environ. Microbiol. 45 (3), 826–829.

Reed, M.L., Glick, B.R., 2005. Growth of canola (*Brassica napus)* in the presence of plant growth-promoting bacteria and either copper or polycyclic aromatic hydrocarbons. Can. J. Microbiol. 51 (12), 1061–1069.

Resosudarmo, B.P., Alisjahbana, A., Nurdianto, D.A., 2012. Energy security in Indonesia. In: Anceschi, L., Symons, J. (Eds.), Energy Security in the Era of Climate Change. Palgrave Macmillan, London, pp. 161–179.

Rodríguez, H., Fraga, R., 1999. Phosphate solubilizing bacteria and their role in plant growth promotion. Biotechnol. Adv. 17 (4-5), 319–339.

Romauld, S.I., Venkataraghavan, R., Yuvaraj, D., Devi, V.I., Hashika, S., 2019. Mycoremediation of hydrocarbon and its products using *Fusarium oxysporum*. Res. J. Pharm. Technol. 12 (9), 4216–4224.

Rossiana, N., Wulandari, A.P., Muthia, A., Indrawati, I., Rahayuningsih, S.R., Mayawati, B., et al., 2020. Growth profile of Penicillium chermesinum Biourge in function of total petroleum hydrocarbon and polycyclic aromatic hydrocarbons (PAH) compounds in oily sludge. World Sci. News 141, 103–114.

Rouphael, Y., Franken, P., Schneider, C., Schwarz, D., Giovannetti, M., Agnolucci, M., et al., 2015. Arbuscular mycorrhizal fungi act as biostimulants in horticultural crops. Sci. Hortic. 196, 91–108.

Ruley, J.A., Amoding, A., Tumuhairwe, J.B., Basamba, T.A., Opolot, E., Oryem-Origa, H., 2020. Enhancing the phytoremediation of hydrocarbon-contaminated soils in the Sudd Wetlands, South Sudan, using organic manure. Appl. Environ. Soil Sci. 2020 (3), 1–8.

Ruley, J.A., Tumuhairwe, J.B., Amoding, A., Opolot, E., Oryem-Origa, H., Basamba, T., 2019. Assessment of plants for phytoremediation of hydrocarbon-contaminated soils in the Sudd Wetland of South Sudan. Plant Soil Environ. 65 (9), 463–469.

Ryu, C.M., Farag, M.A., Hu, C.H., Reddy, M.S., Kloepper, J.W., Paré, P.W., 2004. Bacterial volatiles induce systemic resistance in Arabidopsis. Plant Physiol. 134 (3), 1017–1026.

Sadhana, B., 2014. Arbuscular mycorrhizal fungi (AMF) as a biofertilizer-a review. Int. J. Curr. Microbiol. Appl. Sci. 3 (4), 384–400.

Saha, P., Banerjee, A., Sarkar, S., 2015. Phytoremediation potential of duckweed (*Lemna minor* L.) on steel wastewater. Int. J. Phytoremediation 17 (6), 589–596.

Salt, D.E., Smith, R.D., Raskin, I., 1998. Phytoremediation. Annu. Rev. Plant Biol. 49 (1), 643–668.

Schenker, J.E., 2014. Ecological Remediation Using Bacterial, Fungal, and Plant Microcosms: An Effective Solution for Bunker C Crude Oil Contamination in Waterways. University of Vermont, VT.

Schirawski, J., Perlin, M.H., 2018. Plant–microbe interaction 2017—the good, the bad and the diverse. Int. J. Mol. Sci. 2018 (19), 1374.

Sentchilo, V., Mayer, A.P., Guy, L., Miyazaki, R., Tringe, S.G., Barry, K., 2013. Community-wide plasmid gene mobilization and selection. ISME J. 7 (6), 1173.

Shahab, S., Ahmed, N., Khan, N.S., 2009. Indole acetic acid production and enhanced plant growth promotion by indigenous PSBs. Afr. J. Agric. Res. 4 (11), 1312–1316.

Shahsavari, E., Adetutu, E.M., Ball, A.S., 2015. Phytoremediation and necrophytoremediation of petrogenic hydrocarbon-contaminated soils. In: Ansari, A.A., Gill, S.S., Gill, R., Lanza, G.R., Newman, L. (Eds.), Phytoremediation. Springer, Cham, pp. 321–334.

Shores, A.R., Hethcock, B., Laituri, M., 2018. Phytoremediation of BTEX and naphthalene from produced-water spill sites using Poaceae. Int. J. Phytoremediation 20 (8), 823–830.

Shrestha, P., Bellitürk, K., Görres, J.H., 2019. Phytoremediation of heavy metal-contaminated soil by switchgrass: a comparative study utilizing different composts and coir fiber on pollution remediation, plant productivity, and nutrient leaching. Int. J. Environ. Res. Public Health 16 (7), 1261.

Shukla, K.P., Sharma, S., Singh, N.K., Singh, V., Bisht, S., Kumar, V., 2013. Rhizoremediation: a promising rhizosphere technology. In: Patil, Y.B., Rao, P. (Eds.), Applied Bioremediation-Active and Passive Approaches. IntechOpen, London, UK.

Singha, L.P., Kotoky, R., Pandey, P., 2017. Draft genome sequence of *Alcaligenes faecalis* BDB4, a polyaromatic hydrocarbon-degrading bacterium isolated from crude oil-contaminated soil. Genome Announce. 5 (48), e01346 e01317.

Singleton, D.R., Lee, J., Dickey, A.N., Stroud, A., Scholl, E.H., Wright, F.A., 2018. Polyphasic characterization of four soil-derived phenanthrene-degrading Acidovorax strains and proposal of *Acidovorax carolinensis* sp. nov. Syst. Appl. Microbiol. 41 (5), 460–472.

Smith, S.E., Read, D.J., 2010. Mycorrhizal Symbiosis. Academic Press, MA.

Söhngen, N.L., 1913. Oxidation of petroleum, paraffin, paraffinoil and benzine by microbes., 15, 1145–1151.

Spaepen, S., Vanderleyden, J., 2011. Auxin and plant-microbe interactions. Cold Spring Harb. Perspect. Biol. 3 (4), a001438.

Störmer, K., 1908. Über die Wirkung des Schwefelkohlenstoffs und ähnlicher Stoffe auf den Boden. Zentralblatt fur Bakteriologie, Parasitenkunde und Infektionskrankheiten II 20, 282–286.

Sung, K., Munster, C.L., Rhykerd, R., Drew, M.C., Corapcioglu, M.Y., 2003. The use of vegetation to remediate soil freshly contaminated by recalcitrant contaminants. Water Res. 37 (10), 2408–2418.

Taha, M., Shahsavari, E., Aburto-Medina, A., Foda, M.F., Clarke, B., Roddick, F., 2018. Bioremediation of biosolids with Phanerochaete chrysosporium culture filtrates enhances the degradation of polycyclic aromatic hydrocarbons (PAHs). Appl. Soil Ecol. 124, 163–170.

Tang, J., Wang, M., Wang, F., Sun, Q., Zhou, Q., 2011. Eco-toxicity of petroleum hydrocarbon contaminated soil. J. Environ. Sci. 23 (5), 845–851.

Tanimoto, E., 2005. Regulation of root growth by plant hormones—roles for auxin and gibberellin. Crit. Rev. Plant Sci. 24 (4), 249–265.

Thirkell, T.J., Charters, M.D., Elliott, A.J., Sait, S.M., Field, K.J., 2017. Are mycorrhizal fungi our sustainable saviours? Considerations for achieving food security. J. Ecol. 105 (4), 921–929.

Torres-Farradá, G., Manzano-León, A.M., Rineau, F., Leal, M.R., Thijs, S., Jambon, I., Vangronsveld, J., 2019. Biodegradation of polycyclic aromatic hydrocarbons by native Ganoderma sp. strains: identification of metabolites and proposed degradation pathways. Appl. Microbiol. Biotechnol. 103 (17), 7203–7215.

Truu, J., Truu, M., Espenberg, M., Nõlvak, H., Juhanson, J., 2015. Phytoremediation and plant-assisted bioremediation in soil and treatment wetlands: a review. Open Biotechnol. J. 9 (1), 85–92.

Veerapagu, M., Jeya, K.R., Kalaivani, R., Jeyanthi, K.A., Geethanjali, S., 2019. Screening of hydrocarbon degrading bacteria isolated from oil contaminated soil. Pharma Innov. J. 8 (6), 69–72.

Viesser, J.A., Sugai-Guerios, M.H., Malucelli, L.C., Pincerati, M.R., Karp, S.G., Maranho, L.T., 2020. Petroleum-tolerant rhizospheric bacteria: isolation, characterization and bioremediation potential. Sci. Rep. 10 (1), 1–11.

Wang, D., Lin, J., Lin, J., Wang, W., Li, S., 2019. Biodegradation of petroleum hydrocarbons by *Bacillus subtilis* BL-27, a strain with weak hydrophobicity. Molecules 24 (17), 3021.

Wang, X., Chen, C., Wang, J., 2017. Phytoremediation of strontium contaminated soil by Sorghum bicolor (L.) Moench and soil microbial community-level physiological profiles (CLPPs). Environ. Sci. Pollut. Res. 24 (8), 7668–7678.

Wang, Y., Oyaizu, H., 2009. Evaluation of the phytoremediation potential of four plant species for dibenzofuran-contaminated soil. J. Hazard. Mater. 168 (2-3), 760–764.

Williams, A., de Vries, F.T., 2020. Plant root exudation under drought: implications for ecosystem functioning. New Phytol. 225 (5), 1899–1905.

Wu, C.H., Wood, T.K., Mulchandani, A., Chen, W., 2006. Engineering plant-microbe symbiosis for rhizoremediation of heavy metals. Appl. Environ. Microbiol. 72 (2), 1129–1134.

Xu, X., Liu, W., Tian, S., Wang, W., Qi, Q., Jiang, P., et al., 2018. Petroleum hydrocarbon-degrading bacteria for the remediation of oil pollution under aerobic conditions: a perspective analysis. Front. Microbiol. 9, 2885.

Xun, F., Xie, B., Liu, S., Guo, C., 2015. Effect of plant growth-promoting bacteria (PGPR) and arbuscular mycorrhizal fungi (AMF) inoculation on oats in saline-alkali soil contaminated by petroleum to enhance phytoremediation. Environ. Sci. Pollut. Res. 22 (1), 598–608.

Yang, Y., Han, X., Liang, Y., Ghosh, A., Chen, J., Tang, M., 2015. The combined effects of arbuscular mycorrhizal fungi (AMF) and lead (Pb) stress on Pb accumulation, plant growth parameters, photosynthesis, and antioxidant enzymes in *Robinia pseudoacacia* L. PloS One 10 (12), e0145726.

Yergeau, E., Sanschagrin, S., Maynard, C., St-Arnaud, M., Greer, C.W., 2014. Microbial expression profiles in the rhizosphere of willows depend on soil contamination. ISME J. 8 (2), 344.

Yergeau, E., Tremblay, J., Joly, S., Labrecque, M., Maynard, C., Pitre, F.E., et al., 2018. Soil contamination alters the willow root and rhizosphere metatranscriptome and the root–rhizosphere interactome. ISME J. 12 (3), 869–884.

Yousaf, S., Ripka, K., Reichenauer, T.G., Andria, V., Afzal, M., Sessitsch, A., 2010. Hydrocarbon degradation and plant colonization by selected bacterial strains isolated from Italian ryegrass and birds foot trefoil. J. Appl. Microbiol. 109 (4), 1389–1401.

Yu, Y., Zhang, Y., Zhao, N., Guo, J., Xu, W., Ma, M., et al., 2020. Remediation of crude oil-polluted soil by the bacterial rhizosphere community of Suaeda salsa revealed by 16S rRNA Genes. Int. J. Environ. Res. Public Health 17 (5), 1471.

Yuniati, M.D., 2018. Bioremediation of petroleum-contaminated soil: A ReviewIOP Conference Series: Earth and Environmental Science118. IOP Publishing, 012063.

Zamani, M.M., Fallahpour, M., Harvani, G.Y., Aghmiuni, S.K., Zamani, M., Tehrani, D.M., et al., 2014. Recent proportionate treatment methods for crude oil contamination evaluation of the Tehran refinery site soil. Thrita 3 (1). doi:10.5812/thrita.12113.

Zhang, B., Zhang, L., Zhang, X., 2019. Bioremediation of petroleum hydrocarbon-contaminated soil by petroleum-degrading bacteria immobilized on biochar. RSC Adv. 9 (60), 35304–35311.

Zhou, L., Li, H., Zhang, Y., Han, S., Xu, H., 2016. Sphingomonas from petroleum-contaminated soils in Shenfu, China and their PAHs degradation abilities. Braz. J. Microbiol. 47 (2), 271–278.

Zhu, H., Gao, Y., Li, D., 2018. Germination of grass species in soil affected by crude oil contamination. Int. J. Phytoremediation 20 (6), 567–573.

Zhu, J.Y., Pan, X., Zalesny, R.S., 2010. Pre-treatment of woody biomass for biofuel production: energy efficiency, technologies, and recalcitrance. Appl. Microbiol. Biotechnol. 87 (3), 847–857.

Zou, Y.N., Srivastava, A.K., Wu, Q.S., 2016. Glomalin: a potential soil conditioner for perennial fruits. Int. J. Agric. Biol. 18, 293–297.

Chapter 15

Ecotoxicity of nickel and its possible remediation

Misbah Naz[a], Muhammad Imran Ghani[b], Mohammad Sarraf[c], Mengyun Liu[b], Xiaorong Fan[a,d]

[a]*State Key Laboratory of Crop Genetics and Germplasm Enhancement, Nanjing Agricultural University, Nanjing, China,* [b]*College of natural resource and environment Northwest A&F University, Yangling shaanxi, China,* [c]*Department of Horticulture Science, Shiraz Branch, Islamic Azad University, Shiraz, Iran,* [d]*Key Laboratory of Plant Nutrition and Fertilization in Lower-Middle Reaches of the Yangtze River, Ministry of Agriculture, Nanjing Agricultural University, Nanjing, China*

15.1 Introduction

Nickel (Ni) is a beneficial plant nutrient because of its key role in plant growth (Yusuf, Fariduddin, Hayat, and Ahmad, 2011). Ni is a vital component of some particular enzymes that are participated in the nitrogen (N) metabolism and some involved in nitrogen fixation (NF), (Asher, 1991). If plants suffer from Ni's deficiency, it shows necrosis initiating from the leaf tip (Fig. 15.1). Ni is absorbed through secondary active transport pathway mediated by osmosis and other transport proteins. Insoluble Ni compounds transport to plant root cells mainly via endocrine. Once Ni is sorbed by the roots, then it is transported through the celling via the fermentation flow to the buds and can accumulate in the other parts of plant, such as buds, fruits, and seeds (Georgescu, Georgescu, Dărăban, Bouaru, and Paşcalău, 2011). The Ni transport and reconversion process is strictly regulated by metal ligand complexes such as Nicotianamine, histamine, and organic acids, as well as some proteins specifically combined and transported to Ni. However, the characters of Ni in wheat seedling show that the rise in the actions of the antioxidant enzymes was not appropriate to protect cell membrane in denial of Ni toxicity. Ni toxicity has turned out to be a specific anxiety, due to its enlarged industrial use. Plants evolved various detoxifying mechanisms to combat Ni toxicity problem. These mechanisms include the formation of various complexes such as Ni^{2+} and $NiOH^{+}$ (Krämer, Pickering, Prince, Raskin, and Salt, 2000). With the increasing of world population, the abundance of toxic chemicals in the environment has increased immensely, which raised significant concerns among scientific society (Eisler, 1998). Anthropogenic activities have

Phytoremediation. DOI: https://doi.org/10.1016/B978-0-323-89874-4.00022-4

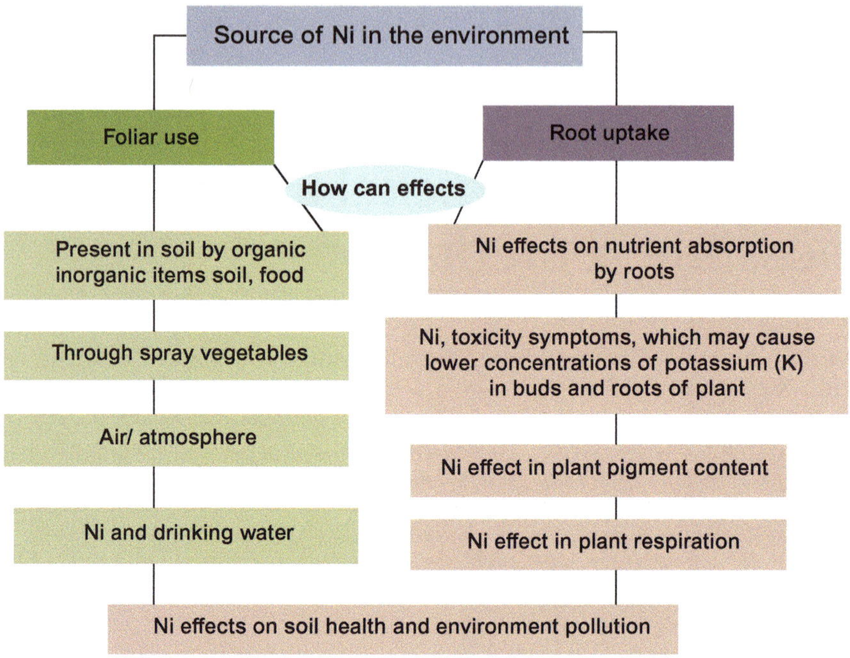

FIG. 15.1 Source of nickel present and its effects on environment.

accelerated the discharge of different organic and inorganic substances such as pesticides, petroleum products, salts, heavy metals acids, and the environment (Correia et al., 2018). Among heavy metal pollutants in different environments, in recent years, Ni has received substantial attention because of the rapid increase in soil, air, and water concentrations in various parts of the world (R. Sharma, Gulati, Pandey, and Adholeya, 2012). Recently, due to the uncontrolled industrial development of the globe, many chemicals have caused serious air, water, and soil pollution, so that environmental pollution has become a serious world problem (Masindi and Muedi, 2018). Ni is one of the most important trace elements that is ubiquitously released into the environment from human activities and natural sources. It has become a major concern due to the release of higher concentrations of Ni particles in agriculture soil and the atmosphere (Bennett, 1982). The burning of coal and oil for power generation is the main source of nitrogen emission and the rest of the emission comes from the Ni-, steel-, or cement-manufacturing process (Moreno et al., 2010; Pacyna, 1987). Ni has strong plant toxicity at high concentrations (Eisler, 1998; Masindi and Muedi, 2018). The role of Ni in the relationship between plant and human by effecting environment and its balance amount suggests for best result. Ni may be involved in the key to plant antioxidant metabolism, especially in stress situations (Fig. 15.1).

However, the phytoremediation is a green plant used to treat and control waste in water, soil, and air, and is an essential part of the new ecological engineering field. In-situ and off-site applications are controlled by the site's soil and water characteristics, nutrient sustainability, meteorology, hydrology, viable ecosystems, and pollutant characteristics. Phytoremediation research requires a proper understanding of the processes and methods that promote phytoremediation. These plants and processes promote tolerance to salty and alkaline and sodium salt soils phytoremediation markets to agricultural products. To make this technology widely commercial, more research and knowledge are needed, ensuring sustainable food security and making the planet more beautiful.

15.2 Hazardous effects of Ni in plants

From germination to productivity plant stage, the concentration of Ni slows down the plant growth, effects branching and development, deforms the different plant parts, produces abnormal flower patterns, also reduces biomass production, induces leaf spots, interferes with the tip of the mitochondrial roots, and produces leaf deficiency that leads to leaf worm disease and leaf necrosis (Ahmad and Ashraf, 2012). A sufficient quantity of Ni plays a vigorous role in a wide range of morphological and physiological functions in the plant (Wood and Reilly, 2007). Besides, an elevated Ni level also influences the nutrient absorption of the root; decelerates plant metabolism, repression photosynthesis, steaming; and leads to hyperstructure modification (Wood and Reilly, 2007). Eventually, when these crops experience a lot of Ni exposure, all these changes in the process reduce crop yields. Hagemeyer (2004) discussed a dual role (i.e., positive and negative) of Ni in plants growth and development).

In plant stages, a high concentration Ni inhibits shoot and root growth, affects the growth of branches, corrects various parts of plants, produces the abnormal shape of flowers, reduces the production of biomass, leaf highlights, disturbs mitotic root signals, and causes Fe deficiency which is a sign of chlorosis and vegetation necrosis (Ahmad and Ashraf, 2012).

15.2.1 Ni effects on nutrient absorption by roots

The literature on the ingestion and accumulation of Ni in plants and their effects on plant physiology and biochemical process are reviewed by Khan, Khan, Khan, Qamar, and Waqas (2015). The review showed the importance of Ni as a nutrient as well as potentially toxic elements in plants and environments. Ni is a heavy metal found ubiquitously in small amounts (Duman and Ozturk, 2010). However, rapid industrialization and urbanization in modern times have resulted in the accumulation of Ni and many other trace elements in different habitats, from plant collection and its further transfer to human and animal populations, which may seriously affect human and animal life (Hammad, 2011). On the other hand, Ni's physiological and toxic effects in higher plants and plant toxicity of metals are well documented (Kopittke, Asher, and Menzies, 2007). The higher concentration above 50 μg/g dry weight has a detrimental effect on plant growth. (Fernandes and Henriques, 1991). Ni is widely spread in the soil throughout the physical and biological world. Metals are extracted from mineral for a variety of industrial, chemical, and biological applications. Ni also occurs in water bodies and the atmosphere, usually in small amounts (Nagajyoti, Lee, and Sreekanth, 2010). Municipal and industrial wastewater contributes significantly to soil and water, but its relative contribution depends on the wastewater source (Kıpçak and Akgün, 2018). In plants Ni uptake relies on a variety of factors, of course, the most critical factor is ions, the concentration of Ni in the medium (Ross et al., 2000). Soil pH below 5.6 seems to facilitate absorption, mainly because the replaceable Ni content of the soil increases with upsurge of soil acidity (Adriano, 2001) and increases absorption by increasing the phosphate content in the soil (Crooke and Inkson, 1955; Li, Shen, Deng, Liu, and Hu, 2009).

15.2.2 Accumulation of Ni in plants

Ni is highly plant toxic at high concentrations. Plants can store Ni, commonly less than 0.1% of the total dryness, although the new Caledonian species can accumulate 25% of its dry weight (Jourand et al., 2010). Metals are mobile and can accumulate in plants and reproductive parts (Schmidt, 2003). During plant growth, most Ni is transferred and accumulated in the leaves. However, during leaf aging, most metals are shipped to seeds such as soybeans (Malan and Farrant, 1998). The study of the chemical form of Ni in plant tissue is limited to the leaves of accumulator (Tovar, 2008). A large number of plants have been identified as the Ni^{-2} plant intermediary. Several plant species and environmental factors govern the degree of accumulation in plants. In terms of numeral accumulation, there are considerable differences among species plants that grow on serpentine soils and are generally considered Ni-resistant species (Tovar, 2008) (Arazi, Sunkar, Kaplan, and Fromm, 1999) studies shown that, if plant medium have manganese (Mn) copper (Cu) and Zinc (zinc) concentrations were also higher at 50 mg Ni/L but under One thousand mg Ni/L induced by Ni toxicity symptoms, which may cause potassium (K) to have lower concentrations of buds and roots in plants when exposed to this concentration, as well as induce high concentrations of shoots (Yang et al., 2014).

15.3 Ni hazardous effects in photosynthesis

In some cases, the decline in plant productivity is attributed to a decrease in the photosynthesis rate due to Ni in the plants. It is well known that this metal inhibits photosynthesis and gas exchange for a sensitive crop such as corn and sunflowers (Atkins and Canvin, 1971; R. K. Sharma and Agrawal, 2005). However, pregraining seeds with Ni salt increases the photosynthesis rate of maize, oats, tomatoes, peas, bean wheat crop, and vegetables, apparently due to higher levels of chlorophyll and other secondary pigments (Atkins and Canvin, 1971). Seneviratne et al., (2019) suggested that the metal suppression Photo

System (PS)II could be more effective at the oxidation site. The activity of photophosphatization, photosynthesis, and PS^{-I} is another possibility of Ni^{-2}. Long-term exposure reduces growth of photochromic pigments of leaves, changes the structure of chloroplasts, and reduces CO_2 assimilating enzyme activity (Vega-Mas et al., 2015). Ni causes harmful effects on the plant's physiological progressions, such as photosynthesis. With the higher level of Ni toxicity in the environment, it is vital to understand functional and toxic effects of the Ni in plants. Ni toxic effects in plants comprise reduction of shoot and root growth and leaf area (Nagajyoti et al., 2010). The growth of *Zea Mays*seedlings decreases as the concentration of Ni (Bhardwaj, Arora, Sharma, and Arora, 2007). However corn and sunflower plants grow in hydroponic cultivation and process different levels of lead, Ni, Cd, and Tl salts. Net photosynthesis, steaming and toxic metal ion concentrations, and total plant biomass were measured. TI was found the most toxic to net photosynthesis and growth (Jadia and Fulekar, 2008).

15.4 Ni effect in plant pigment content

Changes in natural coloring are often obvious signs of the Ni effect. Researchers are trying to define the effect on plant leaves pigments, especially levels of chlorophyll and carotenoid. In the pea plants, carotene is mostly treated with Ni (Ahmad and Ashraf, 2012). An upsurge in chlorophyll content of 100–200 μg per gram Ni was detected in the pretreatment of maize and oat seeds. Pretreatment of potato tubers with a low concentration of Ni solution increases chloramphenicol synthesis, carotene, and flavonoids (Jianxin, 2005; Miri, Shakib, Ebrahimi, and Sharifi-Rad, 2017). The decomposition of chlorophyll in tobacco plates and separatrices is usually accelerated by far-red lights and slowed by red lights (Misra and Fleet, 1973). Surprisingly, however, when the separated rice leaves floated on the 237.7 g/L $NiCl_2$ solution in red and far-red light, the decomposition of chlorophyll accelerated in red light and slowed down at the far red light (Astarita and Santarém, 2008). There have been reports that Ni prevents the demolition of chlorophyll during the aging of wheat and rice leaves separated in the dark (Wani and Khan, 2013). But in the Ni attachment to young wheat leave failed to keep green, it was reported (Rout and Sahoo, 2015) that Ni sheltered chlorophyll from demolition and was more effective in the dark than in the light-separated wheat leaves. Routand Sahoo 2015), reported that higher levels of anthocyanins were preserved in petunia flowers that were incubated in $NiSO_4$ solution for three to four days compared to flowers hatched in the water at the same time. This observation is important in gardening and may be used to extend the life of flowers especially the life of cut flowers.

15.5 Ni effect in plant respiration

Cellular respiration and other metabolic processes are catalyzed by different enzymes that require the metals as cofactors; these metals and other nonessential metals have higher concentrations to inhibit enzyme activity, affecting plant respiration. The respiration rate of wheat leaf healthy tissue increases with the treatment of Ni (Llamas and Sanz, 2008), while treatment inhibits increased breathing associated with rusty infections (Aziz, Gad, and Badran, 2007), which demonstrated that low concentrations of $NiCl_2$ increased the respiration rate of mitochondria of maize. Still, at high meditations, the respiratory response was blocked. Ni produces a maximum breathing reaction at a concentration of 4.7 ppm $NiSO_4$ and a stimulation percentage of 400. The report by Gomes Rocha (2012) further states that, in the absence of phosphates, Ni increased NADH oxidation by about 5% at 5.87 ppm $NiSO_4$. The previous study (Jan, Khan, Asaf, Lee, and Kim, 2019) has exposed that Ni has a harmful effect on plant respiration and photosynthesis. The lack of detailed knowledge of heavy metals and metabolic progressions and other ion interactions hampers the fundamental understanding of toxicity and tolerance, which is especially true in the context of the effects of heavy metals on plant respiration (Hartmann, Skrobankova, and Drozdova, 2013).

15.5.1 Ni oxide nanoparticles

Compared to Ni II oxide (NiO-Bulk), Ni oxide nanoparticles (NiO-NPs) have exceptional properties and can be used in industrial products that promote inventive applications (Oukarroum, Barhoumi, Samadani, and Dewez, 2015). It was recently described that NiO-NPs can be easily transported to biological systems to induce cytotoxic and genotoxic effects. However, the effect of bioaccumulation NiO-NPs on the aquatic plant systems has not been widely studied because there are little data on their toxicity. As metal nanoparticles characterize harmful aquatic pollutants, the toxic effects of NiO-NPs need to be resolved to develop bioassays for toxicity risk valuation. In this respect, the aquatic plant *Lemna gibba* epitomizes a sensitive organism in toxicity analysis (Åkerlund, 2018). In this toxicology survey, we clearly show the potential toxicity of NiO-NPs suspension compared to NiO-Bulk. Ni oxide nanoparticles (NiO-NPs) are used in industries and their release into the environment can lead to contamination in agricultural areas (Debnath, Bansal, Salunke, Sadhu, and Bhattacharyya, 2016).We recommend that NiO-NPs be applied to stimulate toxic effects and increase levels of phytochemicals (glucorate and phenolic compounds) in the cabbage seedlings (Gopiraman et al., 2019; Husen, 2019). So far, there have been few research articles on the use of plants to synthesize NiO-NPs. Different parts of the plant, such as leaves and roots, are used to synthesize NiO-NPs.

15.6 Ni metabolic effects

Physiological aberrations detected during Ni deficiency indicate that Ni may be intricate in many physiological procedures. Some of those are associated with metabolic effects that have been defined earlier (Wood, Reilly, and Nyczepir, 2004). One of the most apparent impacts of N_I pretreatment is metabolism of protein. Ni upsurges the total protein and nitrogen content of the maize and oat plants (Yusuf et al., 2011). The infected plant was sprayed with $NiSO_4$ solution in the five to six leaf phases, increasing the free amino acid content of the leaves (C. Chen, Huang, and Liu, 2009). In the sensitivity of the isolated rice leaves Romeo, V Giofre, and A Chiacchio (2017) reported that the combination of $NiCl_2$ and benzodiazepines was more effective in inhibiting protein breakdown than $NiCl_2$ alone. Asrar, Mozafari, Rezanejad, Pourseyedi, and Yaghoobi (2014) stated that only $NiSO_4$ did not significantly increase the ascorbic acid content in tomato but when combined with complete fertilizer upsurges the ascorbic acid content of albumin lactose in tomato (Tito et al., 2011).

Ni salt increases ascorbic acid concentration (Das and Buchner, 2007), in tomato and corn (Younis, Athar, Malik, Raza Shah, and Mahmood, 2015) deliberate the stabilizing effect of salts from the Ni and the other elements on ascorbic acid's constancy in the water, HCl, H_2SO_4, and NaOH media. In aqueous solution and in HCl, Ni has the most significant stabilization effect against synesatic acid in the elements tested. Ni is beneficial to the oxidation of the phenols in infected plants of wheat (Gajewska et al., 2013). In the pea plants, the treatment by $NiSO_4$ promotes the formation of the western micro (Sreekanth, Nagajyothi, Lee, and Prasad, 2013). The hint of the whole root of the pea incubated with butyric acid solution quickly accumulated acetic acid, but the removal of the root lost this capacity (Rashotte, Poupart, Waddell, and Muday, 2003). Nevertheless, the sucrose, glucose, and fructose presence is nil in the solution, which restores the ability to remove the roots.

15.7 Ni affects enzyme activity

Several works have been carried out for measured the enzyme's response to Ni. Micronutrients play an important role in enzyme catalytic metabolism and are usually cofactored (Kelley, Lu, Haley, Gaddy, and Johnson, 2020). Several possibilities have been raised about the enzyme inhibition of Ni. However, there is a need to describe the molecular mechanism of Ni feat. Although Ni is an important environmental pollutant, there is no sufficient literature to completely understand the mechanism of Ni role in plants (Wellington et al., 2017). The metabolism of several vital elements, such as N (nitrogen), S (sulfur) P (phosphorus), and micro-molecules (proteins, nucleic acids, etc.), was unaffected. However, the literature suggests that plant growth must react with Ni, which of course varies depending on species, the concentration of Ni, and the nutrient composition of the soil (Erol, 2019). Without Ni, plants cannot complete their life cycle. Ni cannot replace Al, Cd for soybean growth. This is in combination with the discovery that Ni is vital to peas and produces beneficial growth reactions in several plants and it is recommended that Ni be categorized as a micronutrient element vital for higher plants (Krebs, 2006).

The test plant yield depends on the various type of plant analyzed, plant portion examined, and the Ni concentration in the nutrient solution. Increasing Ni dose in nutritional solution leads to a decrease in the yield of corn, a field bean. In the case of lettuce, note the opposite dependence, that is increased yield

TABLE 15.1 Nickel effects on enzyme activity in different crops.

Crop/plant	Enzyme	Application	References
Wheat leaves	Oxidative damage SOD and Catalase (CAT)	The leaves after Ni application	(Gajewska and Skłodowska, 2007)
Rice seedlings	SOD (A), ascorbate peroxidase (APX) (B), GR (C), and CAT (D) in roots of rice seedling	Effect of $NiSO_4$ on the activities enzyme	(Lin and Kao, 2005)
Tomato grafting	CAT, APX, and guaiacol peroxidase activities in leaf, proline content, membrane lipid peroxidation and mineral composition	Effect of Ni and grafting, combination on yield, fruit quality	(Kumar, Rouphael, Cardarelli, and Colla, 2015)
Maize root	Oxidative enzyme disorder	Overall expression of Ni toxicity in roots of *Zea mays* of Ni stress	(Baccouch, Chaoui, and El Ferjani, 2001)
Wheat seedling	Peroxidase activity effected	Intracellular peroxidases might act as scavengers of peroxide radicals produced as a result of nickel toxicity	(Pandolfini, Gabbrielli, and Comparini, 1992)
Rice seedlings	Inhibits ribonuclease and protease activities	Ni toxicity inhibits ribonuclease and protease activities in rice seedlings	(Maheshwari and Dubey, 2007)

Brassica napus L. leaves	Antioxidant enzyme activities	Ni-induced toxicity in canola plants and Ni-induced oxidative stress	(Kazemi, Khavari-Nejad, Fahimi, Saadatmand, and Nejad-Sattari, 2010)
Urd (*Vigna mungo* L.) seedlings	Proline content of plants	Source of nitrogen in the form of ammonium nitrate decreased proline content of plants treated with Pb and Ni	(Singh, Agnihotri, Reshma, and Ahmad, 2012)
Aulosira fertilissima	Effect on nitrogen assimilating enzymes	Scavenging of Ni and chromium toxicity in *Aulosira fertilissima* by immobilization	(Banerjee, Mishra, and Chatterjee, 2004)
Soybean seedlings	Antioxidant enzyme activity	Ni stress caused negative impacts on soybean seedlings	(Sirhindi, Mir, Abd-Allah, Ahmad, and Gucel, 2016)
Finger millet plants	Antioxidant enzymes and proline reduction	Oxidative stress observed under Ni treatment.	(Kotapati, Palaka, and Ampasala, 2017)
Seeds of Sunflower (*Helianthus annuus* L.)	α-amylase and protease activities reduced	Toxic effect of Ni on the growth and some key metabolic processes in sunflower, with varying levels of Ni as $Ni(NO_3)^2$	(Ashraf, Sadiq, Hussain, Ashraf, and Ahmad, 2011)

Note: For the proper growth and continuing the other cell function plants need various enzyme activities that are most important for human health and plant, Ni affects enzyme activity in different crops mentioned in this table.

under increased Ni dose effect in the nutritional solution (Yusuf et al., 2011). The Ni content of the field beans and lettuce on the ground is almost double that of the content registered in the corn. Ni accumulates mostly in roots of the experimental plant. The maximum value of TF is for lettuce plants, nearly twice the TF value of corn plant and the lowest field beans. Based on the analysis of NiCI (Kiba et al., 2012), it can be determined from the plants tested that bean revealed the greatest trend of offsetting Ni accumulation in buds. Among the experimental plants analyzed, the highest value of Ni BAF was for field beans, while the lowest was for corn. Taking into account the varieties of the experimental plants, corn shows the largest intake of Ni from a nutritional solution because of the high potential for the plant to produce the species analyzed; lettuce was the least sensitive to increase the Ni dose, while corn was the most sensitive (Broderick, 1999).

15.7.1 Ni effects on soil health

Although Ni is an important environmental pollutant, there is not enough literature to fully understand the mechanism of Ni role in plants. The metabolism of numerous vital elements, such as N (nitrogen), S (sulfur), P (phosphorus), and micro molecules (proteins, nucleic acids, etc.), was unaffected by Ni (dos Reis et al., 2017). However, the literature suggests that plant growth must have a reaction to Ni, which obviously fluctuates depending on species, Ni concentration, and nutrient composition of the soil. Without Ni, plants cannot complete their life cycle. Ni cannot replace Al and Cd for soybean growth (Boyle, 2013). This is in combination with the discovery that Ni is vital to peas and yields valuable growth reactions in numerous plants, and it is recommended that Ni be categorized as a micronutrient element that is vital for higher plants (Yusuf et al., 2011).

The yield of the test plant depends on the type of plant analyzed, the plant portion analyzed, and the Ni concentration in the nutrient solution. Increasing the Ni dose in the nutritional solution leads to a decrease in the yield of corn, a field bean. In the case of lettuce, note the opposite dependence that is increased yield under the influence of increased Ni dose in the nutritional solution (Kaur, Kaur, Siddique, and Nayyar, 2016). The Ni content of the field beans and lettuce on the ground is almost double than that of the content registered in the corn. Ni accumulates mainly in the roots of the experimental plant. Based on the analysis of lettuce, corn, and beans for the sensitivity and tolerance of Ni CI, it can be concluded that from the plants tested, bean revealed the greatest accumulated Ni in buds. Among the experimental plants analyzed, the highest value of Ni was applied in field beans, while the lowest was for corn. Taking into account the varieties of the experimental plants, corn shows the largest intake of Ni from a nutritional solution because of the high potential for the plant to produce. While another study analyzed that lettuce was the least sensitive to increase the Ni amount, while corn was the most sensitive. In addition, the Ni concentration of soybeans obtained from the control treatment is lower than that of the seeds used in the experiment; therefore, according to the results of this study, N metabolism in fixed N-type plants may be impaired or improved by Ni availability (Ahmad and Ashraf, 2012).

This work summarizes the importance of Ni in plant metabolism and its role and importance to crop agronomy. Current research on Ni physiology in plants has shown the importance of this micronutrient to plants, especially in processes associated with nitrogen metabolism. Ni main function is defined by its importance to urea hydration; however, Ni may be important in other physiological processes, such as NF and anthocyanin synthesis (Pais and Jones Jr, 1997). Although nine deficiencies in plants are rare events, crop yields and nitrogen efficiency are positive for the application of different species. Ni is one of the important micronutrient heavy metals used in plants. However, the rise in Ni levels raises serious concerns about plant physiology and survival. Heavy metal pollution of soil is a major environmental problem, which has a negative impact on human health and agriculture (Ahmad and Ashraf, 2012). Rhizosphere, as an important interface between soil and plants, plays an important role in plant restoration of heavy metals that contaminate soils, where microbial populations are known to affect heavy metal flow and plant supply by releasing coating agents, acidification, phosphate dissolution, and oxidation changes; thus enhancing the plant repair process. Plant repair strategies of rhinitis bacteria adapted with appropriate heavy metals have been paid more and more attention. In heavily contaminated soils where metal content exceeds plant tolerance limits, it is possible to treat plants with the bacteria noose that promotes plant growth and increase plant biomass, thereby stabilizing, remedying, and repairing metal-contaminated soil (Hossain, Piyatida, da Silva, and Fujita, 2012). However, there are many areas where there is a lack of understanding or information and more research is needed. To examine changes in heavy metal species in the root layer and to determine whether such changes can alter the accumulation and distribution of heavy metals; the role of bacteria in soil solutions in plant Cd intake is still poorly understood; this knowledge can help us understand the role and mechanisms of soil nasal algae in plant restoration. In summary, although the use of root bacteria with plants provides high efficiency for plant restoration, the microbial ecology in the nasal layer has not been fully understood (Adrees et al., 2015).

15.7.2 Effect of Ni contamination on soil enzymatic activities

In the pot experiment, the effects of soil contamination of 100, 200, 300, and 400 mg of Ni/kg on the activity of dehydrogenase, urease, and acid-base phosphatase were studied. Heavy sand and silt light leaves were used in two series of experiments: spring barley cultivation and plant-free cultivation (Sung, Lee, and Lee, 2011). Activity of enzymes was measured on the 14th, 28th, 42nd, and 56th days of the experiment. On this basis, it is found that soil gets polluted by

Ni as Ni 2.6 H_2O reduces the activity of dehydrogenase, urease, and acid-base phosphatase. This reduction is examined by the applied dosage of the metal. Ni has a greater impact on soil in spring barley cultivation than unseeded soil (Campos, Souza, Coelho, and Gloria, 1996). In the heavy sand sowing with spring barley, the inhibition rate of dehydrogenase activity is more than 50% caused by 300 and 400 mg Ni pollution; and in 200, 300, and 400 mg of soil, the inhibition rate of urease is more than 50%. The activity of dehydrogenase and alkaline phosphatase was observed to be reduced by more than 50% under 400 mg of Ni pollution in silicone flax, which was sown in spring barley. The inhibition of other enzyme activity is no more than 50%. Acidic and alkaline soils showed toxicity of phosphates, while dehydrogenase and acid phosphatase showed higher activity in lighter soils. In Ni-free soil, barley cultivation positively affected enzymatic activity and barley yield was positively correlated with dehydrogenase and phosphatase activity (Gianfreda and Bollag, 1996).

15.8 Cyclic nucleotide-gated channels (CNGCs)

The cyclic nucleotide-gated channels (CNGCs) are a newly well-known family of plant ion channels. This family protein has diverse functions in the switch of ion homeostasis, development, and defense beside biotic and abiotic intimidations. These proteins can sense changes in cNMP levels within cells and regulate many cellular reactions (Pandey, Zhang, and Assmann, 2007). In the current review, we provide an overview of the molecular properties of CNGC structure, regulatory and ion selectivity, subcellular positioning, and describe the signal transduction pathways involved in these channels (Frietsch, 2006). We also review new insights into the role of CNGC proteins in plant growth, development, and coping with stress (Duszyn, Świeżawska, Szmidt-Jaworska, and Jaworski, 2019).

15.9 Ni and human health

If anything has the advantage, it must have a disadvantage, one side Ni is a beneficial element for plant growth and development; and on the other side, it has severe health effects. Because it is transit elements that can be transferred from the plant to the animal than humans; however there is no strong evidence as it is unclear yet (Jaishankar, Tseten, Anbalagan, Mathew, and Beeregowda, 2014). It can affect human health in several ways: pollute soil, affect water, can reduce the normal function of human kidney, when present in air can cause allergy, can also have effects such as cardiovascular diseases, kidney diseases, lung fibrosis, and cancer (Munawer, 2018).

It can mutate the epigenetic effect that may induce the gene mutation due to Ni toxicity. After that, it moves to the environment and can become a reason for environmental pollution, which can be solved using phytoremediation and phytomining. As Ni is ferromagnetic elements so it is naturally present in the environment and have ability to combine other elements such as oxygen(oxides)

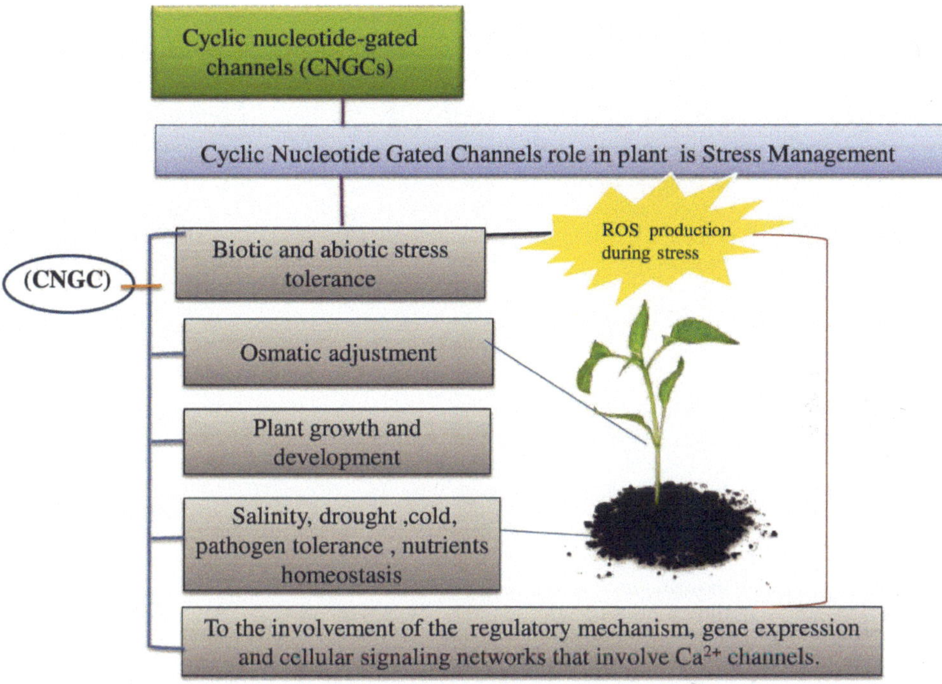

FIG. 15.2 To summarize the Incorporate the mechanism in the recent insights into the role of CNGC proteins in plant growth, development, and coping with stress.

and sulfur (sulfides). Also according Ali, Khan, and Sajad, (2013) the large amount of Ni present the sea water affect life of marine organism. A sufficient amount of Ni is important for microbes and human, plants, and animals. The Ni enzymes associate the use or can production of various gases such as (CO, CO_2, CH_4, H_2, NH_3 and O_2) environment Ni in gas form can be reason of human asthma disease it can effect on the mitochondrial DNA and reduce the up regulation of ATP increase the reactive oxygen species enzyme induce stress hypertension disease in human (Ahmad and Ashraf, 2012). The Ni can affect mental health in humans due to brain gene expression and regulation modifications, which can alter the DNA sequences and final product protein. Sometimes Ni breakdown the cell. It causes the cellular variations that become the reason for cell death. It can induce cancer tumor cell in the brain, skin, and lungs (Burkholder, 2020).

There are some side effects of Ni on human health from food items, soil, and air because of Ni toxicity in different fruits and vegetables and from atmospheric Ni toxicity (Genchi, Carocci, Lauria, Sinicropi, and Catalano, 2020; Giordani, Cecchi, and Zanchi, 2005; Gollhausen and Ring, 1991; Sinicropi, Rovito, Carocci, and Genchi, 2012; Zambelli, Uversky, and Ciurli, 2016) in Table 15.2.

TABLE 15.2 Ni Side effect on human health through food items soil and air.

Sources	Ni side effect on human health	Items	References
Soil and air	Allergy, lung fibrosis, lung and nasal cancer, cardiovascular and kidney diseases	Ordinary sources atmospheric Ni, wind, dust, weathering of rocks and soils, forest fires coal, diesel oil and fuel oil, waste and sewage and some industry fuels	(Genchi et al., 2020; Gollhausen and Ring, 1991; Sinicropi et al., 2012)
Vegetable	Ni exerts both positive and negative effects on human health by Ni toxicity through vegetables allergy and kidney diseases	Cabbage, beans, savoy cabbage, spinach, leeks, lettuce, lentils, peas, and some leafy vegetables	(Giordani et al., 2005)
Grain	Through grains allergy, lung fibrosis, and kidney stones	Buckwheat, sesame seeds, millet, sunflower seeds, whole grain bread, oats, and brown rice	(Genchi et al., 2020; Zambelli et al., 2016)

15.10 Ni and industrial use

The Ni complexes have several industrial and viable uses due to its chemical and physiological capacity and its resistance to high heat. It is most useful in the production of materials, such as cheap some kind jewelry, door keys, students' paper clips, and most important wearing clothing and fasteners (Wolfe, 2012). It combines with other elements such as Ni-cadmium for making batteries and electronic equipment for electronic companies. Moreover, it is useful for food industry, food processing industries, food pigments, artificial color, and catalysts (Dolez, 2015). In medical field, surgical steel instruments are made up of Ni. Finally, the Ni nano-particles (NiNPs), Ni nano-material (NiNMs), and Ni nano- fertilizer (NiNFs) are extensively used nanomaterials in many fields, such as medical field, food industry, catalysts, electronic magnetic materials, plants for soil biological conductive pastes and improver to grease, etc., (Manukyan et al., 2015).

15.10.1 Ni and environment pollution

Ni may occur through soil, air, and water; and in food in different ways but in small amounts. It can be released into environment through urine. It is also a poisonous element; the toxicity in the environment is the main issue of environmental pollution due to its chemical property that affects vegetable and fruits. This kind pollution is known as toxic pollution because of different food and fertilizer industries liquid and solid fuels (Barceloux and Barceloux, 1999; Cempel and Nikel, 2006). Due to toxicity nature affecting wish and beneficial microbes in the sea water, the greenhouse gases in the atmosphere directly affect flora and fauna and can cause skin rashes, skin burning, lead to suffocation, increase the body heat, etc. in humans (Barceloux and Barceloux, 1999).

15.10.2 Ni and drinking water

A few Ni elements salts, such as, acetate, chloride, nitrate, and sulfate, are soluble in water and present in the form of ground water (Nagajyoti et al., 2010). Ni occurs predominantly as the ion Ni $(H_2O)_6^{2+}$.At tap water with pH 5–9,the availability of Ni could depend on the soil use. The soil deep crop rotation heavy tillage, high fertilizer dose (Draycott and Christenson, 2003). Disposable glass pots, dress instrument, food packing exposes of sea water effects on the ability of water.

15.10.3 Ni interaction global warming

In this review, we describe the key allowing resources for environment variation solutions such as solar energy, ethanol production, and carbon capture and storage affecting environment system (Biscéré et al., 2017). As global warming is increasing day-by-day due to increase in greenhouse gases, which affects human health and plant growth and development. The 30% heat is coming from the sun the 70% is coming from the earth rest absorbed by its the environment atmosphere absorbing solar heat of our earth (Bian, Das, Wai, Hongmanorom, and Kawi, 2017). The earth spate more heat which goes back to atmosphere with increasing greenhouse gas come back to earth also increases some gasses like carbon dioxide and methane that affect health (Wang, Liu, Zhong, and Lu, 2019) (Fig. 15.3).

15.11 How to control Ni-based contamination in relation to plants

Two Brassica mustard and Brassica mustard seeds of Indian mustard were planted in artificial Ni-contaminated soil to study their tolerance and Ni accumulation. Specie *Brassica juncea* seems to have higher tolerance to Ni accumulation (X. Chen, Kumari, Cao, Plaza, and Achal, 2019). Consistent with earlier reports, the migration of pollutants from roots to shoots seems to be restricted in Brassica under the effect of higher Ni application rates and EDTA (Zurayk, Sukkariyah, and Baalbaki, 2001). The results of this study indicate that the mustard-type Bacillus may have excessive accumulation of Ni (Burd, Dixon, and Glick, 1998).

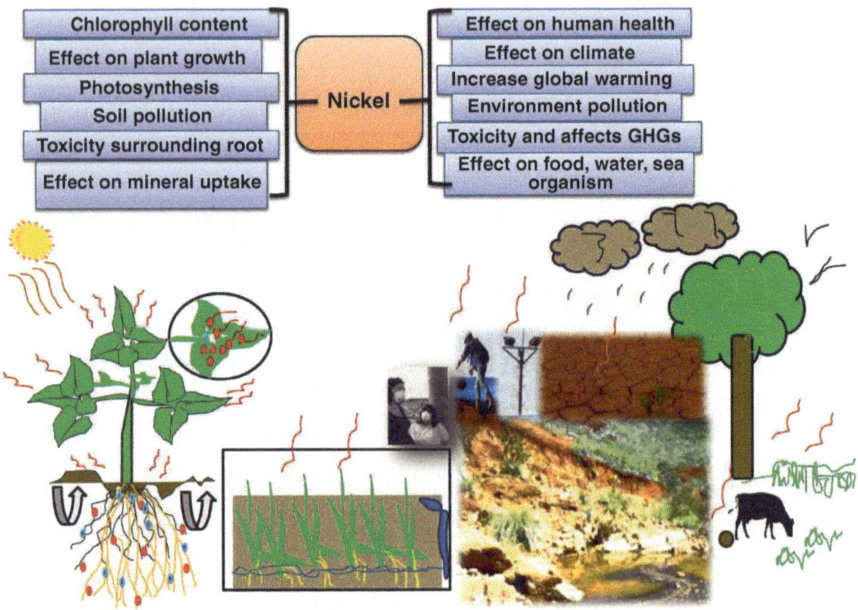

FIG. 15.3 Nickel interaction and effects on plant and then human by atmosphere.

15.11.1 Remediation of heavy metal–contaminated sites

Sites that are highly contaminated with metals can pose a health risk to humans and other organisms in the ecosystem. Phytoremediation is an economical and solar-powered method executed in any condition (Oves, Khan, Zaidi, and Ahmad, 2012). It can be used to remove, stabilize, and detoxify organic and inorganic pollutants, including heavy metals, from air, soil, and liquid reservoirs. The dispersed literature comprehensively reviews the possible sources, chemistry, potential biological hazards, and best available remediation approaches of multiple heavy metals (lead, chromium, arsenic, zinc, cadmium, copper, mercury, and Ni) commonly found in contaminated soil (Marques, Rangel, and Castro, 2011). The principles, advantages, and disadvantages of immobilization, soil washing, and phytoremediation technologies are introduced (Fig. 15.4). These technologies are often listed as the most effective technologies for cleaning heavy metal–contaminated sites (Wuana and Okieimen, 2011). Repairing soil contaminated by heavy metals is necessary to reduce related risks, make land resources available for agricultural production, enhance food security, and reduce land tenure issues caused by changes in land use patterns.

15.11.2 Ni uptake by plants and phytoremediation

Heavy metals are among the most significant types of pollution in the environment. Many approaches were previously used to clean up the environment

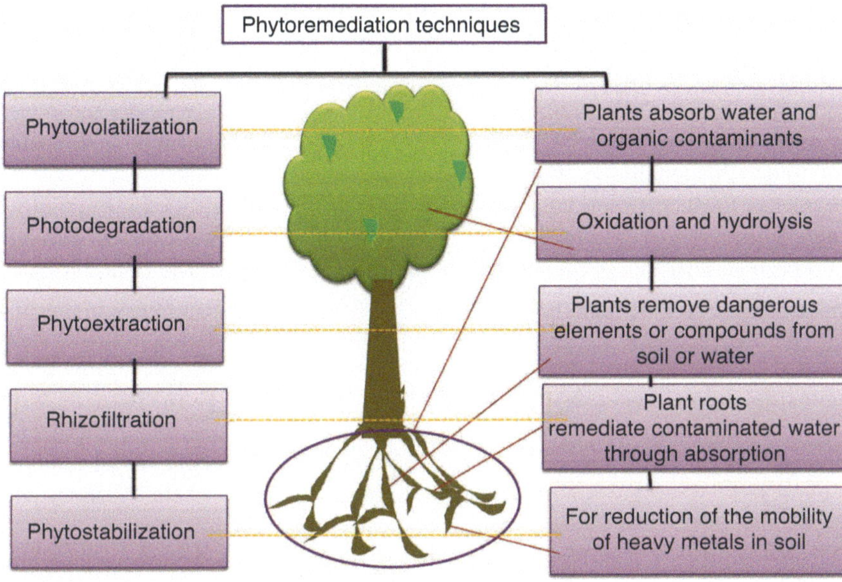

FIG. 15.4 Phytoremediation technology and function.

from this type of pollution, but most of them are expensive and problematic to accomplish maximum outcomes (Ghosh and Singh, 2005). Presently, phytoremediation is an efficient and inexpensive technical solution used to remove inert metals and metal contaminants from polluted soil and water. This technology is environment friendly and possibly cost effective. The advantage of this type of technology is to reduce pollutant factors affecting the optical mechanism and the metal optic mechanism in phytoremediation technology (Cunningham, Shann, Crowley, and Anderson, 1997). Some of the suggested plants commonly used in phytoremediation and their ability to reduce pollution are also reported. Recent concerns about environmental pollution have led to the development of applicable technologies to evaluate the presence and movement of metals in soil, water, and wastewater (Sarma, 2011).

Currently, phytoremediation has become an effective and inexpensive technical solution used to remove inert metals and metal contaminants from contaminated soil. Phytoremediation plants are used to clean soil, sediment, and water pollution. This technology is environmentally friendly and potentially artificial. Plants with extraordinary metal deposits are known as hyperchemulator plants (Awa and Hadibarata, 2020). Root of phytoremediation plants takes advantage of the systems' distinctive and selective yield aptitudes, along with whole plant body displacement, bioaccumulation, and pollution reduction capabilities. Lifting heavy metals by plants using phytoremediation technology seems to be a prosperous way to restore an environment contaminated with heavy metals (Kanamarlapudi, Chintalpudi, and Muddada, 2018). It has some advantages over the more commonly used traditional technologies. There are several factors to consider to achieve high efficiency of remedial results. The most important factor is a suitable breed of plants that can be used to eliminate pollution. Even the phytoremediation technique seems to be a great alternative, there are some limitations. Long-term research is needed to reduce this limitation to use this technique effectively (Sarwar et al., 2017).

At the level of practical technology Interest as with all technologies initially, in the stages of development, however, many mechanisms are considered poor or still unknown. Research and testing will push the boundaries of this powerful technology. The main limitations of the method are relatively shallow depth restriction to the roots of plants, and relatively long time to reach the required level. Also, there are related various issues such as environmental side effects are yet to be resolved, how to keep from changing wild plants, and unfamiliar plants for phytoremediation. In unwanted mourning clothes overall, this approach is very promising. In reasonable cases, the cost is expected to be significantly lower than the traditional system (Vishnoi and Srivastava, 2007). Its energy source is mostly solar (hence renewable), and it promises to be a flexible system that engineer can create to solve multiple tasks, but only one type (such as combining volume reduction with pollution and degradation) at time (because pollution almost never happens). Another benefit of phytoremediation is that the restoration of a soil ecosystem during the cleaning process. In the 1990s, the new method was developed to remove heavy metals from soil contaminated. The costly process of excavating and disposing of contaminated soil has increased, A new way to deal with soil in-situ fixation is a process, in which heavy metals are much less used in biology. This scene cleanup does less damage to people compared to life and environment excavates and dispose of contaminated soil elsewhere. Phytoremediation uses plants in a variety of ways to clean up heavy metals. The advantage of phytoremediation is that environmental problems can be solved naturally at a relatively low cost (Glick, 2003).

15.11.3 Pertaining remediation of Ni from water and soil environments

Environmental pollution affects the quality of the circle, hydrosphere, atmosphere, lithosphere, and biosphere. In the past two decades, people have made great efforts to reduce pollution sources and correct contaminated soil and water resources (Lazarus, 2008). Compared with physical and chemical methods, phytoremediation technology has higher cost-effectiveness and fewer side effects, and is becoming more and more popular in academic and practical circles. More than 400 plants have been identified as having soil and water remediation potential. Among them, *Thlaspi, Brassica, Sedum alfredii H.,* and *Arabidopsis* species are the most studied. It can also be expected that the latest advances in biotechnology will play a promising role in the development of new types of hyperaccumulating plants by transferring metal hyperaccumulator genes from low-biomass wild species to cultivated species that produce higher biomass (Khopkar, 2007). One study attempted to briefly review the latest progress in phytoremediation research and practical application of soil and water resources (De Anil, 2002).

The purpose of this study is to propose an advanced method to remediate Ni pollution in agricultural soils by using canola (*Brassica napus* L.) plants. The conclusion is that the *Consolida ambigua* and *Calendula officinalis* L. ornamental plants can be safely grown for phytoremediation to decrease soil pollution in terms of heavy metal accumulation (Lone, He, Stoffella, and Yang, 2008).

15.12 Conclusion

Ni is a heavy metal, also known as a toxic element. If it is present in excessive amount in the ecosystem, studies are needed to understand the details of biochemistry and molecular levels of the Ni super accumulator's resistance. Over the past few decades, there has been an interest in Ni. Moreover, scientists have shown increased interest because of its progressive industrial and commercial significance and improved electrothermal Ni analysis methods. The concentration of Ni in the environment (air, water, soil, food). However, everyone should remember that, at present, if Ni is widely released into the environment, it may represent the danger of human healing and the ecosystem. The purpose of this review is to describe a chemical feature and its toxicity mechanism in the environment. Besides, people should also pay attention to removing it from the environment because it only affects earth despite or good benefits. The ecotoxicity of Ni in the ecosystem and its possible remediation via phytoremediation is a challenge for nowadays; this chapter focuses on phytoremediation as a way of using plants to remove, transfer, stabilize, and/or destroy contaminants in the soil and groundwater as the plant releases natural substances through its roots, supplying nutrients to microorganisms in the soil.

References

Adrees, M., Ali, S., Rizwan, M., Zia-ur-Rehman, M., Ibrahim, M., Abbas, F., Irshad, M.K., 2015. Mechanisms of silicon-mediated alleviation of heavy metal toxicity in plants: a review. Ecotoxicol. Environ. Saf. 119, 186–197.

Adriano, D.C., 2001. Nickel. In: Adriano, D.C. (Ed.), Trace Elements in Terrestrial Environments. Springer, pp. 677–705.

Ahmad, M.S.A., Ashraf, M, 2012. Essential roles and hazardous effects of nickel in plants. In: Voogt, P. (Ed.), Reviews of Environmental Contamination and Toxicology. Springer, Cham, pp. 125–167.

Åkerlund, E., 2018. Cellular Effects of Nickel and Nickel Oxide Nanoparticles: Focus on Mechanisms Related to Carcinogenicity (Ph.D. thesis). Institute of Environmental Medicine, Stockholm, Sweden.

Ali, H., Khan, E., Sajad, M.A., 2013. Phytoremediation of heavy metals—concepts and applications. Chemosphere 91 (7), 869–881.

Arazi, T., Sunkar, R., Kaplan, B., Fromm, H., 1999. A tobacco plasma membrane calmodulin-binding transporter confers Ni^{2+} tolerance and Pb^{2+} hypersensitivity in transgenic plants. Plant J. *20* (2), 171–182.

Asher, C.J., 1991. Beneficial elements, functional nutrients, and possible new essential elements. Micronutr. Agric. 4, 703–723.

Ashraf, M.Y., Sadiq, R., Hussain, M., Ashraf, M., Ahmad, M.S.A, 2011. Toxic effect of nickel (Ni) on growth and metabolism in germinating seeds of sunflower (Helianthus annuus L.). Biol. Trace Elem. Res. 143 (3), 1695–1703.

Asrar, Z., Mozafari, H., Rezanejad, F., Pourseyedi, S., Yaghoobi, M., 2014. Calcium and L-histidine effects on ascorbate-glutathione cycle components under nickel-induced oxidative stress in tomato plants. Biol. Plant. 58 (4), 709–716.

Astarita, L., Santarém, E., 2008. Plant secondary metabolism in response to the environment. Adv. Plant Physiol. 10, 31–89.

Atkins, C., Canvin, D., 1971. Photosynthesis and CO_2 evolution by leaf discs: gas exchange, extraction, and ion-exchange fractionation of 14C-labeled photosynthetic products. Can. J. Bot. 49 (7), 1225–1234.

Awa, S.H., Hadibarata, T., 2020. Removal of heavy metals in contaminated soil by phytoremediation mechanism: a review. Water Air Soil Pollut. 231 (2), 47.

Aziz, E.E., Gad, N., Badran, N., 2007. Effect of cobalt and nickel on plant growth, yield and flavonoids content of *Hibiscus sabdariffa* L. Aust. J. Basic Appl. Sci. 1 (2), 73–78.

Baccouch, S., Chaoui, A., El Ferjani, E., 2001. Nickel toxicity induces oxidative damage in *Zea mays* roots. J. Plant Nutr. 24 (7), 1085–1097.

Banerjee, M., Mishra, S., Chatterjee, J., 2004. Scavenging of nickel and chromium toxicity in Aulosira fertilissima by immobilization: effect on nitrogen assimilating enzymes. Electron. J. Biotechnol. 7 (3), 13–14.

Barceloux, D.G., Barceloux, D., 1999. Nickel. J. Toxicol. 37 (2), 239–258.

Bennett, B., 1982. Exposure of man to environmental nickel—an exposure commitment assessment. Sci. Total Environ. 22 (3), 203–212.

Bhardwaj, R., Arora, N., Sharma, P., Arora, H.K., 2007. Effects of 28-homobrassinolide on seedling growth, lipid peroxidation and antioxidative enzyme activities under nickel stress in seedlings of *Zea mays* L. Asian J. Plant Sci. 6 (5), 765–772.

Bian, Z., Das, S., Wai, M.H., Hongmanorom, P., Kawi, S., 2017. A review on bimetallic nickel-based catalysts for CO_2 reforming of methane. ChemPhysChem 18 (22), 3117–3134.

Biscéré, T., Lorrain, A., Rodolfo-Metalpa, R., Gilbert, A., Wright, A., Devissi, C., Payri, C., 2017. Nickel and ocean warming affect scleractinian coral growth. Mar. Pollut. Bull. 120 (1-2), 250–258.

Boyle, D., 2013. Surface Complexation Modelling of the Adsorption of Cd (II), Cu (II), and Ni (II) to the Roots of Triticum turgidum. University of Guelph, Canada (Ph.D thesis).

Broderick, C.E., 1999. 403 Promotion of germination and seedling development of lima beans (*Phaseolus* spp.) by nickel chloride. HortScience 34 (3), 513C–5513.

Burd, G.I., Dixon, D.G., Glick, B.R., 1998. A plant growth-promoting bacterium that decreases nickel toxicity in seedlings. Appl. Environ. Microbiol. 64 (10), 3663–3668.

Burkholder, N.T., 2020. Structural basis of RNA Polymerase II C-terminal domain kinase and phosphatase specificity and their impact on transcriptional regulation.

Campos, C.F., Souza, P.E.A., Coelho, J.V., Gloria, M.B.A, 1996. Chemical composition, enzyme activity and effect of enzyme inactivation on flavor quality of green coconut water. J. Food Process Preserv. 20 (6), 487–500.

Cempel, M., Nikel, G., 2006. Nickel: a review of its sources and environmental toxicology. Pol. J. Environ. Stud. 15 (3).

Chen, C., Huang, D., Liu, J., 2009. Functions and toxicity of nickel in plants: recent advances and future prospects. CLEAN–Soil Air Water 37 (4-5), 304–313.

Chen, X., Kumari, D., Cao, C., Plaza, G., Achal, V., 2019. A review on remediation technologies for nickel-contaminated soil. Human Ecol. Risk Assess 26 (3), 571–585.

Correia, L., Marrocos, P., Olivares, D.M., Velasco, F., Luzardo, F., de Jesus, R.M., 2018. Bioaccumulation of nickel in tomato plants: risks to human health and agro-environmental impacts. Environ. Monit. Assess. 190 (6), 317.

Crooke, W., Inkson, R., 1955. The relationship between nickel toxicity and major nutrient supply. Plant Soil 6 (1), 1–15.

Cunningham, S.D., Shann, J., Crowley, D.E., Anderson, T.A., 1997. Phytoremediation of Contaminated Water and Soil. ACS Publications, Washington DC.

Das, K.K., Buchner, V., 2007. Effect of nickel exposure on peripheral tissues: role of oxidative stress in toxicity and possible protection by ascorbic acid. Rev. Environ. Health 22 (2), 157.

De Anil, K., 2002Environmental Chemistry2. *New Age International*, New Delhi, India, p. 232.

Debnath, B., Bansal, A., Salunke, H.G., Sadhu, A., Bhattacharyya, S., 2016. Enhancement of magnetization through interface exchange interactions of confined NiO nanoparticles within the mesopores of $CoFe_2O_4$. J. Phys. Chem. C 120 (10), 5523–5533.

Dolez, P.I., 2015. Nanomaterials definitions, classifications, and applications. In: Dolez, P. (Ed.), Nanoengineering. Elsevier, Amsterdam, pp. 3–40.

dos Reis, A.R., de Queiroz Barcelos, J.P., de Souza Osório, C.R.W., Santos, E.F., Lisboa, L.A.M., Santini, J.M.K., de Figueiredo, P.A.M., 2017. A glimpse into the physiological, biochemical and nutritional status of soybean plants under Ni-stress conditions. Environ. Exp. Bot. 144, 76–87.

Draycott, A.P., Christenson, D.R., 2003. Nutrients for Sugar Beet Production: Soil-plant Relationships. CABI, Oxfordshire, England.

Duman, F., Ozturk, F., 2010. Nickel accumulation and its effect on biomass, protein content and antioxidative enzymes in roots and leaves of watercress (Nasturtium officinale R. Br.). J. Environ. Sci. 22 (4), 526–532.

Duszyn, M., Świeżawska, B., Szmidt-Jaworska, A., Jaworski, K., 2019. Cyclic nucleotide gated channels (CNGCs) in plant signalling—current knowledge and perspectives. J. Plant Physiol. 241, 153035.

Eisler, R., 1998. *Nickel Hazards to Fish, Wildlife, and Invertebrates: A Synoptic Review*. US Department of the Interior. US Geological Survey, Patuxent Wildlife.

Erol, N.O.z., 2019. Genetic Analysis of Nitrogen use Efficiency in *Arabidopsis thaliana*. Wageningen University.

Fernandes, J., Henriques, F., 1991. Biochemical, physiological, and structural effects of excess copper in plants. Bot. Rev. 57 (3), 246–273.

Frietsch, S., 2006. The role of Cyclic Nucleotide-Gated Channels (CNGC) in Plant Development and Stress Responses in *Arabidopsis thaliana*. Universität Ulm.

Gajewska, E., Drobik, D., Wielanek, M., Sekulska-Nalewajko, J., Gocławski, J., Mazur, J., Skłodowska, M., 2013. Alleviation of nickel toxicity in wheat (*Triticum aestivum* L.) seedlings by selenium supplementation. Biol. Lett. 50 (2), 65–78.

Gajewska, E., Skłodowska, M., 2007. Effect of nickel on ROS content and antioxidative enzyme activities in wheat leaves. Biometals 20 (1), 27–36.

Genchi, G., Carocci, A., Lauria, G., Sinicropi, M.S., Catalano, A., 2020. Nickel: human health and environmental toxicology. Int. J. Environ. Res. Public Health 17 (3), 679.

Georgescu, B., Georgescu, C., Dărăban, S., Bouaru, A., Paşcalău, S., 2011. Heavy metals acting as endocrine disrupters. Sci. Pap. Anim. Sci. Biotechnol. 44 (2), 89–93.

Ghosh, M., Singh, S., 2005. A review on phytoremediation of heavy metals and utilization of it's by products. Asian J. Energy Environ. 6 (4), 18.

Gianfreda, L., Bollag, J., 1996. Influence of natural and anthropogenic factors on enzyme activity in soil. Soil Biochem. 9, 123–194.

Giordani, C., Cecchi, S., Zanchi, C., 2005. Phytoremediation of soil polluted by nickel using agricultural crops. Environ. Manag. 36 (5), 675–681.

Glick, B.R., 2003. Phytoremediation: synergistic use of plants and bacteria to clean up the environment. Biotechnol. Adv. 21 (5), 383–393. doi:10.1016/S0734-9750(03)00055-7.

Gollhausen, R., Ring, J., 1991. Allergy to coined money: nickel contact dermatitis in cashiers. J. Am. Acad. Dermatol. 25 (2), 365–369.

Gomes Rocha, A.M., 2012. Calcium Regulation in Chloroplasts and the Role of Calcium-Dependent Phosphorylation of Transketolase in Carbon Metabolism. lmu.

Gopiraman, M., Saravanamoorthy, S., Deng, D., Ilangovan, A., Kim, I.S., Chung, I.M., 2019. Facile mechanochemical synthesis of nickel/graphene oxide nanocomposites with unique and tunable morphology: applications in heterogeneous catalysis and supercapacitors. Catalysts 9 (5), 486.

Hagemeyer, J., 2004. Ecophysiology of plant growth under heavy metal stress. In: Prasad, M.N.V., Hagemeyer, J. (Eds.), Heavy Metal Stress in Plants. Springer, Cham, pp. 201–222.

Hammad, D.M., 2011. Cu, Ni and Zn phytoremediation and translocation by water hyacinth plant at different aquatic environments. Aust. J. Basic Appl. Sci. 5 (11), 11–22.

Hartmann, S., Skrobankova, H., Drozdova, J., 2013. Inhibition of activated sludge respiration by heavy metals, In Proc. 2013 International Conference on Environment, Energy, Ecosystem and Development. Venice, Italy.

Hossain, M.A., Piyatida, P., da Silva, J.A.T., Fujita, M., 2012. Molecular mechanism of heavy metal toxicity and tolerance in plants: central role of glutathione in detoxification of reactive oxygen species and methylglyoxal and in heavy metal chelation. J. Bot. 2012, Article ID 872875.

Husen, A., 2019. Natural product-based fabrication of zinc-oxide nanoparticles and their applications. In: Husen, A, Iqbal, M. (Eds.), Nanomaterials and Plant Potential. Springer, Cham, pp. 193–219.

Jadia, C.D., Fulekar, M.H., 2008. Phytoremediation: the application of vermicompost to remove zinc, cadmium, copper, nickel and lead by sunflower plant. Environ. Eng. Manag. J. 7 (5), 547–558.

Jaishankar, M., Tseten, T., Anbalagan, N., Mathew, B.B., Beeregowda, K.N., 2014. Toxicity, mechanism and health effects of some heavy metals. Interdiscip. Toxicol. 7 (2), 60–72.

Jan, R., Khan, M.A., Asaf, S., Lee, I.-J., Kim, K.M., 2019. Metal resistant endophytic bacteria reduces cadmium, nickel toxicity, and enhances expression of metal stress related genes with improved growth of *Oryza Sativa*, via regulating its antioxidant machinery and endogenous hormones. Plants 8 (10), 363.

Jianxin, L., 2005. Physiological and ecological responses of maize seedlings to cadmium stress. Chin. J. Ecol. 3, 265–268.

Jourand, P., Ducousso, M., Reid, R., Majorel, C., Richert, C., Riss, J., Lebrun, M., 2010. Nickel-tolerant ectomycorrhizal Pisolithus albus ultramafic ecotype isolated from nickel mines in New Caledonia strongly enhance growth of the host plant Eucalyptus globulus at toxic nickel concentrations. Tree Physiol. 30 (10), 1311–1319.

Kanamarlapudi, S., Chintalpudi, V.K., Muddada, S., 2018. Application of biosorption for removal of heavy metals from wastewater. Biosorption 18, 69.

Kaur, S., Kaur, N., Siddique, K.H., Nayyar, H., 2016. Beneficial elements for agricultural crops and their functional relevance in defence against stresses. Arch. Agron. Soil Sci. 62 (7), 905–920.

Kazemi, N., Khavari-Nejad, R.A., Fahimi, H., Saadatmand, S., Nejad-Sattari, T., 2010. Effects of exogenous salicylic acid and nitric oxide on lipid peroxidation and antioxidant enzyme activities in leaves of *Brassica napus* L. under nickel stress. Sci. Hortic. 126 (3), 402–407.

Kelley, B.R., Lu, J., Haley, K.P., Gaddy, J.A., Johnson, J.G., 2020. Metal homeostasis in pathogenic epsilonproteobacteria: mechanisms of acquisition and regulation. Metallomics 13 (1). doi:10.1093/mtomcs/mfaa002.

Khan, A., Khan, S., Khan, M.A., Qamar, Z., Waqas, M., 2015. The uptake and bioaccumulation of heavy metals by food plants, their effects on plants nutrients, and associated health risk: a review. Environ. Sci. Pollut. Res. 22 (18), 13772–13799.

Khopkar, S., 2007. Environmental Pollution Monitoring and Control. New Age International, New Delhi.

Kiba, D.I., Zongo, N.A., Lompo, F., Jansa, J., Compaore, E., Sedogo, P.M., Frossard, E., 2012. The diversity of fertilization practices affects soil and crop quality in urban vegetable sites of Burkina Faso. Eur. J. Agron. 38, 12–21.

Kıpçak, E., Akgün, M., 2018. Biofuel production from olive mill wastewater through its Ni/Al_2O_3 and Ru/Al_2O_3 catalyzed supercritical water gasification. Renew. Energy 124, 155–164.

Kopittke, P.M., Asher, C.J., Menzies, N.W., 2007. Toxic effects of Ni^{2+} on growth of cowpea (*Vigna unguiculata*). Plant Soil 292 (1-2), 283–289.

Kotapati, K.V., Palaka, B.K., Ampasala, D.R., 2017. Alleviation of nickel toxicity in finger millet (*Eleusine coracana* L.) germinating seedlings by exogenous application of salicylic acid and nitric oxide. Crop J. 5 (3), 240–250.

Krämer, U., Pickering, I.J., Prince, R.C., Raskin, I., Salt, D.E., 2000. Subcellular localization and speciation of nickel in hyperaccumulator and non-accumulator Thlaspispecies. Plant Physiol. 122 (4), 1343–1354.

Krebs, R.E., 2006. The History and Use of Our Earth's Chemical Elements: A Reference Guide. Greenwood Publishing Group, CA.

Kumar, P., Rouphael, Y., Cardarelli, M., Colla, G., 2015. Effect of nickel and grafting combination on yield, fruit quality, antioxidative enzyme activities, lipid peroxidation, and mineral composition of tomato. J. Plant Nutr. Soil Sci. 178 (6), 848–860.

Lazarus, R.J., 2008. *The Making of Environmental Law*:. University of Chicago Press.

Li, Z., Shen, B., Deng, Y., Liu, L., Hu, W., 2009. Preparation and microwave absorption properties of electroless Co–P-coated nickel hollow spheres. Appl. Surf. Sci. 255 (8), 4542–4546.

Lin, Y.C., Kao, C.H., 2005. Nickel toxicity of rice seedlings: the inductive responses of antioxidant enzymes by $NiSO_4$ in rice roots. Crop Environ. Bioinform. 2, 330–335.

Llamas, A., Sanz, A., 2008. Organ-distinctive changes in respiration rates of rice plants under nickel stress. Plant Growth Regul. 54 (1), 63–69.

Lone, M.I., He, Z.-l., Stoffella, P.J., Yang, X.-e, 2008. Phytoremediation of heavy metal polluted soils and water: progresses and perspectives. J. Zhejiang Univ. Sci. B 9 (3), 210–220.

Maheshwari, R., Dubey, R., 2007. Nickel toxicity inhibits ribonuclease and protease activities in rice seedlings: protective effects of proline. Plant Growth Regul. 51 (3), 231–243.

Malan, H., Farrant, J., 1998. Effects of the metal pollutants cadmium and nickel on soybean seed development. Seed Sci. Res. 8 (4), 445–453.

Manukyan, K.V., Tan, W., deBoer, R.J., Stech, E.J., Aprahamian, A., Wiescher, M., Weihs, T.P., 2015. Irradiation-enhanced reactivity of multilayer Al/Ni nanomaterials. ACS Appl. Mater. Interf. 7 (21), 11272–11279.

Marques, A.P., Rangel, A.O., Castro, P.M., 2011. Remediation of heavy metal contaminated soils: an overview of site remediation techniques. Crit. Rev. Environ. Sci. Technol. 41 (10), 879–914.

Masindi, V., Muedi, K.L., 2018. Environmental contamination by heavy metals. Heavy Metals 10, 115–132.

Miri, A.H., Shakib, E.s., Ebrahimi, O., Sharifi-Rad, J., 2017. Impacts of nickel nanoparticles on grow characteristics, photosynthetic pigment content and antioxidant activity of *Coriandrum sativum* L. Orient. J. Chem. 33 (3), 1297–1303.

Misra, K., Fleet, M., 1973. The chemical compositions of synthetic and natural pentlandite assemblages. Econ. Geol. 68 (4), 518–539.

Moreno, T., Querol, X., Alastuey, A., de la Rosa, J., de la Campa, A.M.S., Minguillón, M., Gibbons, W., 2010. Variations in vanadium, nickel and lanthanoid element concentrations in urban air. Sci. Total Environ. 408 (20), 4569–4579.

Munawer, M.E., 2018. Human health and environmental impacts of coal combustion and post-combustion wastes. J. Sustain. Min. 17 (2), 87–96.

Nagajyoti, P.C., Lee, K.D., Sreekanth, T., 2010. Heavy metals, occurrence and toxicity for plants: a review. Environ. Chem. Lett. 8 (3), 199–216.

Oukarroum, A., Barhoumi, L., Samadani, M., Dewez, D., 2015. Toxic effects of nickel oxide bulk and nanoparticles on the aquatic plant *Lemna gibba* L. BioMed Res. Int. *2015*, 1–17.

Oves, M., Khan, M.S., Zaidi, A., Ahmad, E., 2012. Soil contamination, nutritive value, and human health risk assessment of heavy metals: an overview. In: Zaidi, A., Wani, P.A, Khan, M.S. (Eds.), Toxicity of Heavy Metals to Legumes and Bioremediation. Springer, Vienna, pp. 1–27.

Pacyna, J.M., 1987. Atmospheric emissions of arsenic, cadmium, lead and mercury from high temperature processes in power generation and industry. In: Hutchinson, TC, Meema, KM (Eds.), *Lead, Mercury, Cadmium and Arsenic in the Environment*. John Wiley & Sons Ltd, USA, pp. 69–87.

Pais, I., Jones Jr, J.B., 1997. The Handbook of Trace Elements. CRC Press, Boca Raton, FL.

Pandey, S., Zhang, W., Assmann, S.M., 2007. Roles of ion channels and transporters in guard cell signal transduction. FEBS Lett. 581 (12), 2325–2336.

Pandolfini, T., Gabbrielli, R., Comparini, C., 1992. Nickel toxicity and peroxidase activity in seedlings of *Triticum aestivum* L. Plant Cell Environ. 15 (6), 719–725.

Rashotte, A.M., Poupart, J., Waddell, C.S., Muday, G.K., 2003. Transport of the two natural auxins, indole-3-butyric acid and indole-3-acetic acid, in Arabidopsis. Plant Physiol. 133 (2), 761–772.

Romeo, R., V Giofre, S., A Chiacchio, M., 2017. synthesis and biological activity of unnatural enediynes. Curr. Med. Chem. 24 (32), 3433–3484.

Ross, C., Chantrell, R., Hwang, M., Farhoud, M., Savas, T., Hao, Y., Humphrey, F., 2000. Incoherent magnetization reversal in 30-nm Ni particles. Phys. Rev. B 62 (21), 14252.

Rout, G.R., Sahoo, S., 2015. Role of iron in plant growth and metabolism. Rev. Agric. Sci. 3, 1–24.

Sarma, H., 2011. Metal hyperaccumulation in plants: a review focusing on phytoremediation technology. J. Environ. Sci. Technol. 4 (2), 118–138.

Sarwar, N., Imran, M., Shaheen, M.R., Ishaque, W., Kamran, M.A., Matloob, A., Hussain, S., 2017. Phytoremediation strategies for soils contaminated with heavy metals: modifications and future perspectives. Chemosphere 171, 710–721.

Schmidt, U., 2003. Enhancing phytoextraction: the effect of chemical soil manipulation on mobility, plant accumulation, and leaching of heavy metals. J. Environ. Qual. 32 (6), 1939–1954.

Seneviratne, M., Rajakaruna, N., Rizwan, M., Madawala, H., Ok, Y.S., Vithanage, M., 2019. Heavy metal-induced oxidative stress on seed germination and seedling development: a critical review. Environ. Geochem. Health 41 (4), 1813–1831.

Sharma, R., Gulati, S., Pandey, A., Adholeya, A., 2012. Novel, efficient and recyclable silica based organic–inorganic hybrid nickel catalyst for degradation of dye pollutants in a newly designed chemical reactor. Appl. Catal. B 125, 247–258.

Sharma, R.K., Agrawal, M., 2005. Biological effects of heavy metals: an overview. J. Environ. Biol. 26 (2), 301–313.

Singh, G., Agnihotri, R.K., Reshma, R.S., Ahmad, M., 2012. Effect of lead and nickel toxicity on chlorophyll and proline content of Urd (*Vigna mungo* L.) seedlings. Int. J. Plant Physiol. Biochem. 4 (6), 136–141.

Sinicropi, M.S., Rovito, N., Carocci, A., Genchi, G., 2012. Acetyl-L-carnitine in Parkinson's disease. In: Dushanova, J. (Ed.), Mechanisms in Parkinson's Disease—Models and Treatments. Book on Demand, Norderstedt, Germany, pp. 367–392.

Sirhindi, G., Mir, M.A., Abd-Allah, E.F., Ahmad, P., Gucel, S., 2016. Jasmonic acid modulates the physio-biochemical attributes, antioxidant enzyme activity, and gene expression in Glycine max under nickel toxicity. Front. Plant Sci. 7, 591.

Sreekanth, T., Nagajyothi, P., Lee, K., Prasad, T., 2013. Occurrence, physiological responses and toxicity of nickel in plants. Int. J. Environ. Sci. Technol. 10 (5), 1129–1140.

Sung, M., Lee, C.-Y., Lee, S.-Z., 2011. Combined mild soil washing and compost-assisted phytoremediation in treatment of silt loams contaminated with copper, nickel, and chromium. J. Hazard. Mater. 190 (1-3), 744–754.

Tito, A., Carola, A., Bimonte, M., Barbulova, A., Arciello, S., de Laurentiis, F., Colucci, G., 2011. A tomato stem cell extract, containing antioxidant compounds and metal chelating factors, protects skin cells from heavy metal-induced damages. Int. J. Cosmetic Sci. 33 (6), 543–552.

Tovar, M.I.P, 2008. Evolutionary Ecology of the Nickel Hyperaccumulator Thlaspi Montanum var. Montanum: The Role of Herbivory. University of California, Berkeley.

Vega-Mas, I., Marino, D., Sanchez-Zabala, J., Gonzalez-Murua, C., Estavillo, J.M., González-Moro, M.B., 2015. CO_2 enrichment modulates ammonium nutrition in tomato adjusting carbon and nitrogen metabolism to stomatal conductance. Plant Sci. 241, 32–44.

Vishnoi, S.R., Srivastava, P., 2007. Phytoremediation–green for environmental clean, In Proc. Taal2007: the 12th World Lake conference, Jaipur, India.

Wang, J.-W., Liu, W.-J., Zhong, D.-C., Lu, T.-B., 2019. Nickel complexes as molecular catalysts for water splitting and CO_2 reduction. Coord. Chem. Rev. 378, 237–261.

Wani, P.A., Khan, M.S., 2013. Nickel detoxification and plant growth promotion by multi metal resistant plant growth promoting Rhizobium species RL9. Bull. Environ. Contamin. Toxicol. 91 (1), 117–124.

Wellington, S., Nag, P.P., Michalska, K., Johnston, S.E., Jedrzejczak, R.P., Kaushik, V.K., Bajrami, B., 2017. A small-molecule allosteric inhibitor of Mycobacterium tuberculosis tryptophan synthase. Nat. Chem. Biol. 13 (9), 943–950.

Wolfe, J.A., 2012. Mineral Resources a World Review. Springer Science & Business Media, Berlin, Germany.

Wood, B.W., Reilly, C., 2007. Nickel and plant diseaseMineral Nutrition and Plant Disease. *The American Phytopathological Society*, St. Paul, MN.

Wood, B.W., Reilly, C.C., Nyczepir, A.P., 2004. Mouse-ear of pecan: a nickel deficiency. HortScience 39 (6), 1238–1242.

Wuana, R.A., Okieimen, F.E., 2011. Heavy metals in contaminated soils: a review of sources, chemistry, risks and best available strategies for remediation. ISRN Ecol. 2011. doi:10.5402/2011/402647 Article ID 402647.

Yang, T., Zhang, S., Hu, Y., Wu, F., Hu, Q., Chen, G., Yu, L., 2014. The role of a potassium transporter OsHAK5 in potassium acquisition and transport from roots to shoots in rice at low potassium supply levels. Plant Physiol. 166 (2), 945–959.

Younis, U., Athar, M., Malik, S., Raza Shah, M., Mahmood, S, 2015. Biochar impact on physiological and biochemical attributes of Spinach (*Spinacia oleracea* L.) in nickel contaminated soil. Glob. J. Environ. Sci. Manag. 1 (3), 245–254.

Yusuf, M., Fariduddin, Q., Hayat, S., Ahmad, A., 2011. Nickel: an overview of uptake, essentiality and toxicity in plants. Bull. Environ. Contamin. Toxicol. 86 (1), 1–17.

Zambelli, B., Uversky, V.N., Ciurli, S., 2016. Nickel impact on human health: an intrinsic disorder perspective. Biochim. Biophys. Acta 1864 (12), 1714–1731.

Zurayk, R., Sukkariyah, B., Baalbaki, R., 2001. Common hydrophytes as bioindicators of nickel, chromium and cadmium pollution. Water Air Soil Pollut. 127 (1-4), 373–388.

Chapter 16

Phytoremediation of pesticides

Manmeet Kaur, Shivani Sharma, Harpreet Singh Sodhi
Department of Microbiology, Punjab Agricultural University, Ludhiana, Punjab, India

16.1 Introduction

As a critical part of agriculture, pesticides have been affordable and have for quite a while anticipated a complete role in protecting harvests and animals from yield decreases. Up to now, the application of pesticides has been considered the safest and most accepted techniques for plant crop ammunition from bugs (Ozkara et al., 2016). However, as pitiful as 1% of an applied pesticide occurs at the target bug and the remaining in air, soil, and water gradually entering our advanced lifestyle and influencing nontarget organisms such as humans (Lozowicka et al., 2016), vegetation, and soil catalytic activity. As many as 80%–90% of pesticides applied to crops specifically hit nontarget vegetation, or can tenor or volatilize off-site and grime soil, air, and nontarget plants from the treated field. Approximately 80% of all pesticides used could be differentiated, with fraction of these innovations being found as transformation products with more than 10 years of industriousness. Pesticides represent a tremendous imminence to hominids in a roundabout way, by means of evolved way of life and adulterant of mineral deposit. Broad horticultural utilization of pesticides can prompt huge menace to nontarget living beings within adjoining terrestrial and aquatic ecosphere. The pollutant of water bodies can straightforwardly or in a roundabout way influence human wellbeing and biological system honesty by prompting a critical danger to oceanic conditions and drinking water assets. The cleanup of water defiled by pesticides is costly and energy devouring. Pesticides, especially the organophosphates and organochlorines, partake in any brackish body through farming run-off (Karunya and Saranraj, 2014).There are different ways by which these synthetic substances can pick up section into water bodies that incorporate surface run-off; float into lakes, streams, and waterway; splash types of gear; and ethereal showers to control water-restraining pests (Singh and Mandal, 2013). In addition, different remediation procedures may be deliberated to bulwark the wellbeing and climate of people as pesticides can surpass the endorsed standard levels. Differentiable chemical, physical, and biological strategies alongside catalytic degradation, adsorption, membrane filtration,

Phytoremediation. DOI: https://doi.org/10.1016/B978-0-323-89874-4.00026-1

oxidation, and biological treatment have been advanced as treatment innovation in the remediation of pesticides (Rani et al., 2017). Ordinarily, biological treatment is best in light of the fact that the chemical and physical treatments are monstrously exorbitant. Bioremediation innovation that includes the use of organisms is generally used to decrease the pesticides. In the previous 10 years, phytoremediation (the utilization of plants for tidying up of xenobiotic mixes) has picked up prominence as an eco-friendly, cost-efficacious, and productive in-situ innovation for an assortment of defilements and among them, numerous pesticides (Eapen et al., 2007).

16.2 Biopesticide scenario

Pesticide consumption in world has reached to 2 million tones and from these 2 million tones Europe utilizes 45% followed by 24% utilization by United States and rest 25% in rest of the world (Abhilash and Singh, 2009). Pesticide consumption in Asia is also alarming. China uses highest percentage followed by Korea, Japan, and India. In India, use of pesticide is about 0.5 kg/hectare and large contribution is from organochlorine pesticides. The usage is because of warm humid climatic conditions. The concept of green revolution has played an important role for utilization of variety of pesticides for high-yield varieties. Currently India is the largest producer of pesticides in Asia and ranks 12th in world. Pesticide residue in several crops has also affected the export of agricultural commodities in the last few years. In this context, pesticide safety, regulation of pesticide use, propel application technologies, and integrated pest management are some of the key strategy for minimizing human exposure to pesticides and to maintain the fertility of the soils for proper productivity. There is a dearth of studies related to these issues in India, Uttar Pradesh is the largest consumer followed by Punjab, Haryana, and Maharashtra. Regarding the pesticide share across agricultural crops, cotton accounts for 45% followed by rice (25%), chillies/vegetables/fruits (13%–24%), plantations (7%–8%), cereals/millets/oil seeds (6%–7%), sugarcane (2%–3%), and other (1%–2%).

16.3 Pesticides: distinct categories and hallmarks

Pesticides is the collective term for all chemicals used to counteract a certain group of organisms. The primary classes of pesticides are insecticides (against insects), herbicides (against plants), and fungicides (against fungi). However, inside these different classes, there still is a wide variety between the compounds and their chemical properties. They are defined by their ionic or nonionic properties, hydrophobic properties, mechanism of action, and their molecular structure (Gavrilescu, 2005).The four main groups of insecticides are: organochlorines, organophosphates, pyrethroids, and carbamate esters. While contemplating about bug spray taints, organochlorines are routinely the engineered materials of most important concern. Customarily known members are DDD (dichlorodiphenyldichloroethane), HCHs (hexacyclochlorohexanes:α, β, γ, δ), DDT (dichlorodiphenyltrichloroethane), DDE (dichlorodiphenyldichloroethylene), and chlordane (Pascal-Lorber and Laurent, 2011).These atoms are profoundly hydrophobic, with log K_{OW} values somewhere in the range of 5.5 and 6.9. At the point when these mixes dwell in soils for quite a long time, a critical enduring effect can be noticed (Lunney et al., 2004). As a normal toxicology procedure, pesticides are also ordered depending on the extreme dermal and oral toxicity of the rodent, introduced as LD_{50} in Table 16.1. LD_{50} is the factual estimate of the quantity of milligrams of poison per kilogram of body weight required to kill half of the huge population of guinea pigs, according to the World Health Organization (2010). While the majority of the characterization of pesticides depends on the intense oral appreciation of LD_{50}, dermal toxicity is constantly taken into account as the dermal liability of pesticides (Strickland et al., 2018).

In addition, as Giusti et al. (2018) suggested, the enthrallment of pesticides in the environment demolishes denizens through inhalation. The use of a median lethal dose (LC_{50}) characterizing the intermediate concentration of chemicals such as gas, smoke, fog, smoke, or residue is capable of executing half of the guinea pigs uncovered by inhalation under explicit test conditions, recorded as milligram (mg) per liter (L) over a given period of time (Deadman, 2017). The articulation of pesticides may include more than one part, such as the wetting expedient of eloquent toxicity, at which point the characterization can compare with the harmfulness of admixtures (World Health Organization, 2010).

Endured DDTs are firmly adsorbed to soil particles, further strengthened by drying and wetting substitutions. DDTs normally accumulate in subcutaneous fat due to their hydrophobic and lipophilic nature and are occasionally exacerbated in the echelons. A few benthic critters emerged to approach pervasive potential impact of pesticides promenade, that is, egg shells in migratory birds and metabolic disruptors in vertebrate species dropped significantly (Sharpe,

TABLE 16.1 Description of biopesticides dependent on acute toxicity.

Class		LD_{50} for rat (mg/kg body weight)	
		Oral	Dermal
IA	Enormously haphazard	<5	<50
IB	Mightily hazardous	~50	~200
II	Partially hazardous	50–2000	200–2000
III	Scantly hazardous	Atop 2000	Above 2000
U	Farfetched to produce acute hazard	Over 5000 or higher	

1995). Another most herbicide tainting occurs with atrazine, which is a photosynthetic herbicide used in agriculture to control herbicides (Garmouna et al., 1998). The above sparked a far-reaching desecration of soil and groundwater supplies. There is far less thought about fungicide filthiness, but hexachlorobenzene (HCB) and pentachlorophenol are the key toxins. HCB is a bioaccumulative organic hydrophobic compound that has been found in environment and aquatic surroundings (Qiu et al., 2004). In light of the dynamic and inactive components and target pestilences, biopesticides have a distinctive mode of activity: meddling with protein synthesis and amino acid, cell division, nervous system, respiration, energy production, photosynthesis, DNA methylation, and damage (Table 16.2).

TABLE 16.2 Pesticides dependent on the chemical types with their mode of activity.

Chemical types	Mode of action and effect on pest	References
Organophosphate	• Acetylcholinesterase inhibitors • Causing rapid twitching of voluntary muscles hence paralysis and death • Leads to nervous impulse failure across the synapses	Aliferis and Jabaji (2011) Heard et al. (2017)
Phenylpyrazoles (Fiproles)	• Gaba-gated chloride channel blockers • Affecting nerves and muscles	Buckingham et al., 2017
Pyrazole	• Affecting mitochondrial respiration	Aliferis and Jabaji, (2011)
Carbamates	• Acetylcholinesterase inhibitors • Affects nervous system and muscles	Aliferis and Jabaji, (2011)
Arsenic compounds	• Effect on DNA methylation • Predominantly DNA damage	Heard et al. (2017)
Nitrophenol derivatives	• Had impact on oxidative phosphorylation	Aliferis and Jabaji, (2011)
Imazamox	Targets: (1) nucleic acid (2) protein synthesis (3) amino acid	Velki and Ecimović, (2015)
Triflumezopyrim	• Nicotinic acetylcholine receptor competitive modulator • Adverse impact on nerves and muscles	Cordova et al. (2016)
Copper compounds	• DNA damage	Bednarska et al., 2017
Avermectins	• Glutamate-gated chloride channel allosteric modulator • leads to paralysis	Sparks and Nauen, (2015)

16.4 Demeanor of biopesticides in the environs

The merging of natural processes affects the execution of pesticides in the climate. In the climate, pesticides are linked to surroundings such as soil, water, and living beings, and coalitions are circumscribed by assortment of physical, chemical, biological reactions (Gavrilescu, 2005). The impact on pesticide defilement can be divided into the following three types:

- *Transport measures*: Dislocation of pesticide from the initial threshold of exordium.
- *Transfer measures*: Pesticides development is controlled through natural compartments, for example, water, soil, climate, and biota.
- *Transformation processes*: Conversion of the structure or alteration in the nature of a pesticide takes place or deterioration into its constituent components.

As well as the properties of the pesticides themselves, the above processes are therefore influenced by different soil and climatic variables. This is followed by the following main soil variables:

- About the surface of the soil and the transportation of soil particles. Soils with a fine texture have a larger surface with a lower porousness and a larger volume. Therefore, pesticides in the soil will diffuse all the water more gradually, allowing the pesticide to sorb soil particles for a longer time. Soils with larger particles can have far fewer accumulations of pesticides.
- The density of soil affects the calculation of the period before the pesticides are washed to deeper layers in conjunction with soil particles.
- Soil pH determines the ability of soils to adsorb and the speed of organic cycles that can reduce the buildup of pesticides. The sorption capacity of pesticide particles can be changed by soil organic matter, as can the proportion of organically dynamic microbes (Fukushima and Tatsumi, 2001). During the plausible degradation of pesticide buildup, organic material appears to provide microbes with a fuel source, and is a deeply unknown medium that can eventually associate with inorganic and organic analytes.
- The soil slant is an inherent geographical factor that can establish the space necessary for the entry of pesticides and water into the soil. Sloping areas have slightly more overflows, making pesticides less of an ideal opportunity for leakage.

The environment can have an effect on pesticide remediation as well. Weathering happens when pesticide derivatives are more impenetrably bound by some means to wet downpour or dew and dry up again to soil particles (White et al., 2005). Endured pesticides are known to be less productive than pesticides recently applied. The material characteristics of the pesticide itself would have an enormous effect on the climatic fate of these mixtures, far apart from soil and climate variables. Essential attributes incorporate water dissolvability, a proclivity to adsorb dirt particles, and a half-life timeframe or climate persistence. Accessibility of the pollutant is dictated by the consolidation of the previously mentioned soil and pesticide characteristics. The utilization of enzyme refers to the rate and degree to which pesticide atoms will be implemented and maintained in the global environment and also had outsized influence on the feasible remediation capability of specific innovations.

16.4.1 Transmit of acaricides

Pesticides may also be transferred between various compartments, despite climate transport: soil, the atmosphere, water, and biota (Gavrilescu, 2005). The majority of pesticides are originally applied to soils and plant surfaces in a solid or fluid state, although volatilization can take place. Volatilization is where the gas stage is transferred to solid and fluid pesticides, resulting in an air exchange (Frazar, 2000). Pesticides can be transported across vast distances until they are detectable all around, and toxins can be distributed across wider areas. Due to the breakup of solid and fluid pesticides, precipitation prospects will mostly change. The sawdust can run off a portion of water in the watersheds where the overflow also accumulates, which can cause pesticide defilement. The amount of run-off depends profoundly on both the characteristics of the pesticide and the topography and occurs when a significant portion of the downpour water reaches the dirt system at a distance from the place of use. It is also possible to transfer pesticides from soil, air, or water to plants, microorganisms, and creatures.

16.4.2 Consign of pesticides

The production of formulations from their place of syntax to the weather patterns is the transport of pesticides (Gavrilescu, 2005). The majority of pesticides are used by spraying, resulting in the analytes being fractionally dissipated in the air. Second, fungicides can vanish from the biological surfaces of the soil. In addition, when pesticides interact with soil, an analyte area that does not adsorb soil particles can filter into aquatic environments through the network and cause desecration there.

16.5 The alarm of pesticides

As agricultural contamination is not certified as a certifiable or dynamic factor, its adverse effects such as air, soil, and water contamination are carried out (Newete and Byrne, 2016). Each part of the general environment and all life forms, including microorganisms, plants and humans, may have an antagonistic effect. Algal bloom (due to eutrophication), rash and other skin disorders, respiratory and neurological problems (due to breathing in dirt air), liver, stomach and kidney problems, and malignancies (due to consuming polluted water & swimming) may be adverse effects of horticultural infection (Edao, 2017).

Therefore, horticultural pesticides are present in all climatic compartments (i.e., water, soil, air) and pose a real risk to the biological system because of their higher levels of poisoning (Moss, 2008). As a consequence, for the wellbeing of nature and human, the elimination of horticultural toxins from the soiled areas is important. Until now, a variety of techniques, including substrate, physical, and organic approaches, have been developed for the demolition of horticultural poisons. Each of the possible methodologies has priorities and disservices of its own. The most popular among each of these methodologies is considered to be organic alternatives (microbially or plant interceded).

16.5.1 Agrarian major pollutants

A few agricultural poisons exist, but they are classified extensively into organic and inorganic toxins. Natural contaminants, such as polycyclic aromatic hydrocarbons (PAHs, polychlorinated biphenyls (PCBs), and phenolic mixtures, include pesticides, herbicides, weedicides, and numerous natural mixtures. Propoxur, acephate, glyphosate metaldehyde, dursban, boric acid, diazinon, malathion, and dichlorodiphenyltrichloroethane are some neonicotinoids used to combat numerous diseases and insect pests (DDT). Heavy metals such as "lead," "mercury," "arsenic," "cadmium," "chromium," and "selenium" are commonly found in inorganic agricultural poisons (Mao et al., 2013).

16.5.2 Mechanisms and wellsprings of contaminations

Generally, through various rural practices and cultivation operations, horticultural pesticides come into the environment. The use of pesticides and the application of compost are significant contributors to rural emissions (Zazai et al., 2018). The application of manures increases dirt richness and complement levels, enhances crop growth and improvement, and ultimately builds on the productivity of crops. Synthetic or mineral fixings can be found in the compost. Though phosphorus, nitrogen, and potassium are readily available in these composts as essential source supplements and have a significant role to play in improving yield quality.

Again, despite the fact that compost, in particular N fertilizer, is introduced into the field, only an insufficient formulated compost criterion (less than half) is used by plants and a large portion is lost by filtering and volatilization steps (Lassaletta et al., 2014). Filtering causes tainting of underground water, whereas air defilement is caused by volatilization (like N oxides). Highly elevated concentrations of potassium and sodium composted manure have an insufficient impact on soil products, such as soil pH reduction, soil structure obliteration, and decrease in field crop productivity (Savci, 2012).

A critical component of surface and groundwater tainting is considered to be nitrates. The prevalence of nitrates in water, soil, and food is improved by the widespread use of compost and pesticides and concentrated farming. Massive concentrations of nitrates in groundwater cause chronic infections in newborns, such as methemoglobinemia. This is a direct result of the transformation of nitrate present in the stomach into nitrite rather than oxygen with hemoglobin and structural safety (L'hirondel et al., 2006). In addition, nitrates and phosphates in aquatic environments cause eutrophication (Pestana et al., 2018). However, the concentration of 10 mg/L nitrogen manure can have detrimental impacts on amphibians (Camargo et al., 2005). Ammonium fertilizers allow ammonia gas to be released, which is also a very toxic gas. Via an oxidation process resulting in acidic rain, ammonia is converted into nitric acid, which has a severe effect on crops and all other life forms.

16.5.3 Remarkable pesticides causing environmental pollution

1. *Organochlorine pesticides*
   Organochlorine pesticides, such as chlordane, lindane, DDT, show climate persistence and robust bioaccumulation in organisms and are thusly referred to as persistent organic pollutants (POPs), mixtures for which climate remediation strategies should be identified regardless of the restriction on their use (Gonzalez et al., 2005). DDT is the key pesticide that has been known since 1945 as the causative specialist for the decrease in the population of feathered creatures. DDT was a significant portion of bug spray used against mosquitoes, vectors of many diseases, including malaria.

   DDT and its metabolites are strongly lipophilic, thus firmly adsorbing soil particles. In plants, these mixtures are generally adsorbed to root frames and are gently absorbed. However, the plants of the Cucurbita family are more successful in taking and moving organochlorine than other plants. Cucumber and pumpkin both display a bioaccumulation factor more important than 1 for S DDT with no isomer-specific accumulation (BAF = ratio of absolute plant focus versus soil fixation) (Lunney et al., 2004). Moreover, perhaps the limitation of courgette to migrate DDT accumulations is more prevalent than those of plant species. Courgette is also recognized for its successful hardened chlordane collection through a mud intake pathway (Mattina et al., 2000). In view of the fact that large-scale DDT or other organochlorine pesticide pollution is a maturing defilement, this is a fundamental perception, with the aftereffect that it is formidable to disrobe toxins from the soil. The part of the root exudates, wherein the nature and extent of the chemicals differ from that of the different plants, may reiterate this more influential limit of the integration of the courgette by the preferable implosion of the soil formations of the DDT (Aigner et al., 1998). The perverting of conditions by DDT and divergent POPs is unavoidable. The frail degradability of these mixtures can explain their high half-life, despite the impotence of accessibility due to soil adsorption. This exemplifies the precondition for climate protection against these pollutants and, furthermore, for the use of appointive strategies for prevalent perforation to reduce the impact of POPs.

2. *Atrazine*
   Atrazine is a photosynthesis-inhibitor herbicide used to intensify the control of maize and sorghum in the pre-and postbeginning control of annual large leafy weeds and fortnightly grass. A substantial impurity in water

was found to be a chemical, depleting both surface water (Garmouna et al., 1998) and groundwater (Davoli et al., 1987), resulting in its restriction in the European Union in 2004. Atrazine's soil nutrient bias is ineffective, with $K_{oc} \sim 100\ cm^3/g$ (K_{oc} is the compound partition coefficient in organic matter versus water) resulting in its exceptional crude portability (Gustafson, 1989). In exposed soil rather than vegetated soil, the reduction in atrazine is low (Anderson and Coats, 1995). Atrazine is broken down by microorganisms into didealkylatrazine (DIDA) or hydroxyatrazine (HO-A), deethylatrazine (DEA), and deisopropylatrazine (DIA). Nevertheless, for quite a long time, the normal use of atrazine for some impurities caused the depravity of the product to shift the bacterial communes; this resulted in faster mineralization and herbicide devaluation (Houot et al., 2000)

Expanded devolution is a metabolic cycle through which atrazine is used by microbial organisms as a solitary fuel source. Other degradation processes, for instance, are cometabolism steps in which microbes use physical matter from the soil as an alternative fuel. Thus, increased oxidation is extremely effective in soils with low organic matter. As a result, although maximum mineralization may be feasible (Barriuso and Houot, 1996), fields are largely weak (Lin et al., 2008). The microbial decolorization pathway with HO-A more adequately reabsorbed in the clay humic complex could provide a limit to the dissipation of the herbicide. Interestingly, the dealkylated metabolites as portable as atrazine finally reach amphibian environments, contributing atrazine to oceanic pollution. A difference in the degradation pathways results in the sensitivity of the plant to atrazine in responsive and resistant plants. Like maize, atrazine is present in the above, in the presence of 2,4-dihydroxy-7-methoxy-1,4-benzoxazine-3-one (DIMBOA), a substance found in the maize roots that can be enzymatically degraded (Raveton et al., 1997).

As in microorganisms, atrazine can go through dechlorination or be dealkylated by means of glutathione formation. Dealkylated metabolites, DIA, or DEA do not lack phytotoxicity, and dealkylation pathways are prevalent in plants. Because of its solvency and deterioration, atrazine quickly disappears from splashed fields, but the ecological consequences persist for quite a while after the end of its use as it is still recognized in water compartments. This is expected to be measured for the use of atrazine or pesticides with a comparative danger to the ocean climate (Edwards and Owen, 1989).

16.6 Plant-associated remediation

Phytoremediation can also be seen as an innovative method that is cost-effective and ecologically sustainable (Hussain et al., 2009). Some techniques (Fig. 16.1) include phytoremediation, such as:

- Phytotransformation
- Phytodegradation

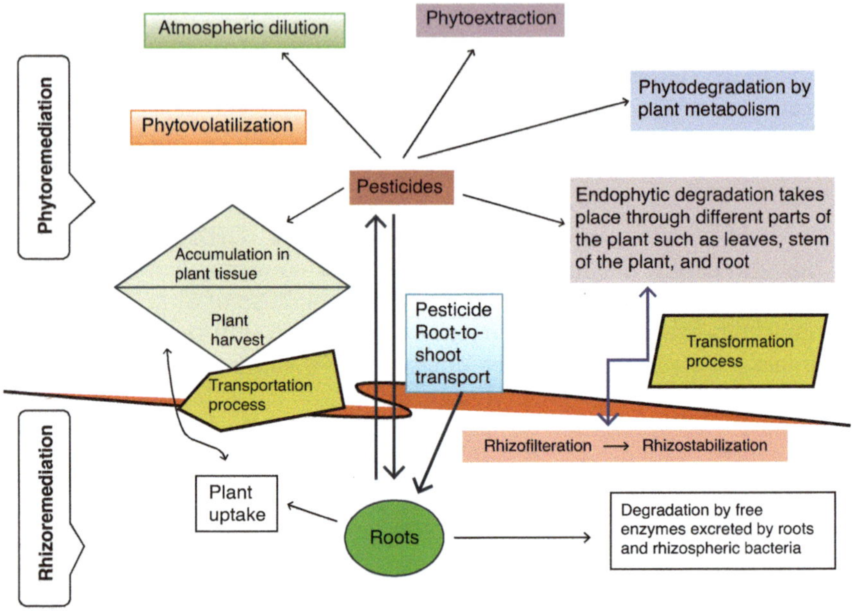

FIG. 16.1 Overview of plant-associated remediation.

- Phytovolatilization
- Rhizoremediation.

For the first three phases, the absorption of the contaminant into the plant tissues is crucial; at the rhizosphere or soil-root interface, rhizoremediation occurs. Plant-associated microorganisms may influence any of the phytoremediation processes described above. By degrading pesticides and promoting plant growth, both endophytic and rhizospheric microorganisms may play a role in remediation (Glick, 2010). At the rhizosphere, plant roots exude organic compounds that support a broad and efficient microbial population, leading to increased sorption of contaminants at the soil root interface.

16.6.1 Rhizoremediation

The amount of soil surrounding the root systems is defined by the term rhizosphere, which is determined by the activities of the growing roots. Rhizosphere is a convoluted climate that retains an immense amount of microbial species that are metabolically dynamic, a few essential things higher than the soil that is not rhizospheric (Capdevila et al., 2004). In rhizosphere soil, pesticide oxidation is frequently more prevalent than in mass soil wonder, perceived as "rhizosphere effect" (Hussain et al., 2009).

The rhizosphere seems to be the amount of soil that immediately surrounds the roots and is strongly affected by the plant's operations. For the rhizosphere, this motion makes it a more chaotic shift. This movement produces a more dynamic atmosphere for the rhizosphere than mass soil, supporting large quantities of microbial ecosystems that are metabolically complex. These estimates might become 10–100 times greater than the sum of soil unvegetated or mass microorganisms, hitting 10^{12} cells for each gram of sample (Pascal-Lorber and Laurent, 2011). The existence of plants with a large rhizospheric network can also produce the amount of microbial cells in mass soil (Leigh et al., 2006).

By developing deterrent or beneficial compounds, such as "1-aminocyclopropane-1-carboxylate (ACC) deaminase" the presence of microbial networks in the field will improve the plant (Glick, 2003). The precursor molecule for ethylene, which is a stress hormone often released by plants in polluted soils, is ACC. Ethylene levels are reduced at the moment when ACC-deaminase denigrates ACC, and the plant species reduce stress from ingredient exposure. These components can decrease the phytotoxicity of plants, speed up plant growth, and increase the phytoremediation potential.

In addition, microbes are able to promote the absorption by plants of indispensable supplements by the production of organic acids (Lunney et al., 2010); microbiota can also protect plants from microorganisms by searching for conditions in the microbiome of the plant and by degrading the pollutants before they are adversely affected by the plant. Rhizodegradation is a natural process, but can be reinforced by choosing the most suitable species of plants or by inoculating pesticide-debasing microorganisms. In the rhizosphere, plant root systems can eradicate pesticide-degrading compounds, but they also dispense photosynthetic products that can be filled in as supplements for rhizobia. Some researchers have secluded rhizosphere microorganisms and growth that display limitations of pesticide degradation (Gerhardt et al., 2009) (Table 16.3).

When the pesticide-degrading bacteria are intensified by inoculums, bacteria-enhanced rhizodegradation is known to be the method. Ahmad et al. (2012) detailed that when *Bacillus pumilus* C2A1 was inoculated in the plants, a quarter of the devaluation of chlorpyrifos is rooted in the *Lolium multiflorum* rhizosphere. Wang et al. (2011) demonstrated the effective decadence of phoxim when green onion (*Allium fistulosum*) and carrot (*Daucus carota*) were impregnated with *Glomus intraradices* and *Glomus mosseae* arbuscular mycorrhizal growths; outstandingly, pollutant deflation in uninoculated plants was imperceptible. The organism *Trichoderma harzianum* was separated from the marine plant and could degrade 50 mg/L of phoxim within 7 days in the fluid medium (Vacondio et al., 2015). A broad investigation into pesticide degradation by fungi was conducted by Jauregui et al. (2003). These analysts tested 17 white decay organisms; 16 had the option of dimming pesticides after a four-day development period, such as "parathion," "azinphos-methyl," "phosmet," "terbufos," and "tribufos."

Rhizoremediation techniques greatly influence the characteristics of soils, plants, and pesticides. Various factors such as soil organic matter, pH, and temperature will affect the bioavailability of the pesticide, as well as the enzymatic

TABLE 16.3 Pesticide degradation by rhizospheric bacteria and fungus with source of isolation.

Pesticide	Rhizospheric bacteria and fungus	Isolated from	Removal efficiency (%)	References
Chlorpyrifos	*Psychrobacter alimentarius*	Bulk soil	93.7	Khalid et al., 2016
Atrazine	*Rhodococcus rhodochrous*	Bulk soil	75	Kolekar et al., 2014
Phoxim	*Glomus sp.*	*Medicago sativa*	70	Wang et al., 2011
Pentachlorophenol	*Aspergillus sp.*	*Didemnum ligulum*	76	Vacondio et al., 2015
Linuron	*Variovorax sp.*	Bulk soil	80	Horemans et al., 2013
Atrazine	*Streptomyces sp.*	*Saccharum officinarum*	98	Mesquini et al., 2015
Sulfentrazone	*Penicillium sp.*	Bulk soil	40	Martinez et al., 2010
Methyl parathion	*Stenotrophomonas sp.*	Sludge	100	Deng et al., 2015
Dieldrin	*Cordyceps* sp.	Bulk soil	45	Xiao and Konodo (2013)

and bacterial degradation capacity in the rhizosphere. However, the existence of the plant roots can endorse a few of these impacts and, in particular, the propensity for deterioration is higher relative to mass soil (Hussain et al., 2009). In phytoremediation processes, such as "rhizostabilization," the rhizosphere may also play a major role. As they escape from agrarian fields and ultimately pollute groundwater, pesticides will pass through the soil. However, root systems and their related microbial species are stabilized in the soil by pesticide molecules (Pascal et al., 2011).

16.6.2 Phytoremediation

"Agro-remediation," "botano-remediation," "green remediation," and "vegetative remediation" have also been referred to as phytoremediation. Phytoremediation is also an advanced, remunerative, and socioeconomically advantageous technology that uses metals, hydrocarbons, pesticides, and chlorinated solvents in polluted sites to degrade, assimilate, metabolize, or detoxify plants (Lewandowski et al., 2006).

TABLE 16.4 Pesticide phytoremediation potential.

Pesticide	Plant species	Removal efficiency (%)	References
Atrazine	*Acorus calamus*	61.8	Wang et al., 2012
Endosulfan sulfate	*Cucumis sativus*	52.3	Somtrakoon et al., 2014
Cyanophos	*Plantago major*	94.7	Romeh, 2014
Methoxychlor	*Ricinus communis*	70	Rissato et al., 2015
Fenpropathrin	*Spirodela polyrhiza*	17.5	Xu et al 2015
Hexachlorobenzene	*Typha latifolia*	90	Zhou et al.,2013

Phytoremediation includes many processes such as "phytotransformation," "phytodegradation," "phytovolatilization," and "rhizoremediation." Plant roots exude organic compounds at the soil-root interface that sustain a large and active microbial community, which contributes to increased biodegradation of pollutants in the rhizosphere. The plant species most widely used in the research paper addressing pesticide phytoremediation are *Cucurbita pepo* and *Zea mays*.

Due to their high number of cultivars because of their significant position in agriculture and gardening, and their strong accumulation capacity of a broad range of organic pollutants, these plants are sometimes considered for phytoremediation (White, 2010). Many plants have been recorded to absorb pesticides effectively; Table 16.4 offers an overview. The plant shown to accumulate the largest number of different pollutants is *Ricinus communis*. The uptake of 11 different pesticides with a broad range of features was identified by Rissato et al. (2015). Many soil and plant characteristics decide the plant's uptake efficiency of these compounds (Hussain et al., 2009).

In addition to pH, soil moisture, temperature, and organic matter, the duration during which pesticide residues in soils can also influence pesticide uptake; time-dependent declines in accessibility are also identified during residue weathering or ageing. Pesticides are primarily absorbed via the soil-to-plant process by plants, while air deposition is often feasible in the form of additional gaseous accumulation or pollutants sorted into potential particles.

Several plant variables, such as water-holding efficiency and root subtlety, can increase the potential for absorption in the soil-to-plant penetration scenario. When pesticides are produced in the root seeds, they can indeed be immobilized in the roots or transferred to the aerial plant, where even analytes can be encoded, volatilized, or metabolized (Fig. 16.1). Often, root pesticide accumulation is inadequate for remediation purposes; while the magnitude of soil contaminants has dropped significantly, root specimens are not culled cost-effectively. Plant-focused aquatic remediation projects are an obvious problem to consider, where the reduction of pollutants from plant roots can be essential. *Eichhornia crassipes* (water hyacinth) can be very effective in accumulating insecticide ethion in its root system because of the fact that the roots provide more than 50% of the plant mass, but also that the total plant can be replanted along with the roots seamlessly.

16.6.2.1 Methods of application

1. *In-situ phytoremediation*
   In-situ phytoremediation requires the protection of live plants in polluted surface and ground water that is already contaminated or in fine sediments in contact with contaminated water and soil for remediation purposes. Polluted material is not removed in this process prior to phytoremediation. If the phytomechanism consisted of hyperaccumulation as applied to in the transformation of the toxic substrate, the plants could be cultivated and removed for treatment or retrieval of pollutants from the area after remediation. The prerequisite of the in-situ process is that reaching the contaminant to the roots should be simple. The least costly approach for phytoremediation is arguably this technique.

2. *In-vivo phytoremediation with relocated contaminants*
   For areas where the contaminant is not accessible to plants, such as deep aquifer contaminants, an alternative phytoremediation solution can be used. The pollutant is collected by a technological action in this process and then transported to a transitory treated area where it is possible to expose the plants selected for optimal phytoremediation. The purified water or soil may be returned to its original site after treatment and the vegetation may, if appropriate, be collected for excavation. Typically, this strategy would be more burdensome than the more passive approach already cited. Except at the site of pollution or at another venue, treatments may happen.

3. *In-vitro phytoremediation*
   Live plants for phytoremediation are used in the first two procedures. Using live plant materials, such as enzymes that are extracted, is the traditional phytoremediation technique. In certain cases, this approach can potentially be implemented in situ, for example by adding plant extracts to a contaminated pond or by using a porous barrier impregnated with enzymes in a polluted groundwater contrail. This method will, most definitely, also be extended to contaminated material that has been relocated to a provisional treatment area. Taking into account the expense of plant enzyme preparation/extraction, this technique may theoretically be the most expensive form of phytoremediation; however, in certain plants, such as tarragon, it may lead to lower overhead costs. The time span in which enzymes are continually involved in pollutant degradation is an important factor to consider in this approach.

16.7 Remediation technologies for pesticides

Phytoremediation of soil pesticides is dependent on plant absorption, volatilization, vegetation degradation, and combined root exudate and microorganism degradation of the rhizosphere (Liu et al., 2016). The rate of soil degradation of "deltamethrin" has been directly linked to soil microbial activity (Ismail et al., 2015). Some of the methods used to extract pesticides from polluted soil are listed in Table 16.5. In bacterial populations with "P450 cytochrome genes," there is a beneficial impact on aerobic pesticide bioremediation. Vetiver (*Chrysopogon zizanioides)* is sufficient for phytostabilizing moderately dioxin-contaminated areas. "Electrokinetic flushing of soils (EKSF)" is used to repair contaminated soils with pollution. The pollutants in the soil can be eliminated or transferred to the flushing fluid when the contaminated soil is exposed to an electric field between anodes and cathodes that can then be extracted or mobilized by electroosmosis. The EKSF approach used to extract pesticides from polluted soils has become a widely discussed subject in the remediation community (Vieira et al., 2016) and has recently been combined with other remediation techniques such as permeable reactive barriers and bioremediation (Trellu et al., 2016).

TABLE 16.5 Technologies for remediation used to extract pesticides from the soil.

Remediation technology	Pesticide	Degradation efficiency (%)	References
Enhanced electrokinetic Fenton	Triazoles	67	Bocos et al., 2015
Phytoremediation	Endosulfan	72	Mitton et al., 2016
Electrokinetic soil flushing	2,4-D	95	Silva et al., 2015
Solar-powered electrokinetic remediation	2,4-D	74	Souza et al., 2016
Phytoremediation	Atrazine	60–80	Khruynk et al., 2017
Electrokinetic soil fences	Oxyfluorfen	26.8	Risco et al., 2016
Electroremediation	2,4-D	57	Risco et al., 2015
Phytoremediation	Aldrin, DDT, hexachlorocyclohexane	25–70	Rissato et al., 2015
Rhizospheric and microbial degradation	Endosulfan	50–60	Singh et al., 2016

16.7.1 Methods of sustainable remediation for agricultural land

For pesticide-polluted agricultural soils, many safe in-situ remediation approaches can be employed. Phytoremediation approaches as well as integrated technologies for use, including "rhizobia remediation," "fertilized-assisted remediation," and "microbial remediation," relevant to rhizoremediation are strongly supported. In the future of rhizobial remediation technology, essential enzymes and fungi for pesticide removal should be further explored, along with potential interactions with many other soil microorganisms (Meharg et al., 2016).

Growing research focuses on innovative methods of remediation, such as "electrokinetics" with microbial permeable reactive barriers to the remediation of electrokinetic remediation. "Ultrasound-assisted soil washing" has been combined with bioaugmentation for heavy metal–contaminated soil and PAHs remediation. To remove polybrominated diphenyl ethers in property dependent soil microcosms, "Tourmaline-catalyzed Fenton-like reaction (TCFR)" coupled with *Phanerochaete chrysosporium* (TCFR + P) was used (Li et al., 2016a).

Nanoremediation "[zero-valent iron nanoparticles (nZVI)]" may be used in combination with electrokinetics to remove PCBs from the soil, with the percentage of removal in polluted soil rising from 20% to >75% (Gomes et al., 2016a). An alternative to PCB-contaminated soil remediation might be Fenton-like oxidation followed by soil washing (Ma et al., 2016). Anaerobic or optional anaerobic microflora and phytoremediation were implemented to contain soil polluted with POPs (DDTs and trifluralin). To biomineralize pesticide-polluted soil, genomic techniques have been used (Douglass et al., 2016). Transgenic plants and microbes can also be used in pesticide-contaminated environmental remediation (Huang et al., 2016).

16.8 Biotechnological advances in phytoremediation

1. *Transgenic plants which express enzyme-degrading pesticides*
   While plants exhibit an intrinsic capacity to dissolve certain xenobiotic compounds, for optimum degradation, there may be a lack of catabolic reactions. As a consequence, the transfer of genetic material involved in the biosynthesis of xenobiotics from microbiota or other eukaryotes to plants would further improve their ability to remediate (Eapen et al., 2007). Sonoki et al. (2005) applied an fungal enzyme, *Coriolus versicolor*, to tobacco plants . The transgenic plant was able to extract "pentachlorophenol" by making laccases and squirting them into the rhizobia. Wang et al. (2008) expressed the microbial "Organophosphorus hydrolase" gene in tobacco plants for organophosphorus pesticide incineration. More than 99% of methyl parathion deteriorated in transgenic plants after 14 days of development.

In plant cells, a range of plant enzymes, including "cytochrome P-450," "N-O-malonyltransferase, peroxygenase," "glutathione-S-transferase," "peroxidase," "N-oglucosyltransferase" and "carboxylesterase" are involved in the phytotransformation of xenobiotics. "Glutathione S-transferases" are a multigene family of herbicide-detoxifying enzymes. For the creation of a binary vector sufficient for genetic modification of tobacco plants, the gene "*gstI-6His*" was used from maize encoding "6His-tagged GST I" (*Nicotiana tabacum*). Transgenic plants have demonstrated significantly higher resistance to alachlorine, indicating their role in the phytoremediation of agricultural areas polluted with herbicides (Karavangeli et al., 2005). GSH overexpression has gained considerable interest in phytoremediation because of its significant function in the plant defense mechanism (May et al., 1998). Peuke and Rennenberg (2005b) used transgenic poplars with increased GSH synthesis in greenhouse experiments and found that transgenic poplars had a great propensity for ingestion and extraction of toxic heavy metals and pesticides. As described in Fig. 16.2, GSH is synthesized in two ATP-dependent phases, catalyzed by "glutamylcysteine synthetase (γ-ECS)"and "glutathione synthetase (GSHS)." GSH is processed in two ATP-dependent steps in the first two phases, which are catalyzed by "γ-ECS" and "GSHS" and combined with pesticide molecules.

Finally, "glutathione s-transferase (GST)" transfers these coupled molecules through cytoplasm into molecules where the pesticide compound is mineralized.

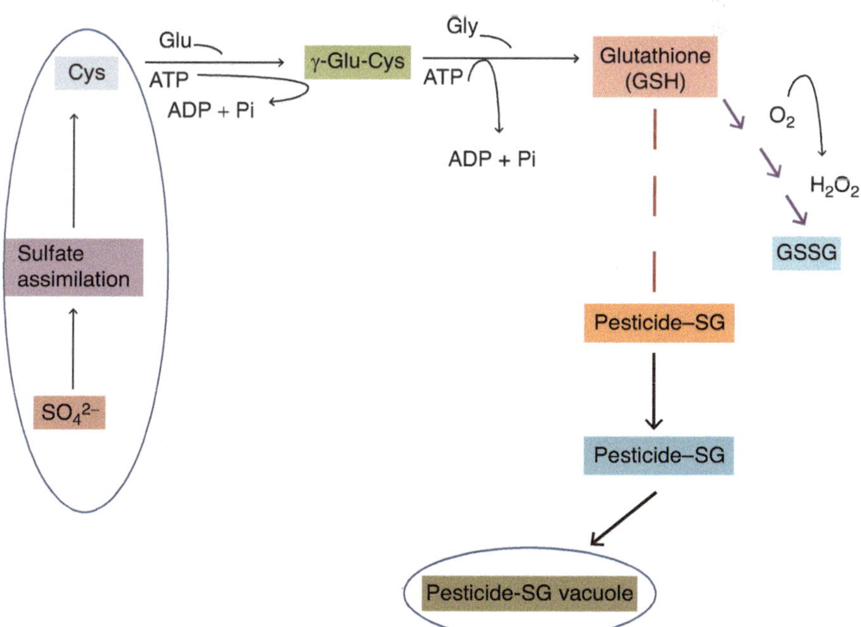

FIG. 16.2 Mechanism of detoxification of pesticides in plant cell.

Both "ECS" and "GSHS" enzymes have been identified in different parts of plants, such as roots, chloroplasts, leaf cell cytosols (May et al., 1998). In rice plants (*Oryza sativa* cv. Nipponbare), the human cytochrome P450 gene "CYP1A1" was introduced by Kawahigashi et al. (2003) to induce resistant herbicides in rice plants. In addition to the improved herbicide resistance and synthesis of transgenic rice plants, it is important to note that the intermediate metabolite levels were higher in the "CYP1A1" crop medium than in plant tissues.

Like many plant enzymes; "arylamine," "N-acetyltransferases (NAT)" are xenobiotic-metabolizing enzymes that catalyze the conversion of the acetyl group from "acetyl CoA (AcCoA) to arylamine N_2 or O_2" based on xenobiotics and their hydroxylated metabolites. They have a personal interest in the detoxification of certain pesticides. NAT isoforms have already been described in several species, but no NAT enzymes have been identified in plants for partial genome sequencing. Rodrigues Lima et al. (2006) recently showed the existence of NAT-dependent pathways in leguminous plants due to the symbiotic relationship with *Mesorhizobium loti*.

Two NAT enzymes cloned by *M. loti* were present. *Mesorhizobium loti* showed that N-acetylation by these two recombinant enzymes of a variety of known "NAT substrates," including aniline-derived pesticides, was catalyzed. They also revealed the existence of a strong "NAT-dependent acetylation pathway" in the *M. loti* inoculated *Lotus japonicus*. "*M. loti* was the first noneukaryotic organism demonstrated to develop two catalytic active NAT isoforms." This research also demonstrates the fact that a plant xenobiotic detoxification pathway is formed by soil microbe symbiotic relationships. Some transgenic plant species are used, as shown in Table 16.6, for advanced phytodegradation of pesticides.

2. *Transgenic plants expressing antibodies for pesticide detoxification*
   By developing transgenic plants that express proteins unique to a specific pesticide culture, plant phytoremediation skills might also be applied to pesticides. For the development of full-length "immunoglobulin G (IgG)" antibodies (Hiatt et al., 1989), "multimeric secretive antibodies"(Ma et al., 1995), and a broad variety of useful fragments of antibodies, transgenic plants have been used (de Neve et al., 1993). Two types of transgenic plants were studied by Drake et al. (2002). In the first plant type ("IgG plants"), the activated full-length antibody was formed and reabsorbed from the roots to the adjacent nutrient broth. The goal was to determine if the secretory antibody would bind to the aqueous phase of "SA I/II" (Vine et al., 2001).

The reactive antibody was rhizosecreted from the very first plant form and was shown to be bound to an antigen to create an immune complex in the extracellular environment. The transmembrane fragment sustained unique antibodies in the second, in the cell membranes plant. Within 24 h, the receptor introduced via the roots of IgG plants to the nutrient agar medium was transmitted to the uppermost leaves and stems, where it was secreted by binding as an immune complex

TABLE 16.6 In plant cells, genes inserted to promote phytoremediation.

Plant	Source of gene	Enzyme	Pesticide	Remarks	Reference
Tobacco	*Homo sapiens*	*P450*	Atrazine	N-dealkylation of atrazine	Bode et al., 2004
Alfalafa	Gram-negative soil bacteria	*Atrazine cholorohydrolase*	Atrazine	Expression of gene leads to phytodegradation	Wang et al., 2005
Rice	Pig	*CYP2B22*	Herbicides	Transgenic	Kawahigashi et al., 2005b
Potato	*Homo sapiens*	*P450*	Sulfonylurea	Higher metabolism of pesticides	Inui., 2005
Tobacco	Maize	*GST 1*	Alachlor	Tolerance towards pesticide	Karavangeli et al., 2005
Tobacco	*Homo sapiens*	*P450*	Metamitron	Plant cell cultures metabolized	Bode et al., 2006

to the plasma membrane antibody. The density of the immune complex increased steadily for 72 h in the young leaves, only after the antigen was extracted from the nutrient broth. Free antigen in wild plant leave was not observed.

Furthermore, Drake et al. (2003) have demonstrated the secretion of a "functional," "full-length chimeric monoclonal matrix" from the genetically modified "*N. tabacum*" roots. Initially, seedlings on the "cellulose acetate membranes" and immune secretions from growing roots were found. Plants were then produced in the "hydroponic culture" over 25 days and estimated in the culture media. "Western blotting" revealed that, along with many other fragments, the concentration depended on the full-length antibody. By binding to the antigen, secreted antibodies have been shown to be active in "enzyme-linked immunosorbent assay (ELISA)" studies.

On the other side, from the recombinant nicotiana, where the same antibody was released as membrane protein in the plasmalemma, no antibody could be detected. According to the results obtained, antibody aggregation is likely caused by rhizosecretion in the cultural media and not by cellular damage. Applying gelatin to the seed germination environment significantly increased the development of antibodies, which meant that the antibodies per plant produced approximately 11.7 μg/g of dry root weight per day. These results strongly recognize that such antibody rhizosecretion can prove to be a novel phenomenon for the removal or neutralization of a toxic substrate in the genome.

Two "rhizosecretion-mediated bonding" and "leaf tissue sequestration" technologies may subsequently be used for the phytoremediation of contaminants, such as pesticides, for which an immune response can be produced, according to several findings. To date, no authentic laboratory findings on just this portion of soil insecticide and phytoremediation of water have been available, but this method may attract widespread interest in the short term.

16.9 Protection of the environment

Prevention of water pollution, the ultimate target for pesticides, is among the main current issues for consumers or authorities of pesticides when it comes to their effectiveness (Tingle et al., 1997). First of all, this consists of the use and careful spraying of better sprayers to prevent the environment from replacing pollutants. The formation of a firewall between sprinkled crops and marine environments is protection of property. This barrier is also a "vegetative filter strip" next to the fields in the water supply network. The second method of safety is the treatment of waste before being released into the water system. Plants can be used for the manipulation of unused industrial effluent spray left in pipes and rinsing water in areas of run-off or drift-interception.

16.9.1 Vegetative filter strips

The creation of "vegetative filter strips" involves the formation between the field crops and the water network of noncropped areas. The unregulated field avoids the treated area, away from groundwater. Next, surface water from drifting pesticides is protected. Second, the primary objective is to restrict the transfer of pesticides in solution to suspended solids or reabsorbed in run-off water. "Phytofiltration" is the philosophy of such facilities. An obstacle to run-off water is the root systems of plants, increasing the time taken to reach the drainage system.

This allows the primed suspended particles, increased water penetration, and decreased leachate to be repositioned. The surplus water that flows into the ground water from the vegetative strip has a reduced pesticide load than the discharge upstream of the strip. The effectiveness of such a device depends on the "pesticide" at issue; "conditions of the geopedological region," "width of the strip," "speed of the flow of penetration" as well as the "species of the plant" used. In addition to these strictly physical phenomena, phenomena of phytoaccumulation and rhizodegradation can also be associated and are causes that have resulted in a reduction of pesticide doses along with soil adsorbate (Mersie et al., 2003).

16.9.2 Constructed wetlands

Contaminated waste from crop spraying also needs environmental protection to be controlled (tank residues, water of sprayers, etc.). Such effluents have already been pumped into water environments, enabling them to purify pollutants (Williams, 2002). Different techniques, using physical and chemical methods (electro-Fenton oxidation, TiO_2) or biological methodologies to minimize effluent concentrations, have already been used to treat them (biobed). The design of engineered reservoirs in a managed and restricted area is a fascinating approach that recognizes the essential environmental process (Moshiri, 1993).

This method is also used to minimize mine waste runoff with water and in urban water treatment. Via a macrophyte-planted engineered river system, waste products flow. With pollutants reabsorbed to periphyton, these plants have been documented to have a filtering effect, reducing the amount of wastewater (Kadlec and Knight, 1996). Plants filter water, increase pollutant sedimentation, and provide habitat for pesticide-spreading microorganisms, as in the case of VFSS (Luckeydoo et al., 2002).

Moore et al. (2006) highlighted the significance of plants for the elimination of most "methyl-parathion" in a built wetland, while only minute quantities of this insecticide trapped in nonplanted sediment. Bulrush (*Scirpus validus)* increased simazine and metolachlor depletion by 30%–50% and achieved 90% purification reliability compared to bare wetlands (Stearman et al., 2003). The system has been shown to be effective, but its environment is necessary for the size of the waste to be handled. There is also a major risk of phytotoxicity due to "herbicide contamination" or "accumulation" of some products, such as Cu.

16.10 Conclusion and future aspects

Pesticide pollution is a big concern for the environment and global protection. Vitrification and other chemical methods are commonly used in polluted areas, while efficient and funded, low-cost alternatives are bioremediation and phytoremediation. The soil is home to a variety of microbial organisms due to extensive "metabolic activities" capable of destroying various types of pesticides. This microbial metabolic ability for in-situ bioremediation techniques could be used. The integration of "catabolic genes" and the "synthesis of appropriate pesticide-degrading mechanisms" in indigenous microflora have expanded the possible use of bioremediation to purge polluted ecosystems.

In different plant tissues, plants can also eliminate, consume, convert, and retain pesticides. In the production of pesticide-contaminated environments and other organic contaminants, recent developments in genetic engineering have made this technique more effective and sustainable. Results indicate that bioremediation and phytoremediation can be used successfully and that pesticide-contaminated sites are currently being considered or used to remediate.

Plants can also eliminate, absorb, convert, and maintain pesticides in various plant tissues. Recent furtherance in genetic engineering has made this technique more efficient and sustainable in the generation of pesticide-contaminated habitats and other organic pollutants. The findings of the study addressed that it is possible to effectively use bioremediation and phytoremediation, and that pesticide-contaminated sites are actually being considered or used for remediation. There are various scenarios for a polluted site, so all site-specific considerations must be taken into consideration, and decisions must be taken as part of the most appropriate technologies available.

To improve our understanding of the consequences and constraints of these innovative techniques, several researchers are conducting pilot-scale phytoremediation and bioremediation research. Further research is based on understanding the capacities of microorganisms under various climatic conditions to expand the use and successful application of bioremediation and phytoremediation. This will help improve the design of built structures for contaminated land remediation. To achieve this objective, efforts should be made to expand work on soil-plant-microbe-pollutant interactions to turn "bench-and pilot-scale" results accurately into "field-scape" results.

References

Abhilash, P.C., Singh, N., 2009. Pesticide use and application: an Indian scenario. J. Hazard. Mater. 165, 1–12.

Ahmad, F., Iqbal, S., Anwar, S., Afzal, M., Islam, E., Mustafa, T., Khan, Q.M., 2012. Enhanced remediation of chlorpyrifos from soil using ryegrass (*Lollium multiflonum*) and chlorpyrifos –degrading bacterium *Bacillus pumilus* C2A1. J. Hazard. Mater., 110–115.

Aigner, E.J., Leone, A.D., Falconer, R.L., 1998. Concentration and enantiomeric ratios of organochlorine pesticides in soil from U.S. corn belt. Environ. Sci. Technol. 32, 1162–1168.

Aliferis, K.A., Jabaji, S., 2011. Metabolomics – a robust bioanalytical approach for the discovery of the modes-of-action of pesticides: a review. Pestic. Biochem. Physiol. 100, 105–117.

Anderson, T.A., Coats, J.R., 1995. Screening of rhizosphere soil samples for the ability to mineralize elevated concentrations of atrazine and metolachlor. J. Environ. Sci. Health B 30, 7473–7484.

Barriuso, E., Houot, S., 1996. Rapid mineralization of the S-triazine ring of atrazine in soils in relation to soil management. Soil Biol. Biochem. 28, 1341–1348.

Bednarska, A.J., Choczynski, M., Laskowski, R., Walczak, M., 2017. Combined effects of chlorpyriphos, copper and temperature on acetylcholinesterase activity and toxicokinetics of the chemicals in the earthworm Eisenia fetida. Environ. Pollut. 220, 567–576.

Bocos, E., Fernandez-Costas, C., Pazos, M., Sanroman, M.A., 2015. Removal of PAHs and pesticides from polluted soils by enhanced electrokinetic-Fenton treatment. Chemosphere 125, 168–174.

Bode, M., Haas, M., Faymonville, T., Thiede, B., Schuphan, I., Schmidt, B., 2006. Biotransformation of metamitron by human P450 expressed in transgenic tobacco cell cultures. J. Environ. Sci. Health 41, 201–222.

Bode, M., Stobe, P., Thiede, B., Schuphan, I., Schmidt, B., 2004. Biotransformation of atrazine in transgenic tobacco cell culture expressing human P450. Pest Manag. Sci. 60, 49–58.

Buckingham, S.D., Ihara, M., Sattelle, D.B., Matsuda, K., 2017. Mechanisms of action, resistance and toxicity of insecticides targeting GABA receptors. Curr. Med. Chem. 24, 27.

Camargo, J.A., Alonso, A., Salamanca, A., 2005. Nitrate toxicity to aquatic animals: a review with new data for freshwater invertebrates. Chemosphere 58, 1255–1267.

Capdevila, S., Martinez-Granero, F.M., Sanchez-Contreras, M., Rivilla, R., Martin, M, 2004. Analysis of *Pseudomonas fluorescens* F113 genes implicated in flagellar filament synthesis and their role in competitive root colonization. Microbiology 150, 3889–3897.

Cordova, D., Benner, E.A., Schroeder, M.E., Holyoke, C.W., Zhang, W., Pahutski, T.F., et al., 2016. Mode of action of triflumezopyrim: a novel mesoionic insecticide which inhibits the nicotinic acetylcholine receptor. Insect Biochem. Mol. Biol. 74, 32–41.

Davoli, E., Benfenati, E., Bagnati, R., Fanelli, R., 1987. Analysis of atrazine in underground waters at part per trillion levels as an early warning method for contamination and for soil distribution studies. Chemosphere 16, 1425–1430.

de Neve, M., de Loose, M., Jacobs, A., Van Houdt, H, Kaluza, B., Weidle, U., Depicker, A., 1993. Assembly of an antibody and its derived antibody fragment in *Nicotiana* and *Arabidopsis*. Transgenic Res. 2, 227–237.

Deadman, M.L., 2017. Sources of pesticide residues in food: Toxicity, exposure, and risk associated with use at the farm level. In Pesticide residue in foods. Cham: Springer International Publishing, pp. 7–35.

Deng, S.Y., Chen, Y., Wang, D.S., Shi, T.Z., Wu, X.W., Ma, X., Li, Q.X., 2015. Rapid biodegradation of organophosphorous pesticides by *Stenotrophomonas* sp.G1. J. Hazard. Mater. 297, 17–24.

Douglass, J.F., Radosevich, M., Tuovinen, O.H., 2016. Biomineralization of atrazine and analysis of 16S rRNA and catabolic genes of atrazine degraders in a former pesticide mixing site and a machinery washing area. J. Soils Sediment 16, 1–12.

Drake, P.M.W., Chargelegue, D.M., Vine, N.D., van Dolleweerd, C.J., Obregon, P., Ma, J.K., 2002. Transgenic plants expressing antibodies: a model for phytoremediation. FASEB J. 16, 1855–1860.

Drake, P.M.W., Chargelegue, D.M., Vine, N.D., van Dolleweerd, C.J., Obregon, P., Ma, J.K., 2003. Rhizosecretion of a monoclonal antibody protein complex from transgenic tobacco roots. Plant Mol. Biol. 52, 233–241.

Eapen, S., Singh, S., D'Souza, S.F, 2007. Advances in development of transgenic plants for remediation of xenobiotic pollutants. Biotechnol. Adv. 25, 442–451.

Edao, H.G., 2017. Heavy metals pollution of soil; toxicity and phytoremediation. Int. J. Adv. Res. Publ. 1, 29–41.

Edwards, R., Owen, W., 1989. The comparative metabolism of the s-triazine herbicides atrazine and terbutryne in suspension cultures of potato and wheat. Pestic. Biochem. Physiol. 34, 246–254.

Frazar, C., 2000. The Bioremediation and Phytoremediation of Pesticide Contaminated Sites. National Network of Environmental studies (NNEMS), Washington.

Fukushima, M., Tatsumi, K., 2001. Functionalities of humic acid for the remedial processes of organic pollutants. Anal. Sci. 17, 821–823.

Garmouna, M., Teil, M.J., Blanchard, M., Chevreuil, M., 1998. Spatial and temporal variations of herbicide (triazines and phenylureas) concentrations in the catchment basin of the Marne river (France). Sci. Total Environ. 224, 93–107.

Gavrilescu, M., 2005. Fate of pesticides in the environment and its bioremediation. Eng. Life Sci. 5, 497 -452.

Gerhardt, K.E., Huang, X.D., Glick, B., Greenberg, B.M., 2009. Phytoremediation and rhizoremediation of organic soil contaminants. potential and challenges. Plant Sci. 179, 20–30.

Giusti, A., Pirard, C., Charlier, C., Petit, J.C.J., Crevecoeur, S., Remy, S, 2018. Selection and ranking method for currently used pesticides (CUPs) monitoring in ambient air. Air Qual. Atmos. Hlth, 1–12.

Glick, B., 2003. Phytoremediation: synergistic use of plants and bacteria to clean up the environment. Biotechnol. Adv. 21, 383–393.

Glick, B., 2010. Using soil bacteria to facilitate phytoremediation. Biotechnol. Adv. 21, 367–374.

Gomes, H.I., Fan, G., Ottosen, L.M., Dias-Ferreira, C., Ribeiro, A.B., 2016. Nanoremediation Coupled to Electrokinetics for PCB Removal from Soil. Springer International Publishing, NY.

Gonzalez, M., Miglioranza, K.S.B., Aizpun, de Moreno, J.E., Moreno, V.J., 2005. Evaluation of conventionally and organically produced vegetables for high lipophilic organochlorine pesticide (OCP) residues. Food Chem. Toxicol. 43, 261–269.

Gustafson, D .I., 1989. Groundwater ubiquity score: a simple method for assessing pesticide leachability. Environ. Toxicol. Chem. 8, 339–357.

Heard, M.S., Baas, J., Dorne, J.L., Lahive, E., Robinson, A.G., Rortais, A., Hesketh, H., 2017. Comparative toxicity of pesticides and environmental contaminants in bees: are honey bees a useful proxy for wild bee species? Sci. Total Environ. 578, 357–365.

Hiatt, A., Cafferkey, R., Bowdish, K., 1989. Production of antibodies in transgenic plants. Nature 342, 76–78.

Horemans, B., Vandermaesen, J., Vanhaecke, L., Smolders, E., Springael, D, 2013. Variovorax sp.-mediated biodegradation of the phenyl urea herbicide linoron at micro-pollutant concentrations and effects of natural dissolved organic matter as supplementary carbon source. Appl. Microbiol. Biotechnol. 97, 9837–9846.

Houot, S., Topp, E., Yassir, A., Soulas, G., 2000. Dependence of accelerated degradation of atrazine on soil pH in French and Canadian soils. Soil Biol. Biochem. 32, 615–625.

Huang, X., He, J., Yan, X., Hong, Q., Chen, K., He, Q., et al., 2016. Microbial catabolism of chemical herbicides: microbial resources, metabolic pathways and catabolic genes. Pestic. Biochem. Physiol. 143, 272–297.

Hussain, S., Siddique, T., Arshad, M., Saleem, M., 2009. Bioremediation and phytoremediation of pesticides: recent advances. Crit. Rev. Environ. Sci. Technol. 39, 843–907.

Inui, H., 2005. Herbicide resistance in transgenic plants with mammalian P450 monooxygenase genes. Pest Manag. Sci. 61, 286–291.

Ismail, B.S., Mazlinda, M., Tayeb, M.A., 2015. The persistence of deltamethrin in Malaysian agricultural soils. Sains Malays. 44, 83–90.

Jauregui, J., Valderrama, B., Albores, A., Vazquez-Duhalt, R., 2003. Microsomal transformation of organophosphorus pesticides by white rot fungi. Biodegradation, 14, 397e406.

Kadlec, R.H., Knight, R.L., 1996. Treatment Wetlands. CRC, Boca Raton, FL.

Karavangeli, M., Labrou, N.E., Clonis, Y.D., Tsaftaris, A., 2005. Development of transgenic tobacco plants overexpressing maize glutathione S-transferase I for chloroacetanilide herbicides phytoremediation. Biomol. Eng. 22, 121–128.

Karunya, S.K., Saranraj, P., 2014. Toxic effects of pesticide pollution and its biological control by microorganisms: a review. Appl. J. Hyg. 3, 1–10.

Kawahigashi, H., Hirose, S., Inui, H., Ohkawa, H., Ohkawa, Y., 2005b. Enhanced herbicide cross-tolerance in transgenic rice plants co-expressing human CYP1A1, CYP2B6, and CYP2C19. Plant Sci. 168, 773–781.

Kawahigashi, H., Hirose, S., Ohkawa, H., Ohkawa, Y., 2003. Transgenic rice plants expressing human CYP1A1 exude herbicide metabolites from their roots. Plant Sci. 165, 373–380.

Khalid, S., Hashmi, I., 2016. Biotreatment of chlorpyrifos in a bench scale bioreactor using *Psychrobacter alimentarius* T14. Environ. Technol. 37, 316–325.

Khrunyk, Y., Schiewer, S., Carstens, K.L., Hu, D., Coats, J.R., 2017. Uptake of C(14)-atrazine by prairie grasses in a phytoremediation setting. Int. J. Phytoremediation 19, 104.

Kolekar, D.P., Phugare, S.S., Jadhav, P.J., 2014. Biodegradation of atrazine by *Rhodococcus* sp. BCH2 to N-isopropylammelide with subsequent assessment of toxicity of biodegraded metabolites. Environ. Sci. Pollut. Res. 21, 2334–2345.

L'hirondel, J.L., Avery, A.A., Addiscott, T., 2006. Dietary nitrate: where is the risk? Environ. Health Perspect. 114, 458–459.

Lassaletta, L., Billen, G., Grizzetti, B., Anglade, J., Garnier, J., 2014. 50 year trends in nitrogen use efficiency of world cropping systems: the relationship between yield and nitrogen input to cropland. Environ. Res. Lett. 9, 105011.

Leigh, M.B., Prouzova, P., Mackov ´ a, M., Macek, T., Nagle, D.P., Fletcher, J.S., 2006. Polychlorinated biphenyl (PCB)-degrading bacteria associated with trees in a PCB-contaminated site. Appl. Environ. Microbiol. 72, 2331–2342.

Lewandowski, I., Schmidt, U., Londo, M., Faaij, A., 2006. The economic value of the phytoremediation function—assessed by the example of cadmium remediation by willow (*Salix* ssp). Agric. Syst. 89, 68–89.

Li, J, Wang, C, Wang, D, Zhou, Z, Sun, H, Zhai, S., 2016a. novel technology for remediation of PBDEs contaminated soils using tourmaline-catalyzed Fenton-like oxidation combined with *P. chrysosporium*. Chem. Eng. J. 296, 319–328.

Lin, C.H., Lerch, R.N., Garrett, H.E., George, M.F., 2008. Bioremediation of atrazine- contaminated soil by forage grasses: transformation, uptake, and detoxification. J. Environ. Qual. 37, 196–206.

Liu, Y., Li, S., Ni, Z., Qu, M., Zhong, D., Ye, C., Tang, F., 2016. Pesticides in persimmons, jujubes and soil from China: residue levels, risk assessment and relationship between fruits and soils. Sci. Total Environ. 542, 620–628.

Lozowicka, B., Jankowska, M., Hrynko, I., Kaczynski, P., 2016. Removal of 16 pesticide residues from strawberries by washing with tap and ozone water, ultrasonic cleaning and boiling. Environ. Monit. Assess. 188, 1–19.

Luckeydoo, L.M., Fausey, N.R., Brown, L.C., Davis, C.B., 2002. Early development of vascular vegetation of constructed wetlands in northwest Ohio receiving agricultural waters. Agric. Ecosyst. Environ. 88, 89–94.

Lunney, A.I., Rutter, A., Zeeb, B.A., 2010. Effect of organic matter additions on uptake of weathered DDT by *Cucurbita pepo* ssp.pepo cv. Howden. Int. J. Phytoremediation 12, 404–417.

Lunney, A.I., Zeeb, B.A., Reimer, K.J., 2004. Uptake of DDT wheathered in vascular plants: potential for phytoremediation. Environ. Sci. Technol. 38, 6147–6154.

Ma, X.H., Zhao, L., Lin, Z.R., Dong, Y.H., 2016. Soil washing in combination with homogeneous Fenton-like oxidation for the removal of 2,4, 4′-trichlorodiphenyl from soil contaminated with capacitor oil. Environ. Sci. Pollut. Res. Int. 23, 7890–7898.

Ma, J.K., Hiatt, A., Hein, M., Vine, N.D., Wang, F., Stabila, P., van Dolleweerd, C., Mostov, K., Lehner, T., 1995. Generation and assembly of secretory antibodies in plants. Science 268, 716–719.

Mao, Y., Sun, M., Yang, X., Wei, H., Song, Y., Xin, J., 2013. Remediation of organochlorine pesticides (OCPs) contaminated soil by successive hydroxypropyl-β-cyclodextrin and peanut oil enhanced soil washing–nutrient addition: a laboratory evaluation. J. Soils Sediment 13, 403–412.

Mattina, M.J.I., Iannucci-Berger, W., Dykas, L, 2000. Chlordane uptake and its translocation in food crops. J. Agric. Food Chem. 48, 1909–1915.

May, M.J., Vernoux, T., Sanchez-Fernandez, R., Van, M.M., Inze, D, 1998. Evidence for posttranscriptional activation of γ -glutamylcysteine synthetase during plant stress responses. Proc. Natl. Acad. Sci. USA 95, 12049–12054.

Meharg, A.A., Cairney, J.W.G., 2016. Ectomycorrhizas—extending the capabilities of rhizosphere remediation? Soil Biol. Biochem. 32, 475–484.

Mersie, W., Seybold, C.A., McNamee, C.M., Huang, J., 2003. Abating endosulfan from runoff using vegetative filter strips: the importance of plant species and flow rate. Agric. Ecosyst. Environ. 97, 215–223.

Mesquini, J.A., Sawaya., A.C.H.F., Lopez, B.G.C., Oliveira, V.M, Miyasaki, N.R.S, 2015. Detoxification of atrazine by endophytic Streptomyces sp. isolated from sugarcane and detection of nontoxic metabolite. Bull. Environ. Contamin. Toxicol. 95, 803–809.

Mitton, F.M., Gonzalez, M., Monserrat, J.M., Miglioranza, K.S.B., 2016. Potential use of edible crops in the phytoremediation of endosulfan residues in soil. Chemosphere 148, 300–306.

Moore, M.T., Bennett, E.R., Cooper, C.M., Smith Jr., S., Farris, J.L., Drouillard, K.G., Schulz, R., 2006. Influence of vegetation in mitigation of methyl parathion runoff. Environ. Pollut. 142, 288–294.

Moshiri, G.A. (Ed.), 1993. Constructed Wetlands for Water Quality Improvement. Lewis Publishers, Boca Raton, FL.

Moss, B., 2008. Water pollution by agriculture. Philos. Trans. R. Soc. B 363, 659–666.

Newete, S.W., Byrne, M.J., 2016. The capacity of aquatic macrophytes for phytoremediation and their disposal with specific reference to water hyacinth. Environ. Sci. Pollut. Res. 23, 10630–10643.

Ozkara, A., Akyıl, D., Konuk, M., 2016. Pesticides, environmental pollution, and health. In: Larramendy, M.L., Soloneski, S. (Eds.), Environmental Health Risk-Hazardous Factors to Living Species. IntechOpen, London, UK.

Pascal-Lorber, S., Laurent, F., 2011. Phytoremediation techniques for pesticide contaminations. In: Lichtfouse, E. (Ed.), Alternative Fanning Systems, Biotechnology, Drought Stress and Ecological Fertilization. Springer.

Pestana, I.A., Meneguelli-Souza, A.C., Gomes, M.A.C., Almeida, M.G., Suzuki, M.S., Vitoria, A.P., Souza, C.M, 2018. Effects of a combined use of macronutrients nitrate, ammonium, and phosphate on cadmium absorption by *Egeria densa* Planch and its phytoremediation applicability. Aquat. Ecol. 52, 51–64.

Peuke, A.D., Rennenberg, H., 2005b. Phytoremediation with transgenic trees. Z. Naturforsch. 60, 199–207.

Qiu, X.H., Zhu, T., Li, J., Pan, H.S., Li, Q.L., Miao, G.F., Gong, G.C., 2004. Organochlorine pesticides in the air around the Taihu Lake, China. Environ. Sci. Technol. 38, 1368–1374.

Rani, M., Shanker, U., Jassal, V., 2017. Recent strategies for removal and degradation of persistent and toxic organochlorine pesticides using nanoparticles: a review. J. Environ. Manag. 190, 208–222.

Raveton, M., Ravanel, P., Kaouadji, M., Bastide, J., Tissut, M., 1997. The chemical transformation of atrazine in corn seedlings. Pestic. Biochem. Physiol. 58, 199–208.

Risco, C., Rodrigo, S., Lopez-Vizcaino, R., Saez, C., Villasenor, J., Navarro, V., et al., 2016. Electrokinetic Remediation of Soils Polluted with Pesticides: Flushing and Fence Technologies. In Proc. Geo-Chicago, Chicago, IL.

Risco, C., Rodrigo, S., Lopez-Vizcaino, R., Yustres, A., Saez, C., Canizares, P., et al., 2015. Electrochemically assisted fences for the electroremediation of soils polluted with 2, 4-D: a case study in a pilot plant. Sep. Purif. Technol. 156, 234–241.

Rissato, S.R., Galhiane, M.S., Fernandes, J.R., Gerenutti, M., Gomes, H.M., Ribeiro, R., Vinicius, de, Almeida, M, 2015. Evaluation of *Ricinus communis* L for the phytoremediation of polluted soil with organochlorine pesticides. Biomed Res. Int., 1–8.

Rodrigues-Lima, F., Dairou, J., Diaz, C.L., Rubio, M.C., Sim, E., Spaink, H.P., Dupret, J.M., 2006. Cloning, functional expression and characterization of Mesorhizobium loti arylamine N-acetyltransferases: rhizobial symbiosis supplies leguminous plants with the xenobiotic N-acetylation pathway. Mol. Microbiol. 60, 505–512.

Romeh, A.A., 2014. Phytoremediation of cyanophos insecticide by *Plantago major* L. in water. J. Environ. Health Sci. Eng. 12, 8.

Savci, S., 2012. An agricultural pollutant: chemical fertilizer. Int. J. Environ. Sci. Dev. 3, 77–80.

Sharpe, R., 1995. Reproductive biology: another DDT connection. Nature 375, 538–539.

Silva, V.P., Moreirasantos, M., Mateus, C., Teixeira, T., Rui, R., Viegas, C.A., 2015. Evaluation of *Arthrobacter aurescens* strain TC1 as bioaugmentation bacterium in soils contaminated with the herbicidal substance terbuthylazine. PLoS One 10, e0144978.

Singh, V., Singh, P., Singh, N., 2016. Synergistic influence of *Vetiveria zizanioides* and selected rhizospheric microbial strains on remediation of endosulfan contaminated soil. Ecotoxicology 25, 1–11.

Singh, B., Mandal, K, 2013. Environmental impact of pesticides belonging to newer chemistry. In: Dhawan, A.K., Singh, B., Brar-Bhullar, M., Arora, R. (Eds.), Integrated Pest Management. Scientific Publishers, Jodhpur, pp. 152–190.

Somtrakoon, K., Krautrachue, M., Lee, H., 2014. Phytoremediation of endosulfan sulfate-contaminated soil by single and mixed plant cultivations. Water Air Soil Pollut. 225, 1886.

Sonoki, T., Kajita, S., Ikeda, S., Uesugi, M., Tatsumi, K., Katayama, Y., Iimura, Y., 2005. Transgenic tobacco expressing fungal laccase promotes the detoxification of environmental pollutants. Appl. Microbiol. Biotechnol. 67, 138–142.

Souza, F.L., Saez, C., Llanos, J., Lanza, M.R.V., Canizares, P., Rodrigo, M.A, 2016. Solar-powered electrokinetic remediation for the treatment of soil polluted with the herbicide 2,4-D. Electrochim. Acta 190, 371–377.

Sparks, T.C., Nauen, R., 2015. IRAC: mode of action classification and insecticide resistance management. Pestic. Biochem. Physiol. 121, 122–128.

Stearman, G.K., George, D.B., Carlson, K., Lansford, S., 2003. Pesticide removal from container nursery runoff in constructed wetland cells. J. Environ. Qual. 32, 1548–1556.

Strickland, J., Clippinger, A.J., Brown, J., Allen, D., Jacobs, A., Matheson, J., Casey, W., 2018. Status of acute systemic toxicity testing requirements and data uses by U.S. regulatory agencies. Regul. Toxicol. Pharmacol. 94, 183–196.

Tingle, C.H., Shaw, D.R., Boyette, M., Palmer, E.W., 1997. Utilizing vegetative filter strips of varying widths to reduce herbicides in runoff water. Proc. South Weed Science Society Meet, Houston, TX 27.

Trellu, C., Ganzenko, O., Papirio, S., Pechaud, Y., Oturan, N., Huguenot, D., et al., 2016. Combination of anodic oxidation and biological treatment for the removal of phenanthrene and tween 80 from soil washing solution. Chem. Eng. J. 306, 588–596.

Vacondio, B., Birolli, W.G., Ferreira, I.M., Seleghim, M.H.R., Goncalves, S., Vasconcellos, S.P., Porto, A.L.M, 2015. Biodegradation of pentachlorophenol by marine derived fungus *Trichoderma harziam* CBMAI 1677 isolated from ascidian Didemmum ligulum. Biocatal. Agric. Biotechnol. 4, 266–275.

Velki, M., Ecimovic, S., 2015. Changes in exposure temperature lead to changes in pesticide toxicity to earthworms: a preliminary study. Environ. Toxicol. Pharmacol. 40, 774–784.

Vieira, D.S.E., Souza, F., Saez, C., Canizares, P., Lanza, M.R., Martinezhuitle, C.A., et al., 2016. Application of electrokinetic soil flushing to four herbicides: a comparison. Chemosphere 153, 205–211.

Vine, N.D., Drake, P., Hiatt, A., Ma, J.K., 2001. Assembly and plasma membrane targeting of recombinant immunoglobulin chains in plants with a murine immunoglobulin transmembrane sequence. Plant Mol. Biol. 45, 159–167.

Wang, F.Y., Tong, R.J., Shi, Z.Y., Xu, X.F., He, X.H., 2011. Inoculations with arbuscular mycorrhizal fungi increase vegetable yields and decrease phoxim concentrations in carrot and green onion and their soils. Plos One 6, e16949.

Wang, Q., Zhang, W.X, Li, C., Xiao, B., 2012. Phytoremediation of atrazine by three emergent hydrophytes in a hydroponic system. Water Sci. Technol. 66, 1282–1288.

Wang, L., Samac, D.A., Shapir, N., Wackett, L.P., Vance, C.P., Olszewski, N.E., Sadowsky, M.J., 2005. Biodegradation of atrazine in transgenic plants expressing a modified bacterial atrazine chlorohydrolase (atzA) gene. Plant Biotechnol. J. 3, 475–486.

Wang, X., Wu, N., Guo, J., Chu, X., Tian, J., Yao, B., Fan, Y, 2008. Phytodegradation of organophosphorus compounds by transgenic plants expressing a bacterial organophosphorus hydrolase. Biochem. Biophys. Res. Commun. 365, 453–458.

White, J.C., 2010. Inheritance of p,p'-DDE phytoextraction ability in hybridized Cucurbita pepo cultivars. Environ. Sci. Technol. 44, 5165–5169.

White, J.C., Parrish, Z., Isleyen, M., Gent, M., Iannucci – Berger, W., Eitzer, B., Mattina, M., 2005. Uptake of weathered p,p'-DDE by plant species effective at accumulating soil elements. Microchem. J. 81, 148–155.

Williams, J.B., 2002. Phytoremediation in wetland ecosystems: progress, problems, and potential. Crit. Rev. Plant Sci. 21, 607–635.

World Health Organization, 2010. The WHO Recommended Classification of Pesticides by Hazard and Guidelines to Classification 2009. WHO Press, Geneva http://www.who.int/ipcs/publications/pesticides_hazard_2009.pdf?ua=1.

Xiao, P.F., Kondo, R., 2013. Biodegradation of dieldrin by Cordyceps fungi and detection of metabolites. Pts 1-4. In: Tang, X.Y., Zhong, W., D.Zhuang, C.Li, Liu, Y. (Eds.), Progress in Environmental Protection and Processing of Resource. Guangdong University, Guangzhou, China, pp. 30–34.

Xu, X.J., Sun, J.Q., Nie, Y., Wu, X.L., 2015. *Spirodela polyrhiza* stimulates the growth of its endophytes but differentially increases their fenpropathrin-degradation capabilities. Chemosphere 125, 33–40.

Zazai, K.G., Wani, O.A., Ali, A., Devi, M., 2018. Phytoremediation and carbon sequestration potential of agroforestry systems: a review. Int. J. Curr. Microbiol. Appl Sci. 7, 2447–2457.

Zhou, Y., Tigane, T., Li, X., Truu, M., Truu, J., Mander, U., 2013. Hexachlorobenzene dechlorination in constructed wetland mecocosms. Water Res. 47, 102–110.

Chapter 17

Nature sucks up explosives

Tanveer Bilal Pirzadah[a], Bisma Malik[a], Khalid Rehman Hakeem[b], Seema Rashid[c]

[a]*University Centre for Research and Development (UCRD), Chandigarh University, Mohali, Punjab, India,* [b]*Department of Biological Sciences, Faculty of Science, King Abdulaziz University, Jeddah, Saudi Arabia,* [c]*Department of Zoology, Amar Singh College, Cluster University, Srinagar, India*

17.1 Introduction

The widespread utilization of munitions globally (explosive compounds) for various applications such as in military ventures, mining, and several industrial endeavors polluted vast area of land, water, and soil and posed health effects due to toxic explosive compounds (Kalderis et al., 2011; Pichtel, 2012; Klodenko et al., 2014). Numerous industrial ventures and time of war endeavors contribute a large portion of contaminants into the environment due to the presence of explosive and the utilization, manufacture, and explosive decommissioning are the largest contributors of this pollution (Just and Schnoor, 2004; Pichtel, 2012; Certini et al., 2013). Sites of production and training centers act as the main areas of contamination due to explosive and along with their byproducts as they have the capability to penetrate deep into the groundwater and ultimately results in the contamination of the total environment. There was history that described the dependency of military actions on the utilization of explosives to strike and safeguard a particular part of the region. Large soil pollutants are located and detected near artillery ranges, ranges, army depots, ammunition evaluation facilities including ordnance disposal sites (composition B), which is usually utilized in the military preparations and is usually consisted of dangerous explosive compounds such as TNT (2,4,6,-trinitrotoluene), HMX (octa hydro-1,3,5,7-teranitro-1,3,5,7-tetrazocine), and RDX (hexahydro-1,3,5-trinitro-1,3,5-triazine) (Douglas et al., 2012). Nowadays, pollution due to explosives is a matter of concern due to their resistance to degradation, toxicity, and complicated health issues results in the overall degradation of environment. For example in United States, the Defense Science Board had identified some closed areas and offshore sites and stated that there could be not less than 15 million acres (6 million hectares) polluted with explosive compounds (Douglas et al., 2012). Other than United States, there are a huge number of polluted sites/areas

Phytoremediation. DOI: https://doi.org/10.1016/B978-0-323-89874-4.00015-7

all over Asia, Africa, and Europe where restricted disputes result in the burying large-scale ordnance in the form of land mines (Kalderis et al., 2011). Apart from this, munitions contain reactive compounds that display their effectiveness as a tool of armies and utilization of gunpowder in battle field dates back in 969 CE in China (Kelly 2004). Currently research has progressively focused on producing environmentally feasible, reliable, and cost-effective rectification approaches (in-situ) and self-cleaning preparations for explosive pollutants. In this approach, one of the important challenges in producing fruitful bioengineered in-situ rectification method lies in the restricted bioavailability of TNT, emerging from the hydrophobicity property of the nitro-aromatic compounds (at 25 °C, solubility of TNT is 101.5 mg/L) (Ro et al., 1996; Rylott and Bruce, 2009). In this context, bioaugmentation utilization provides promising results as this approach exploited the capability of microorganism to degrade different types of xenobiotics; however some research work described that native microbes usually defeat the nonnative microbes, which ultimately leads in the limited achievements (Hannink et al., 2002). Phytoremediation technique (plant-assisted breakdown of organic/inorganic pollutants) provides cost-effective and environmental friendly technique for the degradation of xenobiotic compounds particularly explosives. Some particular plants have evolved a complex degradation mechanism to combat explosive contamination (Douglas et al., 2012) and permit them to withstand in sites polluted with higher concentrations of these pollutants. The plant root system uptakes the toxic compounds (explosives) and simultaneously maintains the soil health and limits the release of polluted dust in the atmosphere (Doty, 2008). Moreover plants provide additional advantages such as they provide nutrients to the rhizospheric bacteria which may symbiotically help in the detoxification mechanism as well as help in the observation of areas simply through collection of tissues and visualization. One of the major limitations of phytoremediation technique is that this process is tedious, as sites polluted with explosives are useless and such pollution commonly expands over a large area of land hence phytoremediation technology could play a critical role in the remediation process of sites polluted with explosive compounds (Rylott and Bruce, 2009). This chapter highlights the phytoremediation process of explosives and how genetic engineering of plants could play an important role in remediating the sites contaminated with explosive compounds.

17.2 Sources and types of explosive compounds

There are numerous sources of explosive compounds that enter into the environment through different ways in which military activities and industrial applications contributed the majority of this pollution. The whole munition approach from development, storage, transportation, distribution, and dismantling can release a number of explosive compounds either in pure solid or liquid forms into the environment through leaching, pollutant spills, trace particles, exploded and unexploded ordnance (UXOs, partially or whole) including discarded UXOs that ranges from bombs to landmines to mortar sounds (Fig. 17.1). Explosive substances either means the pollutants left during an explosion event or the device contained explosive compounds that did not completely get destroyed after explosion since it has been found that trace particles, or residues even whole or partial portion of the munition will remain in the surrounding sites resulting in the release of these harmful compounds in the environment. This confirmed that the areas far away from the actual munition site or conflict zone possessed higher amounts of pollutants. Industrial scale mining is the largest contributor of explosive pollution than the military activities and in the past, heavy mining utilized explosives in large scale and after having shifted away they are still utilized today (Kholodenko et al., 2014). Previously or currently, the functioning of mining operations is still having the potentiality to release a heavy amount of explosive compounds in the environment today (Dick et al., 1983; Tiwary, 2001). However there was a similarity between the pollutants released from mining operations and munition events, although partial device explosives play a lead role in the pollution process (Tiwary, 2001). Moreover, some other sources of explosives such as lacquers, fireworks, paints, personal firearms and dyes also contributed to a less degree of pollution (Hamilton, 1921; Almog and Zitrin 2009).

Traditionally there are three categories of explosive compounds based on the differences in the structure of the compounds associated with it. These

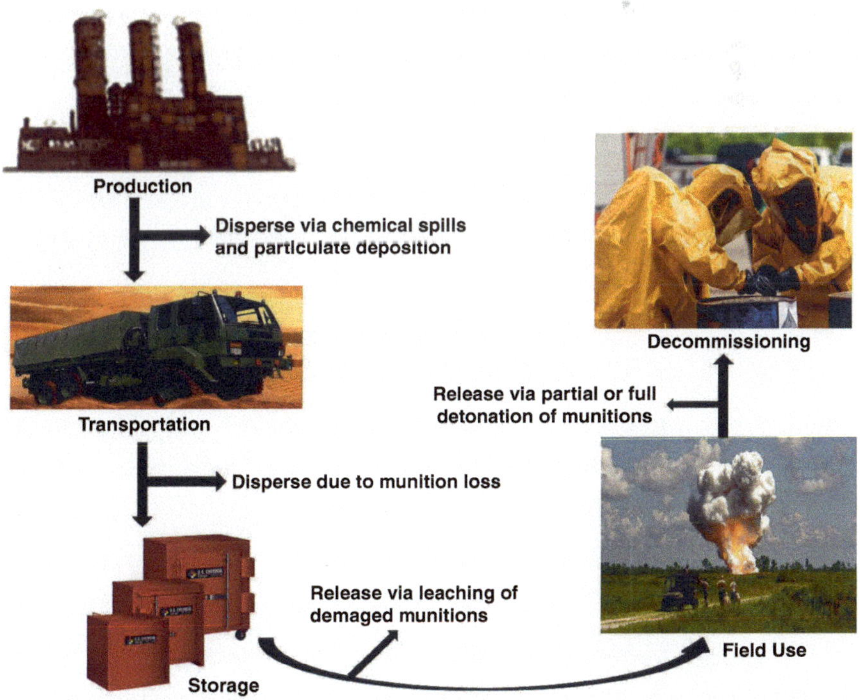

FIG. 17.1 Different sources of explosive contamination.

are as follows: (1) nitroaromatics, (2) nitroamine, and (3) nitrate esters (Fig. 17.2). The characterization of nitroaromatics is done on the basis of aromatic ring in the center and nitro groups connected at different points. TNT among them is the only one that has noticed huge range and everlasting applications. 2, 4-Dinitrotoluene (DNT) and 2-aminodinitrotoluene (ADNT) are the most common other types of nitroaromatic compounds produced during the process of degradation of TNT and are accordingly present in the polluted sites. In addition, nitroamines are a group of minute active compounds consisted of a heterocyclic ring in the center attached with N-nitro group and among them HMX and RDX are the extensively used nitroamines. Besides, nitrate esters are the esters of nitric acid and alcohol and are the smallest harmful explosive compounds utilized till date and the common examples are GTN (glyceryl trinitrate) and PETN (pentaerythritol tetranitrate). Unresponsive munition compounds are usually consisted of heteroaromatic or central aromatic structure with an elevated degree of hydrogen bonding, reduced oxygen stability, and have unconfined area due to effective packing of crystals (Pagoria, 2016), thereby decreasing reactivity to activate the stimuli and directly decrease the human risk during developing, collecting, shipping, and transporting munition compounds compared to others. Recently these compounds drew attention of scientific community toward the ecotoxicological investigations and this approach proves to be a great success as far as human safety is concerned

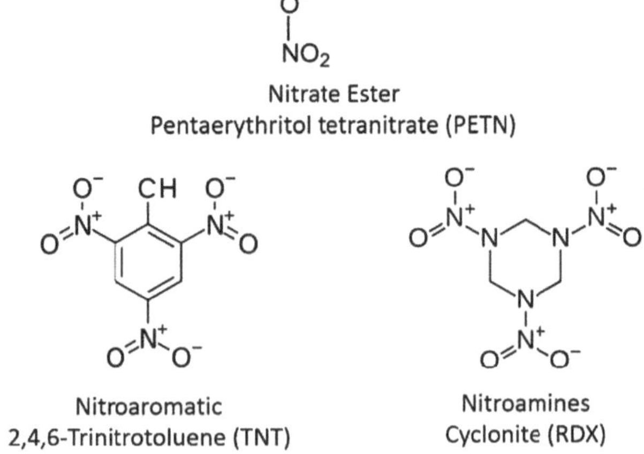

FIG. 17.2 Three categories of explosive compounds based on the differences in the structure of the compounds associated with it.

(Dontsova et al., 2014; Madeira et al., 2018). Currently the most important progress in the field of explosives is the development of "green" explosive compounds, which ultimately results in lowering of environmental risk and health effects. Currently in the environment, nitroamines and nitroaromatics comprised the huge part of explosives due to extended industrial dependence and extensive utilization in critical conflicts of military in the last century (Via and Zinnert, 2016). Munitions containing RDX and TNT in specific amounts have covered the largest part of utilization till date and have assembled much of the attention in the information cited.

17.3 Environmental behavior of explosive compounds

Prediction of environmental behavior of explosive compounds is very tough and hard as it is influenced by different factors such as its mobility in the soil, attachment with the substrate, and how it is detoxified or degraded by the organism it enters (Pichtel, 2012). Explosive pollutants are usually present in soil in the form of residues or particles due to the production and utilization of munitions and released into the environment through soil dissolution (Pennington et al., 2008; Kalderis et al., 2011). The explosives are movable in the soil solution and after their release in the environment they do not persist or adhered to the release point, preferably they dispersed into the environment with the help of pore matrix of the soil (Fig. 17.3) (Pennington and Brannon, 2002; Kiiskila et al., 2015; Taylor et al., 2015). These pollutant compounds have different aqueous solubility rates that result in the protection strategy throughout the soil column (Pichtel, 2012). Irrespective of nitroaromatics and nitrate esters, nitroamines exhibited a high degree of penetration rate deep inside the soil; however, these compounds exist in the higher concentrations at the top surface layer of the soil (15 cm). When there is a chemical release containing explosive compounds through any industrial application, it leads to limited dispersion of pollutants in the area and if there is explosion of munition or UXOs detonation then it can lead to long-lasting or continuing disturbed patterns of release. Comparatively it has been seen that low solubility of several common explosive compounds results in the uninterrupted pulse release into the environment and nearby sites as UXOs and munition residues deteriorate with passage of time (Taylor et al., 2015). Explosive compounds when released come in contact with soil, where they are absorbed and adsorped by soil particles. The existence of sorption and the state of lasting or remaining unchanged indefinitely depend on the structure of the compounds. Nitroaromatics are liable to attach with soil particles and is usually a reversible mechanism. It has been found that nitroaromatics and nitrate esters are not taken up by the soil or get sorbed easily and if it happens then the bond between the two does not break easily (Pennington and Branon, 2002). Phytoremediation technique can be improved or modified successfully if this factors is manipulated and altered; the addition of chemical increases sorption and adsorption and with the addition of

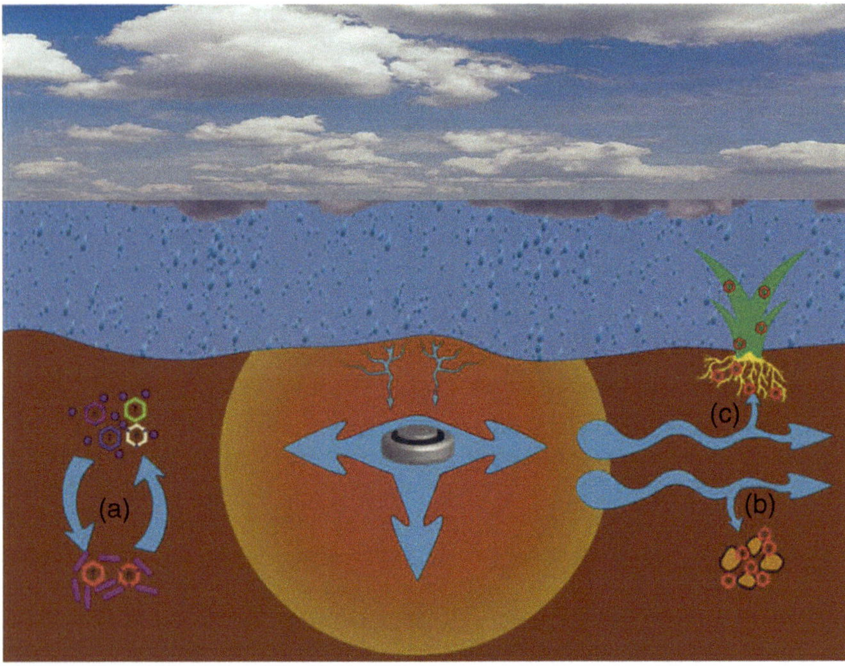

FIG. 17.3 Environmental behavior of explosive compounds in soils, and center icon represents unexploded ordnance (UXO) and the color behind it represents the diffusion of contaminants. Water is denoted by blue arrows and contaminant presence outside of the central diffusion zone is indicated by cultured pentagons. Region (A) represents microbial interactions and metabolism, (B) sorption to soil particles, and (C) uptake and sequestration by above- and below-ground plant tissues. *From Via S.M. (2020) Phytoremediation of Explosives. In: Shmaefsky B. (eds) Phytoremediation. Concepts and Strategies in Plant Sciences. Springer, Cham. https://doi.org/10.1007/978-3-030-00099-8_8.*

surfactants, desorption can be improved (Pennington et al., 1995). Regulating of soil binding can increase or decrease the bioavailability of the compounds in the surrounding environment, which is largely depends on the requirements of the remediation technology.

17.4 Plants and explosives

17.4.1 Pollutant uptake

Interactions of plant-pollutant strategy begin with the pollutant uptake and take place usually through liquid solution present in the pore matrix of the soil. Soil solution containing explosive compounds will penetrate the roots of the plant without any hindrance or freely due to the increased water movement flow aided by transpiration of leaf (Singh and Mishra, 2014). Explosive compounds in the root interiors proceed and move freely between the intermembrane spaces, travel via protective Casparian strip along xylem, and ultimately get deposited completely in the plant (Ghosh and Singh, 2005; Pilon-smits, 2005). Explosive compounds pollution design and segregation are widely preserved across plant species, yet their role and health reactions are not yet defined (Via and Zinnert, 2016).

17.4.2 Phytotoxicity

Explosive compound toxicity in plants is highly dangerous and can give rise to number of stresses in plants ranging from deformation in the morphological patterns and impaired cellular responses to pollutants varying from species to species of plans, exposure time and age, concentration and kind of pollutants (Winfield et al., 2004; Just and Schnoor, 2004; Vila et al., 2008; Ait Ali et al., 2014; Via et al., 2014). In plants, morphological and physiological reactions to TNT emergence will take place at lesser concentration compared to RDX (Peterson et al., 1996; 1998; Pilon-Smits, 2005; Vila et al., 2005; Zinnert 2012; Via et al., 2014). Effects of explosive compounds once exposed mainly displayed localization of compounds in the plants. RDX, one of the most common nitroamines, is primarily attached to the aboveground tissues resulting in considerable leaf impairment (Winfield et al., 2004; Vila et al., 2007; Via et al., 2014) and also damages photosynthetic activities (Thompson et al., 1998; Ait Ali et al., 2006; Zinnert, 2012; Zinnert et al., 2013; Via et al., 2014). RDX induced changes in the morphology of plants ranges from chlorosis, leaf necrosis, impaired leaf margins, atypical bilateral symmetry, and decreased leaf size (Winfield et al., 2004; Vila et al., 2007; Khatisashvili et al., 2009; Singh and Mishra, 2014; Via et al., 2014). TNT one of the most common nitroaromatic compounds primarily attaches to below ground parts (Peterson et al., 1998; Vila et al., 2007; Khatisashvili et al., 2009; Singh and Mishra, 2014) and causing impairment in root growth, altered root structures, and reduced root functions (Peterson et al., 1998; Gong et al., 1999; Krishnan et al., 2000; Vila et al., 2007; Khatisashvili et al., 2009; Singh and Mishra, 2014). On the other hand, nitrate esters have not subjected to considerable analysis methods as other explosives groups undergo; however, nitrate esters have been shown to hinder germination of seeds, development of plants, and seedling formation (French et al., 1999). It is important to note here that morphological responses have been extensively studied than the physiological responses but after subjection to explosive pollution can change carboxylation, assimilation of carbon, electron use efficiency, exchange of gases, and water relations (O'Leary, 1981; Thompson et al., 1998; Dawson et al., 2002; Ait Ali et al., 2006; Zinnert, 2012; Zinnert et al., 2012; Via et al., 2014). Stress conditions result in the activation of some physiological responses in some pathways of plants irrespective of the origin of stress (Chapin III, 1991); natural or man-made. It is recommended that although there are definite dissimilarities between man-made and natural stress reactions and accordingly different types of explosive compounds affect physiology in plants through various approaches. Upon exposure to RDX and TNT, woody plant species have not experienced curvilinear and extremely related reactions in photosynthesis and stomatal conductance; however, this association is conventionally sustained under different types of natural stresses (Flexas et al. 1999; Zinnert et al. 2012; Via et al. 2014). Novel and new reactions have been induced by explosives that may arise from breaking of different physiological processes that are firmly associated under common circumstances. TNT and RDX instigated the same reduction in the fixation of carbon in plants, besides Ait Ali et al., (2006) and Via et al., (2014) indicated that irrespective of some similarities among plants toward the reaction mechanisms to explosives, they play differently under stress conditions. These observations indicated that TNT in plants affects PS II (photosystem-II), while RDX affects the electron transport rate and dark minimal fluorescence ; however, TNT did not induce considerable damage (Ait Ali et al., 2006) but RDX induces impairment upon exposure (Via et al., 2014). Although TNT reduced water relation, RDX results in the considerable elevation in stomatal conductance in affected ones. However, very little is known about the outside effects of RDX and TNT in plant physiology and is yet not fully understood. Toxicity of explosive compounds in plants intensified by the dangerous effects of products induced by light and metabolic degradation of TNT and RDX as these can generate derivatives of amino and other compounds such as oxygen radical, carbon monoxide, nitric oxide, and formaldehyde (Spain, 1995; Hawari et al., 2000; Halasz et al., 2002; Bernstein and Ronen, 2012). Currently, nitrate esters and insensitive explosives have not been yet investigated for physiological responses in plants.

17.5 Environmental clean-up of explosives

Environmental clean-up of soil and water contaminated with explosive compounds appears to be a productive remedy in laboratory conditions and also displays positive results based on nearly less field assessment that has been done in the past. Plants have been utilized for the remediation of contaminated sites with explosive compounds by three main approaches namely; phytoextraction, phytostabilization, and rhizofiltration (Pilon-Smits, 2005). The rhizospheric region of plant roots is referred to as the rhizosphere where rhizofiltration takes place, and it is the stimulation of microbial communities to form and play a critical role in the increasing activity levels around the rhizosphere (Pilon-Smits, 2005). Usually, the rhizospheric zone expands only on a small area from the root structures; however, in plant species having fibrous, broad, and deep roots could expand over a larger area under consideration. Phytoremediation takes place by acting on contaminant near the zone of microbial activity, which is actually the first place where the process of remediation begins. The microbial communities found in rhizosphere of plants have the potential to effectively degrade explosive compounds from the sites. Metabolism of microbes could arise due to the removal of external methyl and nitrate groups and also due to the cleavage of central rings (Hawari et al., 2000; Rylott and Bruce, 2009; Rylott et al., 2011). The wild-type strains of microbes have shown potential to sustain such processes; however, a considerable amount of work has been done on these natural mechanisms for the improvement through the use of genetically modified strains (Chatterjee et al., 2017; Rai et al., 2020). This relationship of plant–microbe interactions goes both ways as several microbes can effectively have impact on growth and health of plant in the same way as the plants have the influence on the microbial and bacterial communities. The rhizospheric region of the plant roots can be altered or modified through addition of microbes, endo- and ectosymbionts initiation, and roots (Kuiper et al., 2004; McGuinness and Dowling, 2009).

Besides, phytostabilization focused on the compound retained in a localized area and restricting their spread and transport to nearby areas. For this purpose, plants play an important role in the degradation of contaminants by the sequestration of pollutants into the cellular systems before it affects the nearby areas. Several plant species have the capability to accumulate higher concentration of toxins in their body than in the nearby sites and this type of plant species is known as hyperaccumulators. Plant species show natural capacity to attach pollutants with their cells and then subjected to detoxification to reduce cellular damage. There are three different mechanisms involved in the degradation of organic pollutants postuptake and these are transformation, conjugation, and ultimately sequestration and often these mechanisms are known as a process called the model of "Green Livers" (Klein and Scheunert, 1982; Sandermann, 1994; Burken and Schnoor, 1997; Hannink et al., 2002). Explosives pollutants experienced transformation by the enzymatic action and conjugation with amino acid, D-glucose, or glutathione resulting in the production of soluble or insoluble byproducts (Sens et al., 1999; Vila et al., 2008). Moreover, soluble products are either attached with the cell wall or can be stored in the vacuole; however, insoluble products are only detoxified in the cellular structures (Burken et al., 2000; Lotufo et al., 2009; Rylott et al., 2011). For the compounds such as TNT, stabilization approaches for rectification show significant potential as these compounds are accumulated in the below-ground tissues and are very hard to collect.

In phytoremediation plants are used as pumps, they uptake the pollutants from the water and the soil, accumulate the pollutants into their tissues, and the pollutants are imported into the ground parts of plants and are removed after the completion of growing season (Pilon-Smits, 2005). In this context, collection should be completed after every growing season, and if plants are not able to behave as a new source for pollutant release, then it takes up pollutants from the soil and accumulates it onto the surface of woody and leafy structures through senescence. Phytoremediation approaches have the potentiality for efficient remediation of nitrate esters and nitroamines that show similar behavior in some ways, although nitroaromatics and its binding actions resulted in more difficulties in the overall process (Pilon-Smits, 2005; Rocheleau et al., 2011). Currently, one approach of phytoremediation that draws and receives a huge attention and response from research communities is phytodegradation, in which plants have the ability to degrade or transform a pollutant or a toxic compound into less harmful one; however, not all but only a few plants are capable of degrading explosive pollutants by permitting the compound transformation and metabolism to happen. Many studies have been undertaken to investigate the phytodegradation approach and many efforts have been applied for its improvement. Reduction of nitro group can be the outcome of nitroreductases (Adamia et al., 2006; Makris et al., 2007; Rylott and Bruce, 2009; Rylott et al., 2011), laccase activities (Schnoor et al., 1995), cytochrome P450 (Rylott et al., 2011), and oxophytodienoate reductases (Beynon et al., 2009). For this process, FMN (flavin mononucleotide), FAD (flavin adenine dinucleotide) as prosthetic groups, and NADH (nicotinamide adenine dinucleotide) or NADPH (nicotinamide adenine dinucleotide phosphate) are reducing agents utilized by the nitroreductases in the presence of TNT (Bryant and DeLuca, 1991). Moreover, in some cases, nitroaromatic detoxification can also be employed with the help of enzyme lactases (Schnoor et al., 1995). Reduction of nitroamines usually involves the enzymes such as cytochrome P450, glutathione-S-transferase, reductases, and peroxidises (Rylott et al., 2011). The phenomenon of phytophotolysis is also documented for the transformation of compound in plant tissues (Just and Schnoor, 2004) and this phenomenon on the plant part provides an extra advantage; however, very less is known about metabolic process of nitrate esters explosive compounds (Rylott and Bruce, 2009) but in the presence of plants nitroglycerin can be converted into isomers of dinitroglycerin (Rocheleau et al., 2011). Plants abilities can be enhanced further to convert explosive pollutants into less or neutral forms, which results in much more effective technology and reduces the requirements for the remediation methods. Table 17.1 presents the group of plants (aquatic, terrestrial, and wetland) for their tolerance and removal capacity for RDX, HMX, and TNT explosive compounds.

17.6 Phytoremediation: prospective and future recommendation

Discussion of sites polluted with a single pollutant permitted for very specific modifications of strategies allowing for nearby climate, soils, land use, as well as type of pollutant and amount to be taken for examination or investigation. One of the most silent hurdles that come in the way of predicting the ecological influences of explosive compounds is the site-specific inconsistency of the concentrations. In this context, for example dry soils, the concentration of TNT ranges from 0.08 to 87,000 ppm and for RDX, the concentration ranges from 0.7 to 74,000 ppm, respectively (Best et al., 2008). From the previous studies, it has been found that huge data were accessible only on rectification technology of explosives focused on a single pollutant. This increases the complexity in the utilization of laboratory results to field applications as explosives are hardly obtained as a single pollutant. Subsidiary or similar pollution that involves different types of explosive compounds, heavy metals, and a diversity of different pollutants is usual and most frequently exists in the other form of explosives. The reason being is detonation and explosions that primarily contain numerous compounds to be verified until preferred outcome is attained

TABLE 17.1 Suggested species for phytoremediation of RDX, HMX, and TNT based on the literature findings. In this case, the removal potential is listed as high (75%–100%), moderate (50%–75%), and low (<50%) based on percent contaminant removed from growing media. Uptake is reported as high (80%–100%), moderate (50%–80%), or low (<50%) depending on relative concentration in plant tissues compared to those of growing media. *From Via S.M. (2020) Phytoremediation of Explosives. In: Shmaefsky B. (eds) Phytoremediation. Concepts and Strategies in Plant Sciences. Springer, Cham. https://doi.org/10.1007/978-3-030-00099-8_8.*

Contaminant	Functional group	Genus	Removal	Uptake	System	References
RDX	Submerged	*Ceratophyllum*	-	High	aquatic	Best et al., (1997b)
RDX	Algae	*Charales*	High	-	Aquatic	Best et al., (1997a)
RDX	Graminoid	*Cyperus*	-	High	wetland	Price et al., (1997)
RDX	Herbaceous	*Lactuca*	-	High	Terrestrial	Price et al., (1997)
RDX	Submerged	*Myriophyllum*	-	High	Aquatic	Best et al., (1997a; b)
RDX	Graminoid	*Oryza*	-	Low	wetland	Vila et al., (2007)
RDX	Graminoid	*Phalaris*	High	Moderate	wetland	Best et al., (1997b); Sikora et al., (1998)
RDX	Woody	*Populus*	-	High	Terrestrial	Thompson et al., (1999)
RDX	Submerged	*Potamogeton*	Low	Low	Aquatic	Best et al., (1997a)
RDX	Woody	*Robina*	-	Moderate	Terrestrial	Schneider et al., (1995)
RDX	Herbaceous	*Sagitaria*	-	Low	Wetland	Schneider et al., (1995) Best et al., (1997b)
RDX	Herbaceous	*Solidago*	-	High	Terrestrial	Schneider et al., (1995)
RDX	Submerged	*Stuckenia*	High	High	Aquatic	Best et al., 1997a
RDX	Graminoid	*Triticum*	-	High	Terrestrial	Vila et al., (2007)
HMX	Herbaceous	*Brassica*	Low	High	Terrestrial	Groom et al., (2002)

(continued)

TABLE 17.1 (Cont'd)

Contaminant	Functional group	Genus	Removal	Uptake	System	References
HMX	Herbaceous	*Medicago*	Low	High	Terrestrial	Groom et al., (2002)
HMX	Graminoid	*Phalaris*	High	-	Wetland	Sikora et al., (1998)
HMX	Graminoid	*Triticum*	Low	High	Terrestrial	Groom et al., (2002)
TNT	Woody	*Abutilon*	High	-	Wetland	Chang et al., (2004)
TNT	Herbaceous	*Alisma*	High	-	Wetland	Best et al., (1997b)
TNT	Graminoid	*Bromus*	Low	Low	Terrestrial	Zellmer et al., (1995)
TNT	Graminoid -	*Carex*	High	-	Wetland	Best et al., (1997b)
TNT	Submerged	*Ceratophyllum*	High	-	Aquatic	Best et al., (1997b)
TNT	Algae	*Charales*	High	-	Aquatic	Best et al., (1997a)
TNT	Herbaceous	*Cicer*	Moderate	-	Terrestrial	Adamia et al., (2006)
TNT	Herbaceous	*Dipsacus*	Low	Low	Terrestrial	Zellmer et al., (1995)
TNT	Submerged	*Egeria*	High	-	Aquatic	Best et al., (1997a)
TNT	Graminoid	*Eleocharis*	High	-	Wetland	Best et al., (1997a)
TNT	Submerged	*Elodea*	High	-	Aquatic	Best et al., (1997a)
TNT	Herbaceous	*Glycine*	High	-	Terrestrial	Adamia et al., (2006)
TNT	Herbaceous	*Helianthus*	Moderate	-	Terrestrial	Adamia et al., (2006)
TNT	Graminoid	*Heteranthera*	High	-	Terrestrial	Best et al., (1997a)

(National Research Council, 2004; Pichtel, 2012). Due to the innate complications in rectifying polluted system and our present interpretation of plant-explosive relationship, there exist huge possibilities for the utilization of phytoremediation technology on polluted areas. The gap that existed in between field applications and laboratory findings specifically that of continuing extensive attempts is the one that needs rectification for the development of optimum phytoremediation technologies as far as both cost and efficiency are concerned. For the rectification or remediation of shallow and surface soil and ground water, phytoremediation of explosive compounds is the need of the hour to remediate the soil and water polluted with explosives. This technique for remediating polluted areas indicated remarkable potentiality for the degradation of nitrate esters and nitroaromatics; however, degradation of nitroamines has been restricted. It is not easy to extract the nitroaromatics from the polluted areas due to their movability but can be decontaminated efficiently by utilizing bioreactors or by establishment of wetlands for waste disposal sites (Sikora et al., 1998; Truu et al., 2015). Based on present knowledge and understanding, phytoremediation technology for explosives (organic/inorganic) holds great potential but need considerable funding for successfully applied into the fields. Moreover, laboratory findings and outcomes are needed for the basic knowledge of any method, but applying on fields several other parameters can exceptionally modify the different responses of plants. It is very crucial to understand these complexities to launch phytoremediation strategies and their application further.

References

Adamia, G., Ghoghoberidze, M., Graves, D., Khatisashvili, G., Kvesitadze, G., Lomidze, E., Ugrekhelidze, D., Zaalishvili, G., 2006. Absorption, distribution, and transformation of TNT in higher plants. Ecotoxicol. Environ. Saf 64 (2), 136–145.

Ait Ali, N.A., Dewez, D., Robidoux, P.Y., Popovic, R., 2006. Photosynthetic parameters as indicators of trinitrotoluene (TNT) inhibitory effect: change in chlorophyll a fluorescence induction upon exposure of Lactuca sativa to TNT. Ecotoxicology 15, 437–441.

Ait, Ali, A., Zinnert, J.C., Muthukumar, B., Peng, Y., Chung, S.M., Stewart, C.N., 2014. Physiological and transcriptional responses of Baccharis halimifolia to the explosive "composition B" (RDX/TNT) in amended soil. Environ. Sci. Pollut. Res 21 (13), 8261–8270.

Almog, J., Zitrin, S., 2009. Colorimetric detection of explosives M. Marshall and J.C. Oxley (Eds.). Aspects of Explosives Detection, 41–58.

Bernstein, A., Ronen, Z., 2012. Biodegradation of the explosives TNT, RDX and HMX. In: Singh, S.N. (Ed.), Microbial Degradation of Xenobiotics. Springer, Berlin, Heidelberg, pp. 135–176.

Best, E.P., Sprecher, S.L., Fredrickson, H.L., Zappi, M.E., Larson, S.L., 1997. Screening Submersed Plant Species for Phytoremediation of Explosives-Contaminated Groundwater from the Milan Army Ammunition Plant, Milan, Tennessee (No. WES-TR-EL-97-24). Army Engineer Waterways Experiment Station Vicksburg MS Environmental Lab.

Best, E.P.H., Tatem, H.E., Geter, K.N., Wells, M.L., Lane, B.K., 2008. Effects, uptake, and fate of 2,4,6-trinitrotoluene aged in soil in plants and worms. Environ. Toxicol. Chem 27 (12), 2539–2547.

Beynon, E.R., Symons, Z.C., Jackson, R.G., Lorenz, A., Rylott, E.L., Bruce, N.C., 2009. The role of oxophytodienoate reductases in the detoxification of the explosive 2,4,6-trinitrotoluene by Arabidopsis. Plant Physiol 151 (1), 253–261.

Bryant, C., DeLuca, M., 1991. Purification and characterization of an oxygen-insensitive NADPH nitroreductase from Enterobacter cloacae. J. Biol. Chem 266 (7), 4119–4125.

Burken, J.G., Schnoor, J.L., 1997. Uptake and metabolism of atrazine by poplar trees. Environ. Sci. Technol 31 (5), 1399–1406.

Burken, J.G., Shanks, J.V., Thompson, P.L., 2000. Phytoremediation and plant metabolism of explosives and nitroaromatic compounds.In: J.C. Spain, J.B. Hughes, H-J. Knackmuss (Eds.) Biodegradation of Nitroaromatic Compounds and Explosives. CRC Press, London, UK, pp. 239–275.

Certini, G., Scalenghe, R., Woods, W.I., 2013. The impact of warfare on the soil environment. Earth. Sci. Rev 127, 1–15.

Chang, Y.Y., Kwon, Y.S., Kim, S.Y., Lee, I.S., Bae, B., 2004. Enhanced degradation of 2, 4, 6-trinitrotoluene (TNT) in a soil column planted with Indian mallow (Abutilon avicennae). J. Biosci. Bioeng 97 (2), 99–103.

Chapin III, F.S., 1991. Effects of multiple environmental stresses on nutrient availability and use. In: Mooney, H.A., Winner, W.E., Pell, E.J. (Eds.), Response of Plants to Multiple Stresses. Academic Press, San Diego, pp. 67–88.

Chatterjee, S., Deb, U., Datta, S., Walther, C., Gupta, D.K., 2017. Common explosives (TNT, RDX, HMX) and their fate in the environment: emphasizing bioremediation. Chemosphere 184, 438–451.

Dawson, T.E., Mambelli, S., Plamboeck, A.H., Templer, P.H., Tu, K.P., 2002. Stable isotopes in plant ecology. Annu. Rev. Ecol. Syst 33 (1), 507–559.

Dick, R.A., Fletcher, L.R., D'Andrea, D.V., 1983. Explosives and Blasting Procedures Manual (No. 8925). US Department of the Interior, Bureau of Mines.

Dontsova, K., Brusseau, M., Arthur, J., Mark, N., 2014. Dissolution of NTO, DNAN and insensitive munitions formulations and their fates in soils, Proc. Jannaf workshop Proceedings—Fate, Transport and Effects of Insensitive Munitions: Issues and Recent Data (May), 32–47.

Doty S.L. 2008 Enhancing phytoremediation through the use of transgenics and endophytes. New Phytol 179:318–333.

Douglas, T.A., Walsh, M.E., Weiss, C.A., McGrath, C.J., Trainor, T.P., 2012. Desorption and transformation of nitroaromatic (TNT) and nitroamine (RDX and HMX) explosive residues on detonated pure mineral phases. Water. Air. Soil. Pollut 223, 2189–2200.

Flexas, J., Escalona, J.M., Medrano, H., 1999. Water stress induces different levels of photosynthesis and electron transport rate regulation in grapevines. Plant Cell Environ 22 (1), 39–48.

French, C.E., Rosser, S.J., Davies, G.J., Nicklin, S., Bruce, N.C., 1999. Biodegradation of explosives by transgenic plants expressing pentaerythritol tetranitrate reductase. Nat. Biotechnol 17 (5), 491.

Ghosh, M., Singh, S., 2005. A review on phytoremediation of heavy metals and utilization of its by-products. Appl. Ecol. Environ. Res 3, 1–18.

Gong, P., Wilke, B., Fleischmann, S., 1999. Soil-based phytotoxicity of 2,4,6-trinitrotoluene (TNT) to terrestrial higher plants. Environ. Contamin. Toxicol, 152–157.

Groom, C.A., Halasz, A., Paquet, L., Morris, N., Olivier, L., Dubois, C., Hawari, J., 2002. Accumulation of HMX (octahydro-1, 3, 5, 7-tetranitro-1, 3, 5, 7-tetrazocine) in indigenous and agricultural plants grown in HMX-contaminated anti-tank firing-range soil. Environ. Sci. Technol 36 (1), 112–118.

Halasz, A., Groom, C., Zhou, E., Paquet, L., Beaulieu, C., Deschamps, S., Corriveau, A., Thiboutot, S., Ampleman, G., Dubois, C., Hawari, J., 2002. Detection of explosives and their degradation products in soil environments. J. Chromatogr. A 963 (1–2), 411–418.

Hamilton, A., 1921. Industrial Poisoning in Making Coal-TAR dyes and Dye Intermediates (No. 280). US Government Printing Office.

Hannink, N.K., Rosser, S.J., Bruce, N.C., 2002. Phytoremediation of explosives. Crit. Rev. Plant. Sci 21 (5), 511–538.

Hawari, J., Beaudet, S., Halasz, A., Thiboutot, S., Ampleman, G., 2000. Microbial degradation of explosives: biotransformation versus mineralization. Appl. Microbiol. Biotechnol 54 (5), 605–618.

Just, C.L., Schnoor, J.L., 2004. Phytophotolysis of hexahydro-1,3,5-trinitro-1,3,5-triazine (RDX) in leaves of reed canary grass. Environ. Sci. Technol 38 (1), 290–295.

Kalderis, D., Juhasz, A.L., Boopathy, R., Comfort, S., 2011. Soils contaminated with explosives: environmental fate and evaluation of state-of-the-art remediation processes (IUPAC Technical Report). Pure. Appl. Chem 83 (7), 1407–1484.

Kelly, J., 2004. Gunpowder: Alchemy, Bombards, and Pyrotechnics: the History of the Explosive that Changed the World. Basic Books, AZ.

Khatisashvili, G., Gordeziani, M., Adamia, G., Kvesitadze, E., Sadunishvili, T., Kvesitadze, G., 2009. Higher plants ability to assimilate explosives. World Acad. Sci. Eng. Technol 33 (9), 256–270.

Kholodenko, T, Ustimenko, Y, Pidkamenna, L, Pavlychenko, A, 2014. Ecological safety of emulsion explosives use at mining enterprises. In: Bondarenko, V., Kovalevs'ka, I., Ganushevych, K. (Eds.), Ecological safety of emulsion explosives use at mining enterprises. Progressive Technologies of Coal, Coalbed Methane, and Ores Mining, 255–260.

Kiiskila, J.D., Das, P., Sarkar, D., Datta, R., 2015. Phytoremediation of explosive-contaminated soils. Curr. Pollut. Rep 1, 23 34.

Klein, W., Scheunert, I., 1982. Bound pesticide residues in soil, plants and food with particular emphasis on the application of nuclear techniquesAgrochemicals: fate in food and environment Vienna. International Atomic Energy Agency, Vienna, Austria, pp. 177 205.

Krishnan, G., Horst, G.L., Darnell, S., Powers, W.L., 2000. Growth and development of smooth bromegrass and tall fescue in TNT-contaminated soil. Environ. Pollut 107 (1), 109–116.

Kuiper, I., Lagendijk, E.L., Bloemberg, G.V., Lugtenberg, B.J., 2004. Rhizoremediation: a beneficial plant-microbe interaction. Mol. Plant. Microbe. Interact 17 (1), 6–15.

Lotufo, G.R., Carr, R.S., Conder, J.M., Nipper, M., 2009. Fate and toxicity of explosives in sediments. In: Lotufo G.R., Nipper M., Carr R.S., Conder J.M. Ecotoxicology of explosives. CRC Press, London UK, pp. 117–134.

Madeira, C.L., Field, J.A., Simonich, M.T., Tanguay, R.L., Chorover, J., Sierra-Alvarez, R., 2018. Ecotoxicity of the insensitive munitions compound 3-nitro-1,2,4-triazol-5-one (NTO) and its reduced metabolite 3-amino-1,2,4-triazol-5-one (ATO). J. Hazard. Mater 343, 340–346.

Makris, K.C., Shakya, K.M., Datta, R., Sarkar, D., Pachanoor, D., 2007. Chemically catalyzed uptake of 2,4,6-trinitrotoluene by Vetiveria zizanioides. Environ. Pollut 148 (1), 101–106.

McGuinness, M., Dowling, D., 2009. Plant-associated bacterial degradation of toxic organic compounds in soil. Int. J. Environ. Res. Public. Health 6 (8), 2226–2247.

National Research Council, 2004. Existing and Potential Standoff Explosives Detection Techniques. The National Academies Press, Washington, DC. https://doi.org/10.17226/10998.

O'Leary, M.H., 1981. Carbon isotope fractionation in plants. Phytochem 20 (4), 553–567.

Pagoria, P., 2016. A comparison of the structure, synthesis, and properties of insensitive energetic compounds. Propellants. Explos. Pyrotech 41 (3), 452–469.

Pennington, J.C., Brannon, J.M., 2002. Environmental fate of explosives. Thermochim. Acta 384, 163–172.

Pennington, J.C., Hayes, C.A., Myers, K.F., Ochman, M., Gunnison, D., Felt, D.R., McCormick, E.F., 1995. Fate of 2,4,6-trinitrotoluene in a simulated compost system. Chemosphere 30 (3), 429–438.

Pennington, J.C., Hayes, C.A., Yost, S., Crutcher, T.A., Berry, T.E., Clarke, J.U., Bishop, M.J., 2008. Explosive residues from blow-in-place detonations of artillery munitions. Soil. Sed. Contamin 17 (2), 163–180.

Peterson, M.M., Horst, G.L., Shea, P.J., Comfort, S.D., 1998. Germination and seedling development of switchgrass and smooth bromegrass exposed to 2,4,6-trinitrotoluene. Environ. Pollut 99 (1), 53–59.

Peterson, M.M., Horst, G.L., Shea, P.J., Cornfor, S.D., Petersonc, R.K.D., 1996. TNT and 4-amino-2,6- dinitrotoluene influence on germination and early seedling development of tall fescue. Environ. Pollut 93 (1), 57–62.

Pichtel, J., 2012. Distribution and fate of military explosives and propellants in soil: a review. Appl. Environ. Soil. Sci. Article ID 617236. https://doi.org/10.1155/2012/617236.

Pilon-Smits, E.A., 2005. Phytoremediation. Annu. Rev. Plant. Biol. 56, 15–39.

Price, R.A., Pennington, J.C., Larson, S.L., Nuemann, D., Hayes, C.A., 1997. Plant Uptake of Explosives from Contaminated Soil and Irrigation Water at the Former Nebraska Ordnance Plant, Mead, Nebraska (No. WES/TR/EL-97-11)Army Engineer Waterways Experiment Station. Vicksburg MS Environmental Lab.

Rai, P.K., Kim, K.H., Lee, S.S., Lee, J.H., 2020. Molecular mechanisms in phytoremediation of environmental contaminants and prospects of engineered transgenic plants/microbes. Sci. Total. Environ 705, 135858.

Ro, K.S., Venugopal, A., Adrian, D.D., Constant, D., Qaisi, K., Valsaraj, K.T., Thibodeaux, D.R., 1996. Solubility of 2,4,6-trinitrotoluene (TNT) in water. J. Chem. Eng. Data 41, 758–761.

Rocheleau, S., Kuperman, R.G., Martel, M., Paquet, L., Bardai, G., Wong, S., Sarrazin, M., Dodard, S., Gong, P., Hawari, J., Checkai, R.T., Sunahara, G.I., 2011. Phytotoxicity and uptake of nitroglycerin in a natural sandy loam soil. Sci. Total. Environ 409 (24), 5284–5291.

Rylott, E.L., Bruce, N.C., 2009. Plants disarm soil: engineering plants for the phytoremediation of explosives. Trend. Biotechnol 27 (2), 73–81.

Rylott, E.L., Lorenz, A., Bruce, N.C., 2011. Biodegradation and biotransformation of explosives. Curr. Opin. Biotechnol 22 (3), 434–440.

Sandermann, H., 1994. Higher plant metabolism of xenobiotics: the "green liver" concept. Pharmacogenetics 4 (5), 225–241.

Schneider, J. F., Zellmer, S. D., Tomczyk, N. A., Rastorfer, J. R., Chen, D., 1995. Uptake of Explosives from Contaminated Soil by Existing Vegetation at the Iowa Army Ammunition Plant. Energy and Environmental Systems Div, Argonne National Laboratory, IL.

Schnoor, J.L., Light, L.A., McCutcheon, S.C., Wolfe, N.L., Carreia, L.H., 1995. Phytoremediation of organic and nutrient contaminants. Environ. Sci. Technol 29 (7), 318A–323A.

Sens, C., Scheidemann, P., Werner, D., 1999. The distribution of14C-TNT in different biochemical compartments of the monocotyledonous Triticum aestivum. Environ. Pollut 104 (1), 113–119.

Sikora, F.J., Almond, R.A., Behrends, L.L., Hoagland, J.J., Kelly, D.A., 1998. Demonstration results of phytoremediation of explosives-contaminated groundwater using constructed wetlands at the Milan Army Ammunition Plant, Milan, TennesseeII. Tennessee Valley Authority, Muscle Shoals, AL.

Singh, S.N., Mishra, S., 2014. Phytoremediation of TNT and RDX. In: Singh, S.N. (Ed.), Biological remediation of explosive residues. Springer, Cham, pp. 371–392.

Spain, J.C., 1995. Biodegradation of nitroaromatic compounds. Annu. Rev. Microbiol 49 (1), 523–555.

Taylor, S., Bigl, S., Packer, B., 2015. Condition of in situ unexploded ordnance. Sci. Total. Environ 505, 762–769.

Thompson, P.L., Ramer, L.A., Schnoor, J.L., 1999. Hexahydro-1, 3, 5-trinitro-1, 3, 5-triazine translocation in poplar trees. Environ. Toxicol. Chem. Int. J 18 (2), 279–284.

Thompson, P.T., Ramer, L.A., Guffey, A.P., Schnoor, J.L., 1998. Decreased transpiration in poplar trees exposed to 2,4,6- trinitrotoluene. Environ. Toxicol. Chem 17 (5), 902–906.

Tiwary, R.K., 2001. Environmental impact of coal mining on water regime and itsmanagement. Water Air Soil Pollut 132 (1-2), 185–199.

Truu, J., Truu, M., Espenberg, M., Nõlvak, H., Juhanson, J., 2015. Phytoremediation and plant-assisted bioremediation in soil and treatment wetlands: a review. Open Biotechnol. J 9 (1), 85–92.

Via, S.M., Zinnert, J.C., 2016. Impacts of explosive compounds on vegetation: a need for community scale investigations. Environ. Pollut 208, 495–505.

Via, S.M., Zinnert, J.C., Butler, A.D., Young, D.R., 2014. Comparative physiological responses of Morella cerifera to RDX, TNT, and composition B contaminated soils. Environ. Exp. Bot 99, 67–74.

Vila, M., Lorber-Pascal, S., Laurent, F., 2008. Phytotoxicity to and uptake of TNT by rice. Environ. Geochem. Health 30 (2), 199–203.

Vila, M., Mehier, S., Lorber-Pascal, S., Laurent, F., 2007. Phytotoxicity to and uptake of RDX by rice. Environ. Pollut 145 (3), 813–817.

Vila, M., Pascal-Larber, S., Rathahao, E., Debrauwer, L., Canlet, C., Laurent, F., 2005. Metabolism of 2,4,6-trinitrotoluene in tobacco cell suspension cultures. Environ. Sci. Technol 39 (2), 663–672.

Winfield, L.E., Rodgers, J.H., D'Surney, S.J., 2004. The response of selected terrestrial plants to short (%3c12 days) and long term (2, 4, and 6 weeks) hexahydro-1,3,5-trinitro-1,3,5-triazine (RDX) exposure Part 1: growth and developmental effects. Ecotoxicology 13, 335–347.

Zellmer, S.D., Schneider, J.F., Tomczyk, N.A., Banwart, W.L., Chen, D., 1995. Plant Uptake of Explosives from Contaminated Soil at the Joliet Army Ammunition Plant. Argonne National Lab, IL.

Zinnert, J.C., 2012. Plants as phytosensors: physiological responses of a woody plant in response to RDX exposure and potential for remote detection. Int. J. Plant. Sci 173 (9), 1005–1014.

Zinnert, J.C., Nelson, J.D., Hoffman, A.M., 2012. Effects of salinity on physiological responses and the photochemical reflectance index in two co-occurring coastal shrubs. Plant Soil 354 (1–2), 45–55.

Zinnert, J.C., Via, S.M., Young, D.R., 2013. Distinguishing natural from anthropogenic stress in plants: physiology, fluorescence and hyperspectral reflectance. Plant Soil 366 (1-2), 133–141.

Chapter 18

Phytoremediation: An alternative approach for removal of dyes

Saksham Garg[b], Arpita Roy[a]
[a]*Department of Biotechnology, School of Engineering & Technology, Sharda University, Greater Noida, India,* [b]*Department of Biotechnology, Delhi Technological University, Delhi, India*

18.1 Introduction

Dyes are the chemical compounds used to impart coloration to the products produced by industries such as but not limited to textile, cosmetics, and pigmentation (Roy and Bharadvaja, 2019; Raina et al., 2020). The rapid growth of some industries such as textiles has resulted in usage hike of these coloration substances. This elevated usage of dyes has led in increase of water pollution, it is estimated that almost 10%–15% of dyes used end up in the environment as waste and even in minute amounts (>1 ppm in few cases) dyes can have adverse effects (Adegoke and Bello, 2015). Many of the synthetically produced dyes have a mutagenic or carcinogenic activity on human, while the wastewater from industries containing dyes can disrupt the aquatic life due to the existence of heavy metals and harmful ions such as chlorides. Along with the direct foreseeable problems, numerous other complications add up to the list, which makes extremely necessary to come up with techniques that can efficiently degrade these compounds. However, the complex chemical organization and xenobiotic nature make them resistant to degradation (Klementova, 2018).

Dyes are also highly stable chemical compounds even under the light and heat exposure (Lellis et al., 2019). Considering the adversities caused by the effluents containing dyes, numerous methods including physical, chemical, and biological have been devised to carry out the degradation. Adsorption, wet air oxidation, photocatalysis, electrocoagulation, jet cavitation, membrane separation, electrochemical methods, and sonolysis are some of examples of chemical and physical methods. All the methods mentioned have some limitation, either they are economically not feasible or end up making toxic byproducts. To overcome these issues, biological systems have been deployed but they are seen to be less efficient against the complex polymers and are time consuming hence their use is limited and less preferred (Lellis et al., 2019). Biological systems use aerobic and anaerobic pathways and processes to degrade the dyes (Sheng

Phytoremediation. DOI: https://doi.org/10.1016/B978-0-323-89874-4.00005-4

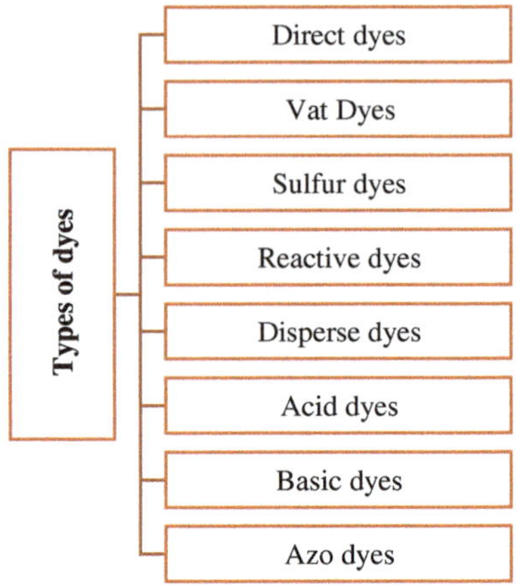

FIG. 18.1 Different types of dyes.

et al., 2018). Therefore, new and innovative methods are needed to overcome the limitation posed by the existing methodologies, which should be highly potent and pose a minimum risk to the environment. Phytoremediation is a biological treatment method for remediation of dyes, which is getting more attention due to its cost-effective and eco-friendly nature. Phytoremediation utilizes living plants for the removal of dye contaminants from the sources of water or soil. Therefore, in this chapter, importance of various plants able to remediate dye pollution has been discussed. Different types of dyes have been mentioned in Fig. 18.1.

18.2 Environmental problems associated with dyes

Mankind is using dyes from ancient times in various industrial activities, it was only after the invention and manufacturing of synthetic ones, the problems started to surface in both environment and health care regions. In many major industries such as leather, printing, textile, cosmetics, plastics, food processing, etc. dyes are extensively used. All these industries utilizing the synthetic dyes contribute toward the exhaustive list of environmental hazards ranging from air pollution to water pollution (Muthu, 2017). The air pollutants involve volatile compounds having oxides of either nitrogen or sulfur, particulate dust, etc. However, other forms of pollutants are produced in the form of solid waste but significant affect is caused on the water bodies. The discharge from the industry houses produced end up untreated in the water bodies (Bhatia, 2017), increasing the biochemical oxygen demand and chemical oxygen demand of the ecosystem denoting the rise in pollution. The high solubility and nonbiodegradable nature of dyes make them an environmental hazard. Effluent produced has a coloration in it due to the presence of dyes, which not only makes the water bodies aesthetically displeasing (Ohgaki et al., 2015) but also leads to reduced photosynthesis (Imran et al., 2015), prevents deep light penetration, and effects the flora and fauna growth (Hassan and Carr, 2018). In terms of nonbiodegradable aspect, they end up entering into the food chain and further get biomagnified causing whole other set of problems (Sandhya, 2010). The soluble organic dyes can lead to accelerated eutrophication of water bodies. Due to contamination in soil, dye can leach to water table and hinder the growth and proliferation of certain microbes essential for the production of food (Imran et al., 2015; Tahir, Sohail and Khan, 2017; Rehman et al., 2018).

18.3 Health impacts caused by dyes

Majority of the synthetically produced dyes are resistant to natural degradation due to their xenobiotic nature. This results in their redundancy in environment for a long period of time impacting the functioning of ecosystems (Rawat et al, 2016). Being the largest industrial pollutants, dyes are also highly harmful to human health (Ito et al., 2016). They are known to have properties such as carcinogenic and mutagenic to both animals and humans. Problems caused by dyes range from skin diseases such as dermatitis to disorders of central nervous system (Khan and Malik, 2018). Oral and respiratory disorders develop due to exposure to dust, causing dermatitis, rhinitis, occupational asthma, and other allergic reactions as well (Hunger, 2007; Clark, 2011). While some dyes have been categorized to have genotoxic effect, others reveal mutagenic and cytotoxic potentials. In all the cases these dyes lead to formation of a long-life ailment (Christie, 2007). At times the structural moiety of dye is not as such harmful rather the intermediates formed by the metabolism of the living system are the more hazardous one. Sudan I dye is one of the examples, which is enzymatically converted into carcinogenic amines (Piątkowska et al., 2018). Another point of notice is that many of the dyes, which show mutagenic and genotoxic nature, are azo dyes (Christie, 2007; Mondal, Purkait and De, 2018). The colored effluent along with dyes also contain heavy metals such as Cu, Zn, As, and Cr and the presence of these trace metals may act as a source to other set of health conditions such as nausea, skin irritation, ulceration, and hemorrhage (Verma, 2021). Different environmental and health impacts associated with dyes have been mentioned in Fig. 18.2.

18.4 Conventional methods of dye removal

The existing techniques employed for the removal of dyes are based on physical and chemical separation of dye from the industry-produced effluent. Conventional methods of dye removal has been mentioned in Fig. 18.3. These methodologies have their own set of advantages and disadvantages. Some of the disadvantages include problems with disposal of the byproducts, cost efficiency, and also, they are less efficient in terms of separation of the dyes. However, their combination may result into more effective and amplified removal system.

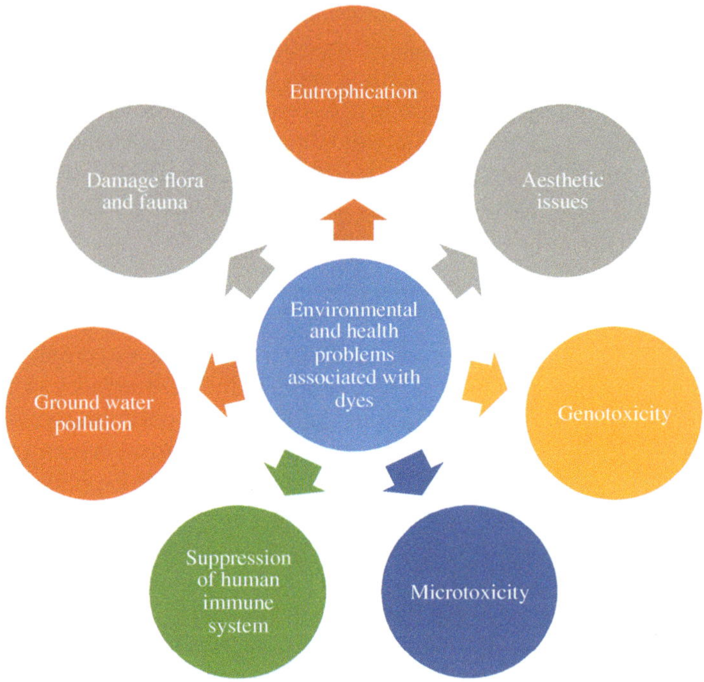

FIG. 18.2 Problem associated with dyes.

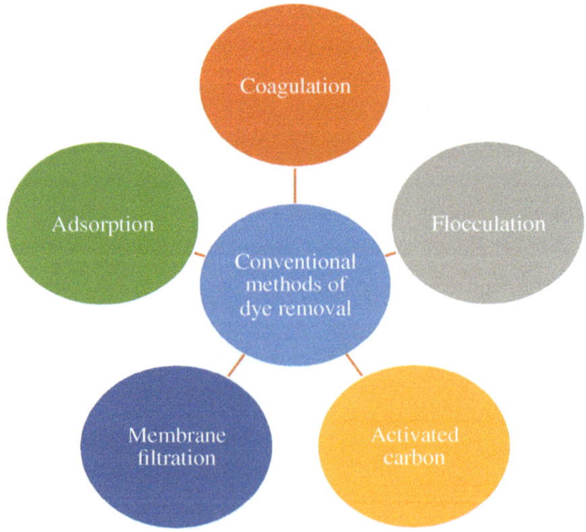

FIG. 18.3 Conventional methods of dye removal.

Adsorption is one of the physical techniques to remove dyes from the wastewater. It is highly efficient process and can be used for a number of dyes. For the removal of dyes from industrial wastewater, liquid phase adsorption is considered to be the best option due to lower cost at the start, less complexity in terms of design, insensitivity to hazardous substances, and ease of operation. Activated charcoal is the most used adsorbent in the industry for its high efficiency of adsorption of different substances. However, activated charcoal loses its efficiency for dispersed and vat dyes. The process yields to loss of the adsorbent and regeneration process is expensive to carry out. Filtration techniques also offer to remove the dyes. The reuse of water is possible by using methods such as ultrafiltration, nanofiltration, and reverse osmosis. These techniques when employed help in lowering significant amount of Biochemical oxygen demand (BOD), Chemical oxygen demand (COD), and color from the wastewater. However, filters tend to create ample of waste product and also installation of filter with every use is not cost-efficient method making filtration process less desirable. Coagulation and flocculation are other physical methods for the separation. While they offer a solution to dispersive and sulfur dyes, they are less effective for the separation of acid, vat, and direct dyes. In case of the latter dyes, the process produces a quantifiable amount of sludge and also not a significant change in color is observed making it less effective. The process is strictly based on the physical separation of dye from the liquid, thereby it tends to produce no toxic intermediates. This property makes the process attractive. The underlying mechanism is complex with lot of unknown aspects as well. Main additive to the coagulation and flocculation is a coagulant; the whole process depends upon the coagulant. Aluminum- and iron-containing salts are usually used as the coagulants as their self-hydrolysis forms hydroxide solids. Passing current through the solution increases the capability of the process. Along with type of coagulant used other factors such as pH, temperature, ions present in the solution also determine the effectiveness. Inorganic flocculants are considered to be the next-generation ones such as polyaluminum chloride. Water quality is retained when these next-generation flocculants are used. They are cost-effective when implemented to a large project, which further makes case for them. However, the demerit of using them is the amount of sludge production and safe disposal of the waste.

Other conventional methods in use are the chemical methods. The most popularly deployed one is oxidation method using oxidizing agents such as ozone, permanganate, and hydrogen peroxide (H_2O_2). The technique comes quite handy in the removal of azo dyes and coloration from the wastewater (Georgiou, 2017). However, in practical it is not cost effective considering high cost of ozone, has a short life-span that is, the oxidizing agent needs to be changed very regularly, chemical reaction is highly susceptible to release byproducts that are further harmful and might cause environmental hazards. It is also observed that the oxidizing method is not as effective as other techniques in lowering the COD value. Dispersed and insoluble dyes are not the substrate for the oxidization reaction. However, some innovative oxidizing methods known as Advances oxidation processes (AOPs) have successfully managed to solve a few demerits. AOPs such as Fenton reaction and H_2O_2/UV process do eliminate the cost factor and have a high COD removal. Harmful sludge production, undesired byproducts formation, and decreased efficiency of color removal in highly contaminated water still remain as the problems to tackle (Brienza and Katsoyiannis, 2017).

18.5 Bioremediation

Usage of biological systems for the treatment or remediation process is known as bioremediation (Kaur and Roy, 2020). It accomplishes removal, reduction, and destruction of hazardous chemical compounds by the usage of biological agents such as algae, bacteria, fungi derived from various sources (Roy and Bharadvaja, 2020) (Singh and Roy, 2021). They are capable of treating soil, water, sludge, and effluents or wastewater from the industries (Saxena, 2015; Das and Dash, 2017). Strategies to carry bioremediation can be implemented in either in situ or ex situ form. In situ is a method to introduce an inoculum of bacteria having desired properties or favoring the growth of certain microorganism in a natural habitat known as bioaugementation and biostimulation, respectively (Saxena, 2015). On the other hand, ex situ strategies include bioremediation by using bioreactors, landfarming, composting, and tools of genetic engineering to improve the microorganisms' activities (Pereira and Alves, 2012; Das and Dash, 2014).

Microorganisms have a unique capability to survive in the harsh environments. It is one of the reasons of why desirability of the biological systems is increasing with continuous addition to the knowledge and processes. The effluents produced by different industries have different chemical properties. Identification of these properties helps us in utilizing the right microorganism. Bacteria are well capable of carrying out the degradation process under both aerobic and anaerobic environments (Telke, Kadam and Govindwar, 2015). Effluents from certain industries such as textile are highly saline in nature due to the presence of varied metal ions in the dyes, in which normal bacteria fail to grow due to plasmolysis (Meng et al., 2012). To overcome these limitations, extremophiles are being deployed which show impressive properties of tolerating highly stressful conditions of pH, salinity, temperature, and toxicity (Amoozegar, Mehrshad and Akhoondi, 2015). A study reported that acid red 27 dye was depigmented by *Shewanella aquimarina* in a NaCl-rich medium (Meng et al., 2012). Another example is of yeast, *Pichia occidentalis* which aerobically discolored acid red B dye by the reduction reaction under highly saline conditions (Song et al., 2017). Thermophiles such as *Anoxybacillus rupiensis,* etc. are also considered to be interesting options as few studies have discovered their potential to degrade the dyes such as reactive black 5 dye and dark red effluent at temperatures above 60 °C (Deive et al., 2010; YH and SG, 2011). Fungi kingdom very well makes their case as they are a reservoir of enzymes;

many of them show high biodegradability (Saratale et al., 2011). Biodegradation process by fungi is mainly dependent and mediated by the enzymes (Kaushik and Malik, 2015; Sen et al., 2016). Azoreductase, Mn peroxidase, laccases, and lignin peroxidase are few of many enzymes facilitating the process of degradation (Saratale et al., 2011). Peroxidases and laccases produces by white rot fungi degrade the dye used in textile industry (Kandelbauer and Guebitz, 2005). *Trametes versicolor* biodegraded reactive red 2 and reactive blue 4 (Nilsson et al., 2006). Fungi such as *Aspergillus* sp. and *Trichoderma* sp. play an important role in discoloration of dyes such as orange G dye, methyl orange, methylene blue by providing a surface for binding and holding the dye in the effluent (Sivasamy and Sundarabal, 2011; Kabbout and Taha, 2014; Takey, 2014).

Bioremediation by microorganism is a cost effective and mostly environment friendly method of treatment of effluents from different industries (Singh, 2006). However, there are other factors that affect its desirability. In the bioreactors, activated sludge cannot act alone, subsidiary compounds are needed. Moreover, the sensitivity of microorganisms to environment conditions provides an operational difficulty. It is also seen that bacterial cultures are not equally effective for every dye. Usage of fungi possesses a different set of problems such as overgrowth of fungi in the bioreactor leading to blockage, contamination of bacteria inhibiting fungal enzyme activity, etc. (Sen et al., 2016). Fungal degradation is also a time-hefty process and it is also retrained by the restrictive environmental factors such as pH, temperature, oxygen-levels, etc. All these physical and chemical parameters need a close monitoring making the bioremediation using microorganism a labor-extensive process as well (Sen et al., 2016).

18.6 Phytoremediation

Phytoremediation is a novel approach that refers to the usage of plants as a medium to carry out the detoxifying process (Alvarez and Illman, 2005). Most of the studies are concentrated on bioremediation and finding microbial communities. It is quite recent that plants have got the recognition and being taken up for the studies. Plants growing in the areas with high levels of contaminants are observed to have advanced metabolic pathways that help them survive in those conditions. This advanced metabolism equips those plants with adaptations that enable them to act as efficient bioreactors to degrade the accumulated pollutants (Campos et al., 2008; Kabra et al., 2011). Major contaminants are metals, pesticides, explosives, landfill leachates, crude oil, solvents, and hydrocarbons, all these pollutants can be detoxified from the environment with the help of plants. Studies have also revealed plants' potential to genetically adapt to grow in these toxic environments with time (Paz-Alberto and Sigua, 2013). Phytoremediation, therefore, has come across as a natural and energy-efficient system for the remediation of pollutants from different sources such as soil, water, air, and other polluted areas with added advantages of planting a plant as well. Plants offer a way out for the disposal

TABLE 18.1 Phytoremediation techniques.

S. No.	Phytotechnique	Definition	Application	Reference
1.	Phytoextraction	A process through which the plants absorb the toxic pollutants from the medium and store in the harvestable plant parts, for example shoots, stem, leaves, etc.	Inorganic contaminants for example, Metals including Pb, Cu, Cr, U, Se, etc.	(Marmiroli, Marmiroli and Maestri, no date; Ali, Khan and Sajad, 2013)
2.	Phytofiltration	Root system of the plant acts as the filter for the removal of contaminant from the water sources	Inorganic pollutants, for example Metals	(Marmiroli, Marmiroli and Maestri, no date; Macek *et al.*, 2009; Al-Baldawi *et al.*, 2013; Ali, Khan and Sajad, 2013)
3.	Phytostabalization	Utilizing the plants and their parts for the immobilization purposes reducing the bioavailability of the pollutants	Metals such as Pb, Cd, Zn, Cu, etc. and hydrophobic organics such as Polychlorinated biphenyl (PCB), Polycyclic aromatic hydrocarbons (PAH), Dichlorodiphenyltrichloroethane (DDT), etc.	(Marmiroli, Marmiroli and Maestri, no date; Macek *et al.*, 2009; Ali, Khan and Sajad, 2013)
4.	Rhizodegradation	Removal of pollutants by the microbes situated in soil by the stimulation of plant released chemicals	Xenobiotics and organic pollutants	(Al-Baldawi *et al.*, 2013; Ali, Khan and Sajad, 2013)
5.	Phytovolatilization	It is the uptake of contaminants from the growth media and converting them into volatile gases and elimination through the transpiration	Organic pollutants, herbicides, xenobiotics, etc.	(Marmiroli, Marmiroli and Maestri, no date; Macek *et al.*, 2009; Ali, Khan and Sajad, 2013)
6.	Phytotransformation	It is the direct degradation of contaminants through metabolic and enzymatic activity of plants	Xenobiotics and organic compounds	(Marmiroli, Marmiroli and Maestri, no date; Al-Baldawi *et al.*, 2013; Ali, Khan and Sajad, 2013)

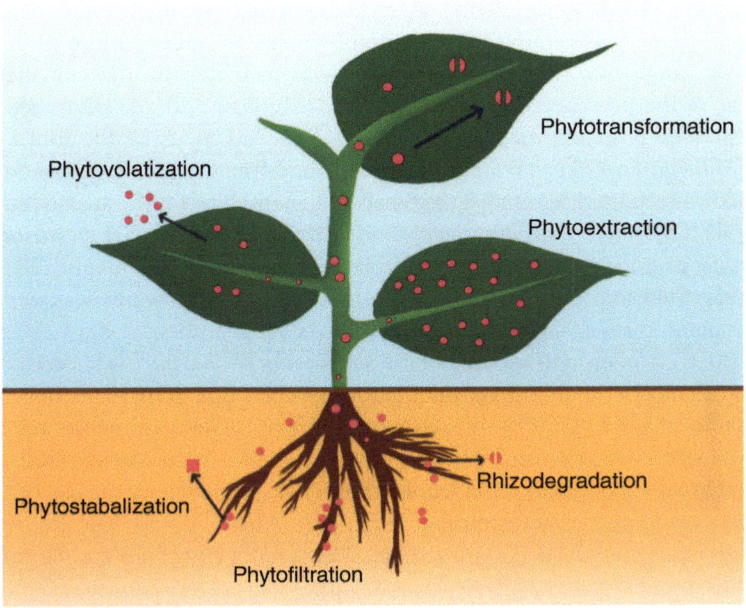

FIG. 18.4 Phytoremediation techniques.

system as well. They are easily harvested, grown, and can be disposed without releasing any further pollutants (Sigua et al., 2004; Etim, 2012). Conventional plants including aquatic plants such as *Phragmites australis* and *Lemna minor* are utilized in phytoremediation (Uysal, Aktas and Caglar, 2014; Hussein and Scholz, 2017). Different phytoremediation techniques have been mentioned in Fig. 18.4 Table 18.1.

18.7 Phytoremediation for removal of dyes

Phytoremediation techniques provide the scientific community an eco-friendly way for the detoxification and revitalization of soil and water bodies with accumulated toxic dye wastage. It is a sustainable alternative carrying out the much needed degradation process through various modes, use of enzymes for the purpose being the most potent. These aspects make the phytoremediation method a desirable and a novel approach (Govindwar and Kagalkar, 2011).

General processes used by plants for the removal of pollutants include dissipation, immobilization, degradation, and accumulation (Wang et al., 2017). Various plants have been reported utilizing these four mechanisms and aiding the process. *Lemna minor* L. in a pond ecosystem was reported to eliminate the basic red 46 particles even if present at low concentrations in the system (Yaseen and Scholz, 2017). It was also reported to have high stress resistance for the dyes such as crystal violet and malachite green. The removal yield of the dyes was about 80% and 90% using phytoextraction and phytodegradation, respectively (Török et al., 2015). A carcinogen and potential teratogenic dye, Brilliant green, was removed by the employment of callus culture of *Tecoma stans* var. *angustata*. Degradation of dye was mediated by the enzyme: peroxidase in the presence of H_2O_2 (Rani and Abraham, 2016). Bioremediation of methylene blue and methyl orange was achieved by using the root system of *Eichhornia crassipes* (Water hyacinth) (Tan, Morad and Ooi, 2016). *Scirpus grossus* was reported to remove the basic dye, methylene blue and also reducing the BOD, COD, and total organic carbon (Almaamary et al., 2017). *Nasturtium officinale* degraded basic red 46 (Torbati, Khataee and Movafeghi, 2014). A study reported that 67% of the dye pollutant in distilled water was degraded by the plant species, *Typhonium flagelliforme* in just about 4 days (Kagalkar et al., 2010). Under hydroponic conditions *Rheum rhabarbarum* showed degradation of sulfur-containing anthraquinones (Aubert and Schwitzguébel, 2004). Reduction of up to 27.33% of color imparted by malachite green and congo red was observed with the implementation of *Aloe barbadensis* (Rai et al., 2014). On the other hand, 95% decolorization was achieved by the use of *Chara vulgaris* using the phytoextractive technique (Mahajan and Kaushal, 2013). Biotransformation method is utilized by *Momordica charantia* for the removal of disperse red 17 and disperse brown 1 dyes (Satar and Husain, 2009). Cell separated from an artificial culture of *Blumea malcolmii* were found 93.41% efficient in decolorizing the waste contaminated with malachite green. Other dyes including Remazol red, red HE8B, methyl orange, red HE7B, etc. were also degraded. Use of enzymes such as indophenol reductase, laccase, and veratryl alcohol oxidase was involved in the process of remediation (Kagalkar et al., 2011).

Some researchers have found that plants working together in a complex system to improve the remediation either with other plant species or a microorganism show potentially useful transformation of organic and inorganic compounds, element cycling, and bioweathering (Deng and Cao, 2017). Wastewater containing acid red B or acid scarlet dye particles was detoxified with the use of a plant microbiome system containing *Medicago sativa* L. or *Sesbania cannabina* as their plant part and *Gracilibacillus* sp. as their microbial partner (Zhou and Xiang, 2013). This type of collective working was also shown by *Petunia grandiflora* and *Gaillardia grandiflora* to reduce the BOD and COD requirements of the textile pollutants (Watharkar and Jadhav, 2014). Other consortiums such as *Ipomea hederifolia* and *Ipomea aquatica* (Rane et al., 2016) and *Paspalum scrobiculatum* and *Typha angutifolia* (Chandanshive et al., 2017) also illustrated biodegradation of brown 5R and congo red, respectively.

18.8 Phytoremediation for removal of azo-dyes

Azo dyes are extensively used in the textile industry contributing as the major pollutant in the effluent by the industry. Structurally, azo dyes contain

TABLE 18.2 Role of different plants in dye degradation.

Plant used	Phytoremediated dye
Lemna minor L	Basic red 46, Crystal violet, malachite green
Tecoma stans var. *angustata*	Brilliant green
Eichhornia crassipes	Methylene blue, methyl orange
Scirpus grossus	Methylene blue
Aloe barbadensis	Malachite green, congo red
Typhonium flagelliforme	Dye pollutants
Blumea malcolmii	Malachite green, remazol red, red HE8B, methyl orange, red HE7B, reactive red 5B dye
Medicago sativa L. or *Sesbania cannabina* with *Gracilibacillus* sp.	Acid red B/acid scarlet dye
Ipomea hederifolia and *Ipomea aquatica*	Brown 5R
Petunia grandiflora with *Gailardia grandiflora*	Textile dyes
Paspalum scrobiculatum and *Typha aguti*	Congo red
Momordica charantia	Disperse red 17, disperse brown 1
Tagetes patula L.	Reactive Red 198, methyl orange, orange M2RL, navy blue HE2R, reactive red M5B

one or more double-bonded nitrogen (-N=N-). Azo dyes are characterized as mutagens and carcinogens to humans. Highly saline conditions provided by the wastewater from the industry and the toxicity introduced by the azo dyes make these dyes a difficult target for any other remediation process especially bioremediation. To which phytoremediation offers some attractive solutions. Plant species including *Sorghum vulgare, Phaseolus mungo,* and *Brassica juncea* have been evaluated for their decolorization (breaking of doubly bound nitrogen atoms) potential and their efficiency was about 57%, 53%, and 79%, respectively (Ghodake et al., 2009). Coco yam plantations showed decolorization of wetlands by 72%–77%. Various studies are being conducted for the appropriate measure of phytoremediation potential of different plants. Use of *Tagetes patula* L. root system degraded the reactive red 198, methyl orange, orange M2RL, navy blue HE2R, and reactive red M5B (Patil et al., 2009).

18.9 Advantages of phytoremediation

Comparing to the other existing methodology and techniques for the removal of dyes, phytoremediation is an interesting approach in terms of cost efficiency. The technique is highly cost-efficient throughout the process. It does not need to have a skilled labor for the maintenance and requires less expensive equipment. It is also easily implemented without disturbing any other aspect of the environment. Growing plants at the places of contamination is also socially acceptable making its implementation way easier. The wide variety of plants also gives the option to find the right plant for the particular type of dye. This is also aesthetically soothing and eco-friendly process also contributing toward the global warming notion. Deployment of the phytoremediation method can also improve the soil fertility and minimize the erosion process as well. Its easy implementation makes it desirable choice for the vast scale uses covering even acres of land (Aken, Correa and Schnoor, 2010; Wuana and Okieimen, 2011; Jacob et al., 2018).

18.10 Disadvantages of phytoremediation

Although phytoremediation solved many existing shortcomings of the conventional detoxifying methodologies, the process in itself comes up with new challenges and disabilities. In some plants the up taken contaminants are plausible to get stored in the edible and consumable parts of the plant entering into food chain, which may result into biomagnification process to occur. Remediation of the soil or water bodies needs an extensive plan for the plantation purposes. The process, no matter how affective, is very slow process taking up to months and even years. Handling of the contaminated plants and disposal still remains an issue (Rostam, 2016).

18.11 Conclusion

Effluents released by the industries using the dyes as their raw material in the manufacturing process are mostly polluted by the dyes that are directly drained into the water bodies. Over the time, accumulation of these particles happens in a local water body, which not pleasing to eye and also possess health hazards. Over the years, a number of techniques and modifications have been implemented to address these concerns and detoxify the water bodies, but all the strategies come with other significant challenges. Bioremediation and phytoremediation are two attractive approaches to deal with dye pollutants in an eco-friendlier and sustainable manner. Of the two, phytoremediation is considered to be a process for the future and researchers are interested in it due to minimalistic sludge formation, high enzymatic activity, and especially due to its cost-effectiveness. Phytoremediation can even be implemented for the wastewater having azo-dye contamination, solving other crisis in hand. Given these factors, field of biotechnology has provided us with a novel, sustainable, and effective approach to carry out detoxification of the soil and water bodies having colored dyes as contaminants.

References

Adegoke, K.A., Bello, O.S., 2015. Dye sequestration using agricultural wastes as adsorbents. Water Resour. Ind. 12, 8–24.

Aken, B.Van, Correa, P.A., Schnoor, J.L, 2010. Phytoremediation of polychlorinated biphenyls: new trends and promises. Environ. Sci. Technol. 44 (8), 2767–2776. doi:10.1021/es902514d.

Al-Baldawi, I.A.W, et al., 2013. Comparative performance of free surface and sub-surface flow systems in the phytoremediation of hydrocarbons using Scirpus grossus. J. Environ. Manag. 130, 324–330. doi:10.1016/j.jenvman.2013.09.010.

Ali, H., Khan, E., Sajad, M.A., 2013. Phytoremediation of heavy metals—concepts and applications. Chemosphere 91 (7), 869–881. doi:10.1016/j.chemosphere.2013.01.075.

Almaamary, E.A.S, et al., 2017. Rawatan metilena biru dalam air sisa menggunakan Scirpus grossus. Malay. J. Anal. Sci. 21 (1), 182–187. doi:10.17576/mjas-2017-2101-21.

Alvarez, P.J.J., Illman, W.A, 2005. Bioremediation and Natural Attenuation. John Wiley & Sons, Inc, Hoboken, NJ, USA. doi:10.1002/047173862X.

Amoozegar, M. A., Mehrshad, M. and Akhoondi, H., 2015 .Application of extremophilic microorganisms in decolorization and biodegradation of textile wastewaterIn: Singh S. (eds) Microbial Degradation of Synthetic Dyes in Wastewaters. Environmental Science and Engineering. Springer, Cham., pp. 267–295. doi: 10.1007/978-3-319-10942-8_12.

Aubert, S., Schwitzguébel, J.-P., 2004. Screening of plant species for the phytotreatment of wastewater containing sulphonated anthraquinones. Water Res. 38 (16), 3569–3575. doi:10.1016/j.watres.2004.04.057.

Bhatia, S. C., 2017. Pollution control in textile industry, Pollution Control in Textile Industry. doi: 10.1201/9781315148588.

Brienza, M., Katsoyiannis, I., 2017. Sulfate radical technologies as tertiary treatment for the removal of emerging contaminants from wastewater. Sustainability 9 (9), 1604. doi:10.3390/su9091604.

Campos, V.M., et al., 2008. Phytoremediation of organic pollutants: a review. Span. J. Agric. Res. 6 (S1), 38. doi:10.5424/sjar/200806S1-372.

Chandanshive, V.V., et al., 2017. Co-plantation of aquatic macrophytes Typha angustifolia and Paspalum scrobiculatum for effective treatment of textile industry effluent. J. Hazard. Mater. 338, 47–56. doi:10.1016/j.jhazmat.2017.05.021.

Christie, R., 2007. Environmental Aspects of Textile Dyeing,. Woodhead Publishing Limited, Cambridge, UK. doi:10.1533/9781845693091.

Clark, M., 2011. Handbook of Textile and Industrial Dyeing: Principles, Processes and Types of Dyes. Woodhead Publishing Limited, Cambridge, UK. doi:10.1533/9780857093974.

Das, S., Dash, H.R., 2014. Microbial bioremediation: a potential tool for restoration of contaminated areas. In: Das, S. (Ed.), Microbial Biodegradation and Bioremediation. Elsevier, Amsterdam, pp. 1–21. doi:10.1016/B978-0-12-800021-2.00001-7.

Das, S., Dash, H.R., 2017. Handbook of Metal-Microbe Interactions and Bioremediation. CRC Press, Boca Raton, FL,. doi:10.1201/9781315153353.

Deive, F.J., et al., 2010. Decolorization of dye reactive black 5 by newly isolated thermophilic microorganisms from geothermal sites in Galicia (Spain). J. Hazard. Mater. 182 (1–3), 735–742. doi:10.1016/j.jhazmat.2010.06.096.

Deng, Z., Cao, L., 2017. Fungal endophytes and their interactions with plants in phytoremediation: a review. Chemosphere 168, 1100–1106. doi:10.1016/j.chemosphere.2016.10.097.

Etim, E.E., 2012. Phytoremediation and its mechanisms: a reviewInt. J. Environ. Bioenergy2, 120–136. http:// www.ModernScientificPress.com.

Georgiou, D., 2017. Destruction of azo-reactive dyes by ozonation and the synergetic effect of a radio-frequency alternating electric field inductance deviceCurr. Trends Fashion Technol. Textile Eng.1. doi:10.19080/CTFTTE.2017.01.555560.

Ghodake, G.S., et al., 2009. Potential of Brassica juncea in order to treat textile—effluent—contaminated sites. Int. J. Phytoremediation 11 (4), 297–312. doi:10.1080/15226510802429518.

Govindwar, S.P., Kagalkar, A.N., 2011. Phytoremediation Technologies for the Removal of Textile Dyes - An Overview and Future Prospects. Nova Science Publishers, Inc., NY, pp. 472–494.

Hassan, M.M., Carr, C.M., 2018. A critical review on recent advancements of the removal of reactive dyes from dyehouse effluent by ion-exchange adsorbents. Chemosphere 209, 201–219. doi:10.1016/j.chemosphere.2018.06.043.

Hunger, K., 2007. Industrial Dyes: Chemistry, Properties, Applications. John Wiley & Sons, NJ. https://books.google.com/books?hl=en&lr=&id=uAzS4Hk2TgwC&oi=fnd&pg=PR5&ots=e6uoCQK4G_&sig=4raJVwlPpztPluci00efjWCiBH0. Accessed: 29 September 2020.

Hussein, A., Scholz, M., 2017. Dye wastewater treatment by vertical-flow constructed wetlands. Ecol. Eng. 101, 28–38. doi:10.1016/j.ecoleng.2017.01.016.

Imran, M., et al., 2015. Microbial biotechnology for decolorization of textile wastewaters. Rev. Environ. Sci. Biotechnol. 14 (1), 73–92. doi:10.1007/s11157-014-9344-4.

Ito, T., et al., 2016. Long–term natural remediation process in textile dye–polluted river sediment driven by bacterial community changes. Water Res. 100, 458–465. doi:10.1016/j.watres.2016.05.050.

Jacob, J.M., et al., 2018. Biological approaches to tackle heavy metal pollution: a survey of literature. J. Environ. Manag. 217, 56–70. doi:10.1016/j.jenvman.2018.03.077.

Kabbout, R., Taha, S., 2014. Biodecolorization of textile dye effluent by biosorption on fungal biomass materials. Phys. Procedia 55, 437–444. doi:10.1016/j.phpro.2014.07.063.

Kabra, A.N., et al., 2011. Phytoremediation of a sulphonated azo dye green HE4B by Glandularia pulchella (Sweet) Tronc. (Moss Verbena). Environ. Sci. Pollut. Res. 18 (8), 1360–1373. doi:10.1007/s11356-011-0491-7.

Kagalkar, A.N., et al., 2010. Studies on phytoremediation potentiality of Typhonium flagelliforme for the degradation of Brilliant Blue R. Planta 232 (1), 271–285. doi:10.1007/s00425-010-1157-2.

Kagalkar, A.N., et al., 2011. Phytodegradation of the triphenylmethane dye malachite green mediated by cell suspension cultures of *Blumea malcolmii* Hook. Bioresour. Technol. 102 (22), 10312–10318. doi:10.1016/j.biortech.2011.08.101.

Kandelbauer, A., Guebitz, G.M., 2005. Bioremediation for the Decolorization of Textile Dyes — A Review. In: Lichtfouse, E., Schwarzbauer, J., Robert, D. (Eds.), Environmental Chemistry. Springer-Verlag, Berlin/Heidelberg, pp. 269–288. doi:10.1007/3-540-26531-7_26.

Kaur, S., Roy, A. 2020. Bioremediation of heavy metals from wastewater using nanomaterials. Environment, Development and Sustainability, 1–24.

Kaushik, P. and Malik, A., 2015. Mycoremediation of synthetic dyes: an insight into the mechanism, process optimization and reactor design, In: S.N. Singh (Ed.) Microbial Degradation of Synthetic Dyes in Wastewaters, Springer, Cham, pp. 1–25. doi: 10.1007/978-3-319-10942-8_1.

Khan, S., Malik, A., 2018. Toxicity evaluation of textile effluents and role of native soil bacterium in biodegradation of a textile dye. Environ. Sci. Pollut. Res. 25 (5), 4446–4458. doi:10.1007/s11356-017-0783-7.

Klementova, S., 2018. Photochemical degradation of organic xenobiotics in natural waters. In: Saha, S. (Ed.), Photochemical degradation of organic xenobiotics in natural waters. Photochemistry and Photophysics: Fundamentals to Applications, 67.

Lellis, B., Fávaro-Polonio, C.Z., Pamphile, J.A., Polonio, J.C., 2019. Effects of textile dyes on health and the environment and bioremediation potential of living organisms. Biotechnol. Res. Innov. *3* (2), 275–290.

Macek, T. et al., 2009. Advances in phytoremediation and rhizoremediation, In: A. Singh, R.C. Kuhad, and O.P. Ward (Eds.) Advances in Applied Bioremediation, Springer, Cham, pp. 257–277. doi: 10.1007/978-3-540-89621-0_14.

Mahajan, P., Kaushal, J., 2013. Degradation of congo red dye in aqueous solution by using phytoremediation potential of *Chara Vulgaris*. Chitkara Chem. Rev. 1 (1), 67–75. doi:10.15415/ccr.2013.11005.

Marmiroli, N., Marmiroli, M., Maestri, E., 2006. Phytoremediation and phytotechnologies: a review for the present and the future. In: Twardowska, I., Allen, H.E., Häggblom, M.M., Stefaniak, S. (Eds.), Soil and Water Pollution Monitoring, Protection and Remediation. Dordrecht: Springer, Netherlands, pp. 403–416. doi:10.1007/978-1-4020-4728-2_26.

Meng, X., et al., 2012. Azo dye decolorization by *Shewanella aquimarina* under saline conditions. Bioresour. Technol. 114, 95–101. doi:10.1016/j.biortech.2012.03.003.

Mondal, S., Purkait, M.K., De, S., 2018. Advances in Dye Removal Technologies, Advances in Dye Removal Technologies. Springer, Cham. doi:10.1007/978-981-10-6293-3.

Muthu, S. S., 2017. Introduction In: S.S. Muthu (Ed.) Sustainability in the Textile Industry, Springer, Cham, pp. 1–8. doi: 10.1007/978-981-10-2639-3_1.

Nilsson, I., et al., 2006. Decolorization of synthetic and real textile wastewater by the use of white-rot fungi. Enzyme Microb. Technol. 38 (1–2), 94–100. doi:10.1016/j.enzmictec.2005.04.020.

Ohgaki, S., et al., 2015. Southeast Asian water environment 1. Water Intelligence Online 4, 9781780402987. doi:10.2166/9781780402987.

Patil, P., et al., 2009. Degradation analysis of reactive red 198 by hairy roots of Tagetes patula L. (Marigold). Planta 230 (4), 725–735. doi:10.1007/s00425-009-0980-9.

Paz-Alberto, A.M., Sigua, G.C., 2013. Phytoremediation: a green technology to remove environmental pollutants. Am. J. Clim. Change 02 (01), 71–86. doi:10.4236/ajcc.2013.21008.

Pereira, L., Alves, M., 2012. 'Dyes—Environmental Impact and Remediation', in Environmental Protection Strategies for Sustainable Development. Dordrecht: Springer, Netherlands, pp. 111–162. doi:10.1007/978-94-007-1591-2_4.

Piątkowska, M., et al., 2018. Absence of evidence or evidence of absence? a transfer and depletion study of Sudan I in eggs. Food Chem. 239, 598–602. doi:10.1016/j.foodchem.2017.06.156.

Rai, M. S. et al., 2014. Original research article degradation of malachite green and congo red using *Aloe barabadensis* Mill. Extract, 3(4), 330–340.

Raina, S., Roy, A., Bharadvaja, N., 2020. Degradation of dyes using biologically synthesized silver and copper nanoparticles. Environmental Nanotechnology, Monitoring & Management, 13, 100278.

Rane, N.R., et al., 2016. *Ipomoea hederifolia* rooted soil bed and *Ipomoea aquatica* rhizofiltration coupled phytoreactors for efficient treatment of textile wastewater. Water Res. 96, 1–11. doi:10.1016/j.watres.2016.03.029.

Rani, N., Abraham, E., 2016. A potential tissue culture approach for the phytoremediation of dyes in aquaculture industry. Biochem. Eng. J. 115, 23–29. doi:10.1016/j.bej.2016.08.001.

Rawat, D., Mishra, V., Sharma, R.S., 2016. Detoxification of azo dyes in the context of environmental processes. Chemosphere 155, 591–605. doi:10.1016/j.chemosphere.2016.04.068.

Rehman, K., et al., 2018. Effect of reactive black 5 azo dye on soil processes related to C and N cycling. PeerJ 6, e4802. doi:10.7717/peerj.4802.

Rostam, F.A., 2016. Advantages and Disadvantages of RAD software development. Int. Res. J. India 1 (VIII), 1–3. http://www.my-project-management-expert.com/the-advantages-and-disadvantages-of-rad-software-development.html.

Roy, A., Bharadvaja, N., 2019. Silver nanoparticle synthesis from Plumbago zeylanica and its dye degradation activity. Bioinspired, Biomimetic and Nanobiomaterials, 8(2), 130–140.

Roy, A., Bharadvaja, N. 2020. Removal of toxic pollutants using microbial fuel cells. In Removal of toxic pollutants through microbiological and tertiary treatment. Elsevier, pp. 153–177.

Sandhya, S., 2010. Biodegradation of azo dyes under anaerobic condition: role of azoreductase. In: Erkurt, H.A. (Ed.), Biodegradation of Azo Dyes. Springer, Cham, pp. 39–57. doi:10.1007/698_2009_43.

Saratale, R.G., et al., 2011. Bacterial decolorization and degradation of azo dyes: a review. J. Taiwan Inst. Chem. Eng. 42 (1), 138–157. doi:10.1016/j.jtice.2010.06.006.

Satar, R., Husain, Q., 2009. Phenol-mediated decolorization and removal of disperse dyes by bitter gourd (*Momordica charantia*) peroxidase. Environ. Technol. 30 (14), 1519–1527. doi:10.1080/09593330903246432.

Saxena, S., 2015. Applied Microbiology. doi:10.1007/978-81-322-2259-0.

Sen, S.K., et al., 2016. Fungal decolouration and degradation of azo dyes: a review. Fungal Biol. Rev. 30 (3), 112–133. doi:10.1016/j.fbr.2016.06.003.

Sigua, G.C., Holtkamp, M.L., Coleman, S.W., 2004. Assessing the efficacy of dredged materials from lake Panasoffkee, Florida: implication to environment and agriculture. Environ. Sci. Pollut. Res. 11 (6), 394–399. doi:10.1007/BF02979659.

Singh, A., Roy, A., 2021. Fungal communities for the remediation of environmental pollutants. In Recent Trends in Mycological Research, Springer, Cham, pp. 127–165.

Singh, H., 2006. Mycoremediation. John Wiley & Sons, Inc, Hoboken, NJ, USA. doi:10.1002/0470050594.

Sivasamy, A., Sundarabal, N., 2011. Biosorption of an azo dye by *Aspergillus niger* and *Trichoderma* sp. fungal biomasses. Curr. Microbiol. 62 (2), 351–357. doi:10.1007/s00284-010-9713-3.

Sheng, S., Liu, B., Hou, X., Wu, B., Yao, F., Ding, X., Huang, L., 2018. Aerobic biodegradation characteristic of different water-soluble azo dyes. International J. Environ. Res. Public Health 15 (1), 35.

Song, L., et al., 2017. Performance of a newly isolated salt-tolerant yeast strain *Pichia occidentalis* G1 for degrading and detoxifying azo dyes. Bioresour. Technol. 233, 21–29. doi:10.1016/j.biortech.2017.02.065.

Tahir, U., Sohail, S., Khan, U.H., 2017. Concurrent uptake and metabolism of dyestuffs through bio-assisted phytoremediation: a symbiotic approach. Environ. Sci. Pollut. Res. 24 (29), 22914–22931. doi:10.1007/s11356-017-0029-8.

Takey, M., 2014. Bioremediation of xenobiotics: use of dead fungal biomass as biosorbent. Int. J. Res. Eng. Technol. 03 (01), 565–570. doi:10.15623/ijret.2014.0301094.

Tan, K.A., Morad, N., Ooi, J.Q., 2016. Phytoremediation of methylene blue and methyl orange using *Eichhornia crassipes*. Int. J. Environ. Sci. Dev. 7 (10), 724–728. doi:10.18178/ijesd.2016.7.10.869.

Telke, A.A., Kadam, A.A., Govindwar, S.P., 2015. Bacterial enzymes and their role in decolorization of azo dyes. In: Singh, S.N. (Ed.), Microbial Degradation of Synthetic Dyes in Wastewaters. Springer, Cham, pp. 149–168. doi:10.1007/978-3-319-10942-8_7.

Torbati, S., Khataee, A.R., Movafeghi, A., 2014. Application of watercress (*Nasturtium officinale* R. Br.) for biotreatment of a textile dye: investigation of some physiological responses and effects of operational parameters. Chem. Eng. Res. Des. 92 (10), 1934–1941. doi:10.1016/j.cherd.2014.04.022.

Török, A., et al., 2015. Biological removal of triphenylmethane dyes from aqueous solution by *Lemna minor*. Acta Chim. Sloven 62 (2), 452–461. doi:10.17344/acsi.2014.1109.

Uysal, Y., Aktas, D., Caglar, Y., 2014. Determination of colour removal efficiency of *Lemna minor* L. from industrial effluents. J. Environ. Protect. Ecol. 15 (4), 1718–1726.

Verma, A., Roy, A., Bharadvaja, N., 2021. Remediation of heavy metals using nanophytoremediation. In Advanced oxidation processes for effluent treatment plants. Elsevier, pp. 273–296.

Wang, L., et al., 2017. A review on in situ phytoremediation of mine tailings. Chemosphere 184, 594–600. doi:10.1016/j.chemosphere.2017.06.025.

Watharkar, A.D., Jadhav, J.P., 2014. Detoxification and decolorization of a simulated textile dye mixture by phytoremediation using *Petunia grandiflora* and, *Gailardia grandiflora*: A plant–plant consortial strategy. Ecotoxicol. Environ. Saf. 103, 1–8. doi:10.1016/j.ecoenv.2014.01.033.

Wuana, R.A., Okieimen, F.E., 2011. Heavy metals in contaminated soils: a review of sources, chemistry, risks and best available strategies for remediation. ISRN Ecol., 1–20. doi:10.5402/2011/402647.

Yaseen, D.A., Scholz, M., 2017. Comparison of experimental ponds for the treatment of dye wastewater under controlled and semi-natural conditions. Environ. Sci. Pollut. Res. 24 (19), 16031–16040. doi:10.1007/s11356-017-9245-5.

YH, G., SG, G., 2011. Decolourization of textile effluent by a thermophilic bacteria *Anoxybacillus rupiensis*. J. Pet. Environ. Biotechnol. 02 (02). doi:10.4172/2157-7463.1000111.

Zhou, X., Xiang, X., 2013. Effect of different plants on azo-dye wastewater bio-decolorization. Procedia Environ. Sci. 18, 540–546. doi:10.1016/j.proenv.2013.04.073.

Chapter 19

Phytoremediation of pharmaceutical wastes

Arpita Banerjee[a#], Sasha Cardozo[a#], Satya Prakash Khuntia[a], Anjani Kumar Upadhyay[b]

[a]*School of Biotechnology, Kalinga Institute of Industrial Technology, Bhubaneswar, Odisha*

[b]*Heredity Lifesciences Pvt. Ltd., Bhubaneswar, Odisha*

[#]*Arpita Banerjee and Sasha Cardozo share equal authorship*

19.1 Introduction

Pharmaceutical compounds are indispensable for maintaining human health and are produced in humongous amounts annually. These compounds, although a boon to mankind, can have varying effects on the environment including disbalancing various ecosystems and threatening natural habitats, and therefore its monitoring and evaluation are of high importance. The pharmaceutical compounds generally released into the environment include therapeutic steroids (Hormones), antibiotics, analgesics, antimicrobials, antidepressants, and many other drugs and their derivatives (Archer et al., 2017; Ebele et al., 2017). Most of these pollutants make their way to water bodies because of the incessant release of untreated waste from hospitals, households, and industries (Loraine and Pettigrove, 2006). The wastewater treatment facilities installed in cities of most developing and developed countries are not designed for the treatment of these compounds and are one of the major reasons for releasing these compounds without being eliminated (Ashton et al., 2004). Pharmaceutical compounds can be very difficult to degrade naturally and can persist in the environment for a long period. Long exposure to antibiotics can even lead to the formation of antimicrobial-resistant pathogens that can lead to deadly outbreaks and claim thousands of lives (Vikesland et al., 2017). Acknowledging the toxicity and adverse effects it can have on the environment, various strategies and technologies are being developed to facilitate the efficient removal of these pollutants, most advanced techniques are chemical based and involve the use of activated carbon for adsorption, ozonization, electrochemical oxidation, nanofiltration, coagulation, chlorination, and reverse osmosis (Quesada et al., 2019). Although these techniques are efficient in removing the pharmaceutical compounds, they are not at all cost-effective, requiring additional setups and

Phytoremediation. DOI: https://doi.org/10.1016/B978-0-323-89874-4.00003-0

expertise (Zhang et al., 2014). The need for a low cost and effective technology for the removal of pharmaceutical compounds can be met by the use of biological systems such as membrane separation systems, photocatalysis, aerobic and anaerobic bioreactors. These technologies, however, are not as efficient as other advanced chemical and physical techniques but can be promising and hold a lot of potential. Especially the emergence of techniques involving plants and their associated rhizospheres that can efficiently remove these pollutants from soil, water bodies, and municipal slurries. The process is called "Phytoremediation" and this process is being used to remove recalcitrant toxic compounds such as pesticides, industrial wastes, organic and inorganic pollutants from contaminated soil and wastewater by use of constructed wetlands (CW) (Li et al., 2020; Cunningham et al., 1995). This is a low-cost and low-maintenance technology and it is also being employed for the elimination of pharmaceutical compounds efficiently (Carvalho, 2020). In this chapter, we will learn about pharmaceutical compounds, their characterization, effects on the environment, and humans. We will understand some key mechanisms, technologies of detection, and elimination of these compounds from the environment. We will also discuss in detail about phytoremediation, its mechanism of action, emphasizing the use of aquatic plants for phytoremediation, its challenges, and future perspectives.

19.2 Emergence of pharmaceutical wastes

In the recent world as the population is emerging the demand for food, land, and water is also increasing. In the same manner, the demand for pharmaceutical products is also at its peak which is one of the most essential commodities for survival. The major complication related to these emerging pharmaceutical products is their disposal and removal from the environment as these products are not easily biodegradable. Waste discharge from hospitals also often carries traces of pharmaceutical compounds. One of the major facts for identifying these pharmaceutical wastes in the environment is the proper channeling of technologies such as gas chromatography and mass spectroscopy (GC-MS), liquid chromatography with MS (LC-MS) (Patel et al., 2020). The primary source of accumulation of these wastes in the environment is through humans. The drugs that the humans take in are not completely metabolized by the body, some remnants of the drug are excreted out in the form of urine and faces which eventually mixes with the water bodies and contaminates them (Boxall, 2012). Other point sources have also been identified which are responsible for the emergence of pharmaceutical waste, such as the sewage treatment plant. The treatment done in these plants is not enough to degrade the hormonal residue or the drug residue and this sewage flows to other water bodies making them contaminated. The sewage from drug manufacturing companies also carries traces of pharmaceutical contaminants, which is a direct cause for emerging pharmaceutical wastes (Kolpin et al., 2002). The emerging production of birth control pills that mimic the natural estrogen and hormonal injections for growth and are given to cattle and other livestock is an alarming cause that increases the level of pharmaceutical products in nature (Valdez-Carrillo et al., 2020). Veterinary medicine when applied to cattle, the remnant of that pharmaceutical products is found in their faces that later is used as a fertilizer, which adds to the pharmaceutical load (Boxall et al., 2003). Besides all these factors there is a risk of the household effluent carrying pharmaceutical contamination from expired drugs.

19.3 Classified pharmaceutical wastes

In general, pharmaceutical waste can be classified into two groups: hazardous pharmaceutical waste and nonhazardous pharmaceutical waste. Hazardous pharmaceuticals are further divided into characteristic waste. These types of waste have a certain characteristic feature such as ignitability compounds that are easily flammable; reactivity compounds that release toxic fumes, gases, or causes an explosion when comes in contact with other compounds; corrosivity compounds that have the ability to burn the skin or corrode any other compound and toxicity. The wastes are generally listed in four groups P, U, F, and K. The P group wastes are acutely hazardous waste that is even a trace amount of the compound is hazardous. The drugs present in this group are added to the criteria of their lethal dose (LD50). LD50 is the quantity of a drug that when administered to any organism causes death in 50% of the organisms. At present, there are eight compounds that come under the P-list. U-listed Pharmaceutical compounds include compounds such as neoplastic waste, phenol, acetone, chloral hydrate, and lindane. These compounds are generally known for their toxicity. At present, there are 21 such compounds. The nonhazardous class of pharmaceutical waste basically deals with the leftover or unused pharmaceutical products, yet this waste should be discarded according to the guidelines of the Drug Enforcement Administration (DEA) and (EPA).

19.3.1 Effects on the environment

Among various other environmental pollutants, pharmaceutically active compounds, belonging to a class of complex molecules, predominantly affect the quality of the environment that encompasses all forms of life. The ionic characteristics of pharmaceutical wastes, along with their varied physicochemical and biological functionalities, make them very hazardous environmental threats. However, systematic chemical classification of these wastes is required to indicate their biologically active components (beta-lactams, cephalosporins, and quinolones). The rampant oral and intravenous administration of pharmaceutical drugs increases their chances of being incorporated into the environment after use. Drugs that are metabolized may lose their potency, but some poorly metabolized drugs such as contrast media are excreted in a near functional and high efficacy state. These inadequately metabolized drugs are disseminated into sewage treatment and wastewater plants. At this stage, notable care is taken to remove these wastes so as to prevent them from entering aquatic ecosystems and finally contaminating drinking water reserves for multitudes of people. Hormones are another disturbing subset of pharmaceuticals that are conventionally nonbiodegradable. These comprise endocrine modulating substances that are frequently found in the effluents of sewage treatment plants. Pharmaceutical compounds used in animal husbandry and veterinary practices also increase the infiltration of drugs and their metabolites into the environment. Various diagnostics and disinfectants seep into the subsoil (Kümmerer, 2004). While we cannot ignore the potential scope of pharmaceutical compounds in the modern world, it is impossible to exempt the toxic nature of the wastes they generate. These pharmaceutical wastes manifest themselves as wastewater runoff, leachate seepages, and contaminants from animal wastes. Limited concentrations of human and animal pharmaceutical wastes have been sourced from various groundwater and surface water reservoirs across the globe. In India, the pharmaceutical companies account for the third-largest industry volumetrically. With an average growth rate of about 8%–9% per year, it is estimated that the net worth of the overall pharmaceutical waste produce is around 4.5 billion dollars. Partially treated pharmaceutical pollutants corrupt the environment and make it a breeding ground for disease and infection. Although most of these pharmaceutical wastes occur in negligible amounts in drinking water sources, it is evident that even trace amounts of these wastes can adversely affect human and animal health. According to the United States Environmental Protection Agency, the major constituents of pharmaceutical wastes consist of birth control substances and antibiotics (Patneedi and Prasadu, 2015). Most of the industrial pharmaceutical wastes arise from the improper handling of manufacturing and formulation units that lead to several challenges while handling, storing, and containing pharmaceutical compounds that ultimately accumulate as waste. The different ports, docks, and pipelines used for the transportation of pharmaceutical compounds are often exposed to breakages, spillages, and mechanical failures that increase the incidence of air pollution by emitting carbon monoxide (CO), Nitrogen dioxide (NO_2), and other volatile organic substances such as toluene, methanol, and acetonitrile-containing compounds. The increased toxicity of effluents that are nonbiodegradable in nature makes them fatal health hazards in drinking water streams and channels (Kapoor, 2015). Pharmaceutical wastes find their way to the environment through the intentional use of growth promoters and animal feed. Personal care products along with pharmaceuticals have aroused alarming rates of concern in the world today. Steroids, cosmetics, antidepressants, disinfectants, and analgesics abound in freshwater reserves, which are sources of drinking water for a large number of people. Anthropogenic activities largely facilitate the dispersion of pharmaceutical wastes from landfill leachate, livestock breeding, irrigation, and sewage discharge, thereby promoting their accumulation in groundwater. The fate of pharmaceutical waste products in groundwater reserves is an area that has not been probed. Being an indispensable source of drinking water for many countries across the globe, it is very challenging to remediate the adverse consequences of pharmaceutical waste contamination in groundwater resources. Emerging organic contaminants, in groundwater have increasingly been reported to have become the primary source of fear and concern for many citizens alike. Additionally, aquatic ecosystems populated by low concentrations of partially degraded antibiotics can promote the occurrence of antimicrobial resistance (AMR), which severely impacts the health of humans. In a study conducted by Fick et al., the detection frequency of antibiotics in a network of 47 groundwater sites was found to exceed 30%, which raises the stakes of AMR in the same areas, contributing to various immune deficiencies and health-related hazards (Fick et al., 2009). Caffeine on the other hand, which is rampantly consumed by almost all demographic sections in a population, also contaminates aquatic ecosystems and thereby drinking water reserves (Sui et al., 2015). The adverse effects of pharmaceutical contamination are frequently observed in organisms that did not intend to consume these products, thus contributing to the ecotoxicity of pharmaceutical wastes. Wastewater dislodged from medical facilities is sometimes directly released into the sewage treatment plants, without being exposed to pretreatment procedures. Municipal wastewater plants are designed with the overall goal to remove all traces of phosphorus, carbon, nitrogen, and pathogenic microbiota from wastewater. However, the efficiency of reducing the percentage of micropollutants in wastewater is highly reduced because of inadequate knowledge of the nature of contaminants and the mechanisms employed to treat wastewater effectively (Cidlinová et al., 2018). The emerging rise in the saturation of pharmaceutical wastes in potable and drinking water should be of paramount concern to us. The clogging of sewage pipes and the inefficient treatment of wastewater plants is a potential health hazard that needs to be addressed regularly. Tube wells and water pumps that source drinking water from groundwater reserves heavily depend on the quality and quantity of toxic elements, which are often neglected. The flora and fauna that are native inhabitants of several polluted aquatic and terrestrial ecosystems, suffer immensely because of the fatal consumption, absorption, and imbibition of pharmaceutical wastes and their emissions in trace amounts (Table 19.1). AMR, a consequence of pharmaceutical waste pollution, is the leading cause of various immunological diseases that chronically impairs the overall health and wellbeing of all organisms alike. It should be the foremost need of the hour to functionalize effective and sustainable treatment and remediation measures to prevent the unforeseen damages that pharmaceutical wastes do to our environment.

19.3.2 Effects on humans

One of the most detrimental effects of pharmaceutical wastes on humans is the persistent accumulation of xenobiotics in plant tissues and organs. Xenobiotics are a class of organic compounds composed of substances such as dyes, solvents, pesticides, and other chemicals that are not naturally produced by any organism or found in the environment. They are the end products

TABLE 19.1 List of Pharmaceutical compounds and their adverse effects on environment.

Pharma compounds	Effects on environment	References
Carbamazepine	Strong growth inhibitory effects, causes damage to the solidity of cell walls, interferes with fatty acid synthesis, and cell division dysfunction in green algae	Mofeed (2020)
Ibuprofen	Stimulation of growth of cyanobacteria and inhibition of growth of aquatic plants	Barbera et al. (2020)
Tylosin	Impacts on the structure of soil microbial communities	Westergaard et al. (2001)
Synthetic steroid	Fish, reptiles and invertebrates face endocrine-disruption	Young et al. (2002)
Methyltestosterone	Lower fecundity, oogenesis, spermatogenesis in snails	Schulte-Oehlmann et al. (2004)
Avermectins	Loss of water balance, impaired mating, and development of physical abnormalities in adult insects	Floate et al. (2005)
Erythromycin	Hampering the growth of cyanobacteria and aquatic plants	Pomati et al. (2004)
Tetracycline	Antibacterial resistance in soil bacteria	Sengelov et al. (2003)
Fenbendazole	Impact on dung decomposition	Sommer & Bibby (2002)
Cocaine	Alteration in the mechanisms of action of the juvenile zebrafish	Parolini et al. (2018)
Oxytetracycline	Interference in neurotransmission mechanisms including physiological disturbance of freshwater fish	Rodrigues et al. (2018)
Diclofenac	Direct adverse effects on fish organs	Mehinto et al. (2010)

of poorly metabolized substances that are endobiotic in nature. When high concentrations of pharmaceutical drugs lie redundant in the environment for long periods of time in semimetabolized states, they become toxic to the surrounding ecosystem. These substances emanating from petrochemical industries enter the environment and accumulate in various plant parts, finally making their way to humans through the food chain. Anthropogenic factors such as metallurgical processes, mining, and fuel production increase the levels of heavy metal contamination in the environment (Vanek et al., 2010). Anthelmintics from veterinary diagnosis and antimicrobial therapeutics are major contributors to the wide-scale assemblage of pharmaceutical wastes in any ecosystem. Antifungals, parasiticides, and antibacterial drugs are constantly percolating into the groundwater table and risking the lives of many people who primarily depend on groundwater for consumption. Residues of certain biologically active compounds remain in surface water and increase the chances of disease and infection in humans. Livestock treatments often cause poorly metabolized drugs to enter the environment through the use of pesticides and fertilizers. The efficacy, administration routes, toxicology, metabolism, nature, and ecotoxicity of these waste compounds should be studied and monitored to improve prevention strategies. The physicochemical properties of these substances largely determine their distribution in the environment. The sorption behavior of materials used in wastewater treatment facilities affects the spread and absorption of these pharmaceutical wastes in soil, subsoil, plants, and water. The breakdown products of these substances often retain the same toxicity as the original pharmaceutical compounds even after treatment, thus affecting the lives of both humans and animals at large. The behavior and longevity of drugs in wastewater, groundwater, plant extracts, and aquatic ecosystems is estimated by analytical techniques such as LC and MS (Boxall, 2004). AMR is one of the most challenging outcomes of pharmaceutical waste accumulation in the environment and can create life-threatening conditions in the lives of human beings. Water sources across the world contain various antimicrobial-resistant bacteria and their genetic components. Wastewater consisting of pharmaceutical wastes and untreated hospital discharge is the primary causative agent of drug resistance in humans and animals. Most of the antibiotics consumed by men are eliminated unaltered through urine and feces, leaving behind trace quantities of toxic residues in the ecosystem. The large efflux of pharmaceutical waste from health and agricultural industries leads to the emergence of multidrug-resistant pathogens in the environment that are carried into surrounding water bodies by rain and floods. Groundwater and surface water, which are both sources of drinking water for more than one-third of the world, are infected with bacterial populations resistant to even third-degree antibiotics sometimes. Pathogenic bacteria, such as those causing malaria, tuberculosis, and cholera are given far more clinical importance than the ever-increasing incidence of drug-resistant bacterial species. Inaccessibility to proper health care and inadequate knowledge about the clinical manifestations of drug resistance in many parts of the world are clear reasons for the reduced prioritization of this challenge (Taneja and Sharma, 2019). Fig. 19.1 shows the percentage prevalence of pharmaceutical waste substances in drinking water, as observed during a study conducted in the United States (Focazio et al., 2008).

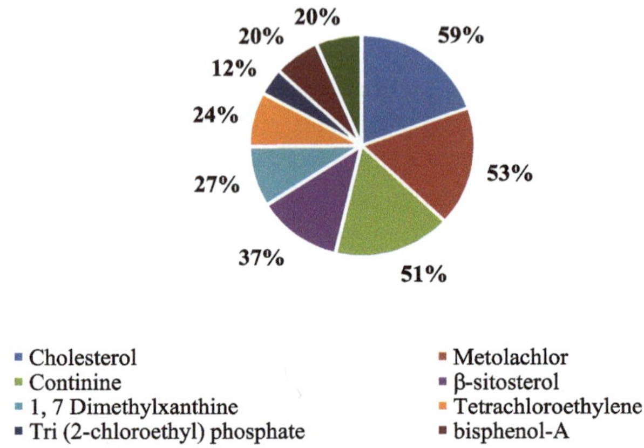

FIG. 19.1 Prevalence of pharmaceutical wastes in drinking water.

19.4 Remediation process

19.4.1 Detection of pharmaceutical compounds

The increase of Pharmaceutical wastes in the environment is a great cause of worry. As these pharmaceutical wastes interfere with human health and their well-being, it is a major concern to detect these pharmaceutical wastes and eliminate them (Fig. 19.2). The pharmaceutical compounds have to be isolated from the environment using processes such as supercritical fluid extraction, microwave-assisted methods, Soxhlet extraction, and ultrasonic waves, which are more convenient for solid compounds and in the case of liquid compounds methods such as solid-phase extraction, stir bar sorptive extraction, liquid-liquid extraction can be used (Ram et al., 2020). Most of the pharmaceutical wastes nowadays are generally enantiomers. The problem lies in distinguishing and separating these enantiomers from one another, but with the application of techniques such as GC and LC, the process of separation of these enantiomers has been made easy. Chiral compounds such as thalidomide, glutamide mandelic acid, and chlortalidone, in wastewater, are generally detected by using HPLC (High-Pressure Liquid Chromatography) (Dogan et al., 2020). Methods such as GC-MS/MS (Gas chromatography-mass spectrometry) and LC-MS/MS (Liquid Chromatography with Mass Spectroscopy) can even identify and detect the pharmaceutical waste even in the range of nanogram per liter. Conventional methods of the water treatment plant are not enough to remove some specific pharmaceutical wastes; hence methods such as the advanced oxidation process that involves photocatalysis, UVH_2O_2 oxidation, ionizing, sonolysis, Fenton, and Phyto Fenton oxidation are used. Recently, it has been found that species of fungi have the presence of enzymes such as oxidoreductase that can degrade pharmaceutical waste. Techniques such as NMR (Nuclear Magnetic Resonance), Raman spectroscopy, and electrochemical methods can also be used to detect pharmaceutical waste in the water (Feier

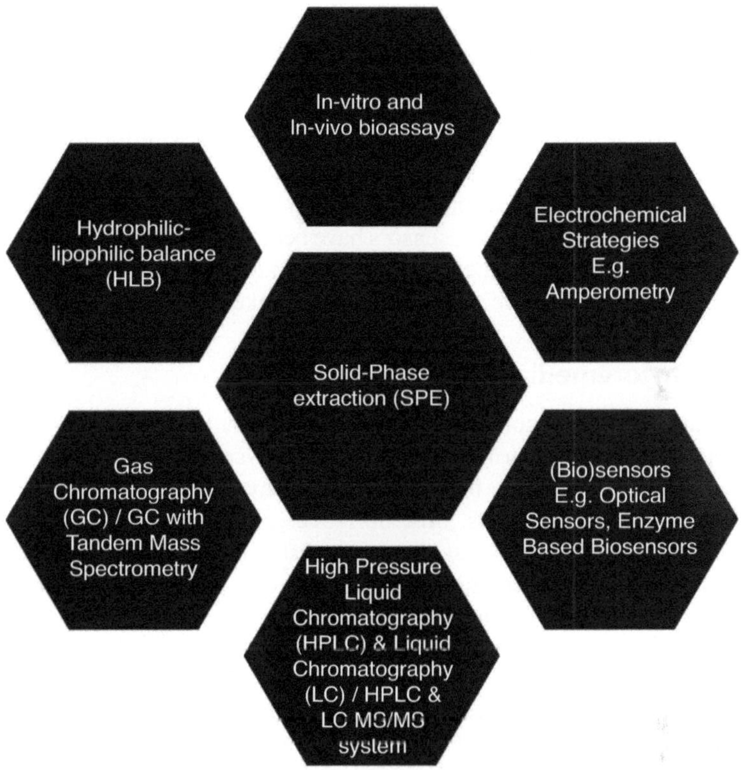

FIG. 19.2 Recent tools for detecting pharmaceutical wastes.

et al., 2018). A number of other detection methods such as ultra-high-performance LC, ultraviolet detection LC with tandem mass spectrometry are being used as a tool to detect wastes, such as phenolic acid, ferulic acid, salicylic acid, potassium permanganate, and their highly sensitive nature makes it more efficient (Chen et al., 2020). In a recent study, a portable optical colorimetric sensor was used to identify pharmaceutical promethazine in lean cocktails (Lantam et al., 2020).

Antibiotics are commonly used as pharmaceutical products. The residue of these antibiotics is found in nature and is a cause of worry as it causes antibiotic resistance in bacteria and makes it more virulent. Modern methods such as screen-printed electrode-based detection systems have been devised to detect the presence of beta-lactam antibiotics. In this process, molecularly imprinted polymers (MIPs) are used for the detection of the beta-lactam antibiotics such as amoxicillin, and these MIPs are highly sensitive. This set-up is a cost-effective (Jamieson et al., 2020). Researchers have also exploited the chemotaxis nature of some bacteria to use them as mobile sensors. These bacteria head toward the region where the concentration of pharmaceutical compounds is high in the water (Lu et al., 2011). Detection of on-site pharmaceutical waste is possible with the use of electrochemical sensors. Recently an electrochemical immunosensor has been designed for the detection of Penicillin-G. Some immunosensors can even detect low levels of antibiotics. H_2O_2 was used as a transducer for the detection of trace amounts of beta-lactam antibiotics in (10–10M) range. The sensor was also sensitive to cefotaxime, amoxicillin, dicloxacillin, and ampicillin (Merola et al., 2014).

Aptamers are a new replacement for antibodies that are being used to develop biosensors. Electrochemical aptasensors have been designed that can effectively sense ampicillin and sulfadimethoxine. It was found that the sensor can detect low levels of drugs in square wave volumetry in lake water (Yang et al., 2017).

19.4.2 Phytoremediation techniques

The decontamination of soil and water ecosystems by traditional methods such as thermal treatment, chlorination, ozonation, physicochemical methods, and autoclaving has increasingly been replaced by phytoremediation methods that use plants to extract effluents, oils, hormones, and drugs from the environment. This typically is a low-cost alternative that does not adversely affect other biotic components of the environment. Some of these processes involve uptake of pharmaceutical wastes from the soil, stabilizing their components, enzyme treatment, preventing leaching of toxic substances into the groundwater, and filtration of hazardous chemicals from wastewater. The chemical components of these contaminants are either degraded by microbial activity or chemically modified to an ecologically safe and stable form. Freshwater is the breeding ground of water-borne diseases because of the accumulation of microquantities of artificial food and pharmaceutical wastes. Thus, a prior environmental analysis is needed to classify these compounds to detect even small amounts of chemicals and drugs in soil and water.

Most of these pharmaceutical waste compounds are highly nonbiodegradable and may exert estrogenic effects on the physiological processes of higher organisms. Suspended pollutants, additives, drugs, and fragrances are discharged into aquatic ecosystems, threatening the life of various aquatic organisms. Through processes such as biomagnification, bioaccumulation, and eutrophication, toxic heavy metals such as Cadmium (Cd), Lead (Pb), Arsenic (As), Copper (Cu), and Zinc (Zn) infiltrate the lives of humans through the food chain. These elements cause health disorders such as gastroenteritis, stomatitis, diarrhea, vomiting, and convulsions. Elaborate and time-consuming techniques such as nanofiltration, electrocoagulation, activated sludge, and ultraviolet disinfection have more recently been substituted by phytoremediation procedures such as phytostabilization, rhizofiltration, rhizodegradation, phytodegradation, phytoextraction, phytoaccumulation, and phytovolatilization. Phytoremediation is relatively less expensive than the other methods and the contaminants can be separated permanently. In the process of phytovolatilization, the pollutants are degraded into their simplest forms that are easy to cast away. Some plants also tend to hyperaccumulate toxic substances in their tissues, thus making them ideal candidates for the detoxification of soil and water deserves. Examples of botanical plants used for phytoremediation include *Scirpus grossus*, *Eleocharis macrostachya*, *Vetiveria zizanioides*, *Pistia stratiotes* (water cabbage), *Typha* sp., *Phragmitis communis*, *Lemna gibba*, *Hydrilla verticillata*, *Spirodela polyrhiza*, *Gracilaria gracilis*, *Ipomoea aquatica*, *Azolla pinnata*, and others that effectively contribute to the overall removal of pharmaceutical wastes from the environment (Jain et al., 2019) (Table 19.2).

TABLE 19.2 List of plants used for phytoremediation of pharmaceutical compounds.

Pharmaceutically active compounds	Plants used for phytoremediation	References
Acetaminophen	*Armoracia rusticana*	Shutcha et al. (2010)
Amoxicillin	Alfaalfa (*Medicago sativa*)	Bartrons & Peñuelas (2017)
Aspirin	*Brassica juncea*	Gahlawat & Gauba (2016)
Azithromycin	*Lactuca sativa, Raphanus sativus, Festuca arundinacea*	Sidhu et al. (2019)
Ciprofloxacin	*Lactuca sativa, Raphanus sativus, Festuca arundinacea*	Sidhu et al. (2019)
Clofibric Acid	*Typha* spp.	Dordio et al. (2009)
Diclofenac	Microalga (*Pseudokirchneriella subcapita*)	Bartrons & Peñuelas (2017)
Diltiazem	*Azolla caroliniana*	Maharjan (2014)
Fluoxetine	*Elodea canadensis*	Carvalho et al. (2014)
Ibuprofen	Great Millet (*Sorghum Bicolor*)	Bartrons & Peñuelas (2017)
Levonorgestrel	*Cyperus alternifolius* and *Eichhornia crassipes*	Li et al. (2014)
Metformin	Carot (*Daucus Carota*)	Bartrons & Peñuelas (2017)
Norfloxacin	*Lythrum salicaria*	Carvalho et al. (2014)
Progesterone	*Lemna Gibba*	Allam et al. (2015)
Sulfadiazine	*Populus nugra*	Carvalho et al. (2014)
Sulfamethoxazole	*Pistia stratiotes*	Maharjan (2014)
Synthetic estrogen	Sandbar Willow (*Salix exigua*)	Franks (2006)
Tetracycline	*Brassica juncea*	Gahlawat & Gauba (2016)

Phytoremediation is a process that uses plants to metabolize and stabilize pharmaceutical contaminants in the soil and groundwater, which has gained popularity because of its noninvasive technical advantages. Some of the most widely used phytoremediation techniques are as follows.

19.4.2.1 Rhizofiltration

In this method, the roots of plants are used for the uptake, absorption, concentration, and precipitation of heavy metal components such as mercury, lead, and cadmium of pharmaceutical wastes sourced from aquatic ecosystems. Hydroponically cultured terrestrial plants such as Indian mustard (*Brassica juncea*) and Sunflower (*Helianthus annuus L.*) can efficiently withdraw toxic substances from aqueous substances. Members of the Brassicaceae family have also been shown to store high levels of pharmaceutical substances in their roots and shoots. In areas of high lead (Pb) concentration, precipitation is more favored as opposed to root absorption. In a study conducted by Dushenkov et al., live roots were better adapted to accumulating lead and drawing it out of its aqueous phase. In the same study, it was inferred that cultured cells of the plant *Datura innoxia*, in live or dead form, can be used to eliminate trace amounts of Ba^{+2} ion from its solution. The high-water content of plants such as the water hyacinth (*Eichhornia crassipes*) and aquatic fern (*Azolla filiculoides*) makes the processes of drying, composting, and extraction more complex. A more efficient alternative approach utilizes the fibrous root system of terrestrial plants to acquire pharmaceutical waste products. These roots offer a seamless surface area for absorption and can be easily air-dried after extraction. Rhizofiltration is economically driven because of its ability to treat large volumes of pharmaceutical wastes, recycle them, and reduce the amount of secondary waste generation (Dushenkov et al., 1995).

19.4.2.2 Phytostabilization

This method entails a series of green approaches to contain pharmaceutical contaminants in situ, thereby withholding them from contaminating other ecosystems. Phytostabilization involves the generation of plant cover on the surface of waste-enriched sites so as to reduce the infiltration of waste materials. The pharmaceutical wastes generated are confined within the rhizosphere through the uptake and accumulation of waste products in roots. The immobilization of wastes prevents leaching and erosion, sustaining an aerobic environment for the waste particles inside the root tissue. Organic contaminants within the plant roots may be further degraded by microbial activity to generate stable and nontoxic hydrocarbons. Phytostabilization facilitates the containment of waste products in the vadose zone of root systems by accumulation and precipitation inside the rhizosphere. Soil amendments can be further procured to enhance the immobilization of pesticides, metalloids, and other contaminants. The nature of metalloids present inside the rhizosphere is altered by the action of organic acids such as acetic acid, citric acid, butyric acid, lactic acid, malic acid, and succinic acid present inside the roots. Lead is precipitated as phosphate, whereas cadmium is available as complexes with sulfide. Phytostabilization comprises steps such as uptake and segregation of pharmaceutical contaminants in root tissues, amendment of soil properties that directly influence the confinement and immobilization of waste products, application of root exudates that control the precipitation of waste substances, incorporation of a vegetation barrier that limits human and animal contact with the soil, mechanical altercations at the site to reduce wind and water erosion, and facilitating evapotranspiration as a means to reduce leaching of contaminants. Agricultural crops induce the phytostabilization of pharmaceutical wastes by generating Zn-rich biomass that can act as nutritious fodder for animals. In calcareous soil containing high volumes of zinc, iron, copper, and manganese, certain plants employ phytosiderophores as a chelating agent. Various ion channels and transport systems such as the K+ ion channels and the Ca^{2+}/Zn^{2+} transport systems further regulate the mechanism of uptake of various ions. Phytostabilization does not necessarily remove waste products from a site, but it effectively reduces the mobility of these wastes and thereby reduces the threat to life in other ecosystems (Bolan et al., 2011).

19.4.2.3 Phytoaccumulation

Micronutrients in plants such as zinc, manganese, magnesium, and iron are essential for their survival. However, when these exceed a specific concentration, they may become toxic for the health of the plant. These micronutrients are an integral part of pharmaceutical wastes along with nonessential heavy metals such as mercury (Hg), cadmium (Cd), and lead (Pb). Plants exhibit different tolerance levels to these ions and must follow a fixed path for the uptake and accumulation of these elements. Phytoaccumulation is a nonintrusive and secure method for the confined storage of toxic substances from pharmaceutical wastes in plant tissues. The uptake of these substances can either be achieved by root exclusion in which wastes are barred from entering the main body of the plant or shoot exclusion, in which translocation of waste materials to the shoot is denied. Hyperaccumulation involves the concentrated accumulation of waste products in the root and shoots tissues. Plants classified as hyperaccumulators can withstand, contain, and translocate high levels of certain pharmaceutical components that could otherwise be toxic for many organisms (Kamal et al., 2004). In an experiment demonstrated by Zavoda et al. (2001) plants whose leaves comprise cadmium (>100 mg/kg), nickel (>1000 mg/kg), zinc (>10,000 mg/kg), and manganese (>10,000 mg/kg) when grown in a nutrient-rich medium show a tendency to hyperaccumulate waste particles. In a study conducted, the pea plant (*Pisum sativum*) was investigated for its role in the uptake of pharmaceutical chemicals. The plant that was first recultivated before being subjected to chemical compounds showed absorption and accumulation of different test pharmaceutical compounds in its leaves, stems, roots, and shoots. The results of the experiment further indicated that the accumulation of Carbamazepine (CBZ) was the highest in the leaves of the pea plant as compared to the net weight of leaves calculated prior to the study. In the same study, the scope of the cucumber plant (*Crocus sativus*), was also examined. The concentration of sap present in the xylem and the test compounds was predetermined. After a period of 24–72 h, the leaves were tested for the uptake of chemicals. The observation inferred at the end of the experiment was that the xylem sap showed no presence of imbibed test chemicals. The concentration ratios of a large number of pharmaceutical compounds were approximately 0.1 in the shoot system. However, the cucumber plant showed a specific affinity for pharmaceutical compounds such as cyclophosphamide, CBZ, and crotamiton and thus favored the uptake of these compounds. The concentration of compounds was evaluated by analytical techniques such as LC-MS/MS during both these experiments. Municipal wastewater plants act as major contributors to pharmaceutical waste pollution, leading to the accumulation of chemicals with decreased absorbability and biodegradability, thus making phytoremediation techniques such as phytoaccumulation a necessity for treating toxic effluents and sewage sludge (Tanoue et al., 2012).

19.4.2.4 Phytodegradation

Organic compounds such as solvents, hydrocarbons, aromatic compounds, and volatile substances emanating from pharmaceutical wastes in air and soil ecosystems need to be removed from the environment periodically. The most convenient strategies involve a protocol consisting of uptake, accumulation, degradation, and volatilization of these toxic substances. These phytoremediation techniques are cost-effective, efficient, and extremely eco-friendly. The kind of plant activity being employed in these techniques depends on the nature of the compounds being remediated and the type of plant parts being used in the process. In a study pursued by Doty et al., the phytodegradation abilities of the tropical leguminous tree *Leucaena leucocephala* were studied to treat pharmaceutical contaminants such as ethylene dibromide and trichloroethylene from groundwater. Phytodegradation is primarily confined to the rhizosphere of the leguminous plant, thus restricting the movement of contaminants. The plants used for the purpose of photodegradation were chosen on the basis of their pollutant removal efficiency and their ability to survive in the climatic conditions of that particular ecosystem. The leguminous plant used in the experiment showed evidence of the metabolization of test compounds present in an unprecedented ethylene bromide (EDB) spill in groundwater. Each plant organ of the chosen plant species was individually capable of debromination of EDB. After drying and freezing the plant parts, the compounds that were analyzed using LC showed a considerable increase in the levels of bromide accumulation, indicating the successful degradation of ethylene (Doty et al., 2003). Evidence of phytodegradation was also seen in a study initiated by Ma and Burken (2002), in which volatile organic compounds (VOCs) were effectively degraded by tree cores. Chlorinated solvents such as tetrachloroethylene and trichloroethylene were used as test compounds in the study, wherein the portioning of these solvents in air and wood biomass was observed. The isotherms were obtained after the study indicated a rise in levels of VOCs in the roots of trees exposed to groundwater contaminated with pharmaceutical wastes (Ma and Burken, 2002).

19.4.2.5 Phytovolatilization

This is a process by which components of pharmaceutical wastes in soil and groundwater are released into the atmosphere as volatile substances through diffusion and transpiration. Plants interact with their surrounding ecosystems, extracting, and releasing chemical compounds throughout their lives. The fate of compounds such as hormones and xenobiotics inside plant organs either results in their accumulation in various plant parts or their translocation to other organs. Phytovolatilization promotes the portioning of waste substances in different air spaces, and their effective diffusion into less contaminated ambient air zones. This process considerably dilutes the concentration of VOCs in pharmaceutical wastes and increases the subsequent chances of photodegradation in the atmosphere. Phytovolatilization promises efficient removal of both organic and inorganic compounds from pharmaceutical waste dumping grounds by lowering the groundwater table and increasing the surface area of contact between the contaminants and the vadose zone of the plant being used. Root activities may increase or decrease the extent of phytovolatilization depending on root-mediated soil modification. Direct phytovolatilization involves a simple pathway in which pharmaceutical chemicals are taken up by the plant and translocated, so as to be volatilized from the stem, trunk, and leaves by transpiration. These compounds being slightly hydrophobic can diffuse across the cutin and epidermis layers. The flux of these volatile contaminants can be enhanced by indirect phytovolatilization, whereby plants are driven to achieve a lower water table, increased chemical mobility mediated by hydraulic redistribution, better soil permeability, and interception of rainfall that would otherwise dilute VOCs (Limmer et al., 2016).

19.4.2.6 Phytoextraction

This technique is primarily used to eliminate residual metal contaminants from pharmaceutical wastes. The soil ecosystem has become both nonarable and toxic to life because of anthropogenic factors that contribute to large-scale dumping of metalloproteins, enzymes, hormones, and cofactors in open areas. These substances block certain essential functional groups that may hinder natural physiological processes in animals and man. Their immutable nature makes them a greater threat to life in the environment. These components of pharmaceutical wastes not only alter biochemical processes but also choke the process of photosynthesis leading to loss of vegetation and habitat for various forms of life (Garbisu et al., 2001). Phytoextraction promotes a steady and cost-effective means for the removal of these hazardous metals from soil and water ecosystems. However, this method is excessively used for the withdrawal of heavy metal wastes such as compounds containing cadmium, mercury, arsenic, lead, and zinc. The scope of phytoextraction in the context of pharmaceutical waste eradication can be pertinent only in the case of poorly metabolized pharmaceutical waste compounds that still contain trace amounts of these heavy metals in their active forms.

19.4.3 Role of aquatic plants in eliminating wastes

Aquatic environments are one of the most polluted environments on the Earth, almost all kinds of pollutants make their way to water bodies and cause ecological disruptions. For a really long time, rivers and oceans have been used as a dump yard, it was believed that the oceans are so vast that the number of pollutants released into them will have an insignificant effect on it. This belief led to purposeful incessant dumping of industrial, household, agricultural, and radioactive wastes into the aquatic environment. No wonder that pharmaceutical wastes are also being released into the water bodies untreated. We have already discussed how dangerous they can be for the aquatic ecosystem and humans, even trace-level concentration of these compounds can be classified as threatening. Aquatic plant–based systems have emerged as a great alternative for bioremediation of wastewater as they are cheap, effective, and easy to set up, and manage. There are few aquatic plants such as *Azolla caroliniana, I. aquatica, E. crassipes, P. stratiotes, Lemna minor, Salix atrocinerea, and Typha latifolia* that have shown considerable efficiency in removing pharmaceutical compounds. Aquatic plants are indispensable for the treatment of wastewater and are also being widely used for all the methods involved in phytoremediation (Mustafa et al., 2020; Herklotz et al., 2010). Aquatic plants capable of performing phytoremediation absorb the pharmaceutical compounds via their roots which is followed by limited distribution in shoots and leaves. Generally, the concentration of the compounds is considerably more in the roots in comparison to the shoot, but in few cases the translocation is very restricted in both roots and shoots as seen in the case of remediation of the compound diclofenac. The pharmaceutical compounds translocated to shoots and leaves also get localized at the subcellular levels in some aquatic plants such as penny grass, water spinach, rice, and purple perilla (He et al., 2017). The translocation of these compounds depends on many factors such as hydrophobicity, dissociation potential of the compound, the ability of the plant to metabolize the compounds, and require many other underlying mechanisms.

Most of the pharmaceutical compounds are ionizable and are either acidic or basic. Acidic compounds are taken by aquatic plants with the help of diffusion, which is facilitated by the presence of pH difference. For the successful uptake of the compounds, the pH of the external solution must be lower than the pH inside the cell. As the acidic pharmaceutical compounds enter into the cell they get dissociated and later translocated. Basic compounds have different fate, they dissociate into cationic or neutral intermediates due to the electrical attraction between cations and negatively charged plasmalemma. Neutral compounds are generally dependent upon their hydrophobicity for their uptake by roots. It has been well established that the uptake of neutral pharmaceutical compounds is dependent on their chemical hydrophobicity. Higher hydrophobicity results in higher uptake of neutral compounds (Wu et al., 2013). Once the compounds are taken up by plants, there are three main phases through which degradation or detoxification can occur in plants. In Phase I, the compounds are converted to more biologically active forms by the action of various enzymes (cytochrome P450). Phase II converts the phase I activated compounds to high molecular weight complexes by subjecting them to conjugation reactions with complex biomolecules such as amino acids and carbohydrates rendering them more water-soluble and biologically less active. Finally, in phase III, the resulting molecules are accumulated inside vacuoles or integrated into the cell wall. These accumulated compounds can still cause toxic effects such as oxidative stress, altered photosynthetic rate, and reduced growth. In some cases, microbes present in the rhizosphere can protect the plant by helping in the metabolism of these compounds. The microbes use these compounds as carbon sources and decompose them to generate energy (Fig. 19.3).

One of the emerging phytoremediation technologies involving aquatic plants is the CW system. CWs are strategically made artificial on-site wetlands propagated with aquatic plants. CWs are regarded as one of the most efficient, sustainable, and cheapest technologies for the removal of pollutants, especially micropollutants such as industrial waste, nutrients, and PPCPs (pharmaceutical and personal care products). It is an eco friendly method that employs plants

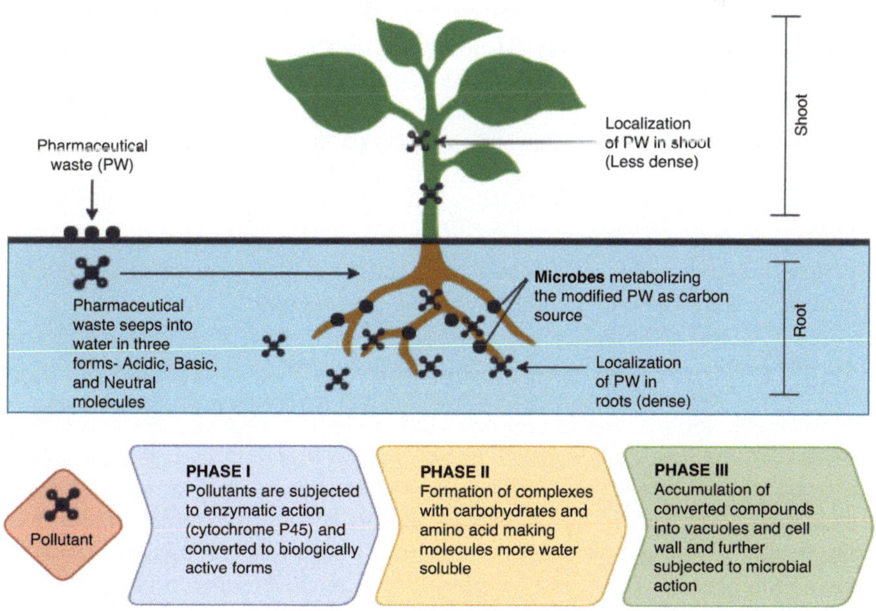

FIG. 19.3 Mechanism of phytoremediation by aquatic plants.

that rely on microbial, physical, and chemical processes for degradation of pollutants from soil or water naturally. CWs do not need the energy to operate and require very little maintenance that generally involves routine water level monitoring and media rinsing to restore adsorption capacity.

Subsurface flow CWs are generally preferred for removing pharmaceutical pollutants. The advantages of using subsurface flow CWs are many, one of them is that it does not generate much biosolids and sludge that require disposal, hence maintenance becomes much convenient; and it also does not have an exposed water surface that prevents direct contact with animals and humans making it ideal for the treatment of toxic compounds. Horizontal flow subsurface CWs are used for anaerobic processes, whereas vertical flow subsurface CWs are employed in aerobic processes because of their capacity to facilitate the high transfer rate of oxygen. The degradation efficiency of the aerobic processes or anaerobic generally varies a lot and can potentially depend on the target compounds desired for removal. As the mechanism through which the CWs detoxify these compounds is actually a plethora of underlying processes carried out by plants and microbes occurring synergistically, it is quite difficult to pinpoint the exact process that leads to the elimination of a particular pollutant (Nguyen et al., 2019). So, we generally compare and study various plants that either directly uptake the compound or provide favorable conditions to the microbes that degrade the compound (Biofilm formation) and then come up with a setup strategy.

Usually, CWs are used for pretreatment and generally are a small part of a waste treatment facility, but in recent years they are also being used as a primary treatment system and even secondary treatment systems. It is now well established that CWs are not only effective in reducing BOD (Biological Oxygen Demand), TDS (Total Dissolved Solids), faecalis coliforms, and nutrients from wastewater but can also remove recalcitrant toxic compounds. Although the efficiency is quite inconsistent and, in some cases, even negative results were found, more than 70% efficiency was observed in the primary treatment of many compounds, which shows the potential of CWs (Liu et al., 2019). If the target pollutants are known and the plant suitable for phytoremediation of that particular compound is well established, then 70%–90% removal of that pollutant can be achieved after configuration and process optimization.

19.5 Phytoremediation challenges

Phytoremediation is an emerging technology to degrade pharmaceutical wastes. In the past decade it has come into the limelight as it is an effective method to remove toxoids from land and water but as every new set up faces challenges there are many challenges associated with phytoremediation too. Phytoremediation is a lengthy process. The plants that are used for phytoremediation generally take time to grow as well as the biomass produced is also less. The toxic elements extracted through phytoremediation also need to be disposed properly. At times it is difficult to accumulate these contaminants in soil. The genetically modified plants used for phytoremediation may affect the growth of the wild-type flora. The weather condition is one of the other factors that affects the phytoremediation process. The plants that are being used for the process must be selected on various criteria such as the type of their root, their ability of transpiration, pH interaction, types of soil in which they grow and other agronomic practices (Mahar et al., 2016). There are many other underlying causes that hinder the process of Phytoremediation such as the location of the contaminants. The plant can only extract these contaminants from the layer where their roots reach if the contaminants lie deep down where the roots of the plants cannot reach the contaminants cannot be extracted through phytoremediation. Another criterion for drawback is the life cycle of the plant as the plant reaches maturity its capability to take up the contaminants decreases because much of the energy is spent in growth. Climate is a big challenging factor that influences phytoremediation. The plants used for phytoremediation must be grown in a suitable climate. At times many plants release its toxins to the environment through transpiration (phytovolatilization), care must be taken so that these contaminants do not spread in the environment. Phytotoxicity is a major problem for the process of phytoremediation because survival of the plants in high toxic conditions becomes difficult hence most of the plants die in high contaminant concentration. Waste accumulated through the process of phytoextraction is accumulated in the plant tissue. Proper care should be taken in harvesting these plants or else the contaminants get exposed to the ecosystem. Genetically modified plants that are used for the phytoremediation process is a cause of worry because it may cause harm to the ecosystem in future. The risk factor that these genetically modified plants possess cannot be identified beforehand and may cause damage to the environment. There is a great difference between the plants being used for field trials and the plants being used for the cause of phytoremediation. Moreover, the chances of cross-pollination are also a reason of worry. Proper care must be taken that only limited people get the access to the place where the phytoremediation is being carried out as these plants are highly toxic in nature. If proper disposal of these plants is not done by a skilled professional then it might raise the risk of contamination and if these biomasses are consumed by the livestock it will enter the food chain, which is proved to be carcinogenic in nature (Orooj et al., 2014). There are other challenging factors such as the availability of land. Phytoremediation of wastewater needs CWs, but countries that have shortage of land cannot have these CWs for phytoremediation. The efficiency of the removal of these pharmaceutical compounds from the wet land decreases over time due to clogging or aging. These pharmaceutical wastes once collected from the wastewater should be treated in a proper manner to eliminate it; several PPCs from these CWs pose threat to aquatic life (Verlicchi et al., 2014; Nguyen et al., 2019). Enzymes belonging to the cytochrome (P450) family plays an important role in phytoremediation and detoxification of soil and water but the major problem with the mammalian

CRP enzyme is that the stability of the enzyme is very low, they have low turnover rate, their expression in plants are also low, and they metabolize a limited number of pollutants (Kumar et al., 2012).

19.6 Enhancement of phytoremediation

19.6.1 Approaches to enhance the phytoremediation of wastewater

Contamination free water is one of the necessities of all living beings but with due course of time we are losing our clean water as the quality of water is getting degraded and the sole reason behind this is that the clean water bodies are getting contaminated with the wastes from industries, hospitals, housing societies, etc. The most challenging task of science here is to remove these kinds of contaminants from the wastewater that are not only toxic to the ecosystem but also very difficult to remove using any other conventional methods. Phytoremediation has been an effective way to remove contaminants such as heavy metals and toxic substances from these water bodies. The process can be enhanced with the help of building wetlands, which has been broadly accomplished in the past decades throughout the globe as a method to treat wastewater (Wang et al., 2017). Enhancement of phytoremediation can also be executed with the help of aquatic plants such as water hyacinth and water lettuce that have a fibrous root system effective for the adsorption of heavy metals and toxic substances (Mays et al., 2001); (Stoltz et al., 2002). Aquatic plants also discard toxic compounds with the help of phytovolatilization. In this process, plants take up the toxic substance through their roots and convert them into a volatile form and release it in the air through transpiration (Wani et al., 2017). Roots play another major role in the remediation of a contaminated site, the toxic compounds get precipitated when external factor such as pH is changed around the rhizosphere and the fluid emits from the roots of the plant (exudates) and when the plant that has taken up the pollutants is harvested and discarded, this whole process falls under the category of rhizofiltration (Zhu et al., 1999). Also, many microorganisms help plants in the process of bioremediation where they assist to gather the toxic substances near the rhizoids of the plant and in turn enhance the plant's ability to take up the toxic element more effectively. Among these microorganisms, endophytic bacteria play a very important role in phytoremediation. These endophytic bacteria enhance the ability of the plant to tolerate both biotic and abiotic stress (Hussain et al., 2018; Ashraf et al., 2018). In a recent approach, transgenic plants were used to enhance the process of phytoremediation. These transgenic plants have specific genes that increase their ability to take up the toxic metals more effectively and also increase their metabolism. Water-containing phenolic or chlorine remnants and even residues of explosives can be treated with these kinds of transgenic plants that have a greater ability to withstand stress conditions and accumulate more of the toxoids (Koźmińska et al., 2018; Macek et al., 2008). There are other effective measures to remove pharmaceutical waste from water bodies. Waste such as clofibric acid with a half-life of 21 years is a pharmaceutical metabolite and can be removed with the use of *Typha* spp. The interaction between the rhizome of the plant and macrophytes enhances the ability to adsorption of clofibric acid from the water. Generally, it is achieved by subsurface flow CW systems, where the wastewater is accumulated in a CW, and plant species such as *Typha* spp are grown to eliminate the pharmaceutical waste that cannot be degraded by other means (Rana et al., 2020).

19.6.2 Approaches to enhance the phytoremediation of waste land

The land is a very important resource for humans. It is often found that a large area of the land is left out because it is unfit for cultivation or use. The main reason behind this is the presence of toxic elements in the soil. This land can be treated and the toxins can be removed from the soil hence making it suitable for agriculture and other usages. One of the ways to remove toxins from the soil is phytoremediation. Plants often act as a barrier and limit the movement of the toxins in the soil. The groundwater moving in the soil carries the contaminants, this movement is restricted to some extent with the help of plants, and the water entering the soil is filtered by the process known as phytostabilization. Large, highly contaminated areas are treated with this method (Mahdavian et al., 2017). Although phytostabilization is not a solution to the cause but a mere process to deactivate the toxic compounds and render it harmless. Another promising means to remove toxins from the soil is phytoextraction, also known as phytoaccumulation. In this process, the toxins are absorbed by the roots of the plants and then transported to the upper parts of the plant (stem, leaves). Once the desired amount of toxins is accumulated by the plants, the plants are harvested and burned as a source of fuel and the heavy metals can be recycled from its ashes (Chandra et al., 2018). Methods such as phytodesalination have shown its efficiency to remove the salinity of the soil. In this process, halophytes absorb salts from the soil and accumulate it in its tissue thus making the soil less saline. Plants such as Suaeda and Salicornia (Fam: Amaranthaceae) and *Sesuvium portulacastrum* (Fam: Aizoaceae) are planted in the saline wasteland to remove the salt and make it suitable for use (Naikoo et al., 2020).

19.6.3 Transgenic approach

The field of biotechnology plays an effective role in the enhancement of phytoremediation. Phytoremediation using microorganisms and recombinant DNA technology are the two most efficient techniques. It helps in enhanced expression of shoot symplast and vacuole, excretion of chelators, level of enzyme modification, and level of leaf chelators. Interaction of plants with microorganisms increases the rate of metal uptake through the plants and hence, in turn, increases the process of phytoremediation. Genetically modified plants have the potential to take up more of the heavy metals. There were successful trials for genetically modified plants that can produce higher biomass. Phytochelatin and glutathione producing gas-gene found in *E. coli* is embedded in transgenic *B. juncea*, which helps in rate constraining. Cd resistance can be increased with the use of γ -ECS seedling that generally contains more amount of phytochelatins, glutathione, and γ -GluCys than the wild type. Anatomical resistance is a restriction factor that limits genetically modified plants. Species such as Indian mustard (*B. juncea*), *T. caerulescens* (hyperaccumulator), maize (*Zea mays*) have been genetically transformed to increase its scavenging properties for heavy metals. The symbiotic relationship of bacteria with plants helps in the in-situ phytoremediation of heavy metals. Modified PGP (Plant growth-promoting) bacteria have genes coding for biodegradation enzymes. Other genes coding for metal uptake regulation have been identified and inserted with the help of phytoremediation (Malik et al., 2014).

19.7 Conclusion

Wastewater treatment is imperative to the health and wellbeing of any community. The infiltration of pharmaceutical wastes from health care industries and the poor handling of old and expired drugs have led to large-scale water pollution. Water, being an indispensable resource for animals, plants, and human beings alike, is an asset to our ecosystem. Regular investigation of water quality, the concentration of chemicals, and redundant pharmaceutical compounds is the foremost investment for any healthy economy. The presence of toxic substances in water and soil degrades life, often causing diseases that affect human beings through the food chain. The consumption of contaminated water can be the source of diseases such as diarrhea, cholera, gastroenteritis, and sometimes even cancer. AMR is the fatal outcome of partially metabolized pharmaceutical wastes in the environment and can cause life-threatening immunological disorders. Infectious pathogens can evade important immune responses if antibiotics prove unresponsive to their pathogenesis. Leaching of such compounds may result in the accumulation of hazardous elements in groundwater, which also acts as a natural reserve of drinking water for a large section of people. Various physicochemical methods and biological techniques have been used to treat contaminated ecosystems. However, most of these treatments are not only exorbitant but also entail a lot of labor and time. Phytoremediation has proved to be a low-cost, and eco-friendly alternative that uses plants to accumulate, translocate and metabolize chemicals, pharmaceuticals, and heavy metals abounding in soil and water. Through processes such as phytoaccumulation, rhizofiltration, phytoextraction, phytostabilization, and phytovolatilization, commercially viable plants have been able to combat the detrimental effects of organic and inorganic wastes in soil and water ecosystems. These plants rooted in the soil are able to take up trace elements and detoxify the underlying groundwater because of their high tolerance and ability to hyperaccumulate toxic substances in their leaves, stems, and shoots. It is this property of certain plants that makes them miracle water purifiers, thus supporting a cleaner, safer, and greener environment.

Acknowledgment

The authors gratefully acknowledge the support of Ms. Mousumi Meghamala Nayak for her valuable suggestions and contribution in editing.

References

Allam, A., Tawfik, A., Negm, A., Yoshimura, C., Fleifle, A., 2015. Treatment of drainage water containing pharmaceuticals using duckweed (Lemna Gibba). Energy Procedia, 74, 973–980.

Archer, E., Wolfaardt, G.M., Van Wyk, J.H., 2017. Pharmaceutical and personal care products (PPCPs) as endocrine disrupting contaminants (EDCs) in South African surface waters. Water Sa, 43 (4), 684–706.

Ashraf, S., Afzal, M., Rehman, K., Naveed, M., Zahir, Z.A., 2018. Plant-endophyte synergism in constructed wetlands enhances the remediation of tannery effluent. Water Sci. Technol. 77 (5), 1262–1270.

Ashton, D., Hilton, M., Thomas, K.V., 2004. Investigating the environmental transport of human pharmaceuticals to streams in the United Kingdom. Sci. Total Environ. 333 (1–3), 167–184.

Barbera, A.C., Leonardi, G., Ferrante, M., Zuccarello, P., Maucieri, C., 2020. Effects of pharmaceuticals (caffeine and ibuprofen) and AMF inoculation on the growth and yield of *Oryza sativa* L. Agric. Water Manag. 232, 106005.

Bartrons, M., Peñuelas, J., 2017. Pharmaceuticals and personal-care products in plants. Trends Plant Sci. 22 (3), 194–203.

Bolan, N.S., Park, J.H., Robinson, B., Naidu, R., Huh, K.Y., 2011. Phytostabilization: a green approach to contaminant containment In Advances in agronomy, Vol. 112. Academic Press, MA, pp. 145–204.

Boxall, A.B., 2004. The environmental side effects of medication: how are human and veterinary medicines in soils and water bodies affecting human and environmental health? EMBO Rep. 5 (12), 1110–1116.

Boxall A.B.A., 2012. New and Emerging Water Pollutants Arising From Agriculture. Paris: OECD.

Boxall, A.B., Kolpin, D.W., Halling-Sørensen, B., Tolls, J., 2003. Are veterinary medicines causing environmental risks? Environ. Sci. Technol., 37 (15), 286A–294A.

Carvalho, P.N., 2020. Constructed wetlands and phytoremediation as a tool for pharmaceutical removal. In: Solsona, S.P., Montemurro, N., Chiron, S., Barceló, D. (Eds.). Interaction and Fate of Pharmaceuticals in Soil-Crop Systems. Handb. Environ. Chem. Springer, Cham, 103, pp. 377–413.

Carvalho, P.N., Basto, M.C.P., Almeida, C.M.R., Brix, H, 2014. A review of plant–pharmaceutical interactions: from uptake and effects in crop plants to phytoremediation in constructed wetlands. Environ. Sci. Pollut. Res. 21 (20), 11729–11763.

Chandra, R., Kumar, V., Tripathi, S., Sharma, P., 2018. Heavy metal phytoextraction potential of native weeds and grasses from endocrine-disrupting chemicals rich complex distillery sludge and their histological observations during in-situ phytoremediation. Ecol. Eng. 111, 143–156.

Chen, X., Mao, J., Wen, F., Xu, X., 2020. Determination of phenolic acids in botanical pharmaceutical products by capillary electrophoresis with chemiluminescence detection. Anal. Lett. 54 (5), 1–13.

Cidlinová, A., Wittlingerová, Z., Zimová, M., Chrobáková, T., Petruželková, A., 2018. Ecotoxicity of wastewater from medical facilities: a review. Sci. Agric. Bohem. 49 (1), 26–31.

Cunningham, S.D., Berti, W.R., Huang, J.W., 1995. Phytoremediation of contaminated soils. Trends Biotechnol. 13 (9), 393–397.

Dogan, A., Płotka-Wasylka, J., Kempińska-Kupczyk, D., Namieśnik, J., Kot-Wasik, A., 2020. Detection, identification and determination of chiral pharmaceutical residues in wastewater: problems and challenges. TrAC Trends Anal. Chem. 122, 115710.

Dordio, A.V., Duarte, C., Barreiros, M., Carvalho, A.P., Pinto, A.P., da Costa, C.T., 2009. Toxicity and removal efficiency of pharmaceutical metabolite clofibric acid by Typha spp.–potential use for phytoremediation? Bioresour. Technol. 100 (3), 1156–1161.

Doty, S.L., Shang, T.Q., Wilson, A.M., Moore, A.L., Newman, L.A., Strand, S.E., Gordon, M.P., 2003. Metabolism of the soil and groundwater contaminants, ethylene dibromide and trichloroethylene, by the tropical leguminous tree, *Leucaena leucocephala*. Water Res. 37 (2), 441–449.

Dushenkov, V., Kumar, P.N., Motto, H., Raskin, I., 1995. Rhizofiltration: the use of plants to remove heavy metals from aqueous streams. Environ. Sci. Technol. 29 (5), 1239–1245.

Ebele, A.J., Abdallah, M.A.E., Harrad, S, 2017. Pharmaceuticals and personal care products (PPCPs) in the freshwater aquatic environment. Emerg. Contamin. 3 (1), 1–16.

Feier, B., Florea, A., Cristea, C., Săndulescu, R., 2018. Electrochemical detection and removal of pharmaceuticals in waste waters. Curr. Opin. Electrochem. 11, 1–11.

Fick, J., Söderström, H., Lindberg, R.H., Phan, C., Tysklind, M., Larsson, D.J., 2009. Contamination of surface, ground, and drinking water from pharmaceutical production. Environ. Toxicol. Chem. 28 (12), 2522–2527. 7-70.

Floate, K.D., Wardhaugh, K.G., Boxall, A.B., Sherratt, T.N., 2005. Fecal residues of veterinary parasiticides: nontarget effects in the pasture environment. Annu. Rev. Entomol. 50, 153–179.

Focazio, M.J., Kolpin, D.W., Barnes, K.K., Furlong, E.T., Meyer, M.T., Zaugg, S.D., Barber, L.B., Thurman, M.E., 2008. A national reconnaissance for pharmaceuticals and other organic wastewater contaminants in the United States—II Untreated drinking water sources. Sci. Total Environ. 402 (2–3), 201–216.

Franks, C.G., 2006. Phytoremediation of Pharmaceuticals with Salix Exigua (Ph.D. Thesis). Faculty of Arts and Science, University of Lethbridge, Alberta, Canada.

Gahlawat, S., Gauba, P., 2016. Phytoremediation of aspirin and tetracycline by *Brassica juncea*. Int. J. Phytoremediation, 18 (9), 929–935.

Garbisu, C., Alkorta, I., 2001. Phytoextraction: a cost-effective plant-based technology for the removal of metals from the environment. Bioresour. Technol. 77 (3), 229–236.

He, Y., Nie, E., Li, C., Ye, Q., Wang, H., 2017. Uptake and subcellular distribution of triclosan in typical hydrophytes under hydroponic conditions. Environ. Pollut. 220, 400–406.

Herklotz, P.A., Gurung, P., Heuvel, B.V., Kinney, C.A., 2010. Uptake of human pharmaceuticals by plants grown under hydroponic conditions. Chemosphere, 78 (11), 1416–1421.

Hussain, Z., Arslan, M., Malik, M.H., Mohsin, M., Iqbal, S., Afzal, M., 2018. Integrated perspectives on the use of bacterial endophytes in horizontal flow constructed wetlands for the treatment of liquid textile effluent: phytoremediation advances in the field. J. Environ. Manag. 224, 387–395.

Jain, P., Andotra, A., Aziz, A., Kaur, P., Mahajan, A., Kumar, A., 2019. Phytoremediation-a miracle technique for waste water treatment. Res. J. Pharma. Technol. 12 (4), 2009–2016.

Jamieson, O., Soares, T.C., Faria, B.A.D., Hudson, A., Mecozzi, F., Rowley-Neale, S.J., et al., 2020. Screen printed electrode based detection systems for the antibiotic amoxicillin in aqueous samples utilising molecularly imprinted polymers as synthetic receptors. Chemosensors, 8 (1), 5.

Kamal, M., Ghaly, A.E., Mahmoud, N., Cote, R., 2004. Phytoaccumulation of heavy metals by aquatic plants. Environ. Int. 29 (8), 1029–1039.

Kapoor, D., 2015. Impact of pharmaceutical industries on environment, health and safety. J. Crit. Rev. 2 (4), 25–30.

Kolpin, D.W., Furlong, E.T., Meyer, M.T., Thurman, E.M., Zaugg, S.D., Barber, L.B., Buxton, H.T., 2002. Pharmaceuticals, hormones, and other organic wastewater contaminants in US streams, 1999– 2000: a national reconnaissance. Environ. Sci. Technol. 36 (6), 1202–1211.

Koźmińska, A., Wiszniewska, A., Hanus-Fajerska, E., Muszyńska, E., 2018. Recent strategies of increasing metal tolerance and phytoremediation potential using genetic transformation of plants. Plant Biotechnol. Rep. 12 (1), 1–14.

Kumar, S., 2012. Challenges and opportunities of cytochrome P450-mediated phytoremediation. J. Pet. Environ. Biotechnol. 4, e001.

Kümmerer, K., 2004. Pharmaceuticals in the environment—scope of the book and introduction. In: Kümmerer, K. (Ed.), Pharmaceuticals in the Environment. Springer, Berlin, Heidelberg, pp. 3–11.

Lantam, A., Limbut, W., Thiagchanya, A., Phonchai, A., 2020. A portable optical colorimetric sensor for the determination of promethazine in lean cocktail and pharmaceutical doses. Microchem. J. 159, 105519.

Li, G., Zhai, J., He, Q., Zhi, Y., Xiao, H., Rong, J., 2014. Phytoremediation of levonorgestrel in aquatic environment by hydrophytes. J. Environ. Sci. 26 (9), 1869–1873.

Li, Y., Lian, J., Wu, B., Zou, H., Tan, S.K., 2020. Phytoremediation of pharmaceutical-contaminated wastewater: insights into rhizobacterial dynamics related to pollutant degradation mechanisms during plant life cycle. Chemosphere, 126681.

Limmer, M., Burken, J., 2016. Phytovolatilization of organic contaminants. Environ. Sci. Technol. 50 (13), 6632–6643.

Liu, X., Guo, X., Liu, Y., Lu, S., Xi, B., Zhang, J., Bi, B., 2019. A review on removing antibiotics and antibiotic resistance genes from wastewater by constructed wetlands: performance and microbial response. Environ. Pollut. 254, 112996.

Loraine, G.A., Pettigrove, M.E., 2006. Seasonal variations in concentrations of pharmaceuticals and personal care products in drinking water and reclaimed wastewater in southern California. Environ. Sci. Technol. 40 (3), 687–695.

Lu, B., Oyekan, J., Gu, D., Hu, H., Nia, H.F.G, 2011. Mobile sensor networks for modelling environmental pollutant distribution. Int. J. Syst. Sci. 42 (9), 1491–1505.

Ma, X., Burken, J.G., 2002. VOCs fate and partitioning in vegetation: use of tree cores in groundwater analysis. Environ. Sci. Technol. 36 (21), 4663–4668.

Macek, T., Kotrba, P., Svatos, A., Novakova, M., Demnerova, K., Mackova, M., 2008. Novel roles for genetically modified plants in environmental protection. Trends Biotechnol. 26 (3), 146–152.

Mahar, A., Wang, P., Ali, A., Awasthi, M.K., Lahori, A.H., Wang, Q., Li, R., Zhang, Z., 2016. Challenges and opportunities in the phytoremediation of heavy metals contaminated soils: a review. Ecotoxicol. Environ. Saf. 126, 111–121.

Maharjan, R. (2014). Phytoremediation of Selected Pharmaceuticals by and their Phytotoxicity to Aquatic Plants. (Ph.D. thesis), University of Toledo, OH.

Mahdavian, K., Ghaderian, S.M., Torkzadeh-Mahani, M., 2017. Accumulation and phytoremediation of Pb, Zn, and Ag by plants growing on Koshk lead–zinc mining area, Iran. J.Soils Sediment. 17 (5), 1310–1320.

Malik, B., Pirzadah, T.B., Tahir, I., Dar, T.H., Rehman, R. (2014). Recent trends and approaches in phytoremediation. Soil Remedation and Plants: Prospects and Challenges, 131–146.

Mehinto, A.C., Hill, E.M., Tyler, C.R., 2010. Uptake and biological effects of environmentally relevant concentrations of the nonsteroidal anti-inflammatory pharmaceutical diclofenac in rainbow trout (Oncorhynchus mykiss). Environ. Sci. Technol. 44 (6), 2176–2182.

Merola, G., Martini, E., Tomassetti, M., Campanella, L., 2014. New immunosensor for β-lactam antibiotics determination in river waste waters. Sens. Actuators B Chem. 199, 301–313.

Mofeed, J., 2020. Effects of three commonly used pharmaceutical products on biochemical parameters of the micro-alga *Pseudokirchneriella subcapitata* (under laboratory conditions). Catrina Int. J. Environ. Sci. 22 (1), 1–10.

Mustafa, H.M., Hayder, G., 2020. Recent studies on applications of aquatic weed plants in phytoremediation of wastewater: a review article. Ain Shams Eng. J. 12 (1), 355–365.

Naikoo, M.I., Kafeel, U., Naushin, F., Khan, F.A., 2020. Halophytes in India and their role in phytoremediation. In: Grigore, M-N. (Ed.), Handbook of Halophytes: From Molecules to Ecosystems towards Biosaline Agriculture. Springer, Cham, pp. 1–21.

Nguyen, P.M., Afzal, M., Ullah, I., Shahid, N., Baqar, M., Arslan, M., 2019. Removal of pharmaceuticals and personal care products using constructed wetlands: effective plant-bacteria synergism may enhance degradation efficiency. Environ. Sci. Pollut. Res., 1–18.

OroojSurriya, S.S.S., Waqar, K., Kazi, A.G, 2014. Phytoremediation of soils: prospects and challenges. In: Hakeem, Khalid Rehman, Sabir, Muhammad, Öztürk, Münir, Mermut, Ahmet Ruhi (Eds.), Phytoremediation of soils: prospects and challenges. Soil Remediation and Plants: Prospects and Challenges, 1–36.

Parolini, M., Bini, L., Magni, S., Rizzo, A., Ghilardi, A., Landi, et al., 2018. Exposure to cocaine and its main metabolites altered the protein profile of zebrafish embryos. Environ. Pollut. 232, 603–614.

Patel, N., Khan, M.D., Shahane, S., Rai, D., Chauhan, D., Kant, C., Chaudhary, V.K., 2020. Emerging pollutants in aquatic environment: source, effect, and challenges in biomonitoring and bioremediation-a review. Pollution, 6 (1), 99–113.

Patneedi, C.B., Prasadu, K.D., 2015. Impact of pharmaceutical wastes on human life and environment. Rasayan J. Chem. 8 (1), 67–70.

Pomati, F., Netting, A.G., Calamari, D., Neilan, B.A., 2004. Effects of erythromycin, tetracycline and ibuprofen on the growth of *Synechocystis* sp. and *Lemna minor*. Aquatic Toxicol. 67 (4), 387–396.

Quesada, H.B., Baptista, A.T.A., Cusioli, L.F., Seibert, D., de Oliveira Bezerra, C., Bergamasco, R, 2019. Surface water pollution by pharmaceuticals and an alternative of removal by low-cost adsorbents: a review. Chemosphere, 222, 766–780.

Ram, S.K., Panidepu, H., Cheernam, V., Tyagi, R.D., et al., 2020. Pharmaceutical metabolites and their by-products in hospital wastewater. In: Tyagi, R.D. et al (Ed.), Current Developments in Biotechnology and Bioengineering: Environmental and Health Impact of Hospital Wastewater. Elsevier, Amsterdam, pp. 43–78.

Rana, V., Maiti, S.K., 2020. Municipal and Industrial Wastewater Treatment Using Constructed Wetlands. In: Shmaefsky, B.R. (Ed.), Phytoremediation. Springer, Cham, pp. 329–367.

Rodrigues, S., Antunes, S.C., Correia, A.T., Nunes, B., 2018. Oxytetracycline effects in specific biochemical pathways of detoxification, neurotransmission and energy production in *Oncorhynchus mykiss*. Ecotoxicol. Environ. Saf. 164, 100–108.

Schulte-Oehlmann, U., Oetken, M., Bachmann, J., Oehlmann, J., 2004. Effects of ethinyloestradiol and methyltestosterone in prosobranch snails. In: Klaus K., (Ed.), *Pharmaceuticals in the Environment*. Springer, Berlin, Heidelberg, pp. 233–247.

Shutcha, M.N., Mubemba, M.M., Faucon, M.P., Luhembwe, M.N., Visser, M., Colinet, G., Meerts, P., 2010. Phytostabilisation of copper-contaminated soil in Katanga: an experiment with three native grasses and two amendments. Int. J. Phytoremediation 12 (6), 616–632.

Sidhu, H., O'Connor, G., Kruse, J., 2019. Plant toxicity and accumulation of biosolids-borne ciprofloxacin and azithromycin. Sci. Total Environ. 648, 1219–1226.

Sommer, C., Bibby, B.M., 2002. The influence of veterinary medicines on the decomposition of dung organic matter in soil. Eur. J. Soil Biol. 38 (2), 155–159.

Stoltz, E., Greger, M., 2002. Accumulation properties of As, Cd, Cu, Pb and Zn by four wetland plant species growing on submerged mine tailings. Environ. Exp. Bot. 47 (3), 271–280.

Sui, Q., Cao, X., Lu, S., Zhao, W., Qiu, Z., Yu, G., 2015. Occurrence, sources and fate of pharmaceuticals and personal care products in the groundwater: a review. Emerg. Contamin. 1 (1), 14–24.

Taneja, N., Sharma, M., 2019. Antimicrobial resistance in the environment: the Indian scenario. Indian J. Med. Res. 149 (2), 119.

Tanoue, R., Sato, Y., Motoyama, M., Nakagawa, S., Shinohara, R., Nomiyama, K., 2012. Plant uptake of pharmaceutical chemicals detected in recycled organic manure and reclaimed wastewater. J. Agric. Food Chem. 60 (41), 10203–10211.

Valdez-Carrillo, M., Abrell, L., Ramírez-Hernández, J., Reyes-López, J.A., Carreón-Diazconti, C., 2020. Pharmaceuticals as emerging contaminants in the aquatic environment of Latin America: a review. Environ. Sci. Pollut. Res. 27 (36), 1–29.

Vanek, T., Podlipna, R., Fialova, Z., Petrova, S., Soudek, P., 2010. Uptake of xenobiotics from polluted waters by plants. In: Fatta-Kassinos, D., Bester, K., Kümmerer, K. (Eds.), Xenobiotics in the Urban Water Cycle. Springer, Dordrecht, pp. 431–444.

Verlicchi, P., Zambello, E., 2014. How efficient are constructed wetlands in removing pharmaceuticals from untreated and treated urban wastewaters? A review. Sci. Total Environ. 470, 1281–1306.

Vikesland, P.J., Pruden, A., Alvarez, P.J., Aga, D., Bürgmann, H., Li, X.D., et al., 2017. Toward a comprehensive strategy to mitigate dissemination of environmental sources of antibiotic resistance. Environ. Sci. Technol. 51(22), 13061–13069.

Wang, M., Zhang, D.Q., Dong, J.W., Tan, S.K., 2017. Constructed wetlands for wastewater treatment in cold climate—a review. J. Environ. Sci. 57, 293–311.

Wani, R.A., Ganai, B.A., Shah, M.A., Uqab, B., 2017. Heavy metal uptake potential of aquatic plants through phytoremediation technique—a review. J. Bioremediat. Biodegrad, 8 (404), 2.

Westergaard, K., Müller, A.K., Christensen, S., Bloem, J., Sørensen, S.J., 2001. Effects of tylosin as a disturbance on the soil microbial community. Soil Biol. Biochem. 33 (15), 2061–2071.

Wu, X., Ernst, F., Conkle, J.L., Gan, J., 2013. Comparative uptake and translocation of pharmaceutical and personal care products (PPCPs) by common vegetables. Environ. Int. 60, 15–22.

Yang, Z., Ding, X., Guo, Q., Wang, Y., Lu, Z., Ou, H., et al., 2017. Second generation of signaling-probe displacement electrochemical aptasensor for detection of picomolar ampicillin and sulfadimethoxine. Sens. Actuators B Chem. 253, 1129–1136.

Young, W.F., Whitehouse, P., Johnson, I., 2002. Proposed Predicted-No-Effect-Concentrations (PNECs) for Natural and Synthetic Steroid Oestrogens in Surface Waters. Environment Agency, Bristol, UK.

Zavoda, J., Cutright, T., Szpak, J., Fallon, E., 2001. Uptake, selectivity, and inhibition of hydroponic treatment of contaminants. J. Environ. Eng. 127 (6), 502–508.

Zhang, D., Gersberg, R.M., Ng, W.J., Tan, S.K., 2014. Removal of pharmaceuticals and personal care products in aquatic plant-based systems: a review. Environ. Pollut. 184, 620–639.

Zhu, Y.L., Zayed, A.M., Qian, J.H., De Souza, M., Terry, N., 1999. Phytoaccumulation of trace elements by wetland plants: II. Water hyacinth. J. Environ. Qual. 28 (1), 339–344.

Chapter 20

Phytoremediation of persistent organic pollutants (POPs)

Bisma Malik[a], Tanveer Bilal Pirzadah[a], Khalid Rehman Hakeem[b]
[a]*University Centre for Research and Development (UCRD), Chandigarh University, Mohali, Punjab, India,* [b]*Department of Biological Sciences, Faculty of Science, King Abdulaziz University, Jeddah, Saudi Arabia*

20.1 Introduction

The unprecedented increase in urbanization, industrialization, intensive agricultural system globally has contributed toward the generation of toxic pollutants and contaminants in the ecosystem and poses great threat to human health and environment. The crucial concern today is the contamination of soil and water due to the discharge of urban wastes by human activities (Gosh and Singh, 2005). Environment becomes polluted by natural as well as by anthropogenic activities; natural sources include urban wastes, soil erosion, volcanic activities, while human factor includes wastes from textile industries, refining, nuclear power plant, electroplating, and several other factors (Akpor et al., 2014). However, manufacturing of several man-made halogenated organic compounds used for different purposes namely, herbicides, pesticides, electrical insulation oil, refrigerants, pants, solvents, and many fire retardants ultimately leads to life-threatening diseases and environmental pollution. This issue is of great concern nowadays as some of these compounds are highly dangerous to human health as they get bioaccumulated in the tissues and highly concentrated and increase progressively in the food web. Natural degradation of these compounds is not fully possible as these compounds are highly tolerant to natural mechanisms and have the ability to transport to long ranges and even to those places where these compounds have never been utilized. Soil microorganisms have the capability to degrade or detoxify some of these contaminants/pollutants except persistent organic pollutants (POPs) (Stockholm convention, UNEP, http://chm.pop.int). According to Stockholm Convention, POPs have been categorized as those substances which are subjected to exclusion, restriction, and accidental production and there has been escalation in the number of these compounds mentioned in the list in Table 20.1. In spite of banning and limiting the use of

Phytoremediation. DOI: https://doi.org/10.1016/B978-0-323-89874-4.00010-8

TABLE 20.1 List of some plants reported for phytoremediation of organic contaminants.

S. No	Plant used for phytoremediation	Pollutant	References
1.	*Festuca arundinacea Schreb.* and *Festuca pratensis Huds*	Water-soluble phenols, total petroleum hydrocarbon (TPH), and polycyclic aromatic hydrocarbon (PAH) contents	Soleimani et al., (2010)
2.	Alfalfa (*Medicago sativa* L.)	Petroleum hydrocarbon	Agnello et al., (2016)
3.	*Kandelia obovata, Bruguiera gymnorrhiza,* and *Avicennia marina*	PAHs	Wang et al., (2014)
4.	Maize (*Zea mays* L.)	Atrazine	Sánchez et al., (2018)
5.	*Scirpus grossus*	Petroleum hydrocarbons	Al-Baldawi et al., (2013)
6.	Fire phoenix (*Festuca arundinacea, Festuca elata, Festuca gigantea*)	PAHs	Liu et al., (2014)
7.	*Achillea millefolium*	Petroleum hydrocarbons	Masu et al., (2014)
8	*Impatiens balsamina* L.	TPHs	Cai et al., (2010)
9.	*Salix miyabeana SX67* and *S. miyabeana SX64*	Mixture of organic pollutants	Guidi et al., (2012)
10.	*Echinacea purpurea, Festuca arundinacea* Schred, *Medicago sativa* L.	PAHs	Liu et al., (2015)
11.	Wild rye (*Elymus angustus)* and alfalfa (*Medicago sativa*)	Hydrocarbon	Phillips et al., (2012)
12.	*Pinus densiflora, Populus tomenti glandulosa,* and *Thuja orientalis*	TPHs	Jagtap et al., (2014)
13.	*Paspalum vaginatum, Tamarix gallica, Spartium junceum*	TPHs	Masciandaro et al., (2014)

14.	*Poa foliosa* (Hook. f.)	Hydrocarbons	Bramley-Alves et al., (2014)
15.	*Vetiveria zizanioides* (L.) Nash	Phenol	Phenrat et al., (2017)
16.	*Melilotus albus*	TPHs	Hernández- Ortega et al., (2012)
17.	*Lolium perenne, Festuca arundinacea, Hordeum vulgare* treated with Plant growth-promoting rhizobacteria (PGPR) *Hordeum vulgare*	Petroleum hydrocarbon	Gurska et al., (2009)
18.	*Helianthus annuus*	POPs	de Almeida et al., (2018)
19.	*Salix matsudana, Populus deltoides, Populus nigra, Eucalyptus camaldulensis, Helianthus annuus*	n-Alkanes, PAHs , and PCBs	Nissim et al., (2018)
20.	*Leptochloa fusca* and *Brachiaria mutica*	Petroleum hydrocarbons	Fatima et al., 2018
21.	*Brassica napus, Sorghum bicolor, drummondii*	Petroleum hydrocarbons	Petrová et al., (2017)
22.	*Zea mays* L.	Trichloroethylene	Moccia et al., (2017)
23.	*Acorus calamus*	PAHs	Jeelani et al., (2017)
24.	*Medicago sativa* cultivar Crioula	Phenanthrene, pyrene, and anthracene	Alves et al., 2018
25.	Ryegrass (*Lolium perenne*), white clover (*Trifolium repens*), and celery (*Apium graveolens*)	PAHs	Meng et al., 2011
26.	*Leucanthemum vulgare*	TPHs	Noori et al., 2018

these compounds, there were reports by several regions in the adulteration of breast milk, soil, and water by POPs globally. However, recently a study conducted by group of researchers reported the presence of POPs in fauna from the huge ocean trenches of >10,000 meters deep (Jameison et al., 2017). Therefore, it is the need of the hour to degrade or eliminate such pollutants from the environment to save human health and integrity of ecosystem worldwide. However, the elimination or degradation of such pollutants was carried out by utilizing mainly energy-concentrated engineering process (Salvato et al., 2003). But these processes required highly sophisticated and heavy equipped machinery that often leads to unsatisfactory results such as incomplete removal of pollutants, green-house gas emission, dismantling of soil integrity, and heavy alteration in the landscape pattern. For the removal of POPs or any other pollutants, viable, eco-friendly and cost-effective technologies are yet to be developed. From the past two decades, abundant data showed the robustness of plants as an alternative technique utilized for the decontamination of organic or inorganic pollutant from the environment (Pilon Smits, 2005). This idea basically depends on the remarkable capabilities of plants for their metabolic and extractive properties that confer them with unique potential for the decontamination of soil and water. In addition to this, several plants have the capability to survive on high metal–accumulated soils that greatly exceeded the normal levels and undergo normal metabolism. Moreover, plants based technology is solar energy–driven processes making them both eco-friendly and cost effective in nature. In the scientific literature, the use of wild or transgenic plants for the degradation, removal, or decontamination of the pollutants from the environment is commonly referred to as phytoremediation (Salt et al., 1998; Dictz and Schnoor, 2001; Mc Cutcheon and Schnoor, 2003) and this technology can be applied to both organic or inorganic pollutants present in water, air, or the soil (Raskin et al., 1994). This chapter will focus on an overview of current knowledge on the phytoremediation of POPs and detoxification mechanism, and current advances in molecular tools and rising challenges to phytoremediation technology will also be discussed.

20.2 Persistence, bioaccumulation, and phytoremediation of POPs

As far as the term persistence is concerned, it refers to how long a pollutant or toxin substance resides in the environment, and bioaccumulation means accumulation of a toxic pollutant overtime in a living organism (Russel, 2005). Toxic pollutants and their residues have been noticed in the soil, water, air, wildlife, aquatic animals, sediments, and in various other types of living organisms globally (WWF, 2005). A toxic substance/pollutant has the ability to stay in the environment for longer periods of time before it was degraded or detoxified into a less suitable form. It has been reported that the concentration of the pollutant will be increased as it was utilized throughout the food web and thus ultimately gets accumulated in the tissue of living organisms that will be proven more dangerous in coming years (Stockholm Convention, 2005). Accumulation of pollutants in biological systems is extremely dangerous as pollutants get absorbed by organic and some other substances and are then ingested by small animals; and POPs get accumulated particularly in the fatty tissues of these animals and in this way each successive predator consume a high amount of toxic pollutant and can be more dangerous than one single dose (FAO, 2005). These properties of pollutants make them highly detrimental and are one of the main reasons that they are such a great threat to human health and ecosystem. Therefore, environmental cleanup is the need of the hour and plants used for this purposes are highly recommendable. Utilization of green plants and their associated microbiota for the removal of pollutants from the environment are collectively referred to as phytoremediation (Cunningham et al., 1996). The term phytoremediation was coined in early 1900s to define the role of plants in the remediation or removal of pollutants from the environment (Raskin et al., 1994; Salt et al., 1995). This technology can be applied to both organic and inorganic pollutants present in the soil, water, and air (Pilon Smits, 2005). As this technology is powered by solar energy that ultimately makes it cost effective and eco-friendly than any other method of detoxification (washing, excavating, burning soil, or pump-and-treat system) (Schnoor, 1997). Another positive result of phytoremediation is that it rejuvenates the soil health after treatment and also helps in the balancing of various soil properties. When compared to other chemical methods, phytoremediation is suitable for in-situ reduction of pollutants over a large area, which makes this technology sustainable and cost effective. The transport mechanisms of pollutants to terrestrial plants are distinguished into sorption of pollutant by the roots through xylem and dispersal with shoots by phloem transport. The root concentration factor to the transpiration stream concentration factor indicates the flow of organic pollutants from soil to the plants (Briggs et al., 1982). For the remediation process, bioavailability of pollutants is an important factor for its removal procedure and the bioavailability depends on the properties of soil, pollutant, and edaphic factors including bioactivities. The mobility of a pollutant in soil is affected by its hydrophobicity that is one of the most important chemical properties and is usually expressed as log K_{ow} (the octanol-water partition coefficient) and value of K_{ow} is directly proportional to the pollutant hydrophobicity value. POPs having value of K_{ow} >3 are highly hydrophobic molecules that are adhered tightly with soil organic material and never diffused into pore water of soil (Briggs et al., 1982). This absence of bioavailability restricts their phytoremediation capability and results in their distribution as intractable pollutants. Besides, volatility of pollutants indicated by Henry's Law Constant (H_i) is one more key chemical property that affects the overall properties of pollutants to be phytoremediated. Besides, POPs are often categorized as semivolatile compounds.

Usually phytoremediation technology is divided into different types depending upon the type of mechanism involved in the pollutant removal

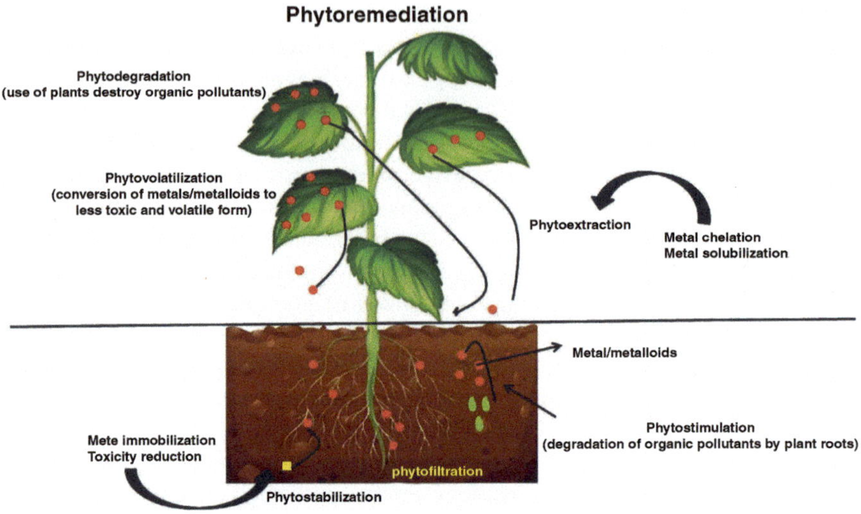

FIG. 20.1 Phytoremediation technology is divided into different types depending upon the type of mechanism involved in the pollutant removal process.

process: (1) eliminating pollutants from the soil by accumulating them in aboveground plant parts (phytoextraction), (2) liberation of pollutants in the environment through transpiration mechanism of plants (phytovolatilization), (3) limiting the movement of pollutant in the soil (phytostabilization) (4) increasing the activities of soil microbiota for the efficient removal of pollutants by plants in the rhizospheric region (phytostimulation), and (5) detoxification of pollutants inside the tissues of plants (phytodegradation) (Fig. 20.1). However, these technologies are not completely interactive because sometimes stabilization, volatilization, and accumulation take place concomitantly. The capability to take up higher amounts of pollutants is necessary for the operative elimination of pollutants from the soil. The plants species that accumulate large amount of inorganic/organic metals are called as hyperaccumulator plants (Brooks et al., 1977; Peer et al., 2005). Several hyperaccumulator plants or metallophyte species (*Elsholtzia splendens, Silene vulgaris, Thlaspi caerulescens, Alyssum species, Arabidopsis halleri, Sedum alfredii, Brassica juncea,* and *Pteris vittata*) have been studied for different heavy metals such as Cu, Zn, Ni, Cd, Pb, Se, and As (Ma et al., 2001; Song et al., 2004; Weng et al., 2005; Liu et al., 2007; Singer et al., 2007; Milner and Kochian, 2008; Ueno et al., 2008). Besides, several plants species have been studied for the efficient removal of toxic organic pollutants such as explosives and polycyclic aromatic hydrocarbons (Hannink et al., 2001; Harms et al., 2002; Tang and Mo, 2007) as well as there are also reports for normal organic pollutants describing the uptake of or resistance of POPs (Dichlorodiphenyltrichloroethane (DDT), Polychlorinated biphenyls (PCBs), and hexachlorocyclohexanes (HCHs)) in several plant species (Table 20.1) (Calvelo Pereira et al., 2006; Mo et al., 2008; Mikes et al., 2009; Abhilash and Singh, 2010; Abhilash et al., 2013; Dubey et al., 2014; Rissato et al., 2015). The limiting factor for the phytoremediation of POPs from the soil is the inadequate consumption by plants and is due to their highly hydrophobic value of K_{ow} >3, which makes them intractable for their uptake from the soil by the plant species. However, there is no standard criterion fixed for organic pollutant hyperaccumulator, although *Cucurbita* species is commonly regarded as POPs hyperaccumulator and has been widely investigated for the translocation and uptake POPs (DDT, PCBs, dieldrin, endrin, and dioxine) (Hulster et al., 1994; White, 2001; Lunney et al., 2004; Mattina et al., 2006; Otani et al,. 2007; Inui et al., 2008; Namiki et al., 2013; Sugiyama et al., 2013). As the molecular mechanism for the uptake of highly hydrophobic organic pollutants are not yet studied however, the key characteristics for the POPs uptake and solubilization have been reported in many recent studies (Murano et al., 2010; Inui et al., 2013; 2015).

20.3 Technological outlook

Organic pollutants have the possibility to get metabolized by living organisms in contrast with heavy metals. On these compounds, the plants responded in a multiple ways such as they can be stored, immobilized, volatilized, and even mineralized, or a mixture of them, based on the type of plant genotype involved, particular compound, and the overall environmental conditions. As a result, several phytotechnologies have been developed to reduce the pollution caused by organic pollutants (Fig. 20.1). Different phytotechnologies developed for the reduction of organic pollution from the environment are phytoextraction (Singh et al., 2014), rhizofiltration (Ensley, 2000), phytodegradation (Etim, 2012), phytovolatilization, and phytostabilization (Etim, 2012). Bioavailability of pollutants is clearly as one of the crucial factors for the phytoremediation completion. The mobility of an organic pollutant in the soil mainly depends on the physical and chemical properties (water solubility, molecular size, charge, and vapor pressure) and its relationship with nearby molecules; however, soil characteristics such as structure, pH, texture, and organic matter content including the rhizosphere also play a significant role in the process (Dzantor, 2007). As per the previous reports, it is very important that different factors must be taken into consideration before utilizing the full-scale phytoremediation technology to a contaminated site and the most important one's are: (1) characterization of the polluted site in detail (pollution level), (2) selection of site-specific plant species, (3) determining the total cultivation costs (including soil modification), and (4) evaluating the predetermined state of the pollutant. Besides, time required for the process of remediation is another important factor that largely depends on the uptake and removal rate of the pollutant and small-scale preliminary studies was usually conducted to evaluate these factors (McCutcheon and Schnoor, 2003). In addition to this, other important factors to be addressed after the harvesting stage are the plant biomass fate and collection as well as pollution level of the remaining material (particularly the below-ground parts that are difficult to remove and costly). Currently, the area of phytoextraction and phytodegradation draws more attention by the most of the scientific and commercial communities and the plant species selected for the same is being a crucial factor as various countries still apply strict rules and regulations for the use of genetically modified plants. Usually the plant biomass is collected and either used in the beneficial way (as energy sources or in the fiber production) or discarded once the pollutant was taken up and accumulated by the plants after approximately improved by metabolic activities. The overall technique can be extremely effective and is cost effective (average tenfold cheaper) than any other physical technique.

20.4 Physiological and biochemical outlook

It is very important to make significant attempts to elucidate the mechanism underlying the uptake, allocation, and decontamination of the xenobiotic compounds once absorbed by the plants and these compounds are absorbed primarily in the plants through leaves and roots (Wang and Liu, 2007). Usually leaf absorption is a result of organochemical spraying in the agriculture system; however immediate uptake of volatile compounds can also be noteworthy (Burken et al., 2005). The absorption into the roots takes place by simple diffusion method and xenobiotic compounds enter into the xylem channels through unsuberized cell walls. It is evident that there are no particular transporters in plants for these xenobiotic compounds so the mobility rate of these compounds depends mostly on their physical and chemical characteristics into and through the plants, as this mobility rate is basically a passive physical procedure, it is rather expected and responsive to moderate modeling system (Fujisava, 2002). Some of the agronomic methods may increase the efficacy of the absorption activity for example, roots of the plants can be managed unnaturally in the polluted areas by supplementing irrigation, fertilization, fertilization, and oxygenation of roots and in the same way plants can be chosen or modified for suitable root structure (Wang et al., 2006). Certain plant species such as *Popular spp*, *Salix spp* with deep root system and huge rate of transpiration showed high potentiality for phytoremediation approaches (Jonsson and Douglas, 2007). Xenobiotic compounds are improved through reduction, oxidation, or by hydrolysis reaction accompanied by the conjugation with organic acids, sugar, and glutathione (GSH) after penetrating into the symplast. After that the next step provides them more dissolvability and likely their next binding to transporters, enzymes, and other applicable proteins (Dietz and Schnoor, 2001; Pilon-Smits, 2005). In the cellular compartments (vacuoles), conjugated xenobiotics are detoxified as a part of insoluble cell wall polymers where they are further metabolized preferably to carbon dioxide (Pilon-Smits, 2005). Moreover removal of pollutants from sensitive areas in the cytosol, cell compartmentation, emerges to be one of the essential steps for the removal of various pollutants (Mezzari et al., 2005). GSH conjugate synthesis has been largely investigated in all living beings, besides it is arbitrated by a broad ranges of GSTs (Glutathione-S-transferase) majority of them have required place for hydrophobic ligands (reviewed by Edwards and Dixons, 2005). For the consecutive detoxification process, the resulting complexes communicated with downstream enzymes and gamma glutamyl transpeptidases (GGTs). Currently, Heynh vacuoles have been determined in which a particular GGT catalyzes the required initial step for conjugate GSH detoxification in *Arabidopsis thaliana* (Ohkama-Ohtsu et al., 2007). Moreover, there is no information regarding other alternatives for conjugation such as those arbitrated by malonyl transferases or glucosyl transferases (Bruzier-Hicks and Edwards, 2005). Despite the particular system, many research works have indicated that ATP-binding cassette (ABC) transporters play a critical function in the conjugate transformation from cytosol to apoplast or the vacuole and therefore are the key possible elements for the modification/ genetic transformation strategies (Klein et al., 2006). It has been found that various plant enzymes have a significant role in the degradation of xenobiotics such as phosphatises, carboxylesterases, nitrilases, mono-and dihydrogenases, dehalogenases, hydrolases, nitroreductases, dehydrogenases, and many more (Dietz and Schnoor, 2001; Singer et al., 2003; Wolfe and Hoehamen, 2003; Pilon-Smits, 2005). However, biotechnologists progressively know the capability of these enzymes to enhance the potentiality of specific species of plants. These enzymes emerge naturally and then liberated into the soil and are able to degrade the myriads of organic contaminants varying from explosives to solvent (Singer et al., 2003). Wang et al., (2004) described the feasible approach of transgenic *Arabidopsis* with the ability of increased liberation of *laccase* enzyme into the soil by the overexpression of genes that lead to the subsequent degradation and remediation process. The function of various enzymes and transporters of plants toward xenobiotics or any other man-made compounds may be regarded basically as a phenomenon of chance as the actual function of these enzymes are well known in the plant system. In these circumstances, cytochrome P450 monooxygenases might be regarded as an appropriate enzyme for the removal of environmental pollutants (Morant et al., 2003), as oxidation reaction catalyzed by these enzymes has been described in secondary metabolism in which a network of reactions allowed plants to interact with its living environment, that is, to captivate soil microorganism and pollinators and avoiding pathogens and herbivores. Hydroxylation and epoxidation are the usual reactions catalyzed by plant P450 and are also able to catalyze other reactions such as ring development, phenol coupling, and decarboxylation or modification of suitable substrates. It has been also described that there is a possible role of P450s in the survival of the plant at higher level of herbicide and its detoxification from the polluted soil. The first biological catalysts of plant source involving in the effective herbicide detoxification process are *soybean* [*Glycine max* (L) Merr.] CYP71A10 and *Helianthus tuberosus* CYP76B1 (Robineau et al., 1998; Siminszky et al., 1999) besides, several P450s of plant source were also mentioned that have an active role in the mineralization of various organic pollutants particularly POPs. Interestingly in most of the cases, transgenic plants (overexpression of P450 gene) used for effective remediation procedure belong to mammalian source. A study conducted by Doty et al., (2000) described the development of transgenic tobacco plants having the capability to synthesizing mammalian CYP2E1 with remarkable potential in mineralization and absorption of trichloroethylene and ethylene bromide. In addition to this, many studies have reported that the genetically modified *Solanum tuberosum* L. (potato) and *Oryza sativa* L. (rice) have significant increase in their resistance against dangerous herbicide application and accumulation (Inui et al., 2001; Kawahigashi et al., 2007).

However, in this direction considerable research must be promoted so that the process of phytoremediation of transgenic plants results in effective and attainable process through the insertion of P450 gene at a large scale. Several reports have described that P450s acquired from mammalian system have the greater ability to degrade huge number of xenobiotic compounds compared to P450s obtained from plant sources. Didierjean et al., (2002) investigated transgenic *Arabidopsis* and *Nicotiana tabaccum* expressing the P450 gene (plant origin) for herbicide resistance and revealed that both the plant species indicated higher resistance to herbicides and this was the first extensive study in transgenic plants. Sandermann, (1994) put forward the idea of "green livers" to emphasize the resemblances between plants and mammals to handle a particular type of organic contaminant.

20.5 Current molecular tools

Novel molecular tools along with current developments in our apprehension of plant genetics and biochemistry to investigate different plant mechanisms (plant metabolism, protein function and gene regulation) led to the unraveling of new and unique approaches for xenobiotics degradation process. After the utilization of these tools, more in-depth information emerged regarding different characteristics along with additional attributes of the associated mechanisms. In the coming years, three research areas at the molecular level are believed to unravel mechanisms and can cause significant effect in upcoming times.

20.6 Protein modification

From any sources, suitable transporters or enzymes can be improved through progressive approaches of protein modification before overexpressed in genetic engineered plants. In this regard, enzymes are of specific interest in rate-limiting steps in remediation process possibly such as reductases and oxygenases. Previously in environmental biotechnology, the potentiality of P450 cytochrome monooxygenases is well described (see also Hannemann et al., 2007). Some current results are also promising with aromatic ring-fission dioxygenases and ring-hydroxylating dioxygenases that actually take parts in the initial steps of xenobiotics degradation (polycyclic aromatic hydrocarbon , PCBs, pesticides and others) (Furukawa, 2006). The suitable design of novel or improved characteristics needs an in-depth information on structure-activity association and with this aim, the 3D structure of various enzymes has been established and mutagenesis approaches have been produced to locate certain residues and enhance the activity of enzymes or specificity of substrate (Barriault and Sylvestre, 2004; Labrou et al., 2004; Vardar et al.,

2005; Leungsakul et al., 2006; Zielinski et al., 2006; Camara et al., 2007). The physiochemical properties have also been utilized to collect specific knowledge on the structure of certain pollutants and unravel their charge attributes and chemical reactivity (Pacious Gomez, 2006). Biphenyl dioxygenase BphA enzyme is an outstanding candidate to be modified as it catalyzed the first step in the well-studied pathway for PCBs aerobic degradation. The modification of this enzyme has been done through site-directed mutagenesis and recently it has been expressed in transgenic tobacco and showed the capability of accomplish the addition of oxygen to 4-chlorobiphenyl (Mohammadi et al., 2007). Moreover, comprehensive and detailed information on the specific genes with effective pollutant degradation potential and techniques involved in their modification has been precisely demonstrated by Whiteley and Lee, (2006).

20.7 Omics approaches

Various techniques are now accessible to investigate extensively genomes, proteomes, metabolomes, and transcriptomes for different applications in cellular systems. Utilizing these techniques in phytoremediation studies has enormous capabilities irrespective of our limited knowledge regarding various cellular systems. Although these techniques have been employed usually for determining hyperaccumulation (Hooda, 2007) , limited research have focused on the genomic variation appeared due to the presence of explosives or any other xenobiotics (Ekman et al., 2003; 2005; Patel et al., 2004; Mentewab et al., 2005) or herbicide (Baerson et al., 2005; Castro et al., 2005; Mezzari et al., 2005; Zhang et al., 2007). Various proteins assumed to be involved in the detoxification of PCBs have been recognized recently by employing the proteomic approach. However, GSTs and ABC transporters induction mechanism have been frequently described in such studies suggesting an important function of these proteins in the degradation process. In response to organic contaminants, enzymes involving in oxidative stress are usually up-regulated and sequestered the dangerous contaminants from the cellular systems. The combination of various omics approach with earlier molecular information and designs will definitely increase the length and breadths of gene discovery in this area of research and besides it, regulatory networks also hold great promise and open different novel means for genetic modification strategies.

20.8 Genetic engineering

Several studies have been conducted and investigated for genetic transformation in plants and have revealed the usefulness of manipulating higher degradative and or extractive capabilities in plants through genetic engineering. Much work so far done in this area of research has concentrated on the genes whose protein products play an important role in the accumulation and uptake of metals (Hooda, 2007). The natural hyperaccumulators such as *A. halleri*, *Alyssum lesbiacum, Thlaspi caerulescens*, and *Thlaspi goesingense* after performing the systematic analyses have been certainly perceptive and lead the plant biotechnologists to focus on metal transport and metal chelation process as the two main mechanisms for the achievement of phytoremediation strategy (Song et al., 2003; Becher et al., 2004). However, there are no reports on hyper-degradators for the degradation of organic pollutants (may be they exist but hard to pinpoint), that is why well-organized strategies are less expected to conduct research on such fields. Another limitation is that we have limited knowledge on degradation mechanism as there are actually hundreds or thousands of varied organic contaminants associated to a broad range of structural families. In spite of these limitations which clearly describe why the xenobiotics phytoremediation is not fully explored and falls behind metals, some encouraging outcomes in the previous years have been achieved. French et al., (1999) provided a significant proof of transgenic capabilities by increasing considerably the capability of tobacco to detoxify explosives such as 2,4,6-trinitrotoluene (TNT) and glyceryl trinitrate (GTN) by the overexpression of bacterial NADPH-dependent nitroreductases. Other studies have successfully observed that genetically engineered plants have the ability to decontaminate herbicides (Noctor et al., 1998; Karavangeli et al., 2005), phenolic compounds (Wang et al., 2004), organomercurials (Bizily et al., 2000), nitroaromatics (Hannink et al., 2001; Rylott et al., 2006), and PCBs (Mohammadi et al., 2007). Nowadays manipulation or genetic modification of plant endophytic bacteria also results in the improvement of organic pollutant degradation mechanisms (Mastretta et al., 2006).

20.9 Emerging challenges

The first and foremost limitation of phytoremediation technology such as ex-situ approach and excavation is moderate measure of remediation process. Contaminated soil restricts the plant growth and survival and thereby restricting the catabolism of pollutants (Harvey et al., 2002; Huang et al., 2004; 2005). Numerous biotic (herbivory by insects or animals, plant pathogens, competition with herbs and grass) and abiotic (light, temperature, nutrient, atmospheric CO_2) aspects may also limit the growth and survival of plant in fields used for phytoremediation process (Nedunuri et al., 2000). These aspects (biotic/abiotic) eventually limits the growth of the plants and affect the phytoremediation mechanism. For example, the phytoremediation attempt for contaminated ground water and soil at a hydrocarbon burn facility (NASA Kennedy space Centre in Florida) was stopped due to water stress condition and struggle with herbs. Moreover, different investigations revealed that herbs/weeds may also degrade organic contaminants and in this project, degradation of organic pollutants by weeds was not studied and in-depth weed culmination was announced as the failure reason for this project. Besides, another limitation of phytoremediation is physical challenge that actually limits the efficacy of remediation process. For example, in the organic pollutant–contaminated site increasing moisture availability might increase the difficulty of issue after being moistureless and finally results in the hampering of growth and development of seedlings. In addition to this, another limitation of phytoremediation strategy is associated with a specific pollutant of interest as in majority of cases; general soil depth of usually 50 cm is acquired by the plant roots (Pilon-Smits, 2005) and before phytoremediation, such type of hurdles are important to scrape out before process. Moreover, it is important to mention that roots of trees (deeper) promote phytoremediation at higher depth (Pilon-Smits, 2005). In case of ground and soil remediation process, dendroremediation of explosives indicates greater ability for the degradation of explosives (Meagher, 2000; Susarla et al., 2002; Doucette et al., 2003). However, growth, survival, and development are the important issues encountered in the polluted sites and the trees require years together to achieve enough biomass for successful phytoremediation strategy and disposal of contaminated plants/trees (roots/woods) are another problematic issues after the commencement of remediation process. In green house and laboratory experiments soil is properly mixed to homogeneity in their attributes whereas at field sites there is heterogeneity in contaminants distribution, which is crucial factor for the phytoremediation success. In field experiments, it is not possible to attain consistency in the conditions of soil even if the site is tilled properly before planting and thus this inconsistency in the soil conditions makes phytoremediation data dispersed and disintegrated. Besides, components such as structure of roots, morphology of soil, pH, composition of nutrients, and availability of moisture and soil, microorganism activities usually show significant structural and material distinction (Nedunuri et al., 2000; Corwin et al., 2006). For the removal of organic pollutants, utilization of transgenic organisms has received a lot of attention and showed successful results under laboratory conditions; however, the issue related to regulatory mechanism has limited their estimation under normal field conditions (Pilon-Smits and Freeman, 2006). If there is no appropriate pollutant system present, then there is high probability of gene transfer from transgenic organism to local one. In addition, the improved strains possessing the antibiotic genes could not play a role in their release in the natural environments (Villacieros et al., 2005). Furthermore, if there is any solution to the problem of transfer of genes from genetically modified organism to local one then the fewer acceptances by public may arise as a main hindrance for the release of transgenic organism as stated for Europe (INRA, 2000). Although due to irrelevant fear, public rejected the transgenic organisms for their application instead of having some relevant and scientific knowledge, regulatory laws in majority of such cases during the development of policy requires response of public for the final execution and implementation.

20.10 Outlook and conclusion

When compared with traditional methods of pollutant degradation from the environments, phytoremediation is reasonably the latest technology that provides several benefits for environment cleanup. This technology is widely applied on the

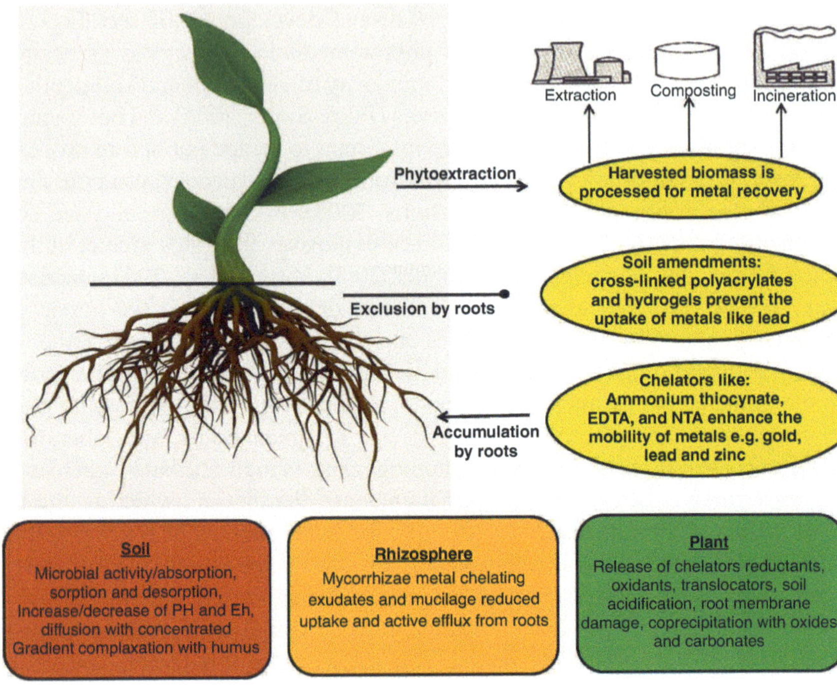

FIG. 20.2 Phytoremediation for sustainable development.

contaminated soil and water as this technology is cost effective and eco-friendly in relation to traditional remediation techniques. In this technology, plants reduce, degrade, breakdown, and remove pollutants from soil and wastewater and transfer them into their biomass. However, its lesser utilization has been tested in laboratory and green-house experiments, whereas some of its applications have been adequately tested in fields to permit the complete effectiveness of this technology. It is important to note here that the basic research is in its infancy stage and lacks many aspects to utilize effectively the enormous benefits promised by these technologies. In this context, the combination of new multiple omics tools with traditional information on the plant physiology, biochemistry, and genetics is predicted to facilitate considerably our knowledge and understanding of the appropriate mechanism for the degradation of contaminants. Genetic engineering approach and the information gathered will be utilized to produce high-level varieties and this approach is previously proven applicable, and selection of relevant species for this technology is also a crucial step for the successful implementation. Besides, there are open challenges to phytoremediation technology which could be defeated through genetic engineering of plants and endophytic microbes that have the ability to overexpress the enzyme for efficient pollutant degradation and sustainable environment (Fig. 20.2).

References

Abhilash, P.C., Singh, B., Srivastava, P., Schaeffer, A., Singh, N., 2013. Remediation of lindane by Jatropha curcas L: utilization of multipurpose species for rhizoremediation. Biomass. Bioenerg. 51, 189–193.

Abhilash, P.C., Singh, N., 2010. Withania somnifera Dunal-mediated dissipation of lindane from simulated soil: implications for rhizoremediation of contaminated soil. J. Soils. Sediments. 10, 272–282.

Agnello, A.C., Huguenot, D., Hullebusch Van, E.D., Esposito, G., 2016. Citric acid-and Tween® 80-assisted phytoremediation of a co-contaminated soil: alfalfa (*Medicago sativa* L.) performance and remediation potential. Environ. Sci. Pollut. Res. 23 (9), 9215–9226.

Akpor, O.B., Ohiobor, G.O., Olaolu, T.D., 2014. Heavy metal pollutants in waste water effluents: sources, effects and remediation. Adv. Biosci. Bioengineering. 2 (4), 37–43.

Al-Baldawi, I.A., Abdullah, S.R.S., Suja, F., Anuar, N., Mushrifah, I., 2013. Effect of aeration on hydrocarbon phytoremediation capability in pilot sub-surface flow constructed wetland operation. Ecol. Eng. 61, 496–500.

Almeida, M.V.D., Rissato, S.R., Galhiane, M.S., Fernandes, J.R., Lodi, P.C., Campos, M.C.D., 2018. In vitro phytoremediation of persistent organic pollutants by Helianthus annuus L. plants. Química. Nova 41 (3), 251–257.

Alves, W.S., Manoel, E.A., Santos, N.S., Nunes, R.O., Domiciano, G.C., Soares, M.R., 2018. Phytoremediation of polycyclic aromatic hydrocarbons (PAH) by cv. Crioula: a Brazilian alfalfa cultivar. Int. J. Phytoremediation 20 (8), 747–755.

Baerson, S.R., Sánchez-Moreiras, A., Pedrolbonjoch, N., Schulz, M., Kagan, I.A., Agarwal, A.K., Reigosa, M.J., Duke, S.O., 2005. Detoxification and transcriptome response in Arabidopsis seedlings exposed to the allelochemical benzoxazolin-2(3H)-one. J. Biol. Chem. 280, 21867–21881.

Barriault, D., Sylvestre, M., 2004. Evolution of the biphenyl dioxygenase BphA from *Burkholderia xenovorans* LB400 by random mutagenesis. J. Biol. Chem. 279, 47480–47488.

Becher, M., Talke, I.N., Krall, L., Kramer, U., 2004. Cross-species microarray transcript profiling reveals high constitutive expression of metal homeostasis genes in shoots of the zinc hyperaccumulator *Arabidopsis halleri*. Plant. J. 37, 251–268.

Bizily, S.P., Rugh, C.L., Meagher, R.B., 2000. Phytodetoxification of hazardous organomercurials by genetically engineered plants. Nat. Biotechnol. 18, 213–217.

Bramley-Alves, J., Wasley, J., King, C.K., Powell, S., Robinson, S.A., 2014. Phytoremediation of hydrocarbon contaminants in subantarctic soils: an effective management option. J. Environ. Manag. 142, 60–69.

Brazier-Hicks, M., Edwards, R., 2005. Functional importance of the family 1 glucosyltransferase UGT72B1 in the metabolism of xenobiotics in *Arabidopsis thaliana*. Plant. J. 42, 556–566.

Briggs, G.G., Bromilow, R.H., Evans, A.A., 1982. Relationship between lipophilicity and root uptake and translocation of nonionized chemicals by barley. Pestic. Sci. 13, 495–504.

Brooks, R.R., Lee, J., Reeves, R.D., Jaffre, T., 1977. Detection of nickeliferous rocks by analysis of herbarium specimens of indicator plants. J. Geochem. Explor. 7, 49–57.

Burken, J.G., Ma, X.M., Struckhoff, G.C., Gilbertson, A.W., 2005. Volatile organic compound fate in phytoremediation applications: natural and engineered systems. J. Biosci. 60, 208–215.

Cai, Z., Zhou, Q., Peng, S., Li, K., 2010. Promoted biodegradation and microbiological effects of petroleum hydrocarbons by Impatiens balsamina L. with strong endurance. J. Hazard. Mater. 183 (1–3), 731–737.

Calvelo-Pereira, R., Camps-Arbestain, M., Rodríguez Garrido, B., Macías, F., Monterroso, C., 2006. Behaviour of α-, β, γ-, and δ-hexachlorocyclohexane in the soil-plant system of a contaminated site. Environ. Pollut. 144, 210–217.

Camara, B., Seeger, M., González, M., Standfussgabisch, C., Kahl, S., Hofer, B., 2007. Generation by a widely applicable approach of a hybrid dioxygenase showing improved oxidation of polychlorobiphenyls. Appl. Environ. Microbiol. 73, 2682–2689.

Castro, A.J., Carapito, C., Zorn, N., Magne, C., Leize, E., Van Dorsselaer, A., Clement, C., 2005. Proteomic analysis of grapevine (*Vitis vinifera* L.) tissues subjected to herbicide stress. J. Exp. Bot. 56, 2783–2795.

Corwin, D.L., Hopmans, J., De Rooij, G.H., 2006. From field-to landscape-scale vadose zone processes: scale issues, modeling, and monitoring. Vadose. Zone. J. 5 (1), 129–139.

Cunningham, S.D., Ow, D.W., 1996. Promises and prospects of phytoremediation. Plant. Physiol. 110, 715–719.

Didierjean, L., Gondet, L., Perkins, R., Lau, S.M., Schaller, H., O'Keefe, D.P., Werck-Reichhart, D., 2002. Engineering herbicide metabolism in tobacco and Arabidopsis with CYP76B1, a cytochrome P450 enzyme from Jerusalem artichoke. Plant. Physiol. 130, 179–189.

Dietz, A.C., Schnoor, J.L., 2001. Advances in phytoremediation. Environ. Health. Perspect. 109, 163–168.

Doty, S.L., Shang, T.Q., Wilson, A.M., Tangen, J., Westergreen, A.D., Newman, L.A., Strand, S.E., Gordon, M.P., 2000. Enhanced metabolism of halogenated hydrocarbons in transgenic plants containing mammalian cytochrome P450 2E1. Proc. Natl. Acad. Sci. USA. 97, 6287–6291.

Doucette, W.J., Bugbee, B.G., Smith, S.C., Pajak, C.J., Ginn, J.S., 2003. Uptake, metabolism, and phytovolatilization of trichloroethylene by indigenous vegetation: impact of precipitation. Phytoremediation:. Transformation. and. Control. of. Contaminants, 561–588.

Dubey, R.K., Tripathi, V., Singh, N., Abhilash, P.C., 2014. Phytoextraction and dissipation of lindane by *Spinacia oleracea* L. Ecotoxicol. Environ. Saf. 109, 22–26.

Dzantor, E.K., 2007. Phytoremediation: the state of rhizosphere engineering for accelerated rhizodegradation of xenobiotic contaminants. J. Chem. Technol. Biotechnol. 82, 228–232.

Edwards, R., Dixon, D.P., 2005. Plant glutathione transferases. Methods. Enzymol. 401, 169–186.

Ekman, D.R., Lorenz, W.W., Przybyla, A.E., Wolfe, N.L., Dean, J.F., 2003. SAGE analysis of transcriptome responses in Arabidopsis roots exposed to 2,4,6-trinitrotoluene. Plant. Physiol. 133, 1397–1406.

Ekman, D.R., Wolfe, N.L., Dean, J.F., 2005. Gene expression changes in *Arabidopsis thaliana* seedling roots exposed to the munition hexahydro-1,3,5-trinitro-1,3,5- triazine. Environ. Sci. Technol. 39, 6313–6320.

Ensley, B.D., 2000. Rationale for use of phytoremediation. In: Raskin, I., B.D., Ensley (Eds.), Phytoremediation of toxic metals. Using plants to clean up the environment. J. Wiley and Sons., New York, USA, pp. 3–11.

Etim, E.E., 2012. Phytoremediation and its mechanism: a review. Int. J. Environ. Bioenerg. 2 (3), 120–136.

Fatima, K., Imran, A., Amin, I., Khan, Q.M., Afzal, M., 2018. Successful phytoremediation of crude-oil contaminated soil at an oil exploration and production company by plants-bacterial synergism. Int. J. Phytoremediation 20 (7), 675–681.

Food and Agricultural Organization (FAO), 2005. Pesticide disposal series: prevention and disposal of obsolete pesticides. Food and Agricultural Organization of the United Nations, New York, NY.

French, C.E., Rosser, S.J., Davies, G.J., Nicklin, S., Bruce, N.C., 1999. Biodegradation of explosives by transgenic plants expressing pentaerythritol tetranitrate reductase. Nat. Biotechnol. 17, 491–494.

Fujisawa, T., 2002. Model of the uptake of pesticides by plants. J. Pest. Sci. 27, 279–286.

Furukawa, K., 2006. Oxygenases and dehalogenases: molecular approaches to efficient degradation of chlorinated environmental pollutants. Biosci. Biotechnol. Biochem. 70, 2335–2348.

Gosh, M., Singh, S.P., 2005. A review on phytoremediation of heavy metals and utilization of its by-products. Appl. Ecol. Environ. Res. 31 (1), 1–8.

Guidi, W., Kadri, H., Labrecque, M., 2012. Establishment techniques to using willow for phytoremediation on a former oil refinery in southern Quebec: achievements and constraints. Chem. Ecol. 28 (1), 49–64.

Gurska, J., Wang, W., Gerhardt, K.E., Khalid, A.M., Isherwood, D.M., Huang, X.D., Glick, B.R., Greenberg, B.M., 2009. Three year field test of a plant growth promoting rhizobacteria enhanced phytoremediation system at a land farm for treatment of hydrocarbon waste. Environ. Sci. Technol 43 (12), 4472–4479.

Hannemann, F., Bichet, A., Ewen, K.M., Bernhardt, R., 2007. P450 systems: biological variations of electron transport chains. Biochem. Biophys. Acta. 1770, 330–344.

Hannink, N., Rosser, S.J., French, C.E., Basran, A., Murray, J.A., Nicklin, S., Bruce, N.C., 2001. Phytodetoxification of TNT by transgenic plants expressing a bacterial nitroreductase. Nat. Biotechnol. 19, 1168–1172.

Harms, H., Bokern, M., Kolb, M., Bock, C., 2002. Transformation of organic contaminants by different plant systems. In: McCutcheon, S.C., Schnoor, J.L. (Eds.), Phytoremediation: transformation and control of contaminants. Wiley, New York, pp. 285–316.

Harvey, P.J., Campanella, B.F., Castro, P.M., Harms, H., Lichtfouse, E., Schäffner, A.R., Smrcek, S., Werck-Reichhart, D., 2002. Phytoremediation of polyaromatic hydrocarbons, anilines and phenols. Environ. Sci. Pollut. Res 9 (1), 29–47.

Hernández-Ortega, H.A., Alarcón, A., Ferrera-Cerrato, R., Zavaleta-Mancera, H.A., López-Delgado, H.A., Mendoza- López, M.R., 2012. Arbuscular mycorrhizal fungi on growth, nutrient status, and total antioxidant activity of Melilotus albus during phytoremediation of a diesel-contaminated substrate. J. Environ. Manage. 95, S319–S324.

Hooda, V., 2007. Phytoremediation of toxic metals from soil and waste water. J. Environ. Biol. 28, 367–376.

Huang, X.D., El-Alawi, Y., Gurska, J., Glick, B.R., Greenberg, B.M., 2005. A multi-process phytoremediation system for decontamination of persistent total petroleum hydrocarbons (TPHs) from soils. Microchem. J 81 (1), 139–147.

Huang, X.D., El-Alawi, Y., Penrose, D.M., Glick, B.R., Greenberg, B.M., 2004. A multi-process phytoremediation system for removal of polycyclic aromatic hydrocarbons from contaminated soils. Environ. Pollut 130 (3), 465–476.

Hülster, A., Müller, J.F., Marschner, H., 1994. Soil-plant transfer of polychlorinated dibenzo-p-dioxins and dibenzofurans to vegetables of the cucumber family (*Cucurbitaceae*). Environ. Sci. Technol. 28, 1110–1115.

Inra, (Europe)-ECOSA, March 15, 2000. Eurobarometer 52.1, the Europeans and Biotechnology Report. http://ec.europa.eu/public_opinion/archives/ebs/ebs_134_en.pdf.

Inui, H., Hirota, M., Goto, J., Yoshihara, R., Kodama, N., Matsui, T., Yamazaki, K., Eun, H., 2015. Zinc finger protein genes from *Cucurbita pepo* are promising tools for conferring non-*Cucurbitaceae* plants with ability to accumulate persistent organic pollutants. Chemosphere, 123, 48–54.

Inui, H., Sawada, M., Goto, J., Yamazaki, K., Kodama, N., Tsuruta, H., Eun, H., 2013. A major latex-like protein is a key factor in crop contamination by persistent organic pollutants. Plant. Physiol. 161, 2128–2135.

Inui, H., Shiota, N., Motoi, Y., Ido, Y., Inoue, T., Kodama, T., Ohkawa, Y., Ohkawa, H., 2001. Metabolism of herbicides and other chemicals in human cytochrome P450 species and in transgenic potato plants co-expressing human CYP1A1, CYP2B6 and CYP2C19. J. Pest. Sci. 26, 28–40.

Inui, H., Wakai, T., Gion, K., Kim, Y.S., Eun, H., 2008. Differential uptake for dioxin-like compounds by zucchini subspecies. Chemosphere. 73, 1602–1607.

Jagtap, S.S., Woo, S.M., Kim, T.S., Dhiman, S.S., Kim, D., Lee, J.K., 2014. Phytoremediation of diesel-contaminated soil and saccharification of the resulting biomass. Fuel. 116, 292–298.

Jamieson, A.J., Malkocs, T., Piertney, S.B., Fujii, T., Zhang, Z., 2017. Bioaccumulation of persistent organic pollutants in the deepest ocean fauna. Nat. Ecol. Evol. 1 (3), 51.

Jansson, S., Douglas, C.J., 2007. *Populus*: a model system for plant biology. Annu. Rev. Plant. Biol. 58, 435–458.

Jeelani, N., Yang, W., Xu, L., Qiao, Y., An, S., Leng, X., 2017. Phytoremediation potential of Acorus calamus in soils cocontaminated with cadmium and polycyclic aromatic hydrocarbons. Sci. Rep. 7 (1), 8028.

Karavangeli, M., Labrou, N.E., Clonis, Y.D., Tsaftaris, A., 2005. Development of transgenic tobacco plants overexpressing maize glutathione S-transferase I for chloroacetanilide herbicides phytoremediation. Biomol. Engineer. 22, 121–128.

Kawahigashi, H., Hirose, S., Ohkawa, H., Ohkawa, Y., 2007. Herbicide resistance of transgenic rice plants expressing human CYP1A1. Biotechnol. Adv. 25, 75–84.

Klien, M., Burla, B., Martinoia, E., 2006. The multidrug resistance-associated protein (MRP/ABCC) subfamily of ATP-binding cassette transporters in plants. FEBS. Lett. 580, 1112–1122.

Labrou, N.E., Kotzia, G.A., Clonis, Y.D., 2004. Engineering the xenobiotic substrate specificity of maize glutathione S-transferase I. Prot. Engineer. Design. Select. 17, 741–748.

Leugsakul, T., Johnson, G.R., Wood, T.K., 2006. Protein engineering of the 4-methyl-5-nitrocatechol monooxygenase from *Burkholderia* sp strain DNT for enhanced degradation of nitroaromatics. Appl. Environ. Microbiol. 72, 3933–3939.

Liu, D., Li., T., Yang, X., Islam, E., Jin, X., Mahmood, Q., 2007. Enhancement of lead uptake by hyperaccumulator plant species Sedum alfredii hance using EDTA and IAA. Bull. Environ. Contam. Toxicol. 78, 280–283.

Liu, R., Dai, Y., Sun, L., 2015. Effect of rhizosphere enzymes on phytoremediation in PAH-contaminated soil using five plant species. PLoS. One. 10 (3), e0120369.

Liu, R., Xiao, N., Wei, S., Zhao, L., An, J., 2014. Rhizosphere effects of PAH-contaminated soil phytoremediation using a special plant named Fire Phoenix. Sci. Total. Environ. 473, 350–358.

Lunney, A.I., Zeeb, B.A., Reimer, K.J., 2004. Uptake of weathered DDT in vascular plants: potential for phytoremediation. Environ. Sci. Technol. 38, 6147–6154.

Ma, L.Q., Komar, K.M., Tu, C., Zhang, W., Cai, Y., Kennelley, E.D., 2001. A fern that hyperaccumulates arsenic. Nature. 409, 579.

Masciandaro, G., Di Biase, A., Macci, C., Peruzzi, E., Iannelli, R., Doni, S., 2014. Phytoremediation of dredged marine sediment: monitoring of chemical and biochemical processes contributing to sediment reclamation. J. Environ. Manage. 134, 166e174.

Mastretta, C., Barac, T., Vangronsveld, J., Newman, L., Taghavi, S., Van Der Lelie, D., 2006. Endophytic bacteria and their potential application to improve the phytoremediation of contaminated environments. Biotechnol. Genet. Engineer. Rev. 23, 175–207.

Masu, S., Albulescu, M., Balasescu, L.C., 2014. Assessment on Phytoremediation of Crude Oil Polluted Soils with Achillea millefolium and Total Petroleum Hydrocarbons Removal Efficiency. Rev. Chimie. 65, 1103–1107.

Mattina, M.I., Isleyen, M., Eitzer, B.D., Iannucci-Berger, W., White, J.C., 2006. Uptake by cucurbitaceae of soil-borne contaminants depends upon plant genotype and pollutant properties. Environ. Sci. Technol. 40, 1814–1821.

Mc-Cutcheon, S.C., Schnoor, J.L., 2003. Phytoremediation: transformation and control of contaminants. Wiley, NY USA. p. 987.

Meagher, R.B., 2000. Phytoremediation of toxic elemental and organic pollutants. Curr. Opin. Plant. Biol 3 (2), 153–162.

Meng, L., Qiao, M., Arp, H.P.H., 2011. Phytoremediation efficiency of a PAH-contaminated industrial soil using ryegrass, white clover, and celery as mono-and mixed cultures. J. Soils. Sediments 11 (3), 482–490.

Mentewab, A., Cardoza, V., Stewart, C.N., 2005. Genomic analysis of the response of *Arabidopsis thaliana* to trinitrotoluene as revealed by cDNA microarrays. Plant. Sci. 168, 1409–1424.

Mezzari, M.P., Walters, K., Jelinkova, M., Shih, M.C., Just, C.L., Schnoor, J.L., 2005. Gene expression and microscopic analysis of Arabidopsis exposed to chloroacetanilide herbicides and explosive compounds: a phytoremediation approach. Plant. Physiol. 138, 858–869.

Mikes, O., Cupr, P., Trapp, S., Klanova, J., 2009. Uptake of polychlorinated biphenyls and organochlorine pesticides from soil and air into radishes (*Raphanus sativus*). Environ. Pollut. 157, 488–496.

Milner, M.J., Kochian, L.V., 2008. Investigating heavy-metal hyperaccumulation using *Thlaspi caerulescens* as a model system. Ann. Bot. 102, 3–13.

Mo, C.H., Cai, Q.Y., Li, H.Q., Zeng, Q.Y., Tang, S.R., Zhao, Y.C., 2008. Potential of different species for use in removal of DDT from the contaminated soils. Chemosphere. 73, 120–125.

Moccia, E., Intiso, A., Cicatelli, A., Proto, A., Guarino, F., Iannece, P., Castiglione, S., Rossi, F., 2017. Use of *Zea mays* L. in phytoremediation of trichloroethylene. Environ. Sci. Pollut. Res. 24 (12), 11053–11060.

Mohammadi, M., Chalavi, V., Novakova-Sura, M., Laliberté, J.F., Sylvestre, M., 2007. Expression of bacterial biphenyl-chlorobiphenyl dioxygenase genes in tobacco plants. Biotechnol. Bioeng. 97, 496–505.

Morant, M., Baky, S., Linderg, B., Werckreichhart, D., 2003. Plant cytochromes P450: tools for pharmacology, plant protection and phytoremediation. Curr. Opin. Biotechnol. 14, 151–162.

Murano, H., Otani, T., Seike, N., 2010. Dieldrin-dissolving abilities of the xylem saps of several plant families, particularly Cucurbita pepo L. Environ. Toxicol. Chem. 29, 2269–2277.

Namiki, S., Otani, T., Seike, N., 2013. Fate and plant uptake of persistent organic pollutants in soil. Soil. Sci. Plant. Nutr. 59, 669–679.

Nedunuri, K.V., Govindaraju, R.S., Banks, M.K., Schwab, A.P., Chen, Z., 2000. Evaluation of phytoremediation for field-scale degradation of total petroleum hydrocarbons. J. Environ. Eng. 126 (6), 483–490.

Nissim, W.G., Cincinelli, A., Martellini, T., Alvisi, L., Palm, E., Mancuso, S., Azzarello, E., 2018. Phytoremediation of sewage sludge contaminated by trace elements and organic compounds. Environ. Res. 164, 356e366.

Noctor, G., Arisi, A., Jouanin, L., Kunert, K., Rennenberg, H., Foyer, C., 1998. Glutathione: biosynthesis, metabolism and relationship to stress tolerance explored in transformed plants. J. Exp. Bot. 49, 623–647.

Noori, A., Zare Maivan, H., Alaie, E., Newman, L.A., 2018. Leucanthemum vulgare Lam. crude oil phytoremediation. Int. J. Phytoremediation 20 (13), 1292–1299.

Ohkama-Ohtsu, N., Zhao, P., Xiang, C.B., Oliver, D.J., 2007. Glutathione conjugates in the vacuole are degraded by gamma-glutamyl transpeptidase GGT3 in Arabidopsis. Plant. J 49, 878–888.

Otani, T., Seike, N., Sakata, Y., 2007. Differential uptake of dieldrin and endrin from soil by several plant families and Cucurbita genera. Soil. Sci. Plant. Nutr. 53, 86–94.

Pacios, L.F., Gómez, L., 2006. Conformational changes of the electrostatic potential of biphenyl: a theoretical study. Chem. Phys. Lett. 432, 414–420.

Patel, N., Cardoza, V., Christensen, E., Rekapalli, B., Ayalew, M., Stewart, C.N., 2004. Differential gene expression of *Chlamydomonas reinhardii* in response to 2,4,6-trinitrotoluene (TNT) using microarray analysis. Plant. Sci. 167, 1109–1122.

Peer, W.A., Baxter, I.R., Richards, E.L., Freeman, J.L., Murphy, A.S., 2005. Phytoremediation and hyperaccumulator plants. In: Martinoia, E., Tamás, M.J. (Eds.), Molecular biology of metal homeostasis and detoxification. Springer/Berlin/Heidelberg, New York, pp. 299–340.

Petrová, S., Rezek, J., Soudek, P., Van_ek, T., 2017. Preliminary study of phytoremediation of brownfield soil contaminated by PAHs. Sci. Total. Environ. 599, 572–580.

Phenrat, T., Teeratitayangkul, P., Prasertsung, I., Parichatprecha, R., Jitsangiam, P., Chomchalow, N., Wichai, S., 2017. Vetiver plantlets in aerated system degrade phenol in illegally dumped industrial wastewater by phytochemical and rhizomicrobial degradation. Environ. Sci. Pollut. Res. 24 (15), 13235–13246.

Phillips, L.A., Greer, C.W., Farrell, R.E., Germida, J.J., 2012. Plant root exudates impact the hydrocarbon degradation potential of a weathered-hydrocarbon contaminated soil. Appl. Soil. Ecol. 52, 56–64.

Pilon-Smits, E., 2005. Phytoremediation. Annu. Rev. Plant. Biol. 56, 15–39.

Pilon-Smits, E.A., Freeman, J.L., 2006. Environmental cleanup using plants: biotechnological advances and ecological considerations. Front. Ecol. Environ. 4 (4), 203–210.

Raskin, I., Kumar, P.N., Dushenkov, S., Salt, D.E., 1994. Bioconcentration of heavy metals by plants. Curr. Opin. Biotechnol. 5, 285–290.

Rissato, S.R., Galhiane, M.S., Fernandes, J.R., Gerenutti, M., Gomes, H.M., Ribeiro, R., de Almeida, M.D., 2015. Evaluation of *Ricinus communis* L. for the Phytoremediation of Polluted Soil with Organochlorine Pesticides. Biomed. Res. Int, 1–8.

Robineau, T., Batard, Y., Nedelkina, S., Cabellohurtado, F., Leret, M., Sorokine, O., Didierjean, L., Werck-Reichhart, D., 1998. The chemically inducible plant cytochrome P450 CYP76B1 actively metabolizes phenylureas and other xenobiotics. Plant. Physiol. 118, 1049–1056.

Russel, K., 2005. The use and effectiveness of phytoremediation to treat persistent organic pollutants. US Environmental Protection Agency Office on Solid Wastes Emergency Response Technology Innovation and Field Service Division, Washington DC.

Rylott, E.L., Jackson, R.G., Edwards, J., Womack, G.L., Seth-Smith, H.M.B., Rathbone, D.A., Strand, S.E., Bruce, N.C., 2006. An explosive-degrading cytochrome P450 activity and its targeted application for the phytoremediation of RDX. Nat. Biotechnol. 24, 216–219.

Salt, D.E., Blaylock, M., Nanda, Kumar, P.B.A., Dushenkov, V., Ensley, B.D., Raskin, I., 1995. Phytoremediation: a novel strategy for the removal of toxic metals from the environment using plants. Biotechnol 13, 468–474.

Salt, D.E., Smith, R.D., Raskin, I., 1998. Phytoremediation. Annu. Rev. Plant. Physiol. Plant. Mol. Biol. 49, 643–668.

Salvato, J.A., Nemerow, N.L., Agardy, F.J., 2003. Environmental engineering. Wiley, NY, USA, p. 1584.

Sánchez, V., López-Bellido, F.J., Cañizares, P., Rodríguez, L., 2018. Can electrochemistry enhance the removal of organic pollutants by phytoremediation? J. Environ. Manage 225, 280–287.

Sanderman, H., 1994. Higher plant metabolism of xenobiotics: the «green liver» concept. Pharmacogenet 4, 225–241.

Schnoor, J.L., 1997. Phytoremediation. Ground-water remediation technologies analysis center, series E. http://www.clu-in.org/download/toolkit/phyto_e.pdf. Accessed 21 Feb 2018.

Siminsky, B., Corbin, F.T., Ward, E.R., Fleischmann, T.J., Dewey, R.E., 1999. Expression of a soybean cytochrome P450 monooxygenase cDNA in yeast and tobacco enhances the metabolism of phenylurea herbicides. Proc. Natl. Acad. Sci. USA. 96, 1750–1755.

Singer, A.C., Bell, T., Heywood, C.A., Smith, J.A., Thompson, I.P., 2007. Phytoremediation of mixed-contaminated soil using the hyperaccumulator plant Alyssum lesbiacum: evidence of histidine as a measure of phytoextractable nickel. Environ. Pollut. 147, 74–82.

Singer, A.C., Crowley, D.E., Thompson, I.P., 2003. Secondary plant metabolites in phytoremediation and biotransformation. Trends. Biotechnol 21, 123–130.

Singh, N.K., Rai, U.N., Verma, D.K., Rathore, G., 2014. Kocuria flava induced growth and chromium accumulation in Cicer arietinum L. Int. J. Phytorem 16, 14–28.

Soleimani, M., Hajabbasi, M.A., Afyuni, M., Mirlohi, A., Borggaard, O.K., Holm, P.E., 2010. Effect of endophytic fungi on cadmium tolerance and bioaccumulation by Festuca arundinacea and Festuca pratensis. Int. J. Phytoremediation 12 (6), 535–549.

Song, W.Y., Sohn, E.J., Martinoia, E., Lee, Y.J., Yang, Y.Y., Jasinski, M., Forestier, C., Hwang, I., Lee, Y., 2003. Engineering tolerance and accumulation of lead and cadmium in transgenic plants. Nat. Biotechnol 21, 914–919.

Song, J., Zhao, F.J., Luo, Y.M., McGrath, S.P., Zhang, H., 2004. Copper uptake by *Elsholtzia splendens* and *Silene vulgaris* and assessment of copper phytoavailability in contaminated soils. Environ. Pollut. 128, 307–315.

Stockholm Convention on Persistence Organic Pollutants, 2005. Ridding the World of POPs: A Guide to the Stockholm Convention on Persistent Organic Pollutants.

Sugiyama, K., Kami, D., Muro, T., Otani, T., Seike, N., Namiki, S., 2013. Identification and inheritance of a winter squash cultivar (*Cucurbita maxima*) with low heptachlor epoxide uptake ability. Sci. Hortic. 161, 35–42.

Susarla, S., Medina, V.F., McCutcheon, S.C., 2002. Phytoremediation: an ecological solution to organic chemical contamination. Ecol. Eng. 18 (5), 647–658.

Tang, S., Mo, C., 2007. Phytoremediation in China. In: Willey, N (Ed.), Phytoremediation: Methods and Reviews. Humana Press, New York, pp. 381–391.

Ueno, D., Iwashita, T., Zhao, F.J., Ma, J.F., 2008. Characterization of Cd translocation and identification of the Cd form in xylem sap of the Cd-hyperaccumulator *Arabidopsis halleri*. Plant Cell Physiol 49, 540–548.

Vardar, G., Ryu, K., Wood, T.K., 2005. Protein engineering of toluene-o-xylene monooxygenase from *Pseudomonas stutzeri* OX1 for oxidizing nitrobenzene to 3-nitrocatechol, 4-nitrocatechol, and nitrohydroquinone. J. Biotechnol 115, 145–156.

Villacieros, M., Whelan, C., Mackova, M., Molgaard, J., Sánchez-Contreras, M., Lloret, J., de Cárcer, D.A., Oruezábal, R.I., Bolanos, L., Macek, T., Karlson, U., 2005. Polychlorinated biphenyl rhizoremediation by *Pseudomonas fluorescens* F113 derivatives, using a *Sinorhizobium meliloti* nod system to drive bph gene expression. Appl. Environ. Microbiol 71 (5), 2687–2694.

Wang, C., Liu, Z.Q., 2007. Foliar uptake of pesticides: present status and future challenge. Pest. Biochem. Physiol. 87, 1–8.

Wang, G.D., Li, Q.J., Luo, B., Chen, X.Y., 2004. *Ex planta* phytoremediation of trichlorophenol and phenolic allelochemicals via an engineered secretory laccase. Nat. Biotechnol 22, 893–897.

Wang, H., Inukai, Y., Yamauchi, A., 2006. Root development and nutrient uptake. Crit. Rev. Plat. Sci. 25, 279–301.

Wang, Y., Fang, L., Lin, L., Luan, T., Tam, N.F., 2014. Effects of low molecular-weight organic acids and dehydrogenase activity in rhizosphere sediments of mangrove plants on phytoremediation of polycyclic aromatic hydrocarbons. Chemosphere 99, 152–159.

Weng, G., Wu, L., Wang, Z., Luo, Y., Christie, P., 2005. Copper uptake by four Elsholtzia ecotypes supplied with varying levels of copper in solution culture. Environ. Int. 31, 880–884.

White, J.C., 2001. Plant-facilitated mobilization and translocation of weathered 2,2-bis (p-chlorophenyl)-1,1-dichloroethylene (p,p'-DDE) from an agricultural soil. Environ. Toxicol. Chem. 20, 2047–2052.

Whiteley, C.G., Lee, D.J., 2006. Enzyme technology and biological remediation. Enzyme. Microb. Technol. 38, 291–316.

Wolfe, N.L., Hoehamer, C.F., 2003. Enzymes used by plants and microorganisms to detoxify organic compounds. In: McCutcheon, S.C., Schnoor, J.L. (Eds.), Phytoremediation. Transformation and Control of Contaminants. Wiley, NY, USA, pp. 159–187.

World Wildlife Fund (WWF), 2005. Toxic fact sheets. Available at http://www.worldwildlife.org/toxics/pubs.cfm. 3 May, 2005.

Zhang, Q., Xu, F.X., Lambert, K.N., Riechers, D.E., 2007. Safeners coordinately induce the expression of multiple proteins and MRP transcripts involved in herbicide metabolism and detoxification in *Triticum tauschii* seedling tissues. Proteomics 7, 1261–1278.

Zielinski, M., Kahl, S., Standfuss-Gabisch, C., Cadmara, B., Seeger, M., Hofer, B., 2006. Generation of novel-substrate-accepting biphenyl dioxygenases through segmental random mutagenesis and identification of residues involved in enzyme specificity. Appl. Environ. Microbiol. 72, 2191–2199.

Chapter 21

Role of biotechnology in phytoremediation

Riya Sharma, Rajesh Kumar, Younis Ahmad Hajam, Raksha Rani
Department of Biosciences, Division Zoology, Career Point University, Hamirpur, Himachal Pradesh, India

21.1 Introduction

Phytoremediation is a bioremediation method by which different plants and their related microorganisms get accumulate, depollute, and stabilize in a nature-friendly and balanced manner that helps to remediate contaminated soil and water. However, "phyto" word puts up with plant henceforth the remediation facilitated by the plant system. Phytoremediation is one of the best processes of bioremediation. It is an actual biological phenomenon to degrade dreadful xenobiotics and unmanageable compounds dependable for environmental effluence. Phytoremediation is a manageable, cost-effective application of using green plants for the purpose of reestablishing, cleaning, and rescuing areas polluted with different levels of environmental pollutants. To remediate affected sites, plant should have ability of survival in toxic and stressful environment. During phytoremediation, plants absorb a contaminant from the soil and also mineralized it, consequently preventing parallel contamination of groundwater and holding ecosystem safety for human inhabitance. In phytoremediation, carefully chosen or genetically improved plants are used to perform the process of uptake of contaminants from the soil. Plant should have fast growth and increased biomass to fulfill the criteria of phytoremediation because a plant with proper growth has the potential of uptake and accumulate large number of heavy metals from the soil into its upper harvested parts. Recent studies showed major improvement in understanding the mechanism of uptake of metals and passage of pollutants in plants performing phytoremediation. Excess in contaminants affects the biomass and growth of metal hyperaccumulator plants and negatively impacts the phytoremediation. To overcome this problem, genetically modified plants should be used. Plants with unique gene sequencing generally show particular properties to perform the phytoremediation. Generally, these plants are present in the area where there is an existence of metal ores. Genes are isolated from the particular plant that provides resistance, and then

Phytoremediation. DOI: https://doi.org/10.1016/B978-0-323-89874-4.00021-2

these genes are injected into several other species for widespread of these kinds of plant species. This elevates the number of plant species that may be used for this kind of purposes. Biotechnological modifications are beneficial in promoting gene expression for providing the maximum resistance to plants. Certain microbes also promote phytoremediation as they promote resistance in different plants. Probably, their separation is possible through biotechnology as genetically engineered microbes enrich the soil and promote phytoremediation by the particular plant species. Phytoremediation can be used for both inorganic and organic pollutants present in solid and liquid substrate. Construction of wetlands or swamps is also an approach to develop phytoremediation, as it has been capable to remediate wastewater with dilute concentrations of toxins such as "petroleum, hydrocarbons, chlorinated solvents, pesticides, explosives, heavy metals, and radio nucleoids. Plants can significantly contribute to remediate toxins from the environment. Some of the reactions are phytoextraction, phytovolatilization, rhizofiltration, phytotransformation, rhizosphere-bioremediation, and phytostabilization. Biotechnology compromises the prospect to transfer hyperaccumulator phenotypes into firm developing, high biomass plants that possibly will be operational in phytoremediation. Hence, hyperaccumulator plants have mainly been reported from family *Brassicaceae*, *Cunoniaceae*, *Caryophyllaceae*, *Asteraceae*, *Euphorbiaceae*, *Cyperaceae*, *Fabaceae*, *Lamiaceae*, *Violaceae*, *Poaceae*, etc. (Padmavathiamma and Li, 2007). Various contaminants have various fates in plant-substrate systems; thus, they have various rate-limiting factors for phytoremediation that may be targeted using genetic engineering. Economic value of phytoremediation is much more than other alternatives of remediation. Phytoremediation has various advantages over microbial bioremediation and considered as an essential process in the present time due to the rapid increase in industrialization, and human population makes this a necessity of today's world. Hence, to overcome the problem of fixation of several toxins, phytoremediation come forth with different kinds of techniques to remediate our environment.

21.2 Role of biotechnology in phytoremediation

Absolute phytoremediator may have various properties such as huge tolerance with respect to impure matter, power to deteriorate or reduce impurities in concentrated biofuel, a large root structure, huge absorption of water, rapid growth level, or excessiveness of biofuel (Cherian and Oliveira, 2005). However, there are many species that grow or develop in a polluted area. They generally develop tardy, lesser energy level or are modified to a particular environmental state. Traditional plants are terminating to reach the demands to efficient phytoremediation (Gratao and Braz, 2005). The medicinal function of flora may enhance through utilization or some floral-modified technique (Kraomer, 2005). The absorption or increase in contaminants within large biofuel flora forms a well-developed system for the occurrence of better phytoremediators (Martanez et al., 2006).

21.3 Transgenic plant–associated bacteria for phytoremediation

21.3.1 Transgenic rhizosphere bacteria

Plants increase the activity of microbes and their movement in the soil to enhance biodegradation action (Anderson et al., 1993; Limbert and Betts, 1996). Granulation of root or root change enhances carbon activities for the effective development of microbes or cometabolic biodegradation. Root granules may suppress organic acids, alcohols, or phenolic compounds, which enhance "bioavailability" or "biodegradation by enduring hydrophobic pollutants (Chaudhry et al., 2005). Some particular compounds present in root granules are recommended to incite microbial enzyme or develop biodegradation capabilities (Fletcher and Hegde, 1995). In the rhizosphere, the presence of O_2 is essential for the "oxidative biodegradation" through "oxygenase" or "laccases" (Anderson et al., 1993; Shimp et al., 1993; Chaudhry et al., 2005). Some changes occur in distinct "Rhizobium strains" and bacteria from the rhizosphere of "*Polymnia Canadensis*" or shrub of Southern California through the initiation of toluene o-monooxygenase gene (TOM) from aerobic Gram-negative bacillus (*Burkholderia cepacia*). These bacteria show the ability to degenerate trichloroethylene (TCE) or the ability to take over or take apart through "poplar," "wheat," or "barley" bacterial rhizosphere. A pseudomonas recombinant bacterium retains the ability to express TCE (TOM) for 29 days. Such a procedure forms "PCP genes" that encrypt pentachlorophenol monooxygenases with familiar "PCP degrader" and *Sphingobium chlorophenolicum* within "*Pseudomonas gladioli*" that forms homesteader rhizosphere of *Allium tuberosum* (Nakamura et al., 2004). The metamorphosis of bacteria has the ability to find *A. tuberosum* or widely degenerate "PCP" in solvent. TCC (tetrachlorocatechol) is formed in tissues of flora having functional metabolic activities of "PCP."

21.4 Endophytic bacteria

Some bacterial species associated with the rhizosphere or living on the surface of the leaf are grouped into floral tissues without showing any symptoms (Ryan et al., 2008). Nearby floral species form some beneficial bacteria in tissue that show a "symbiotic" and "commensal" correlation, for example bacteria supply advance developmental compounds such as "plant hormone" or "siderophores," and diazotrophic bacteria stabilize nitrogen and stop the pathogen effect to trigger Isr (Van aken et al., 2004; Doty et al., 2007). Numerous beneficial bacteria (endophytes) participate in the metabolic process of contaminants such as "toxic xenobiotics" to increase "phytoremediation." Methylobacterium separated from poplar tree shows many metabolic activities such as nitro substitution or questions the over-importance of bacteria in the flora for "biodegradation" (Van aken et al., 2004). Moore et al. (2006) reported that many poplar bacteria separated from "BTEX-contaminated site," also show liberality to "BTEX and TCE." The removal of bacteria from the rhizosphere or tissue of plants shows enhancement in TCE degrades. Barac et al. (2004) reported that the modification of bioendophytes enhances the floral degeneration of C7H8 utilized for remediation of chlorinated solvents. Change in *B. cepacia* occurs due to interlinking with organic-endophytes of "yellow lupine." Studies analyzed that the significant degeneration of C7H8 through endophytic bacteria may decrease the C7H8 "phytotoxicity" to host plant. Approximately, 50%–70% demotion of C7H8 has been studied by the researchers, which indicates that increased motility in tissues may cause a reduction in danger to the environment.

21.5 Plant response mechanisms to heavy metals

There are various methods, such as the utilization of a particular gene, discharge inflates, or chelator, required for the assimilation or growth of heavy rock (metals). The classification of detoxification of heavy metals by various floral Species indicates liberality due to root cells multiplying along with amino acids and organic acid Mkumbo et al. (2012). Some methods utilized, such as obstruct assimilation of toxic metals within root tissues, require "anionic groups of cell walls" (Okem, 2014). There are three methods for the development of metal-contaminated soil: metal excluders, metal indicators, and hyperaccumulator.

21.5.1 Metal excluders

Metal excluders are the plants (flora) that maintain the constant metal concentration in the shoot and lower metal concentration of soil over a broad range. They can also have a huge number of metals in their roots. They maintain a huge accumulation of metals in soil and also limit metals in the rhizome. Flora can change the permeability of the membrane also alters holding capability or apart from other chelators.

21.5.2 Metal indicators

These are the species that increase metal in airy cells or concede red metal position in the soil. They also permit the consolidation of metal by making a bond between the chelators and transformed metals partitions to keep metals in an insensitive portion (Singh et al., 2015; Chukwuma, 2014; Mkumbo et al., 2012;Shu and Xia, 2012; Okem, 2014; Ghosh and Singh, 2005).

21.5.3 Hyperaccumulators

These are the floral species that can permit, assimilate, or change dense congregation of some toxic metals in the tissue. These floras are able to extract the toxic metals from the soil, which are utilized in phytoremediation (Mganga et al., 2011). By stratifying the flora as a "hyperaccumulator," a stranded combination is determined on the basis of metal types. Cadmium may become poisonous when it reaches a concentration of more than 100 mg/kg and for nickel, chromium, lead, cobalt, arsenic (As), or copper the concentration may be >1000 mg/kg; and for zinc and manganese it may be >10,000 mg/kg when they are found in metal-rich soil (Okem, 2014; Ghosh and Singh, 2005; Boularbah et al., 2006).

21.6 Biotechnological approach in phytoremediation for different pollutants

It is clear that different toxics have a distinct end product in plant–substrate system and they have distinct rate-limiting factors to phytoremediation by genetic evolutionary techniques (Dhankher et al., 2012). In any case, transgenic approaches having phytoremediation properties are essential to recognize the process. If any time, "potential rate-limiting steps" are recognized during "physiochemical" or biochemical tests, it may be overexpression of a particular membrane transporter. If genes encoding for these properties are present in organisms, then these organisms can be recognized as transgenics and differentiate from nontransgenic in terms of pollutants remediation (Elizabeth et al., 2006).

21.7 Inorganic pollutants

Chemical impurities are found to be a primitive constituent in soil or in the environment, and also in the actions of humans, such as industry, mining, motorized traffic, agriculture, lagging, and military actions encourage concentration of the majority of contaminants (Elizabeth et al., 2006). Inanimate contaminants comprise metals such as As, cadmium, copper, mercury (Hg), manganese, selenium (Se), or zinc; radionuclides such as cesium, phosphorus (P), uranium; or some plant fertilizers. These all are present in surroundings with either in a negative or positive charge, which is determined by a floral transporter to assimilate or variations. Inanimate concentration may reduce or proceed inside flora (plant), or Hg and Se, however, do not deteriorate. Thus, phytoremediation systems are accessible for inorganics' phytostabilization, phytoextraction, or phytovolatilization. The bioremediation strategy that may change the capability of the plant to phytoremediate inanimate concentration is tolerance or accumulation. Selected genes comprise "metal transporter genes" or also those genes that ease the chelate compound exhibition.

21.7.1 Arsenic

It is mainly found in the slab or soil-free zones within groundwater. The use of this polluted water may cause some skin diseases such as gangrene or cancer of kidney and bladder. A large quantity of As in contaminated crops ultimately enters the "food chain." In Bangladesh, about 40 million people are suffering from As infections, making it the largest poisoning of a population in history as reported by the WHO (Nordstrom, 2002). Some "nonbiodegradable" species are found in or on the surface of agriculture soil. During immersion, As is formed as P analog, and appropriated in plants by a P-assimilated method (Meharg and Acnair, 1992). Phosphate transporter (PHT1) can form connection with AsV assimilation in *Arabidopsis thaliana*. Moreover, AsV clamp down genes are formed in P appetite reaction, suggesting AsV obstruction along with P reorganization or change in the P communication process (Catarecha et al., 2007). Genetically engineered plants are produced to enhance the liberality or inflation of As. The overexpression of genes elaborated in the integration of phytochelatin (PCs) and its harbinger reduced glutathione (GSH) accordingly increases the liberality of As but not like inflation (Li et al., 2005; Gasic and Korban, 2007). Overall related work stated that both "GSH" and "PC" are solitary and are not able to increase As inflation in the shoot. Equivalency of "γ-glutamylcysteine synthetase (γ-ECS)" or "phytochelatins (PCs)" in *Arabidopsis* shows a good result in As liberality or inflation subsequently "overexpression" of one or the other gene isolated. Improving "GSH" or "PCs" in flora is a productive proposition to enhance As liberality of flora, which gives rise to a better flora having powerful "phytoremediation" capability. The liberality or inflation of As increases because of "coexpressing two bacterial genes" (Dhankher et al., 2002). Moreover, *Escherichia coli* synthesizes Gamma-glutamylcysteine, in roots or shoot, through effective constitutive "Actin2 promoter" (Dhankher et al., 2002) (Fig 21.1).

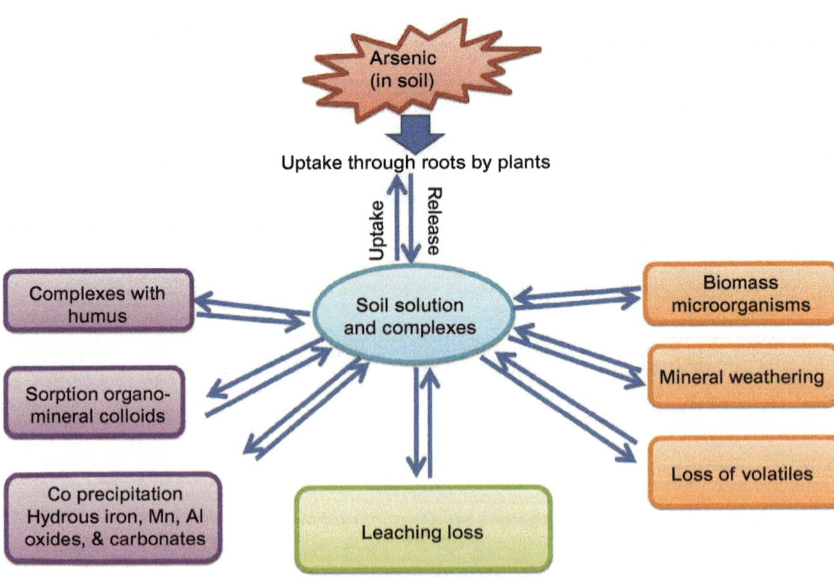

FIG. 21.1 Effect of arsenic on soil and plants.

21.7.2 Selenium

Se is introduced in the soil having greater consolidation. Methylation of Se at a particular place may result in volatilization of Se compound (Nriagu and Pacyna, 1988). Se is important supplement in some creatures such as man, but at a high concentration it may be poisonous. Se insufficiency or poisonousness is a big issue all over the world. There is no study to show that Se is necessary in "higher plant"; however, in some cases Se is willingly absorbed or uptaken by plants through "sulfur transporters or biochemical pathways." Flora absorbed Se in whole parts as well as "seeds" or may evaporate it in the environment. In some species, increase in the Se concentration is about 1% of its dry weight. Flora is able to assemble or evaporate the Se, which can be used in "phytoremediation." In mustard ATP-sulfurylase-overexpressing transgenic plants (APs) converts selenate inro selenite leads to the increased reduction of selenite (Pilon-smits et al., 1999).Transgenic APS assembles have more Se folds than undomesticated assembles that have one and half folds of Se. *Brassica juncea* indicated "cystathionine gamma synthase" exhibits two- to three-fold increased turbulence level than that of unprocessed flora (Van Huysen et al., 2003).

21.7.3 Mercury

Hg is a heavy metal found in the environment, and it causes human health diseases or atmospheric pollution (Dean et al., 1972; Kraemer and Chardonnens, 2001). Hg is present in the atmosphere in an inanimate form may be in Hg(0) or Hg(II) structure. Hg(II) has a high binding capacity with the soil elements and may decrease the accessibility or immersion. Animate forms of Hg such as methylmercury, dimethylmercury, or phenylmercury are very poisonous. CH_3Hg is especially highly contaminated and causes major effect on man or atmosphere due to systematic "biomagnifications" in "food web" (Patra and Sharma, 2000). In 1950 a diaster in Minamata Bay (Japan), methylmercury toxicity was observed in humans (Harada, 1995). Of all forms of Hg, due to distinct compounds of Hg, some are bound with the organic matter but others also remain in contact with the soil and cause human health hazard at present as well as in future. It is not possible to directly change the toxic elements into the less harmful structure. It is reported earlier that "gene encoding bacterial Hg" generated well-maintained features (Smith et al., 1998) that increase the liberality of Hg in flora. In 1990, Richard Meagher and coworkers reported that accumulation of Hg decreases after the introduction of bacteria in which mer operan (merA or merB) were inserted (Rugh et al., 1996, 1998; Bizily et al., 1999,2000). Various floral species such as *A. thaliana* (Rugh et al., 1996), "tobacco" (Heaton et al., 2005), "yellow poplar" (Rugh et al., 1998), "cottonwood" (Che et al., 2003), or "rice" (Heaton et al., 2003) fundamentally indicate altered "merA" is more effective then Hg(II) that damages "nontransgenic commands". To standardize the level of Hg(0) in genetically modified flora, the potential of transgenic yellow poplar or "cottonwood" for development and to increase the ability of "Hg(II)" can signify the capability of "phytovolatilization processes of Hg remediation in wetlands." The study of pot's soil revealed that genetically engineered flora shows that "merA" outperformed undomesticated flora on Hg-polluted area (Heaton et al., 1998). However, the movement of Hg(0) from the roots of these plants to the soil or environment was not determined. Hence, methylmercury is very poisonous than "ionic or metallic Hg" and is able to accumulate into the food chain, whereas "ionic and metallic" Hg species. are not accumulated. Therefore, methylmercury can cause a major threat to human beings. Single merA gene can only transform Hg(II) to Hg(0), and hence is not able to prevent or detoxify plants from the harmful "methylmercury"; whereas both the genes "merA" and merB" are needed together to remediate "methylmercury." In Hg(II), Hg species is very much reactive and less mobile, "MerB" promotes the cleavage of the "carbon-Hg bond" removing the organic ligand and releasing "Hg(II)." Plants were engineered by overexpressing a modified bacterial "organomercury lyase" gene to change "methylmercury" to "ionic Hg "(Bizily et al., 1999, 2000). *Arabidopsis thaliana* in engineered plants expressing "merB" grew vigorously and widely within distinct ranges of "monomethylmercuric chloride" and "phenylmercuric acetate" compared to nontransformed plants that were severely inhibited or died at the same "organomercurial" ranges (Bizily et al., 1999). In all plants presence of merB is sufficient to confer "methylmercury tolerance," due to high contaminants of "methylmercury in eukaryotic cells. These Endangered floras manage to outgrow the environmentally relevant concentrations of methylmercury" to change "methylmercury" to Hg(II), present in the flora. "Methylmercury" is highly toxic, plants were engineered to coexpress "merA" and "merB." The resulting engineered flora carries out the two-step conversion of "methylmercury" to "volatile Hg0" and is tolerant to 50 times higher concentrations of "methylmercury," which is able to kill control plants, and five times greater than the concentrations that kill "merB" plants (Bizily et al., 2000). The "engineered eastern cottonwood trees" showing merA and merB genes were also greatly tolerant to "organic Hg" (Lyyra et al., 2007). These results showed that flora (trees, shrubs, and grasses) can be engineered to detoxify the most abundant forms of ionic and organic Hg found at polluted sites, and it is likely that a number of "phytoremediation strategies" that block its flow into the environment can be adopted. The "chloroplast" and "endoplasmic reticulum" (ER) play a significant role in Hg poisoning (Bernier and Carpentier, 1995; Bizily et al., 2003). Therefore, a transgenic form of Hg detoxification systems in "chloroplasts" or "ER" may offer high levels of Hg tolerance and detoxification. Ruiz (2003) used chloroplast engineering for Hg detoxification by integrating the merA and merB genes into the chloroplast genome. Engineered tobacco plants exhibited high levels of tolerance to the organomercurial compound, phenylmercuric acetate, and accumulated 100- and 4-fold more Hg in the shoot in the presence of phenylmercuric acetate (PMA) and $HgCl_2$, respectively, than untransformed plants (Hussein et al., 2007). Therefore, chloroplast engineering can be declared a significant approach for Hg "phytoremediation" also.

21.7.3.1 Hg hyperaccumulation

Hg is one of the most common heavy metals and closely associated with many diseases in humans and other animals. Therefore, immediate remediation of Hg is needed which is possible through its decontamination and complete volatilization. But this technology bears public resistance due to the deposition of Hg(0) that is less toxic on the earth's surface. Further strategies come into use that can trap and hyperaccumulate Hg(II) in the air in various plant parts, which avoids the release of Hg(0) in the air. The use of genetically modified plants should be the first step in the direction of remediation of metal. Engineered plants with expression of the merB gene may enhance the conversion of methylmercury to Hg(II). These plants accumulate more Hg and show up more resistance toward methylmercury in comparison to transgenic plants that lack gene expression. Future genetic manipulations to improve the phytoremediation of Hg need to include enhanced Hg uptake into roots, transport to shoot, and adoption into aboveground tissues. Hence, Hg is a major neurotoxin, and its extensive contamination simulations have various human health concerns. Reclamation of highly dangerous methylmercury into less harmful volatile form namely Hg(0) can be done through modification of various plant species by bacterial genes, merB, and merA. Even, if these modified plants showed significant prospective for Hg decontamination in laboratory and greenhouse conditions, the efficacy of these plants in the natural environment needs to be proven. Although genetically modified plants showed a significant range of Hg decontamination in greenhouse and laboratory conditions, the potential of these plants in a natural environment still needs to be proven. Because Hg is very contagious, further improvements are carried out in engineering plants with respect to climatic changes. Presently, there is a need to develop techniques for the purpose of finding alternatives to Hg(0) volatilization and reduction of methylmercury to Hg(II). Additionally, enzymes present in subcellular compartments particularly in the ER and chloroplast will be beneficial in Hg detoxification.

21.8 Organic pollutant

Addition of organic pollutants in the environment is mainly man-made and xenobiotic. The maximum pollutants are harmful and carcinomatous. The unintentional release of industrial waste, including chemicals and petrochemicals, agricultural waste such as pesticides and herbicides, explosives and chemical weapons are the foremost sources of organic pollution in the environment. Also, a polluted site comprises a mixture of both organic and inorganic pollutants. Phytoremediation compromises the prospective for the complete degradation of organic pollutants. For this, plants should have specific genes to biodegrade the chemicals by taking up through soil (Schwitzguébel, 2017). To increase the effectiveness of phytoremediation, phytotoxic properties of organic pollutants must be astounded. A number of organic pollutants are present in the environment. Some of these are solvents such as TCE; explosives such as trinitrotoluene

(TNT), cyclotrimethylenetrinitramine, or research department explosive; polycyclic aromatic hydrocarbons such as naphthalene, pyrene; petroleum products namely benzene, toluene, ethylbenzene, and xylene polychlorinated biphenyls; and pesticides such as atrazine and chlorpyrifos.

21.9 Solvents

The constant removal of numerous solvents from the ground is directly linked with the environmental contamination, ultimately causing pollution of the groundwater. One of the major organic pollutants is TCE. Air sparing or in-situ stripping is one of the biotechnological procedure to remediate TCE, in which a cylinder filled with air is pumped into the polluted water from ground, making TCE volatile to pass into the atmosphere. Accumulation of oxidants for example potassium permanganate or hydrogen peroxide as they react with TCE is one of the common engineering methods. Bioremediation by means of anaerobic bacteria is one of the common approaches to remediate TCE. The site is first made CO_2 rich by the addition of huge amounts of substrates called molasses, vegetable oil, or lactate. These substrates then convert the land rich in bacterial associations to degrade TCE to cis dichloroethene (DCE), vinyl chloride, and finally to nontoxic ethane. Sites with narrow groundwater within the range of tree roots prove effective for the phytoremediation of various polluting solvents including TCE. Poplar trees played a very important role in phytoremediation of "TCE" because of their deep-root system. Many herbs namely tobacco, *Leucaena leucocephala*, Arabidopsis also have the genomic ability to vitiate "TCE" (Shang et al., 2001; Doty et al., 2003; Doty, 2008). Both plants and mammals having similarities in TCE degradation pathway; hence, both have the same end product that is, metabolite trichloroethanol (Shang et al., 2001). However, phytoremediation of TCE is limited by the apparent low expression of the cytochrome P450 enzyme. The TCE metabolism within plants is usually considered too slow, which may result in phytovolatilization of pollutants. There are several approaches to increase phytoremediation of TCE with the help of genetic modification and endophyte-assisted phytoremediation (Doty, 2008; James and Strand, 2009). Inoculation or injection of the mammalian cytochrome P450 CYP2E1 in transgenic tobacco (Doty et al., 2000) and poplar plant (Doty et al., 2007) results rapid increase in the "TCE" uptake. Genetically engineered poplar plants showed an increase in the rate of removal of TCE both from the air and liquids. Although only the first gene in the pathway was overexpressed in the transgenic plant, many other genes showing equivalence to pollutant degradation genes were also upregulated in response to TCE in the transgenic poplar (Kang et al., 2010). Pollutant activation, conjugation to sugars, and transport is activated by these genes. Previous studies reported that CYP2E1-engineered plants in comparison to vector-control plants had more TCE metabolites called trichloroethanol. Genetically engineered plants show elevated levels of chloride ions, which results in dechlorinating TCE. However, the concentration of TCE in the effluent was not further reduced by the transgenic plants compared to the controls. Therefore, the plants may be more suitable for the remediation of contaminated ground water (Doty et al., 2007) rather than for pump and treat systems that have a continuous source of TCE. Phytoremediation through endophytes is another strategy to remediate TCE as they have the ability to metabolize it. It is supposed that both bacteria and plant work mutually to promote the effective uptake of toxins by plant, where the bacteria is well known to colonize various vascular tissues and intercellular spaces which may help in decontamination of the pollutants (Doty, 2008; Weyens et al., 2009). Burkholderia strain G4 is a well-studied aerobic bacterium having large, self-spreadable, degradative plasmid, and TOM genes to degrade TCE (Shields et al., 1995). The plasmid or a mutant version with constitutive expression of the TOM genes can be transferred to native endophytes of poplar (Taghavi et al., 2005). Poplar trees were injected in situ with the improved endophytic strain containing pTOMBu61 to improve phytoremediation of TCE (Weyens et al., 2009a). As we know, organic and inorganic toxins highly exist in contaminated sites, so the purpose of phytoremediation through endophytes is to ensure the tolerance of inoculated plant species toward these different pollutants. Studies reported that the inoculation results in no increase in shoot mass of plants, while 30% of root mass of inoculated plant has increased vulnerability to Ni and TCE. A decline in phytotranspiration of TCE was observed in inoculated plants, which is statistically insignificant. Usually, endophyte-aided phytoremediation of TCE ensures some progresses hence, no further changes have been reported (Fig 21.2).

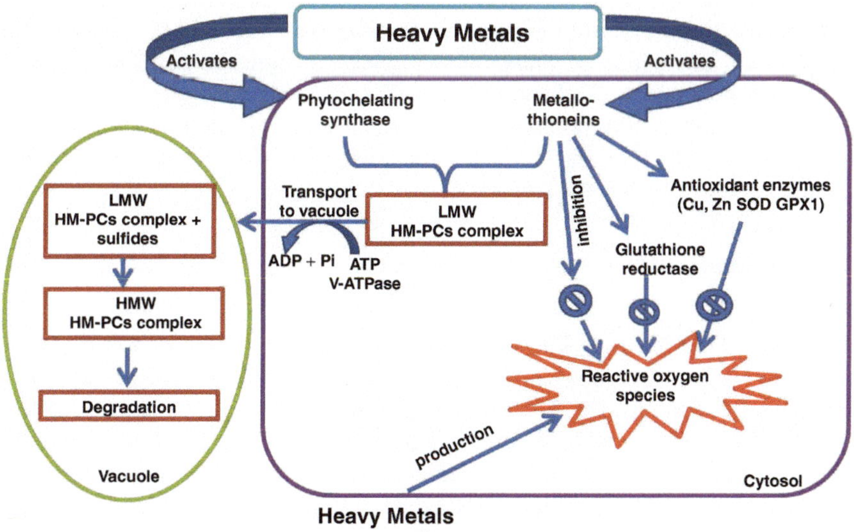

FIG. 21.2 Response of plant enzymes to heavy metals.

21.10 Explosives

The immense spread of ground and water pollution has been occurring due to the release of tons of explosives into the environment. During World War 2, research department explosive was the chief explosive used. These days its newer byproducts are still in use by many agencies. These explosives and their deprivation products are highly toxic and destructive. Remediation of "nitro aromatic explosive" and research department explosive is required at military training ranges to prevent the spreading to nearby populations. Liver damage and anemia are the major impacts of TNT , while research department explosive affects the central nervous system, initiating convulsions (Bizily et al., 2000).There are plant species that can tolerate comparatively less levels of TNT by converting it into amino dinitrotoluene that is conjugated to sugars or glutathione and then stored in the vacuole of cells. Different classes of enzyme complex have been discovered with the help of DNA array and other gene expression techniques to improve the plants reaction toward several nitro aromatics. (Gandia-Herrero et al., 2008). Tobacco plants are produced by the bacterial gene for an NADPH-dependent nitro-reductase tolerance and degrade high levels of TNT (Rylott et al., 2006). Genetic manipulation of *Arabidopsis* plants with xplA genes of *Rhodococcus* bacteria provides resistance to RDX (Cherian and Oliveira,2005). There are a number of bacterial strains isolated from polluted sites that degrade RDX and used it as a source of nitrogen (Crocker et al., 2006). The xplA gene translates a unique, bonded flavodoxin-cytochrome P450 enzyme capable for biodegradation of "RDX" (Jackson et al., 2007). *Arabidopsis* a genetically modified plant with introduced xp1A (CYP177) genes extracted from *Rhodococcus rhodochrous* 11Y can tolerate and remediate significant amounts of RDX, whereas nonmodified plants cannot uptake effectively. Studies reported that the presence of both the xplA and xplB genes in engineered plants shows greater progresses in RDX removal than xplA gene alone (Jackson et al., 2007). The enhanced genetic ability of plants to remediate the pollutants is needed because of the contamination of military sites with TNT and RDX. Poplar plants with nfsI and xplA genes show augmentation of both TNT and RDX.

21.11 Pesticide

Pesticides cause various chronic diseases in human beings and other living organisms. Pesticides are known as pollutants because they degrade the environment. Various methods such as ignition and land filling have been used to eliminate these toxins. But these are costly and not in use due to their low effectiveness toward pollutants. Microorganisms having the potential to detoxify the toxic pollutants used in bioremediation and transgenic plants are proved effective in phytoremediation of pesticides. Both are effective solutions against the pollutants (Hussein et al., 2007). Biotechnology plays a very important role in remediation of pesticides with the help of genetically modified plants.

Wang et al. (2005) reported that atrazine chlorohydrolase, the first enzyme encoded by the atzA gene, expressed a six-step pathway in genetically engineered plants. Plant organs are capable for the dechlorination of pesticides into hydroxyl atrazine. In tobacco cell culture, mammalian cytochrome P450 having genes CYP1A1 and CYP1A2, was expressed to amplify atrazine metabolism (Bode et al., 2004). Transgenic rice plants express genes CYP1A1, CYP2B6, and CYP2C19, which results in typical improvement in metabolism of various herbicides including atrazine and metolachlor (Kawahigashi et al., 2006). Biotechnological approaches are providing tolerance in various plants against the variety of herbicides. The genetically modified plants significantly remediate atrazine from soil than normal plants. Methods have been developed to enhance the remediation of chlorpyrifos using mammalian CYP2B6 and PON1 in transgenic poplar (Lee et al., 2012).

21.12 Oil spills (petroleum products)

Industrialization results in the production of different kinds of pollutants usually petroleum products that are harmful to many communities of the environment. It reduces species diversity and habitats of aquatic and terrestrial organisms, and finally leads death. As oil-spills promote the environmental degradation, it negatively impacts the collective and commercial value of oil-producing companies. Because of this, several water bodies are not able to maintain their aquatic lives. Benzene, toluene, ethylbenzene, and xylene also called (BTEX) are some petroleum derivatives. These petroleum derivatives can be remediated using plants. Petroleum-degrading bacteria present in the roots or rhizosphere of plant are used to detoxify the contaminated land on which plant has been grown (Siciliano et al., 2003). A number of poplar trees grown on BTEX-polluted sites give shelter to many endophytes. These endophytes have pollutant-degradation abilities that are helpful in phytoremediation (Moore et al., 2006). Further, improvements in phytoremediation of BTEX pollutants can be achieved through insertion of particular component degrading genes, that is, toluene-degrading genes into an endophytic strain and this strain is then inoculated into lupin (Barac et al., 2004). These modified plants then become ten times more capable to remediate toluene than the normal plant. When the original toluene-degrading strain was inoculated into the more suitable remediation plant such as poplar, the strain conjugatively transferred the plasmid in plants to the native endophytes, resulting in increased tolerance against the toluene (Taghavi et al., 2005). Additionally, the expression of mammalian cytochrome P450 2E1 in modified plants had prominently improved the remediation of petroleum derivatives, that is, toluene and benzene (Doty, 2008). Removal of toluene is ten times more quick in hydroponic solution that is, within two days in comparison to the regulators. Benzene is one of the derivatives removed completely within three days by CYP2E1 transgenic tobacco. (Cunningham and Berti 1993; Guo et al., 2008; Schnoor et al., 1995; Suresh and Ravishankar, 2004).

21.13 Future perspective

A number of sources like urbanization and more or less anthropogenic activities are polluting the environment. Nowadays, the environmental pollution has become a problem that needs to be resolved. Different techniques are being used against this problem. Among the recent approaches, phytoremediation is one of the frequently used, low-priced, and active methods in comparison to the traditional methods for the elimination of toxins from the contaminated areas. However, this method is time taking to remove contaminants and produces the lower concentrations of biomass. Currently, the bioavailable chemical are augmented more quickly from the environment compared with other waste products. Genetically modified plants may be a good approach to improve metal uptake capacity, movements of chemicals, and then detoxification of the pollutants. Numbers of studies are ongoing for the removal of bioavailable impurities from the environment. The growth of transgenic plants with improved plant–microbe interaction or rhizosphere microbial activity may prove useful. The use of transgenic plants in phytoremediation should also focus on the problem of mixed contamination occurring in many of the contaminated sites. A multigene technique concerning the synchronized transfer of various genes into appropriate candidate plants may help to eliminate poisons of mixed or complex nature. For the better acceptance in the remediation industry, it is important that new transgenic studies should be continued to be tested in the field. Phytoremediation knowledge is presently accessible for only a small subgroup of waste product. Hence, several sites are still polluted with different type of toxins. Thus, to improve phytoremediation, plants with efficiency should be modified with improved selected genes in the direction to fulfill the needs of the current time.

References

Anderson, T.A., Guthrie, E.A., Walton, B.T., 1993. Bioremediation in the rhizosphere. Environ. Sci. Technol. 27 (13), 2630–2636.

Barac, T., Taghavi, S., Borremans, B., Provoost, A., Oeyen, L., Colpaert, J.V., Vangronsveld, J., van der Lelie, D., 2004. Engineered endophytic bacteria improve phytoremediation of water-soluble, volatile, organic pollutants. Nat. Biotechnol. 22 (5), 583–588.

Bernier, M., Carpentier, R., 1995. The action of mercury on the binding of the extrinsic polypeptides associated with the water oxidizing complex of photosystem II. FEBS Lett 360 (3), 251–254.

Bizily, S.P., Kim, T., Kandasamy, M.K., Meagher, R.B., 2003. Subcellular targeting of methylmercury lyase enhances its specific activity for organic mercury detoxification in plants. Plant Physiol 131 (2), 463–471.

Bizily, S.P., Rugh, C.L., Meagher, R.B., 2000. Phytodetoxification of hazardous organomercurials by genetically engineered plants. Nat. Biotechnol. 18, 213–217.

Bizily, S.P., Rugh, C.L., Summers, A.O., Meagher, R.B., 1999. Phytoremediation of methylmercury pollution: merB expression in Arabidopsis thaliana confers resistance to organomercurials. Proc. Natl. Acad. Sci. USA 96 (12), 6808–6813.

Bode, M., Stöbe, P., Thiede, B., Schuphan, I., Schmidt, B., 2004. Biotransformation of atrazine in transgenic tobacco cell culture expressing human P450. Pest Manag. 60 (1), 49–58.

Boularbah, A., Schwartz, C., Bitton, G., Aboudrar, W., Ouhammou, A., Morel, J.L., 2006. Heavy metal contamination from mining sites in South Morocco: 2. Assessment of metal accumulation and toxicity in plants. Chemosphere 63 (5), 811–817.

Catarecha, P., Segura, M.D., Franco-Zorrilla, J.M., García-Ponce, B., Lanza, M., Solano, R., Paz-Ares, J., Leyva, A., 2007. A mutant of the Arabidopsis phosphate transporter PHT1; 1 displays enhanced arsenic accumulation. Plant Cell 19 (3), 1123–1133.

Chaudhry, Q., Blom-Zandstra, M., Gupta, S.K., Joner, E., 2005. Utilising the synergy between plants and rhizosphere microorganisms to enhance breakdown of organic pollutants in the environment. Environ. Sci. Pollut. Res. 12 (1), 34–48.

Che, D., Meagher, R.B., Heaton, A.C., Lima, A., Rugh, C.L., Merkle, S.A., 2003. Expression of mercuric ion reductase in Eastern cottonwood (Populusdeltoides) confers mercuric ion reduction and resistance. Plant Biotechnol. J. 1 (4), 311–319.

Cherian S., Oliveira M.M., 2005, Transgenic plants in phytoremediation: recent advances and new possibilities. Environm. Sci. Technol. 39(24), 9377–9390.

Chukwuma, C., 2014. Concerns in the sustainable management of heavy metals in plants and soils. Int. Intervention J. Agric. Soil Sci. 2 (9), 143–152.

Crocker, F.H., Indest, K.J., Fredrickson, H.L., 2006. Biodegradation of the cyclic nitramine explosives RDX, HMX, and CL-20. Appl. Microbiol. Biotechnol. 73 (2), 274–290.

Cunningham, S.D., Berti, W.R., 1993. Remediation of contaminated soils with green plants: an overview. In Vitro Cell. Dev. Biol. Plant 29 (4), 207–212.

Dean, J.G., Bosqui, F.L., Lanouette, K.H., 1972. Removing heavy metals from wastewater. Environ. Sci. Technol. 6 (6), 518–522.

Dhankher, O.P., Li, Y., Rosen, B.P., Shi, J., Salt, D., Senecoff, J.F., Sashti, N.A., Meagher, R.B., 2002. Engineering tolerance and hyperaccumulation of arsenic in plants by combining arsenate reductase and γ-glutamylcysteinesynthetase expression. Nat. Biotechnol. 20 (11), 1140–1145.

Dhankher, O.P., Pilon-Smits, E.A., Meagher, R.B., Doty, S., 2012. Biotechnological approaches for phytoremediation. Plant biotech. agric. Academic Press, 309–328.

Doty S.L., James C.A., Moore A.L., Vajzovic A., Singleton G.L., Ma C., Khan Z., Xin G., Kang J.W., Park J.Y., Meilan R., 2007, Enhanced phytoremediation of volatile environmental pollutants with transgenic trees. Proc. Natl. Acad. Sci. 104 (43), 16816–16821.

Doty, S.L., 2008. Enhancing phytoremediation through the use of transgenics and endophytes. New Phytol. 179 (2), 318–333.

Doty, S.L., Shang, T.Q., Wilson, A.M., Moore, A.L., Newman, L.A., Strand, S.E., Gordon, M.P., 2003. Metabolism of the soil and groundwater contaminants, ethylene dibromide and trichloroethylene, by the tropical leguminous tree, *Leuceana leucocephala*. Water Res. 37 (2), 441–449.

Doty, S.L., Shang, T.Q., Wilson, A.M., Tangen, J., Westergreen, A.D., Newman, L.A., Strand, S.E., Gordon, M.P., 2000. Enhanced metabolism of halogenated hydrocarbons in transgenic plants containing mammalian cytochrome P450 2E1. Proc. Natl. Acad. Sci. USA 97 (12), 6287–6291.

Elizabeth A.H., Pilon-Smith, Freeman J.L., 2006, Environmental cleanup using plants Biotechnological advances and ecological consideration. Front Ecol. Environ. 4, 203–210.

Fletcher, J.S., Hegde, R.S., 1995. Release of phenols by perennial plant roots and their potential importance in bioremediation. Chemosphere 31 (4), 3009–3016.

Gandia-Herrero, F., Lorenz, A., Larson, T., Graham, I.A., Bowles, D.J., Rylott, E.L., Bruce, N.C., 2008. Detoxification of the explosive 2, 4, 6-trinitrotoluene in Arabidopsis: discovery of bifunctional O-and C-glucosyltransferases. Plant J 56 (6), 963–974.

Gasic, K., Korban, S.S., 2007. Transgenic Indian mustard (Brassica juncea) plants expressing an Arabidopsis phytochelatin synthase (AtPCS1) exhibit enhanced As and Cd tolerance. Plant Mol. Biol. 64 (4), 361–369.

Ghosh, M., Singh, S.P., 2005. A review on phytoremediation of heavy metals and utilization of it's by products. Asian J. Energy Environ. 6 (4), 18.

Gratão, P.L., Prasad, M.N.V., Cardoso, P.F., Lea, P.J., Azevedo, R.A., 2005. Phytoremediation: green technology for the cleanup of toxic metals in the environment. Braz. J. Plant Physiol. 17 (1), 53–64.

Guo, J., Dai, X., Xu, W., Ma, M., 2008. Overexpressing GSH1 and AsPCS1 simultaneously increases the tolerance and accumulation of cadmium and arsenic in Arabidopsis thaliana. Chemosphere 72 (7), 1020–1026.

Harada, M., 1995. Minamata disease: methylmercury poisoning in Japan caused by environmental pollution. Crit. Rev. Toxicol. 25 (1), 1–24.

Heaton, A.C., Rugh, C.L., Wang, N.J., Meagher, R.B., 1998. Phytoremediation of mercury-and methylmercury-polluted soils using genetically engineered plants. J. Soil Contamin. 7 (4), 497–509.

Heaton, A.C., Rugh, C.L., Wang, N.J., Meagher, R.B., 2005. Physiological responses of transgenic merA-tobacco (*Nicotiana tabacum*) to foliar and root mercury exposure. Water Air Soil Pollut 161 (1), 137–155.

Heaton, A.C., Rugh, C.L., Kim, T., Wang, N.J., Meagher, R.B., 2003. Toward detoxifying mercury-polluted aquatic sediments with rice genetically engineered for mercury resistance. Environ. Toxicol. Chem. 22 (12), 2940–2947.

Hussein, J., Matekwe-phoya, A.N.N., Tornui, J.A., Okiwelu, T., 2007. Midwifery practice in Ghana and Malawi: influences of the health system. J. Midwifery Pract., 75.

Jackson, R.G., Rylott, E.L., Fournier, D., Hawari, J., Bruce, N.C., 2007. Exploring the biochemical properties and remediation applications of the unusual explosive-degrading P450 system XplA/B. Proc. Natl. Acad. Sci. USA 104 (43), 16822–16827.

James, C.A., Strand, S.E., 2009. Phytoremediation of small organic contaminants using transgenic plants. Curr. Opin. Biotechnol. 20 (2), 237–241.

Kang, J.W., Wilkerson, H.W., Farin, F.M., Bammler, T.K., Beyer, R.P., Strand, S.E., Doty, S.L., 2010. Mammalian cytochrome CYP2E1 triggered differential gene regulation in response to trichloroethylene (TCE) in a transgenic poplar. Funct. Integr. Genomics 10 (3), 417–424.

Kawahigashi, H., Hirose, S., Ohkawa, H., Ohkawa, Y., 2006. Phytoremediation of the herbicides atrazine and metolachlor by transgenic rice plants expressing human CYP1A1, CYP2B6, and CYP2C19. J Agric. Food Chem. 54 (8), 2985–2991.

Krämer, U., 2005. Phytoremediation: novel approaches to cleaning up polluted soils. Curr. Opin. Biotechnol. 16 (2), 133–141.

Krämer, U., Chardonnens, A., 2001. The use of transgenic plants in the bioremediation of soils contaminated with trace elements. Appl. Microbiol. Biotechnol. 55 (6), 661–672.

Lee K.Y., Strand S.E., Doty S.L., 2012, Phytoremediation of chlorpyrifos by Populus and Salix. Int. J. Phytoremediation. 14, 48–61.

Limbert, E.S.B., Betts, W.B., 1996. Influences of substrate chemistry and microbial metabolic diversity on the bioremediation of xenobiotic contamination. Genetic Eng. Biotechnol 16 (3), 159–180.

Li, Y., Dhankher, O.P., Carreira, L., Balish, R.S., Meagher, R.B., 2005. Arsenic and mercury tolerance and cadmium sensitivity in Arabidopsis plants expressing bacterial γ-glutamylcysteines ynthetase. Environ. Toxicol. Chem. 24 (6), 1376–1386.

Lyyra, S., Meagher, R.B., Kim, T., Heaton, A., Montello, P., Balish, R.S., Merkle, S.A., 2007. Coupling two mercury resistance genes in Eastern cottonwood enhances the processing of organomercury. Plant Biotechnol. J. 5 (2), 254–262.

Martínez, M., Bernal, P., Almela, C., Vélez, D., García-Agustín, P., Serrano, R., Navarro-Aviñó, J., 2006. An engineered plant that accumulates higher levels of heavy metals than Thlaspicaerulescens, with yields of 100 times more biomass in mine soils. Chemosphere 64 (3), 478–485.

Meharg, A.A., Acnair, M.R., 1992. Suppression of the high affinity phosphate uptake system: a mechanism of arsenate tolerance in Holcuslanatus L. J. Exp. Bot. 43, 519–524.

Mganga, N., Manoko, M.L.K., Rulangaranga, Z.K., 2011. Classification of plants according to their heavy metal content around North Mara gold mine, Tanzania: implication for phytoremediation. Tanz. J. Sci. 37 (1), 109–119.

Mkumbo, S., Mwegoha, W., Renman, G., 2012. Assessment of the phytoremediation potential for Pb, Zn and Cu of indigenous plants growing in a gold mining area in Tanzania. Int. J. Environ. Sci. 2 (4), 2425–2434.

Moore, M.N., Allen, J.I., McVeigh, A., 2006. Environmental prognostics: an integrated model supporting lysosomal stress responses as predictive biomarkers of animal health status. Mar. Environ. Res. 61 (3), 278–304.

Nakamura, T., Motoyama, T., Suzuki, Y., Yamaguchi, I., 2004. Biotransformation of pentachlorophenol by Chinese chive and a recombinant derivative of its rhizosphere-competent microorganism, Pseudomonas gladioli M-2196. Soil Biol. Biochem. 36 (5), 787–795.

Nordstrom, D.K., 2002, PUBLIC HEALTH: Enhanced: Worldwide Occurrences of Arsenic in Ground Water. Sci. 296 (5576), 2143–2145.

Nriagu, J.O., Pacyna, J.M., 1988. Quantitative assessment of worldwide contamination of air, water and soils by trace metals. Nature 333 (6169), 134–139.

Okem, A., 2014. Heavy metals in South African medicinal plants with reference to safety, efficacy and quality Ph.D. thesis). University of KwaZulu-Natal, Pietermaritzburg, South Africa.

Padmavathiamma, P.K., Li, L.Y., 2007. Phytoremediation technology: hyper-accumulation metals in plants. Water, Air, Soil Pollut 184 (1-4), 105–126.

Patra, M., Sharma, A., 2000. Mercury toxicity in plants. Bot. Rev. 66 (3), 379–422.

Pilon-Smits, E.A., Hwang, S., Lytle, C.M., Zhu, Y., Tai, J.C., Bravo, R.C., Chen, Y., Leustek, T., Terry, N., 1999. Overexpression of ATP sulfurylase in Indian mustard leads to increased selenate uptake, reduction, and tolerance. Plant Physiol 119 (1), 123–132.

Rugh, C.L., Senecoff, J.F., Meagher, R.B., Merkle, S.A., 1998. Development of transgenic yellow poplar for mercury phytoremediation. Nat. Biotechnol. 16 (10), 925–928.

Rugh, C.L., Wilde, H.D., Stack, N.M., Thompson, D.M., Summers, A.O., Meagher, R.B., 1996. Mercuric ion reduction and resistance in transgenic *Arabidopsis thaliana* plants expressing a modified bacterial merA gene. Proc. Natl. Acad. Sci. USA 93 (8), 3182–3187.

Ruiz, C.B.B., 2003. Osparadoxos do imaginário. Unisinos, Brazil.

Ryan, R.P., Germaine, K., Franks, A., Ryan, D.J., Dowling, D.N., 2008. Bacterial endophytes: recent developments and applications. FEMS Microbiol. Lett. 278 (1), 1–9.

Rylott, E.L., Jackson, R.G., Edwards, J., Womack, G.L., Seth-Smith, H.M., Rathbone, D.A., Strand, S.E., Bruce, N.C., 2006. An explosive-degrading cytochrome P450 activity and its targeted application for the phytoremediation of RDX. Nat. Biotechnol. 24 (2), 216–219.

Schnoor, J.L., Light, L.A., McCutcheon, S.C., Wolfe, N.L., Carreia, L.H., 1995. Phytoremediation of organic and nutrient contaminants. Environ. Sci. Technol. 29 (7), 318A–323A.

Schwitzguébel, J.P., 2017. Phytoremediation of soils contaminated by organic compounds: hype, hope and facts. J. Soils Sediments 17 (5), 1492–1502.

Siciliano, S.D., Germida, J.J., Banks, K., Greer, C.W., 2003. Changes in microbial community composition and function during a polyaromatic hydrocarbon phytoremediation field trial. Appl. Environ. Microbiol. 69 (1), 483–489.

Shang, T.Q., Doty, S.L., Wilson, A.M., Howald, W.N., Gordon, M.P., 2001. Trichloroethylene oxidative metabolism in plants: the trichloroethanol pathway. Phytochemistry 58 (7), 1055–1065.

Shimp, J.F., Tracy, J.C., Davis, L.C., Lee, E., Huang, W., Erickson, L.E., Schnoor, J.L., 1993. Beneficial effects of plants in the remediation of soil and groundwater contaminated with organic materials. Crit. Rev. Environ. Sci. Technol. 23 (1), 41–77.

Shu, Wensheng, Xia, Hanping, 2012. Technique for remediation of heavy metal contamination: potential and practice. Chin. J. Integr. Med. 2, 406–408.

Singh, D., Vyas, P., Sahni, S., Sangwan, P., 2015. Phytoremediation: a biotechnological intervention. In: Kaushik, G. (Ed.), Applied Environmental Biotechnology: Present Scenario and Future Trends. Springer, New Delhi, pp. 59–75.

Smith, T., Pitts, K., McGarvey, J.A., Summers, A.O., 1998. Bacterial oxidation of mercury metal vapor, Hg (0). Appl. Environ. Microbiol. 64 (4), 1328–1332.

Suresh, B., Ravishankar, G.A., 2004. Phytoremediation—a novel and promising approach for environmental clean-up. Crit. Rev. Biotechnol. 24 (2–3), 97–124.

Taghavi, S., Barac, T., Greenberg, B., Borremans, B., Vangronsveld, J., van der Lelie, D., 2005. Horizontal gene transfer to endogenous endophytic bacteria from poplar improves phytoremediation of toluene. Appl. Environ. Microbiol. 71 (12), 8500–8505.

Van Aken, B., Yoon, J.M., Just, C.L., Schnoor, J.L., 2004. Metabolism and mineralization of hexahydro-1, 3, 5-trinitro-1, 3, 5-triazine inside poplar tissues (Populusdeltoides× nigra DN-34). Environ. Sci.Technol. 38 (17), 4572–4579.

Van Huysen, T., Abdel-Ghany, S., Hale, K.L., LeDuc, D., Terry, N., 2003. Overexpression of cystathionine-gammasynthase enhances selenium volatilization in *Brassica juncea*. Planta, 71–78.

Wang, L., Samac, D.A., Shapir, N., Wackett, L.P., Vance, C.P., Olszewski, N.E., Sadowsky, M.J., 2005. Biodegradation of atrazine in transgenic plants expressing a modified bacterial atrazine chlorohydrolase (atzA) gene. Plant Biotechnol. J. 3 (5), 475–486.

Weyens, N., Van Der Lelie, D., Artois, T., Smeets, K., Taghavi, S., Newman, L., Carleer, R., Vangronsveld, J., 2009. Bioaugmentation with engineered endophytic bacteria improves contaminant fate in phytoremediation. Environ. Sci. Technol. 43 (24), 9413–9418.

Chapter 22

Engineering plants for metal tolerance and accumulation

Amauri Ponce-Hernández[a], Angel Josabad Alonso-Castro[b], Ramón Fernando García-De La Cruz[a], Candy Carranza-Alvarez[c]

[a]Research and Postgraduate Center of the Faculty of Chemical Sciences, Universidad Autónoma de San Luis Potosí, S.L.P., Mexico, [b]Department of Pharmacy, Division of Natural and Exact Sciences, University of Guanajuato, Guanajuato, Mexico, [c]Faculty of Professional Studies Zona Huasteca and Center for Research and Postgraduate Studies of the Faculty of Chemical Sciences, Autonomous University of San Luis Potosí, Cd.Valles, S.L.P., Mexico

22.1 Introduction

Heavy metals (HM) are found naturally in the earth's crust at different levels and concentrations (Singh et al., 2016). Some HM, at low concentrations, are necessary for plant development (e.g., Zn, Cu, Mn, Ni, and Co) and others have unknown biological functions (e.g., Cd, Pb, Hg) (Jagtap and Bapat, 2015). Although the term HM has not yet been well defined, it generally refers to metals or metalloids that have a density (>5 g/cm^3) (Duffus, 2001; Jarup, 2003). In recent decades there has been an increase in anthropogenic activities such as industrialization, urbanization, energy generation, agricultural practices, and the burning of fossil fuels (Chibuike and Obiora, 2014; Sadiq et al., 2017), which has resulted in an increase from HM concentrations in soils and waters to amounts that are harmful to plants, animals, and humans. The HM can present changes in their chemical forms and bioavailability; however, unlike organic pollutants, HM cannot be degraded, and they can persist in the environment for a long time (Wuana and Okieimen, 2011). The above coupled with the toxicity and the nature of bioaccumulation makes HM potentially dangerous elements for the environment and human health (Sadiq et al., 2017).

To deal with or mitigate these problems of the terrestrial and aquatic environment, various physical, chemical, and biological methods have been implemented and developed. Within the focus of physical and chemical methods are electrokinetics, encapsulation, stabilization, solidification, steam extraction and washing, soil washing, vitrification, filtration, flocculation, activated carbon, reverse osmosis, and chemical precipitation or coagulation. (Marques et al., 2009; Dada et al., 2015; Derakhshan-Nejad et al., 2017). However, this

Phytoremediation. DOI: https://doi.org/10.1016/B978-0-323-89874-4.00012-1

approach is generally expensive, it is inefficient when the HM concentration is low, it presents changes in the physicochemical and biological properties of the soil and water, and it can cause secondary contamination (Ali et al., 2013). On the other hand, the biological approach or bioremediation is a way where nature is used to clean up the polluted environment. Bioremediation includes processes and actions where living organisms act to detoxify pollutants from the environment. Within bioremediation are technologies such as microbial remediation, mycoremediation, zooremediation, vermiremediation, phycoremediation, and phytoremediation (Dada et al., 2015). Of these, phytoremediation is considered a very promising technology to recover contaminated sites, it is more environmentally friendly, acceptable, and economical (60%–80% or less expensive) than physicochemical approaches. It is a technique that uses plants to remove, degrade, metabolize, or immobilize many organic and inorganic compounds that pollute the environment (Garbisu and Alkorta, 2001; Njoku et al., 2009). According to Wuana and Okieimen (2011), so far, approximately 400 species of plants belonging to at least 45 families have been reported that have the ability to hyperaccumulate metals.

Although phytoremediation is a technique that has many advantages and is known to have been widely used in the restoration of contaminated sites, this technique also has several limitations, among which are a slow speed of metal removal, the amount of biomass produced, depth of roots, incomplete metabolism of compounds, and sometimes low tolerances to high concentrations of contaminants (Van Aken, 2008; Tangahu et al., 2011). To overcome the existing limitations of phytoremediation, the development of modified plant genotypes began. The objective of this approach was to develop plants that present high biomass, withstand different climatic conditions, easy to grow, and that allow the elimination of high concentrations of pollutants (Maestri and Marmiroli, 2011). The main properties of plants for phytoremediation are tolerance and accumulation of metals, which are dependent on metal absorption capacity, translocation, intracellular sequestration, chemical modification, and general resistance to stress.

22.2 Plant responses to HM

The response to HM is different for each plant species, and depends on the exposure time and the concentration of the metal to which the plants are exposed. In Fig. 22.1, the different types of response that occur in plants are presented.

22.2.1 Growth responses

HM can affect plant growth in various metabolic and physiological processes. The damage can be at different levels of cell structures and plant function (Ahmad et al., 2011). Some of the main effects of HM are to inhibit germination, leaf and shoot growth, and root elongation (Munzuroglu and Geckil,

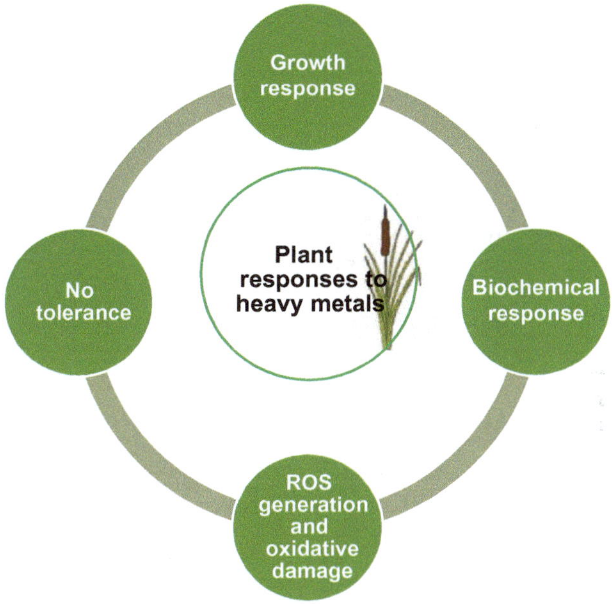

FIG. 22.1 Mechanism of plant responses to heavy metals.

2002). In addition, HM affect cell division, which causes a reduction in roots and shoots. On the other hand, it has been reported that exposure to HM inhibits the mobilization, uptake, and assimilation of nutrients (e.g., nitrogen, phosphorus, and sulfur) and water, inhibiting plant development (Wang and Greger, 2004). reported that the stress caused by HM affects the diameter of the xylem vessel, the area of the parenchyma cells, the medulla, the root cortex, the dimensions of the stem, the vascular bundles, the number of xylem arms at the root, the frequency of stomata on the axial leaf surface, and the reduction of grain yield.

22.2.2 Biochemical responses

Regarding the biochemical responses of plants to the stress caused by HM, a decrease in photosynthetic pigments has been reported, which lead to the affectation of the photosynthetic apparatus and consequently to the respiration process. Lipids and proteins, in the same way, show a decrease, as they are denatured or oxidized by proteolytic activity and/or production of H_2O_2 (Anjum et al., 2015). The accumulation of carbohydrates generally decreases when the content of HM in the medium is progressively increased (Singh et al., 2007). In contrast, proline accumulation increases under HM stress, and can maintain an osmotic setting, provide enzymatic protection, and stabilize the structure of organelles and macromolecules (Ahmad et al., 2011).

22.2.3 ROS generation and oxidative damage

HM can alter homeostasis in plants because they induce the formation of reactive oxygen species (ROS) and methylglyoxal (Anjum et al., 2012). The main ROS production sites are chloroplasts, mitochondria, and peroxisomes (Hameed et al., 2016). The increase in ROS in cells produces oxidative damage on cell structures, nucleic acids (DNA), lipids, and proteins (Tuteja and Sopory, 2008). ROS-produced DNA damage includes single- or double-stranded DNA breaks, purine, pyridine, or deoxyribose base modifications. Lipid peroxidation caused by ROS decreases membrane fluidity, increases filtration, and damages membrane proteins. Oxidative damage to proteins causes effects on cell physiology, such as causing enzymatic oxidation and inhibiting their activities (Hameed et al., 2016).

22.3 Mechanisms of uptake and tolerance to HM

Plants have various mechanisms at the physiological, biochemical, and genomic levels that provide tolerance when exposed to environments with HM (Fig. 22.2). Some mechanisms are unique and are activated only in a certain exposure to HM.

First, when HM induces oxidative stress by ROS production, plants activate the antioxidant defense system, which integrates the action of (1) enzymatic compounds, such as catalase, ascorbate peroxidase, monodehydroascorbate reductase, dehydroascorbate reductase, glutathione peroxidase, glutathione-S-transferase, glutathione reductase, peroxiredoxin, and (2) nonenzymatic compounds, such as ascorbic acid, glutathione (GHS), α-tocopherol, and flavonoids. These components regulate the concentration of ROS and alleviate the damage caused in the cellular structures and macromolecules of the plant (Kapoor et al., 2015).

On the other hand, plants protect themselves from HM through the release of root exudates Fig. 22.3. These exudates bind to HM to limit their absorption

FIG. 22.2 Mechanisms of uptake and tolerance to HM.

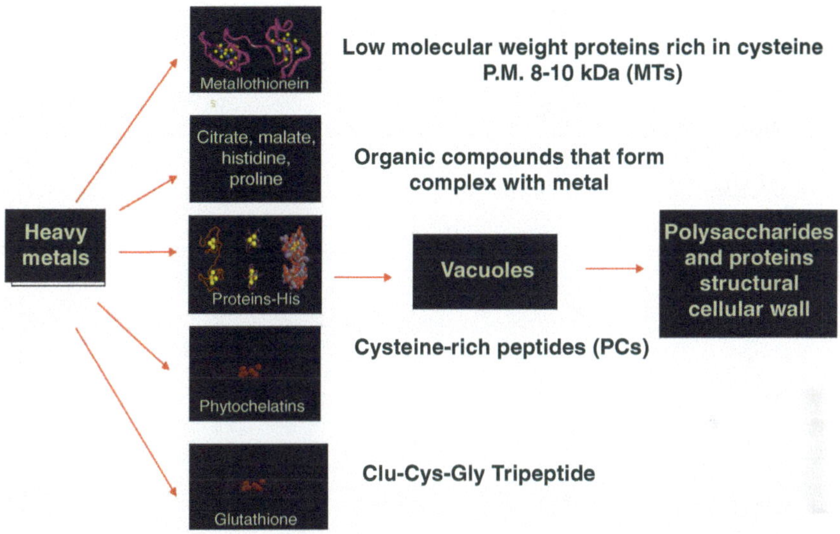

FIG. 22.3 Biochemical uptake and bioaccumulation mechanisms of heavy metals in plants. Figure based on the information from Cobbett and Golsbrougth, 2000.

and increase biosorption in cell walls. To tolerate the stress caused by HM, plants also have mechanisms to sequester metal ions in compartments, in this way they protect the sensitive components of cells from metal interactions. Compounds such as amino acids, GHS, phytochelatins, metallothioneins (MTs); and enzymes such as superoxide dismutase are involved in these mechanisms (Ghori et al., 2019). When the HM are absorbed by the root, they are taken to the xylem and transferred to the shoots. Various transport proteins are responsible for transporting HM from the root to the shoot. Transport proteins are classified as ZIP protein family, ATP-binding cassette transporters (ABC), P-type metal ATPases, natural resistance-associated macrophage protein family, mitochondria ABC transporters, family of proteins that facilitate cation diffusion, proteins of the copper transporter family among others (Kramer et al., 2007; Hameed et al., 2016). Uptake and transport mechanisms help alleviate HM toxicity in plants.

22.4 Main transgenic modifications in plants to improve tolerance and accumulation of metals

To improve the performance of plants in environments contaminated by HM, the transgenic plants approach was introduced with the ability to improve phytoremediation using genetic manipulation and plant transformation technologies (Ellis et al., 2004). Transgenic plants are plants that have been genetically modified, which can express genes that are not native to it. To do this, hyperaccumulative genes are identified that are expressed in plants and are subsequently transferred to plant species that have the capacity for rapid growth, deep roots, and tolerance to HM and various climatic conditions. Table 22.1 shows

TABLE 22.1 Transgenics plants in heavy metals/metalloids tolerance and accumulation.

Recipient plant	Introduced gene	Source	Metal tolerance and or accumulation	Reference
Beta vulgaris	γ-glutamyl-cysteine synthetase glutathione synthetase GCS-GS operon	*Streptococcus thermophilus*	Increased in the tolerance and accumulation of Cd, Zn, and Cu.	Liu et al. (2015)
Populus tremula x *P. alba*	γ-glutamyl-cysteine synthetase (γ-ECS)	*Escherichia coli*	Increase in the tolerance of Cd	He et al., 2014
Brassica juncea (L.) Czern.	γ-ECS, glutathione synthetase (GS), and triphosphate sulfurylase	*Escherichia coli*	Increase in the tolerance and accumulation of Se	Bañuelos et al. (2005)
Brassica juncea L.	γ-ECS and GS	*Escherichia coli*	Increase in the accumulation and tolerance of mixtures of (As, Cd, Cr), (As, Cd, Zn), and (As, Pb, Zn)	Reisinger et al. (2008)
Brassica juncea L	γ-ECS and GS	*Escherichia coli*	Increase in the accumulation of Cd, Cr, Cu, Pb, and Zn	Bennett et al. (2003)
Arabidopsis thaliana	γ-ECS and arsenate reductase (*arsC*)	*Escherichia coli*	Increase in the accumulation and tolerance of As	Dhankher et al. (2002)
Arabidopsis thaliana and *Nicotiana tabacum*	*arsC*	*Escherichia coli*	Increase in the accumulation and tolerance of Cd	Dhankher et al. (2003)
Nicotiana tabacum	Arabidopsis phytochelatin synthase (*AtPCS1*)	*Arabidopsis thaliana*	Increase in the tolerance and accumulation of Cd	Guo et al. (2008)
Brassica juncea	*AtPCS1*	*Arabidopsis thaliana*	Increase in the tolerance of As and Cd	Gasic and Korban (2007a)

Brassica juncea	*AtPCS1*	*Arabidopsis thaliana*	Increase in the tolerance of Cd and Zn	Gasic and Korban (2007b)
Nicotiana tabacum	*AtPCS1*	*Arabidopsis thaliana*	Increases in the tolerance and accumulation of As and As + Cd mix	Zanella et al. (2016)
Astragalus sinicus	*AtPCS1* and *MTL4*	*Arabidopsis thaliana* and *Homo sapiens*, respectively	Increase in the accumulation of Cd	Ike et al. (2007)
Arabidopsis thaliana	*AsPCS1* and GS	*Allium sativum* and *Saccharomyces cerevisiae*, respectively	Increase in the accumulation and tolerance of Cd and As	Guo et al. (2008)
Nicotiana glauca	*TaPCS1*	*Triticum aestivum*	Increase in the accumulation of Cd and Pb	Martínez et al. (2006)
Brassica juncea	ATP sulfurylase (*APS1*)	*Arabidopsis thaliana*	Increase in the concentration de As, Cd, Cr, Cu, Hg, Mo, V, and W in shoot	Wangeline et al. (2004)
Brassica juncea L.	*APS* and selenocysteine methyltransferase (*SMT*)	*Arabidopsis thaliana* and *Astragalus bisulcatus*, respectively	Increase in the tolerance and accumulation of Se	
Brassica juncea L.	*SMT*	*Astragalus bisulcatus*	Increase in the tolerance and accumulation of Se	LeDuc et al. (2004)
Brassica juncea (L.) Czern.	*SMT* and selenocysteine lyase (SL)	*Astragalus bisulcatus* and *Mus musculus*, respectively	Increase in the accumulation of Se	Bañuelos et al. (2007)
Arabidopsis thaliana	SL	*Mus musculus*	Increase in the tolerance and accumulation of Se	Pilon et al. (2003)

(continued)

TABLE 22.1 (Cont'd)

Recipient plant	Introduced gene	Source	Metal tolerance and or accumulation	Reference
Nicotiana tabacum	Cysteine synthase *RCS1*	*Spinacia oleracea*	Increase in the accumulation of Cd in shoot	Kawashima et al. (2004)
Nicotiana tabacum	Cysteine synthase *RCS1*	*Oryza sativa*	Increase in the tolerance of Cd, Se, and Ni and Increase in the accumulation of Cd	Harada et al. (2001)
Nicotiana tabacum	NA synthase *NAS1*	*Arabidopsis thaliana*	Increase in the accumulation and tolerance of Ni	Douchkov et al. (2005)
Nicotiana tabacum	NA synthase *NAS1*	*Hordeum vulgare*	Increase in the tolerance of Ni and increase in the accumulation of Mn	Kim et al. (2005)
Arabidopsis thaliana	NA synthase *NAS1*	*Hordeum vulgare*	Increase in the tolerance of Cd, Cu, Fe, Mn, Zn, and Ni	Kim et al. (2005)
Arabidopsis thaliana	NA synthase *NAS1*	*Thlaspi caerulescens*	Increase in the accumulation and tolerance of Ni	Pianelli et al. (2005)
Arabidopsis thaliana	*mer*Ape9	*Escherichia coli* modified	Increase in the accumulation Hg and Au	Rugh et al. (1996)
Nicotiana tabacum	*mer*Ape9	*Escherichia coli* modified	Increase in the tolerance and volatilization of Hg	He et al. (2001)
Arabidopsis thaliana	*mer*Ape9 and *mer*B	*Escherichia coli* modified	Increase in the tolerance and degradation of organic Hg	Bizily et al. (2003)

Nicotiana tabacum	*merA* and *merB*	*Escherichia coli* Tn21	Increase in the resistance of phenyl mercuric acetate	Ruiz et al. (2003)
Populus deltoides	*merA*	*Escherichia coli* modified	Increase in the tolerance of As and volatilization of Hg	Che et al. (2003)
Oryza sativa	*merA*	*Escherichia coli* modified	Increase in the tolerance and volatilization of Hg	Heaton et al. (2003)
Liquidambar styraciflua × L. formosana	*merA*	*Escherichia coli* modified	Increase in the tolerance of Hg	Dai et al. (2009)
Nicotiana tabacum	*merA*	*Escherichia coli*	Increase in the volatilization of Hg ion	Haque et al. (2010)
Arabidopsis thaliana	*merC*	*Acidithiobacillus ferrooxidans*	Increase in the accumulation of Hg	Sasaki et al. (2006)
Arabidopsis thaliana	*merC and* SYP121	-	Increase in the accumulation Hg	Uraguchi et al. (2019)
Arabidopsis thaliana	Hg-binding protein, *merP*	*Bacillus megaterium*	Increase in the tolerance and accumulation of Hg, Cd, and Pb	Hsieh et al. (2009)
Nicotiana tabacum	Glutathione S-transferase	*Arabidopsis thaliana*	Increase in the tolerance of Al, Cu, and Na	Ezaki et al. (2000)
Medicago sativa L.	Glutathione S-transferase and human CYP2E1	-	Increase in the tolerance of Cd and trichloroethylene	Zhang and Liu (2011).
Medicago sativa L.	Glutathione S-transferase and human CYP2E1	-	Increase in the accumulation of Hg and trichloroethylene mix	Zhang et al. (2013)
Nicotiana tabacum and *Carica papaya*	Citrate synthase	*Pseudomonas aeruginosa*	Increase in the tolerance of Al	Dela Fuente et al. (1997)

(continued)

TABLE 22.1 (Cont'd)

Recipient plant	Introduced gene	Source	Metal tolerance and or accumulation	Reference
Medicago sativa L.	Citrate synthase	*Pseudomonas aeruginosa*	Increase in the tolerance of Al	Barone et al. (2008)
Nicotiana benthamiana	Mitochondrial citrate synthase	*Citrus Junos* Sieb. Ex Tanaka	Increase in the tolerance of Al	Deng et al. (2009)
Populus alba	Metallothionein MT_{A1}	*Pisum sativum*	Increase in the tolerance of Cu and Zn	Balestrazzi et al. (2009)
Pisum sativum	Metallothionein MT_{A1}	*Populus alba*	Increase in the tolerance and accumulation of Cu and Zn.	Turchi et al. (2012)
Nicotiana tabacum	Metallothionein MT1	*Mus musculus*	Increase in the tolerance of Cd	Pan et al. (1994)
Arabidopsis thaliana	Metallothionein MT1	*Cajanus cajan*	Increase in the tolerance and accumulation of Cu and Cd	Sekhar et al. (2011)
Nicotiana tabacum	Metallothionein MT1	*Elsholtzia haichowensis*	Increase in the tolerance and accumulation of Cu	Xia et al. (2012)
Nicotiana tabacum	Metallothionein MT2	*Homo sapiens*	Increase in the tolerance of Cd	Misra and Gedamu (1989)
Nicotiana tabacum	Metallothionein MT2	*Saccharomyces cerevisiae*	Increase in the accumulation of Zn and Cd	Daghan et al. (2013)
Arabidopsis thaliana	Metallothioneins MT1 and MT2	*Brassica campestris*	Increase in the tolerance of Cu and Cd and Increase in the accumulation of Cu	Lv et al. (2013)

Arabidopsis thaliana	Metallothionein MT2b	*Iris lactea* var. *chinensis*	Increase in the tolerance of Cd	Gu et al. (2014)
Arabidopsis thaliana	Metallothionein MT2b	*Iris lactea* var. *chinensis*	Increase in the tolerance and accumulation of Cu	Gu et al. (2015)
Arabidopsis thaliana	Metallothionein MT4	*Arabidopsis thaliana*	Increase in the tolerance of Cu and Zn and Increase in the accumulation of Cu	Rodríguez-Llorente et al., (2010)
Arabidopsis thaliana	Metallothionein MTA	*Pisum sativum*	Increase in the accumulation of Cu	Evans et al. (1992)
Nicotiana tabacum	Metallothionein CUP-1	Levadura	Increase in the accumulation of Cu	Thomas et al. (2003)
Nicotiana tabacum	Metallothionein *His*CUP1	Recombinant fusion	Increase in the accumulation and tolerance of Cd	Pavlíková et al. (2004)
Arabidopsis thaliana	Acetyltransferase (*SAT*)	*Noccaea goesingense*	Increase in the tolerance of Ni	Freeman et al. (2004)
Nicotiana tabacum	Serine acetyltransferase from tobacco (*NtSAT4*)	*Brassica napus* L	Increase in the tolerance of Cd	Rajab et al. (2020)
Vicia sativa	Vicia sativa Caffeoyl-CoA O-methyltransferase (VsCCoAOMT)	*Arabidopsis thaliana*	Increase in the absorption and tolerance of Cd	Xia et al. (2018)
Nicotiana tabacum	Encoding Fe(III) reductases (*FRE-1* y *FRE-2*)	*Saccharomyces cerevisiae*	Increase in the concentration of Fe	Samuelsen et al. (1998)
Nicotiana tabacum	Ferretín	*Glycine max*	Increase in the accumulation of Fe	Goto et al. (1998)
Nicotiana tabacum	Protein *Nt CBP4*	*Nicotiana tabacum*	Increase in the tolerance of Ni and accumulation of Pb	Arazi et al. (1999)

(continued)

TABLE 22.1 (Cont'd)

Recipient plant	Introduced gene	Source	Metal tolerance and or accumulation	Reference
Arabidopsis thaliana	Protein YCF1	*Saccharomyces cerevisiae*	Increase in the tolerance and accumulation of Cd and Pb	Song et al. (2003)
Populus alba × P. tremula	Protein YCF1	*Saccharomyces cerevisiae*	Increase in the tolerance and accumulation of Cd	Shim et al. (2013)
Arabidopsis thaliana	Zn transporters ZAT (At MTP1)	*Arabidopsis thaliana*	Increase in the accumulation of Zn	Van der Zaal et al. (1999)
Nicotiana tabacum	Transporter *At*MHX	*Arabidopsis thaliana*	Increase in the tolerance of Mg and Zn	Shaul et al. (1999)
Arabidopsis thaliana	Mitochondrial transporter *ATM3*	*Arabidopsis thaliana*	Increase in the tolerance of Cd, Pb and increase in the accumulation of Cd	Kim et al. (2006)
Nicotiana tabacum	Phosphate transporter *PHT1* and *PHT7* + vacuolar Cd transporter YCF1	*Arabidopsis thaliana* and *Saccharomyces cerevisiae*	Increase in the tolerance and accumulation of As	LeBlanc et al. (2013)
Nicotiana tabacum	Vacuolar ABC transporter MRP7	*Arabidopsis thaliana*	Increase in the tolerance and accumulation of Cd	Wojas et al. (2009)
Nicotiana tabacum	Vacuolar antiporter CAX2	*Arabidopsis thaliana*	Increase in the accumulation of Cd, Ca, Mn	Korenkov et al. (2007)
Petunia × hybrida	Vacuolar antiporter CAX1	*Arabidopsis thaliana*	Increase in the tolerance and accumulation of Cd	Wu et al. (2011)
Arabidopsis thaliana	As(III) vacuolar antiporter ACR3	*Pteris vittata*	Increase in the tolerance and translocation of As	Chen et al. (2013)

Arabidopsis thaliana	HMA4	*Arabidopsis thaliana*	Increase in the tolerance of Cd, Zn, Co and increase in the accumulation of Cd and Zn	Verret et al. (2004)
Nicotiana tabacum	HMA4 + MT2	*Arabidopsis thaliana*	Increase in the tolerance of Cd	Grispen et al. (2011)
Nicotiana tabacum	HMA3	*Sedum alfredii*	Increase in the tolerance and accumulation of Cd	Zhang et al. (2016)
Arabidopsis thaliana	ZntA	*Escherichia coli*	Increase in the tolerance of Cd and Pb	Lee et al. (2003)
Nicotiana tabacum	Vacuolar H+-pyrophosphatase VP1	*Triticum aestivum*	Increase in the tolerance and accumulation of Cd	Khoudi et al. (2012)
Arabidopsis thaliana	MTP1	*Noccaea goesingense*	Increase in the tolerance and accumulation of Zn	Gustin et al. (2009)
Nicotiana tabacum	MTP1	*Oryza sativa*	Increase in the tolerance and accumulation of As and Cd	Das et al. (2016)
Nicotiana tabacum	Polyphosphate kinase ppk	*Klebsiella aerogenes*	Increase in the tolerance and accumulation of Hg	Nagata et al. (2006a)
Modified *Arabidopsis thaliana*	Adenosine 5′-phosphosulfate reductase 2 (APR2)	Modified *Arabidopsis thaliana*	Regulates accumulation and tolerance of Cd	Xu et al. (2019)
Arabidopsis thaliana	ZAT6 Transcription factor	*Arabidopsis thaliana*	Increase in the tolerance and accumulation of Cd	Chen et al. (2016)
Arabidopsis thaliana	*GmWRKY142* transcription factor	*Glycine max* (L.) Merr	Decreases Cd absorption and positively regulates Cd tolerance.	Cai et al. (2020)

(continued)

TABLE 22.1 (Cont'd)

Recipient plant	Introduced gene	Source	Metal tolerance and or accumulation	Reference
Solanum lycopersicum	ACC deaminase	*Enterobacter cloacae*	Increase the tolerance and accumulation of Cd, Co, Cu, Ni, Pb, and Zn	Grichko et al. (2000)
Brassica napus	ACC deaminase	*Enterobacter cloacae*	Increase in the tolerance and accumulation of As	Nie et al. (2002)
Brassica napus	ACC deaminase	*Pseudomonas putida*	Increase in the tolerance and accumulation of Ni	Stearns et al. (2005)
Petunia × hybrida and *Nicotiana tabacum*	iaaM and ACC deaminase	*Agrobacterium tumefaciens* and *Pseudomonas putida*, respectively	Increase in the accumulation of Co and Cu	Zhang et al. (2008)
Arabidopsis thaliana	Apoplastic endo-β-mannase MAN3	*Arabidopsis thaliana*	Increase in the tolerance and accumulation of Cd	Chen et al. (2015)
Arabidopsis thaliana	G protein *DEP1*	*Oryza sativa*	Increase in the tolerance and accumulation of Cd	Kunihiro et al. (2013)
Nicotiana tabacum	*rhl*A and *rhl*B	*Pseudomonas aeruginosa*	Increase in the tolerance of Al	Brichkova et al. (2007)
Arabidopsis thaliana	Acyl-CoA- binding protein ACBP1	*Arabidopsis thaliana*	Increase in the tolerance and accumulation of Pb	Xiao et al. (2008)
Nicotiana tabacum	Class 1 hemoglobin Hb1	*Nicotiana tabacum*	Increase in the tolerance Cd and reduces accumulation of Cd	Lee and Hwang (2015)

the main classes of genes that have been used to design plants with improved tolerance and accumulation characteristics of HM.

A defense mechanism in plants against metal stress is the synthesis of GHS, which plays an important role in the detoxification of ROS. As in plants, genes for GHS biosynthesis are also expressed in bacteria (Bañuelos et al., 2005; Liu et al., 2015; He et al., 2014). Therefore, in transgenic plants such as *Beta vulgaris*, *Populus tremula* × *Populus alba* and *Brassica juncea* L., genes that express GHS synthesizing enzymes (γ-Glutamyl-cysteine synthetase, (γ-GCS) and Glutathione synthetase (GS) have been inserted, from *Streptococcus thermophilus* and *Escherichia coli*. In these transformed plants it was observed that the expression of foreign genes increased the tolerance and accumulation of metals (Cd, Cr, Cu, Pb, and Zn) (Bennett et al., 2003; Reisinger et al., 2008). These results are attributed to the increase in GHS production due to the increase in the expression of γ-GCS and GS enzymes. In Fig. 22.4, the main stages for the creation of transgenic plants of environmental interest are presented.

Like GHS, phytochelatins are a class of posttranslationally synthesized peptides, playing an important role in tolerance to HM, by chelating these contaminants and reducing their concentrations. The gene that expresses phytochelatin synthase *AtPCS1*, from *Arabidopsis*, was introduced into tobacco and mustard plants. The transformed plants showed an increase in the tolerance and accumulation of metals such as As, Cd, Pb, and Zn (Guo et al. 2008; Gasic and Korban,

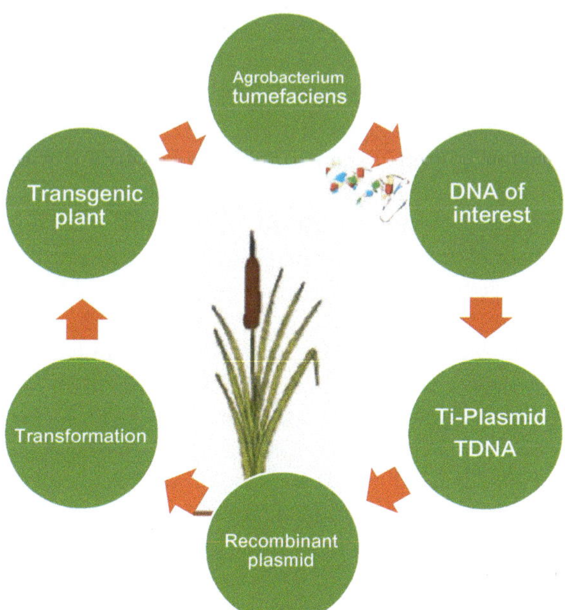

FIG. 22.4 Main stages for the creation of transgenic plants.

2007a, b; Zanella et al., 2016). In the work of Guo et al. (2008), it was showed that in transformed *Arabidopsis thaliana* plants that overexpress the *AsPCS1* and GS genes, from *Allium sativum* and *Saccharomyces cerevisiae,* respectively, led to a high production of phytochelatins, consecutively increasing the accumulation and tolerance of As and Cd. On the other hand, the phytoremediation potential of the *Nicotiana glauca* plant transformed with the *TaPCS1* gene, from *Triticum aestivum*, was evaluated. The overexpression of the gene gave the plant up to 9 and 36 times more accumulation of Cd and Pb under hydroponic conditions, and 3 and 6 times increase in soils compared to their respective controls (Martínez et al., 2006). In summary, it can be stated that transformed plants that synthesize phytochelatins are suitable to be used in phytoremediation.

Due to the sequestration with peptides rich in cysteine, plants tolerate HM (e.g., phytochelatins). Cysteine synthase catalyzes the final reaction for cysteine synthesis. Therefore, in the work of Kawashima et al. (2004), the tolerant capacity of transformed tobacco plants that overexpress cysteine synthase cDNA against HM such as Cd, Se, Ni, Pb, and Cu, was evaluated. The results of the present study showed that transgenic plants increased tolerance toward these metals and the accumulation of Cd in shoot was improved. In addition, previous studies demonstrated that in tobacco plants transformed with the *RCS1* gene, the cytosolic gene for rice cysteine synthase, the activity of cysteine synthase was three times higher than in control plants when exposed to cadmium. Consequently, the transgenic plants exhibited tolerance and greater growth than the control plants (Harada et al., 2001). Both studies show that the capacity for tolerance and accumulation to HM improves in transformed plants that synthesize a greater quantity of compounds rich in cysteine.

MTs are cysteine-rich metal-binding proteins. These proteins have also been tested for phytoremediation purposes. It has been shown that in transformed plants, the overexpression of genes encoding MTs produced an increase in tolerance to Cu. Lv et al. (2013) isolated two MT genes, *BcMT1* and *BcMT2,* from the *Brassica campestris* plant. Subsequently, they induced the ectopic expression of *BcMT1* and *BcMT2* in *A. thaliana.* The expression of the *BcMT1* gene was induced by Cd and Cu, while the *BcMT2* gene was induced by only Cd. The expression of both genes increased the tolerance of Cd and Cu and increased the Cu concentration in shoots. Finally, transformed *Arabidopsis* accumulated less ROS than wild plants. Similarly, yeast MT (*CUP1*) has been reported to increase metal sequestration in plants. In *CUP1*-transformed plants, an accumulation of up to seven times more copper was observed in older leaves than in younger leaves during Cu stress (Thomas et al., 2003). Previous studies support that the expression of genes for the synthesis of MT in transgenic plants could be a strategy to improve the phytoremediation of metallic contaminants.

Studies have shown the efficacy of the enzymes MerB (organomercurial lyase) and MerA (mercury reductase) in detoxifying highly toxic mercury compounds. The native bacterial enzyme mercury reductase (*mer*A) is responsible for the transport of Hg^{2+} to the cell and its subsequent reduction to Hg0 and

MerB transforms methylmercury into less toxic nonbiomagnified ionic mercury. An example of this is the work by Ruiz et al. (2003), who integrated a native operon that contained the *mer*A and *mer*B genes in the chloroplast genome of tobacco plants. The transgenic tobacco plants resulted with a high increase in tolerance to the compound phenylmercuric acetate (PMA), when grown in 400 μM of PMA. Furthermore, Haque et al. (2010), in their work evaluated the viability of transgenic tobacco designed to express *mer*A. The transformed tobacco plants produced a high amount of merA protein in leaves and showed a phenotype more resistant to $HgCl_2$ than wild plants. Furthermore, they indicate that transgenic tobacco expressing merA volatilized more Hg than wild-type plants. On the other hand, it is also known that the *mer*C gene from *Acidithiobacillus ferrooxidans* works as an Hg absorption pump. This is supported by the work carried out by Sasaki et al. (2006), who transformed *A. thaliana* and tobacco plants incorporating the *mer*C gene. Overexpression of the *mer*C gene made transformed plants hypersensitive to Hg^{2+} and accumulated approximately twice as many ions as wild plants. Considering the results shown in the previous works, it can be shown that transgenic plants that overexpress genes for *mer*A and *mer*B enzymes and *mer*C transporter genes may be reasonable candidates for phytoremediating sites contaminated by Hg.

As a primary defense mechanism, plants release metal-binding compounds that counteract the toxic damage of metals. It is well known that nicotianamine (NA) is a nonproteinogenic amino acid, chelator of HM, it is synthesized from the trimerization of S-adenosylmethionine catalyzed by NA synthase (*NAS*). In the work of Kim et al. (2005) established transgenic *Arabidopsis* plants and tobacco overexpressed the *NAS* gene from *Hordeum vulgare*. The overexpression of the *NAS* gene resulted in an increase in NA in the transgenic plants, conferring on the plants an increase in tolerance to high levels of HM, especially Ni. On the other hand, the study carried out by Pianelli et al. (2005) expressed a TcNAS1 cDNA from the hyperaccumulator *Thlaspi caerulescens* in *A. thaliana*. Transformed plants that expressed TcNAS1 accumulated up to 100 times more NA than control plants. The increase in NA production was correlated with an increase in tolerance to Ni. Finally, in transgenic tobacco plants an *NAS* gene from *A. thaliana* was ectopically expressed. The transformed tobacco plants showed an increase in NA levels, consequently leading to an increase in the accumulation of Zn and Mn, and an improvement in the tolerance of Ni (Douchkov et al., 2005). Together, these studies support that NA plays an important role in the detoxification of HM and may be a suitable tool for phytoremediation.

Finally, metal transporters are necessary for the uptake, translocation, and compartmentalization of metals. An example of this are the transporters *PHT1* and *PHT7*, which are membrane phosphate transporters and caused hypersensitivity to As by being overexpressed in tobacco due to increased absorption and accumulation of As (LeBlanc et al., 2013). The *MPR7* transporter present in the tonoplast and in the plasma membrane is overexpressed in tobacco plants, increasing the tolerance and concentration of Cd in the vacuoles of the leaves (Wojas et al., 2009). In the same way, CAX2 and CAX4 antiporters of the tonoplast were expressed in tobacco plants showing an increase in tolerance to high levels of toxic Cd and Zn (Korenkov et al., 2007). Taken together, these studies indicate that gene expression for HM transporters is a reasonable strategy for use in phytoremediation of HM-contaminated sites.

22.5 Conclusion

Plants are able to remove and accumulate in their plant tissues a wide variety of HM, to tolerate high concentrations; plants develop various biochemical mechanisms of tolerance or accumulation. To facilitate phytoremediation plants develop various biochemical mechanisms to tolerate high concentrations of HM. However, to improve phytoremediation, genetic engineering tools can be used to enhance the capacity of plants in phytoremediation mechanisms.

References

Ahmad, P., Nabi, G., Ashraf, M., 2011. Cadmium-induced oxidative damage in mustard [*Brassica juncea* (L.) Czern. & Coss.] plants can be alleviated by salicylic acid. S. Afr. J. Bot. 77, 36–44. https://doi.org/10.1016/j.sajb.2010.05.003.

Ali, H., Khan, E., Sajad, M.A., 2013. Phytoremediation of heavy metals-concepts and applications. Chemosphere 91, 869–881. doi:10.1016/j.chemosphere.2013.01.075.

Anjum, N.A., Ahmad, I., Mohmood, I., Pacheco, M., Duarte, A.C., Pereira, E., et al., 2012. Modulation of glutathione and its related enzymes in plants responses to toxic metals and metalloids—a review. Environ. Exp. Bot. 75, 307–324.

Anjum, N.A., Sofo, A., Scopa, A., Roychoudhury, A., Gill, S.S., Iqbal, M., Lukatkin, A.S., Pereira, E., Duarte, A.C., Ahmad, I., 2015. Lipids and proteins–major targets of oxidative modifications in abiotic stressed plants. Environ. Sci. Pollut. Res. 22, 4099–4121. doi:10.1007/s11356-014-3917-1.

Arazi, T., Sunkar, R., Kaplan, B., Fromm, H.A., 1999. Tobacco plasma membrane calmodulin binding transporter confers Ni_+ tolerance and $Pb2_+$ hypersensitivity in transgenic plants. Plant J. 20, 171–182. doi:10.1046/j.1365-313x.1999.00588.x.

Balestrazzi, A., Botti, S., Zelasco, S., Biondi, S., Franchin, C., Calligari, P., Racchi, M., Turchi, A., Lingua, G., Berta, G., Carbonera, D., 2009. Expression of the PsMTA1 gene in white poplar engineered with the MAT system is associated with heavy metal tolerance and protection against 8-hydroxy-2′-deoxyguanosine mediated-DNA damage. Plant Cell Rep 28, 1179–1192. doi:10.1007/s00299-009-0719-x.

Bañuelos, G., LeDuc, D.L., Pilon-Smits, E.A.H., Terry, N, 2007. Transgenic Indian mustard overexpressing selenocysteine lyase or selenocysteine methyltransferase exhibit enhanced potential for selenium phytoremediation under field conditions. Environ. Sci. Technol 41 (2), 599–605. https://doi.org/10.1021/es061152i.

Bañuelos, G., Terry, N., LeDuc, D.L., Pilon-Smits, E.A.H., Mackey, B, 2005. Field trial of transgenic Indian mustard plants shows enhanced phytoremediation of selenium-contaminated sediment. Environ. Sci. Technol 39 (6), 1771–1777. doi:10.1021/es049035f.

Barone, P., Rosellini, D., Lafayette, P., Bouton, J., Veronesi, F., Parrott, W., 2008. Bacterial citrate synthase expression and soil aluminum tolerance in transgenic alfalfa. Plant Cell Rep. 27 (5), 893–901. doi:10.1007/s00299-008-0517-x.

Bennett, L.E., Burkhead, J.L., Hale, K.L., Terry, N., Pilon, M., Pilon-Smits, E.A., 2003. Analysis of transgenic Indian mustard plants for phytoremediation of metal-contaminated mine tailings. J. Environ. Qual 32, 432–440. doi:10.2134/jeq2003.4320.

Bizily, S.P., Kim, T., Kandasamy, M.K., Meagher, R.B., 2003. Subcellular targeting of methylmercury lyase enhances its specific activity for organic mercury detoxification in plants. Plant Physiol 131, 463–471. doi:10.1104/pp.010124.

Brichkova, G.G., Shishlova, A.M., Maneshina, T.V., Kartel, N.A., 2007. Tolerance to aluminum in genetically modified tobacco plants. Cytol. Genet 41, 151–155. https://doi.org/10.3103/S0095452707030036.

Cai, Z., Xian, P., Wang, H., Lin, R., Lian, T., Cheng, Y., Ma, Q., Nian, H., 2020. Transcription factor GmWRKY142 confers cadmium resistance by up-regulating the cadmium tolerance 1-like genes. Front. Plant Sci. 3 (11), 724. doi:10.3389/fpls.2020.00724.

Che, D., Meagher, R.B., Heaton, A.C., Lima, A., Rugh, C.L., Merkle, S.A., 2003. Expression of mercuric ion reductase in Eastern cottonwood (*Populus deltoides*) confers mercuric ion reduction and resistance. Plant Biotechnol. J 1, 311–319. https://doi.org/10.1046/j.1467-7652.2003.00031.x.

Chen, J., Yang, L., Gu, J., Bai, X., Ren, Y., Fan, T., Han, Y., Jiang, L., Xiao, F., Liu, Y., Cao, S., 2015. MAN3 gene regulates cadmium tolerance through the glutathione-dependent pathway in *Arabidopsis thaliana*. New Phytol 205, 570–582. doi:10.1111/nph.13101.

Chen, J., Yang, L., Yan, X., Liu, Y., Wang, R., Fan, T., Ren, Y., Tang, X., Xiao, F., Liu, Y., Cao, S., 2016. Zinc-finger transcription factor ZAT6 positively regulates cadmium tolerance through the glutathione-dependent pathway in Arabidopsis. Plant Physiol 171, 707–719. doi:10.1104/pp.15.01882.

Chen, Y., Xu, W., Shen, H., Yan, H., Xu, W., He, Z., Ma, M., 2013. Engineering arsenic tolerance and hyperaccumulation in plants for phytoremediation by a PvACR3 transgenic approach. Environ. Sci. Technol 47, 9355–9362. doi:10.1021/es4012096.

Chibuike, G.U., Obiora, S.C., 2014. Heavy metal polluted soils: effect on plants and bioremediation methods. Appl. Environ. Soil Sci. 2014, 752708. Article ID. https://doi.org/10.1155/2014/752708.

Cobbett, C.S., Golsbrougth, P.B., 2000. Mechanisms of metal resistance: phytochelatins and metalothioneins. In: Raskin, I., Ensley, B.D. (Eds.), Phytoremediation of Toxic Metals: Using Plants to Clean Up the Environment. John Wiley, pp. 247–269.

Dada, E., Njoku, K., Osuntoki, A., Akinola, M., 2015. A review of current techniques of physico-chemical and biological remediation of heavy metals polluted soil. Ethiop. J. Environ. Stud. Manag. 8 (5), 606. doi:10.4314/ejesm.v8i5.13.

Daghan, H., Arslan, M., Uygur, V., Koleli, N., 2013. Transformation of tobacco with ScMTII gene-enhanced cadmium and zinc accumulation. Clean Soil, Air Water 41, 503–509. https://doi.org/10.1002/clen.201200298.

Dai, J., Balish, R., Meagher, R.B., Merkle, S.A., 2009. Development of transgenic hybrid sweetgum (*Liquidambar styraciflua* × *L. formosana*) expressing γ-glutamylcysteine synthetase or mercuric reductase for phytoremediation of mercury pollution. New Forests 38, 35–52. https://doi.org/10.1007/s11056-008-9130-7.

Das, N., Bhattacharya, S., Maiti, M.K., 2016. Enhanced cadmium accumulation and tolerance in transgenic tobacco overexpressing rice metal tolerance protein gene OsMTP1 is promising for phytoremediation. Plant Physiol. Biochem 105, 297–309. doi:10.1016/j.plaphy.2016.04.049.

dela Fuente, J.M., Ramirez-Rodriguez, Y., Cabrera-Ponce, J.L., Herrera Estrella, L., 1997. Aluminium tolerance in transgenic plants by alteration of citrate synthesis. Science 276, 1566–1568. doi:10.1126/science.276.5318.1566.

Deng, W., Luo, K., Li, Z., Yang, Y., Hu, N., Wu, Y., 2009. Overexpression of *Citrus junos* mitochondrial citrate synthase gene in *Nicotiana benthamiana* confers aluminum tolerance. Planta 230 (2), 355–365. doi:10.1007/s00425-009-0945-z.

Derakhshan-Nejad, Z., Chae Jung, M., Ki-Hyun, K, 2017. Remediation of soils contaminated with heavy metals with an emphasis on immobilization technology. Environ. Geochem. Health 40, 1–2. doi:10.1007/s10653-017-9964-z.

Dhankher, O.P., Li, Y., Rosen, B.P., Shi, J., Salt, D., Senecoff, J.F., Sashti, N.P., Meagher, R.B., 2002. Engineering tolerance and hyperaccumulation of arsenic in plants by combining arsenate reductase and γ-glutamylcysteine synthetase expression. Nat. Biotechnol 20, 1140–1145. https://doi.org/10.1038/nbt747.

Dhankher, O.P., Shasti, N.A., Rosen, B.P., Fuhrmann, M., Meagher, R.B., 2003. Increased cadmium tolerance and accumulation by plants expressing bacterial arsenate reductase. New Phytol 159, 431–441. https://doi.org/10.1046/j.1469-8137.2003.00827.x.

Douchkov, D., Gryczka, C., Stephan, U.W., Hell, R., Bäumlein, H., 2005. Ectopic expression of nicotianamine synthase genes results in improved iron accumulation and increased nickel tolerance in transgenic tobacco. Plant Cell Environ 28, 365–374. https://doi.org/10.1111/j.1365-3040.2005.01273.x.

Duffus, J.H., 2001. Heavy metals – a meaningless term. (IUPAC Technical Report).Pure Appl. Chem.74, 793–807.

Ellis, D.R., Sors, T., Brunk, D.G., Albrecht, C., Orser, C., Lahner, B., Wood, K.W., Harris, H.H., Pickering, I.G., Salt, D.E., 2004. Production of Se-methylselenocysteine in transgenic plants expressing selenocysteine methyltransferase. BMC Plant Biol 4 (1), 1. doi:10.1186/1471-2229-4-1.

Evans, K.M., Gatehouse, J.A., Lindsay, W.P., Shi, J., Tommey, A.M., Robinson, N.J., 1992. Expression of the pea metallothionein like gene PsMTA in *Escherichia coli* and *Arabidopsis thaliana* and analysis of trace metal ion accumulation: implications of PsMTA function. Plant Mol Biol 20, 1019–1028. doi:10.1007/BF00028889.

Ezaki, B., Gardner, R.C., Ezaki, Y., Matsumoto, H., 2000. Expression of aluminium induced genes in transgenic Arabidopsis plants can ameliorate aluminium stress and/or oxidative stress. Plant Physiol 122, 657–665. doi:10.1104/pp.122.3.657.

Freeman, J.L., Persans, M.W., Nieman, K., Albrecht, C., Peer, W., Pickering, I .J., Salt, D.E., 2004. Increased glutathione biosynthesis plays a role in nickel tolerance in Thlaspi nickel hyperaccumulators. Plant Cell 16, 2176–2191. https://doi.org/10.1105/tpc.104.023036.

Garbisu, C., Alkorta, I., 2001. Phytoextraction: a cost effective plant-based technology for the removal of metals from the environment. Bioresour. Technol 77 (3), 229–236. doi:10.1016/S0960-8524(00)00108-5.

Gasic, K., Korban, S.S., 2007a. Transgenic Indian mustard (*Brassica juncea*) plants expressing an Arabidopsis phytochelatin synthase (AtPCS1) exhibit enhanced As and Cd tolerance. Plant Mol. Biol. 64 (4), 361–369. doi:10.1007/s11103-007-9158-7.

Gasic, K., Korban, S.S., 2007b. Expression of Arabidopsis phytochelatin synthase in Indian mustard (*Brassica juncea*) plants enhances tolerance for Cd and Zn. Planta 225 (5), 1277–1285. doi:10.1007/s00425-006-0421-y.

Ghori, N.-H., Ghori, T., Hayat, M.Q., Imadi, S.R., Gul, A., Altay, V., Ozturk, M., 2019. Heavy metal stress and responses in plants. Int. J. Environ. Sci. Technol 16, 1807–1828. doi:10.1007/s13762-019-02215-8.

Goto, F., Yoshihara, T., Saiki, H., 1998. Iron accumulation in tobacco plants expressing soybean ferritin gene. Transgen. Res 7, 173–180. https://doi.org/10.1023/A:1008836812714.

Grichko, V.P., Filby, B., Glick, B.R, 2000. Increased ability of transgenic plants expressing the bacterial enzyme ACC deaminase to accumulate Cd, Co, Cu, Ni, Pb, and Zn. J. Biotechnol 81, 45–53. doi:10.1016/s0168-1656(00)00270-4.

Grispen, V.M., Hakvoort, H.W., Bliek, T., Verkleij, J.A., Schat, H., 2011. Combined expression of the *Arabidopsis metallothionein* MT2b and the heavy metal transporting ATPase HMA4 enhances cadmium tolerance and the root to shoot translocation of cadmium and zinc in tobacco. Environ. Exp. Bot 72, 71–76. https://doi.org/10.1016/j.envexpbot.2010.01.005.

Gu, C.S., Liu, L.Q., Deng, Y.M., Zhu, X.D., Huang, S.Z., Lu, X.Q., 2015. The heterologous expression of the *Iris lactea* var. chinensis type 2 metallothionein IlMT2b gene enhances copper tolerance in *Arabidopsis thaliana*. Bull. Environ. Contamin. Toxicol 94, 247–253. doi:10.1007/s00128-014-1444-x.

Gu, C.S., Liu, L.Q., Zhao, Y.H., Deng, Y.M., Zhu, X.D., Huang, S.Z., 2014. Overexpression of *Iris lactea* var. chinensis metallothionein llMT2a enhances cadmium tolerance in *Arabidopsis thaliana*. Ecotoxicol. Environ. Saf 105, 22–28. doi:10.1016/j.ecoenv.2014.04.002.

Guo, J., et al., 2008. Overexpressing GSH1 and AsPCS1 simultaneously increases the tolerance and accumulation of cadmium and arsenic in *Arabidopsis thaliana*. Chemosphere 72 (7), 1020–1026.

Gustin, J.L., Loureiro, M.E., Kim, D., Na, G., Tikhonova, M., Salt, D.E., 2009. MTP1-dependent Zn sequestration into shoot vacuoles suggests dual roles in Zn tolerance and accumulation in Zn-hyperaccumulating plants. Plant J 57, 1116–1127. https://doi.org/10.1111/j.1365-313X.2008.03754.x.

Hameed, A., Rasool, S., Azooz, M.M., Hossain, M.A., Ahanger, M.A., Ahmad, P., 2016. Heavy metal stress: plant responses and signaling. In: Ahmad, P. (Ed.), Plant Metal Interaction. Elsevier, Amsterdam, pp. 557–583. doi:10.1016/b978-0-12-803158-2.00024-2.

Haque, S., Zeyaullah, M., Nabi, G., Srivastava, P.S., Ali, A., 2010. Transgenic tobacco plant expressing environmental *E. coli* merA gene for enhanced volatilization of ionic mercury. J. Microbiol. Biotechnol 20 (5), 917–924. https://doi.org/10.4014/jmb.1002.02001.

Harada, E., Choi, Y.E., Tsuchisaka, A., Obata, H., Sano, H., 2001. Transgenic tobacco plants expressing a rice cysteine synthase gene are tolerant to toxic levels of cadmium. J. Plant. Physiol 158 (5), 655–661.

He, J., Li, H., Ma, C., Zhang, Y., Polle, A., Rennenberg, H., Cheng, X., Zhi-Bin, L., 2014. Overexpression of bacterial γ-glutamylcysteine synthetase mediates changes in cadmium influx, allocation and detoxification in poplar. New Phytol 205 (1), 240–254. https://doi.org/10.1111/nph.13013.

He, Y.K., Sun, J.G., Feng, X.Z., Czakó, M., Márton, L., 2001. Differential mercury volatilization by tobacco organs expressing a modified bacterial merA gene. Cell Res 11, 231–236. doi:10.1038/sj.cr.7290091.

Heaton, A.C.P., Rugh, C.L., Kim, T., Meagher, R.B, 2003. Toward detoxifying mercury-polluted aquatic sediments with rice genetically engineered for mercury resistance. Environ. Toxicol. Chem 22, 2940–2947.

Hsieh, J.L., Chen, C.Y., Chiu, M.H., Chein, M.F., Chang, J.S., Endo, G., Huang, C.C., 2009. Expressing a bacterial mercuric ion binding protein in plant for phytoremediation of heavy metals. J. Hazard. Mater 161, 920–925. doi:10.1016/j.jhazmat.2008.04.079.

Jagtap, U.B., Bapat, V.A., 2015. Genetic engineering of plants for heavy metal removal from soil. In: Sherameti, I., Varma, A. (Eds.), Heavy Metal Contamination of Soils. Soil Biology. Springer, Cham, pp. 443–470. doi:10.1007/978-3-319-14526-6_22.

Jarup, L., 2003. Hazards of heavy metal contamination. Br. Med. Bull 68, 167–182. doi:10.1093/bmb/ldg032.

Kapoor, D., Sharma, R., Handa, N., Kaur, H., Rattan, A., Yadav, P., Gautam, V., Kaur, R., Bhardwaj, R., 2015. Redox homeostasis in plants under abiotic stress: role of electron carriers, energy metabolism mediators and proteinaceous thiols. Front. Environ. Sci 3, 13. https://doi.org/10.3389/fenvs.2015.00013.

Kawashima, C.G., Noji, M., Nakamura, M., Ogra, Y., Suzuki, K.T., Saito, K., 2004. Heavy metal tolerance of transgenic tobacco plants over-expressing cysteine synthase. Biotechnol. Lett. 26 (2), 153–157.

Khoudi, H., Maatar, Y., Gouiaa, S., Masmoudi, K., 2012. Transgenic tobacco plants expressing ectopically wheat H+-pyrophosphatase (H+-PPase) gene TaVP1 show enhanced accumulation and tolerance to cadmium. J. Plant Physiol 169, 98–103. https://doi.org/10.1016/j.jplph.2011.07.016.

Kim, D.Y., Bovet, L., Kushnir, S., Noh, E.W., Martinoia, E., Lee, Y., 2006. AtATM3 is involved in heavy metal resistance in Arabidopsis. Plant Physiol. 140, 922–932.

Kim, S., Takahashi, M., Higuchi, K., Tsunoda, K., Nakanishi, H., Yoshimura, E., Mori, S., Nishizawa, N.K., 2005. Increased nicotianamine biosynthesis confers enhanced tolerance of high levels of metals, in particular nickel, to plants. Plant Cell Physiol 46, 1809–1818. https://doi.org/10.1093/pcp/pci196.

Korenkov, V., Hirschi, K., Crutchfield, J.D., Wagner, G.J., 2007. Enhancing tonoplast Cd/H antiport activity increases Cd, Zn, and Mn tolerance, and impacts root/shoot Cd partitioning in *Nicotiana tabacum* L. Planta 226, 1379–1387. doi:10.1007/s00425-007-0577-0.

Kramer, U., Talke, I.N., Hanikenne, M., 2007. Transition metal transport. FEBS Lett. 581, 2263–2272. https://doi.org/10.1016/j.febslet.2007.04.010.

Kunihiro, S., Saito, T., Matsuda, T., Inoue, M., Kuramata, M., Taguchi-Shiobara, F., Youssefian, S., Berberich, T., Kusano, T., 2013. Rice DEP1, encoding a highly cysteine-rich G protein γ subunit, confers cadmium tolerance on yeast cells and plants. J. Exp. Bot 64, 4517–4527. https://doi.org/10.1093/jxb/ert267.

LeBlanc, M.S., McKinney, E.C., Meagher, R.B., Smith, A.P., 2013. Hijacking membrane transporters for arsenic phytoextraction. J. Biotechnol. 163, 1–9. doi:10.1016/j.jbiotec.2012.10.013.

LeDuc, D.L., Tarun, A.S., Montes-Bayon, M., Meija, J., Malit, M.F., Wu, C.P., AbdelSamie, M., Chih-Yuan, C., Tagmount, A., deSouza, M., Neuhierl, B., Böck, A., Caruso, J., Terry, N., 2004. Overexpression of selenocysteine methyltransferase in Arabidopsis and Indian mustard increases selenium tolerance and accumulation. Plant Physiol 135, 377–383. https://doi.org/10.1104/pp.103.026989.

Lee, B.R., Hwang, S., 2015. Over-expression of NtHb1 encoding a non-symbiotic class 1 hemoglobin of tobacco enhances a tolerance to cadmium by decreasing NO (nitric oxide) and Cd levels in *Nicotiana tabacum*. Environ. Exp. Bot 113, 18–27. doi:10.1016/j.envexpbot.2015.01.003.

Lee, J., Bae, H., Jeong, J., Lee, J.Y., Yang, Y.Y., Hwang, I., Martinoia, E., Lee, Y., 2003. Functional expression of a bacterial heavy metal transporter in Arabidopsis enhances resistance to and decreases uptake of heavy metals. Plant Physiol 133, 589–596. doi:10.1104/pp.103.021972.

Liu, D., An, Z., Mao, Z., Ma, L., Lu, Z., 2015. Enhanced heavy metal tolerance and accumulation by transgenic sugar beets expressing *Streptococcus thermophilus* StGCS-GS in the presence of Cd, Zn and Cu alone or in combination. PLoS One 10 (6). https://doi.org/10.1371/journal.pone.0128824.

Lv, Y., Deng, X., Quan, L., Xia, Y., Shen, Z., 2013. Metallothioneins BcMT1 and BcMT2 from *Brassica campestris* enhance tolerance to cadmium and copper and decrease production of reactive oxygen species in *Arabidopsis thaliana*. Plant Soil 367, 507–519. https://doi.org/10.1007/s11104-012-1486-y.

Maestri, E., Marmiroli, N., 2011. Transgenic plants for phytoremediation. Int. J. Phytoremediation 13, 264–279. https://doi.org/10.1080/15226514.2011.568549.

Marques, A.P.G.C., Rangel, A.O.S.S., Castro, P.M.L, 2009. Remediation of heavy metal contaminated soils: phytoremediation as a potentially promising clean-up technology. Crit. Rev. Environ. Sci. Technol. 39 (8), 622–654.

Martínez, M., Bernal, P., Almela, C., Vélez, D., García-Agustín, P., Serrano, R., Navarro-Aviñó, J., 2006. An engineered plant that accumulates higher levels of heavy metals than *Thlaspi caerulescens*, with yields of 100 times more biomass in mine soils. Chemosphere 64 (3), 478–485. https://doi.org/10.1016/j.chemosphere.2005.10.044.

Misra, S., Gedamu, L., 1989. Heavy metal tolerant transgenic *Brassica napus* L and *Nicotiana tabacum* L plants. Theor. Appl. Genet 78, 16–18. https://doi.org/10.1007/BF00288793.

Munzuroglu, O., Geckil, H., 2002. Effects of metals on seed germination, root elongation, and coleoptile and hypocotyl growth in *Triticum aestivum* and *Cucumis sativus*. Arch. Environ. Contam. Toxicol 43, 203–213. doi:10.1007/s00244-002-1116-4.

Nagata, T., Ishikawa, C., Kiyono, M., Pan-Hou, H., 2006a. Accumulation of mercury in transgenic tobacco expressing bacterial polyphosphate. Biol. Pharma. Bull 29, 2350–2353. doi:10.1248/bpb.29.2350.

Nie, L., Shah, S., Rashid, A., Burd, G.I., Dixon, D.G., Glick, B.R., 2002. Phytoremediation of arsenate contaminated soil by transgenic canola and the plant growth-promoting bacterium *Enterobacter cloacae* CAL2. Plant Physiol. Biochem 40, 355–361. https://doi.org/10.1016/S0981-9428(02)01375-X.

Njoku, K.L., Akinola, M.O., Oboh, B.O., 2009. Phytoremediation of crude oil contaminated soil: the effect of growth of Glycine max on the physico-chemistry and crude oil contents of soil. Nat. Sci 12, 22–30.

Pan, A., Yang, M., Tie, F., Li, L., Chen, Z., Ru, B., 1994. Expression of mouse metallothionein-1-gene confers cadmium resistance in transgenic tobacco plants. Plant Mol. Biol 24, 341–351. doi:10.1007/BF00020172.

Pavlíková, D., Macek, T., M. Macková, M., Száková, J., Balík, J, 2004. Cadmium tolerance and accumulation in transgenic tobacco plants with a yeast metallothionein combined with a polyhistidine tail. Int. Biodeterior. Biodegrad 54 (2-3), 233–237. https://doi.org/10.1016/j.ibiod.2004.03.008.

Pilon, M., Owen, J.D., Garifullina, G.F., Kurihara, T., Mihara, H., Esaki, N., Pilon-Smits, E.A.H, 2003. Enhanced selenium tolerance and accumulation in transgenic Arabidopsis expressing a mouse selenocysteine lyase. Plant Physiol 131, 1250–1257. doi:10.1104/pp.102.014639.

Rajab, H., Khan, S.M., Wirtz, M., Malagoli, M., Qahar, F., Hell, R., 2020. Sulfur metabolic engineering enhances cadmium stress tolerance and root to shoot iron translocation in *Brassica napus* L. Plant Physiol. Biochem. 152, 32–43. https://doi.org/10.1016/j.plaphy.2020.04.017.

Reisinger, S., Schiavon, M., Terry, N., Pilon-Smits, E.A.H, 2008. Heavy metal tolerance and accumulation in Indian mustard (*Brassica Juncea* L.) expressing bacterial γ-glutamylcysteine synthetase or glutathione synthetase. Int. J. Phytoremediation 10 (5), 440–454. doi:10.1080/15226510802100630.

Rodríguez-Llorente, I.D., Pérez-Palacios, P., Doukkali, B., Caviedes, M.A., Pajuelo, E., 2010. Expression of the seed-specific metallothionein MT4a in plant vegetative tissues increases Cu and Zn tolerance. Plant Sci 178, 327–332. https://doi.org/10.1016/j.plantsci.2010.01.011.

Rugh, C.L., Wilde, H.D., Stack, N.M., Thompson, D.M., Summers, A.O., Meagher, R.B., 1996. Mercuric ion reduction and resistance in transgenic *Arabidopsis thaliana* plants expressing a modified bacterial merA gene. Proc. Natl. Acad. Sci. USA 93, 3182–3187. doi:10.1073/pnas.93.8.3182.

Ruiz, O.N., Hussein, H.S., Terry, N., Daniell, H., 2003. Phytoremediation of organomercurial compounds via chloroplast genetic engineering. Plant Physiol 132, 1344–1352. doi:10.1104/pp.103.020958.

Samuelsen, A.I., Martin, R.C., Mok, D.W.S., Machteld, C.M, 1998. Expression of the yeast FRE genes in transgenic tobacco. Plant Physiol 118, 51–58. doi:10.1104/pp.118.1.51.

Sasaki, Y., Hayakawa, T., Inoue, C., Miyazaki, A., Silver, S., Kusano, T., 2006. Generation of mercury-hyperaccumulating plants through transgenic expression of the bacterial mercury membrane transport protein MerC. Transgen. Res. 15, 615–625. https://doi.org/10.1007/s11248-006-9008-4.

Sekhar, K., Priyanka, B., Reddy, V.D., Rao, K.V., 2011. Metallothionein 1 (CcMT1) of pigeonpea (*Cajanus cajan*, L.) confers enhanced tolerance to copper and cadmium in *Escherichia coli* and *Arabidopsis thaliana*. Environ. Exp. Bot 72, 131–139. https://doi.org/10.1016/j.envexpbot.2011.02.017.

Shaul, O., Hilgemann, D.W., de Almedia-Engler, J., Van Montagu, M., Inze, D., Galili, G., 1999. Cloning and characterization of a novel $Mg2_+/H_+$ exchanger. EMBO J. 18, 3973–3980. doi:10.1093/emboj/18.14.3973.

Shim, D., Kim, S., Choi, Y.I., Song, W.Y., Park, J., Youk, E.S., Soon-Chun, J., Martinoia, E., Eun-Woon, N., Lee, Y., 2013. Transgenic poplar trees expressing yeast cadmium factor 1 exhibit the characteristics necessary for the phytoremediation of mine tailing soil. Chemosphere 90, 1478–1486. https://doi.org/10.1016/j.chemosphere.2012.09.044.

Singh, D., Nath, K., Sharma, Y.K., 2007. Response of wheat seed germination and seedling growth under copper stress. J. Environ. Biol 28, 409–414.

Singh, S., Parihar, P., Singh, R., Singh, V.P., Prasad, S.M., 2016. Heavy metal tolerance in plants: role of transcriptomics, proteomics, metabolomics, and ionomics. Front. Plant Sci. 6, 1143. doi:10.3389/fpls.2015.01143.

Song, W., Sohn, E.J., Martinoia, E., Lee, Y.J., Yang, Y., Jasinski, M., Forestier, C.I., Lee, Y, 2003. Engineering tolerance and accumulation of lead and cadmium in transgenic plants. Nat. Biotechnol 21, 914–919. https://doi.org/10.1038/nbt850.

Stearns, J.C., Shah, S., Greenberg, B.M., Dixon, D.G., Glick, B.R., 2005. Tolerance of transgenic canola expressing 1-aminocyclopropane-1-carboxylic acid deaminase to growth inhibition by nickel. Plant Physiol. Biochem 43, 701–708. https://doi.org/10.1016/j.plaphy.2005.05.010.

Tangahu, B.V., Sheikh Abdullah, S.R., Basri, H., Idris, M., Anuar, N., Mukhlisin, M., 2011. A review on heavy metals (As, Pb, and Hg) uptake by plants through phytoremediation. Int. J. Chem. Eng 2011. https://doi.org/10.1155/2011/939161.

Thomas, J.C., Davies, E.C., Malick, F.K., Endreszi, C., Williams, C.R., Abbas, M., Petrella, S., Swisher, K., Perron, M., Edwards, R., Osenkowski, P., Urbanczyk, N., Wiesend, W.N., Murray, K.S., 2003. Yeast metallothionein in transgenic tobacco promotes copper uptake from contaminated soils. Biotechnol. Prog 19, 273–280. doi:10.1021/bp025623q.

Turchi, A., Tamantini, I., Camussi, A.M., Racchi, M.L., 2012. Expression of a metallothionein A1 gene of *Pisum sativum* in white poplar enhances tolerance and accumulation of zinc and copper. Plant Sci 183, 50–56. doi:10.1016/j.plantsci.2011.11.008.

Tuteja, N., Sopory, S.K., 2008. Chemical signaling under abiotic stress environment in plants. Plant Signal Behav 3 (8), 525–536. doi:10.4161/psb.3.8.6186.

Uraguchi, S., Sone, Y., Kamezawa, M., Tanabe, M., Hirakawa, M., Nakamura, R., Takanezawa, Y., Kiyono, M., 2019. Ectopic expression of a bacterial mercury transporter MerC in root epidermis for efficient mercury accumulation in shoots of Arabidopsis plants. Sci. Rep. 9, 4347. doi:10.1038/s41598-019-40671-x.

Van Aken, B., 2008. Transgenic plants for phytoremediation: helping nature to clean up environmental pollution. Trends Biotechnol 26 (5), 225–227. doi:10.1016/j.tibtech.2008.02.001.

Van der Zaal, B.J., Neuteboom, L.W., Pinas, J.E., Chardonnen, A.N., Schat, H., Verkleij, J.A.C., Hooykaas, P.J.J, 1999. Overexpression of a novel Arabidopsis gene related to putative zinc transporter genes from animals can lead to enhanced zinc resistance and accumulation. Plant Physiol 119, 1047–1055. https://doi.org/10.1104/pp.119.3.1047.

Verret, F., Gravot, A., Auroy, P., Leonhardt, N., David, P., Nussaume, L., Vavasseur, A., Richaud, P., 2004. Overexpression of AtHMA4 enhances root-to-shoot translocation of zinc and cadmium and plant metal tolerance. FEBS Lett 576 (3), 306–312. https://doi.org/10.1016/j.febslet.2004.09.023.

Wang, Y., Greger, M., 2004. Clonal differences in mercury tolerance, accumulation, and distribution in willow. J. Environ. Qual 33 (5), 1779–1785. doi:10.2134/jeq2004.1779.

Wangeline, A.L., Burkhead, J.L., Hale, K.L., Lindblom, S.D., Terry, N., Pilon, M., Pilon-Smits, E.A.H, 2004. Overexpression of ATP sulfurylase in Indian mustard. J. Environ. Qual 33, 54–60. https://doi.org/10.2134/jeq2004.5400.

Wojas, S., Hennig, J., Plaza, S., Geisler, M., Siemianowski, O., Skłodowska, A., Ruszczyńska, A., Bulska, E., Antosiewicz, D.M., 2009. Ectopic expression of Arabidopsis ABC transporter MRP7 modifies cadmium root-to-shoot transport and accumulation. Environ. Pollution 157, 2781–2789. doi:10.1016/j.envpol.2009.04.024.

Wu, Q., Shigaki, T., Williams, K.A., Han, J.S., Kim, C.K., Hirschi, K.D., Park, S., 2011. Expression of an Arabidopsis $Ca2_{+}/H_{+}$ antiporter CAX1 variant in petunia enhances cadmium tolerance and accumulation. J. Plant Physiol 168, 167–173. doi:10.1016/j.jplph.2010.06.005.

Wuana, R.A, Okieimen, F.E, 2011. Heavy metals in contaminated soils: a review of sources, chemistry, risks and best available strategies for remediation. ISRN 2011. doi:10.5402/2011/402647.

Xia, Y., Liu, J., Wang, Y., Zhang, X., Shen, Z., Hu, Z., 2018. Ectopic expression of Vicia sativa Caffeoyl-CoA O-methyltransferase (VsCCoAOMT) increases the uptake and tolerance of cadmium in Arabidopsis. Environ. Exp. Bot. 145, 47–53. https://doi.org/10.1016/j.envexpbot.2017.10.019.

Xia, Y., Qi, Y., Yuan, Y., Wang, G., Cui, J., Chen, Y., Zhang, H., Shen, Z., 2012. Overexpression of *Elsholtzia haichowensis* metallothionein 1 (EhMT1) in tobacco plants enhances copper tolerance and accumulation in root cytoplasm and decreases hydrogen peroxide production. J. Hazard. Mater. 233-234, 65–71. https://doi.org/10.1016/j.jhazmat.2012.06.047.

Xiao, S., Gao, W., Chen, Q.F., Ramalingam, S., Chye, M.L., 2008. Overexpression of membrane-associated acyl-CoA-binding protein ACBP1 enhances lead tolerance in Arabidopsis. Plant J 54, 141–151. doi:10.1111/j.1365-313X.2008.03402.x.

Xu, Z., Wang, M., Xu, D., Xia, Z., 2019. The Arabidopsis APR2 positively regulates cadmium tolerance through glutathione-dependent pathway. Ecotoxicol. Environ. Saf 187, 109819. doi:10.1016/j.ecoenv.2019.109819.

Zanella, L., Fattorini, L., Brunetti, P., Roccotiello, E., Cornara, L., D'Angeli, S., Rovere, F.D., Cardarelli, M., Barbieri, M., di Toppi, L.S., Degola, F., Lindberg, S., Altamura, M.M., Falasca, G., 2016. Overexpression of AtPCS1 in tobacco increases arsenic and arsenic plus cadmium accumulation and detoxification. Planta 243 (3), 605–622. doi:10.1007/s00425-015-2428-8.

Zhang, J., Zhang, M., Shohag, M.J.I., Tian, S., Song, H., Feng, Y., Yang, X, 2016. Enhanced expression of SaHMA3 plays critical roles in Cd hyperaccumulation and hypertolerance in Cd hyperaccumulator *Sedum alfredii* Hance. Planta 243, 577–589. https://doi.org/10.1007/s00425-015-2429-7.

Zhang, Y., Liu, J., 2011. Transgenic alfalfa plants co-expressing glutathione S-transferase (GST) and human CYP2E1 show enhanced resistance to mixed contaminates of heavy metals and organic pollutants. J. Hazard. Mater. 189 (1–2), 357–362. https://doi.org/10.1016/j.jhazmat.2011.02.042.

Chapter 23

Economic feasibility of phytoremediation

Umair Riaz[a], Tabinda Athar[b], Umair Mustafa[c], Rashid Iqbal[d]
[a]*Soil and water Testing Laboratory for Research, Agriculture Department, Government of Punjab, Bahawalpur, Pakistan,* [b]*Institute of Soil and Environmental Sciences, University of Agriculture, Faisalabad, Pakistan,* [c]*Department of Soil Science, Faculty of Agriculture and environmental sciences, The Islamia university of, Bahawalpur, Pakistan,* [d]*Department of Agronomy, Faculty of Agriculture and environmental sciences, The Islamia university of Bahawalpur, Pakistan*

23.1 Introduction

Remediation of contaminated soil is a burning issue around the globe and has gained significant attention of researchers and scientific communities. Phytoremediation is an ecologically responsible alternate to the expansive and destructive physical and chemical methods (Hou et al., 2019). Use of short rotation coppice and energy crops is rapidly developing in many countries. The produced biomass is being used for energy supply and targeting renewable energy resources (Mleczek et al., 2010).

In the United States, there are currently $6 \times 3 \times 10^5$ brownfields that are polluted with a variety of toxic elements; within them $1 \times 3 \times 10^5$ ha of croplands have 55,000 ha of pasture, and 50,000 ha of forests that are currently considered as severely polluted, thus, land reclamation is urgently required. According to estimation, more than $150 \times 3 \times 10^6$ km^2 of farmlands have been contaminated with heavy metals in China, triggering the damage of about 10 million metric tons of grain per year, and producing a direct economic loss of approximately US$2.99 billion. In New Zealand and Australia, Cd accumulation in the offal of grazing animals made it unacceptable for human consumption, complicating the access of meat products to foreign markets. Unfortunately, pollution assessment is not a priority in many developing countries and therefore, available data regarding the area or the number of polluted sites are rarely available. Phytoremediation is offering novel opportunities to produce quality biomass in multiple land use systems and ecological zones, and is significantly beneficial to gain economic value. Phytoremediation removes heavy metals, and toxic pollutants from soil, thus physical, chemical, and biological activities of soil are greatly improved that in turn support excellent crop growth and development (Lewandowski et al., 2006). Besides the extraction rates, the economic

Phytoremediation. DOI: https://doi.org/10.1016/B978-0-323-89874-4.00025-X

feasibility of phytoremediation technique is greatly determined by the profitability. The economic value can be improved by production of renewable energy along with a good concern of phytoremediation (Yang et al., 2013). If the contamination level of soil with heavy metals is not severe, the benefits of phytoremediation can be evident in few years. The total cost and economic feasibility of phytoremediation are greatly dependent on various factors such as operational costs, initial capital, level of contamination, prevailing climatic conditions, environmental factors, and management practices (Wan et al., 2016). This technique is an ideal solution to mitigate the soil and landfill contamination problems. Cultivation of perennial energy grasses and short-rotation coppice is an excellent source of power and heat generation that can significantly improve the economic feasibility of phytoremediation. Moreover, recent developments in the bacterial species and transgenic plants are also significantly improving the economic feasibility of phytoremediation in the contaminated soils (Van Aken, 2008). The majority of the metal(loids) in the contaminated soil sites are valuable and useful natural resources that are naturally dispersed in the soils and their concentration is totally dependent on the extent of commercial and industrial activities and geographical conditions. Recovery of these elements is the best approach to increase economic feasibility for sustainable industrial and agricultural development. Phytoremediation offers low-cost soil cleaning and elemental recovery. Moreover, a larger amount of biomass containing a mixture of lignin, organic substances, cellulose, and hemicellulose containing substantial levels of calorific values is also helpful to generate significant profit by proper utilization of these substances for various purposes (Jiang et al., 2015).

Currently, high values and demands of valuable lands in the urban areas is a major driving force for the developers of residential areas, and landscaping sites to remediation of contaminated sites for the agricultural systems and urban development. Integrated phytoremediation is essentially important for addressing both renewable energy demands and land contamination. Biomass production by the hyperaccumulators having fewer metal contents (1%) have achieved around 500 USD from the recovered metals. Even the average annual biomass production of about 30 tons/ha containing noble metals, manganese, cadmium, tin, nickel, and cobalt is a financially viable approach (Robinson et al., 1997). Although phytoremediation techniques are mature technologies, still some uncertainties exist due to integrated approaches. Therefore, economic risks and challenges always exist with the integrated remediation techniques, but can be properly appraised by following efficient technologies and economic motivations (Danh et al., 2014). Table 23.1.

Phytoremediation technologies and approaches must be modulated by the involvement of conceptual changes about fundamental economic disciplines. The marketing potential of phytoremediation technologies is reported to billions of USD across the globe. Unfortunately, there has been no virtualization of phytoremediated materialized recoveries and products in the subsequent decades. The quantification of land capability and land use functioning in the biophysical

TABLE 23.1 Some important website addresses for phytoremediation companies/organizations all over the world.

S. No	Companies/organization	Country	Website addresses
01	OverVerde Ltd	Flintshire, United Kingdom	http://www.oververde.co.uk/
02	Hoskin Scientific Limited	Burnaby, BC, Canada	http://www.hoskin.ca/
03	Earthwork	Rome, Italy	https://www.earthworksaction.org/
04	CH2M Hill Companies, Ltd.	Englewood, Colorado, United States	https://www.ch2m.com/
05	Global Innovation Solution SRL	Bucharest, Romania	http://global-innovation.com.ro/?page_id5751&lang5en
06	Lemna Technologies, Inc.	Vadnais Heights, Minnesota 55110, United States	http://www.lemnatechnologies.com/
07	Treefree Biomass Solutions Inc	Seattle, Washington, United States	www.treefreebiomasssolutions.com
08	Edenspace Systems Corporation	Purcellville, Virginia, United States	www.edenspace.com
09	AquaBio Environmental Technologies, Inc.	Hawthorne, California, United States	http://aquabio.us/
10	Weston Solutions, Inc.	West Chester, Pennsylvania, United States	http://www.westonsolutions.com/
11	DWC DecRen Water Consult	Trier, Germany	http://www.dwc-water.com/
12	Edenspace systems corporation	Virginia, United States	http://www.edenspace.com/our-team/
13	Microhumus	54140 Jarville, France	www.microhumus.fr
14	Applied PhytoGenetics, Inc.	Athens, Georgia, United States	www.appliedphytogenetics.com
15	Clemmens Consult bvba	Watou, Belgium	http://www.clemmens-consult.be/
16	Waterloo Environmental Biotechnology Inc.	Hamilton, Ontario, Canada	http://www.waterlooenvironmentalbiotechnology.com/

(continued)

TABLE 23.1 (Cont'd)

S. No	Companies/organization	Country	Website addresses
17	Applied Natural Sciences, Inc	Ohio, United States	http://www.treemediation.com/
18	Greenvironment Innovation & Marketing	Taramani, Chennai, India	http://www.greenvironmentindia.com
19	Clean Biotec	Spain	http://www.cleanbiotec.com/en/
20	Tolunay-Wong Engineers, Inc. (TWE)	Houston, Texas, United States	https://www.tweinc.com/
21	EnviroLogek Technologies	Culver City, California, United States	http://envirologek.com/
22	The Association for Environmental Health and Sciences (AEHS)	Amherst, Massachusetts, United States	http://www.aehs.com/
23	Abydoz Environmental Inc.	Canada	https://www.abydoz.com/
24	OP-TECH Environmental Services, Inc	East Syracuse, New York, United States	http://www.op-tech.us
25	Wolverton Environmental Services, Inc.	Picayune, Mississippi, United States	www.wolvertonenvironmental.com
26	Ecolotree	United States	https://www.ecolotree.com/
27	Phytorem	Miramas, France	http://en.phytorem.com/
28	Symbio Greentech	Kolkata, India	http://www.symbiogreentech.com
29	ERIN Consulting Ltd	Regina, Saskatchewan, Canada	http://www.erinconsulting.com/
30	The Vetiver Network International (TVNI)	San Antonio, Texas, United States	http://www.vetiver.org/
31	Civil & Environmental Consultants, Inc. (CEC)	Pittsburgh, Pennsylvania, United States	http://www.cecinc.com/
32	Microbe Inotech Laboratories, Inc.	St. Louis, Missouri, United States	http://www.microbeinotech.com/

33	BioRemed AB	Sweden	http://bioremed.se
34	Remedy Engineering, Inc	Redding, California, United States	www.remedyengineering.com
35	Bioclear earth	Groningen, Netherlands	http://bioclearearth.nl/
36	Ecological Engineering Group	Massachusetts 024930003 United States	http://ecological-engineering.com
37	TEA, Inc	Baton Rouge, Louisiana, United States	http://www.teainconline.com/
38	Agua Inc	Boulder, Colorado, United States	http://aguainc.com/
39	Viridian Environmental consultants	Montclair, New Jersey, United States	http://www.viridianinc.com/
40	Oceans-ESU Ltd.	Barnsley, United Kingdom	http://www.oceans-esu.com/
41	Cox-Colvin & Associates, Inc. Environmental Services	Plain City, Ohio, United States	https://www.coxcolvin.com/
42	Argonne National Laboratory	Argonne, Illinois, United States	https://www.anl.gov
43	Trident Environmental Services & Technologies Inc.	Mississippi, United States	http://tridentenviro.com/about-us/
44	D. Glass Associates Inc	Massachusetts, United States	http://www.dglassassociates.com/
45	Thomas consultants, Inc.	Cincinnati, Ohio 45254, United States	http://thomas-consultants.com
46	Living Technologies, Inc.	Charlottesville, Virginia 22911 United States	http://www.livingmachines.com/Home.aspx
47	HydroQual Laboratories Ltd	Calgary, Alberta, Canada	http://www.hydroqual.ca/
48	Verdant Technologies, Inc.	Yardville, New Jersey, United States	www.verdanttech.com
49	Blue Planet Environmental Inc. (BPE)	Ontario, Canada	http://www.blueplanetenviro.com/
50	Sustainable Strategies DC	Washington DC, United States	http://www.strategiesdc.com/
51	Ceres	Boston, United States	https://www.ceres.org/

terms for economic feasibility of bioremediation techniques needs site-specific information about site conditions, landscaping information, and plant species to opt beneficial phytoremediation technologies for industries, authorities, and farmers (Ali et al., 2013). The economic costs are greatly dependent on the specific site conditions, thus economic assumptions about yield rates can only be determined by proper and in-depth studies of site-specific conditions. Using bioenergy-producing plants is relatively a new technology to clean the contaminated soils and offers multiple benefits from the remediation activities. A multitude of plant species are being used for concurrent phytoremediation and obtaining useful products such as bioplastics, alkaloids, charcoal, wood, fiber, biodiesel, and bioethanol. Obtaining bioenergy from phytoremediation plants has gained significant attention of scientists and is being considered as a versatile and clean source for energy production (Tripathi et al., 2016). Usually, the plants used for phytoremediation have high levels of nitrogen and phosphorus and produce good weight of biomass that can be used for composting and for the production of bioenergy and various important byproducts. Phytoremediation targeting the nitrogen and phosphorus will produce nutrient-rich compost and fertilizer that in turn can be used to improve soil fertility and productivity of the low yielding soils. Therefore, there is a good justification of further policy development and research for improved feasibility of phytoremediation on the contaminated soils (Burges et al., 2018). Despite some problems and limitations of phytoremediation, numerous effective techniques still exist for the efficient management of produced biomass from phytoremediation. Simply harvesting can efficiently reduce the reentry of 75% of contaminants. The absence of cost evaluation and detailed economic studies are major hurdles for the real-time application of phytoremediation technologies. This process is real time consuming and demands additional cost, but the results are significantly astonishing in terms of benefits. Therefore, there should be a proper site evaluation for phytoremediation technologies to maximize the profit on a sustainable basis (Ashraf et al., 2019). Table 23.2.

23.2 Emerging phytoremediation technologies

23.2.1 Buffers

Phytoremediation buffers are used to clean up low to moderately contaminated sites, mine waste, landfill chelate, and brownfields. The concept of buffer strips along with the riparian corridors is a suitable technique that employs planting suitable plant species in the stripes near the stream ranks for removing the contaminants in the surface water, and underground water streaming toward rivers. This process is extremely helpful to prevent the further spreading and interfusion of water contaminants to the groundwater resources. This system causes the removal of 90% eroded soil particles and 40%–72% of herbicides residues. This process is being used in many countries to reduce contamination caused by the pesticides and fertilizers. Growing Populus tree has been reported as the

TABLE 23.2 An indicative list of highly valued plant-based commercial phytoremediation for the phytoremediation business world.

S. No	Name of crop	Family	Tolerance	Economic use	Type of contaminated sites	References
01	*Lavandula vera* L.	Lamiaceae	Tolerant	Essential oil	Pb, Cd, and Zn contaminated soil	Angelova et al. (2015)
02	*Jasminum sambac*	Oleaceae	Tolerant	Floriculture	Cr-contaminated soil, tannery effluent irrigation	Ramasamy (1997), Anandhkumar (1998) and Mahimairaja et al. (2011)
03	*Leucena leucocephala*	Fabaceae	Tolerant	Pulpwood	Fly ash amended soil; dye-contaminated site	Rai et al. (2004), Pandey and Kumar (2013) and Jayanthy et al. (2014)
04	*Vetiveria zizanioides*	Poaceae	Most tolerant	Essential oil	Pb/Zn mine tailings; Fly ash dumps; Pb, Cu, and Zn sites; chromite-asbestos mine waste dumps	Wu et al. (2010), Chakraborty and Mukherjee (2010), Danh et al. (2014), Verma et al. (2014), Srivastava et al. (2014), Das et al. (2013) and Kumar and Maiti (2015)
05	*Azadirachta indica*	Meliaceae	Tolerant	Timber-wood	Sewage, filth drainage systems	Pandey and Prakash (2014)
06	*Chrysanthemum indicum*	Asteraceae	Tolerant	Floriculture	Cd-contaminated soil	Lal et al. (2008)
07	*Eucalyptus tereticornis*	Myrtaceae	Tolerant	Eucalyptus oil, timber-wood	Tannery sludge dumps; fly ash dumps	Shukla et al. (2011), Juwarkar and Jambhulkar (2008) and Ram et al. (2008)
08	*Nelumbo nucifera*	Nelumbonaceae	Hypertolerant	Floriculture	Domestic wastewater treatment, ex-mining area, mined out shallow pond	Kanabkaew and Puetpaiboon (2004) and Ashraf et al. (2011)
09	*Jasminum grandiflorum*	Oleaceae	Most tolerant	Floriculture	Cr-contaminated soil	Anandhkumar (1998) and Mahimairaja et al. (2011)

(continued)

TABLE 23.2 (Cont'd)

S. No	Name of crop	Family	Tolerance	Economic use	Type of contaminated sites	References
10	*Cymbopogon citratus*	Poaceae	Tolerant	Essential oil	Chromite-Asbestos mine waste dumps	Kumar and Maiti (2015)
11	*Salix spp.*	Salicaceae	Tolerant	Timber-wood	Cadmium contaminate soil; zinc, and cadmium contaminated soil	Lewandowski et al. (2006) and Konlechner et al. (2013)
12	*Miscanthus*	Poaceae	Most tolerant	Biodiesel	PAH-polluted soils; fly ash dumps	Didier et al. (2012) and Techer et al. (2012)
13	*Lavandula angustifolia Mill.*	Lamiaceae	Tolerant	Essential oil	Heavy metals (i.e., Cd, Pb, Cu, Mn, Zn, and Fe) contaminated soil	Zheljazkov and Nielsen (1996a)
14	*Chrysopogon zizanioides*	Poaceae	Tolerant	Essential oil	Chromite-asbestos mine waste dumps	Kumar and Maiti (2015)
15	*Calendula officinalis*	Asteraceae	Tolerant	Floriculture	Cd-contaminated soil	Ramana et al. (2013)
16	*Ocimum basilicum L.*	Lamiaceae	Tolerant	Essential oil	Heavy metals (i.e., Cd, Cu, and Pb) contaminated soil	Zheljazkov and Nielsen (1996a,b); Zheljazkov et al. (2006)
17	*Ricinus communis*	Euphorbiaceae	Most tolerant	Biodiesel	Fly ash–contaminated sites; mine tailings; DDTs and cadmium (Cd) cocontaminated soil; Cd-contaminated soil; heavy metal–contaminated soil	Pandey (2013), Olivares et al. (2013), Huang et al. (2011), Bauddh and Singh (2012), Rajkumar and Freitas (2008) and Pandey and Prakash (2014)
18	*Mentha arvensis*	Lamiaceae	Tolerant	Essential oil	Fly ash dumps, heavy metals (i.e., Cd, Pb, Cu, Mn, and Zn	Srivastava et al. (2014) Das et al. (2013) and Zheljazkov and Nielsen (1996b)
19	*Anethum graveolens L.*		Tolerant	Essential oil	Heavy metals (i.e., Cd, Cu, and Pb) contaminated soil	Zheljazkov et al. (2006)

20	*Mentha piperita L*	Lamiaceae	Tolerant	Essential oil	Fly ash and organic waste, heavy metals (i.e., Cd, Pb, Cu, Mn, and Zn)	Kumar and Patra (2012), Zheljazkov and Nielsen (1996b) and Zheljazkov et al. (2006)
21	*Jatropha curcas*	Euphorbiaceae	Most tolerant	Biodiesel	Heavy metal–contaminated soil; arsenic-, chromium-, and zinc-polluted soil; soil contaminated with used lubricating oil; lead- and cadmium-polluted soil; hexavalent chromium–polluted soil	Juwarkar et al. (2008), Yadav et al. (2009), Kumar et al. (2008), Agamuthu et al. (2010), Mangkoedihardjo and Surahmaida (2008) and Mangkoedihardjo et al. (2008)
22	*Populus spp.*	Salicaceae	Tolerant	Pulpwood	Tannery sludge dumps; boron-contaminated sites; fly ash dumps	Shukla et al. (2011), Robinson et al. (2007) and Das et al. (2013)
23	*Gladiolus grandiflorus*	Iridaceae	Most tolerant	Floriculture	Cd-contaminated soil	Lal et al. (2008)
24	*Dendrocalamus strictus*	Poaceae	Tolerant	Pulpwood	Tannery sludge dumps	Shukla et al. (2011) and Pandey and Prakash (2014)
25	*Gmelina arborea*	Verbenaceae	Tolerant	Timber-wood	Crude oil sites, heavy metal sites	Agbogidi et al. (2007) and Ogbonna and Ukiwe (2010)
26	*Nerium oleander*	Apocynaceae	Tolerant	Floriculture	Tannery effluent irrig	Anandhkumar (1998)
27	*Holoptelea integrifolia*	Ulmaceae	Tolerant	Timber-wood	Sewage, filth drainage systems	Pandey and Prakash (2014)
28	*Tagetes erecta*	Asteraceae	Tolerant	Floriculture	Cd-contaminated soil	Lal et al. (2008)
29	*Tectona grandis*	Lamiaceae	Tolerant	Timber wood	Crude oil sites, iron (Fe) ore tailings	Yenn et al. (2014) and Maiti et al. (2005)
30	*Jasminum auriculatum*	Oleaceae	Most tolerant	Floriculture	Cr-contaminated soil	Ramasamy (1997)

most efficient for phytoremediation buffers (Gabor, 2001).The process of phytoremediation by buffers can be accelerated by following some specific techniques and management practices. Some of the important management practices include selection of fast-growing and easy-to-maintain vegetation that can easily cause pollutants transformation to the nontoxic or relatively fewer toxic forms (DalCorso et al., 2019). Monocultures should always be avoided to reduce the associated risks of insects and disease attack. Moreover, field pot trials, and screening studies should also be conducted to check the suitability of plant species and mitigation potential. Furthermore, proper and efficient removal can only be ensued by timely harvesting and proper disposal. Thus, all these factors must be considered and managed for the effective removal of soil contaminants.

23.2.2 Phytoremediation vegetation

Vegetation cover has also been significantly important to remove soil contaminants by following self-growing and long-term vegetative system. Vegetation cover can be easily expanded over the contaminated sites and can be easily supported by minimum care. Mainly this system works by setting up barriers that cause prevention of further expansion of contaminants and pollutants. Usually, these vegetative covers are applied to the sites that are active hotspots for the further spreading of contaminants. This is economically feasible system due to self-regeneration and can control soil erosion (Vanlı, 2007). Accumulation of heavy metals in plants and their removal from soils is supported by various metabolic activities, cellular functions, and plant processes. These activities include mobilization of heavy metals, uptake by the roots, loading in xylem, transport from roots to shoots, cellular compartmentation, and heavy metal's sequestration. Mostly, the heavy metals are present in the insoluble form in the soil and generally plants are not able to uptake their high concentrations (Chen et al., 2019). Their bioavailability and uptake is increased by releasing different kinds of root exudates from specific plants that can alter the pH of rhizosphere and thus increases their solubility and bioavailability for effective remediation.

23.2.3 Controlling vegetation caps

Containment of soil pollutants by capping is a widely implemented remedial technique and is being used by different regulatory authorities to reduce the risks of different contaminants. Phytoremediation caps can potentially reduce the infiltration, and precipitation along with improved integrity of caps. These caps prevent the potential entries of chelates to the natural resources. These caps were tested in 1990 at the waste sites and are now being used across the globe. Implementation of already existing and high maintenance caps with the vegetation caps consists of minor cap repair or grading, adding high organic matter contents on the upper surface of cap with perforated and interbedded piping network, and plantation of specific plants and trees on all the cap area. Once this vegetation is established, its rooting system will perform two main functions: precipitation uptake and treatment of leachates. This precipitation generation is helpful to reduce leachate generation and cap integrity is maintained by the adjustment of landfill subsidence (Arnao, and Hernández-Ruiz, 2019). Furthermore, reduced volume of leachate is pumped through perforated pipes and is discharged to the cap system. Here this leachate is treated by bioremediation and phytoremediation. Bioremediation mainly occurs in the anaerobic and aerobic zones created by associations between bacteria and rooting system. However, phytoremediation is favored by the rooting system uptake and it causes treatment of leachate by various chemical and physical processes in the poplar trees (Xie, and Van, 2020). These vegetation caps are cost effective and permanent solutions that exactly meet the criteria of Environmental Protection Agency of United States. This technique is being considered as green and sustainable remediation process against other alternates.

23.2.4 Phytoremediation plantings

Phytoremediation plantings have been efficiently used for remediation of organic and inorganic contaminants in ground water and soil. Various plants are helpful to reduce the contaminants; however, all plants are not specialized for removing all types of pollutants. Instead some species are good at removing some specific contaminants, so phytoremediation plantings should be carefully selected according to prevailing climatic conditions, geography, land use capability, and types of contaminants present. *Avena sativa*, *Brassica napus L.*, *Viola calaminaria*, *Thlaspi caerulescens*, and *Hordeum Vulgare* can easily accumulate and tolerate the metals such as zinc, cadmium, nickel, copper, and selenium, whereas *Panicum virgatum* is well known for accumulating Sr-90, and Cs-137 from nuclear fallout weapons. Hybrid poplar trees have been reported to reduce the nitrates concentrations in the superficial groundwater and cause the degradation of atrazine herbicide in the contaminated soils. Forage grasses with compatible bacterial inoculations cause the degradation of chlorinated benzoic acids, chlorinated herbicides, and polychlorinated biphenyls (Furini et al., 2015).

Transportation mechanisms and chelators activities of plants have been greatly modified and enhanced by the application of genetic engineering and biotechnological approaches. In this way tolerance, capacity, and effectiveness of phytoremediation is greatly improved. An association between microbial strains and metallophyte plants significantly improves the redox potential in soil, absorption, and precipitation of heavy metals. These microbial strains are known as metallophyte growth–promoting rhizobacteria. These microorganisms are resistant to heavy metals and greatly protect plants from toxic effects of metals by increasing the metal absorption in the hyperaccumulating plants (Cioica et al., 2019). Cultivation of transgenic plants for the phytoremediation provides novel opportunities for effective bioremediation. Incorporation of specific genes to these plants increases the uptake, accumulation and metabolism of these plants. Ideally, the plantation approaches for bioremediation should provider higher biomass yield according to prevailing environmental and climatic conditions and should be able to perform efficient transformation protocol (Koźmińska et al., 2018). These plants enhance the detoxification process by the multiple times and prevent the potential entry of toxic pollutants to the food chain. First, these plants were only used for the removal of inorganic pollutants but now these are also used to remediate the organic contaminants. *Arabidopsis thaliana* and *Nicotiana tabacum* are greatly efficient in the removal of mercury and cadmium (Misra and Gedamu, 1989; Rugh et al., 1996).

23.2.5 Percolation and phytoremediation

Deep percolation gradually removes the risk of food and environmental contamination by removing and percolating contaminants from the polluted soils. Percolation along with phytoremediation is a clean and sustainable approach from ecological and environmental point of view and is a viable economic constituent compared to chemical approaches. This process is intended to remove soil contaminants by following infiltration and percolation by the vegetative covers. However, the efficiency of system can only be ensured by following management measures according to soil types, concentration of pollutants in soil, and their physicochemical characteristics (Mandi et al., 2009).

This technique is important from a stabilization point of view and thereby greatly improves the physical, chemical, and biological properties of soil. The presence of dense vegetation cover and efficient rooting system helps to reduce the dispersion and mobilization of contaminants by promoting percolation and infiltration instead of runoff. Therefore, bioavailability of contaminants in the rhizosphere will be decreased.

23.2.6 In-situ and phytoremediation

In-situ phytoremediation techniques are most commonly used as they cause significant reduction in the contaminant's multiplication in the airborne waste and water, which ultimately reduces the risk of contaminants movement to the environment. It allows treating many pollutants simultaneously on specific site and also reduces the needs for management of disposal sites. Moreover, it also reduces the further spread of contaminants by preventing leaching and soil erosion (Ali et al., 2020).

However, in-situ phytoremediation is being used in many parts of the globe, but full-scale application of this technique is still limited. Generally, phytoremediation removes pollutants by various ways, as plants are being used for different purposes such as degradation, stabilization, and accumulation of pollutants by acting as traps or filters. Some of the important processes in this regard include phytoextraction, rhizofiltration, phytostabilization, phytodegradation, rhizodegradation, and phytovolatilization.

Phytoremediation can also be promoted by the use of chemicals such as ethylenediaminetetraacetic acid, diethylenetriaminepentaacetic acid, ethylenediamine-N,N′-bis(2-hydroxyphenylacetic acid), and N-(2-hydroxyethyl) ethylenediaminetriacetic acid. However, the concentrations of applied chemicals should be regularly checked to avoid the increased concentrations of heavy metals in the soil systems (Vangronsveld et al., 2009).

- *Phytoextraction*
  This technique is also known as phytoaccumulation and is referred to uptake and translocating metals contaminants to the aboveground portions of plants. Hyperaccumulators are efficient plants for phytoextraction and absorb larger proportions of metal contaminants in their body parts. Planting a combination of these plant species on a specific site is depends on nature and concentration of metal contaminants that can offer significant removal efficiency (Jeke et al., 2019). These plants are harvested after several months of growth and are composted or incinerated to recycling the metals. The ash obtained after incineration should be carefully disposed in the landfill sites, as otherwise it can create serious environmental and health concerns.
- *Rhizofiltration*
  Rhizofiltration is precipitation or adsorption of contaminants present in the rhizosphere on the plant roots or in the plant roots. The plants for this purpose are grown in the greenhouse conditions in the hydroponics culture rather than the soil. These plants are acclimatized with the contaminated water after the development of efficient rooting system. Then these plants are transplanted in the contaminated areas and roots can easily uptake the contaminants and water in the surrounding conditions. When the roots are completely occupied by the contaminants these plants are harvested and can be either composted or incinerated for recycling the contaminants. In this technique both roots and shoots are used to remove contaminants from wasterwaters and surface waters. In this process pollutants are removed by both absorption and adsorption. Modification in the rhizospheric pH due to release of exudates causes heavy metals precipitation on the roots and greatly minimizes the movement of contaminants to the ground water (Javed et al., 2019). Ideally, these plants have dense and efficient rooting system, good tolerance toward heavy metals, and are capable to produce high biomass yield. Both aquatic and terrestrial plants are used for this purpose and offer best removal efficiencies. Aquatic plant species for wetland remediation includes poplar, cattail, duckweed, Azolla, and hyacinth. Terrestrial plants such as sunflower and Indian mustard also have good potential for remediation of contaminated lands by offering good capacities for heavy metals accumulation (Dhanwal et al., 2017).
- *Phytostabilization*
  This technique is focused to use plant species for immobilization of contaminants in the ground water and soil by the process of absorption to the roots, accumulation by the roots, and precipitation in the root zone. In this way,

mobility of contaminants and their migration to the air and ground water is greatly reduced. Moreover, it also helps to reduce the potential entry of contaminants to the food chain and protects animals and humans from serious health consequences. Most of the selected species for phytoremediation have fewer limitations for removal efficiencies. These limitations include slow growth and adaptation to the good range of environmental conditions. However, these limitations can be significantly minimized by the modification and improvement of certain traits for the phytoremediation (Sarwar et al., 2017).

- *Phytodegradation*
  This technique is also known as phytotransformation and it causes breakdown of pollutants and contaminants that are taken up by the plants through different metabolic approaches within the plant bodies. Also, there is a breakdown of pollutants on the external surface of plants by the plant-produced enzymatic actions. These degraded contaminants are incorporated to the plant tissues and are being used as an essential source of nutrients (Jeevanantham et al., 2019).

- *Rhizodegradation*
  Rhizodegradation is also known as phytostimulation, plant-assisted biodegradation, and enhanced rhizospheric biodegradation. In this process, breakdown of contaminants is accelerated in the soil by the microbial activities. Microorganisms such as bacteria, fungi, and yeast consume and digest most of the organic substances to obtain energy and nutrition. Some microorganisms are specialized to digest hazardous solvents and fuels and break them to the less toxic substances by the process of biodegradation. However, plants release and provide natural substances (alcohols, organic acids, sugars) to the microorganisms. Moreover, the presence of additional nutrients in the rhizospheric soil promotes the activities of microorganisms. Resultantly, it causes significant improvement in the physical, chemical, and biological soil activities that favor good plant growth and food production (Cheng et al., 2019).

- *Phytovolatilization*
  Phytovolatilization is related to the contaminants uptake and transpiration followed by the release of modified contaminants to the environment. This is a common process in the growing trees as plants take up the organic contaminants and water. Some of the contaminants can easily pass to the plant leaves and their lower concentrations are volatilized to the environment.

  One of the best advantages of this technique is that it efficiently causes the removal of contaminants from the site and disperses them to the atmosphere as gaseous compounds; thereby minimizing the needs for disposal and harvesting (Mahar et al., 2016). However, the direct release of pollutants to environment is still a concern that should be resolved on a sustainable basis. Moreover, these pollutants can also be redeposited by precipitation; therefore, there should be proper testing and evaluation of risk assessment before applying this technique in the field conditions.

23.3 Potential economic opportunities for biomass produced by phytoremediation

Heavy metals are present as persistent pollutants in the environment, and they must be completely eliminated for complete remedial objectives. Phytoremediation is environmentally friendly, economical, and less disruptive cleanup technology. Proper and effective disposal of produced plant biomass can be used for feeding animals and for biogas production.

Phytoremediation offers cost-effective opportunities in both in-situ and ex-situ application and improves financial benefits by the recovery and reuse of metals. This is least harmful method than the other methods as it involves the use of biological activities to reduce the hazardous contaminants. In-situ phytoremediation can be either used alone or in combination with other chemical and mechanical methods to clean contaminants and pollutants such as landfill chelates, polyalicyclic hydrocarbons, crude oil, explosives, solvents, and synthetic agricultural chemicals (Gatliff et al., 2016).

Phytoremediation offers potential economic and social benefits through the prism of ecosystem functioning and improved ecological balance. Selling the produced biomass for anaerobic digestion and for the production of renewable energy is potentially important to increase farmer's income by multiple folds. The produced biomass is used for anaerobic digestion to convert organic matter contents to methane by the bacterial actions and in the absence of oxygen. The end products digestate and biogas are highly useful due to the presence of high-energy contents. The biogas can be utilized for functioning and running of machines and engines that are working on natural gas. This gas can also be injected in the distribution network of natural gas and can also be used as a transportation fuel.

Some of the reclaimed metals such as Se, Mn, Pb, Zn, Cd, U, Cu, Ni, Co, and Au, can generate huge revenue by Phytomining. Many mining companies are generating profit by recovery and utilization of metals from biomass and generating energy from biomass. Ash is a good source of potassium and carbon, and good profit can be obtained by selling carbon dioxide credits (Sheoran et al., 2009). Extracted nickel from the nickel-contaminated soils is being used as an important substitute for nickel fertilizers and is helpful to save fertilizers costs. Numerous mining companies are working in Canada, United States, Eastern and Western Europe, Australia, Japan, China, and other parts of Asia.

The economic benefits associated with phytoremediation are greatly dependent on metal contents in plants and soil, production of biomass on annual basis, and generation of energy by the combustion of biomass. Production of biomass is most important aspect of phytoremediation and is an ultimate approach for future agrofarming systems and remediation disciplines. Metal values are greatly variable between \$1.793 and \$39,368.59 kg-1 for ordinary and expensive metals (Mahar et al., 2016). Phytoremediation offers great protection of environment and reduces the potential entry of contaminants to the ground water and environment. Therefore, the costs associated with ground water purification, and environmental remediation are greatly reduced. This amount can be used for educational opportunities, health care system, and for other developmental activities. Moreover, production of safe food is also helpful to reduce the potential negative effects on human health, and thus the burden on pharmaceutical industries and health care system is greatly reduced.

Based on the agroecological zones and prevailing climatic conditions, different types of energy crops such as wheat, maize, and rape seed can be grown followed by the short-rotation coppice. The phytoremediation potential of energy-efficient crops, fate of contaminants in these plants, and production of bioenergy by different processes such as incineration, pure plant oil production, anaerobic digestion, and gasification has been explored by various studies (Saxena et al., 2019). Cultivation of 1 ha of maize for phytoremediation can generate 33,000–46,000 kWh thermal and electrical energy per year. Thus it can easily substitute the energy production by fossil fuels in the coal-fed power plants. Studies have also supported that concentration of metals in the grains is very low due to natural exclusion mechanism of plants, so these grains can be used as animal feed. Moreover, the recovery of metals from digestate is a viable opportunity from a technological route. An external benefit of about € 55–501 per hectare has been reported by CO_2 abatement using phytoremediation crops (Witters et al., 2012).

Some studies have also shown that carbon system of traditional and integrated coppicing system is more favorable than the short coppicing rotations. The utilization of land surface is as a functional unit, thus intensively managed bioenergy plantations produce significantly higher yields. The economic feasibility and viability of phytoremediation to produce heat and electricity is also impacted by different production methodologies and technologies. Biofuel remediation significantly reduces the damage to environment. However, legislation, and management according to policies is a major concern around the globe, and economic value of phytoremediation can be greatly enhanced by following rights measures and practices.

Moreover, phytoremediation significantly improves the quality of degraded lands and helps to produce food. Thus, it is directly related to improved cultural and socioeconomic conditions of deprived communities. Resultantly, improved health conditions and availability of opportunities will motivate them to play their role for the development of countries and regions (Gomes, 2012).

However, good economic potential is being achieved by the phytoremediation of heavy metals–contaminated soils, but this profit and value of phytoremediation products can be increased by multiple fold by identification and better understanding of mechanisms of heavy metals binding in the soil, their release, bioavailability, uptake, translocation to different plant parts, detoxification processes, and characterization of various molecular and signaling pathways. Moreover, genetic studies and molecular engineering approaches can be used for enhancing the heavy metals uptake, sequestration, translocation, and tolerance. Furthermore, microorganisms are chelating agents that can also be used for enhancing the bioavailability of metals, which in turn directly facilitates the accumulation of heavy metals in the plants and therefore improves soil health. Practically, any single approach is never enough for complete cleanup of heavy metals and contaminants from the soil, thus a combination of different effective approaches should be used for exhaustive and highly effective phytoremediation (Yan et al., 2020).

References

Agamuthu, P., Abioye, O.P., Aziz, A.A., 2010. Phytoremediation of soil contaminated with used lubricating oil using *Jatropha curcas*. J. Hazard. Mater. 179, 891894.

Agbogidi, O.M., Dolor, E.D., Okechukwu, E.M., 2007. Evaluation of *Tectona grandis* (Linn.) and *Gmelina arborea* (Roxb.) for phytoremediation in crude oil contaminated soils. Agric. Conspec. Sci. 72 (2), 149152.

Ali, H., Khan, E., Sajad, M.A., 2013. Phytoremediation of heavy metals—concepts and applications. Chemosphere, 91 (7), 869–881.

Ali, S., Abbas, Z., Rizwan, M., Zaheer, I.E., Yavaş, İ., Ünay, A., …and Kalderis, D., 2020. Application of floating aquatic plants in phytoremediation of heavy metals polluted water: a review. Sustainability, 12 (5), 1927.

Anandhkumar, S.P., 1998. Studies of Treated Tannery Effluent on Flower Crops and Its Impact on Soil and Water Quality (M.Sc. thesis). Tamil Nadu Agricultural University, Coimbatore, India.

Angelova, V.A., Grekov, D.F., Kisyov, V.K., Ivanov, K.I., 2015. Potential of Lavender (*Lavandula* vera L.) for phytoremediation of soils contaminated with heavy metals. Int. J. Biol. Agric. Food Biotechnol. Eng. 9 (5), 522529.

Ashraf, M.A., Maah, M.J., Yusoff, I., 2011. Heavy metals accumulation in plants growing in ex tin mining catchment. Int. J. Basic Appl. Sci. 8 (2), 401416.

Ashraf, S., Ali, Q., Zahir, Z.A., Ashraf, S., Asghar, H.N., 2019. Phytoremediation: environmentally sustainable way for reclamation of heavy metal polluted soils. Ecotoxicol. Environ. Saf. 174, 714–727.

B Arnao, M., Hernández-Ruiz, J., 2019. Role of melatonin to enhance phytoremediation capacity. Appl. Sci. 9 (24), 5293.

Bauddh, K., Singh, R.P., 2012. Cadmium tolerance and its phytoremediation by two oil yielding plants *Ricinus communis* (L.) and *Brassica juncea* (L.) from the contaminated soil. Int. J. Phytoremediation. 14, 772785.

Burges, A., Alkorta, I., Epelde, L., Garbisu, C., 2018. From phytoremediation of soil contaminants to phyto management of ecosystem services in metal contaminated sites. Int. J. Phytoremediation, 20 (4), 384–397.

Chakraborty, R., Mukherjee, A., 2010. Technical note: vetiver can grow on coal fly ash without DNA damage. Int. J. Phytoremediation. 13 (2), 206214.

Chen, F., Yang, Y., Mi, J., Liu, R., Hou, H., Zhang, S., 2019. Effects of vegetation pattern and spontaneous succession on remediation of potential toxic metal-polluted soil in mine dumps. Sustainability. 11 (2), 397.

Cheng, L., Zhou, Q., Yu, B., 2019. Responses and roles of roots, microbes, and degrading genes in rhizosphere during phytoremediation of petroleum hydrocarbons contaminated soil. Int. J. Phytoremediation, 21 (12), 1161–1169.

Cioica, N., Tudora, C., Iuga, D., Deak, G., Matei, M., Nagy, E.M., Gyorgy, Z., 2019. A review on phytoremediation as an ecological method for in situ clean up of heavy metals contaminated soils, E3S Web of Conferences, 112. EDP Sciences.

DalCorso, G., Fasani, E., Manara, A., Visioli, G., Furini, A., 2019. Heavy metal pollutions: state of the art and innovation in phytoremediation. Int. J. Mol. Sci. 20 (14), 3412.

Danh, L.T., Truong, P., Mammucari, R., Foster, N., 2014. A critical review of the arsenic uptake mechanisms and phytoremediation potential of *Pteris vittata*. Int. J. Phytoremediation, 16 (5), 429–453.

Das, M., Agarwal, P., Singh, R., Adholeya, A., 2013. A study of abandoned ash ponds reclaimed through green cover development. Int. J. Phytoremediation. 15 (4), 320329.

Dhanwal, P., Kumar, A., Dudeja, S., Chhokar, V., Beniwal, V., 2017. Recent advances in phytoremediation technology. In: Kumar, R., Sharma, A.K., Ahluwalia, S.S. (Eds.), Advances in Environmental Biotechnology. Springer, Singapore, pp. 227–241.

Didier, T., Philippe, L.G., Sonia, H., Amar, B., D'Innocenzo Marielle, M.-C.C., Jairo, F., 2012. Prospects of *Miscanthus* × *giganteus* for PAH phytoremediation: a microcosm study. Indian Crop. Prod. 36, 276281.

Furini, A., Manara, A., DalCorso, G., 2015. Environmental phytoremediation: plants and microorganisms at work. Front. Plant Sci. 6, 520.

Gabor, T. S. (2001). Beyond the pipe: the importance of wetlands and upland conservation practices in watershed management: functions and values for water quality and quantity.

Gatliff, E., Linton, P.J., Riddle, D.J., Thomas, P.R., 2016. Phytoremediation of soil and groundwater: economic benefits over traditional methodologies. In: Prasad, M.N.V. (Ed.), Bioremediation and Bioeconomy. Elsevier, Amsterdam, pp. 589–608.

Gomes, H.I., 2012. Phytoremediation for bioenergy: challenges and opportunities. Environ. Technol. Rev. 1 (1), 59–66.

Hou, D., 2019. Sustainable Remediation of Contaminated Soil and Groundwater: Materials, Processes, and Assessment. Butterworth-Heinemann, Oxford, UK.

Huang, H., Yu, N., Wang, L., Gupta, D.K., He, Z., Wang, K., et al., 2011. The phytoremediation potential of bioenergy crop Ricinus communis for DDTs and cadmium contaminated soil. Bioresour. Technol. 102, 1103411038.

Javed, M.T., Tanwir, K., Akram, M.S., Shahid, M., Niazi, N.K., Lindberg, S., 2019. Phytoremediation of cadmium-polluted water/sediment by aquatic macrophytes: role of plant-induced pH changes. In: Hasanuzzaman, M., Prasad, M.N.V., Fujita, M. (Eds.), Cadmium Toxicity and Tolerance in Plants: From Physiology to Remediation. Academic Press, MA, pp. 495–529.

Jayanthy, V., Geetha, R., Rajendran, R., Prabhavathi, P., Sundaram, S.K., Kumar, S.D., et al., 2014. Phytoremediation of dye contaminated soil by *Leucaena leucocephala* (subabul) seed and growth assessment of Vigna radiate in the remediated soil. Saudi J. Biol. Sci. 21, 324333.

Jeevanantham, S., Saravanan, A., Hemavathy, R.V., Kumar, P.S., Yaashikaa, P.R., Yuvaraj, D., 2019. Removal of toxic pollutants from water environment by phytoremediation: a survey on application and future prospects. Environ. Technol. Innov. 13, 264–276.

Jeke, N.N., Zvomuya, F., Cicek, N., Ross, L., Badiou, P., 2019. Nitrogen and phosphorus phytoextraction by cattail (Typha spp.) during wetland-based phytoremediation of an end-of-life municipal lagoon. J. Environ. Qual. 48 (1), 24–31.

Jiang, Y., Lei, M., Duan, L., Longhurst, P., 2015. Integrating phytoremediation with biomass valorisation and critical element recovery: a UK contaminated land perspective. Biomass Bioenergy, 83, 328–339.

Juwarkar, A.A., Jambhulkar, H.P., 2008. Restoration of fly ash dump through biological interventions. Environ. Monit. Assess. 139, 355365.

Kanabkaew, T., Puetpaiboon, U., 2004. Aquatic plants for domestic wastewater treatment: lotus (*Nelumbo nucifera*) and hydrilla (*Hydrilla verticillata*) systems. Songklanakarin J. Sci. Technol. 26 (5), 749756.

Konlechner, C., Türktą s, M., Langer, I., Vaculı ́k, M., Wenzel, W.W., Puschenreiter, M., 2013. Expression of zinc and cadmium responsive genes in leaves of willow (*Salix caprea* L.) genotypes with different accumulation characteristics. Environ. Pollut. 178, 121127.

Koźmińska, A., Wiszniewska, A., Hanus-Fajerska, E., Muszyńska, E., 2018. Recent strategies of increasing metal tolerance and phytoremediation potential using genetic transformation of plants. Plant Biotechnol. Rep. 12 (1), 1–14.

Kumar, A., Maiti, S.K., 2015. Effect of organic manures on the growth of *Cymbopogon citratus* and *Chrysopogon zizanioides* for the phytoremediation of Chromite-Asbestos mine waste: a pot scale experiment. Int. J. Phytorem. 17 (5), 437447.

Kumar, G.P., Yadav, S.K., Thawale, P.R., Singh, S.K., Juwarkar, A.A., 2008. Growth of *Jatropha curcas* on heavy metal contaminated soil amended with industrial wastes and Azotobacter—a greenhouse study. Bioresour. Technol. 99, 20782082.

Kumar, K.V., Patra, D.D., 2012. Alteration in yield and chemical composition of essential oil of *Mentha piperita* L. plant: effect of fly ash amendments and organic wastes. Ecol. Eng. 47, 237241.

Lal, K., Minhas, P.S., Chaturvedi, R.K., Yadav, R.K., 2008. Extraction of cadmium and tolerance of three annual cut flowers on Cd-contaminated soils. Bioresour. Technol. 99, 10061011.

Lewandowski, I., Schmidt, U., Londo, M., Faaij, A., 2006. The economic value of the phytoremediation function—assessed by the example of cadmium remediation by willow (Salix spp. Agric. Syst. 89, 68–89.

Mahar, A., Wang, P., Ali, A., Awasthi, M.K., Lahori, A.H., Wang, Q., ... and Zhang, Z., 2016. Challenges and opportunities in the phytoremediation of heavy metals contaminated soils: a review. Ecotoxicol. Environ. Saf. 126, 111–121.

Mahimairaja, S., Shenbagavalli, S., Naidu, R., 2011. Remediation of chromium-contaminated soil due to tannery waste disposal: potential for phyto and bioremediation. Pedologist 54, 175181.

Maiti, S.K., Nandhini, S., Das, M., 2005. Accumulation of metals by naturally growing herbaceous and tree species in iron ore tailings. Int. J. Environ. Stud. 62 (5), 593603.

Mandi, L., Tiglyene, S., Jaouad, A., 2009. Depuration of tannery effluent by phytoremediation and infiltration percolation under arid climate. Option Mediterr. 88, 199–205.

Mangkoedihardjo, S., Ratnawati, R., Alfianti, N., 2008. Phytoremediation of hexavalent chromium polluted soil using *Pterocarpus indicus* and *Jatropha curcas* L. World Appl. Sci. J. 4, 338342.

Mangkoedihardjo, S.Surahmaida, 2008. *Jatropha curcas* L. for phytoremediation of lead and cadmium polluted soil. World Appl. Sci. J. 4, 519522.

Misra, S., Gedamu, L., 1989. Heavy metal tolerant transgenic *Brassica napus* L. and *Nicotiana tabacum* L. plants. Theor. Appl. Genet. 78 (2), 161–168.

Mleczek, M., Rutkowski, P., Rissmann, I., Kaczmarek, Z., Golinski, P., Szentner, K., … and Stachowiak, A., 2010. Biomass productivity and phytoremediation potential of *Salix alba* and *Salix viminalis*. Biomass Bioenergy 34 (9), 1410–1418.

Ogbonna, P.C., Ukiwe, E.O., 2010. Metals (Hg, As, Fe, Cr) uptake in selected woody species in Umuahia. Nigeria. J. Appl. Sci. Environ. Sanit. 6 (1), 2328.

Olivares, A.R., Carrillo-Gonza ́lez, R., Gonza ́lez-Cha ́vez, Mdel, C.A., Herna ́ndez, R.M.S., 2013. Potential of castor bean (*Ricinus communis* L.) for phytoremediation of mine tailings and oil production. J. Environ. Manage. 114, 316323.

Pandey, D.N., Prakash, N.P., 2014. Sequential restoration of urban green space in tropical dry forest region: Kulish Smriti Van, Jaipur Rajasthan, India. In: RSPCB Occasional Paper No. 8/2014. Climate Change and CDM Cell.

Pandey, V.C., Kumar, A., 2013. *Leucaena leucocephala*: an underutilized plant for pulp and paper production. Genet. Resour. Crop. 60, 11651171.

Rai, U.N., Pandey, K., Sinha, S., Singh, A., Saxena, R., Gupta, D.K., 2004. Revegetating fly-ash landfills with *Prosopis juliflora* L.: impact of different amendments and Rhizobium inoculation. Environ. Int. 30, 293300.

Rajkumar, M., Freitas, S.H., 2008. Influence of metal resistant-plant growth promoting bacteria on the growth of *Ricinus communis* soil contaminated with heavy metals. Chemosphere 71, 834842.

Ram, L.C., Jha, S.K., Tripathi, R.C., Masto, R.E., Selvi, V.A., 2008. Remediation of fly ash landfills through plantation. Remediation 18, 7190.

Ramana, S., Biswas, A.K., Singh, A.B., Ajay, Ahirwar, N.K., Rao, A.S., 2013. Phytoremediation ability of some floricultural plant species. Indian J. Plant Physiol. 18 (2), 187190.

Ramasamy, K., 1997. Tannery effluent related pollution on land and water ecosystems, Proc. Extended Abstracts from the 4th International Conference on the Biogeochemistry of Trace Elements. Berkeley, CA, USA, 771772.

Robinson, B.H., Chiarucci, A., Brooks, R.R., Petit, D., Kirkman, J.H., Gregg, P.E.H., De Dominicis, V, 1997. The nickel hyperaccumulator plant *Alyssum bertolonii* as a potential agent for phytoremediation and phytomining of nickel. J. Geochem. Explor. 59 (2), 75–86.

Robinson, B.H., Green, S.R., Chancerel, B., Mills, T.M., Clothier, B.E., 2007. Poplar for the phytomanagement of boron contaminated sites. Environ. Pollut. 150, 225233.

Rugh, C.L., Wilde, H.D., Stack, N.M., Thompson, D.M., Summers, A.O., Meagher, R.B., 1996. Mercuric ion reduction and resistance in transgenic *Arabidopsis thaliana* plants expressing a modified bacterial merA gene. Proc. Natl. Acad. Sci. USA 93 (8), 3182–3187.

Sarwar, N., Imran, M., Shaheen, M.R., Ishaque, W., Kamran, M.A., Matloob, A.…and Hussain, S., 2017. Phytoremediation strategies for soils contaminated with heavy metals: modifications and future perspectives. Chemosphere, 171, 710–721.

Saxena, G., Purchase, D., Mulla, S.I., Saratale, G.D., Bharagava, R.N., 2019. Phytoremediation of heavy metal-contaminated sites: eco-environmental concerns, field studies, sustainability issues, and future prospects. Rev. Environ. Contamin. Toxicol. 249, 71–131.

Sheoran, V., Sheoran, A.S., Poonia, P., 2009. Phytomining: a review. Miner. Eng. 22 (12), 1007–1019.

Shukla, O.P., Juwarkar, A.A., Singh, S.K., Khan, S., Rai, U.N., 2011. Growth responses and metal accumulation capabilities of woody plants during the phytoremediation of tannery sludge. Waste Manage. 31, 115123.

Srivastava, N.K., Ram, L.C., Masto, R.E., 2014. Reclamation of overburden and lowland in coal mining area with fly ash and selective plantation: a sustainable ecological approach. Ecol. Eng. 71, 479489.

Técher, D., Laval-Gilly, P., Bennasroune, A., Henry, S., Martinez-Chois, C., D'Innocenzo, M., et al., 2012. An appraisal of *Miscanthus* ×*giganteus* cultivation for fly ash revegetation and soil restoration. Indian. Crop. Prod. 36, 427433.

Tripathi, V., Edrisi, S.A., Abhilash, P.C., 2016. Towards the coupling of phytoremediation with bioenergy production. Renew. Sust Energy Rev. 57, 1386–1389.

Van Aken, B., 2008. Transgenic plants for phytoremediation: helping nature to clean up environmental pollution. Trends Biotechnol. 26 (5), 225–227.

Vangronsveld, J., Herzig, R., Weyens, N., Boulet, J., Adriaensen, K., Ruttens, A.…and van der Lelie, D., 2009. Phytoremediation of contaminated soils and groundwater: lessons from the field. Environ. Sci. Pollut. Res. 16 (7), 765–794.

Vanlı, Ö., 2007. Pb, Cd, B Elementlerinin Topraklardan Şelat Destekli Fitoremediasyon Yöntemiyle Giderilmesi Ph.D. thesis). Fen Bilimleri Enstitüsü.

Verma, S.K., Singh, K., Gupta, A.K., Pandey, V.C., Trivedi, P., Verma, R.K., et al., 2014. Aromatic grasses for phytomanagement of coal fly ash hazards. Ecol. Eng. 73, 425428.

Wan, X., Lei, M., Chen, T., 2016. Cost–benefit calculation of phytoremediation technology for heavy-metal-contaminated soil. Sci. Total Environ. 563, 796–802.

Witters, N., Mendelsohn, R., Van Passel, S., Van Slycken, S., Weyens, N., Schreurs, E., ...and Vangronsveld, J., 2012. Phytoremediation, a sustainable remediation technology? II: Economic assessment of CO_2 abatement through the use of phytoremediation crops for renewable energy production. Biomass Bioenergy 39, 470–477.

Wu, S.C., Wong, C.C., Shu, W.S., Khan, A.G., Wong, M.H, 2010. Mycorrhizo-remediation of lead/zinc mine tailings using vetiver: a field study. Int. J. Phytoremediation 13 (1), 61–74.

Xie, L., van Zy, D., 2020. Distinguishing reclamation, revegetation and phytoremediation, and the importance of geochemical processes in the reclamation of sulfidic mine tailings: a review. Chemosphere, 126446.

Yadav, S.K., Juwarkar, A.A., Kumar, G.P., Thawale, P.R., Singh, S.K., Chakrabarti, T., 2009. Bioaccumulation and phyto-translocation of arsenic, chromium and zinc by *Jatropha curcas* L.: impact of dairy sludge and biofertilizer. Bioresour. Technol. 100, 46164622.

Yan, A., Wang, Y., Tan, S.N., Yusof, M.L.M., Ghosh, S., Chen, Z, 2020. Phytoremediation: a promising approach for revegetation of heavy metal-polluted land. Front. Plant Sci., 11. doi:10.3389/fpls.2020.00359.

Yang, F., Li, X., Li, Y., Wei, H., Yu, G., Yin, L....and Pu, Y., 2013. Lysing activity of an indigenous algicidal bacterium *Aeromonas* sp. against *Microcystis* spp. isolated from Lake Taihu. Environ. Technol. 34 (11), 1421–1427.

Yenn, R., Borah, M., Boruah, H.P., Roy, A.S., Baruah, R., Saikia, N., et al., 2014. Phytoremediation of abandoned crude oil contaminated drill sites of Assam with the aid of a hydrocarbon-degrading bacterial formulation. Int. J. Phytorem. 16 (7–12), 909925.

Zheljazkov, V.D., Craker, L.E., Baoshan, X., 2006. Effects of Cd, Pb and Cu on growth and essential oil contents in dill pepper mint, and basil. Environ. Exp. Bot. 58, 916.

Zheljazkov, V.D., Nielsen, N.E., 1996a. Studies on the effect of heavy metals (Cd, Pb, Cu, Mn, Zn and Fe) upon the growth, productivity and quality of lavender (*Lavandula angustifolia* Mill) production. J. Essent. Oil Res. 8, 259274.

Zheljazkov, V.D., Nielsen, N.E., 1996b. Effect of heavy metals on peppermint and cornmint. Plant Soil. 178, 5966.

Index

Page numbers followed by "*f*" and "*t*" indicate, figures and tables respectively.

Q

R

S

FSC
www.fsc.org
MIX
Papier | Fördert
gute Waldnutzung
FSC® C083411

Kolibri 360 GmbH
Ferdinand-Jühlke-Straße 7
99095 Erfurt, Deutschland
produktsicherheit@kolibri360.de